W9-CHU-001

SCIENCEPLUS®

Technology and Society

Project Directors

International: **Charles McFadden**
Professor of Science Education
The University of New Brunswick
Fredericton, New Brunswick

National: **Robert E. Yager**
Professor of Science Education
The University of Iowa
Iowa City, Iowa

This new United States edition has been adapted from prior work by the Atlantic Science Curriculum Project, an international project linking teaching, curriculum development, and research in science education.

Holt, Rinehart and Winston

Harcourt Brace Jovanovich

Austin • Orlando • San Diego • Chicago • Dallas • Toronto

ACKNOWLEDGMENTS

Project Advisors

Herbert Brunkhorst
Director, Institute for Science Education
California State University,
San Bernardino
San Bernardino, California

David L. Cross
Science Consultant
Lansing School District
Lansing, Michigan

Jerry Hayes
Associate Director, Scientific Outreach
Teacher's Academy, Mathematics
and Science
Chicago, Illinois

William C. Kyle, Jr.
Director, School Mathematics
and Science Center
Purdue University
West Lafayette, Indiana

Mozell Lang
Science Education Specialist
Michigan Department of
Education
Lansing, Michigan

Project Authors

Earl S. Morrison (Author in Chief)
Alan Moore (Associate Author in Chief)
Nan Armour
Allan Hammond
John Haysom
Elinor Nicoll
Muriel Smyth

Project Associates

We wish to thank the hundreds of science educators, teachers, and administrators from the scores of universities, high schools, and middle schools who have contributed to the success of *SciencePlus*.

Acknowledgments: See page S158, which is an extension of the copyright page.

Printed in the United States of America

ISBN 0-03-074963-8

6 7 8 9 071 98 97 96

Contents

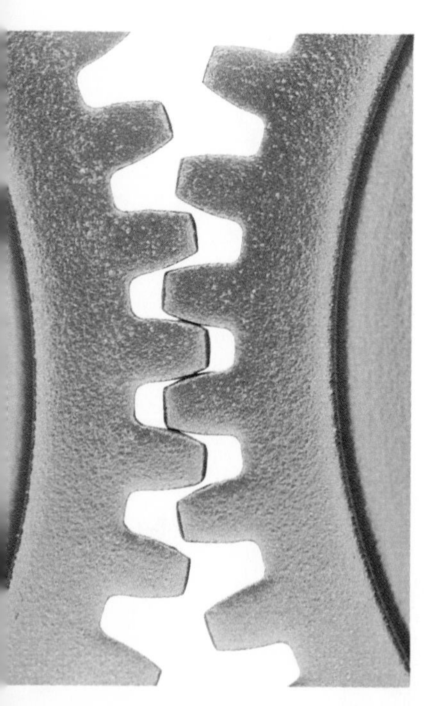

SourceBook

To The Student

This book was written with you in mind! There are many things to try, to create, and to investigate — both in and out of class. There are stories to be read, articles to think about, puzzles to be solved, and even games to play.

GET INVOLVED!

The best way to learn is by doing. In the words of an old Chinese proverb:

Tell me — I will forget.

Show me — I may remember.

Involve me — I will understand.

The authors of this book want ***you*** to ***get involved*** in science.

The activities in this book will allow you to make some basic and important scientific discoveries on your own. You will be acting much like the early investigators in science who, without expensive or complicated equipment, contributed so much to our knowledge.

What these early investigators had, and had in abundance, was ***curiosity*** and ***imagination***. If you have these qualities, you are in good company! And if you develop sharp scientific skills, who knows, you might make your own contributions to science someday.

THE LEADING EDGE

Scientists are usually interested in understanding things that happen in ***nature***. However, the discoveries that scientists make are often used by inventors and engineers. The end result is our most sophisticated ***technology***, including such things as computers, laser discs, nuclear reactors, and instant global communication.

There is an interaction between science and technology. Science makes technology possible. On the other hand, the products of technology are used to make further scientific discoveries. In fact, much of the scientific work that is done today has become so technically complicated and expensive that no one person can do it entirely alone. But make no mistake, the creative ideas for even the most highly technical and expensive scientific work still come from ***individuals***.

GO FOR IT!

Science is a process of discovery: a trek into the unknown. The skills you develop as you do the activities in this book — like observing, experimenting, and explaining observations and ideas — are the skills you will need to be a part of science in the future. There is a universe of scientific exploration and discovery awaiting those who take the challenge.

Keep a Journal

A Journal is an important tool in creative work. In this book, you will be asked to keep a Journal of your thoughts, observations, experiments, and conclusions. As you develop your Journal, you will see your own ideas taking shape over time. This is often the way scientists arrive at new discoveries. You too may log some discoveries as you develop your own Journal.

About Safety

Science investigations and experiments should be both enjoyable and safe. If you follow the safety guidelines listed here, as well as any others mentioned in the explorations, you should have no problems. You should ***always*** follow these guidelines, even when you think that there is little or no danger.

The major causes of laboratory accidents are carelessness, lack of attention, and inappropriate behavior. These all spring from a person's ***attitude***. With a proper attitude and ***consistent*** safety habits, you should be able to feel quite comfortable and at home in a science laboratory.

Safety Guidelines

Eye Safety

Wear goggles when handling acids or bases, using an open flame, or performing any other activity that could harm the eyes. If a substance gets into your eyes, wash them with plenty of water and notify your teacher at once. Never place a chemical substance near your unprotected eyes. Never use direct sunlight to illuminate a microscope.

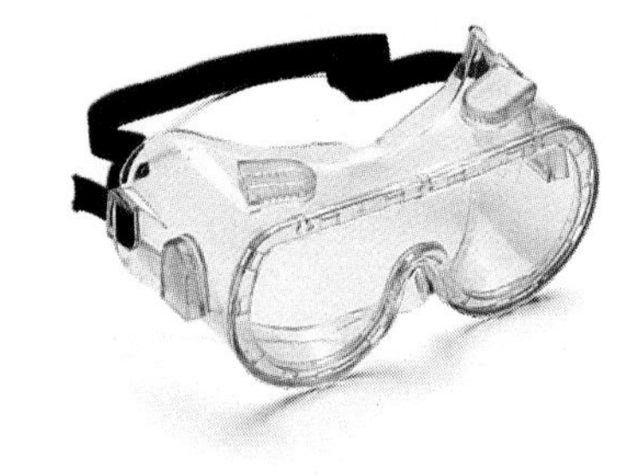

Safety Equipment

Know the location of all safety equipment, such as fire extinguishers and first aid kits.

Neatness

Keep work areas free of all unnecessary books and papers. Tie back long, loose hair and button or roll up loose sleeves when working with chemicals or near a flame.

Chemicals

Chemicals and other dangerous substances can be dangerous if they are handled carelessly. When handling certain chemicals, such as acids or bases, you should protect your eyes and clothes with safety glasses and an apron.

Heat

Whenever possible, use an electric hot plate instead of an open flame. If you must use an open flame to heat a glass container, shield the flame with a wire screen that has a ceramic center. When heating chemicals in a test tube, do not point the test tube toward anyone.

Electricity

Be cautious around electrical wiring. When using a microscope with a lamp, do not place its cord where it can cause someone to trip. Do not let cords hang over a table edge in a way that will cause equipment to fall if a cord is pulled. Do not use equipment with frayed cords.

Glassware

Examine all glassware before using. Glass containers for heating should be made of heat-resistant material. Never use cracked or chipped glassware. Use caution when inserting glass tubing into a rubber stopper. Hold the tubing and stopper with a towel to prevent puncture wounds.

Never taste chemicals, unless you are specifically instructed to do so.

Never pour water into a strong acid or base. The mixture produces heat. Sometimes the heat causes splattering. The correct procedure is to pour the acid or base slowly *into the water*. This way the mixture will stay cool.

If any solution is spilled, wash it off with plenty of water. If a strong acid or base is spilled, neutralize it first with an agent such as baking soda (for acids) or boric acid (for bases).

If you are instructed to note the odor of a substance, wave the fumes toward your nose with your hand rather than putting your nose close to the source.

Sharp/Pointed Objects

Use knives, razor blades, and other sharp instruments with care. Do not use double-edged razor blades in the laboratory.

Cleanup

Wash your hands immediately after handling hazardous materials. Before leaving the laboratory, clean up all work areas. Put away all equipment and supplies. Make sure water, gas, burners, and electric hot plates are turned off.

The instructions for the explorations in this book will include warning statements where necessary. In addition, you will find one or more of the following safety symbols when a procedure requires specific caution.

Wear Safety Goggles

Wear a Laboratory Apron

Flame/ Heat

Sharp/ Pointed Object

Dangerous Chemical/ Poison

Corrosive Substance

Unit 1 Life Processes

For Your Journal

Write a paragraph summarizing your thoughts about each of the following:

1. Where does the food we eat come from?
2. Why are plants green?
3. How is the sun connected to food production?
4. What life processes are taking place in this photograph?

The Source of Food

As you probably already know, three main characteristics of living things are:

- motion or locomotion
- growth
- response to stimuli

In this unit you will study the internal processes that make these activities possible.

Have you ever wondered where living things get the energy they need to carry out life processes? Perhaps it has occurred to you that most of the characteristics of living things are somehow connected with food. In fact, all life is dependent on food.

Now consider the following idea:

Without plants there would be no food.

Do you think this statement is true? To help you decide, think about a simple meal—a ham sandwich. The basic components of the sandwich might be described as "ground-up seeds, green leaves, and pig muscle"! But where did the pig muscle come from? It didn't just appear. Where did the lettuce come from? How about the wheat from which the bread was made? How did it grow? It seems that humans and animals alike eat green plants either directly or indirectly. But where do plants get their food?

Here's another example. Suppose that your lunch consisted of a hot dog with mustard, ketchup, and relish, and a tall glass of milk. Try to trace each of these foods back to plants.

You can see that other living things must depend, directly or indirectly, upon food made by plants. Therefore, the first life process you will look at deals with the question asked earlier: Where do plants get their food?

The raw materials for a ham sandwich.

Photosynthesis: The Food-Making Process

As you might already have guessed, plants have to make their own food. Scientists call the process by which plants make food **photosynthesis**. Like many scientific terms, this one is a combination of two Greek words: *photo,* which means "light," and *synthesis,* which means "putting together." This term will soon have more meaning for you.

By now you should realize that if green plants were unable to make food, all other living things would perish. What is this food that plants make?

The answer is **starch,** a major component of many different kinds of food. In the form of rice, potatoes, wheat, and other cereal products (which are almost completely composed of starch), this important food material represents over 70 percent of the world's food supply. Starch is produced by green plants and stored in their leaves. As you will see in Exploration 1, the presence of starch is easily detected.

This electron micrograph is of a starch-storage cell in a potato. The sphere-shaped objects are granules of pure starch.

EXPLORATION 1

The Search for Starch

Part 1

A Simple Test

You Will Need

- 2 containers
- cornstarch
- an eyedropper
- iodine solution
- sugar solution
- water

What to Do  

1. Mix some cornstarch in water, then add several drops of iodine solution (iodine dissolved in alcohol). What happens? **Be Careful:** *Iodine is poisonous.*
2. Repeat the test, first with plain water and then with a sugar solution. How do the results of these tests differ from the result of the test with cornstarch?
3. When testing for a substance, a *positive* result signals that the substance is present; a *negative* result means it is not. What signaled the presence of starch? For each of your tests for starch, was the result positive or negative?

Part 2

Starch in Green Leaves

You Will Need

- oven mitts
- water
- a geranium plant
- a large beaker
- a small beaker
- methanol
- a hot plate
- iodine solution
- an eyedropper
- white paper
- a watch or clock

Add iodine to each container to test for starch.

What to Do

1. Remove two leaves from a healthy plant that has been growing in a sunny spot.
2. Remove the green color from the leaves. To do this, first heat water in a large beaker until it boils.
3. Turn off the heat source. Move the beaker to a place away from any heat source.
4. Put the leaves in a small beaker and cover them with methanol. **Be Careful:** *Methanol is poisonous and highly flammable. Keep methanol away from open flames and other heat sources.*
5. Put the small beaker into the large beaker containing hot water, as shown in the diagram on the following page. The methanol will become warm enough to boil.

Be sure to save iodine solution and methanol for later Explorations.

A Safe Way to Heat Methanol

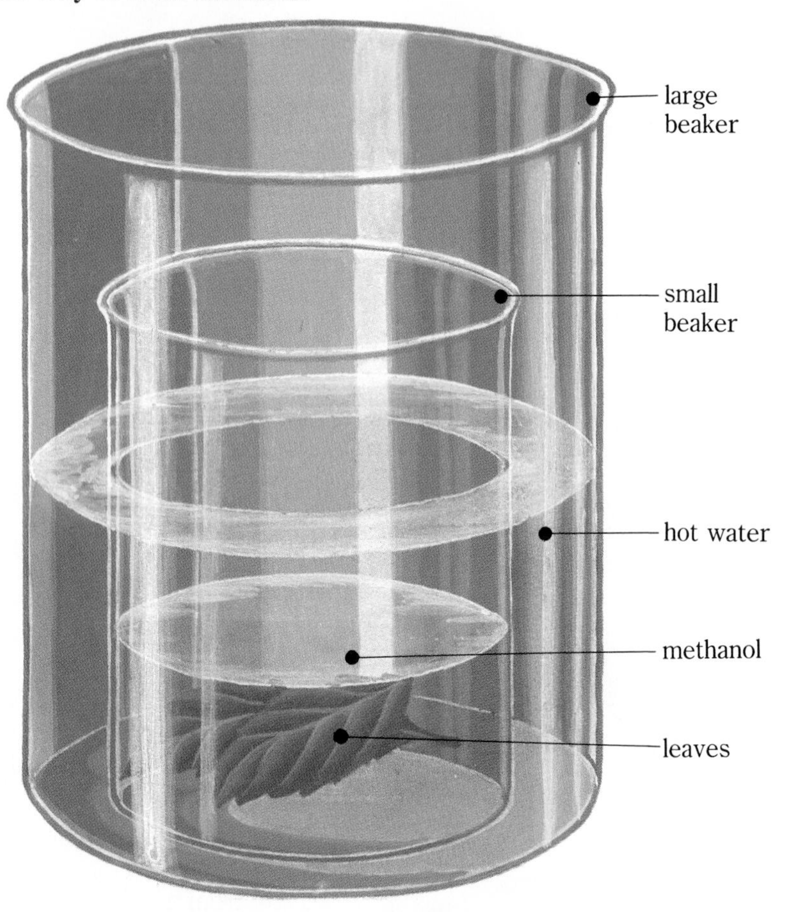

6. Soak the leaves in the methanol for 5 to 10 minutes. The methanol should become green as it dissolves the coloring from the leaves. Simultaneously, the leaves should become pale.
7. Remove the leaves from the methanol and rinse them in cool water. Spread them face up on a piece of white paper.
8. Treat one leaf with drops of iodine solution and place it in water to *fix* the color (make it permanent). Compare the treated leaf with the untreated leaf. Look for any slight change in color.

Questions

1. Sketch the two leaves in your Journal, noting any changes observed.
2. What evidence shows you that starch is present in a plant leaf?
3. Why did you heat the methanol you used to soak the leaves?
4. What was the purpose of using two leaves if only one was treated with iodine?

Starch—Essential for Life

The ability of plants to make starch is critical to all other living things. Even organisms that do not feed on plants depend on them indirectly for food. Carnivorous (meat-eating) animals, for example, prey upon other animals. But the preyed-upon animals either ate plants themselves or ate animals that ate plants. Somewhere in the food chain is a connection to plants.

Consider the statement "All flesh is ultimately grass." What does it mean? Write your explanation in your Journal.

A Formula for Food

You now know that plants make starch using a process called photosynthesis. At this point you might be wondering how photosynthesis works. What ingredients are needed for photosynthesis? What special conditions must be met for it to take place? The next three Explorations will introduce you to this vital process, and in doing so answer these and other fundamental questions.

EXPLORATION 2

Is Light Necessary for Photosynthesis?

The leaf you tested for starch in Exploration 1 was from a plant that had received plenty of light. If a plant receives no light, will it also have starch in its leaves?

You Will Need

- a geranium plant (grown outdoors)
- thin cardboard
- 3 straight pins
- materials and equipment to test leaves for starch

What to Do   

1. Put a geranium plant in a dark but warm place for four days. The soil should be moist but not too wet.
2. After the four days have passed, test some of the plant's leaves for starch as in Exploration 1. Record your findings in your Journal.
3. Now, cover one leaf with cardboard, as shown in the illustrations. (Important—the leaf must still be attached to the plant!) To the top surface of the leaf, fasten a design such as a letter or a number. To the bottom surface, fasten a cardboard square.
4. Put the plant in bright light such as a grow lamp for 24 hours. Then remove the cardboard pieces and test the covered and uncovered parts of the leaf for starch as you did in Part 2 of Exploration 1. Record your findings.

Questions

1. Do you have any indication that putting the plant in the dark caused a change in the leaf's production of starch? If so, what was this change?
2. Which part of this Exploration, steps 1 and 2 or steps 3 and 4, suggests that green plants need light to produce starch? Why?
3. Could you perform steps 3 and 4 without first doing steps 1 and 2?
4. In light of what you have just learned, why do you think the food-making process is called "photosynthesis"?

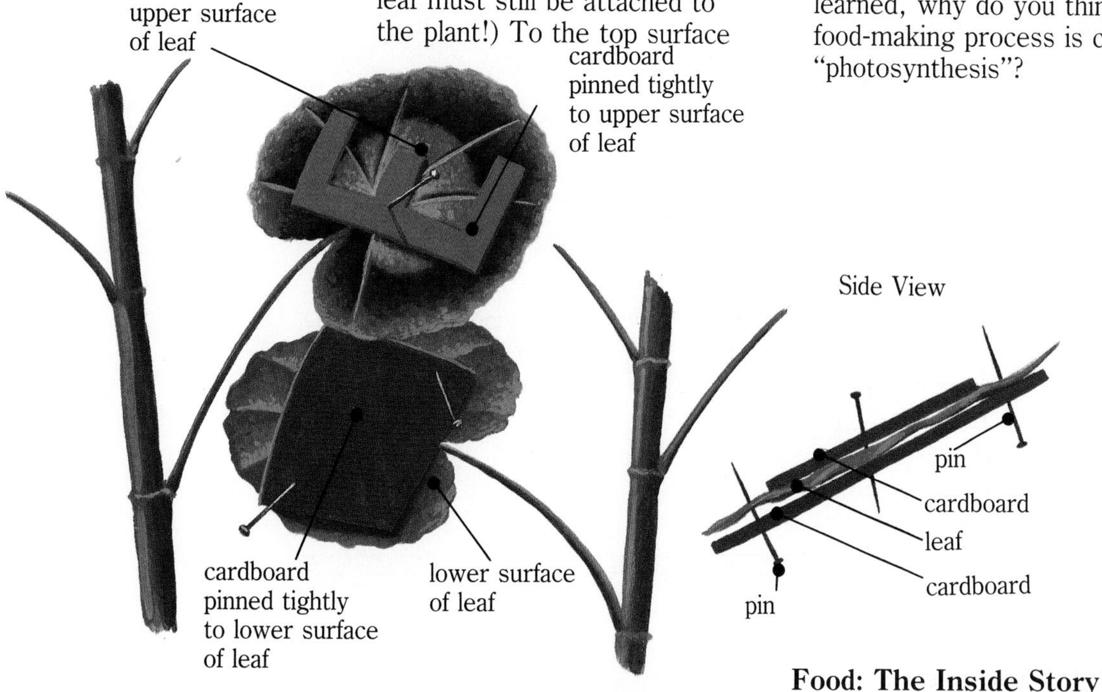

Water—How Essential Is It?

You have seen what happens when a plant is deprived of light. But what happens when a plant is deprived of water? You may have seen this before, perhaps when you or somebody else forgot to water a houseplant. Plants that don't get enough water wilt. If they continue to be deprived of water they eventually die. Clearly, water is important, but why?

One of the great scientists of the past, Jan Baptist van Helmont of Belgium, performed an important experiment in the middle of the seventeenth century. Van Helmont was investigating the growth of plants. He came to the conclusion that water was completely responsible for the great change in mass that plants undergo. But let Jan tell you about his experiment in his own words. (The masses have been changed to metric units.)

The corn shown above is healthy, having received adequate rainfall. The corn on the facing page grew during a drought; as a result, it is stunted.

From the Journal of van Helmont

I have learned from this handicraft operation that all vegetables do immediately and materially proceed out of the element of water only. For I took an earthen vessel in which I put ninety kilograms of earth that had been dried in a furnace, which I moystened with rainwater, and I implanted therein the trunk, or stem, of a Willow tree, weighing two kilograms and about two hundred and fifty grams. At length, five years being finished, the tree sprung from thence did weigh seventy-six kilograms. But I moystened the earth vessel with rainwater, or distilled water (always when there was need) . . . and lest the dust should be co-mingled with the earth, I covered the lip or mouth of the vessel with an iron plate covered with tin, and easily passable with many holes. I computed not the leaves that fell off, in the four autumnes. At length, I again dried the earth of the vessel, and there were found the same ninety kilograms, wanting a few grams. Therefore seventy-four kilograms of wood, bark, and roots arose out of water only.

Here is a summary of van Helmont's experiment. Fill in the missing data in your Journal.

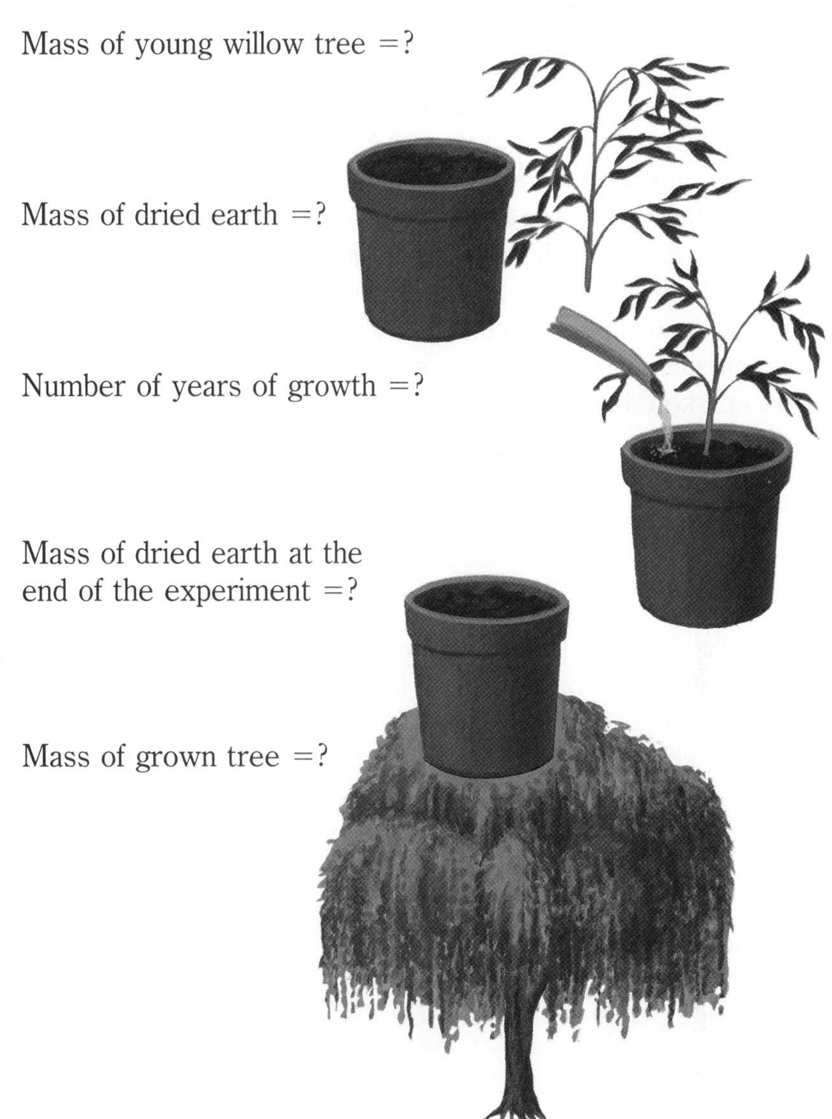

Van Helmont's experiment indicated that water was an important raw material for plants. Somehow, plants used water to build new tissues and structures. It was left to others, however, to find out exactly how the conversion of water into plant matter takes place.

Van Helmont was a careful experimenter. Can you give some examples that show this? Even though he was careful, van Helmont made a major mistake when forming his conclusion. What was it?

In the next lesson you will perform an investigation involving materials about which van Helmont knew nothing. You may also discover that more than just water and light are necessary to make starch.

Gases and Photosynthesis

Scientists know that starch is composed of three elements: carbon, hydrogen, and oxygen. You cannot see hydrogen or oxygen (they are colorless gases), but you can see carbon. If you cook carrots too long, they eventually turn black. What happens if bread gets stuck in a toaster? It, too, turns black. In fact, any food that is cooked too long will turn black because food is made from living things and all living things contain carbon.

Where do plants get their carbon? The answer is: from the air. As you will learn in a later unit, air is made up of different gases. One of these gases is carbon dioxide, which is produced whenever carbon-containing substances are burned. A burning candle produces carbon dioxide, for example. You can test for the presence of carbon dioxide with the chemical *limewater,* which consists of calcium hydroxide dissolved in water. Limewater turns milky when it comes in contact with carbon dioxide.

Testing for Carbon Dioxide

Try the following activity. Take two beakers and put a candle in each one. Light only one candle. Then cover both containers as shown.

Pour limewater into both beakers after the candle flame goes out. Remove the candles. Gently swirl the liquid, keeping the covers on the beakers. What happens to the limewater? Why is a beaker with an unlit candle also used?

EXPLORATION 3

Do Green Plants Use Carbon Dioxide to Make Food?

You Will Need

- 2 beakers or containers with plastic covers
- sodium hydroxide solution (10%)
- baking soda (sodium bicarbonate) solution (5%)
- 2 test tubes
- 2 leaves
- a lamp with 100-W bulb
- materials and equipment to test leaves for starch

What to Do

Arrange the two beakers as shown in the illustration below. Beaker 1 contains a 10% solution of sodium hydroxide, which absorbs carbon dioxide, thus removing it from the air. **Caution:** *Handle sodium hydroxide solution carefully—it can burn your skin!*

The second beaker holds a 5% baking soda solution. To make a 5% solution, add 1 part baking soda to 19 parts water. Baking soda contains carbon; its chemical name is *sodium bicarbonate.* If all of the carbon dioxide in the air is used up by the plant, the baking soda releases more carbon dioxide into the air.

Put the containers in strong light for three to five days. Then test each leaf for starch. During the three to five days, how has the air in Beaker 1 become different from that in Beaker 2? Suggest a way to use limewater to find out whether carbon dioxide is present in either container.

So what do you conclude about the role of carbon dioxide in the production of starch by plants?

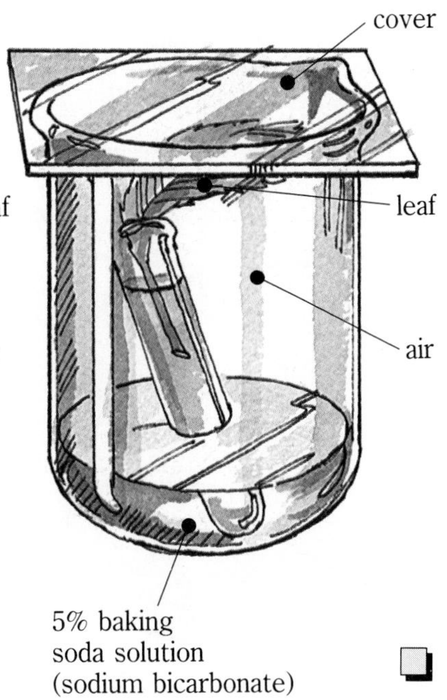

Gases Made by Greenery

You are now aware that when plants make starch they need light energy plus raw materials: water and carbon dioxide gas. Do plants give off anything during the food-making process? The Exploration that follows will help answer this question.

Guess Which Gas!

You Will Need

- a lamp with 100-W bulb
- 2 test tubes
- a wooden splint
- baking soda
- 2 beakers
- 2 funnels
- an elodea sprig
- water
- matches

What to Do

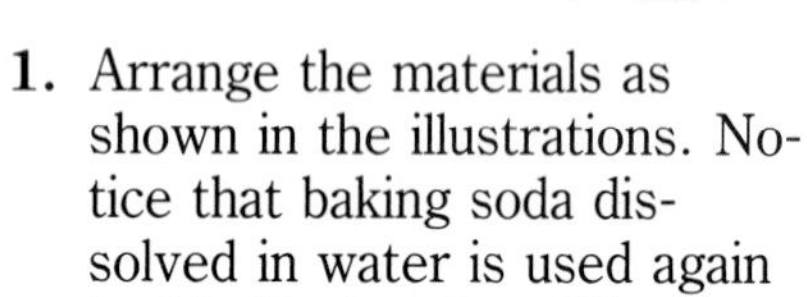

1. Arrange the materials as shown in the illustrations. Notice that baking soda dissolved in water is used again in this Exploration. What purpose does this serve?

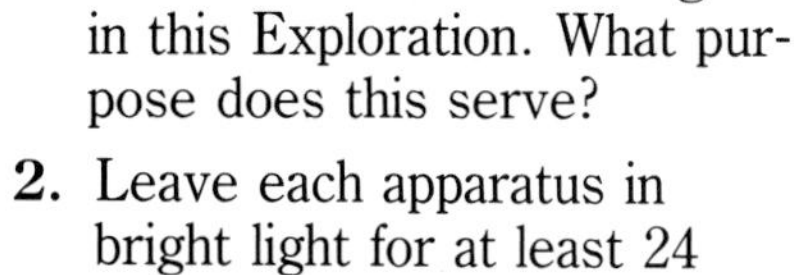

2. Leave each apparatus in bright light for at least 24 hours. What happens?

(1)

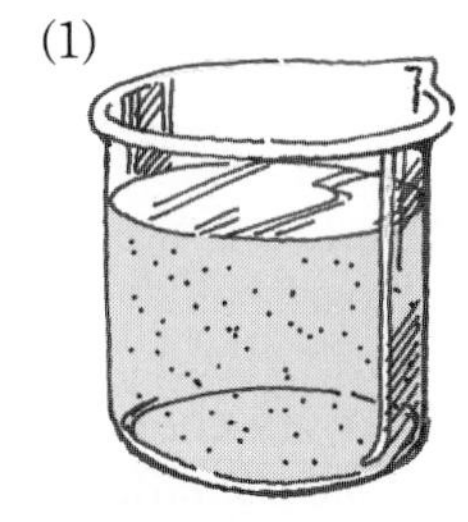

baking soda solution

(2)

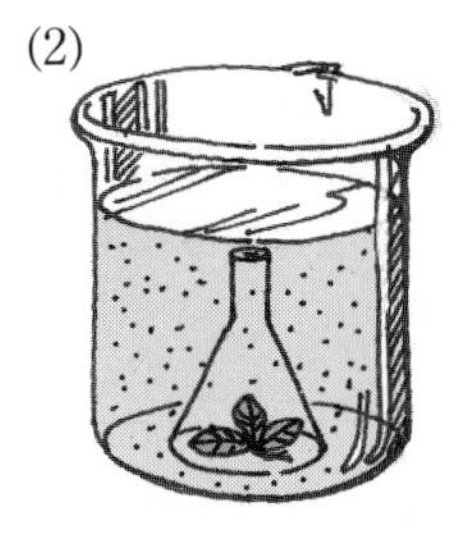

Place plant and funnel in beaker.

(3)

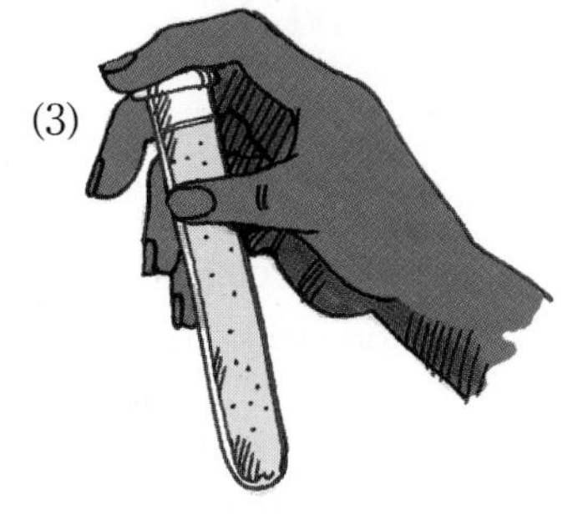

test tube filled with baking soda solution

(4)

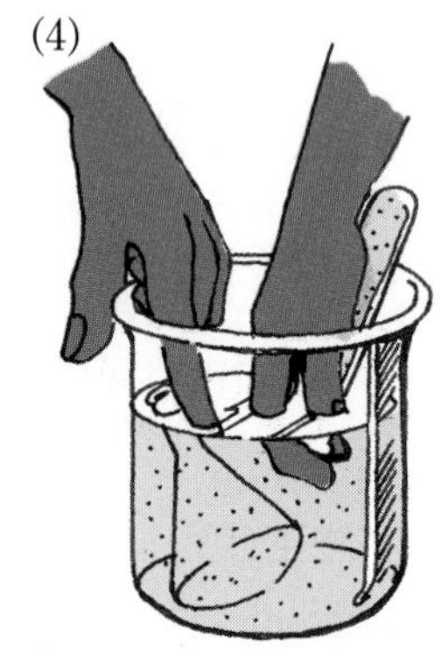

Invert the test tube while holding your finger over its mouth.

(5)

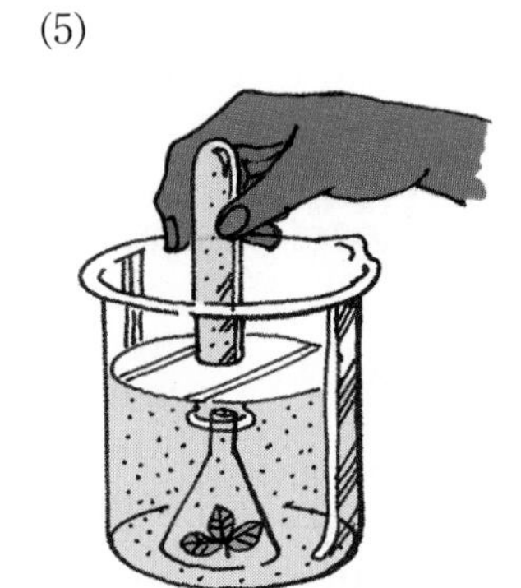

Lower the filled test tube over the funnel.

(6)

Let sit for at least 24 hours.

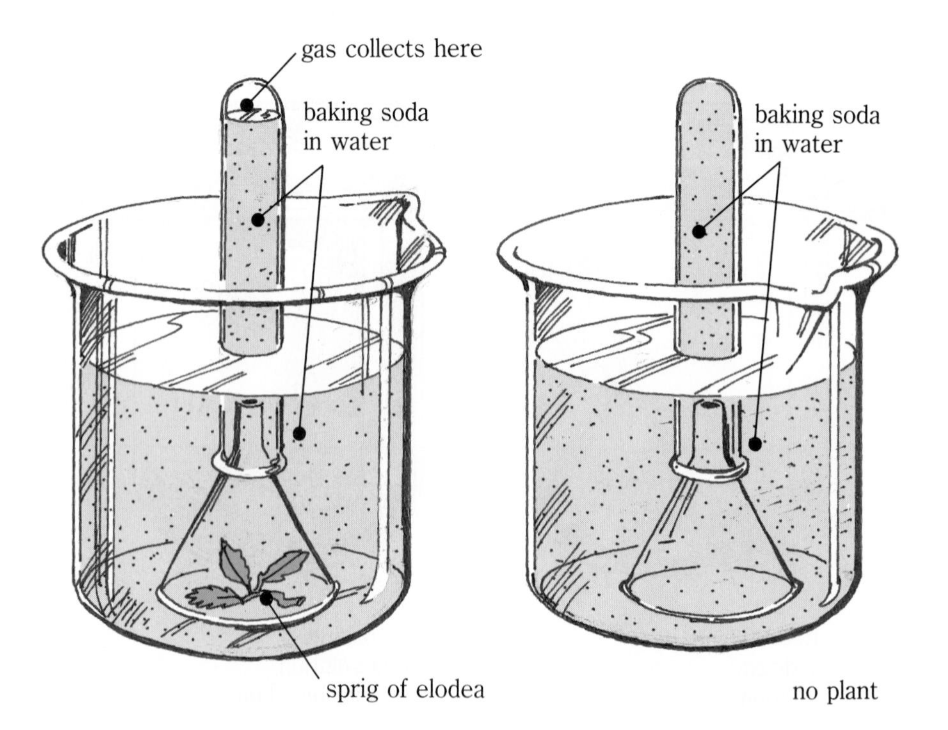

3. Remove the test tube from the beaker, making sure to hold your finger over the end of the tube so the gas does not escape.
4. Insert a glowing (not burning) wooden splint into each of the test tubes that collected gas. If the gas is carbon dioxide, the splint will go out immediately. If the gas is oxygen, the splint will burst into flames.

Testing for Gas

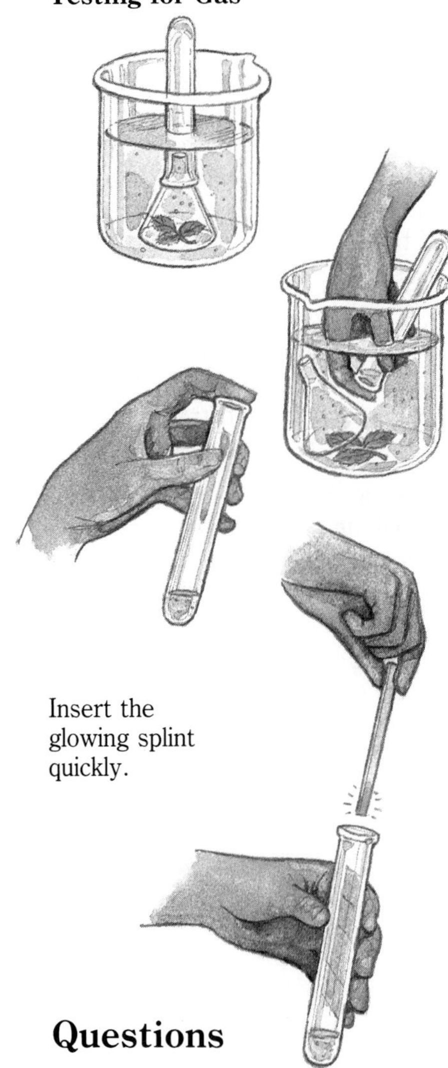

Insert the glowing splint quickly.

Questions

1. What discoveries did you make?
2. Why is a setup without a plant sprig also used?

Putting It All Together

Photosynthesis can be likened to the processes that occur in a factory: raw materials come in, energy is used, a finished product results, and wastes are produced. Now, check what you have learned about leaves as "food factories."

1. What important product results from photosynthesis?
2. What is one essential raw material needed for photosynthesis?
3. What other raw material is required?
4. What kind of energy is required?
5. What is a waste product of photosynthesis?

Now, write a description of photosynthesis based on what you have learned so far in this unit. Use the following terms in your description: *light, oxygen, waste, starch, water, raw material, carbon dioxide, food, manufacture,* and *product.*

Your description of photosynthesis could turn out to be quite long. A much shorter description, which still conveys a great deal of information, is the following simple equation:

carbon dioxide + water ⟶ starch + oxygen
(?) (?) light (?) (?)
(?)

Copy this equation into your Journal. Then replace each question mark with the number of the question above that relates to it. What does the arrow (→) mean?

A Final Word on Photosynthesis

You have learned that the food which results from photosynthesis is starch. However, when it is first made, the food in the leaf is actually in a form that is chemically less complex than starch—that of a simple sugar. Many sugar particles join together in a specific way to form the more complex starch particle. The sugar is changed into starch when the plant stores food for future use. When the plant needs food, it changes some of its stored starch back into sugar. If this is so, then why didn't you simply test for sugar? Mostly because it is easier to test for starch than for sugar, and the presence of starch is still evidence that food is being manufactured by the leaf.

You can therefore rewrite the photosynthesis word equation more accurately as follows:

carbon dioxide + water ⟶ simple sugar + oxygen
↑
light energy

3 Leaves: Food Factories

A well-designed structure is a thing of beauty. Architects and engineers strive to design buildings, bridges, machines, and other structures that are elegant at the same time that they are efficient and functional. Structures are also found in nature. Take leaves, for example. How do they rate in terms of their efficiency? their functionality? their "ingenuity" of design?

Over millions of years, plants have adapted to many different environments. In the following Exploration, you will examine some leaves up close to see some of the different features they have developed—features that help them produce food and grow.

EXPLORATION 5

A Good Look at Leaves

Activity 1

An Outside View

You Will Need

- a magnifying glass or binocular microscope
- a variety of fresh leaves

What to Do

Choose a leaf. Look carefully at both its upper and lower surfaces. Which surface has more features? Look for patterns of veins, hairlike projections, and other structures. Compare the leaf to leaves from several other plants.

Choose one type of leaf and write descriptions of both of its surfaces. Let a classmate read your description and try to identify the leaf you described.

To get a closer look at a leaf, you will need to use a microscope. You will also need to prepare a slide of leaf tissue thin enough for light to pass through. This you will do in the next Activity.

Activity 2

A Close-up View

You Will Need

- a microscope
- a lettuce leaf
- a single-edged razor blade
- a microscope slide and cover-slip
- a clear plastic ruler
- an eyedropper
- water

What to Do 

You are going to make a wet-mount slide of the cells of the lower surface of a lettuce leaf. To do this, use a single-edged razor blade to gently scrape the upper part of the leaf away so that only the thin lower layer remains. Or you might be able to peel a small section of the lower surface from the leaf with the blade. Mount the thin leaf section on a slide, as shown in the numbered illustrations.

Look for the kinds of cells shown in the diagram below. Do you see any slit-like openings?

1

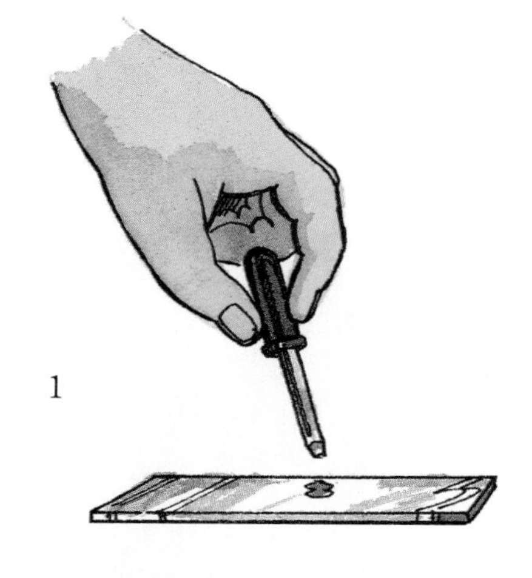

2

3

These openings are called **stomata** (singular: stoma). Study them closely. Then draw and label one stoma and the cells around it. You should see two kinds of cells: guard cells and regular epidermal (surface) cells. For comparison, you may want to look at the cells of the upper surface as well.

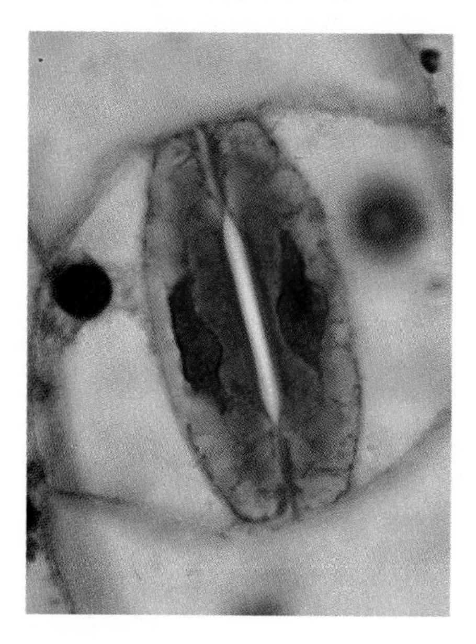

A stoma (with guard cells stained purple)

Interpreting Your Observations

Working with a classmate, find the answers to the following questions.

1. On which leaf surface (top or bottom) did you observe more stomata?
2. Estimate the number of stomata on one square centimeter of the lower leaf surface. (*Hint:* How much magnification are you getting with the microscope? What do you estimate to be the area of the field of view in the microscope?)
3. What purpose do you think the stomata might serve?
4. Can you devise an explanation for how the guard cells operate to open and close the stomata?
5. Do you think the stomata close at night? Why or why not?
6. What do you predict would happen if grease was smeared on the lower surface of the leaves of a healthy plant and left there for several days?

In case you still have questions about the purpose the stomata serve, their function will be explained further, beginning on the following page.

Stomata Data

As you saw in Exploration 5, the stomata are located mainly on the underside of a leaf. Stomata allow water vapor and other gases to pass into and out of the leaf. The thickness of this page is about 10 times the width of a stoma. Each stoma is flanked by a pair of guard cells, which control the size of the opening. When the guard cells absorb water and swell, the stoma is opened. The guard cells relax when water leaves them—this closes the stoma. Usually stomata open during the day and close at night. Considering what you know about photosynthesis, why does this make sense?

An Inside View

The drawing shows a cross section of a typical leaf.

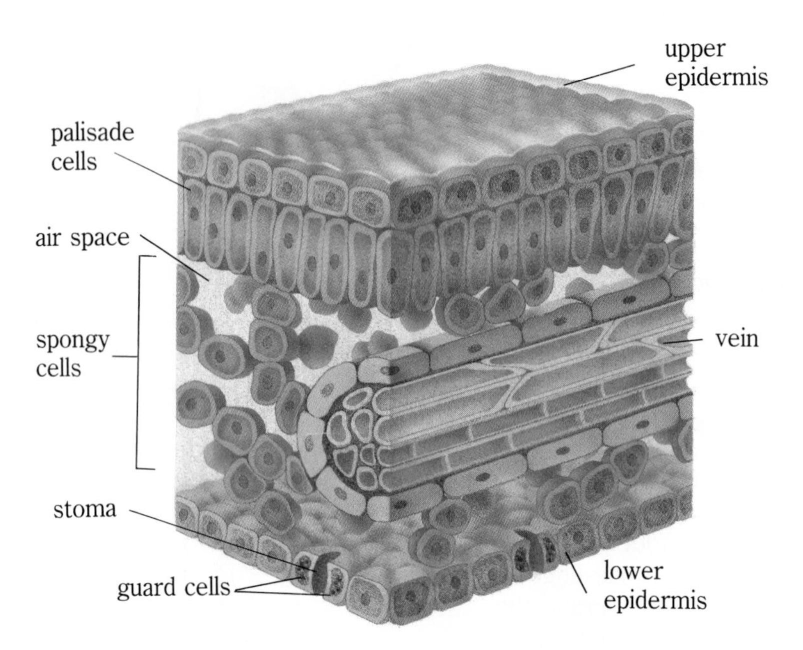

Suppose you could shrink to the size of a water or carbon dioxide particle. You could journey into a leaf and make some interesting discoveries about this "food factory." First, study the diagram of the leaf above, then answer these questions.

1. What are two ways you could enter the factory?
2. Where would you wait, along with large numbers of other gas particles, until you could be used?
3. Where would you find a pipeline system to carry liquids such as water?
4. Where would you find huge, bulging "doors"?
5. Where would you leave the factory?
6. Can you find the elongated cells near the top of the leaf? Would these cells be well-lit? Might these be good places to make food?
7. Food is also made in the closely-packed, somewhat rounded cells farther inside the leaf. Would these be as well-lit as the elongated cells? Could as much food be made there?

A Microbiology Activity

You Will Need

- a leaf
- 2 microscope slides and cover-slips
- a single-edged razor blade
- a microscope
- an eyedropper
- water
- unlined paper

What to Do 

In this activity, you will use a thin cross section of a leaf to make a wet mount for study under a microscope.

1. Hold a leaf between two glass slides so that a very thin edge of the leaf protrudes. Slice this edge off carefully, using a single-edged razor blade. Move the leaf up by the smallest amount and shave off another slice. With practice, you will be able to make a section thin enough for light to pass through.

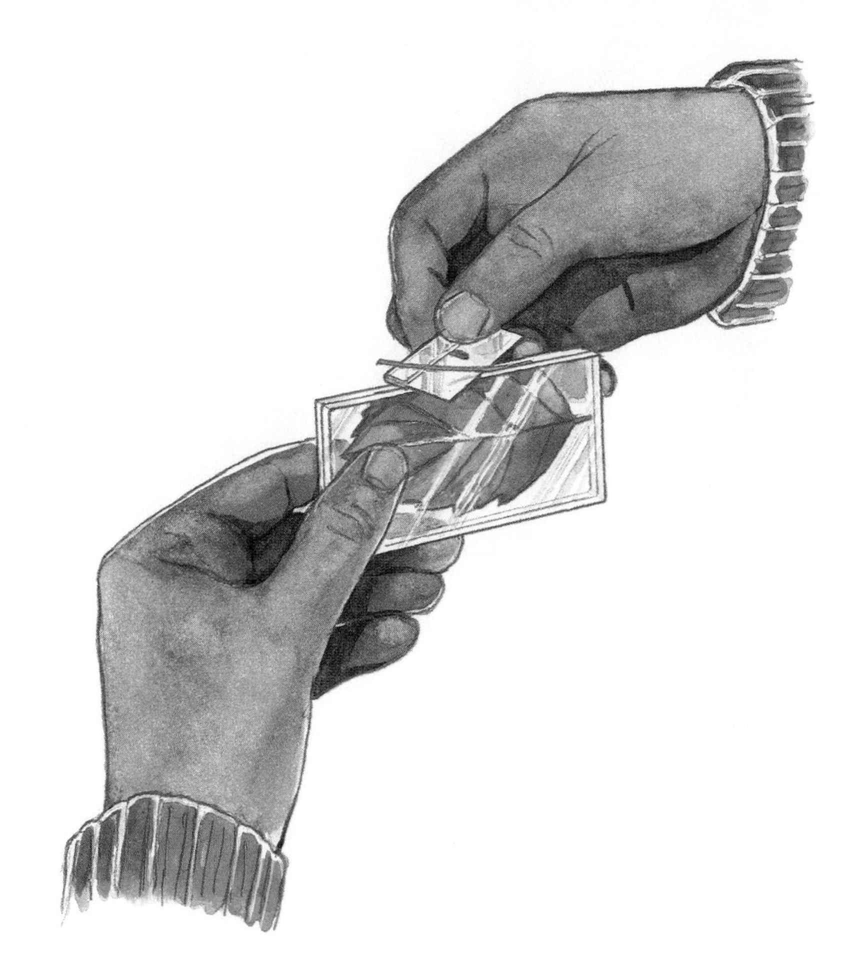

2. Use the cross-sectional view of a leaf on page 18 to help you identify the kinds of cells present. Then draw a detailed diagram of each type of cell on unlined paper. Label the various kinds of cells.
3. Consider the cell types in the leaf. Each is specialized; that is, each type has its own particular function.
   (a) Which cells protect the leaf, preventing the inside from drying up?
   (b) In which cells is food manufactured?
   (c) Which of the cells mentioned in (b) can also function to give the leaf strength as a result of the way they are arranged?
   (d) Which structure looks like it might be a transportation system for water and manufactured food?
   (e) Which cells control the flow of gases into and out of the leaf?

4 Why Green?

You have been investigating the process by which plants produce food. You now know about the raw materials plants use, their source of energy, and the products of photosynthesis. You have no doubt noticed that most living plants are green. Obviously this color must be important, but why?

First of all, this green color is due to a substance called *chlorophyll*. When you tested for starch in Exploration 1, you dissolved chlorophyll and removed it from the leaf so that you could see the results of the test for starch. Now you will see whether chlorophyll is necessary for photosynthesis to occur.

EXPLORATION 7

Chlorophyll and Photosynthesis

To find out whether chlorophyll is required for photosynthesis, you need leaves, such as those from a coleus, that have some white areas (indicating a lack of chlorophyll).

You Will Need

- materials and equipment to test leaves for starch

What to Do   

Give the plant a good chance to make food (you know the conditions), then remove a couple of leaves. Test for starch as before. (Remember to heat the methanol safely!)

Make a brief report in your Journal, including a diagram showing the areas with and without starch.

A Crash Course in Colors

Light from the sun seems to have no particular color. But in fact sunlight is a mixture of many different colors blended together. We can see the colors that make up sunlight by passing it through a prism, or by looking at a rainbow. If sunlight is white, then why do the things around us appear to have color?

Suppose that a substance absorbed all of the colors of sunlight except red, which it reflected. What color would you perceive the substance to be? Red, of course. Substances appear to have color because they absorb some colors and reflect others.

Pigments are substances that absorb certain colors but reflect the rest. In this way pigments give substances their color. Chlorophyll is a pigment. What color do you think chlorophyll reflects?

As you know, sunlight has energy. Therefore, in absorbing sunlight, chlorophyll absorbs energy. How do you think plants put this energy to work?

The process of photosynthesis puts the energy absorbed by chlorophyll to work, chemically combining carbon dioxide and water to form sugars. Further reactions change these sugars into starches. Now that you know this, you can answer the following questions.

Questions

1. State any evidence you have observed that suggests that chlorophyll acts to capture energy for the process of photosynthesis.
2. Write a word equation for photosynthesis, placing chlorophyll in the equation.
3. Study the detailed diagram of the cross section of a leaf on page 18, and find the food-making bodies that contain chlorophyll (or look at a leaf under a microscope). What are they called?
4. In which leaf cells is chlorophyll found?
5. Which type of leaf cell would you expect to be the chief food-producer? Support your inference with at least two observations you have made.

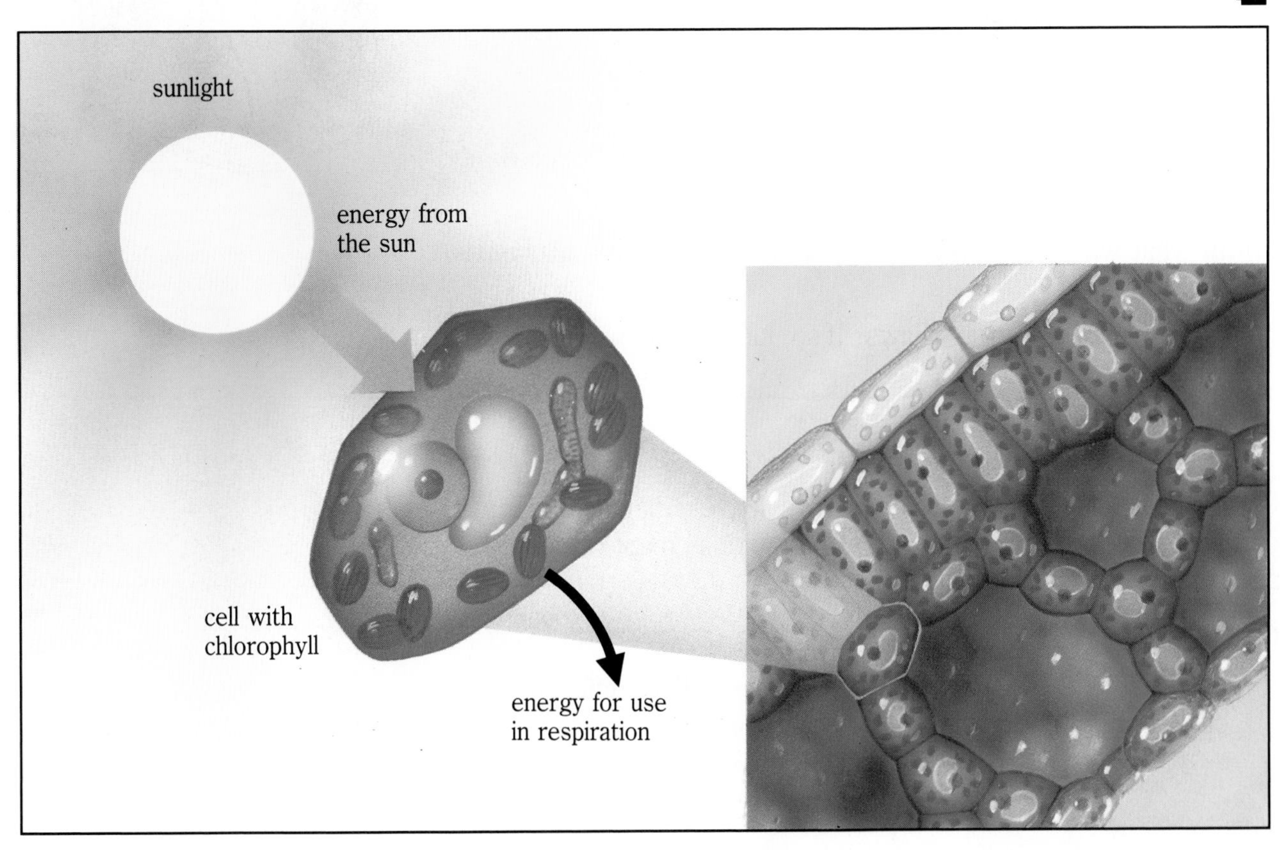

Problems to Discuss

Form small groups and discuss any two of the following topics. You may need to do some research first. Then share your conclusions with the whole class.

Topic 1

A tree releases water through its stomata. How, and from where, did the tree get this water? How do you think the water reaches the leaves?

Topic 2

(a) Why are leaves often broader at the bottom of a tree?

(b) How do plants without green leaves (such as the purple velvet vine) make food?

(c) How does a mushroom, which does not have chlorophyll, manage to obtain food? (Although it resembles a plant in some ways, a mushroom is a fungus.)

(d) How do the leaves of evergreen trees survive the winter?

Topic 3

(a) How do plants in closed terrariums manage to get raw materials for photosynthesis?

(b) How can plants get enough light to grow in deep water?

Topic 4

Discuss the design of a leaf, taking into consideration the following questions:

(a) In what ways is the leaf well-designed for its particular environment?

(b) Can you find any design flaws? If so, identify them and suggest improvements.

Topic 5

Coniferous trees, such as pines, have leaves that are radically different in design from those of plants such as maple trees, sycamore trees, or banana plants.

(a) Identify some of the differences.

(b) How might these structural differences reflect differences in each plant's mode of life?

Challenge Your Thinking

1. You have been asked to help settle the issue raised in the cartoon below. What is the underlying question? How would you respond? How would you go about providing evidence to support your response?

2. The diagram to the right shows the *absorption spectrum* (pattern of light absorption) for a typical plant. Light that is not absorbed is reflected. Study the diagram and then answer the following questions. Remember that white light (such as that from the sun) is a mixture of all the different colors of light.

   - What causes plants to appear green?
   - If a plant received only green light, how might it grow compared to a plant grown in normal light?
   - Would a plant grown in red light do well? blue light?

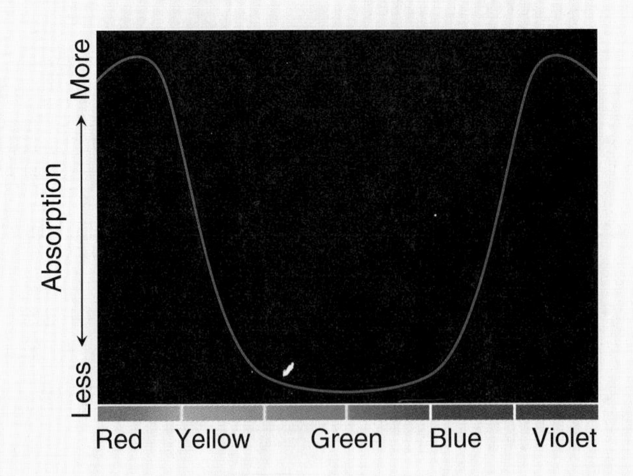

3. A friend of yours has a theory. She believes that the starch in plants comes either from the soil they grow in or the water they use. "Plants won't grow where the soil or water isn't starchy," she says. How might you test her theory?

A World of Water

If you looked at the Earth from space, what would strike you about it more than anything else—its smallness in the vast emptiness of space? the lack of national boundaries? How about the colors? Blues and whites totally dominate the view. What is the significance of these colors? These colors signal the presence of liquid water. Look around you. Water is everywhere. One of the remarkable things about our planet is that conditions are just right for liquid water to exist. If our planet were a little colder or a little warmer, or had a weaker gravitational field, this would not be the case. Without an abundant supply of liquid water, life as we know it could not exist.

Water Facts

What is this stuff we call water, and why is it so important? Here are just a few facts about this remarkable chemical.

- About 70 percent of the Earth's surface is covered by water; that's a total of 1.4 billion cubic kilometers. Three percent of this water is fresh (and 75 percent of that is ice!)
- Most living things are about two-thirds water. A chicken is about 75 percent water, an earthworm is about 80 percent water, and a tomato is about 95 percent water. You are about 65 percent water.
- Water can dissolve more substances than any other substance. It is the closest thing there is to a universal solvent.
- Water has a higher *heat capacity* (absorbs more heat without increasing in temperature) than almost any other substance.
- Water is the only substance on Earth that is naturally present in solid, liquid, and gaseous states.
- About 265,000 cubic kilometers of rain or snow fall every year (enough to cover Arizona to a depth of 1 km).
- A person will drink about 60,000 L of water in an average lifetime.

A few of the many things that are made mostly of water

Points to Ponder

1. What functions do you think water has in living things?
2. Why is it important to living things that water can dissolve so many different substances?
3. Why is it important to living things that water has such a high heat capacity?
4. Where did all of the water come from? Can it ever be used up?
5. What are some ways in which life on Earth depends on water?

6 Transpiration—One Process Involving Water

So far in this unit, you have been looking at various aspects of the food-making process in leaves—photosynthesis. There are, however, other leaf processes besides photosynthesis, as the following demonstration will show.

Transpiration: A Demonstration

Use a geranium or some other hardy plant. Give the plant some water and then cover it with a plastic bag or glass jar. If you use a bag, seal it carefully around the stem. Observe the setup over a period of 24 hours. Record any changes you see. What do you think causes the change(s)?

The process you observed in the previous demonstration, in which water evaporates from the plant's leaves, is called **transpiration.** Transpiration occurs continually. From which part (type of cell) do you think the water vapor exits a leaf?

You might be surprised to learn that a mature tree gives off as much as 15 L of water daily through its stomata. Imagine how much water an entire forest gives off! The Brazilian rainforest transpires so much water that rain clouds form almost every day.

Transpiration taking place on a large scale

7 Spreading It Around

Where does a plant obtain the water it uses? As you may already know, this water enters the plant through its roots and travels upward. The water contains dissolved minerals that combine with sugar to make other kinds of materials used by the plant. The plant uses these materials to grow and maintain itself.

A seedling needs a generous supply of water. Its *root hairs* ensure that this need is met. Just look at the many hairs on a sprouting radish seed. You can sprout radish seeds yourself between two damp pieces of paper towel. Keep the paper towels slightly moist at all times. Once the seeds have sprouted, estimate the number of root hairs on a single seedling. But don't just guess. Use a systematic method.

Now look at the close-up drawing of a section of a root. Note that each root hair is a projection of one of the cells that makes up the outside of the root tip.

Root hairs grow among the tiny particles of soil. The spaces between soil particles are often filled with moisture. This is where plants get most of their water.

How do the root hairs work to provide a plant with its water and mineral requirements? The answer is, through the process of **diffusion.** Under normal conditions, water *diffuses* inward from the soil, through the walls of the root hair, and into the other cells. Now you know *what* happens. However, you still do not know *how* it happens. What causes water and minerals to enter root hairs? To understand this you will need to look more closely at the process of diffusion.

Defining Diffusion

Here are some other situations in which diffusion occurs. Thinking through what happens in each will help you find a definition for diffusion.

(a) A sugar cube is left in a beaker of water for a while. What happens?

(b) Several drops of food coloring are placed into a beaker of water at room temperature. What happens? The experiment is repeated using very hot water. What do you predict will happen?

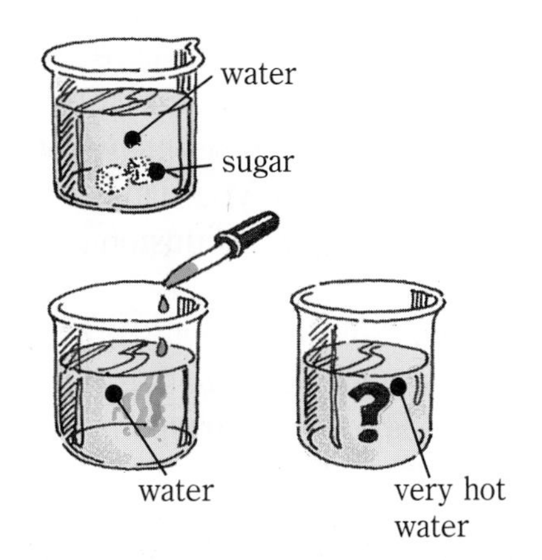

(c) Invisible fumes of ammonia gas rise from a container full of concentrated ammonium hydroxide with the stopper removed. A person stands about 3 m away. How long is it before she smells the ammonia gas?

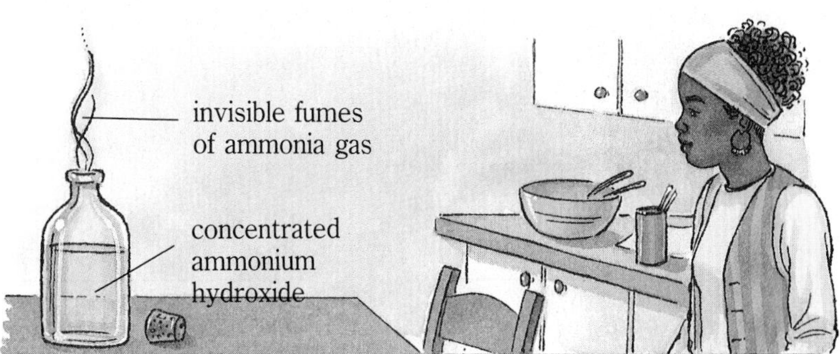

How is diffusion at work here?

In situation (a) the sugar diffuses throughout the water and seems to disappear, yet you can taste the sugar in every part of the water. In situation (b) the red food coloring diffuses throughout the water until the mixture is the same color throughout. In situation (c) the ammonia gas diffuses through the air, eventually reaching your nose. In a short while it can be smelled in every part of the room. Now can you supply a definition for *diffusion?* Can you suggest other examples of diffusion?

Additional information that might help you define diffusion more precisely comes later in this lesson.

A Theory of Diffusion

As you probably already know, scientists have found that all matter is composed of unimaginably tiny particles. (You will take this up in more detail later in this book.) These particles are in constant motion. The hotter the temperature, the faster the particles move. These particles are far too small to be seen, therefore you cannot directly observe their motion. But you can see indirect evidence of their motion. Do you know how? Diffusion is evidence of this particle motion.

A diagram may help you understand the connection between particles and diffusion. Here is a drawing of situation (a) from the previous page. This is only a model of what happens. You can't really see the sugar and water particles.

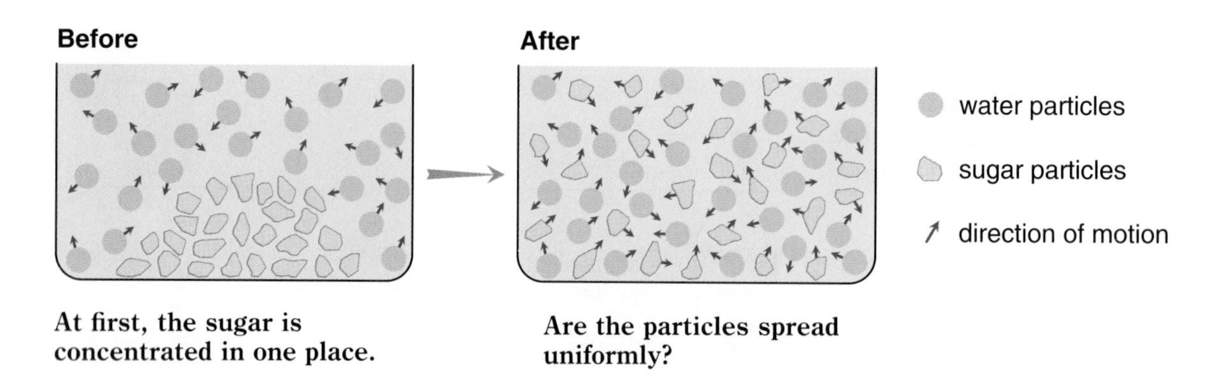

At first, the sugar is concentrated in one place.

Are the particles spread uniformly?

Refining the Definition of Diffusion

Once you have seen the connection between particles and diffusion in situation (a), you can work in small groups to try to explain how diffusion occurs in each of the other two situations. Make use of the ideas suggested by the theory of particles. After you have finished, flex your creative muscles a little bit with the following exercise: Write a short account of the experiences of a particle undergoing diffusion.

A Law of Diffusion

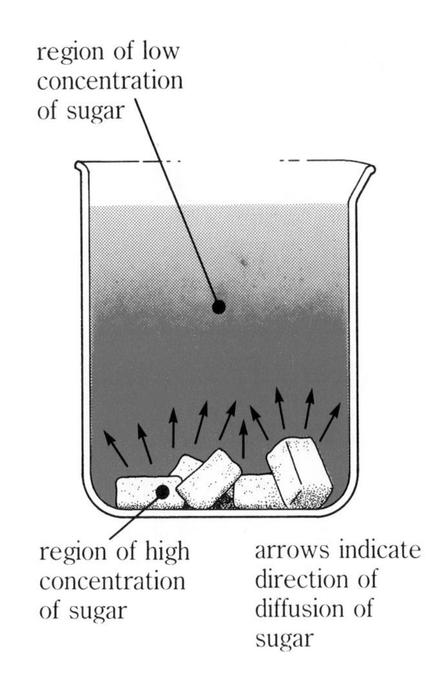

The most important law governing the process of diffusion states:

The particles of a substance always diffuse from a region of high concentration to a region of low concentration.

Do you understand why the diffusion process results in solvent and solute becoming uniformly mixed? This is because diffusion continues until the particles of the substances involved are completely and randomly scattered.

Turn back to the situations illustrated earlier. In your Journal, identify for situation (b) the areas of high and low concentration of red food coloring and its direction of diffusion. For situation (c), identify the regions of high and low concentrations of ammonia gas, and its direction of diffusion.

A Further Refinement

Here is a quick summary of what you have learned so far about diffusion:

- One substance intermingles with another.
- The intermingling seems to be uniform.
- The two substances that intermingle do not need to be mixed mechanically; mixing seems to occur naturally.

Using these observations, we can formulate a good working definition of the process of diffusion:

Diffusion is the uniform intermingling of the particles of one substance with those of another substance. This occurs naturally because of the motion of both types of particles.

Compare this definition to your own. Do you need to refine yours in any way?

The Root of the Matter

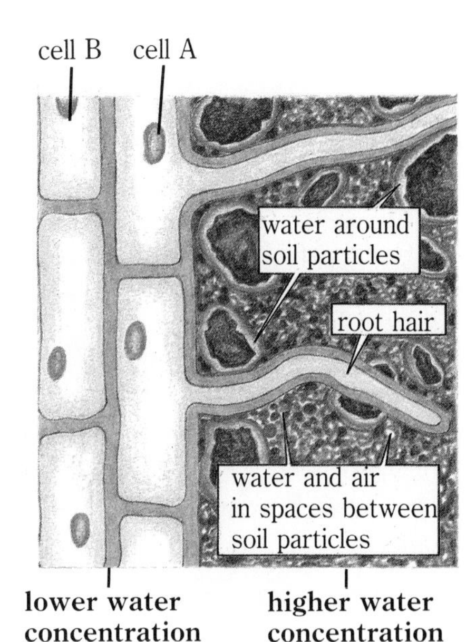

Look at the illustration of a plant's root hair. Where is an area of high concentration of water particles? low concentration? In what direction will diffusion take place?

If water is to diffuse from the soil into the root-hair cells, the water must pass through both the cell wall and the cell membrane, which enclose the contents of each root-hair cell. Water does, in fact, pass easily through both. After this it must go from cell A to cell B, again through the membrane and wall of each cell. Diffusion through a membrane occurs in living things. In the Exploration that follows, you will study this special kind of diffusion.

Diffusion Through a Membrane

Here are two models of diffusion through root hairs. Experiment with one or both.

Activity 1

You Will Need

- a ring stand and clamp
- an egg • vinegar
- a wide-mouthed bottle
- a needle
- a thin glass tube (10 cm)
- candle wax or silicone sealant
- Petri dish • water

An egg is a single cell—a very large one. This cell, like all cells, is protected from its surroundings by membranes.

What to Do

1. Find a way to support the egg so that only the small end rests in the vinegar. The shell will dissolve in about 24 hours. With the shell gone, you can see the inner shell membrane.

dissolving the shell

2. Fill the bottle with water and place the egg (membrane end down) on the mouth of the bottle as shown in the drawing. Use the needle to poke a hole through the top of the egg large enough to insert the glass tube.

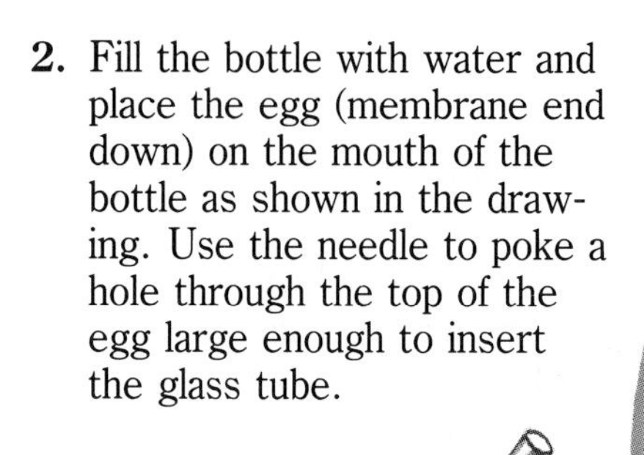
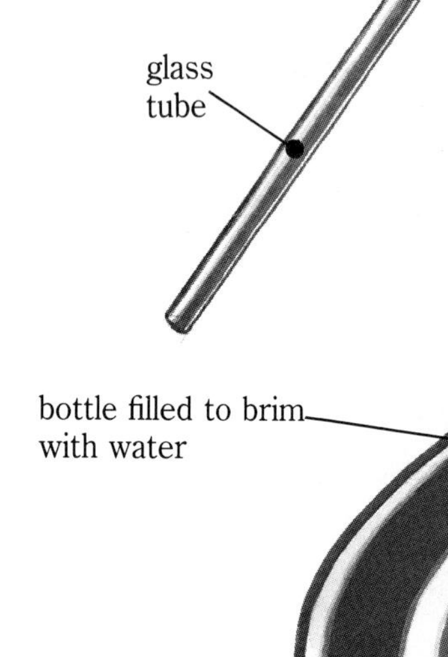

3. Insert the glass tube. It should just penetrate the membrane. Seal the glass tube in place with candle wax or silicone.

4. Leave the egg in place for several hours.
5. Record your observations.

Activity 2

You Will Need

- a ring stand and clamp
- a thin glass tube (20 cm)
- an animal membrane (such as those used in sausage casings)
- sugar solution (5%)
- a rubber band
- a large beaker or similar container

What to Do

Arrange the equipment as shown in the illustration, then leave it for several hours.

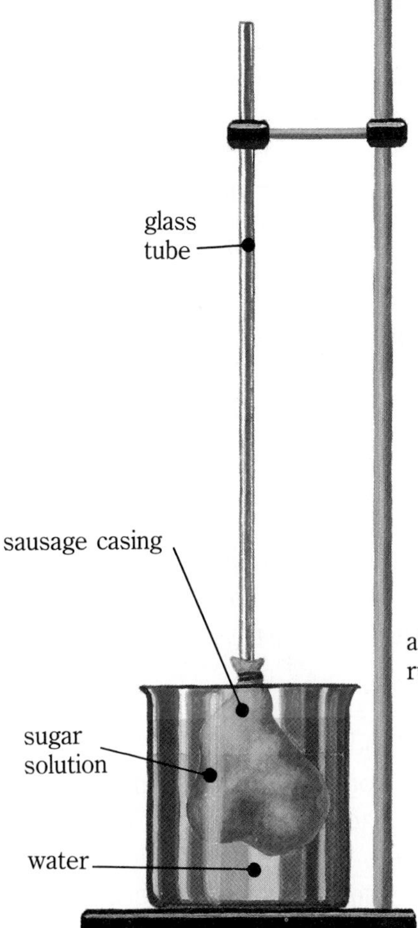

Questions

1. What evidence is there in Activities 1 and 2 that water has passed through the membranes?
2. For each of Activities 1 and 2, identify in your Journal the regions of high and low concentrations of water. Identify the direction of diffusion of the water.
3. What can you say about the concentration of the egg contents in Activity 1? the sugar in Activity 2? Did these substances also diffuse through the membranes? How could you find out?
4. Develop an explanation for the results of this Exploration.

A Model of Cell Membranes

Exploration 8 demonstrated that not all particles can pass through every type of cell membrane. To understand the selection process better, consider what happens when some everyday substances are used as shown below:

- Does water, salt, or sand pass through glass?
- Does water, salt, or sand pass through a fine-meshed screen?
- Does water, a solution of salt and water, or sand pass through filter paper?

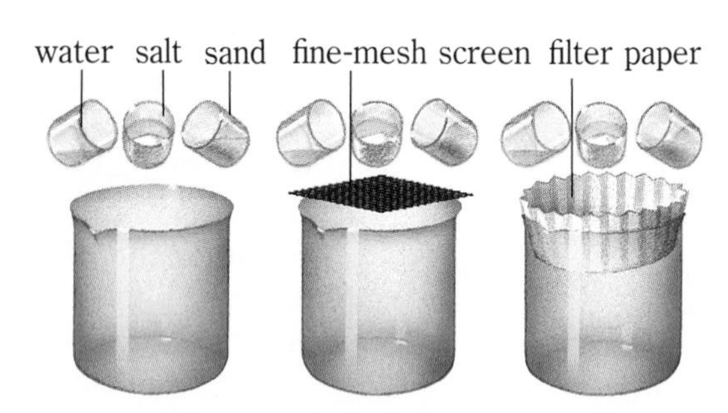

The word **permeable** means "open to passage," while **impermeable** means "closed to passage." Of the glass, screen, and filter paper, which is permeable to all three—water, salt, and sand? Which is impermeable to all three? Which is permeable to some substances but impermeable to others? What might you call a material with such a property?

The screen is a permeable material, while the glass is an impermeable material. The filter paper is a **semipermeable** material. In general, what do you think determines whether materials are permeable, semipermeable, or impermeable? (Think about what you have learned so far in this unit.)

Semipermeable membranes allow some particles to pass through them, depending on the size and shape of the particles. In general, semipermeable membranes prevent larger particles from passing through them.

Suppose, for a moment, that the egg and the animal membrane in Exploration 8 are accurate models for the way diffusion takes place in cells. What kind of membranes do cells have: permeable, semipermeable, or impermeable? What does this model suggest about the way a cell interacts with the surrounding environment?

The action of the cell membrane is similar to that of the filter paper, which allowed passage of salt and water particles, but prevented passage of the sand. In other words, the membranes allow some particles to pass through, while acting as a barrier to other particles. Cell membranes are, therefore, semipermeable.

The next Exploration will reveal more about semipermeable membranes.

EXPLORATION 9

Semipermeable Membranes

You Will Need

- two 14-cm to 16-cm lengths of dialysis tubing
- two 250-mL beakers
- 20 mL of cornstarch and water mixture (5%)
- 20 mL of 5% sugar solution (use corn syrup and water)
- iodine solution
- Benedict's solution
- 2 paper clips
- a hot plate
- a graduated cylinder
- water
- a test tube
- an eyedropper
- thread

What to Do

Arrange the materials as shown in the illustration on the facing page.

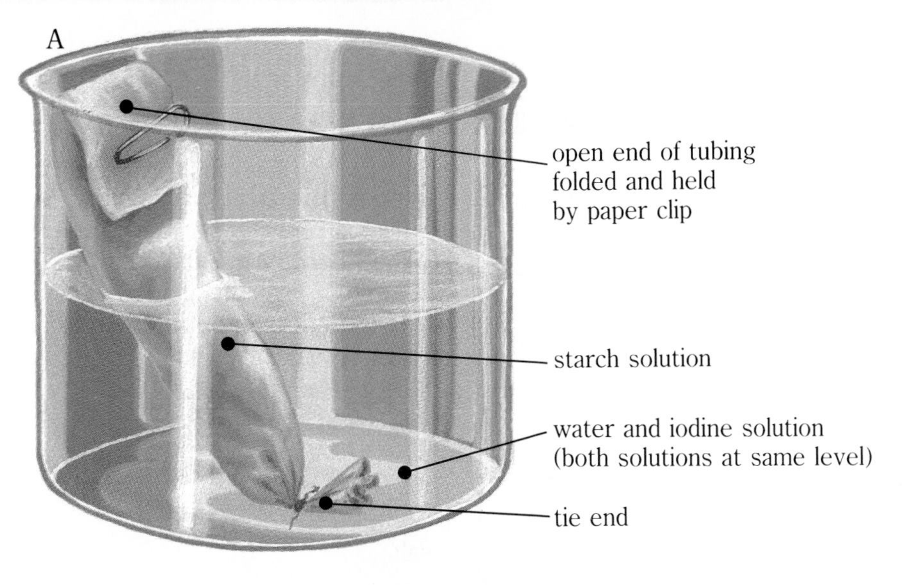

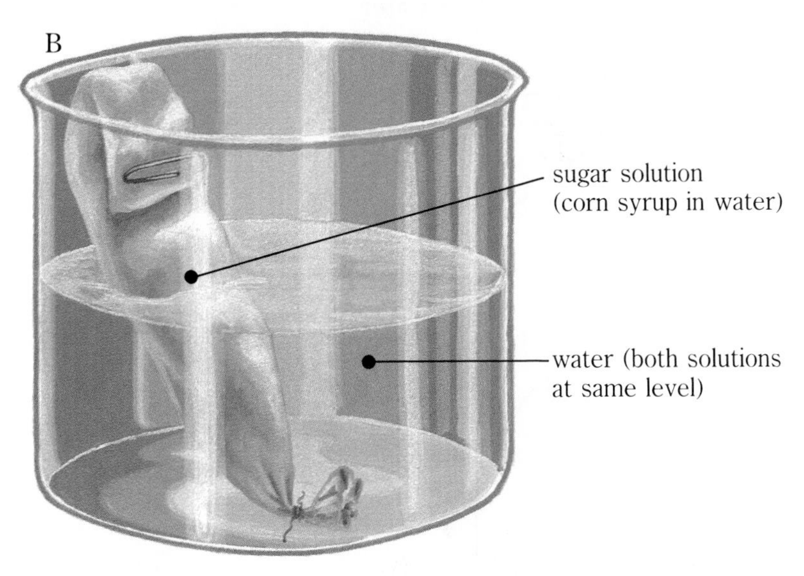

Check the liquids in the beakers after about 20 minutes and then again after one full day has passed. To determine whether starch particles have passed through the tubing in A and whether sugar particles have diffused in B, you will have to apply the chemical test for each of these substances to the liquid in the beakers. You already know how to test for starch; the test for sugar follows. Report your results in your Journal, then answer the questions at the end of the Exploration.

A Test for Sugar

Place 2 mL of the solution to be tested for sugar in a test tube. Add an equal amount of Benedict's solution. Put the test tube in boiling water for 5 minutes. Watch what happens.

When heated, Benedict's solution by itself is blue. Benedict's solution heated with a concentrated sugar solution is brick-red. Benedict's solution heated with weak sugar solution is greenish-yellow.

Questions

1. Is the dialysis tubing selective in what it allows to pass? Which substance(s) does it permit to pass through?
2. What conclusions would you draw about the relative size of starch particles and sugar particles?
3. What property do cell membranes and dialysis tubing have in common?

8 Osmosis: Controlled Diffusion

Diffusion of water (or another liquid) through membranes is called **osmosis** from the Greek word *osmos,* meaning "push." Osmosis takes place continuously in living things. The cells of all plants and animals are surrounded by membranes that allow smaller particles to pass through but stop larger ones. Even when two different kinds of particles are both small enough to pass through a cell membrane, the smaller ones are likely to get through more quickly. You can see this for yourself in Exploration 10.

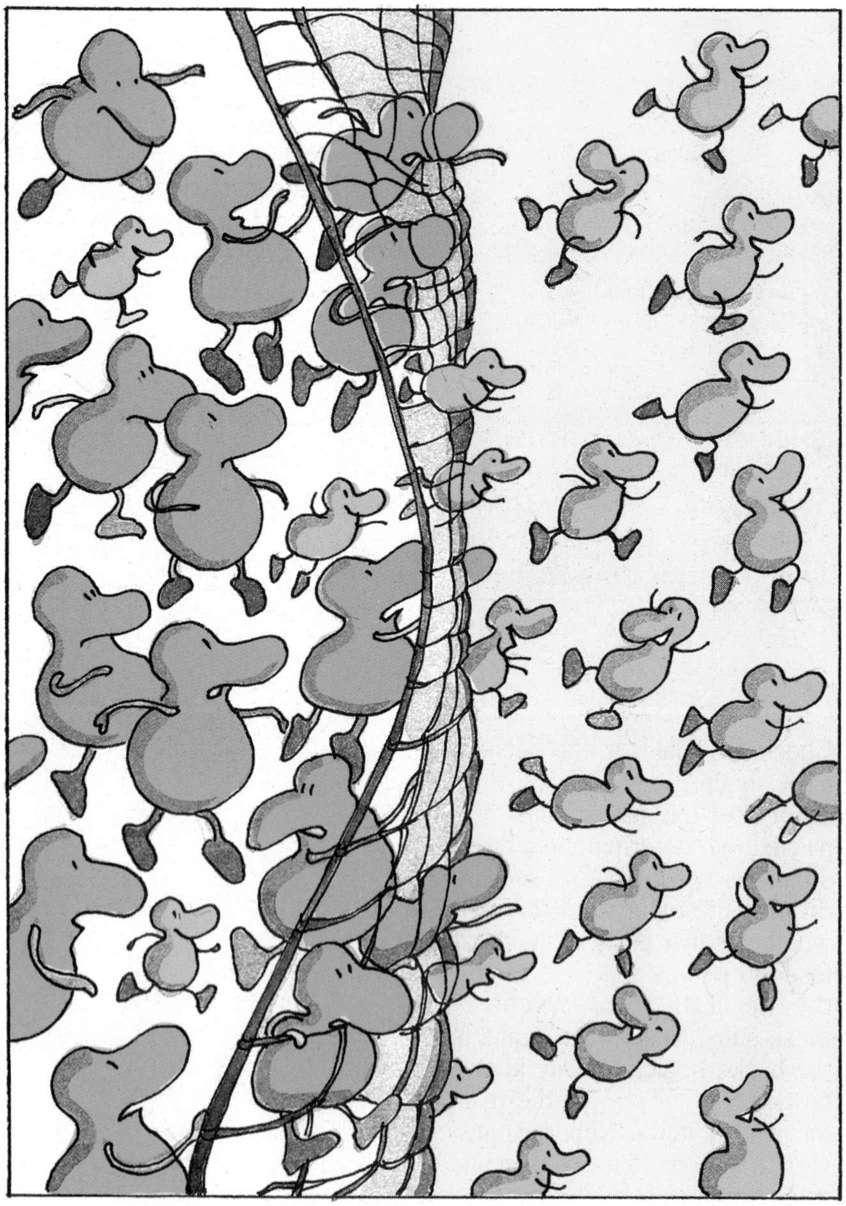

EXPLORATION 10

Observing Osmosis

Potatoes have a very high concentration of water. It is for this reason that they are used in this Exploration. You are going to study the effect of putting potato slices into different liquids.

You Will Need

- a potato
- 4 beakers
- salt
- water
- a potato peeler
- a knife

What to Do

Peel a potato and cut four slices of the same size. Place each slice into a beaker containing either air, tap water, dilute salt solution (20 g/L), or concentrated salt solution (150 g/L). After about 30 minutes, remove the slices and examine them. Record all observations in your Journal.

Follow Up

1. Describe the size and rigidity of each of the four slices before and after being placed in a beaker.
2. What seems to be the effect of tap water on a slice?
3. What effect does a strong salt concentrate have on a slice?
4. Compare the concentration of water in each potato slice and in each of the beakers. Remembering the rule for diffusion, state the direction of water flow through the cell membranes (*i.e.,* into or out of the cells). Does this explain your results?

5. Plant cells become less rigid (more spongy) when water leaves them. How does this principle explain what happens to plants when they are deprived of water (they wilt)?
6. Before you started, you made predictions about how the rigidity of each potato slice would change. How did your predictions compare with the results you observed? Did any slice seem to become more rigid? Which liquid was it in? Explain what happened to this slice. Which slice seems to have lost some water? a lot of water? How do you explain this?

In the beakers containing salt solutions, water particles move out of the cells and into the beaker because initially, the water concentration is greater inside the cells than outside. There is a kind of pressure forcing the water toward the outside until the solution outside has the same concentration as that inside. Diffusion actually always occurs in both directions, from the inside of the cell outward and from the outside of the cell inward. Diffusion occurs fastest, though, where the difference in concentration is greatest. So even though some water diffuses from a salty solution into a cell, it is more than offset by the water diffusing outward from the cell into the salt water. The reverse situation occurs when cells are placed in plain water.

Osmosis: How Fast?

Purpose: to obtain information about how fast osmosis can occur

You Will Need

- a carrot
- a coring knife
- a 20-cm long glass tube
- glycerine
- a towel
- molasses
- a one-holed rubber stopper
- candle wax or silicone sealant
- materials and equipment to test for sugar (see Exploration 9)
- a ring stand and clamp
- a beaker
- water
- graph paper

What to Do

1. With a coring knife, cut a hole in the carrot about 3/4 down its length. The hole should be just large enough so that a one-holed stopper will fit and close it tightly.
2. Lubricate the glass tube with glycerine and hold the tubing and rubber stopper with a towel. Gently insert the glass tube into the rubber stopper.
3. Fill the hole in the carrot with molasses.
4. Push the stopper and glass tube into the hole in the carrot. Seal any openings between the stopper and the carrot with candle wax or silicone sealant.
5. Mark the initial level of the molasses in the glass tube. This is the level at time zero.
6. Record the level of the molasses in the glass tube several times over the next three days, at the same time each day; *e.g.*, before school, during lunch hour, and after school.

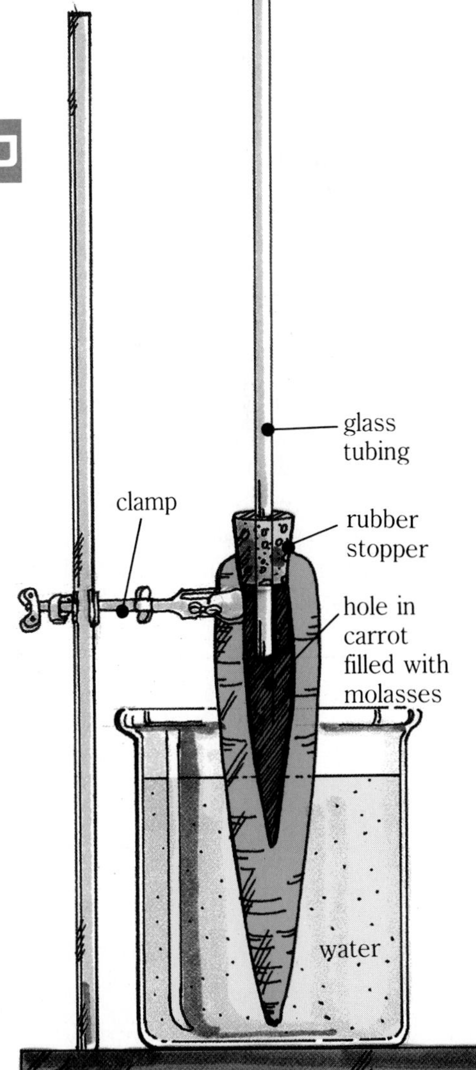

EXPLORATION 11—CONT.

Graph your findings. Interpret your results using the questions that follow.

Questions

1. What was the rate of rise of the molasses in millimeters per hour:
   (a) during day 1?
   (b) during day 2?
   (c) during day 3?
2. Use your graph to indicate the level of the molasses at the following points after time zero:
   (a) 12 hours
   (b) 18 hours
   (c) 30 hours
   (d) 60 hours
3. Use the sugar test to determine whether any molasses particles went through the carrot into the water.
4. Which do you think is made up of smaller particles, water or molasses? Justify your answer.
5. What happened to the molasses concentration inside the carrot during osmosis?
6. What causes the process of osmosis to slow down? Under what conditions do you think it would seem to stop?

Applied Osmosis

Six examples of applied osmosis are presented here. See if you can figure out how osmosis explains each of these common occurrences.

1. When old, spongy potatoes are being prepared for cooking, which would it be better to soak them in, tap water or salt water? Explain.
2. Why is salt sometimes used to kill weeds?
3. To examine blood cells under a microscope, a technologist puts them in an *isotonic* salt solution, which has the same concentration of substances as the blood. Explain why water alone is not used. Predict what would happen if blood cells were dropped in water.
4. Placing sugar on a grapefruit makes it become moist very quickly; a sweet syrup forms over its surface. Explain why this happens.

5. The plants below were damaged by "fertilizer burn." How does osmosis help explain what happened?

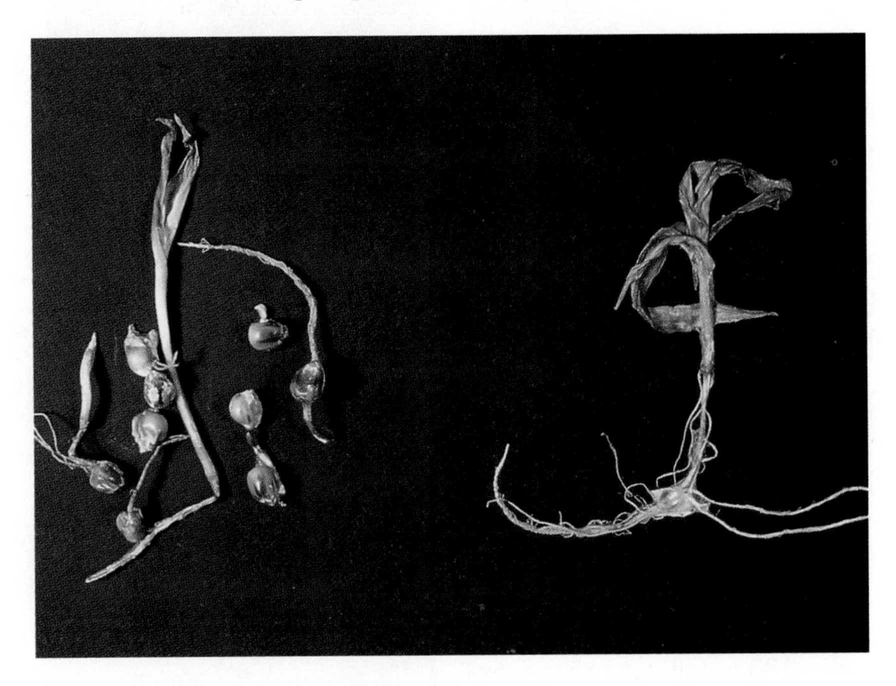

6. Celery becomes limp when it dries out. What could you do to restore its original crispness?

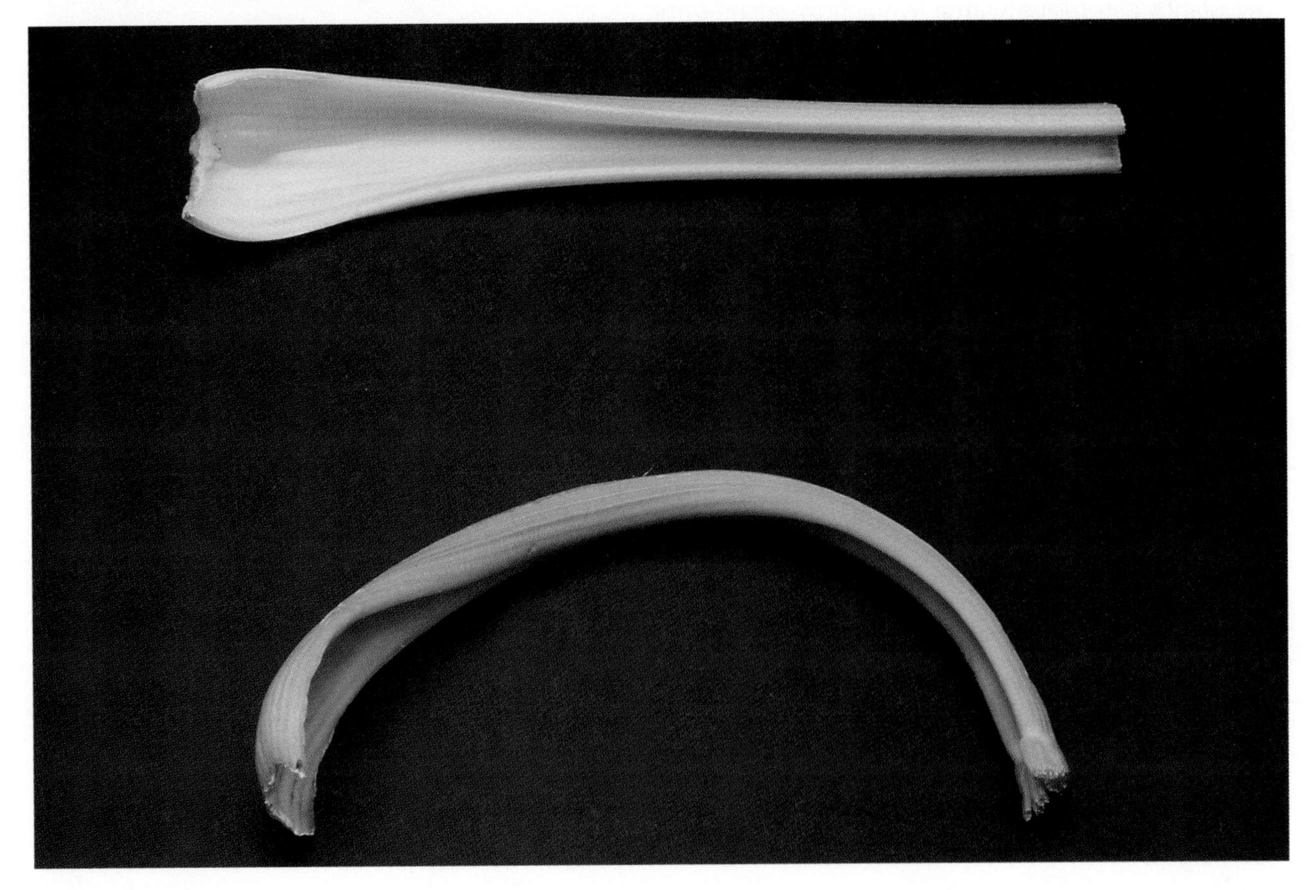

Putting Water Where It Counts

As you know, water enters the roots of plants by diffusion. But what happens then? How does the water get from the root hairs to the leaves where it is needed?

The concentration of water around the root hairs is normally higher than the concentration of water inside the root hair. This difference in concentration results in the diffusion of water into the plant, causing a type of phenomenon known as *root pressure.* Root pressure forces a steady stream of water into the plant. At the same time, the membranes surrounding the root-hair cells prevent the vital contents of the cell from leaking out. The root hairs pass the water they absorb from cell to cell. Eventually the water reaches special cells that function very much like pipes. These cells carry water to the leaves.

In a living plant there is an unbroken "pipeline" of water from roots to leaves. This becomes pretty impressive when you realize that there are some trees that are over 100 m tall. Can root pressure alone push the water that high (this would require more than 100 N/cm^2 of pressure!), or are there other forces at work? What do you think? Come up with as many ways as you can to explain how plants can move water to such great heights.

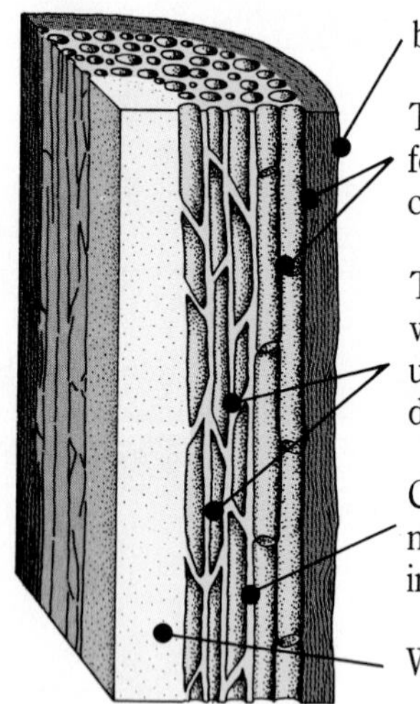

Cutaway view of a woody stem

Study the ideas presented below. At one time or another, each has been suggested as the mechanism responsible for delivering water to the tops of the highest trees. Which, if any, seem logical to you? Which, if any, do not?

Root pressure caused by osmosis.

Example A

Water rising through the process of *capillary action.* Why does it do so? Can you think of other situations in which this occurs? In each case, water rises through fine tubes.

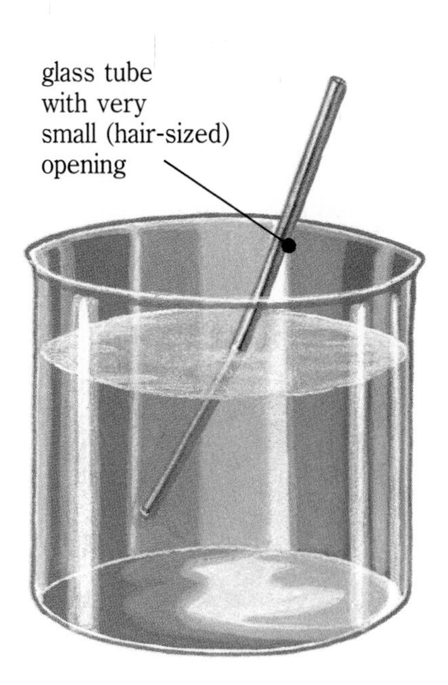

Example B

Here atmospheric pressure forces the water up the glass tube of the dropper. The partial vacuum created by water evaporating from the leaves creates a similar situation.

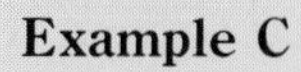

Particles of water have a strong attractive force for one another. Notice how droplets of water cling to each other. (This is called *cohesion.*)

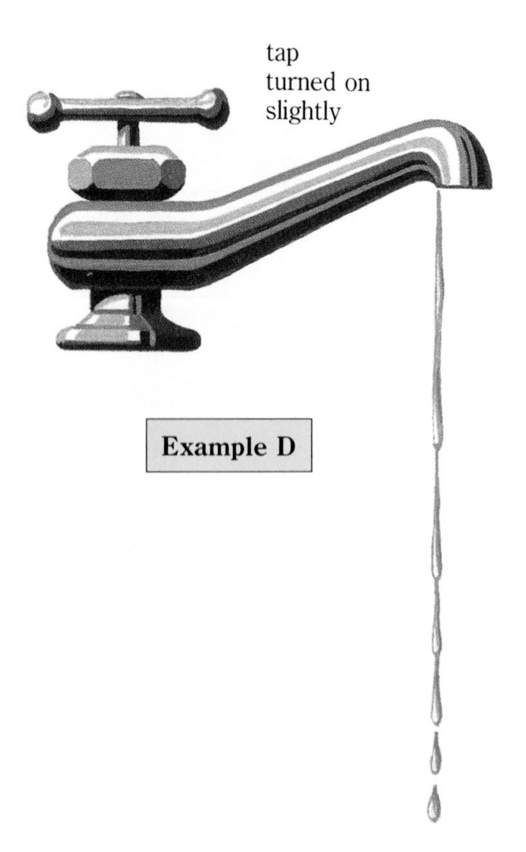

The strong attraction of water to itself allows water particles to pull each other along. As each water droplet evaporates or drips from the leaf, it "tugs" on the droplet next to it.

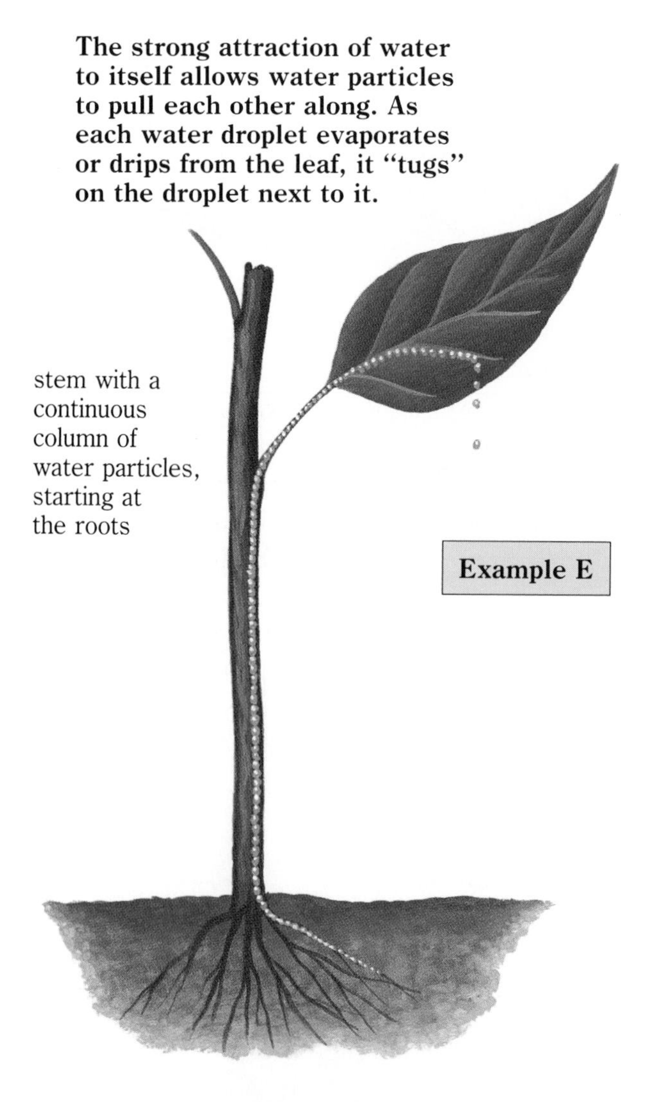

Theories of Water Movement

In looking at the examples on pages 41 and 42, you discovered four kinds of forces that could account for the rise of water in the stems of plants. Here are four explanations in more detail:

- Water particles enter the root cells at a rate greater than they leave the cell, so pressure in the cells increases, pushing water upward. This is called root pressure.
- The narrow passages inside the plant stem cause the water particles to rise by capillary action. Capillary action results from *adhesion* (the attraction between the water and the walls of the passages), and cohesion.
- Water is constantly evaporating through the stomata of the leaves (transpiration). This lowers the pressure in the cells at the top of the plant, allowing atmospheric pressure to push more water upward.
- The force of cohesion keeps the water column intact. Between the pull of water caused by transpiration and the push caused by root pressure, the water column is able to move upward in a plant.

Which explanation do scientists accept today? It turns out that, alone, the forces outlined in the first three explanations are inadequate. Root pressure can only move water upward a few meters. Capillary action is slow and cannot push water very high at all. Evaporation creates only a slight vacuum, so atmospheric pressure would not be able to push the water very far. The fourth explanation, made possible by the great attraction between water particles, is the only one that can, by itself, account for the movement of water to the tops of the tallest trees. Actually, all four phenomena assist the upward movement of water in plants.

In 1894 it was first suggested that transpiration and cohesion worked together to move water up the stems of plants. A simple experiment, such as the one described below and depicted at right, illustrates the way that evaporation and cohesion work together.

First, a continuous column of water is made in the glass tube by forcing steam from the boiling water through the porous cup. Water condenses in the glass tube, filling it. Notice how an unbroken column of water exists between the two containers.

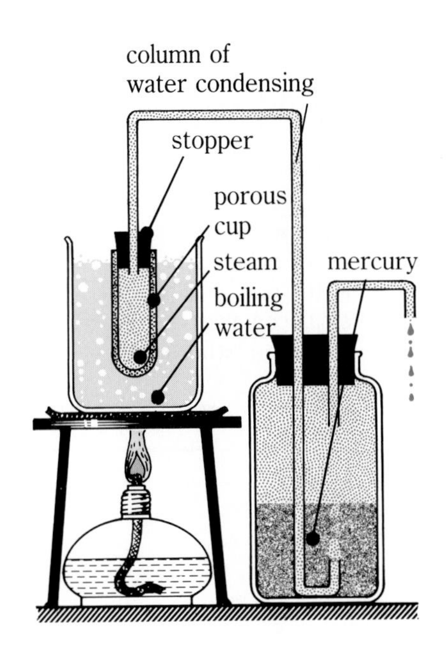

Then the boiling water is removed. (See the illustration at right.) Water now evaporates from the porous cup, just as it does from leaves. As the water evaporates, the water particles attract and pull on each other, drawing the mercury from the lower container upward in the glass tube to a level much higher than in the container. Measurements of this pulling force show that it is great enough to account for the lifting of water to a height of over 100 m.

In 1894 the *cohesion-tension hypothesis* was proposed. This hypothesis states that transpiration and cohesion work together to move water up the stems of plants. It has been suggested that *cohesion-adhesion-tension* might be a better name for the theory of how water rises in plants. Why would this name be more descriptive?

Defying Gravity

Each of the illustrations below shows liquids overcoming gravity. Match each picture with the explanation for the upward movement given below.

- attraction between water particles (cohesion)
- atmospheric pressure
- capillary action
- pressure caused by osmosis
- some other force

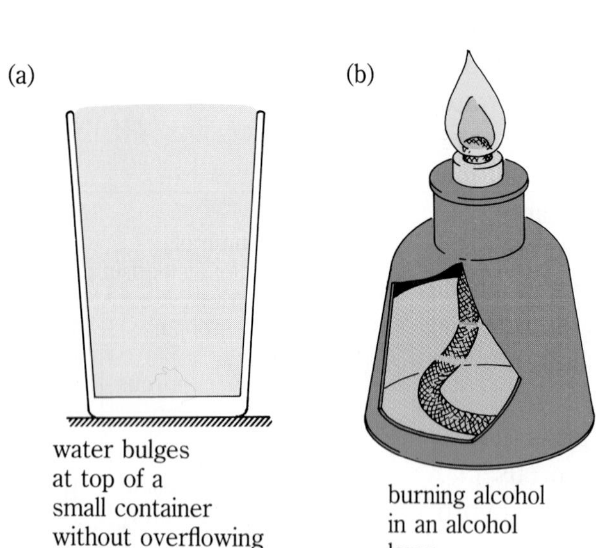

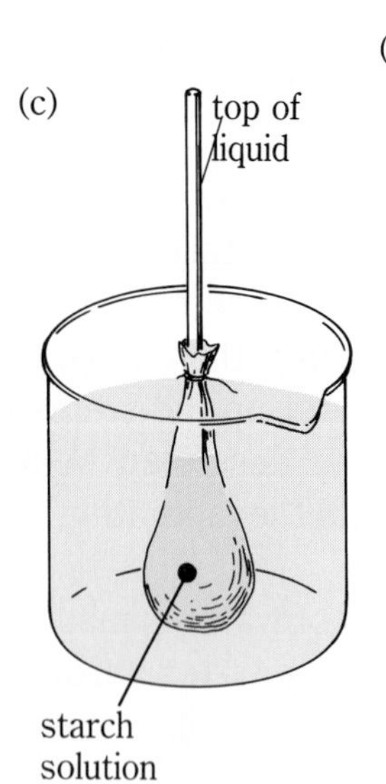

Challenge Your Thinking

1. The diagram below shows a cell under three different conditions: *hypotonic, hypertonic,* and *isotonic.* Use the information in the diagram to answer the following questions:
   - What do *hypertonic, hypotonic,* and *isotonic* mean?
   - In which direction are water particles moving in each example?
   - What might eventually happen to each of the cells?

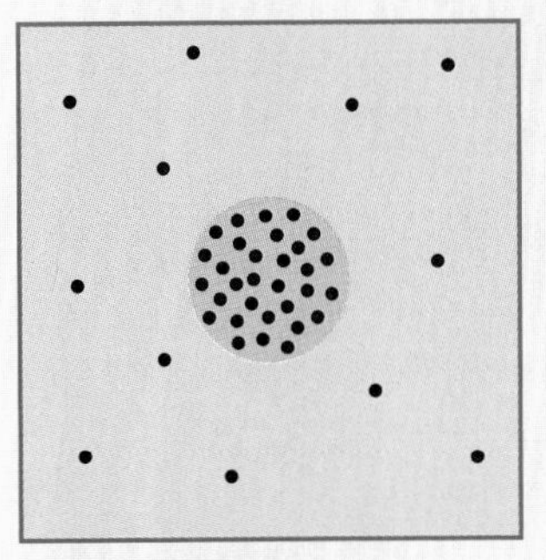

Hypotonic solution

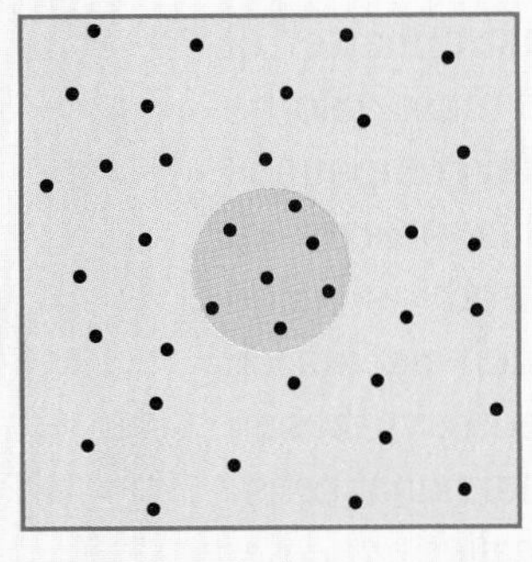

Isotonic solution

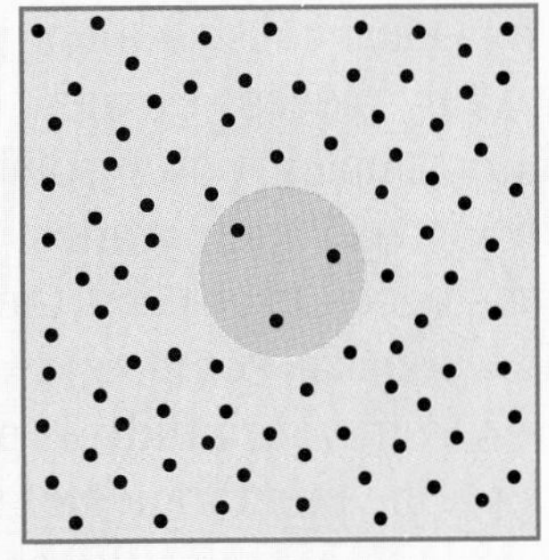

Hypertonic solution

cell

• dissolved particle

2. Honey is basically an extremely concentrated solution of sugar and water. It is also remarkably durable; it has been known to last for centuries without spoiling. How does the principle of osmosis help to explain honey's ability to resist spoilage?

3. Your 9-year-old cousin has just read about diffusion and he has a few questions. Help him answer them.

   How does the stuff know when to start diffusing?

   How does it know when to stop diffusing?

   When the stuff diffuses, why does it always go from a concentrated area to a less-concentrated area? Why doesn't it ever go in the other direction?

4. How does the principle of osmosis help explain why plants wilt when they don't get enough water?

5. Why will drinking salt water make you thirstier than not drinking any water at all?

Turning Food into Fuel

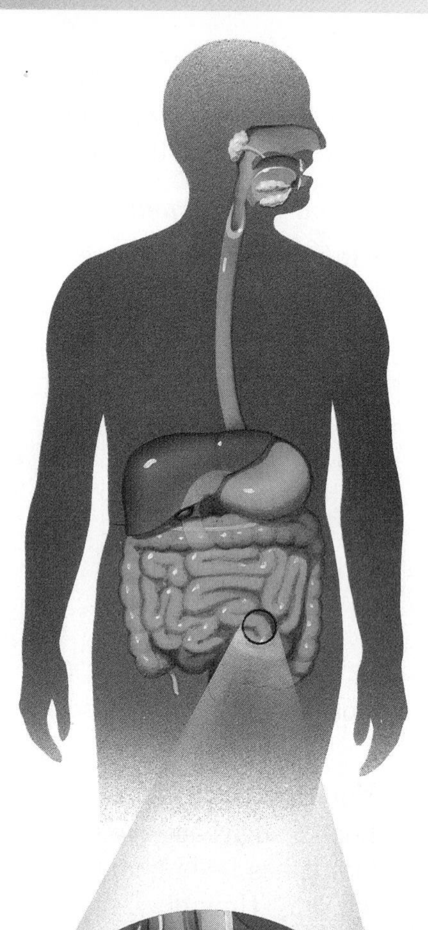

You know that the energy living things need to carry out life functions comes from food. But exactly how do living things—people for example—get energy from food?

Food in its raw form is not usable by our bodies. Before our bodies can put food to work it must be converted into a form suitable for absorption by individual cells. It's not just a matter of breaking down the food into tiny pieces, though. It must also be broken down chemically into simpler water-soluble compounds that can pass by diffusion through cell membranes. The process of breaking down food into substances usable by the body is called **digestion.** Even a simple chemical compound such as starch (which you should remember from your study of photosynthesis) is too complex to pass through the membranes of individual cells. Digestion breaks starch down into a simple sugar called *glucose,* which cells can absorb.

Where does digestion take place in the body? It starts in the mouth, as you will see for yourself if you do the following. Put an unsalted soda cracker into your mouth and chew it slowly, letting it soften thoroughly. Hold the food in your mouth and describe any change in taste. What substance do you think has formed?

You will do a more precise laboratory experiment on this phenomenon in Exploration 12.

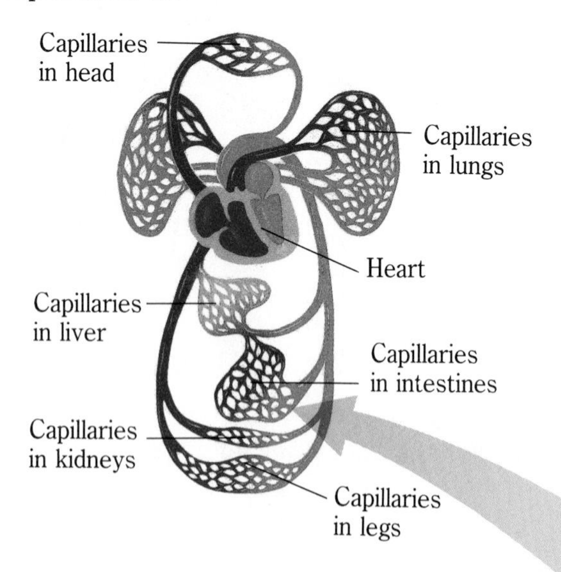

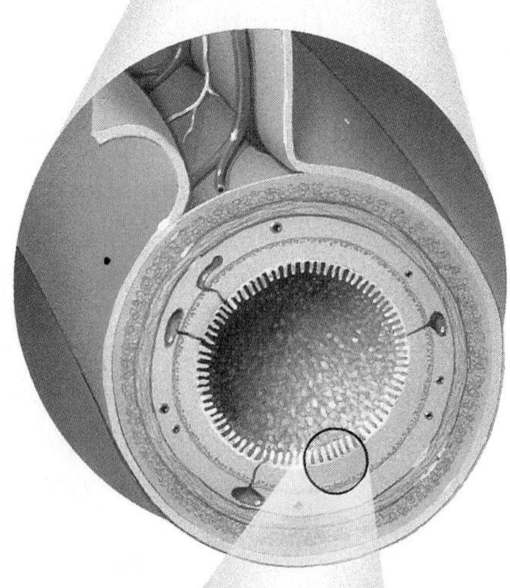

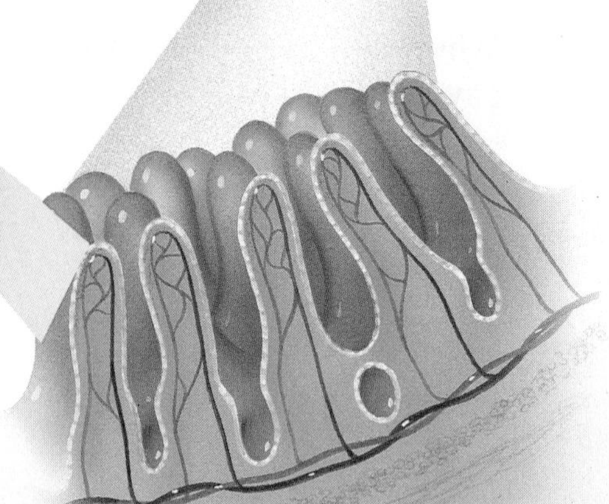

The small intestine contains many tiny projections. Digested food is absorbed through these into the bloodstream. The food is then carried by the bloodstream throughout the body. Food reaches individual cells through a network of *capillaries,* tiny blood vessels that reach every part of the body.

Digestion of Starch

You Will Need

- a small container
- 3 test tubes
- a beaker
- water
- a hot plate
- Benedict's solution
- starch (a crushed, unsalted soda cracker)
- an eyedropper
- a test-tube rack

To gather saliva, first be sure you are not chewing gum or candy. Collect some saliva in your mouth, take a small amount of water and rinse your mouth with it, then deposit this rinse material containing the saliva in a clean container. (Be sure to dispose of this liquid properly at the end of the Exploration.)

What to Do  

1. Number the test tubes 1, 2, and 3.
2. Into test tube 1, put a pinch of starch and enough saliva solution to cover the starch. Shake the tube gently to mix the substances.
3. Into test tube 2, put a pinch of starch and enough water to cover the starch. Mix the substances.

Be sure to work only with your own saliva.

4. Into test tube 3, put the saliva and water solution.
5. After 15 minutes, test the substance in each test tube for sugar, and record your results in a table similar to this one:

Test Tube Number	Substances in Test Tube	Results of Sugar Test
1	Starch, saliva, and water	
2	Starch and water	
3	Saliva and water	

The Rest of the Story

Once food is completely digested, it is able to diffuse through cell membranes. In the small intestine, most of the digested food diffuses into the blood vessels. The bloodstream then carries the food to all parts of the body. The small particles of digested food are able to pass from the blood vessels into the body cells by means of diffusion through particularly thin-walled vessels called capillaries. (How does diffusion determine the direction that the food particles move?) Once the digested food is in the body cells, it is ready for the next process.

Unlocking the Energy in Food

A Matter of Energy

Perhaps in earlier studies you burned a peanut (or some other food) to find out how much energy it contained. Where did the energy in the peanut come from? Where does it go when you eat the peanut? How is the energy released? The illustration might give you some ideas. Study it closely, then write a brief caption of about two or three sentences for it.

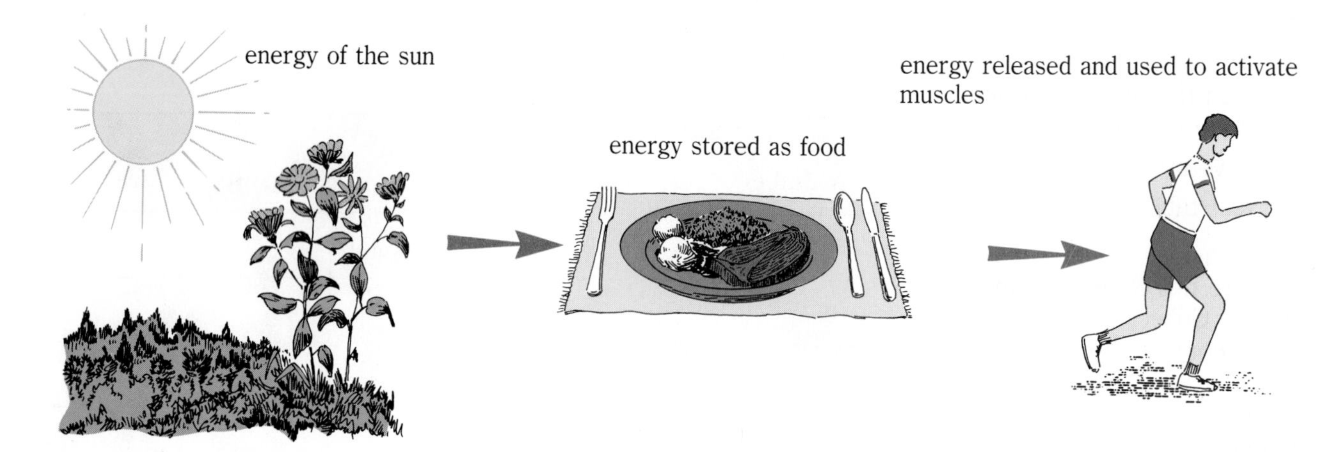

How is the energy in gasoline released in an engine? How is the energy in candle wax released as heat and light energy? The energy within digested food is released in a somewhat similar way. Look at the following diagrams. In what ways is the release of energy similar in each case? How does it differ?

Energy being released

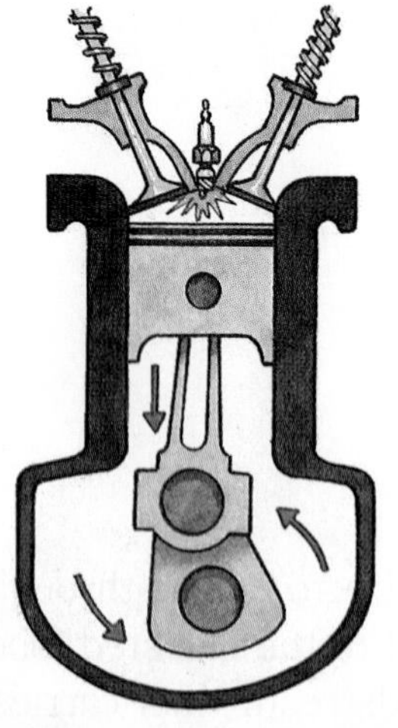

Exploding gasoline pushes the piston down, turning the crank-shaft.

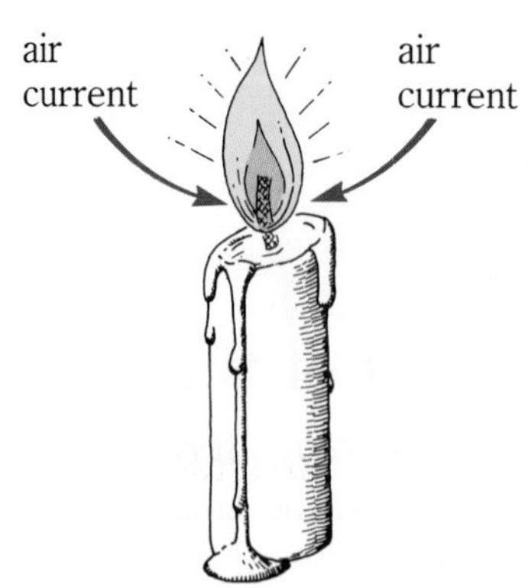

Oxygen combines with candle wax, releasing heat energy.

digested food + oxygen → Energy

"burning"

cells of living things

Food is "burned" in living cells to release energy.

The release of energy that occurs when some types of digested food combine with oxygen in the cells of living things—both plants and animals—is called **respiration.**

The Purpose of Breathing

Air, which contains oxygen, is necessary for the combustion (explosive burning) of gasoline in a car engine. Likewise, oxygen is essential for respiration in the cells of living things. But how does oxygen reach the cells?

In animals, the process begins with breathing. While you were reading the last few sentences, were you conscious of the fact that you were breathing? How many times a minute do you breathe? Do you have to think about it for it to happen? Unless something goes wrong with your "breathing system," such as a stuffed nose, you are not even aware of the flow of air into and out of your lungs. In winter, why are you more likely to be aware of your breathing?

How do you think the following living things obtain the oxygen they need for their cells: you, a dog, a fish, a bird, a tadpole, a frog, an earthworm, a lobster, a grasshopper, a whale, a mosquito? How about a tree? It may surprise you, but plants also "breathe." Their cells must carry on respiration just as the cells of animals do.

The illustrations on page 50 show the breathing structures of some of the living things just mentioned. Which animal has gills when it is young but lungs when it is an adult? Which animals get oxygen from the air? from water?

From Breathing Structures to Cells

Most animals have an internal transportation system that carries materials to and from the cells, just as plants do. Also, as in plants, the medium that does the carrying consists mainly of water.

In the margin is a diagram of the human transportation system, better known as the *circulatory system.*

- What acts as the pump in the system?
- What is the liquid being pumped through the system?
- What are the tubes leading from the heart called?
- What are the tubes leading back to the heart called?
- Where do materials get in and out of the transportation system?

In human blood, red blood cells are the special carriers of oxygen. The oxygen diffuses out of minute air sacs in the lungs into tiny, thin-walled capillaries. In the capillaries, oxygen is absorbed by red blood cells and is taken by the blood to cells in all other parts of the body. Wherever there are capillaries, the oxygen can diffuse through their thin walls to the body cells near them.

The circulatory system carries blood to and from all body parts. Red, moving away from the heart; blue, moving towards the heart.

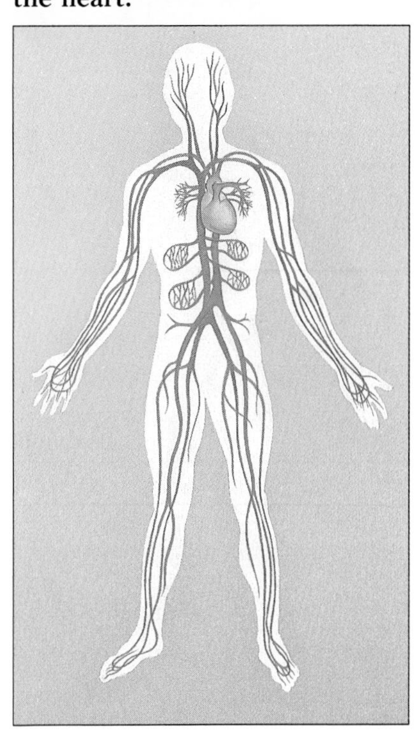

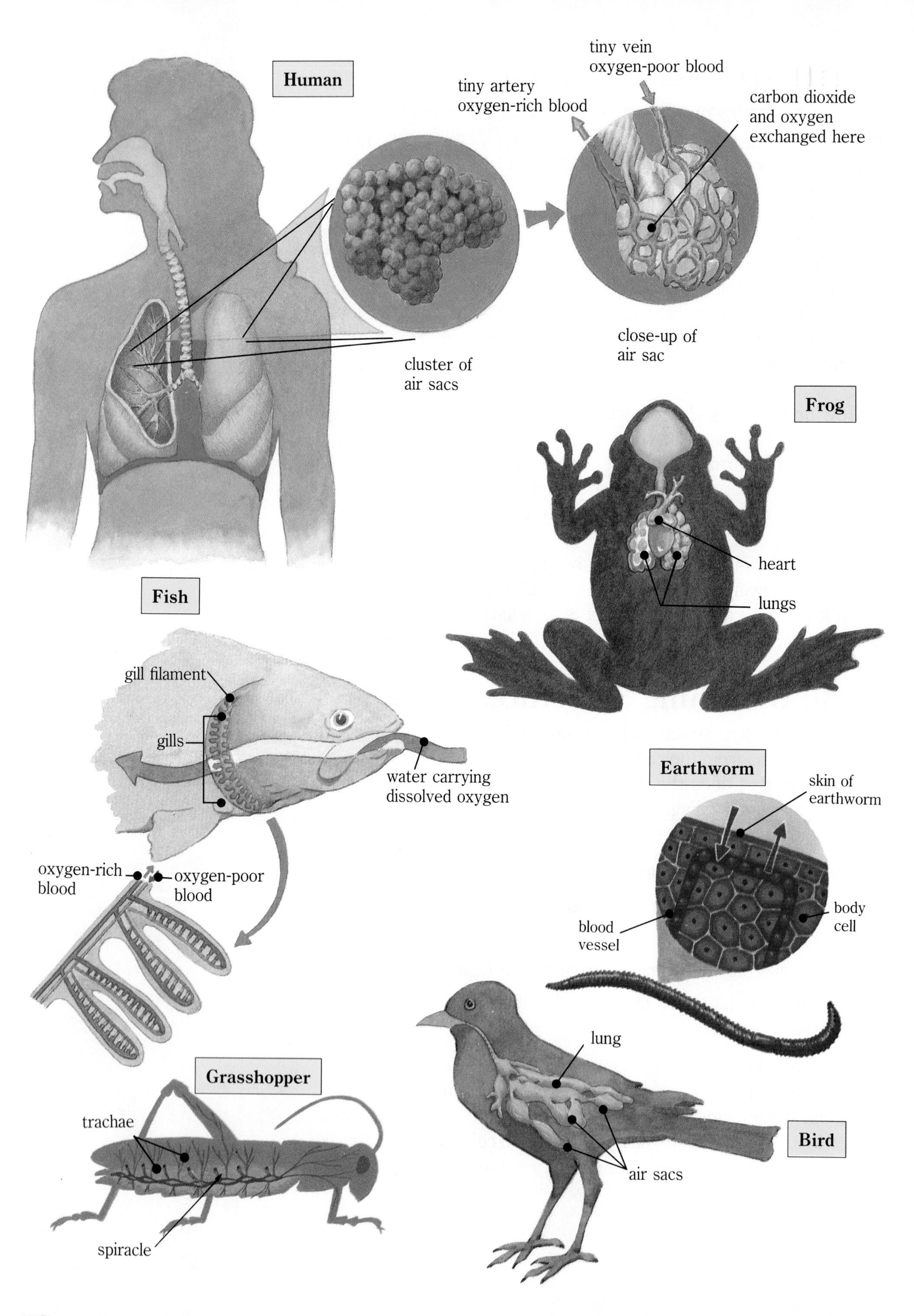
Human
tiny vein
oxygen-poor blood
tiny artery
oxygen-rich blood
carbon dioxide
and oxygen
exchanged here
cluster of
air sacs
close-up of
air sac
Frog
heart
lungs
Fish
gill filament
gills
water carrying
dissolved oxygen
oxygen-rich
blood
oxygen-poor
blood
Earthworm
skin of
earthworm
blood
vessel
body
cell
lung
Grasshopper
trachae
spiracle
air sacs
Bird

Breathing Rate

Does your breathing rate stay constant? If not, what causes it to change? Why do you sometimes breathe with your mouth open to take in large amounts of air quickly? Exploration 13 will help test your answers to these questions.

EXPLORATION 13

Testing Your Breathing and Pulse Rates

You will need a partner to help you count and measure time. (Use a stopwatch or a watch with a second hand to measure the time accurately.)

First find out your breathing rate while sitting quietly. Have a partner count the number of breaths you take in one minute. Try not to think about your breathing. Repeat the procedure five times and average the results. Then take your pulse using the following methods. Count the number of beats in 30 seconds and multiply by 2 to get your resting pulse rate. Repeat the procedure five times and average the results.

Find out your breathing rate while walking around for 2 minutes. Again average five readings. Calculate your pulse immediately following this activity using the method outlined above.

Find out how many times you breathe in one minute while running in place. Again, calculate your pulse rate immediately after the activity.

Record all results in your Journal in a data table of your own design.

Interpreting Your Data

1. What effect does exercise have on your breathing rate? on your pulse rate?
2. Is there any relationship between your breathing and pulse rates? If so, what is it? Why do you think this relationship exists?
3. Why did you repeat your measurements to find average breathing and pulse rates?
4. Are the breathing and pulse rates the same for everybody? If they are different, can you suggest some reasons for the difference?
5. How are breathing and blood circulation related to respiration in the cells?
6. What part of the body do you think regulates the rate at which respiration occurs?
7. Did you find that you got hot as you exercised? What does this suggest about the relationship between respiration and burning?

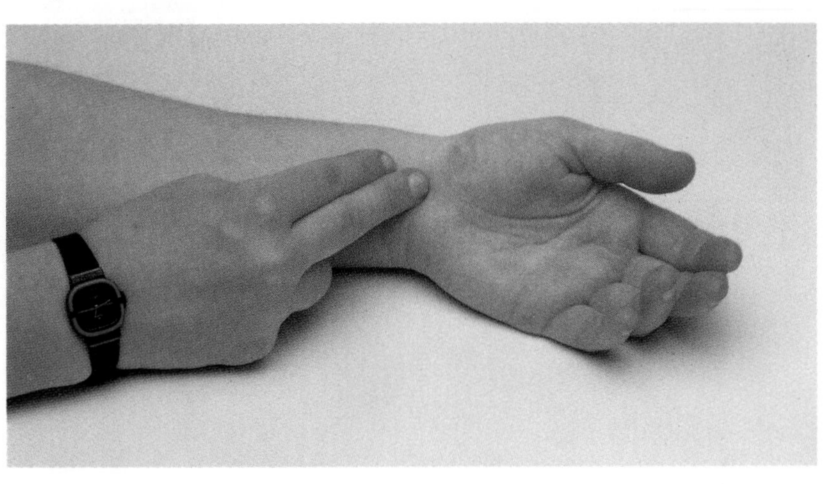

The Ins and Outs of Respiration

The process of respiration uses oxygen and releases energy, just as the combustion process in a car engine or the burning of a candle do. Each of these processes also produces wastes. In Exploration 14 you will learn more about the waste products of respiration.

EXPLORATION 14

Products of Respiration

Test 1

You Will Need

- a drinking straw
- limewater
- a test tube

What to Do

Blow *gently* through a straw placed below the surface of clear limewater in a test tube. What happens? Is there carbon dioxide in the air you breathe out? Might carbon dioxide be a product of the respiration process that takes place in the cells of living things?

How can you be certain that the carbon dioxide you breathed out was not already in the air you breathed in? Here is a way to find out.

Test 2

You Will Need

- a stopwatch or watch with a second hand
- 2 flasks
- glass tubing (as in the illustration)
- 2 two-holed stopppers
- limewater

What to Do

Set up the equipment as shown. Inhale through the tube at X. Air from the room will bubble into the flask of limewater on the left. Exhale into the tube at X. Your breath will bubble through the flask of limewater on the right. Compare the time required for the limewater on the right to turn milky with the time required for the limewater on the left to turn milky. Which has more carbon dioxide in it, the air in the room or the air you exhale? What conclusion can you make regarding one possible product of respiration?

The next experiment will reveal another product of respiration.

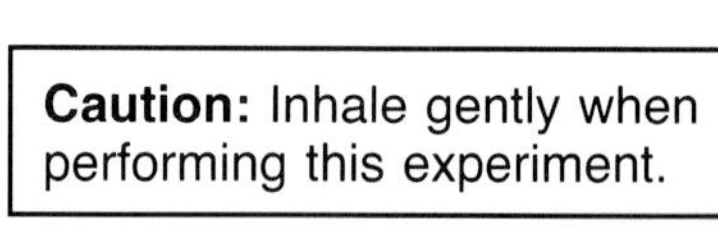

Caution: Inhale gently when performing this experiment.

Test 3

You Will Need

- a mirror or a shiny piece of metal
- cobalt chloride paper

What to Do

Exhale so that your breath strikes the surface of the mirror or piece of metal. Examine the shiny surface closely for any change in appearance. Repeat this process several times. Wipe the surface with a piece of cobalt chloride paper. This paper is blue when dry but changes to pink when wet. This color change signals the presence of water.

Questions

1. What was the appearance of the mirror when you exhaled on it?
2. What color was the cobalt chloride paper before you wiped the mirror? After you wiped the mirror?
3. What was the substance on the mirror?
4. Where did it come from?
5. What conclusion can you draw regarding another product of respiration?

Water and carbon dioxide are produced by the burning of most fuels, for example gasoline or candle wax. Living things, in the process of respiration, give off these same waste products. Carbon dioxide diffuses into the blood from the cells and is carried back to the lungs to be breathed out. (This is the reverse of how oxygen reaches the cells.) What do you think happens to water, the other product of respiration?

Now you are ready to devise a definition of respiration that takes into consideration the substances and the energy required, the products, and the location at which the process takes place. When you have written down your definition, complete this word equation describing respiration.

digested foods + ? →
energy + ? + ?

Now look at the following simple chemical equation:

fuel (gasoline, coal, wood, etc.) + oxygen → carbon dioxide + water + energy

What does this equation tell you about the process of respiration?

Plants Also Respire

Plants must use some of the food they make to carry out their own life processes. Respiration is therefore just as essential to their existence as is photosynthesis. However, it is difficult to show that a green plant takes in oxygen and gives out carbon dioxide because photosynthesis is simultaneously taking place, with its own exchange of gases. To demonstrate respiration in plants, it is easier to use germinating (sprouting) seeds, since they have not yet begun to make their own food through photosynthesis. Instead, they are still using food stored inside them. You will do this in the following Exploration.

Plant Respiration

You Will Need

- 20 seeds (such as radish or bean seeds)
- 4 test tubes
- 4 stoppers
- a wooden splint
- limewater
- a graduated cylinder

What to Do

1. Place 20 moist seeds in two test tubes. (Water helps seeds carry on their life processes.)
2. Stopper the test tubes tightly and leave them alone for two days.

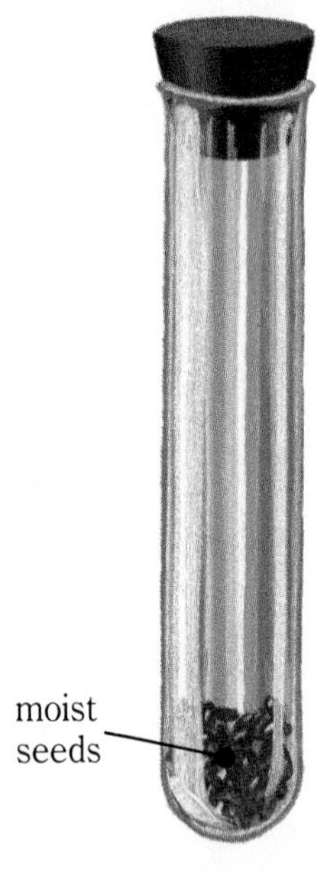

3. After two days, remove the stopper from one test tube and quickly insert a lit wooden splint. Did the splint go out? What does this suggest that the seeds might have done?

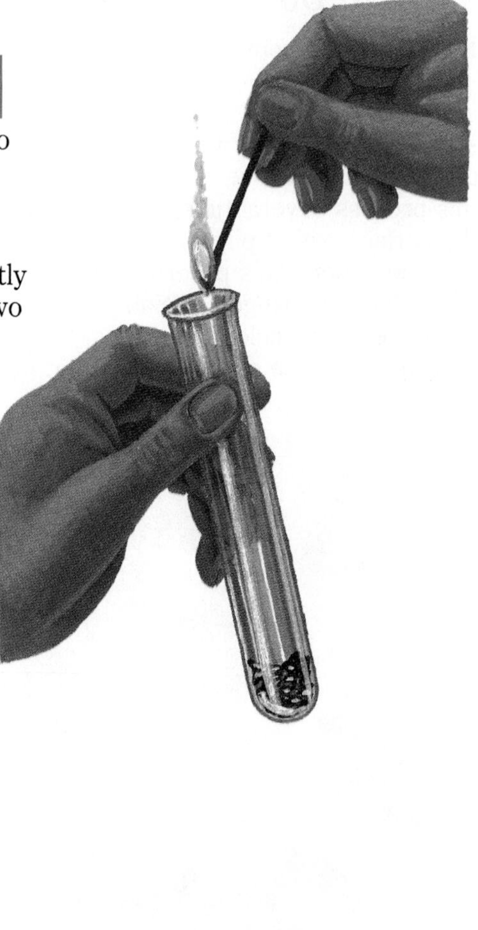

4. Unstopper the second test tube and quickly add 5 mL of limewater.

5. Stopper the tube again, and gently swirl it for a minute. Make note of your observations in your Journal.

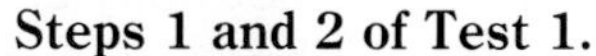

Steps 1 and 2 of Test 1.

6. Repeat steps 2–5 with stoppered test tubes containing only air (why?), and compare the results with those obtained in steps 3 and 5. How can you explain your observations?

What do the results of this Exploration suggest concerning respiration in seeds? Write a summary of your results, along with your conclusions, in your Journal.

Going Further

Design an activity to find out whether respiring seeds produce heat energy.

You have just verified the existence of a similarity between plants and animals: both depend upon the process called respiration. From your studies so far in this unit, make a list of *differences* between plant and animal processes.

Stored Energy

What happens to the energy released in respiration—where does it go, and how is it used? Can it be stored?

The energy provided by respiration is used throughout the body for all bodily processes and functions. But this energy is not used immediately by the body. It must first be stored in a usable form. The storage process is very complicated, and is not yet fully understood. In brief, this is what happens. The energy released by the breakdown of food is used by cells to make a chemical known as **ATP** (short for *adenosine triphosphate*). In this process, the energy is stored in the ATP. To make the ATP, the body uses **ADP** (short for *adenosine diphosphate*), which is present in all cells. Then, as energy is needed by the body, the ATP is converted back to ADP. This releases the energy that was stored in the ATP.

Controlling Respiration

While reading about respiration, did you wonder how your body knows where energy is needed, how much is needed, and so on? The answer lies in the glands, the "process controllers" of the body. The glands produce chemicals that regulate or control different functions in the body. One of the most important, the **thyroid gland,** regulates the rate of respiration in body cells; that is, it controls how fast food is "burned" in the cells. It does this by means of a chemical substance, *thyroxin,* which diffuses from the thyroid gland into the blood and is thus carried to other body cells. The respiration process is monitored and regulated by the nervous system, which consists of the brain and nerves.

Another vital gland is the **pancreas.** It is tucked between the stomach and the first part of the small intestine. The pancreas makes *insulin,* a chemical that enables sugar to be used as "fuel" for respiration.

These two glands, the thyroid and the pancreas, are directly connected with the process of respiration. There are many other glands in the body as well, not all of which regulate respiration. You will study these in later courses.

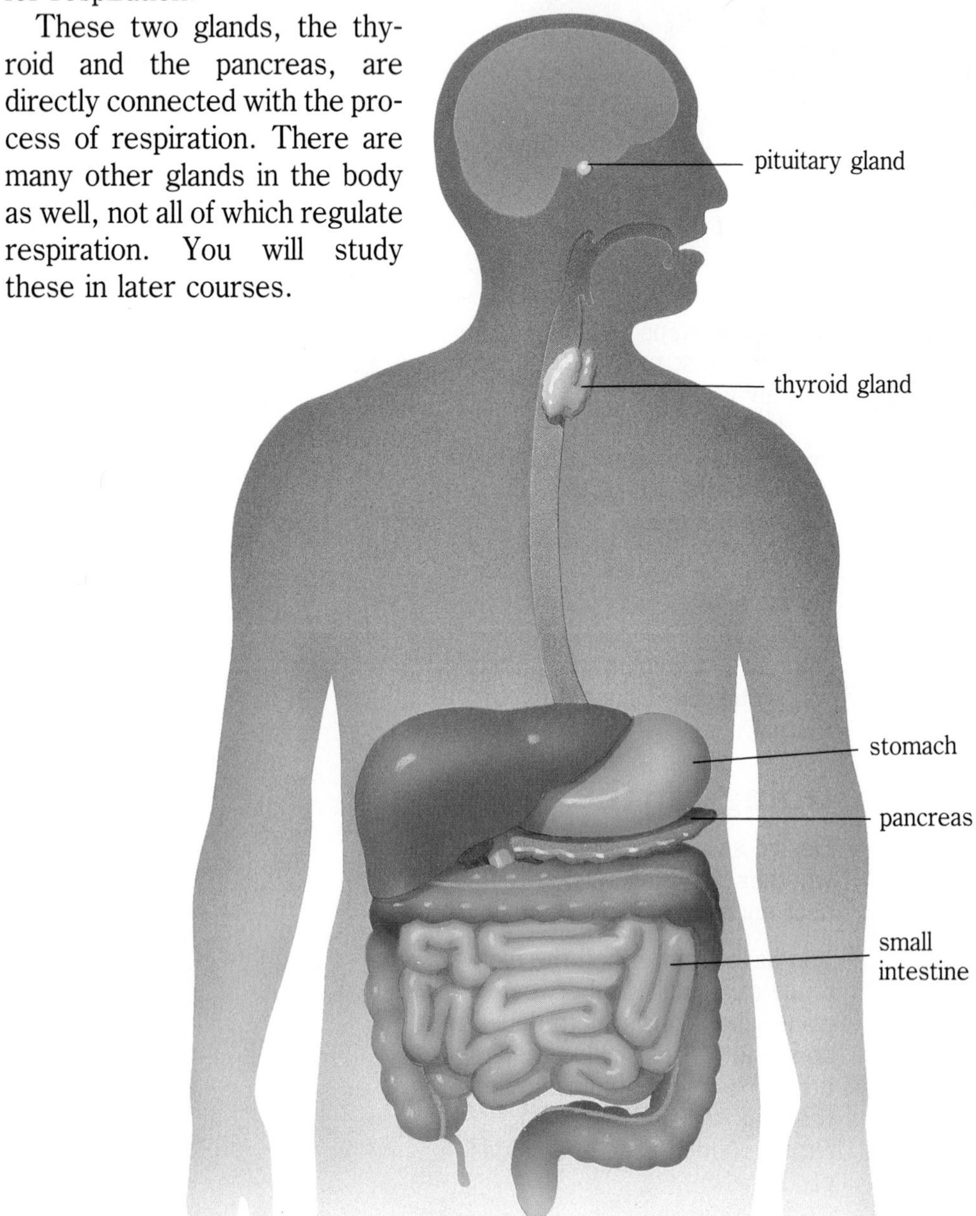

Respiration: A Summary

Imagine an individual cell in your body carrying on respiration.

1. What does the cell need? How do these materials reach the cell?
2. What does the cell give off? How do these materials leave first the cell and then the body?
3. How do substances get into and out of the cell?
4. Why does the cell carry on respiration? Why is this process so important?
5. Many parts of the body are either directly or indirectly involved with respiration. How do each of the following contribute to respiration at the level of individual cells: teeth, heart, skin, kidneys, lungs, salivary glands, thyroid gland, ATP, ADP, brain, capillaries, pancreas, and stomach?

Body Systems

Question 5 in the summary above named parts of many systems in the body. For example, the teeth and stomach are part of the **digestive system.** The kidneys and skin are two organs in the **excretory system.** The heart and capillaries are part of the **circulatory system.** The lungs are part of the **respiratory system.** The brain is part of the **nervous system.** All of these systems in your body work together to accomplish respiration and digestion, as well as other life processes.

This unit is intended to introduce you to a few fundamental life processes. It is not meant to teach you everything you need to know about life processes or about the human body. You will learn much more about these subjects in later biology, health, or physiology classes.

Maintaining the Balance

So far in this unit, we have talked about life processes carried out on a small scale, within a single plant or animal. Let us now turn our attention to life processes carried out on a much larger scale, that of the entire planet.

Our Earth is a very special place. It is the only world we know of where life exists. No other planet in our solar system is even close to being habitable. Life is able to exist on Earth for many reasons, not the least of which is the delicate balance that exists between plants and animals. Animals take in oxygen and give off carbon dioxide, and plants take in carbon dioxide and given off oxygen. Plants and animals require each other.

Consider a sealed terrarium, in which plants and animals live in balance with each other. Once the terrarium is sealed, no additional materials enter and no materials leave. What kinds of plants and animals would you put in a terrarium to make it self-sustaining? What additional materials from outside the terrarium are required? How could the terrarium environment be damaged so that everything in it died?

The Earth is much like a sealed terrarium, only on a much larger scale, of course. Balance must be maintained in order for life to survive.

A balance between plants and animals, and between oxygen and carbon dioxide, has existed on Earth for over one thousand million years. Can human beings upset this balance? If so, how? Read the following article to help you answer these questions. In your Journal, create headings for each paragraph. Each heading should be fairly brief and to the point, but should also capture the essence of each paragraph.

A Critical Balance

The present way of life of human beings requires the use of much fuel. What does burning this fuel do to the amount of carbon dioxide in the air?

(title) ________

Carbon dioxide in the amount that is normally found in air (a little more than 3000 parts carbon dioxide per million parts of air) is absolutely necessary. As you know, it is essential for photosynthesis, the food-making process in green plants. You may not know that it also stimulates the breathing centers in the brains of animals.

(title) ________

At present, it appears that green plants cannot remove the carbon dioxide from the air as fast as humans produce it by burning fossil fuels to heat homes, generate electricity, run factories, and power vehicles. In the mid-1980s, the carbon dioxide concentration in the air was about 3400 parts per million. That concentration could double by the year 2050 if the present rate of burning some types of fuels continues.

(title) ________

Why is an excess of carbon dioxide a threat? The air in the atmosphere is transparent to the sun's radiation; that is, it allows the sun's radiant light and heat energy to reach the Earth's surface. This energy warms the Earth. As the Earth cools at night, it tends to radiate the energy back into space, but at a lower energy level. Carbon dioxide blocks this lower-energy radiation from getting through; heat becomes trapped, causing the atmosphere to heat up. This process is commonly called the greenhouse effect. Water vapor also causes this effect.

(title) ________

The normal greenhouse effect is beneficial. As long as there is a normal concentration of carbon dioxide in the air, the Earth is neither too warm nor too cold. However, a higher concentration of carbon dioxide could result in a significant increase in the average temperature of the Earth. What could some of the possible consequences be, in your opinion?

(title) ________

In fact, it is difficult to predict just what would happen. There has been a slight increase of about 0.5 °C in the average global temperature since 1900, but it is not possible yet to identify the increased levels of carbon dioxide as the cause. It is possible that excess heat due to the greenhouse effect is being absorbed by the Earth's oceans. In that case, the heat could be released gradually over a period of 30 to 50 years. Some predicted effects of a continued rise in global temperature over the next 50 years include drastic changes in climates as well as a rise in sea level, due to melting polar ice.

(title) ________

What can we do about this potential problem? First of all, there may be ways of limiting the use of fuels such as coal and oil. Can you suggest some ways? How might you alert people and governments to this problem, and encourage people to maintain the delicate balance on our remarkable earth?

Points for Discussion

Choose one or more of the following topics and discuss them in small groups. Record your findings for presentation to the class.

1. Some people have suggested that in the future, human energy needs could be met by *biomass* alone—that is, energy would come solely from vegetation. What are the possible advantages and disadvantages of this approach? Would this be a viable solution? Why or why not?
2. As cities expand, trees often have to be cut down to permit the widening of streets and the building of houses and more industrial space. Do we need to conserve trees in our cities and industrial areas? Consider these facts: One large tree can absorb 2300 g of carbon dioxide in one hour. This is the amount given off by ten single-family houses. During the same hour, that tree can give out about 1700 g of oxygen. How might large trees help control the level of carbon dioxide in the air around your town or city?
3. So far, you have learned that carbon dioxide gets into the air through cellular respiration and the burning of fuels. Are there other ways? If so, what are some of them? Do a little research. How long a list of specific examples of the emission of carbon dioxide into the air can you generate? How many of these carbon dioxide sources might you find in a typical home?
4. What are some of the pollutants found in the air? How do they get into the atmosphere? Do they affect breathing? Find out some of the breathing difficulties faced by the inhabitants of such large cities as London, Tokyo, and Los Angeles. What are some of the solutions proposed for these problems?

Processes such as these may contribute to worsening the greenhouse effect.

Challenge Your Thinking

A Magic Square

Match the number of each structure in the illustration with the letter of the best description, and place the number in the appropriate box. If your choices are correct, the sum of the numbers in each row, column, and diagonal will be the same. The mathematical term for such a box is Magic Square. (Some questions refer to earlier sections.)

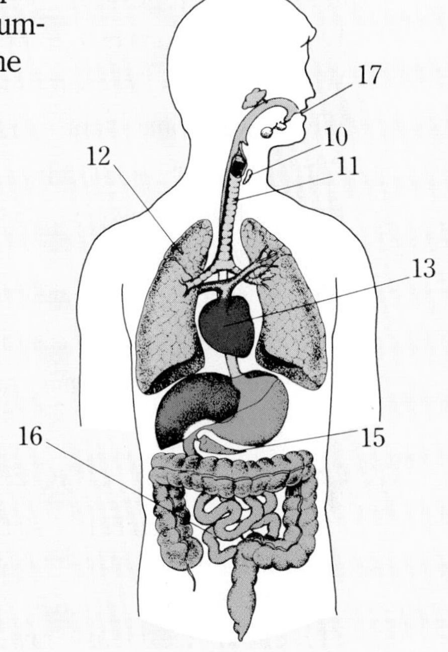

A where diffusion of digested food into the blood primarily occurs

B gland that controls the rate of respiration in body cells

C water pipes in the food factory

D protective waterproof covering for the factory

E the food factory

F air space in the factory

G air pipe with strong supporting rings

H where insulin is made

I center of blood circulation

J where digestion of starch begins

K the regulators for opening and closing the food factory doors

L cell layer where most of the factory doors are located

M site of the actual manufacture of food

N the service doors of the factory

O capillaries, where diffusion of digested food from the blood into the body cells occurs

P where the blood exchanges gases with the air

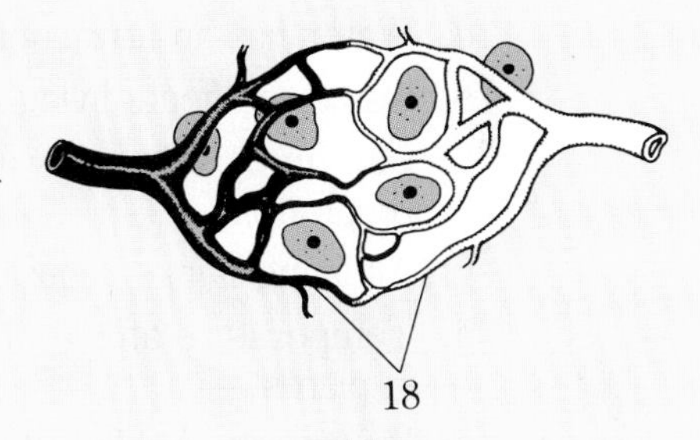

A	B	C	D
E	F	G	H
I	J	K	L
M	N	O	P

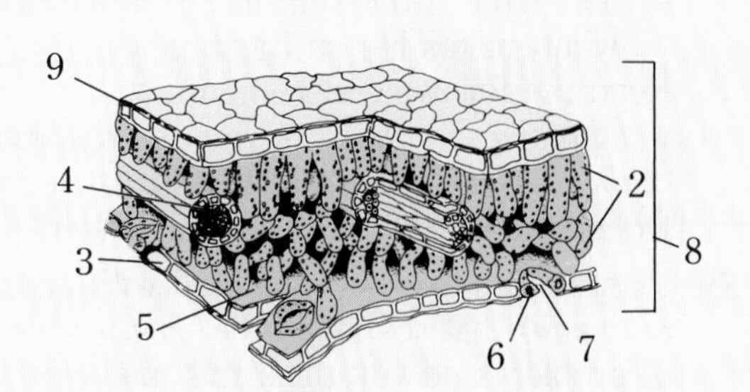

Unit 1 Making Connections

The Big Ideas

In your Journal, write a summary of this unit, using the following questions as a guide.

- Where does food come from?
- Why are leaves like food factories?
- Why is water important to living things?
- What role does osmosis play in living things?
- How do living things get energy from food?
- Why do we breathe?
- What is respiration? How is it similar to burning?
- How do plants and animals depend on each other?
- How are humans changing the natural environment?

Checking Your Understanding

1. Below are a number of statements about topics you studied in this unit. Suggest ways in which each statement could be scientifically verified.

a. Plants give off oxygen only during daylight.

b. Carbon dioxide is a raw material in the manufacture of food.

c. Plants exchange gases through their stomata.

d. Water has an attraction to itself.

e. Osmosis affects living cells.

f. Carbon dioxide is a waste product of respiration.

g. All living things require water.

h. Temperature affects the rate of respiration in plants.

i. Plants need sunlight to produce food.

j. Osmosis creates pressure.

k. The light of the sun, not its warmth, is what makes life on Earth possible.

2. Read the following statement.

Diffusion begins when you add a substance to water, and it ends when the substance is completely mixed in the water.

Does it really tell the whole story, or is there more to it? Explain.

3. Plants are sometimes afflicted with an inherited abnormality (called *albinism*) that results in the plants having no color whatsoever. What kind of problems do you think this might cause for a plant? What would the chances be for this plant to survive and reproduce?

4. Look at the illustrations below. Each tells a story. It is your job to figure out what that story is.

   a. After a few days the leaves were mottled with blue and red patches.

   b. After a few hours, the water seemed to disappear. What happened?

5. Make a concept map using the following terms: *carbon dioxide, breathing, water, respiration, oxygen,* and *energy.*

Reading *Plus*

Now that you have been introduced to some basic life processes, let's take a closer look at what makes living things *living.* By reading pages S1–S14 in the *SourceBook* you will learn about the basic chemistry of life. You will be introduced to the primary organic particles of life and examine water's role as the medium for life's chemical processes. You will also learn about how living things carry out the complicated activities that make life possible.

Updating Your Journal

Reread the paragraphs you wrote for this unit's "For Your Journal" questions. Then rewrite them to reflect what you have learned in studying this unit.

Spotlight on Zoology

Eric Pianka has spent his life studying the role living organisms play in their environment. This branch of science is called **ecology.** The study of animals is called **zoology.**

Eric is a world-famous professor of zoology at the University of Texas and has published around 100 books and papers! Eric's specialty is the ecology of desert lizards. His education is extensive, including a B.A. in Biology and both a D.Sc. and a Ph.D. in Zoology.

Q: How did you get involved in the study of lizards?

Eric: I became a biologist when I was 4 or 5 years old! On a trip across the country with my family, I saw a big green lizard at a roadside park. I tried to catch it, but all I got was the tail. At that moment, as I stood there amazed by every move of that fascinating creature, I knew I had to find out everything I could about the kind of life it led.

Jackson's chameleon

Eric Pianka with a perentie in Australia's Great Victoria Desert. Perenties are some of the largest lizards in the world, ranging up to 2.4 meters in size.

Q: How would you describe the importance of your work to high school students?

Eric: Everyone always asks, "Why lizards?" I turn that around and say, "Why you?" We are a very human-oriented society—the general attitude is that everything on Earth has to somehow serve humans.

Take, for example, the story of a boy who went for a walk in the forest with his father. The boy said, "Dad look at that beautiful tree!" The father responded by saying, "Oh son, that's not a beautiful tree—it's too crooked to be used for lumber and too far away from anything to be used for firewood!"

The fact is that wildlife is not irrelevant to our existence. As products of natural selection that have adapted to their natural environments for millions of years, wild plants and animals have a right to this planet too. They were here long before us and I'm sure they'll be here long after us. We have a lot to learn from them!

By looking at how species have lived and died and changed over millions of years, we can gain an understanding of the world we live in. This understanding may lead to the wisdom that will save the human species from extinction, like so many other species that have gone before us.

Q: How do you carry out your research?

Eric: Well, most of my research has been on the ecology of desert lizards, so I go to a desert, collect lizards, and examine and classify them. Then I compile my data and interpret it into books or papers.

I try to answer questions like: Why are there more lizards in one place than in another? How do they live through a wildfire? How do they interact with each other and with other species? How have they adapted to their environment? In order to do this, I pay close attention to the different varieties of lizards and to the unique features they each possess.

Q: What's the most exciting aspect of your work?

Eric: There are lots of little things that are exciting about my job—like being in the wilderness and seeing things no one has ever seen before. One time I saw two huge lizards fighting in a life-or-death struggle. It was a once-in-a-lifetime event for me, and I play it back in my mind time and time again.

Q: Do you spend much time traveling?

Eric: I've been almost everywhere! I get invitations to attend meetings and lectures all over the world, and I've done actual field work in most of the world's deserts. Initially, I spent a lot of time studying deserts in the western U.S. Since then, I've been to deserts in southern Africa, India, and Chile. My most current (and oldest) interest is in the deserts of Australia. I haven't had a chance to study the Brazilian Amazon yet, but that's my goal for the future!

Q: Is most of your work done independently or in cooperation with others?

Eric: Dozens of people help me to collect and compile data from the field. I also collaborate with other scientists on projects. For example, right now I am working on a project funded by NASA to use satellite imagery to view sections of deserts. We will use computers to analyze images that are beamed to us from the satellite. The project will allow us to see huge expanses of land that were previously difficult, if not impossible, to study.

Q: Do you have any interesting pets you could tell us about?

Eric: I have a 120-cm lizard, called a Savannah Monitor (from southern Africa), that a graduate student of mine found wandering through a movie-studio parking lot. He thinks she wandered off the set of a horror movie. We named her Basil. She has been very helpful to me in training people how to use radio equipment on and in lizards. We do this to track them and learn what they do. I also have five beautiful baby bison that graze freely on my 40-hectare ranch.

A Project Idea

Select an animal that you are really interested in and place yourself in its shoes! Find a spider, a lizard, a frog, or some other animal in a natural environment. It will take a lot of patience, but watch what it eats, what it does, and where it goes. Ask yourself: Why does it do these things? What other animals does it interact with? Why does it live in this particular environment? What specific features of this animal allow it to be well-suited to this environment? Carefully document everything you observe.

The collared lizard lives in rocky regions of the Southwestern United States.

Science and Technology

Far Out Cuisine

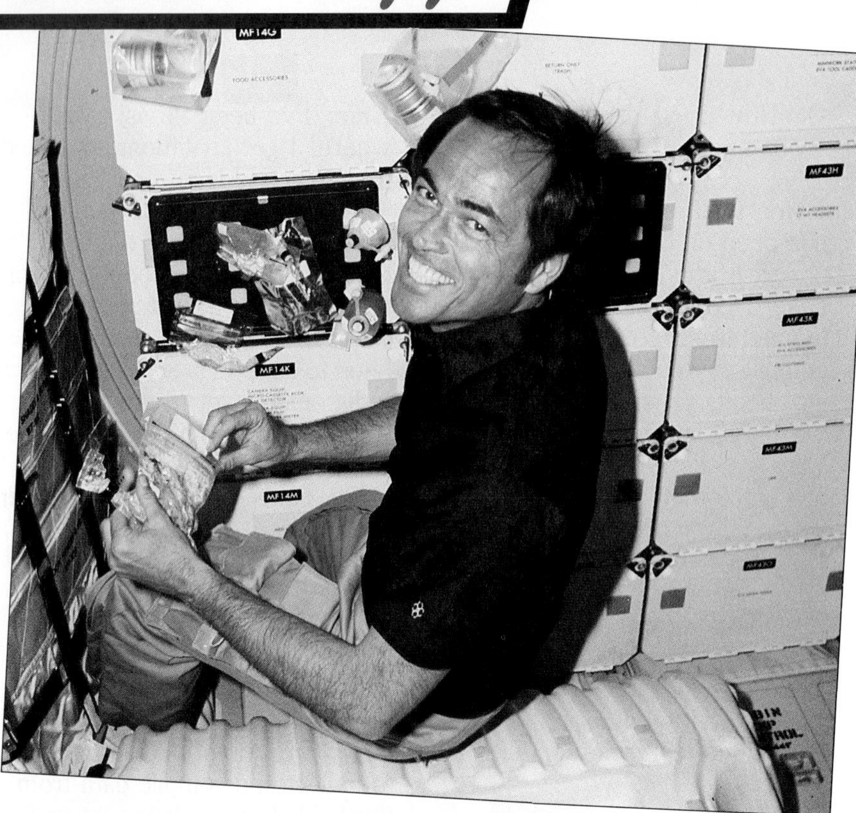

Robert Crippen preparing food on the Space Shuttle Columbia

Do you think gravity is important when you swallow? Would your food go down if you were in a weightless environment? Before the first space flights, scientists were asking these questions. So, one of the things John Glenn did on the first American orbital space flight in 1962 was to eat some food in space.

Fortunately, it turns out that swallowing food in space is no problem. But getting the food to the mouth can be quite another matter. Anything that is not fastened down may float away. How can you have a glass of water with your meal if the glass won't stay put and the water won't stay in the glass?

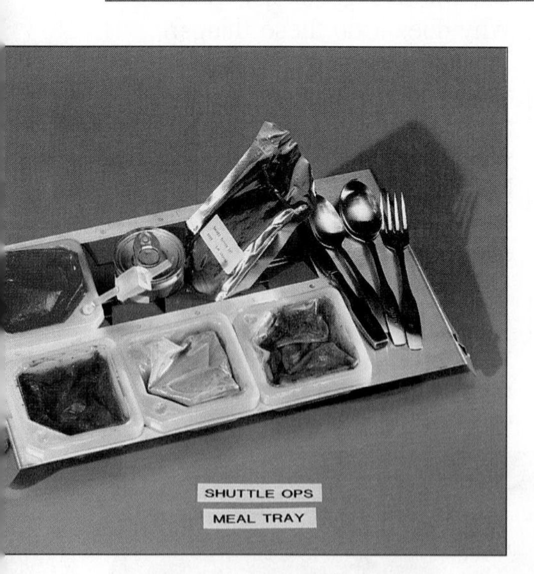

A complete meal for space travelers

Yuck!

Early space food consisted of small dry food cubes and food paste in toothpaste-type tubes. As you might suspect, astronauts quickly found these unappetizing and irritating. Imagine the stress of working and living in space—with the added stress of being hungry with nothing good to eat!

That's More Like It!

Luckily, the menus in space have improved since the early days of space travel. The meal on the left, for example, consists of smoked turkey (foil bag), mixed Italian vegetables, mushroom soup, strawberries, butterscotch pudding (can), and tropical fruit punch. Sound good? Notice that there is a straw attached to the container of fruit punch in the upper left-hand corner. This permits an astronaut to drink the punch without opening the container and allowing the liquid to escape. The silverware is held in place by magnets when it isn't being used.

No Refrigeration Required

All the food the astronauts eat must be in a nonperishable form. This is because there are no refrigerators on the U.S. space shuttles. Each category of food in the list on page 67 will keep for quite some time without refrigeration.

Thermostabilized (T) Foods that, like canned foods, are heat processed. They are stored in cans or pouches.
Intermediate Moisture (IM) Dried foods that have a low moisture content. Dried apricots are an example.
Freeze Dried (FD) These may be eaten as is, or they may need to have hot or cold water added to them.
Rehydratable (R) Dried foods that are prepared by adding water. They are packed in semi-rigid containers with a special provision for water injection.
Natural Form (NF) Includes foods like nuts, cookies, and crunch bars.
Beverages (B) Powdered drinks packed in rehydration containers.

A Typical Menu for a Day in Space

Breakfast	Lunch	Dinner
Dried apricots (IM)	Ground beef w/ pickle sauce (T)	Tuna (T)
Breakfast roll (NF)	Noodles and chicken (R)	Macaroni and cheese (R)
Granola w/ blueberries (R)	Stewed tomatoes (T)	Peas w/butter sauce (R)
Vanilla instant breakfast (B)	Pears (FD)	Peach ambrosia (R)
Grapefruit drink (B)	Almonds (NF)	Chocolate pudding (R) (T)
	Strawberry drink (B)	Lemonade (B)

Space shuttle food is easy to carry on backpacking trips.

Backpacking Tips From the Space Pros

You may think that the space shuttle menu reads like a backpacker's menu. This should not surprise you. When you go backpacking, the food you take must be lightweight, compact, easy to prepare, and nonperishable. These are the same things that are considered when selecting food for space missions.

On some packpacking trips, there are safe, reliable sources of water. In these cases you won't have to carry much water with you. This is important, because water is very heavy! On space-shuttle missions, the water supply comes from the fuel cells. These produce electricity by combining hydrogen and oxygen. Therefore, a by-product of the fuel cell is water, H_2O.

Find Out for Yourself

When possible, NASA astronauts use food that is available in stores. How many of the different types of space foods can you find in stores? Plan a menu for a 3-day backpacking trip.

Do some research to find out about backpacking trips in your area. How would you provide for any special nutritional needs you'd have there? How much water would you need to carry? How much would it weigh?

Unit 2 Machines, Work, and Energy

For Your Journal

Write a paragraph summarizing your thoughts about each of the following:

1. What is work?
2. How are work and energy related?
3. What are machines?
4. How are machines using energy to do work in this photograph?

Machines for Work and Play

Can you think of a time when people did not use machines of some sort? If you went back in time 5500 years, you would discover the first plow. How was this invention beneficial? How did it change peoples' lives?

There were also great construction projects long ago. During the construction of Egypt's Great Pyramids 4500 years ago, huge earthen ramps were constructed to help raise enormously heavy stone blocks into position. A ramp is a kind of machine. A pulley is another kind of machine. Pulleys have been in use for over 3000 years. Archimedes is said to have invented the first combination of pulleys. These were used to haul ships ashore.

Estimate when each of the following machines was invented: motorcycle, internal-combustion engine, jet aircraft, steam locomotive, electric motor, weighing scales, gears, mechanical clocks.

A sampling of machines

Machines in Your Life

Now compare your dates with the actual dates of invention given below.

- weighing scales—3500 B.C.
- gears—100 B.C.
- mechanical clocks—1300
- steam locomotive—1804
- electric motor—1830
- internal-combustion engine—1860
- motorcycle—1885
- jet aircraft—1939

Were you surprised by some of the early dates? Notice those of the 1800s. At that time, in the wake of the Industrial Revolution, machines were developed quickly and in large numbers.

Take a look at the machines shown on this and the facing page. What does each do? How does each use energy?

Now take a moment to consider the following questions.

- How would you define *machine?*
- How many machines have you used so far today?
- Where do machines get their energy?
- What are some hand-operated machines?
- Do they make life easier?
- Do machines save work?
- What is work anyway?
- What is energy?

Some of the questions may stump you now, but by the end of this unit you will be a regular expert in the subject of machines, work, and energy.

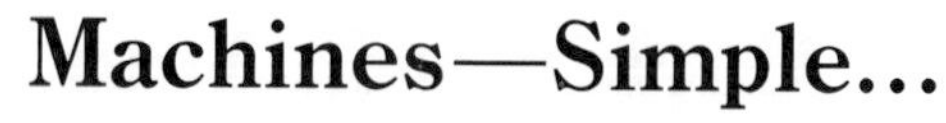

Machines—Simple...

When many people think of machines, they think of complex devices such as cars, photocopiers, or computers. But many machines are quite simple. Just look at the devices below. What does each one do?

...and Not So Simple

The machines pictured below are not simple. They are complex **mechanical systems.** Each of these complicated machines is composed of many simpler machines—*subsystems*—all working together to perform some overall function. What does each of these mechanical systems do?

Familiar Mechanical Systems

You use mechanical systems all the time. Some, like a bicycle, are complex. Others, like a ballpoint pen, are simpler. Even the pen has a surprising number of parts and subsystems, each of which performs functions that contribute to its overall function. The subsystems of a bicycle are more obvious. The braking subsystem is just one example. How many more subsystems of a bicycle can you identify? What does each do? What parts make up each subsystem?

You are able to move faster on a bicycle than you could by walking. How does the bicycle accomplish this increase in speed? As you progress through this unit, you will be investigating many mechanical systems, their subsystems, and their functions. You will also discover how they can increase either force or speed.

Work and Energy

Some machines are hand-powered—you have to do some work to use them. A shovel is a very simple kind of machine that requires you to do the work. If you have to dig a trench, a shovel helps to make your work easier. Imagine clearing away the dirt without it—using your hands only! Of course, it's still easier to use a power trencher. In this case, though, you don't do the real work. The energy to dig up the dirt and to move the machine forward comes from the gasoline burned by the trencher's engine. You do just a little work in operating the machine's controls.

Notice the words "work" and "energy" in the preceding paragraph. What do they mean? You probably have a fairly clear idea of what work and energy are, whether you realize it or not. The lessons that follow will help you define both terms in a scientific sense.

The Idea of Work

What Is Work?

The word *work* is used several different ways in the statements below. How many different meanings can you find?

"Tom *works* part-time after school."
"What kind of *work* does he do?"
"He shovels snow. At least that's not as much *work* as mowing grass."
"Yes—pushing the mower back and forth sure is hard *work*."
"It's not as hard as home*work*."
"Thinking is hard *work* too!"
"Our next unit in science is on machines, *work,* and energy."
"Machines sure save a lot of *work*."
"I hope I have the energy to do the *work*!"
"Get to *work*!" interrupts a voice from the front.
"All *work* and no play makes for a very dull day!"

As you can see from the statements on the previous page, the word *work* can be used in many different ways with very different meanings. Look at the statements again. Now, in your Journal, group the statements by their meanings. Which statements reflect what you would consider to be a scientific definition of work? Which statements suggest that work is:

- a vocation or a job?
- anything that occupies our time?
- the opposite of leisure or recreation?
- a way to earn money?
- connected with forces?
- connected with motion?
- an assignment?

Some statements fit into more than one category.

In science, *work* has a very precise meaning. Thinking would not be called work. Preparing for an exam would not be work. Work, in the scientific sense, requires that a force be applied to an object and that the object move. Another way to define work is "force applied through a distance." Try to group the sentences on the previous page and the pictures on these two pages into examples of work in a scientific sense and a non-scientific sense.

Another word that is often used with work is *energy*. Work and energy are obviously related, but do you really know what energy is? "Energy," too, can have different meanings. Construct some sentences using the different meanings of "energy". One way to start would be to look up "energy" in a dictionary.

Work and energy—how are they related? In this unit, you will explore both.

The Scientific Idea of Work

Work in Progress

It's moving day in the library. The new shelves have arrived and the librarian has recruited ten volunteers to help put the books back. "Here's your job," the librarian announces. "The books are stacked on the floor by the shelves. Use the call numbers on the books' spines to place each book on the proper shelf."

Each volunteer has a different task. Look at the illustrations on this page showing the work of five of the volunteers at one specific moment. Then answer these questions.

- Who is doing the least amount of work?
- Who is doing the greatest amount of work?
- Are some volunteers doing the same amount of work?
- Can you place the volunteers in order according to the amount of work each is doing?

Assume that each book has the same weight.

Lifting a book 2 m obviously requires twice the work needed to lift a book 1 m. Therefore, Kyle must do twice as much work as Marie. What about Mark? Did you conclude that the amount of work depends on the distance through which the force is exerted?

To lift two books 1 m requires twice the work needed to lift one book 1 m. So Roberta does twice as much work as Marie. When you lift two books, you exert twice the force compared to when you lift one book. Therefore, the amount of work also depends on the size of the force exerted.

Do you see why Jean does more work than Roberta and Kyle? The order of the volunteers, from least to most work done, is Marie, Kyle/Roberta, Jean, and Mark.

On the next page is a second set of illustrations to study. Answer the same questions as before. Also, answer this question: What is a formula for determining how much work is done?

Calculating Work

Did you find a good method for calculating how much work each volunteer does? If you multiply the force exerted by the distance through which the force is exerted, you will obtain a quantity that indicates the amount of work done:

work done = force × distance
(when the force exerted is constant)

Suppose that each book weighs 10 newtons (N). How much work does Marie do in lifting one book up to the second shelf? She exerts an upward force of 10 N to lift the book. She exerts this force for 1 meter.

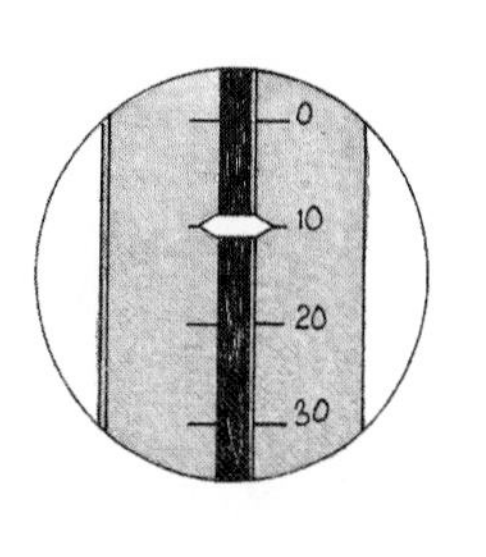

work = 10 N × 1 m = 10 N·m = 10 J

The unit for work is the *joule* (J). It is named after the English scientist James Prescott Joule.

> **The Joule Rule**
> One *joule* of work is done when a force of 1 N is exerted for a distance of 1 m:
> 1 J = 1 N × 1 m

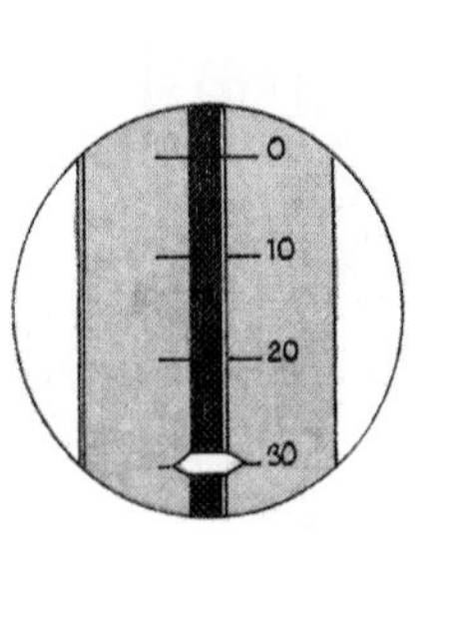

How much work has Kyle done?

Now that you know the scientific formula for work, calculate the work being done by each set of volunteers. Do your results verify your earlier rankings of the volunteers in terms of how much work they are doing?

What Work Feels Like

Here are five tasks to help you understand what work really means, in scientific terms.

Task 1

A Work-Experience Display

You Will Need

- 1 sheet of paper
- scissors
- measuring tape
- 100-g, 1-kg, 5-kg masses
- masking or transparent tape

What to Do

1. Make a set of labels like the ones below.

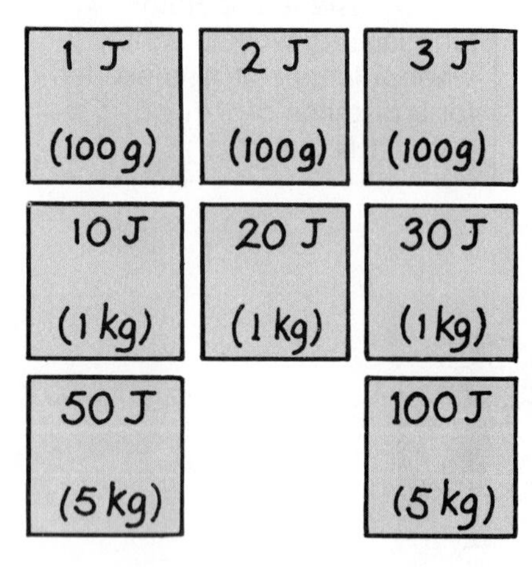

2. Choose one label. What height above the floor must you raise the mass indicated to experience the amount of work shown on the label?
3. Lift the mass to this height and tape the label on the wall.
4. Repeat for each of the labels.

You are also lifting part of your body when you do this work. Can you calculate how many extra joules of work are needed?

Task 2

You're Pushing It!

You Will Need

- a chair
- a bathroom scale

What to Do

1. Hold the bathroom scale up against the back of the chair as shown in the illustration. As smoothly as you can, push the chair with a person seated in it for a distance of 30 cm. Does it matter where you place the bathroom scale as you push?

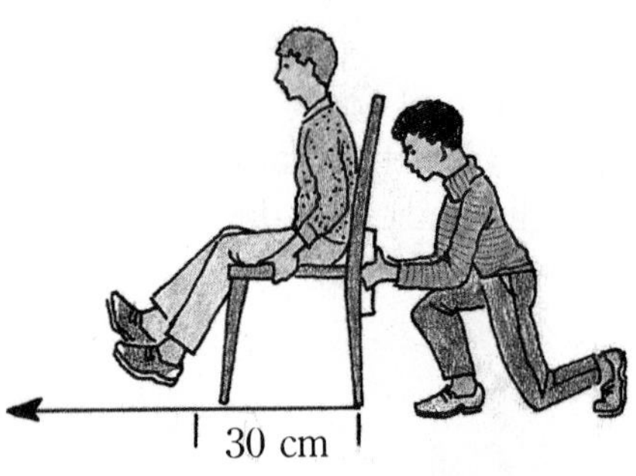

2. Determine the force needed to move the chair the measured distance. Calculate the work done.
3. Would you do the same amount of work if you lifted the chair and the person to a height of 30 cm? Explain.
4. Will the amount of work done change if a different person sits in the same chair? Explain.

If the scale measures in kilograms, multiply by 10 to get the force in newtons; if it measures in pounds, multiply by 4.5.

Task 3

Comparing Different Examples of Work

You Will Need

- a force meter
- various objects

What to Do

Try these activities:

(a) Open a door.
(b) Pull out a drawer.
(c) Lift this book from the floor to directly over your head.
(d) Pull a 1-kg mass along the floor for 2 m.
(e) Pull down a screen or blind.

Find out how many joules of work are required to do each task. To do so, first use a spring scale or force meter to determine the size of the force needed for each task. Then measure the distance through which the force is applied.

Task 4

Work While You Exercise

You Will Need

- a meter stick
- a pull-up bar

What to Do

1. Calculate your weight in newtons.
2. Do a pull-up.
3. Have someone measure how high you lift your weight.
4. Calculate the work you do in one pull-up and in five pull-ups. What muscles are doing this work?
5. Repeat this task, this time doing a vertical jump. Have someone measure the distance from your heels to the floor, at the highest point of your jump with your legs straight.

Task 5

You Will Need

- 1-kg bag of gravel (or other 1-kg masses)
- 2 pulleys, with cord
- a force meter or spring scale

Machines Also Work

Answer the following questions about the diagram below.

(a) How much upward force must the pulley machine put on 1 kg of gravel to lift it?

(b) If the gravel is raised 0.25 m how much work does the machine do?

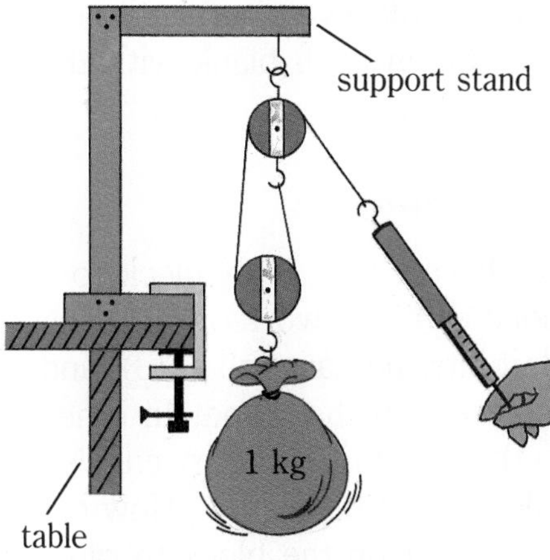

What to Do

1. Arrange the setup as shown.
2. Measure with a force meter the force needed to operate the machine.
3. Measure how far you need to pull the spring to lift the gravel 0.5 m.
4. Calculate the work you put on the machine. Compare your work with the work the machine puts on the gravel.

Why is it helpful to use the machine?

A Powerful Idea

By now you should have a good feel for the scientific meaning of work. But what is the scientific meaning of *power,* an energy-related word we often use? Take a moment to write down your own definition of power. Using your definition, which is more powerful, a horse or a car? a diesel locomotive or a jet aircraft? How powerful are you? How would you find out?

What to Do

Find a staircase or a steep hill that isn't too tall. Measure the vertical distance to the top. Find out how much work you would do in propelling yourself to the top. As fast as you can, run to the top of the hill or staircase. (Don't stumble!) Time how long it takes. To find your *power* in joules per second, or *watts,* divide the amount of work you did by the time it took to do it.

How did the definition of power you wrote earlier compare with the definition of power that you just learned?

A horse can do about 750 J of work in 1 second, or has a *power* of 750 W. The engine in a 6-cylinder compact car can do about 100,000 J in 1 second. That is, it has a power of 100,000 W, or 100 kW.

Work and Energy

What Do You Need to Do Work?

Look at the diagram below. What do you think its message is? Write a caption for this diagram.

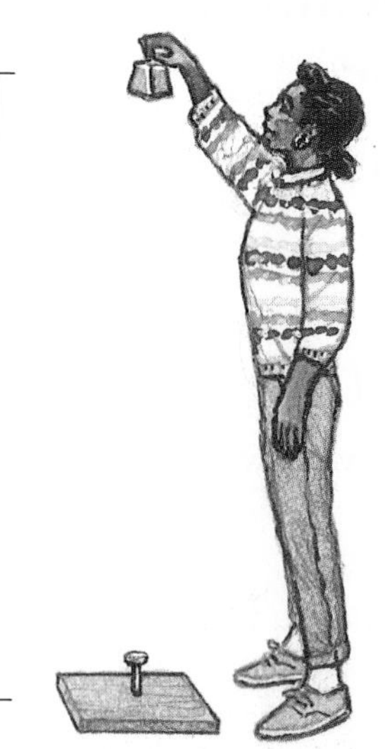

The Work/Energy story that follows will give you many new ideas that relate work with energy. Fill in each blank with the correct term or terms.

The Work/Energy Story

Energy enables us to do (1). If you lift a 10-N block to a height of 2 m, you will have done (2) J of work on it. If you now let the block fall on a nail, it hits the nail with a (3) and the nail is pushed a short distance into the wood. In other words, the block does (4) on the nail. Since energy enables work to be done, the raised block must have (5). How did the block get it? By the (6) you put on the block to raise it.

The energy of raised objects is called potential (stored-up) energy. Since you did 20 J of work on the block, the potential energy of the raised block must be (7) J. You, of course, got your energy from the (8) you ate. If you had felt the top of the nail, it would be warm. As the block did work on the nail, some of the kinetic energy was changed into (9) energy. Note that in every case when work was done, (10) was transferred.

Answers: [1]work, [2]20, [3]force, [4]work, [5]energy, [6]work, [7]20, [8]food, [9]heat, [10]energy

Potential and Kinetic Energy

Alexandra has a lot of *potential* as a hockey player. What does *potential* mean? It doesn't mean that she's a good player yet. But she does have the ability to become one.

A raised object has **potential energy.** This does not mean that the raised object is doing work in that position. Rather, it means that the object can perform work on something else, if it is released.

When you lift the object, you do *work* to overcome the gravitational force on the object. The object now has potential energy. When the object is released, gravitational force acts on it, causing it to accelerate downward. The faster the object moves, the more kinetic energy the object has.

A raised object is only one example of potential energy. Potential energy also takes other forms. Work can be done on objects by overcoming other forces, and in so doing give the objects potential energy. For example, if you pull back on a bowstring, you are overcoming opposing forces—the tensions of the bow and the bowstring. The work you do against the opposing forces gives potential energy to the bow. When the bowstring is released, the elastic forces of the bow and bowstring send the arrow flying. In turn, the arrow does work on any object it hits.

In the above example, the potential energy of the drawn bow is converted into **kinetic energy** in the form of the flying arrow. Kinetic energy is the energy of motion. Kinetic energy takes many different forms. Can you think of some? Can you think of examples in which potential energy is converted to kinetic energy?

Potential-Energy Study

The illustrations on these pages show objects that have potential energy. Working with a partner, answer the following questions for each situation.

1. Which objects or substances have potential energy?
2. How is the potential energy stored?
3. What opposing force is overcome to store the potential energy?
4. How is the potential energy put to work?

Consider the bow and arrow shown on page 81. When the archer pulls back on the string, the bow bends to store potential energy. When the archer releases the string, the bow snaps back to its original position, forcing the arrow forward.

Note that the energy represented by (f) and (l) is stored in food and fuel, respectively. This energy is released by "chemical changes." For this reason, the potential energy in food and fuel is often called **chemical energy.**

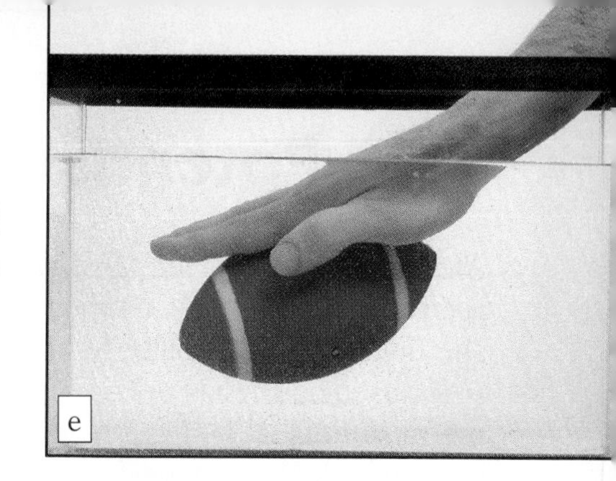

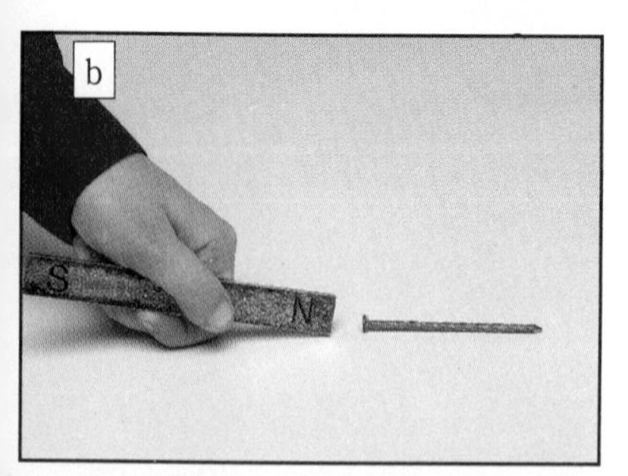

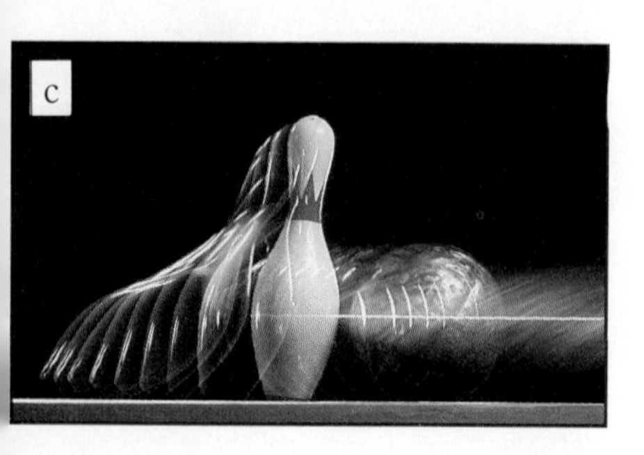

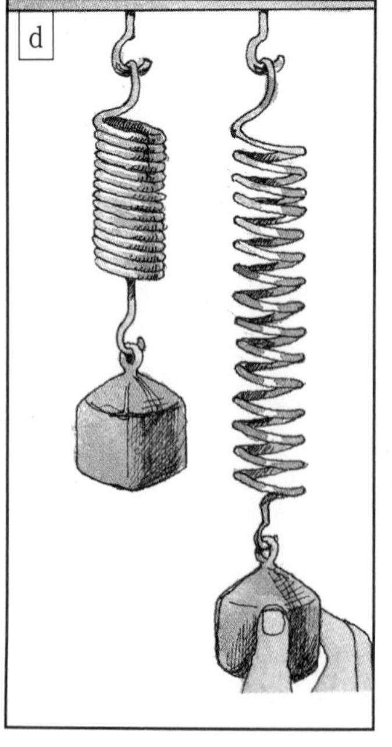

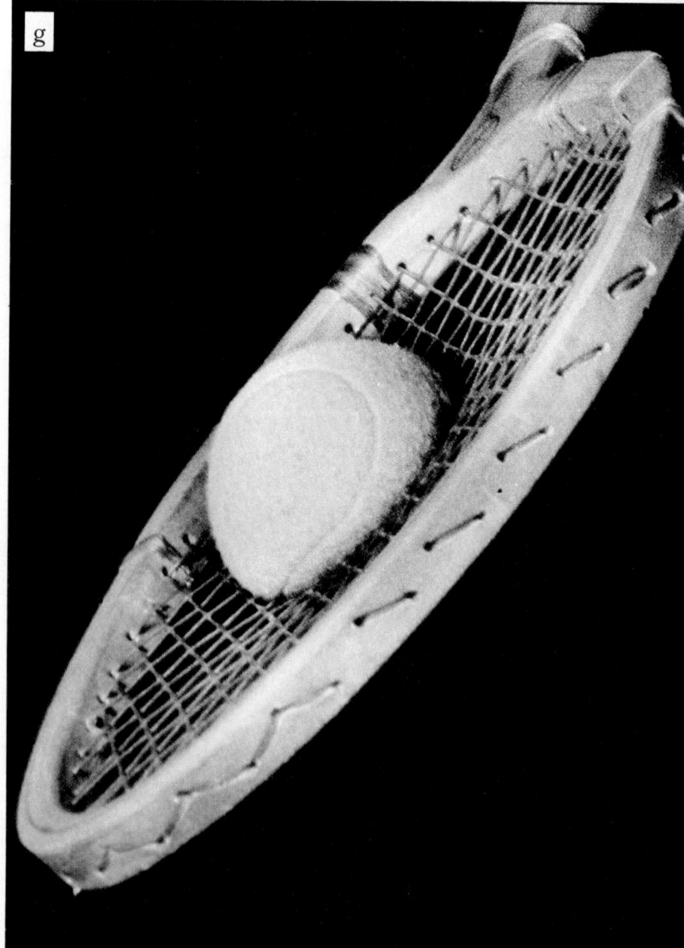

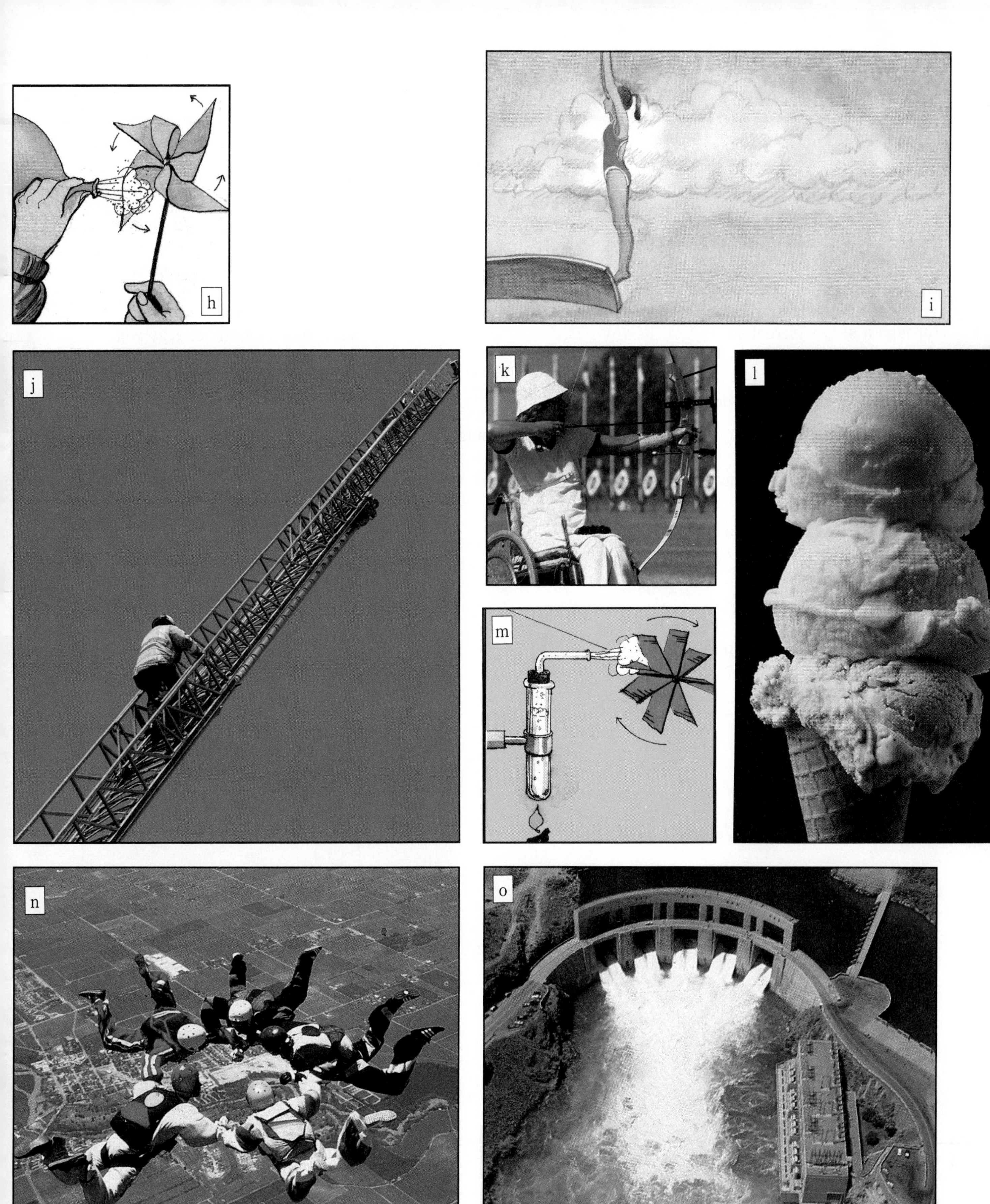
h
i
j
k
l
m
n
o

Another Form of Energy: An Investigation

In this Exploration, you will investigate the questions below through a series of real and imagined experiences.

- Do moving objects do work?
- Do moving objects have energy?
- If moving objects have energy, where do they get it?
- What factors affect the amount of energy moving objects might have?

The first two experiences are thought experiments; only the last one is a laboratory experience. Put on your thinking cap!

Experience 1

Does the moving bowling ball in the illustration below do work when it hits the pins? If the bowling ball does work, what must the moving ball have? Where does this energy come from?

Rodney puts the bowling ball in motion by applying force to it. He does this by swinging the ball forward. If Rodney puts 100 N of force on the ball through a swing of 1 m, how much work does he do? How much kinetic energy does the bowling ball have? How much work will the bowling ball be able to do when it strikes the pins?

Experience 2

A 2-kg brick is raised to a height of 3 m. How much potential energy does it have? When it is dropped, it moves downward with increasing speed. What kind of energy does it have now? How much kinetic energy will it have as it just begins to strike the nail? Where does this energy come from? Does the moving brick do work on the nail? How much work does it do?

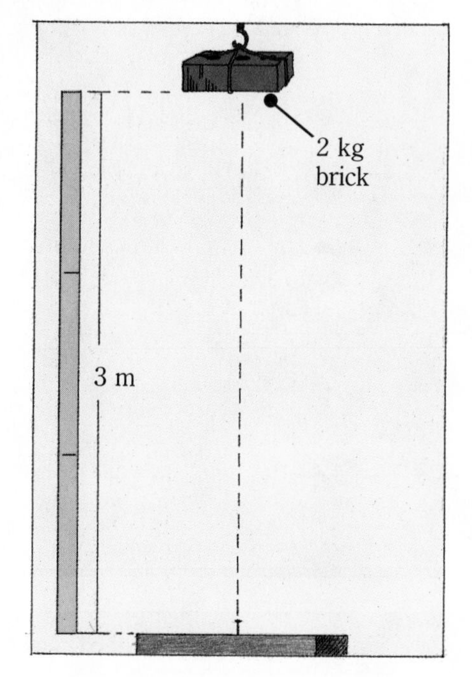

- In Experience 1, the source of the kinetic energy was Rodney. Where did Rodney get the energy to do this work?
- In Experience 2, the source of the brick's kinetic energy was the potential energy it had before it was dropped. Whenever energy is being transformed from one form to another, work is being done. Therefore, work must have been done as the potential energy of the brick was changed into kinetic energy. What did work on the brick?

Experience 3

What factors affect the amount of kinetic energy a moving body has?

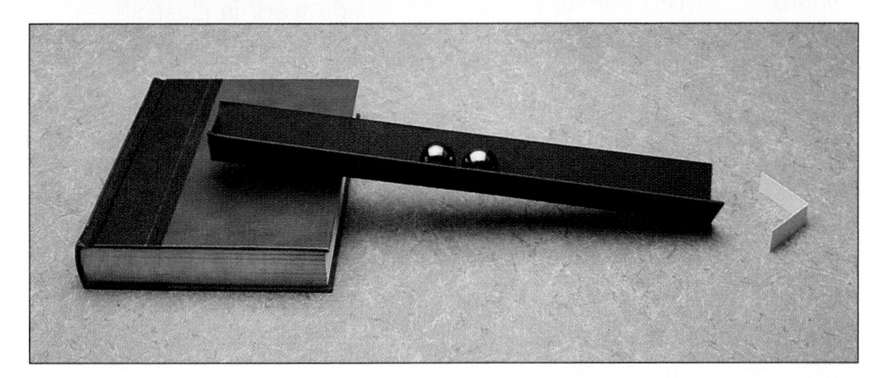

You Will Need

- a trough (made of metal or cardboard approximately 60 cm long)
- 3 identical books
- 3 marbles
- a piece of cardboard (approximately 2 cm × 15 cm)

What to Do

1. Raise the trough by placing a book under it. Let a marble roll down the trough to strike a piece of cardboard folded as in the drawing. What happens? Is work done by the marble? What kind of energy does the marble have before it starts moving? after it starts rolling?

   The distance moved by the cardboard is an indication of the amount of kinetic energy. Roll the marble again, and measure the distance the cardboard moves from the starting point.
2. Add another book to raise the trough more. Roll the marble from the same spot on the trough. How does the speed of this marble compare with that of the marble in step 1? Measure and record the distance the cardboard moves. How does the kinetic energy compare in each case? What is one characteristic of the marble that affects how much kinetic energy it has?
3. Try using three books. Record the distance that the piece of cardboard moves.
4. Place the trough on one book again. Roll two marbles down the trough. Compare the distance the cardboard moves with that in step 1. What causes the difference? Would you expect the same result if the two marbles were made into a single large marble? What is another characteristic that affects the amount of kinetic energy?
5. Check the conclusion you drew in step 4 by using three marbles.
6. Where does the kinetic energy of the marble come from? How does the potential energy of the marble (before rolling) compare in steps 1, 2, and 3? How does its kinetic energy (just before striking the cardboard) compare in steps 1, 2, and 3?

Summary: Which factors (variables) do you think affect the amount of kinetic energy a moving body has? Make a list in your Journal.

❑

4 Energy Changes

How many different types of energy have you seen so far? Can one type of energy change into another type of energy? Look over the last few pages and discover some examples where this happens.

The following Exploration features a number of investigations. Some are thought investigations that can be performed best by talking about them; others you may actually perform.

Below, the major forms of energy are defined. As you do this lesson, try to identify these forms of energy as they are mentioned.

Potential: Stored energy; gained by lifting an object, stretching or compressing a spring, and so on.

Chemical: A form of potential energy stored in foods and fuels.

Kinetic: The energy in moving things.

Heat: A form of energy released when substances are burned, or when surfaces are rubbed together.

Electrical: The kinetic energy of electrically charged particles.

EXPLORATION 4

An Energetic Discussion

Working in groups, discuss question 1 and any two others that interest you. Your group should be prepared to discuss the results with the class.

1. Refer to the "Potential-Energy Study" on pages 82 and 83 to answer the following questions.

(a) How is potential energy changed to kinetic energy in each situation in the "Potential Energy Study" on pages 82–83?

(b) When energy is transferred, work is done. What agent is doing the work as potential energy changes into kinetic energy?

(c) How does kinetic energy do work in illustrations (a) and (h)? Any moving part of a machine has kinetic energy, sometimes called **mechanical energy.** Can you think of other examples? In which illustration(s) is kinetic energy eventually changed into **electrical energy?**

2. Hold a marble at the rim of a large bowl. Release the marble. What happens? You know that potential energy can change into kinetic energy. Can kinetic energy change into potential energy? When does the marble have the most potential energy? the most kinetic energy?

3. Marta lets go of a ball that is attached to the end of a rope suspended from the ceiling. What happens to the ball? Do you think Marta would be safe as long as she stood in the same place? What would happen if she moved forward one pace? Is there a "reversible" change of energy going on here?

4. You are at the circus watching an acrobatic act. One acrobat is standing on a raised platform. A second acrobat is standing on the end of the seesaw. Their initial positions are both labeled *1*. The illustration shows two other positions for each acrobat after the one on the platform jumps off (positions 2 to 5). What happens? Can you describe the potential and kinetic energy of each acrobat in each position? How do you explain the fact that the second acrobat ends up on a platform higher than the platform used by the first acrobat?

5. Diagram (a) represents the lifting of a brick through a vertical distance of 6 m, in two steps. The brick has a mass of 2 kg. Diagram (b) represents the releasing of the same brick.

In your Journal, calculate the amount of potential and kinetic energy that you think the brick has in each position in the two diagrams.

Diagram (a)

Height	Position	Potential Energy	Kinetic Energy
6 m	Step 2	?	?
3 m	Step 1	60 J	0 J
0 m		0 J	0 J

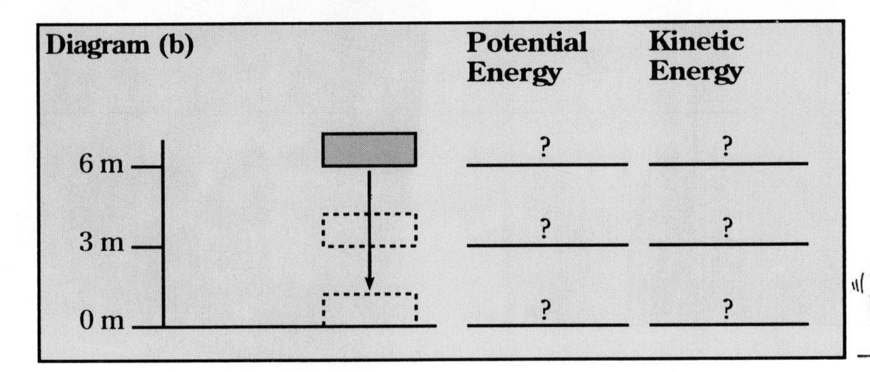

6. Think of dragging a table across the floor. What kind of energy does the table have? When you stop putting a force on the table, it stops moving. Where does the energy go?

7. Touch a nickel to your chin. Place this nickel flat on a piece of paper. While pressing down hard on it, move it back and forth briskly 10 to 20 times. Now touch the nickel to your chin. What do you observe? What happened to the nickel?

8. Look at the illustration below. Where have you seen this before? When is the potential energy the smallest? Why does the ball stop bouncing? What other type of energy might be present?

Discovering New Forms of Energy

You have become acquainted with some of the ways in which energy can be converted from one form to another. The photos on this page show a number of energy-conversion devices. In your Journal, make a table like the one below. For each energy conversion device, indicate the form of energy that goes into the device, the form or forms of energy that come out, and any steps in between.

Example	Energy Converter	Energy Change
(e)	toaster	electrical to . . . .

(c)

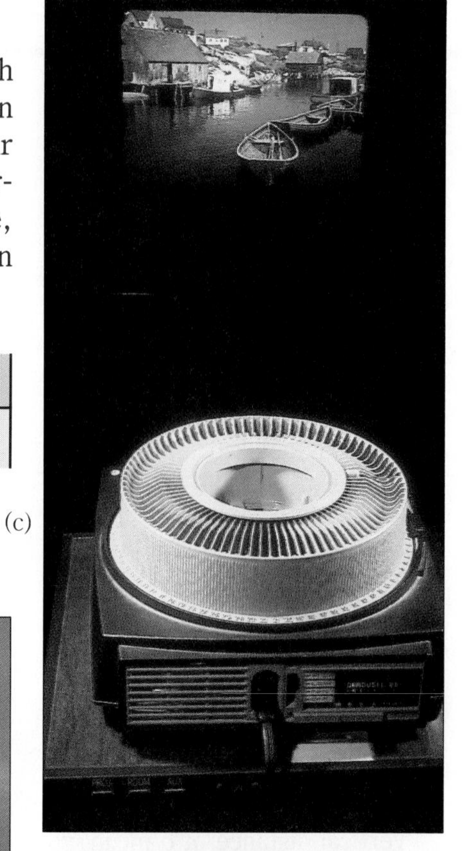

(a)

(b)

(d)

(e)

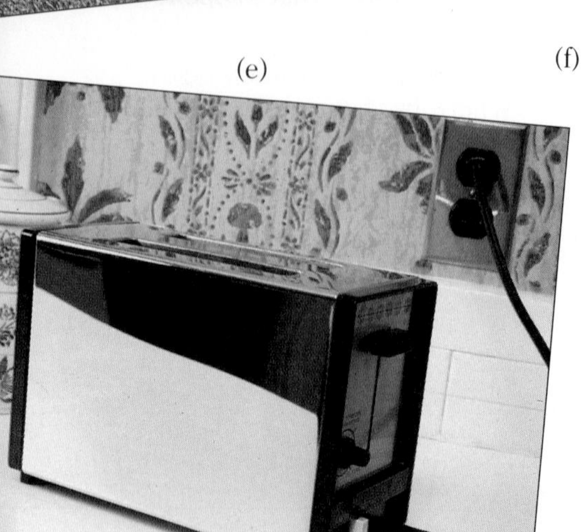

(f)

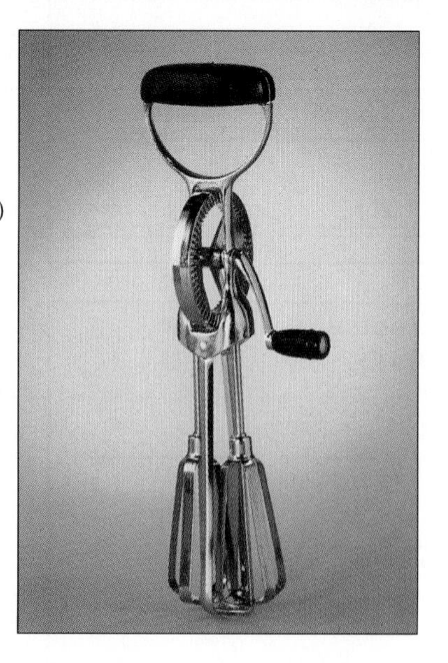

(g)

Challenge Your Thinking

1. The illustration below shows a fanciful machine, a type of imaginary mechanical system. This machine is an absurdly complicated device that is designed to accomplish a simple task.

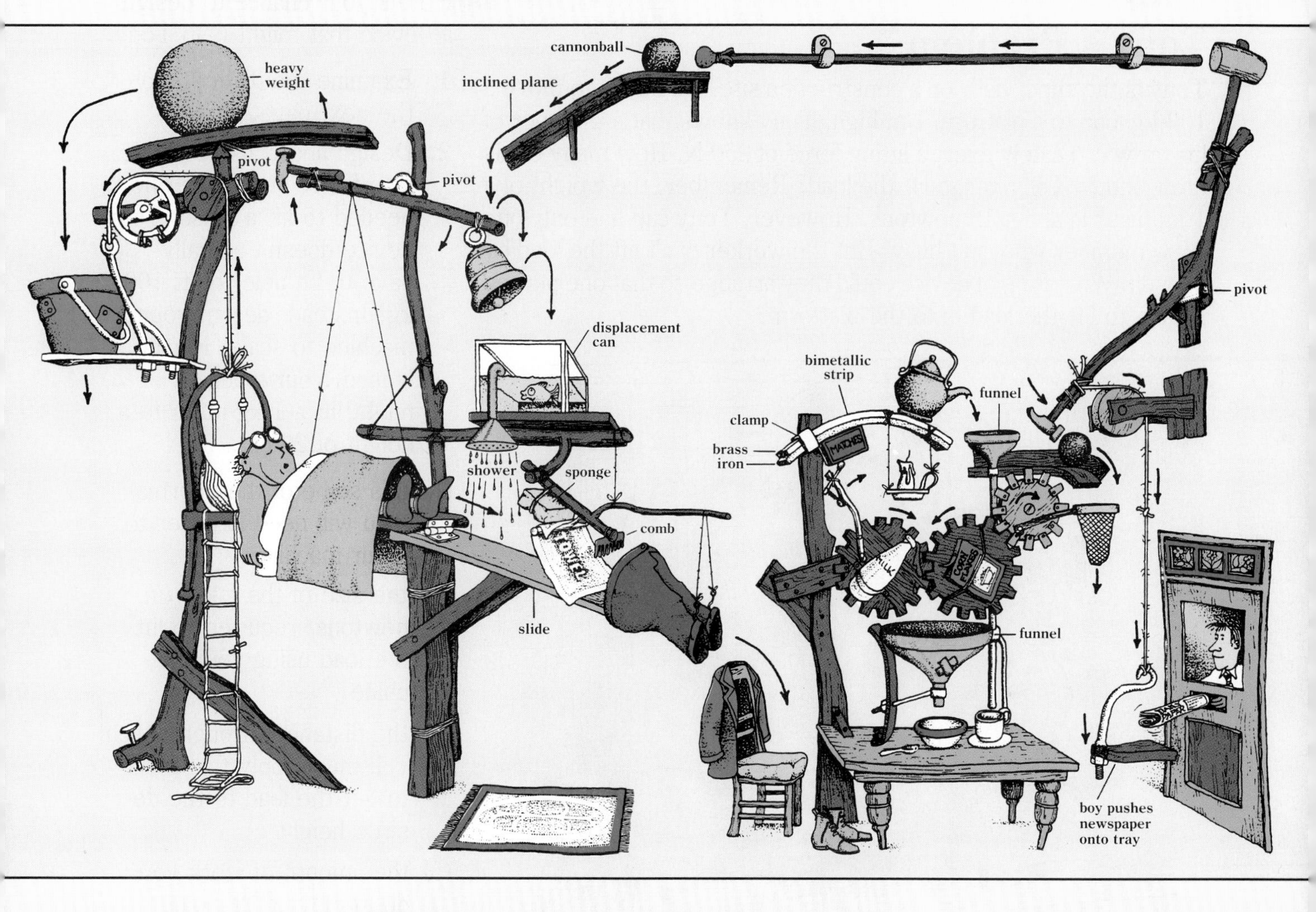

(a) What is this mechanical systems's overall function?
(b) Identify the subsystems.
(c) Where is energy being converted?
(d) Could a machine like this exist in real life? Why or why not?

2. Describe how this machine works.

3. Design your own fanciful machine, for example, to squeeze toothpaste onto a toothbrush, or open and close a door.

4. Do you agree or disagree with the following statements? Explain your answers.
   (a) A machine is a device for converting or transferring energy.
   (b) If work is being done, then energy is being converted.
   (c) Work is a form of energy.

Lightening the Load

Tony's Problem

Tony is the supervisor on a construction site. He needs to move a 100-kg load to a platform 1 m high. Tony knows that a member of his crew can safely exert a lifting force of 250 N. How many of his crew should Tony get to lift the load? Remember, the weight of a 1-kg mass is about 10 newtons. However, Tony can find only one crew member who isn't busy. But the worker can't lift the load by himself. What type of device could they arrange so that one person is able to lift the load onto the platform?

Helping Tony

Here's your chance to design a device that might help Tony.

1. Examine the diagrams on the next page.
2. Design and construct a device for reducing the effort needed to lift a mass. Your device doesn't actually have to be able to lift 100 kg! Instead, design your machine to scale. For instance, your machine might lift a 1 kg-mass to a height of 20 cm.

Here is some of the information you will need in order to test your machine:

(a) the size of the force (in newtons) required to lift the load using your machine

(b) the distance through which you must apply that force to get the load to the desired height

(c) the amount of work you do in moving the load

(d) the comparison of the amount of work you do with your machine to the amount of work you would do without the machine

What might you do to improve your machine? Can you make the load move with an even smaller force? If you are able to do this, is there a trade-off somehow? Does your machine actually *save* work?

Some possible devices to help Tony.

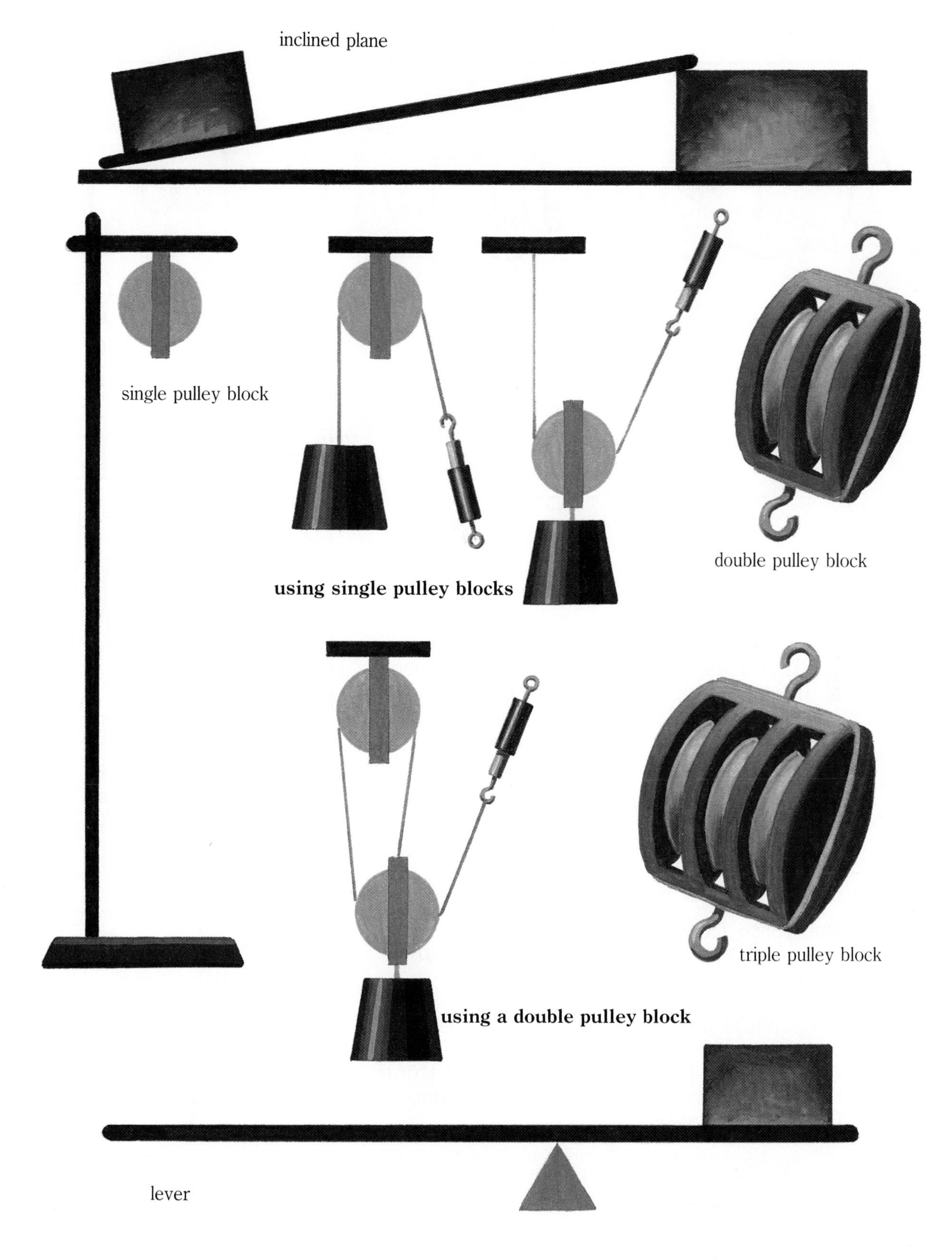

A Look at Two Solutions

Leslie's Machine

Leslie used an arrangement of pulleys to solve Tony's problem. Here's Leslie's report about the machine she made. From the information in her report, solve the problems on the facing page.

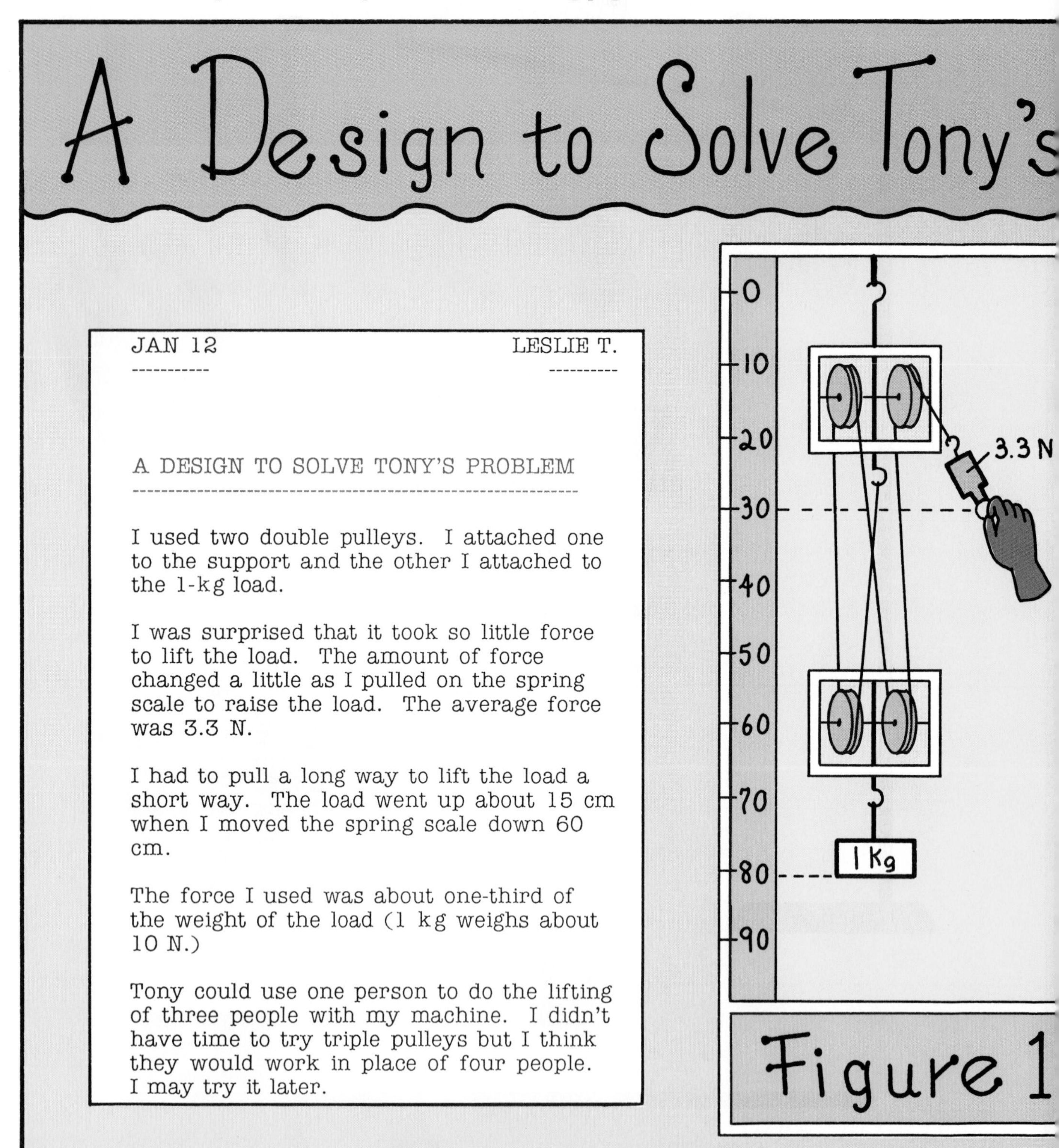

JAN 12 LESLIE T.

A DESIGN TO SOLVE TONY'S PROBLEM

I used two double pulleys. I attached one to the support and the other I attached to the 1-kg load.

I was surprised that it took so little force to lift the load. The amount of force changed a little as I pulled on the spring scale to raise the load. The average force was 3.3 N.

I had to pull a long way to lift the load a short way. The load went up about 15 cm when I moved the spring scale down 60 cm.

The force I used was about one-third of the weight of the load (1 kg weighs about 10 N.)

Tony could use one person to do the lifting of three people with my machine. I didn't have time to try triple pulleys but I think they would work in place of four people. I may try it later.

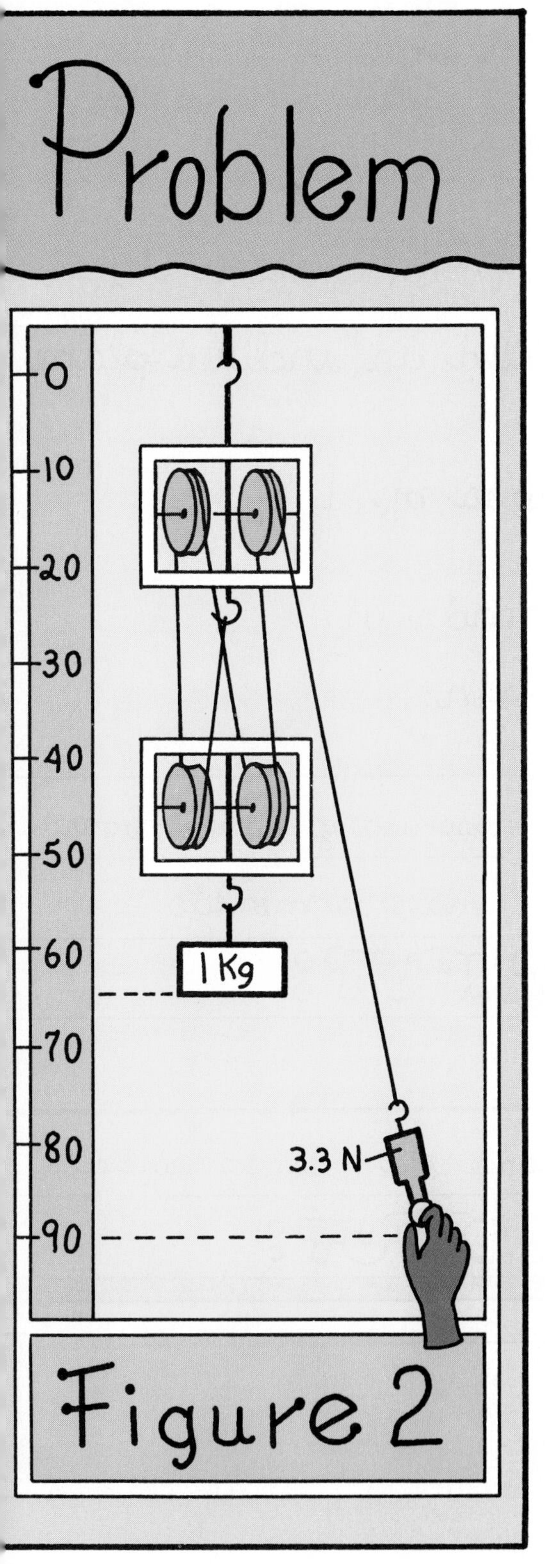

Problem 1

(a) How much force did Leslie put on the string?

(b) How far did she move the string? How far did the load rise?

(c) How many joules of work did Leslie do in using the machine?

Machines are commonly used to make work easier. The work put into a machine is called **work input.**

Problem 2

(a) What upward force does the machine (pulley and strings) place on the load? (This would be the same as the force you would need to lift the load without any machine at all.)

(b) How far is the load raised by the machine?

(c) How much work was done on the load by the machine?

The work done by a machine on a load is commonly called **work output;** that is, it is the work you get out of the machine, regardless of the work you put into it.

Problem 3

How does the work output in 2(c) compare with the work input in 1(c)? What do you think accounts for these results?

Pam's Machine

Pam had her own solution. She tried three different versions of her slanted-board idea. A slanted board is a type of simple machine called an *inclined plane*. Pam tried three boards of different lengths.

Complete her table in your Journal, then write some conclusions that she might draw from her solution. The questions below will help you.

Questions

1. Does the length of the board affect the force needed to pull the load? If so, how?
2. Compare the *work output* (the amount of work actually done by the machine) to the *work input* (the amount of work put into the machine). What do you notice? Compare the work input and work output of Leslie's machine. Which of the two machines was better at converting work input to work output?
3. Would a long inclined plane and a short inclined plane be equally good at converting work input to work output? How and why might they differ?
4. Which solution, Pam's or Leslie's, is better? Why?

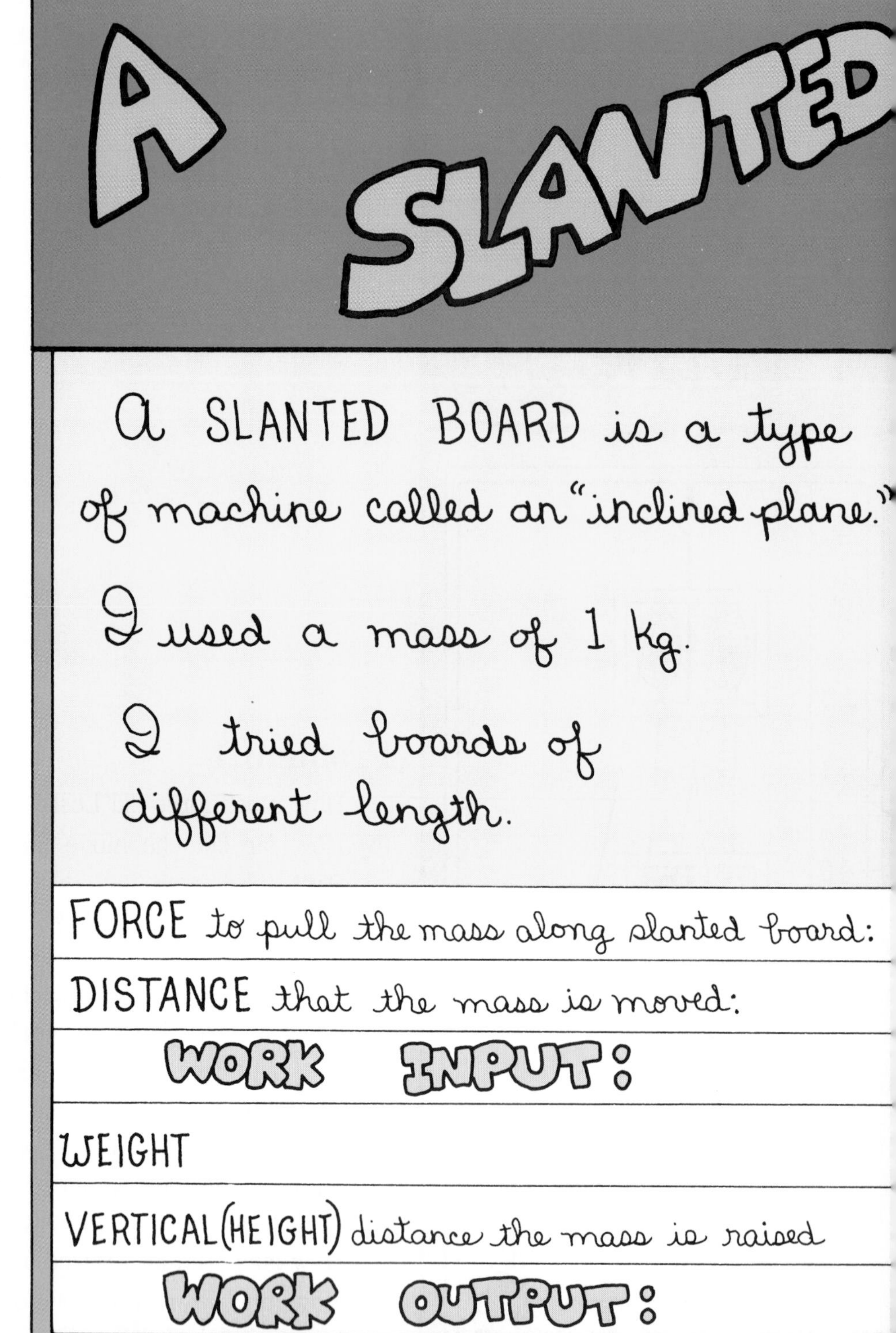

BOARD MACHINE

by Pam S.

Force to pull the mass along the board = 2.5 N	Force to pull = 3.2 N	Force = 2.1 N
1 m 1 Kg	0.7 m	1.3 m
Weight = 10 N		
2.5 N		
1 m		
2.5 N × 1 m = 2.5 J		
10 N (Weight of 1 Kg mass = 10 N)		
0.15 m		
0.15 m × 10 N = 1.5 J		

Some Machines Are Better Than Others

Which of the following machines would you prefer for a job?

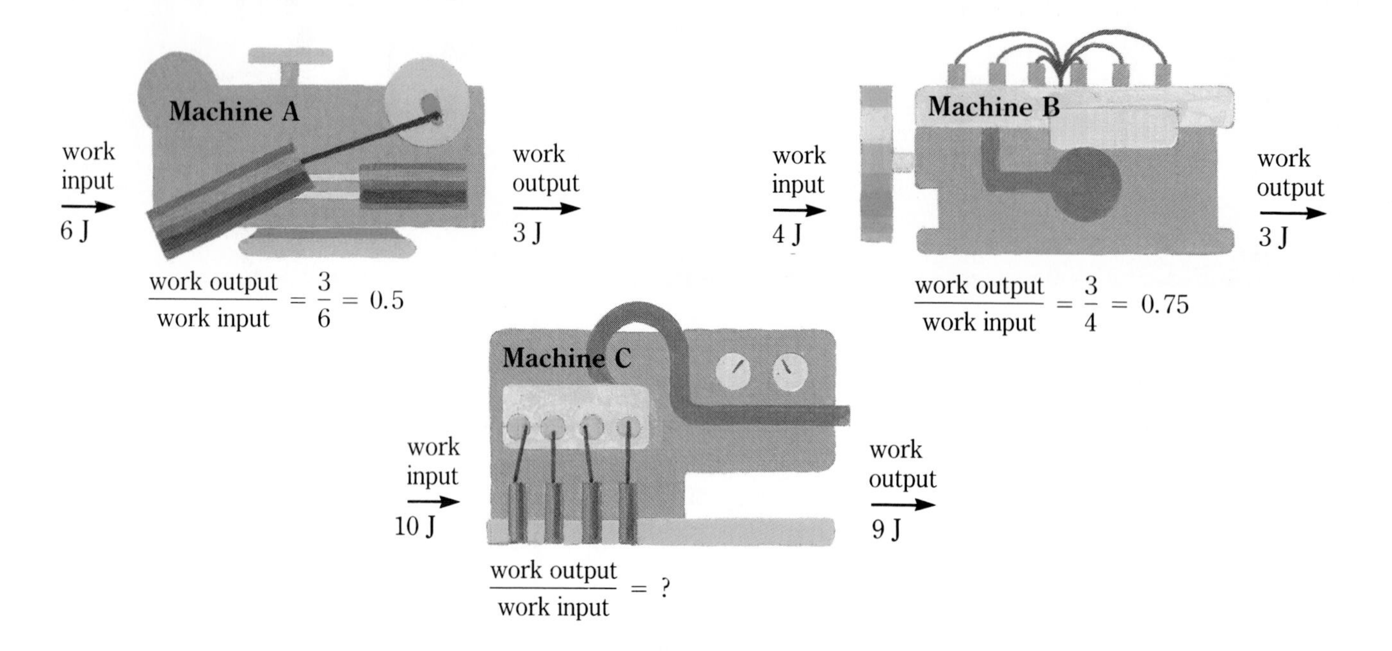

If you chose Machine C, you probably did so because it is better at converting work input to work output. Machine C is more *efficient.* There is a way to measure how efficient a machine is. The fraction:

$$\frac{\text{work output}}{\text{work input}}$$

tells you how much of the work put into a machine becomes useful work done by the machine. This fraction is called the **efficiency** of the machine. Efficiency can be expressed as a percentage by multiplying by 100. For instance, Machine A is $0.5 \times 100 = 50$ percent efficient.

Machine B is 75 percent efficient. This means that 75 percent of the work put into Machine B emerges as useful work done by the machine. What is the efficiency of Machine C? What is the efficiency of Leslie's machine? of Pam's machine?

Now consider the machines your class designed to solve Tony's problem. Compute the efficiency of your machine. Compare your machine's efficiency with that of each of the other designs. What is the highest efficiency? How do you think your machine might be made more efficient?

What Do You Think?

Look at the machine below. Do you think that such a machine is possible? Why or why not? How can Pam and Leslie's results help you reach a conclusion?

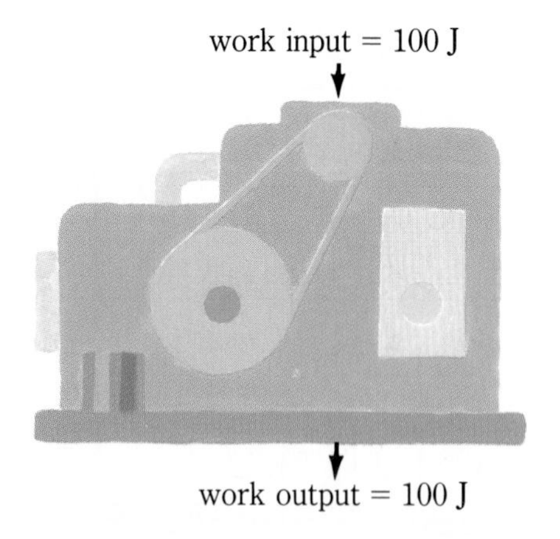

In fact, no machine can be 100 percent efficient. Some of the energy is "wasted" and does not go into work output. What do you think happens to this energy?

The Advantage of Machines

If machines do not *save* work, is there any advantage in using them? Indeed there is! With a machine, one person can do the work of many. A small force can be turned into a larger force.

Leslie, using her machine, can lift the load of 10 N with a force of only 3.3 N. She increases her force three times! You can see this by using the following fraction:

$$\frac{\text{force exerted by the machine}}{\text{force exerted on the machine}} = \frac{10}{3.3} = 3$$

The number of times a force exerted on a machine is increased by the machine is called the **mechanical advantage** of the machine. It is a distinct advantage to the user of the machine. Of course, you never get something for nothing. The smaller force put on the machine must be moved through a longer distance, while the larger force that the machine exerts moves a shorter distance.

$$\textbf{mechanical advantage} = \frac{\textbf{force exerted by machine}}{\textbf{force exerted on machine}}$$

A "Trade-Off" – Force for Distance

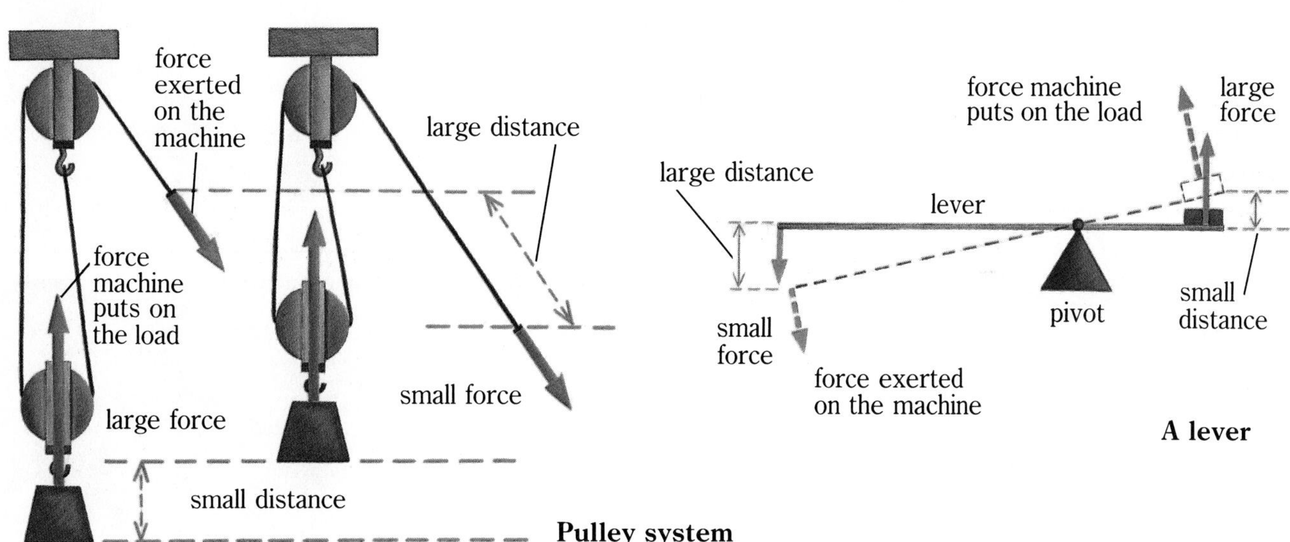

Pulley system

A lever

Some machines work differently. They multiply distance at the expense of the force. You will see some of these on pages 98–99.

1. What is the mechanical advantage of Pam's machine? Leslie's machine? the machine you constructed?
2. If Leslie used triple pulleys and found that a 2-N force could lift the 1-kg mass, what would be the mechanical advantage?
3. Now suppose Tony used a triple-pulley system to raise the 100-kg load to a platform 1 m high. How much force, in newtons, would be needed to raise the load? How many people would be needed for the job?

Which Machines?

Your task is to analyze each of these simple machines. They are all simple machines, although in one or two cases you might not believe it! *Simple* machines make up *complex* machines.

1. Classify each machine as one of the following:
   - lever • pulley • inclined plane • other type of machine
2. Which machines
   - multiply (increase) the force put into them?
   - multiply (increase) the distance put into them?
   - change the direction of the force put into them?

   The same type of machine may multiply either force or distance or change the direction of the force, depending on its use.
3. Where do you exert a force on each machine?
4. Where does the machine exert a force on something else?

A sketch showing the forces might be the best way of answering Questions 3 and 4. Here are the answers for Example (g).

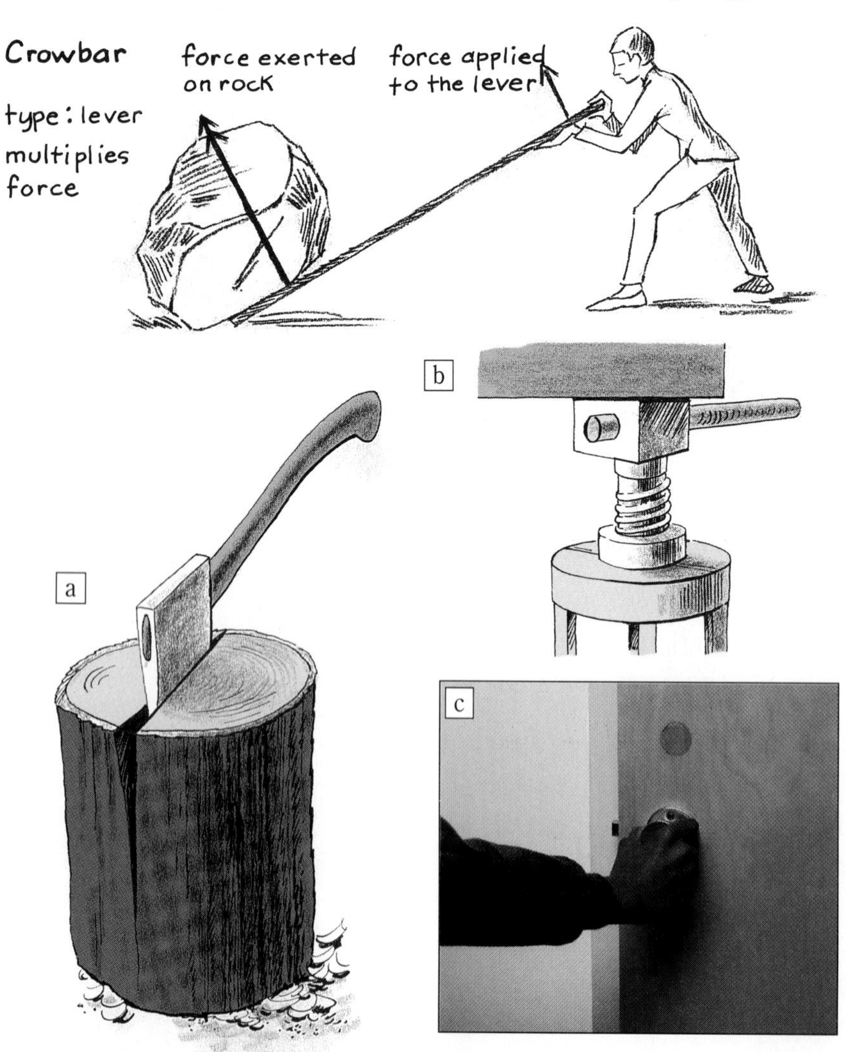

g
h
i
j
k
l
m
n
o
p

Machines and Energy

You have seen that machines can convert energy from one form to another. You have also seen that machines can multiply force or distance, but never both. Increases in force always come at the expense of distance, and increases in distance come at the expense of force. What accounts for this?

As you know, no machine is 100 percent efficient, but where does the "lost" energy go? Does it simply disappear, or is it converted into some unseen form?

In this lesson you will tackle these questions and others. You will also learn a bit more about the ways in which machines convert or convey energy.

EXPLORATION 5

Miguel's Machine

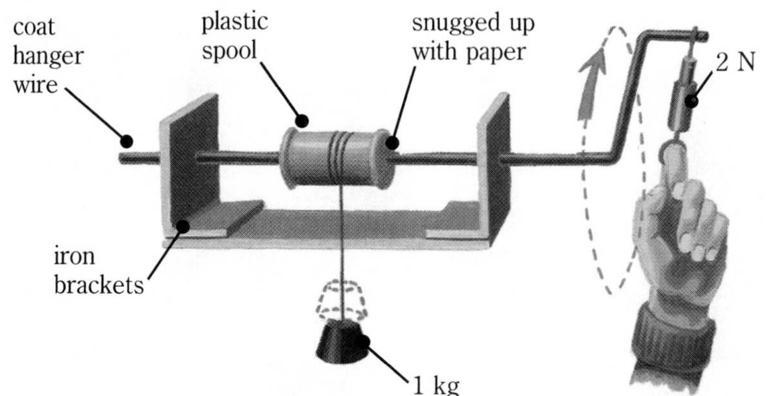

Miguel devised a machine that would lift loads easily. With the machine, he found that he could lift a load of 10 N with a force of only 2 N.

Miguel realized that he had to turn the handle a long way to lift the load just a little. In fact, when he turned the handle all the way around once, the load came up by just the distance around the spool.

Evaluate Miguel's machine by completing the evaluation report in your Journal. Could Miguel's Machine be used to solve Tony's problem on page 90?

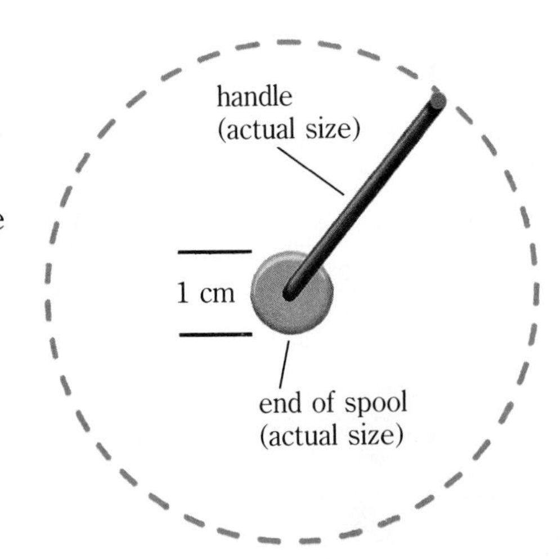

Evaluation Report—Miguel's Machine	
Distance handle moves in one turn = ? Force put on handle = ? Miguel's work input = ?	Distance load moves up when handle makes one turn = ? Upward force put on the load by the cord = ? Machine's work output = ?
Efficiency = ? Mechanical advantage = ?	
Conclusions: Recommendation: Miguel can get an even greater mechanical advantage if he were to . . .	

Count Rumford

James Joule

A Closer Look at Efficiency

Miguel put 0.38 J of work into his machine. How much energy did he transfer to the machine? His machine did 0.30 J of work on the load. How much energy did the machine transfer to the load? Was some energy lost? It seemed so. Where did the 0.08 J of energy go?

Situations like the above puzzled scientists such as Count Rumford (1753-1814) and James Joule (1818-1889). Through many observations and experiments, the two scientists came to theorize separately that the energy apparently lost in machines was actually only diverted; the energy was turned into heat by friction. Their work established that heat was a form of energy, not a form of matter as was commonly believed at that time. Rumford and Joule reasoned that the amount of work actually done by a machine is equal to the amount of energy put into the machine minus the energy lost to friction. Each scientist concluded that energy is *conserved* in nature. In other words: *Energy can be transferred and transformed, but the total amount of energy in a system always stays the same.*

How would you apply Rumford's and Joule's findings to Miguel's machine?

Energy transferred to the machine (?J)	=	Potential energy gained by the mass raised by the machine (?J)	+	Heat energy caused by friction (?J)

Compute the energy values for one turn of the handle. What would the energy values be for 10 turns of the handle? If the machine had less friction, how would these values be altered? How would the value for the efficiency be changed? Consider these principles.

- A machine is a device for transferring or converting energy.
- Some of the work (energy) input to a machine is used to overcome friction. This produces heat energy.
- Energy output is equal to energy input minus energy turned into heat by friction.

Do you understand now why the energy you put into a machine (work input) is always greater than the energy you get out of the machine (work output)?

How much heat energy was produced as the load was lifted 15 cm by Leslie's machine (pages 92-93)? by Pam's machine (pages 94-95)? How could both of these machines be made more efficient?

Wheels and Axles

Miguel's machine is commonly called a **wheel and axle.** Why? Did Miguel apply his force to the wheel or to the axle of the machine? Where does the machine apply its force on the load?

Wheels and axles are simple machines that are normally found together—neither is very useful without the other. Now turn back to the pictures on pages 98 and 99. Which two pictures show examples of a wheel and axle? For each of these examples, determine where the force is applied and where the machine exerts a force on something else.

EXPLORATION 6

Solving a Technological Problem

You have already encountered a wheel and axle machine. Miguel designed such a machine. What steps might he have followed in designing and building it? Try to approach the problem as he might have.

As you follow the suggestions below, you will be tracing out the steps commonly used in solving technological problems.

You Will Need

Choose from the following readily available materials and tools

- spools of various sizes
- mailing tubes
- film cannisters with lids
- wire coat hangers
- knitting needles
- craft sticks
- milk cartons
- shoe boxes
- plastic-foam or paper cups
- cardboard
- plywood
- a hand saw
- white glue
- cans
- wire
- straws
- string
- pliers
- corks
- scissors

What to Do

1. First, review Miguel's problem. He wanted to use simple materials to make a hand-powered wheel and axle strong enough to lift a load of 10 N with a much smaller force.
2. Observe Miguel's machine again. Identify each of its parts:
   (a) the enlarged "spool-like" part of the axle, the cord holding the mass

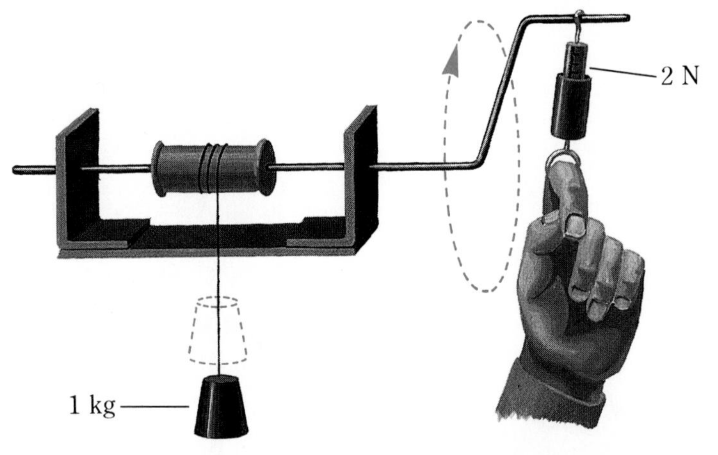

(b) the "wire-like" part that is bent at one end to form a handle—the wheel of the wheel and axle
(c) the frame to support the wheel and axle

3. Make decisions regarding the following:
   (a) What will I choose for the "spool" part of the axle?
   (b) What will I use for the "wire" part of the axle?
   (c) How will I construct the wheel?
   (d) How can I attach the spool to the wire so that both turn together?
   (e) What size would be best for each of these parts?
   (f) What will I use for a supporting frame?
   (g) How strong should each part be?
4. Make a simple diagram showing your design solution. Name all the materials you plan to use.
5. Construct a model of the wheel and axle according to your design.
6. Test its operation. Are there some improvements that are needed so that it will operate better? If so, make them.
7. Determine your machine's mechanical advantage. Also determine its efficiency and the amount of energy it changes into heat due to friction with every turn.
8. Think of other ways you might judge your machine critically: for example, appearance, strength, maximum load it can carry.
9. Compare your wheel and axle with those of your classmates. Identify strengths and weaknesses in their wheels and axles as well as your own.
10. List suggestions to make your machine better.

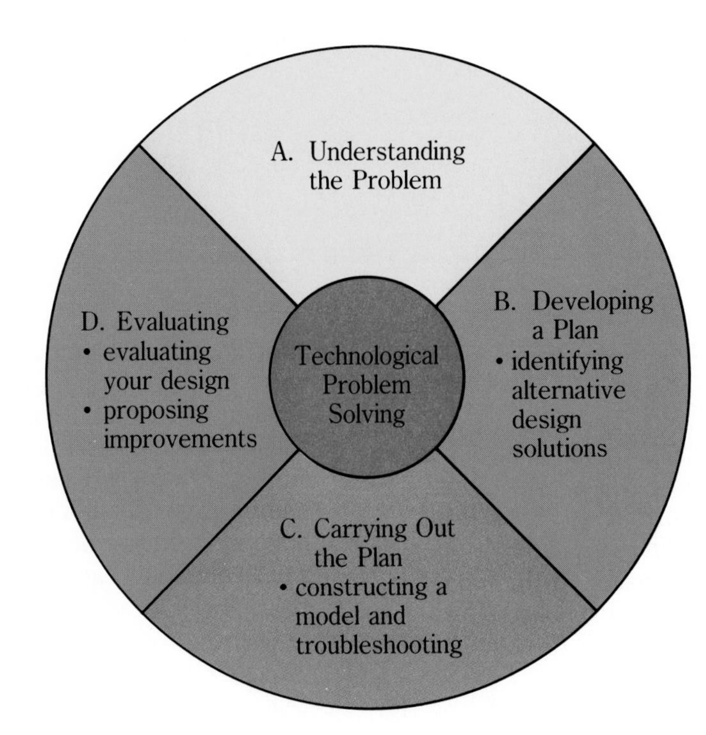

Congratulations! You have a workable hand-powered wheel and axle of sufficient mechanical advantage to do the job you wanted. You did this by "technological problem solving." The diagram to the left shows the steps. Match what you did in the Exploration (steps 1–10) with the steps in the diagram (A, B, C, and D). More than one item of "what you did" may fit into one of these steps.

Exploration 7 extends the problem. Your task will be to adapt your machine so that it can be powered by some source of energy besides your hand.

EXPLORATION 7

Powering Your Machine

Can you suggest another way to power your wheel and axle machine, other than by hand? You might consider using any of the following sources of energy:

- water
- steam
- wind
- gasoline
- electricity
- a rubber band
- a raised weight

How would you connect your energy source to your machine?

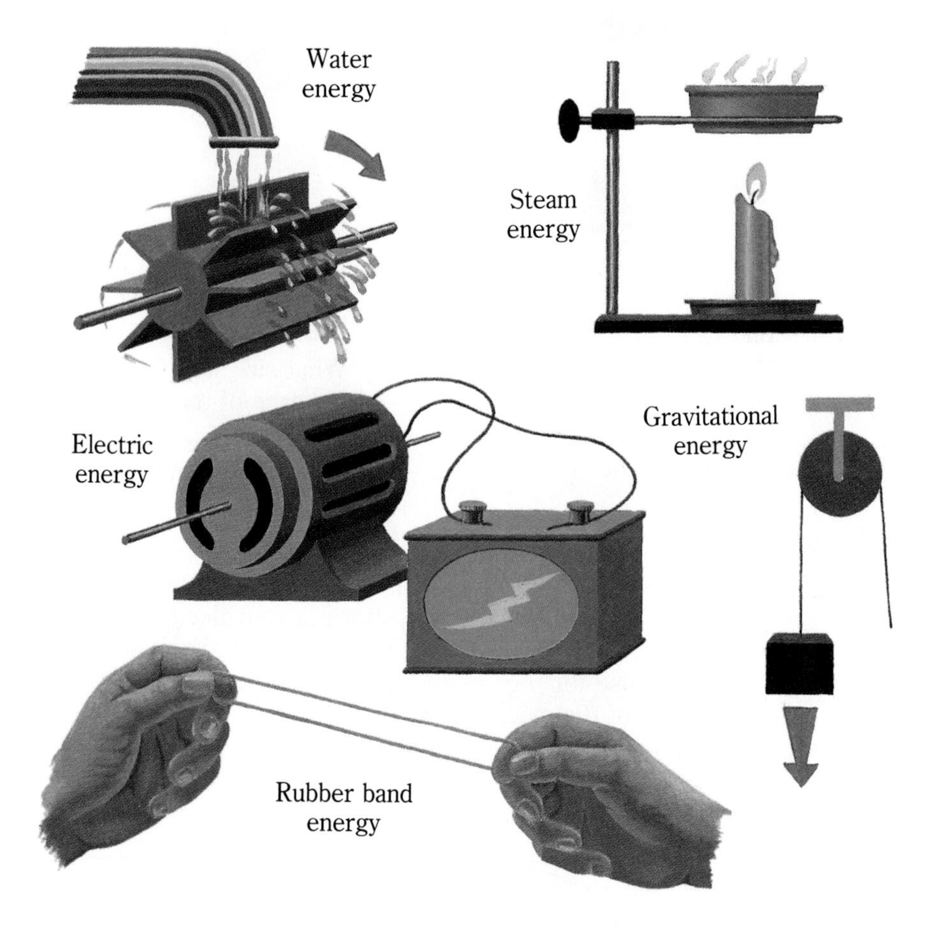

Part 1

What to Do

1. Choose an energy source.
2. Decide on a possible way to connect the source to your machine. If you need to modify the machine, determine how you will do it.
3. Check your design. Ask yourself if you are using simple and readily available materials and equipment. Modify your design if necessary.
4. Draw your design.
5. Show your design to others and ask for their opinions and advice.

Part 2

What to Do

1. Obtain the necessary materials and equipment and construct a model of the powered wheel and axle according to your design.
2. Connect the machine to the source of energy you have chosen.
3. Evaluate your model. Does it work? If not, why did it not work? If so, what load will it lift? How does it operate compared to the models of others? Propose some improvements to your model.
4. How good a problem solver were you? Exploration 6 showed you a model for solving technological problems. Draw a chart like the one on page 103 and complete it by adding details of what you did at each of the steps A–D.

Other Wheels and Axles

Let's look at some more examples of wheel-and-axle machines. As you will see, the following machines are different in a key way from those you have seen so far.

Look carefully at diagrams (a) and (b). How are these wheels and axles different from Miguel's machine? In each of the diagrams, does the operator apply force to the wheel or to the axle? Is the output force of the machine being exerted by the wheel or by the axle?

(a)

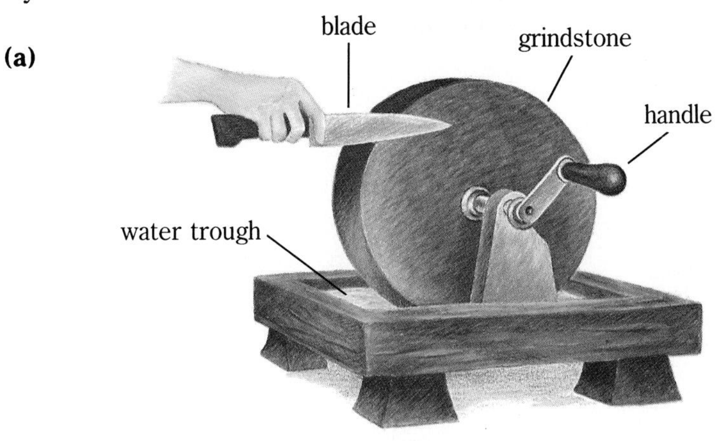

Hand-operated grindstone

(b)

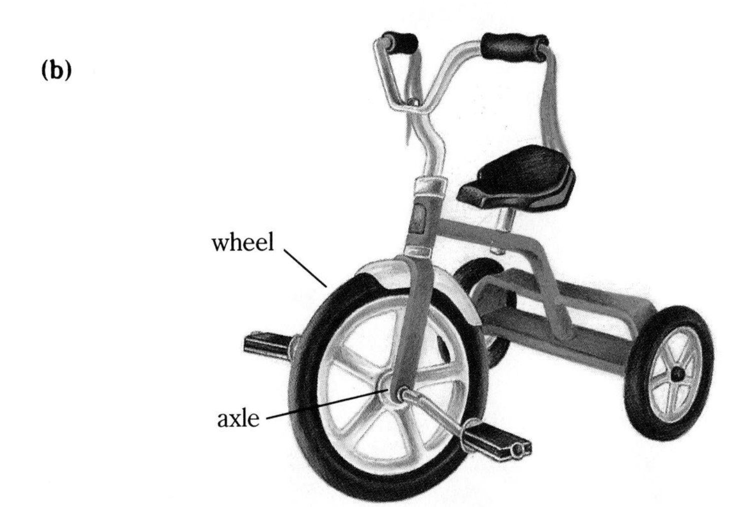

Using the grindstone, a person applies force to the handle, which turns through a smaller circumference than does the rim of the grindstone. How does this wheel and axle affect speed, distance, and force?

Analyze the operation of the tricycle in the same way. Which of these statements is true for these two examples of a wheel and axle?

1. Force is multiplied at the expense of distance.
2. Distance is multiplied at the expense of force.

Suppose you turn the pedal of the unicycle shown below through one complete revolution.

1. Through what distance does each pedal move?
2. Through what distance would the wheel move along the road?

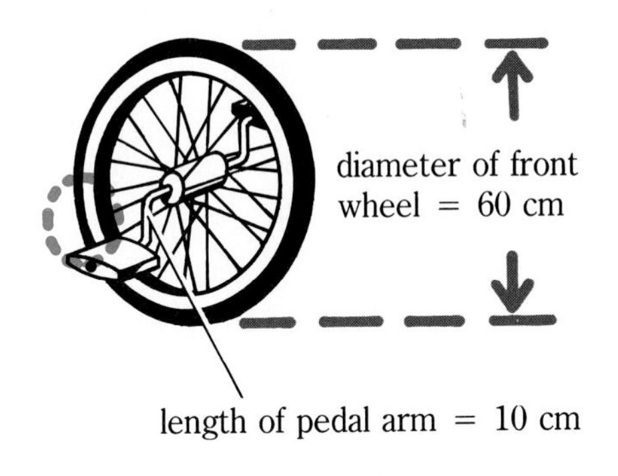

3. How much has the pedal distance been multiplied?
4. Suppose you move the pedals through one full revolution in 1 second. What is the speed of the pedal in centimeters per second? What is the speed of the unicycle?
5. The speed of the pedals has been multiplied. How did this happen? What's the trade-off?

Transferring Energy

Getting into Gear

Examine the following pictures of machines. Identify how the energy is transferred from the energy source to the operating parts of each machine.

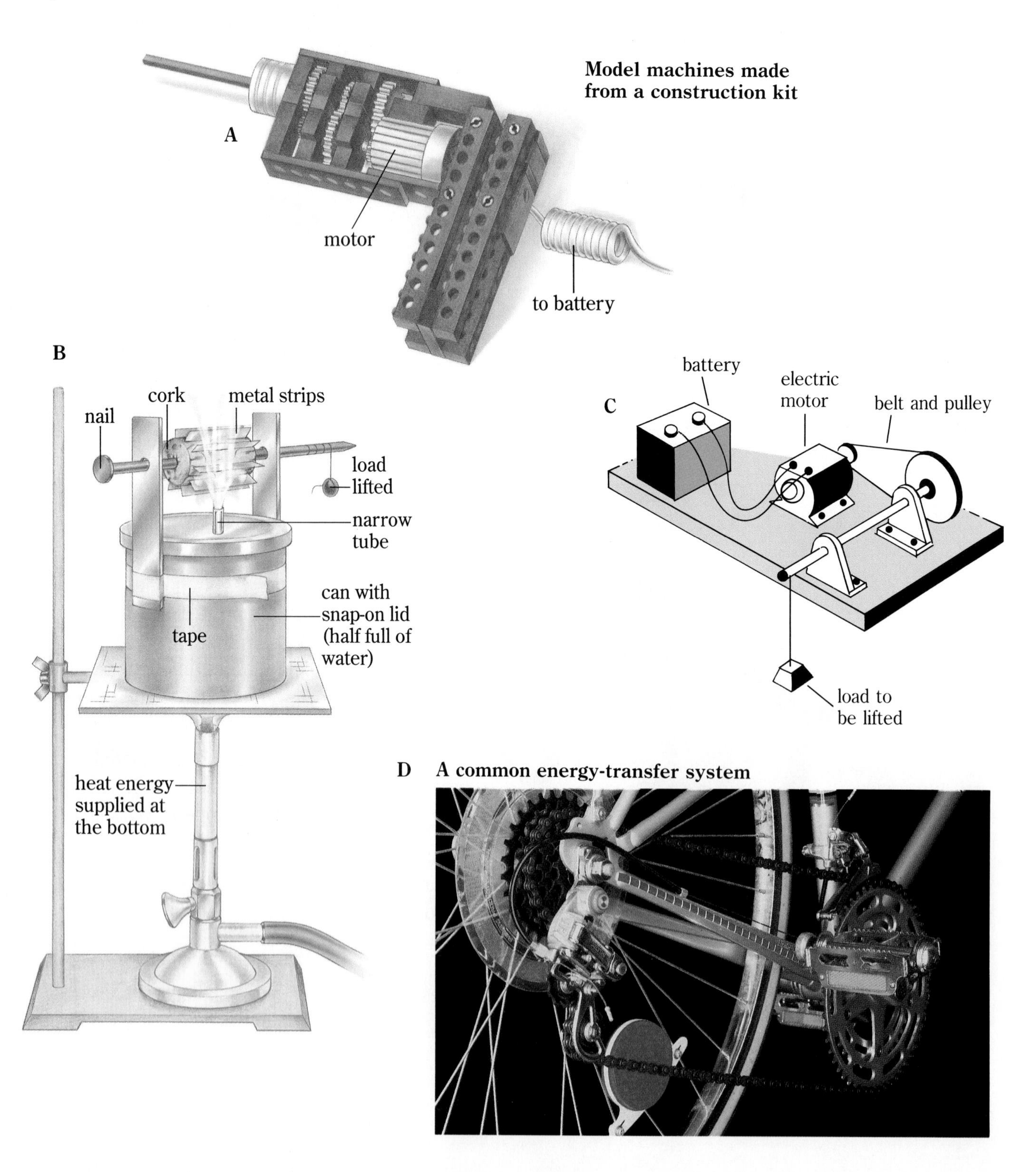

D A common energy-transfer system

What Makes It Tick?

You Will Need

- unused mechanical devices—such as toys, clocks, or watches—powered by some form of energy.

What to Do

1. Working in small groups, first select an object to observe in detail.
2. Disassemble the object to examine the parts necessary for its operation.

For example, If you are examining a toy car, does it have a large wheel that spins faster and faster, depending on how much energy you put into the car? This wheel is a *flywheel*. Of what use is the flywheel?

Can you find toothed wheels or cylinders that fit into one another? These are *gears*. What do they do?

3. Identify the path of energy through your machine, from source to output.
4. Sketch your machine, and describe how it works.

Gears ready for action

Joy's Construction Kit

Most hobby and toy shops have technology construction kits, for example, LEGO or Meccano brand. Joy had such a kit. It contained a motor, wheels, propellers, and all the other items needed to get energy from the battery to these moving parts.

Joy put together four assemblies to see how everything worked. She also asked herself questions about each. Joy's assemblies are shown at right.

Assembly 1

1. If I turn the top gear, in what direction will the lower gear turn?
2. How many turns does the lower gear make for one complete turn of the top gear? Is this related to the number of teeth in each gear?
3. What differences would I note if I turned the lower gear instead?

Assembly 2

1. As I turn the small gear, what happens to the large gear?
2. For one turn of the small gear, how far does the large gear move?
3. What differences would I observe if I turned the large gear instead?

Assembly 3

Joy included a worm gear, another gear, and a wheel in this assembly.

1. As I turn wheel A through one complete turn, how far does the thread of the worm gear appear to move to the left or right?
2. How far does gear B move at the same time?
3. If this system were attached to a motor, how many turns of the motor would be needed to make one turn of gear B?

Assembly 4

1. How does the energy get from the battery to the moving parts?
2. For one revolution of the wheel on the motor, how much does each of the successive parts turn?
   - the wheel attached to the motor by a belt
   - the small wheel on the other side of the short axle
   - the wheel on the long axle
   - the tractor wheel

Assembly 1

Assembly 2

Assembly 3

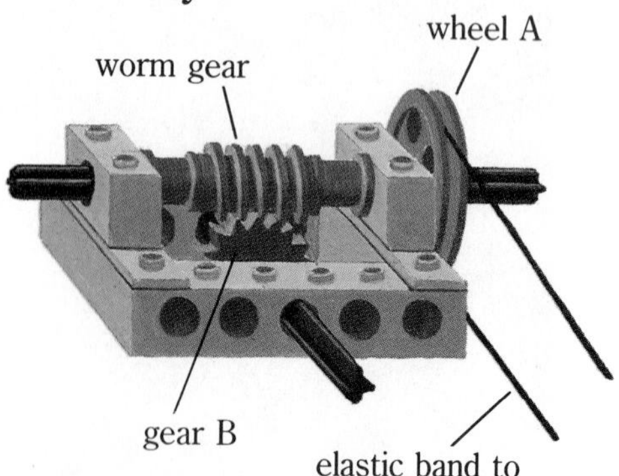

Joy first turned the connecting wheels by hand. She marked the top of each wheel to keep track of its motion.

3. How many turns of the axle of the motor are needed to produce one turn of the tractor wheel?
4. How does the speed of the motor compare with the speed of the tractor wheels?

Assembly 4

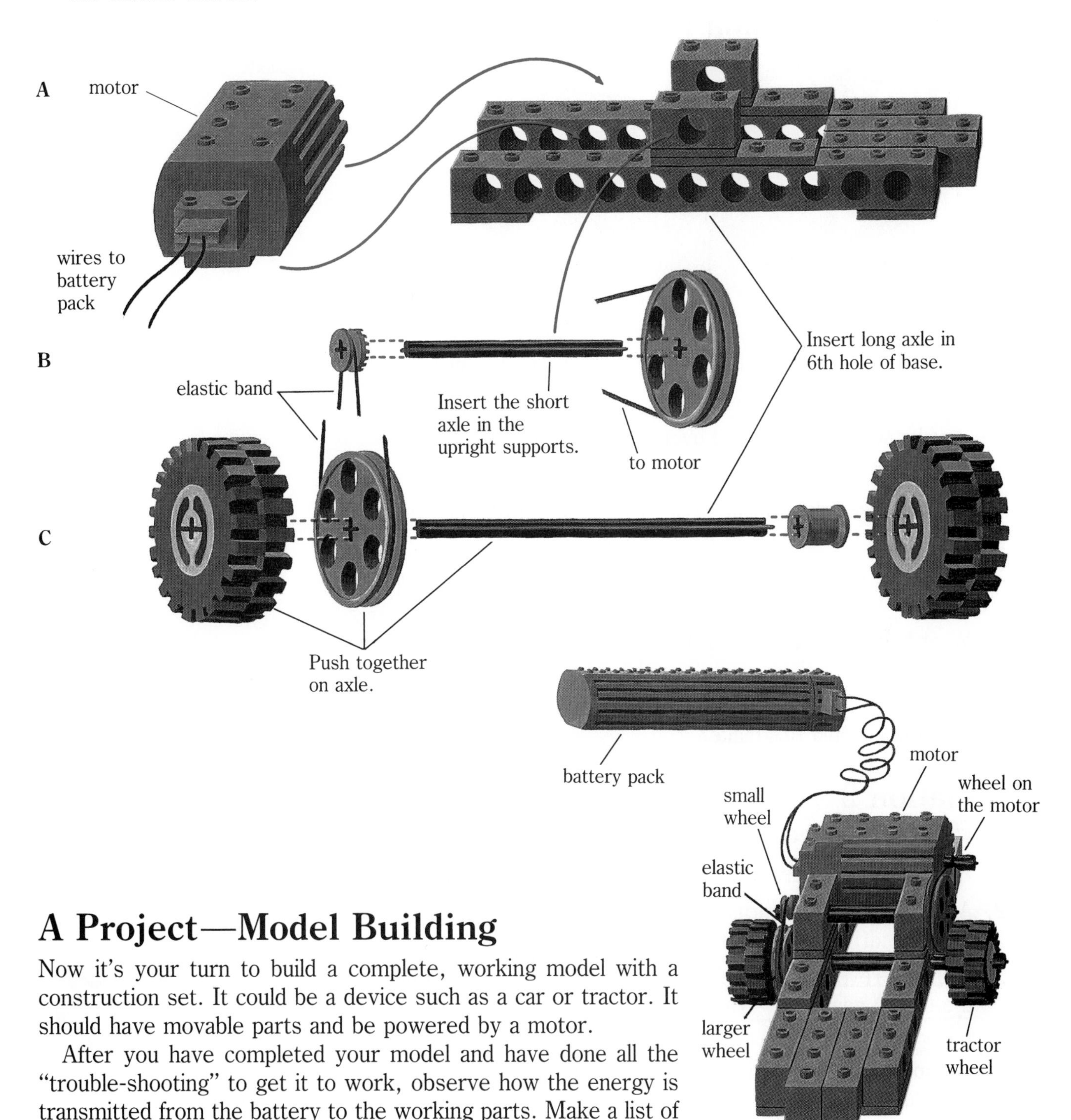

Assembled Model

A Project—Model Building

Now it's your turn to build a complete, working model with a construction set. It could be a device such as a car or tractor. It should have movable parts and be powered by a motor.

After you have completed your model and have done all the "trouble-shooting" to get it to work, observe how the energy is transmitted from the battery to the working parts. Make a list of all the "links" in the "energy train." Describe how they function and where changes in speed and direction occur.

A Closer Look at Wheels, Belts, and Gears

Here is a chance to see what you have learned about wheels and gears. Below are four situations to analyze.

Situation 1
Driving Wheels and Driven Wheels

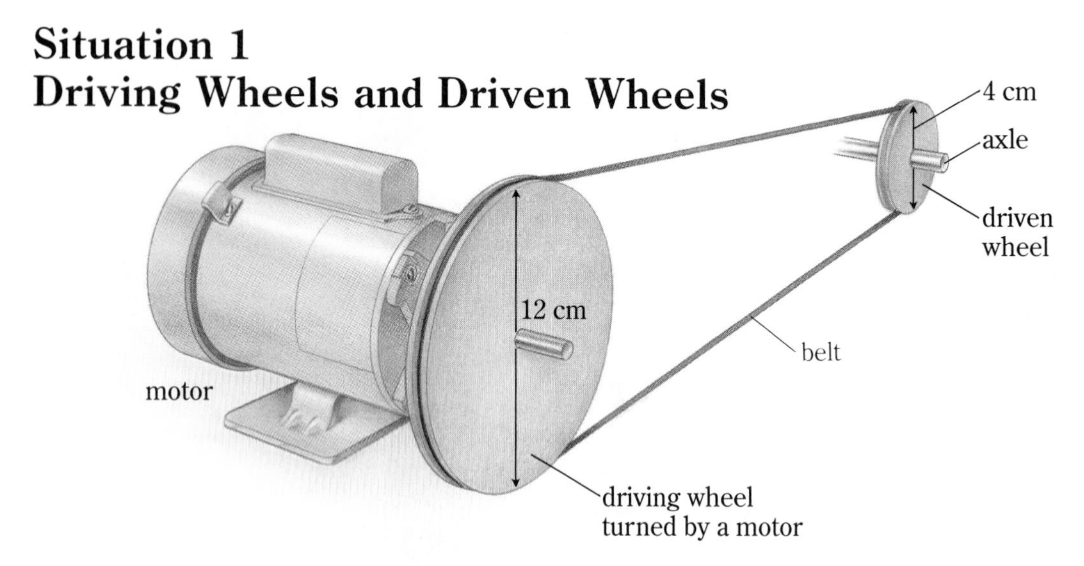

Here the larger wheel is called the *driving* wheel because it drives the smaller wheel. It is turned by a motor. The belt transfers the energy to the smaller wheel, which is called the *driven* wheel.

1. As the large wheel makes one revolution (turn), how many turns does the smaller wheel make?
2. How does the speed of the axle of the driven wheel compare with the speed of the axle of the driving wheel?
3. What has been increased in the driven wheel? Of what value could this be?
4. If the driven wheel were to become the driving wheel, what difference would this make?

Situation 2
Different-Sized Gear Pairs

1. How many teeth are in each gear?
2. As the driving gear turns counterclockwise, in what direction does the driven gear turn?
3. For each revolution of the driving gear, how many turns does the driven gear turn? Which gear axle moves faster?
4. If the speed of the driving gear is 20 turns per second, what is the speed of the driven gear? Which gear exerts less force at its axle?
5. What changes would you make to the gears to increase the speed of the driven gear? to decrease its speed?

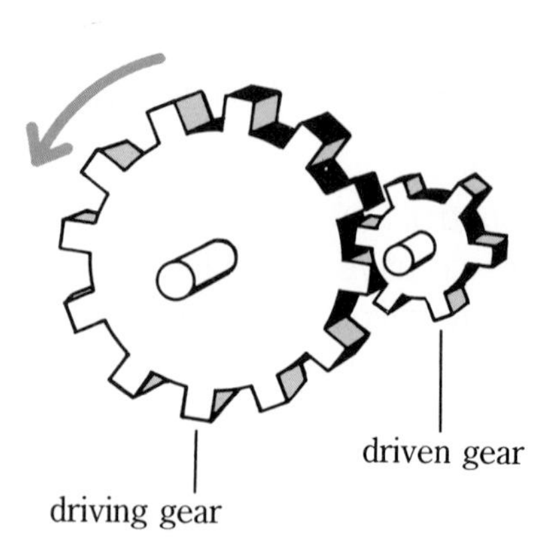

Situation 3
Different Kinds of Gears

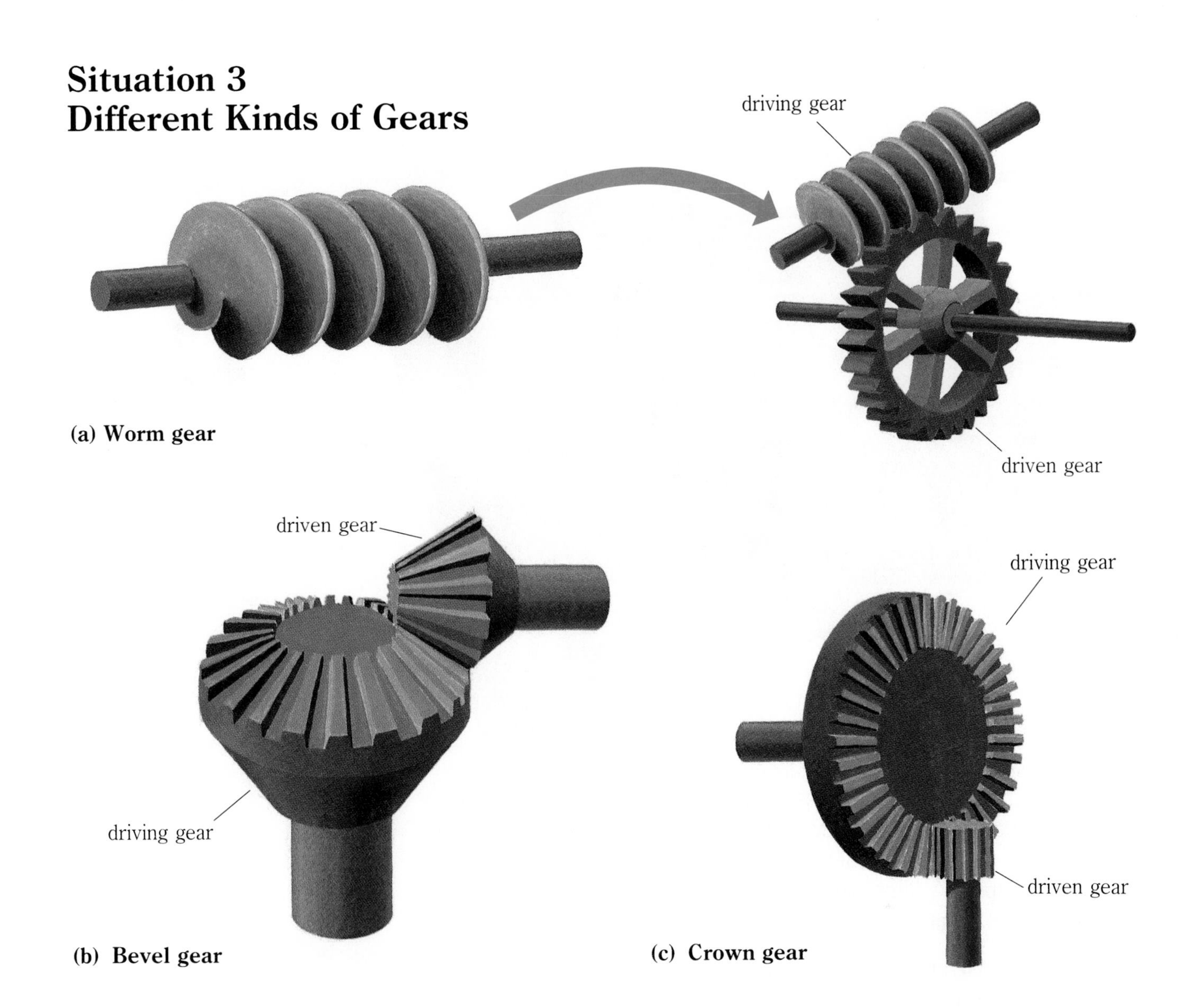

(a) Worm gear

(b) Bevel gear

(c) Crown gear

1. The diagrams above show three different kinds of gear pairs that perform in a similar way. How do they work?
2. In diagram (a), when the worm gear moves one turn, how far do you estimate the driven gear will turn? Is the speed of the driven gear greater than, the same as, or smaller than the driving gear? Which gear will exert the greater force? The worm gear is almost always the driving gear. Can you figure out why?
3. Estimate the number of teeth in each of the bevel gears shown in diagram (b). Suppose the driving bevel gear makes seven turns in one second, approximately how many turns will the driven gear make?
4. In diagram (c), when the driving gear makes one turn, how many turns will the smaller driven gear make? Has speed or force been multiplied?

Situation 4

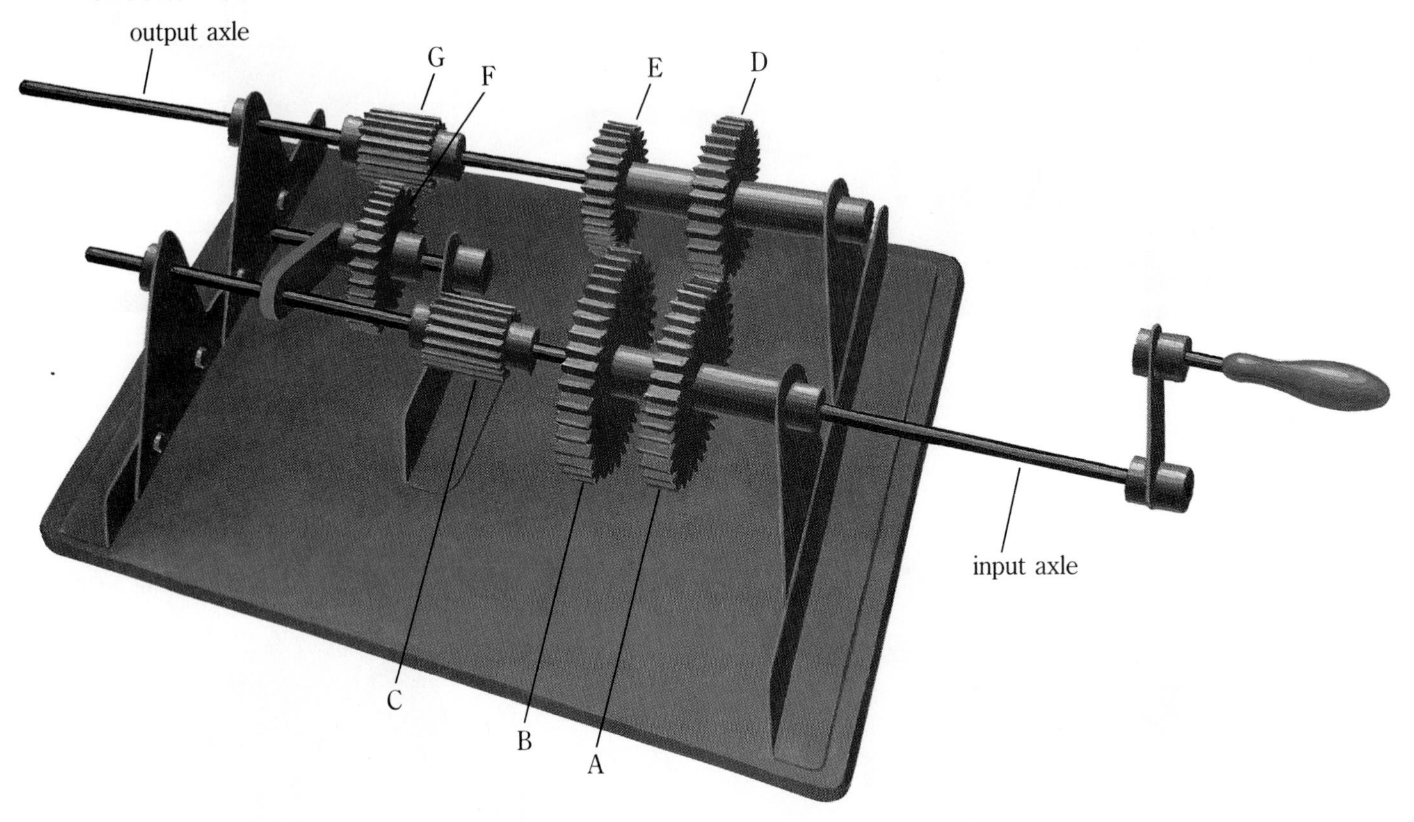

A student made this transmission model of a car (gear box) from a construction set. The handle can move in and out, causing different gears to engage. There are two "forward speeds" and one "reverse speed". The handle is attached to the input axle. Gears on this axle engage with the gears on the output axle.

1. In the illustration, gear A on the input axle is engaged with gear D on the output axle. Suppose the handle is turned counterclockwise. In what direction will gear D on the output axle turn? Which gear, A or D, will move with greater speed?
2. Suppose the handle is pushed to the left so that gear B now engages with gear E. Suppose again that the handle is turned counterclockwise at the same speed. How does the speed of the output axle now compare with its speed in question 1? Why?
3. Suppose now that the handle is pushed farther to the left so that gear C engages with gear F. As the handle is turned counterclockwise, in what direction will gear G and the output axle move?
4. What pair or group of gears in the model corresponds to low gear? high gear? reverse gear? (In an actual car there would be an additional set of gears between the gear box and the wheels.) Which gear gives more force to the forward motion of the car? gives more speed to the forward motion?

Mechanical Systems

Most machines that you see and use every day could be properly called *mechanical systems*. This is because they are made up of two or more simpler machines working together. Look back at the mechanical systems throughout this unit. What simple machines do you recognize?

A can opener is one example of a mechanical system that contains several subsystems. Try to identify the following subsystems.

(a) a lever that operates like a nutcracker

(b) a kind of gear

(c) a wheel and axle

(d) a wedge like the blade of an axe

You remember that a mechanical system is designed to do a single task and is made up of groups of parts called *subsystems*. What function does each of the parts of the can opener system perform? Can you group these four parts into two subsystems? What has been added to one part so that it can serve an additional purpose?

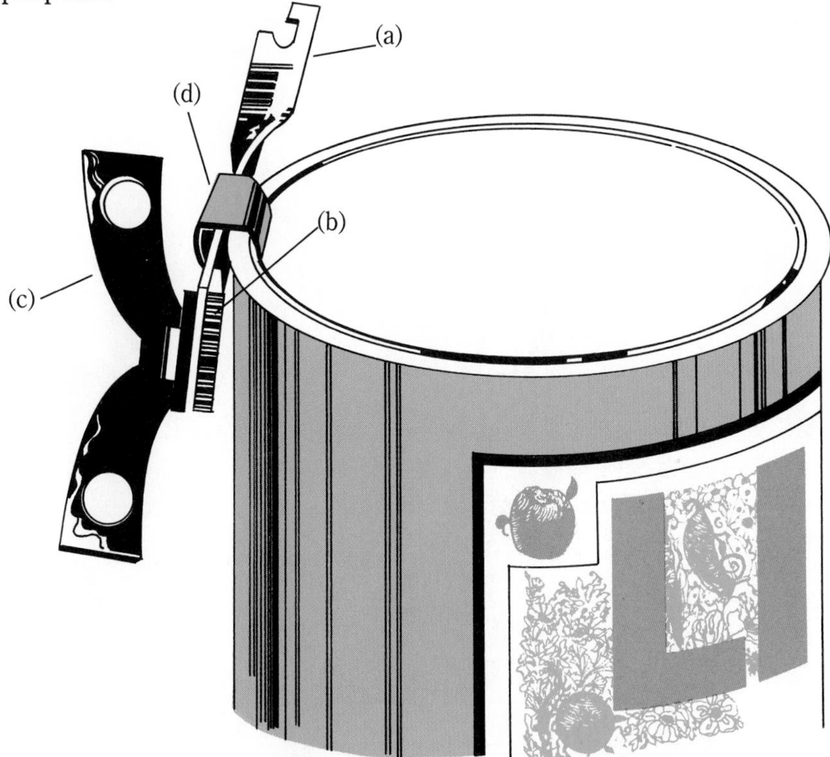

How do you operate the kind of can opener illustrated here? The letters in this diagram correspond to the subsystems listed above. What simple steps of operation would you describe for a much younger person to follow? Write them in a language that he or she could understand.

Analyzing Mechanical Systems

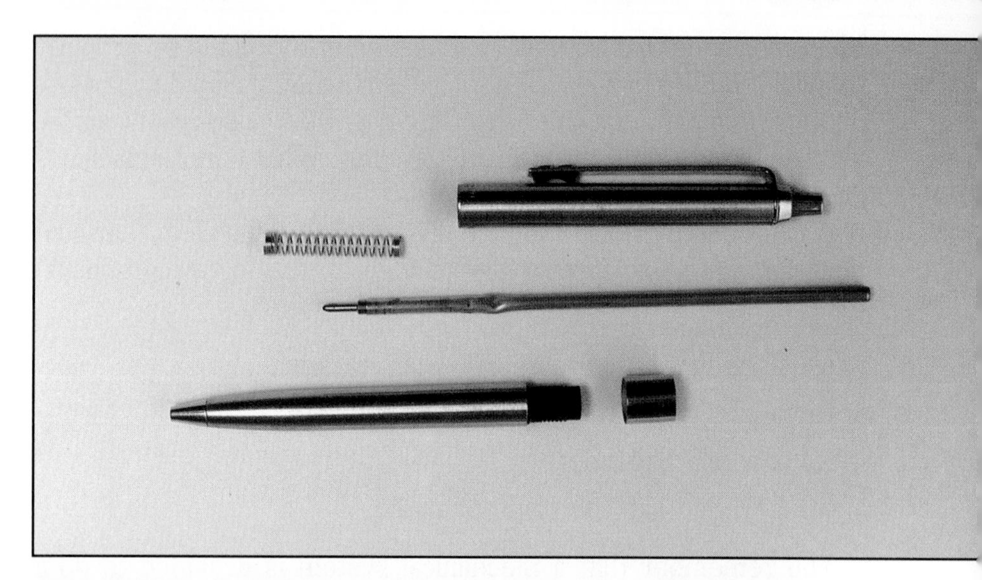

Spend a few minutes at each of the following stations. Record all your answers in your Journal.

Station A The Write Stuff

1. Check the operation of the pen. How does it write? Is it retractable? Is it refillable? How?
2. Take apart the pen to see its contents.
3. How do the parts operate? A drawing of the parts may help you explain it.
4. Examine another type of pen. How is it similar? How is it different? Is it a better design? Why or why not?

Station B A Fasten-ating Device

1. Staple two pieces of paper together. As you do this, carefully observe each aspect of the stapler's operation.
2. Open the stapler and identify each main part. What is the function of each part?
3. Compare the staple before and after stapling. What causes the staple to bend? Could it be bent in another direction?

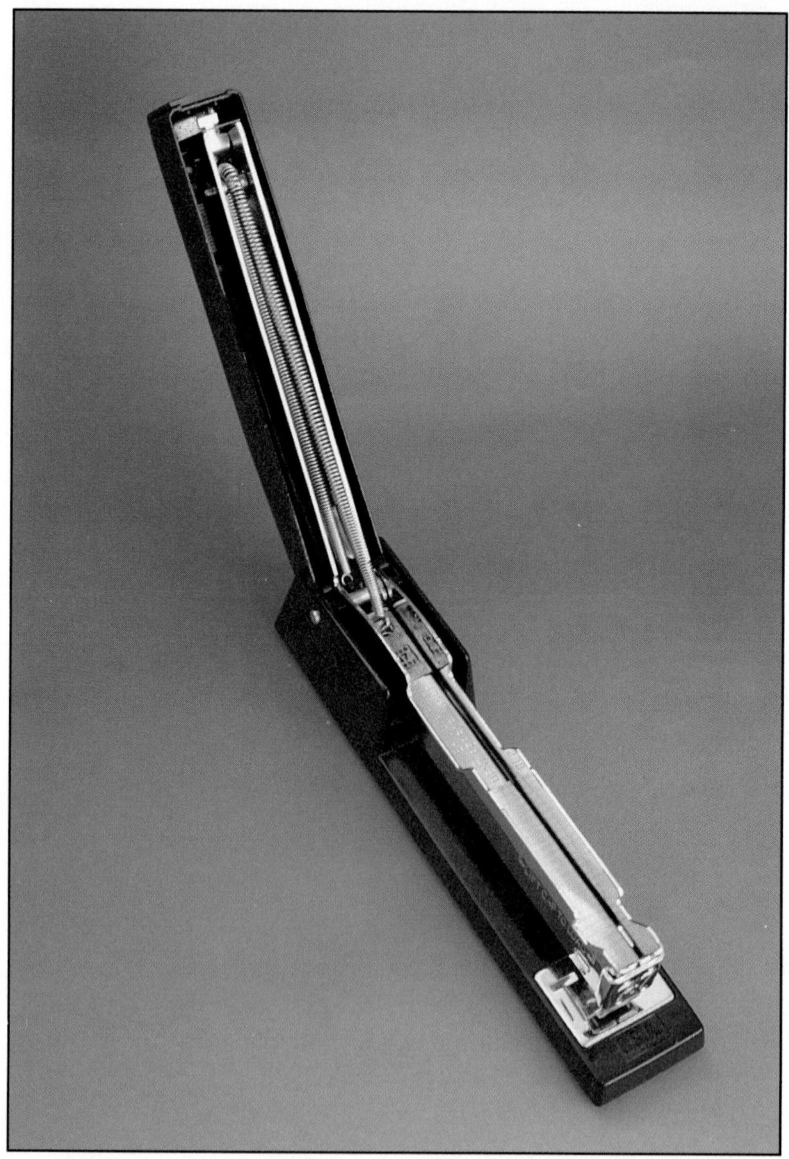

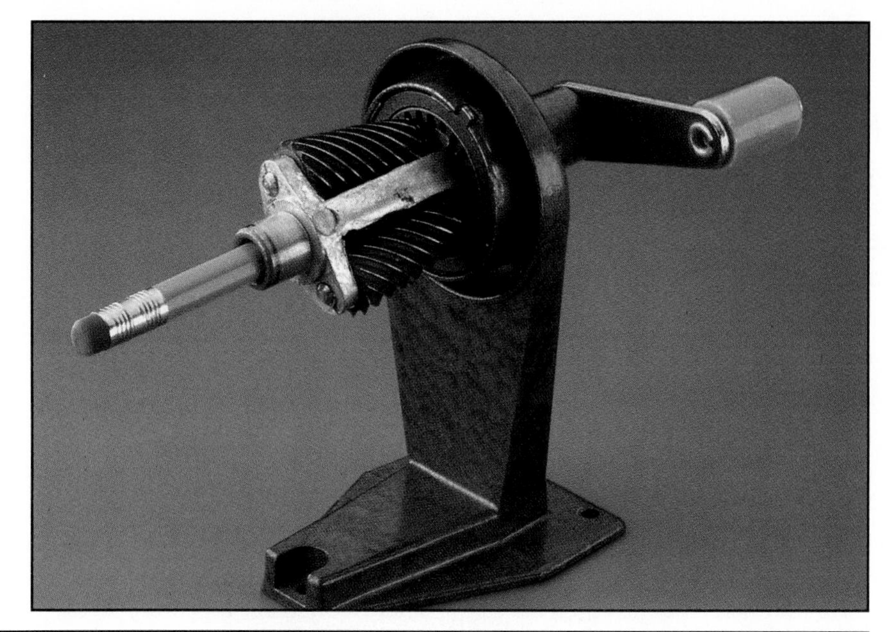

Station C
What a Grind!

1. How does this pencil sharpener operate?
2. Identify all of the pencil sharpener's subsystems. Which of these subsystems are simple machines that you studied earlier?
3. Can you suggest an alternative design that accomplishes the same purpose?
4. Suggest a design for a pencil sharpener that is not powered by hand.

Station D
Wheel You Look at That!

1. In the introduction you considered the bicycle as a mechanical system. Examine a bicycle closely. Observe how each part and subsystem operate.
2. Where do you apply energy to the bicycle? How is the energy transferred?
3. Compare the mechanical systems of different makes and models of bicycles.

Station E
A Hole-some Machine!

1. What is the overall function of a hand drill? How does it work?
2. Identify at least four simple subsystems in this mechanical system.
3. What causes the speed of the drill to be so great?
4. How is the drill powered? In what other ways could it be powered?

One Purpose—Many Designs

The can opener on page 113 is just one device for opening cans. Other devices also serve the same purpose.

1. What designs of can openers have you used or seen?
2. Various designs may have different features. What are they?
3. Which designs work best? Which are most practical?

Collect as many can opening devices as you can. Examine each closely. Identify subsystems and any features that make them useful and effective.

Constructing Your Own Mechanical System

The following three pages outline a traveling mechanical system. Your task is to develop a design for the system and test it. The outline provides information to get you started.

EXPLORATION 10

Constructing a Model Car

Auto manufacturers are researching different energy sources to power the cars of the future. In this Exploration, you will construct your own car and power it with an unusual alternative energy source.

Here is a challenge for you! Design a model car powered by a rubber band. The car must be able to move in a straight line for a distance of at least 5 m. (Slingshot propulsion is not allowed.) Then, construct a prototype of the new car.

You have been presented with a problem. How are you going to solve it? First re-read the steps in solving technological problems on page 103. The following suggestions may also help you.

What You Might Want to Use

- rubber bands
- masking tape
- plastic-foam or paper cups
- cup lids
- drinking straws
- index cards
- cardboard tubes
- wooden dowels
- a drawing compass
- straight pins
- a pencil
- corks
- pliers
- scissors
- paper
- paper clips
- white glue
- thumbtacks
- anything else you can think of

What to Do

1. Keep a record of your work in your Journal. Keep track of what you do, why you do it, what problems you encounter, and how you solve the problems.
2. Work with one or two other people. This will help you get ideas flowing. Start by discussing the project.
3. Try to think of as many ideas for solving the problem as you can. Sketch them to help you refine the concepts.
4. Review all your possible designs. Choose the best one and start to work. Decide basic questions such as what materials to use and how large the car should be. Draw a blueprint of your car, showing exact details, identifying subsystems and individual parts, and indicating the scale. (A blueprint should be detailed enough that someone could use it to build what the blueprint represents.)
5. Make a list of the materials you will need. Collect the parts needed.
6. Build, test, and evaluate your model. Use the results of your tests to make improvements.
7. Give your product a name. Prepare a promotional brochure designed to convince people to buy the product. Be sure to include technical data in your brochure.

Examining One Model Car Design

Jim Louviere, a science teacher, constructed an "air car" that worked surprisingly well. To make it very light he used nothing but paper, index cards, rubber bands, and paper clips. Examine the diagram of the assembled model below. Discuss the questions that follow with two other students.

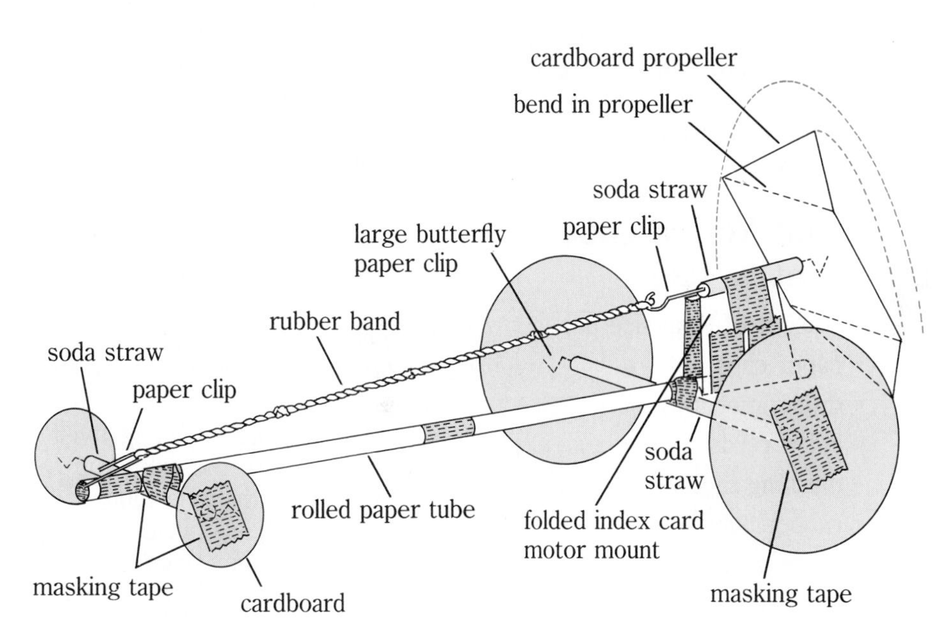

1. How does Jim's model car work? How do you get it to move?
2. What is the source of energy that powers the car?
3. What energy changes take place during its operation?
4. What force moves the car?
5. What causes the car to slow down and eventually stop?
6. How might you measure the energy input into the car?
7. What variables might be important in getting the car to work? You might consider the length of the body, the length of the axles, and other characteristics.

Exploration 11 gives you the opportunity to build your own version of Jim's air car.

Constructing an Air Car

Here are some suggestions to help you construct an air car by following Jim Louviere's general design. However, you may wish to make an original design of your own.

What You Might Want to Use

- a sheet of legal size paper
- paper clips
- drinking straws
- index cards
- masking tape
- rubber bands
- white glue
- a pencil
- a drawing compass
- pliers to bend and cut wire
- scissors
- anything else you can think of

What to Do

1. Working with two other people, first detail the design of the air car you want to build. Make decisions about the variables you identified in Question 7 of Exploration 10.
2. Collect, prepare, and assemble all the parts.
3. Test and trouble-shoot the model. See the trouble-shooting suggestions provided next.

Trouble-Shooting

If your car does not move, don't despair. First, compare your car to those of other groups. You may notice some construction flaws in your car. Here are some common problems and possible causes that you might need to investigate.

1. *The propeller is not turning.* Is it hitting something? Do you need to change its position?
2. *The propeller is spinning but the car is not moving.* Are the propeller blades bent properly? Is there sufficient twist in them? Are the axles rubbing inside the axle tubes? Are the wheels perfectly round? Do they need trimming? You may have to investigate other possibilities such as using model airplane propellers.
3. *Your car is going in circles.* Do the axle tubes and motor mount need slight turning? Are the wheels straight?
4. *The car is flipping over.* This might be related to the size or weight of the propeller, to the length of the rear axle, or to how tightly you wound the rubber band.
5. *The body tube of the car folds up.* Are you winding the rubber band too tightly? Does the body tube need strengthening, perhaps by inserting a pencil into it?
6. *The wheels and/or propeller are flipping around.* Are they firmly attached? Are they sturdy enough?

Extending Your Design

You may want to build a car of more durable materials and power it in other ways. What materials might you use? How would you power the car? Draw your design with possible specifications for its construction. Submit it to others for their critique. You might construct this car as a science project, test it, modify it, and evaluate it.

Challenge Your Thinking

1. On the right is a simplified diagram of a common mechanical clock. Use the labels and diagram to answer the following questions. This should be a real challenge!
   (a) How does this device work?
   (b) Where does the energy come from to power this device?
   (c) What is the path of energy through this device?
   (d) Why do the hour and minute hands move at different speeds?
   (e) The hairspring and balance rock back and forth to mark off the seconds. How does this work?
   (f) Does this device measure *time,* or is it really measuring something else? Explain.
2. The diagram below shows a device for powering a model boat.
   (a) How does it work?
   (b) What is the source of energy?
   (c) How is the energy transferred?
   (d) How is friction reduced?
   (e) Is a trade-off taking place here? If so, what is it?

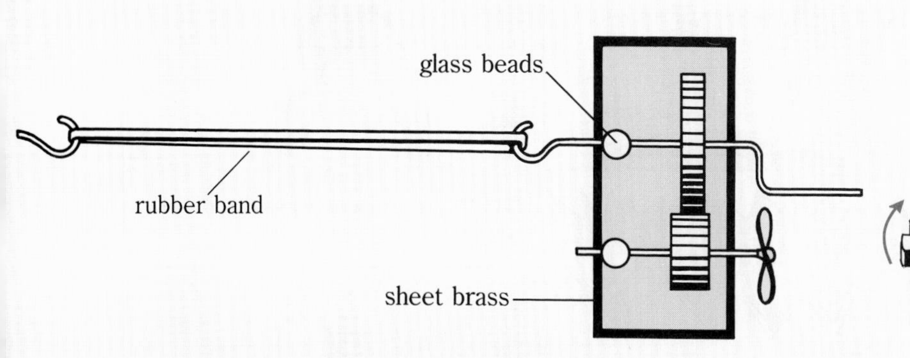

3. The distributor in a car engine sends a pulse of electric current to each of the spark plugs. It must turn at *exactly* 1/2 the speed of the engine. Would gear pairs or a wheel-and-belt system work better? Why?

Mechanical system of a basic clock

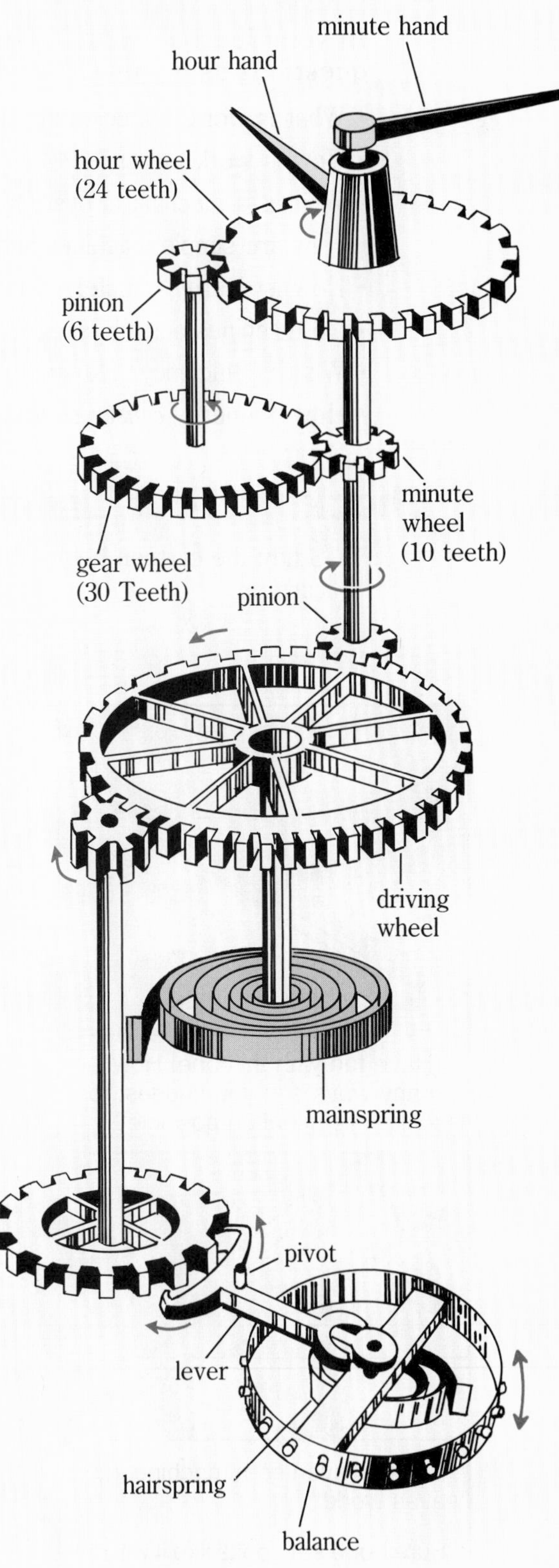

Unit 2 Making Connections

The Big Ideas

In your Journal, write a summary of this unit, using the following questions as a guide.

- What is work in a scientific sense? How is it related to energy?
- How is work measured?
- What is a mechanical system?
- How are simple machines and mechanical systems related?
- How is energy transferred in mechanical systems?
- Do machines actually save work? Explain.
- What is efficiency? How is efficiency related to friction?
- How is energy *conserved* in machines?

Checking Your Understanding

1. Study the cartoon below. Identify the error or errors in each panel of the cartoon.

2. The highest mountain on Earth, measured by how far it rises above its base, is the island of Hawaii. It rises about 9500 m above the surrounding sea floor. Imagine that you could walk from the bottom of the ocean to the highest point on the island. How much work would it take to get you to the top? (Assume you weigh 60 kg.) At 1400 kJ per hamburger, how many hamburgers would you have to eat to get enough energy to go all the way to the top?
3. Here is an interesting collection of gears.
   (a) Trace the motion of the components.
   (b) Where does the device change the direction of motion?
   (c) Where does a trade-off take place?
   (d) Which way does energy probably flow through this system? Explain.
   (e) What might you use this system for?
4. Look at the device pictured below. It is a so-called *perpetual motion machine.*

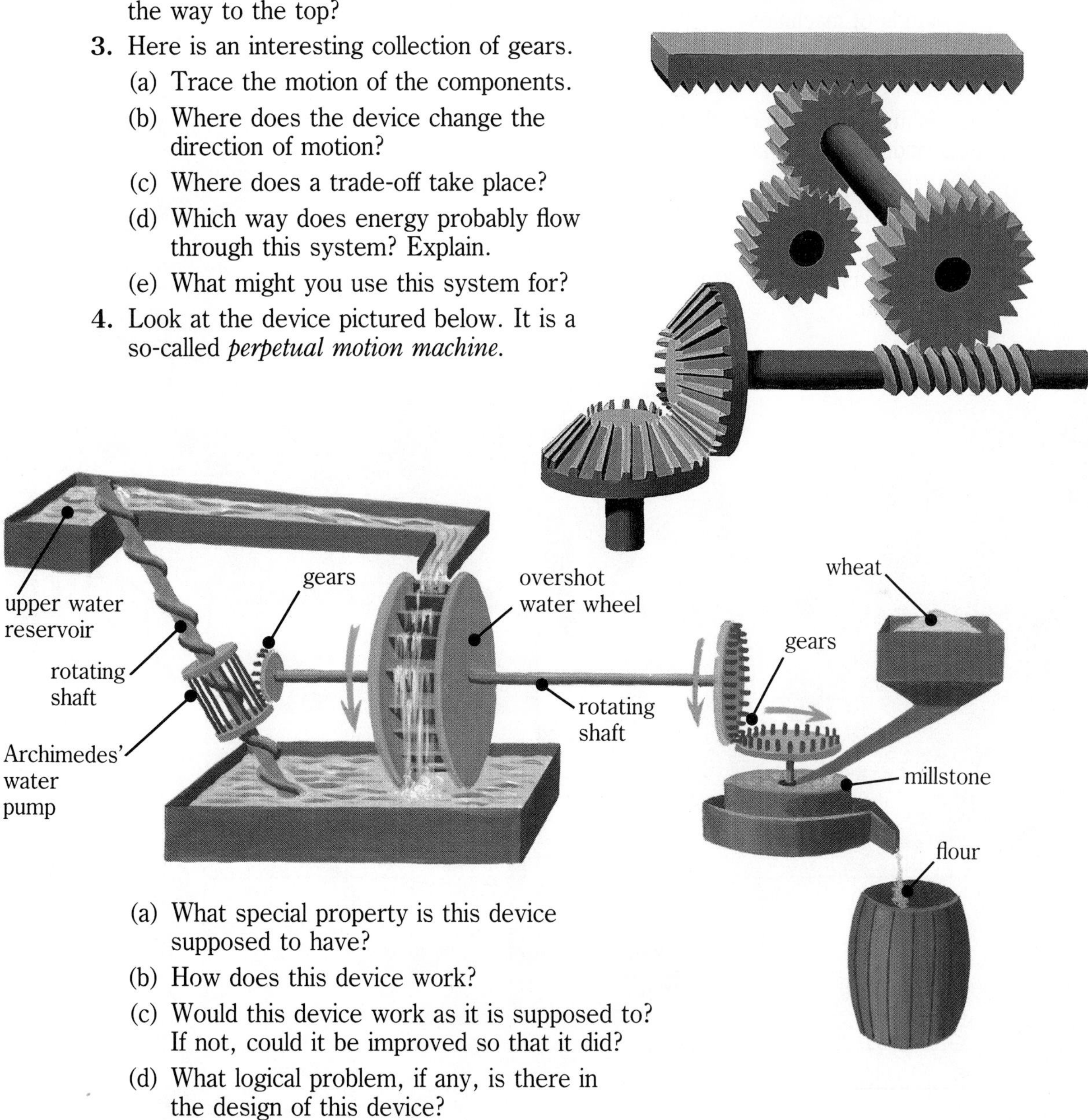

   (a) What special property is this device supposed to have?
   (b) How does this device work?
   (c) Would this device work as it is supposed to? If not, could it be improved so that it did?
   (d) What logical problem, if any, is there in the design of this device?
5. Design your own perpetual motion machine (something that looks like one anyway).

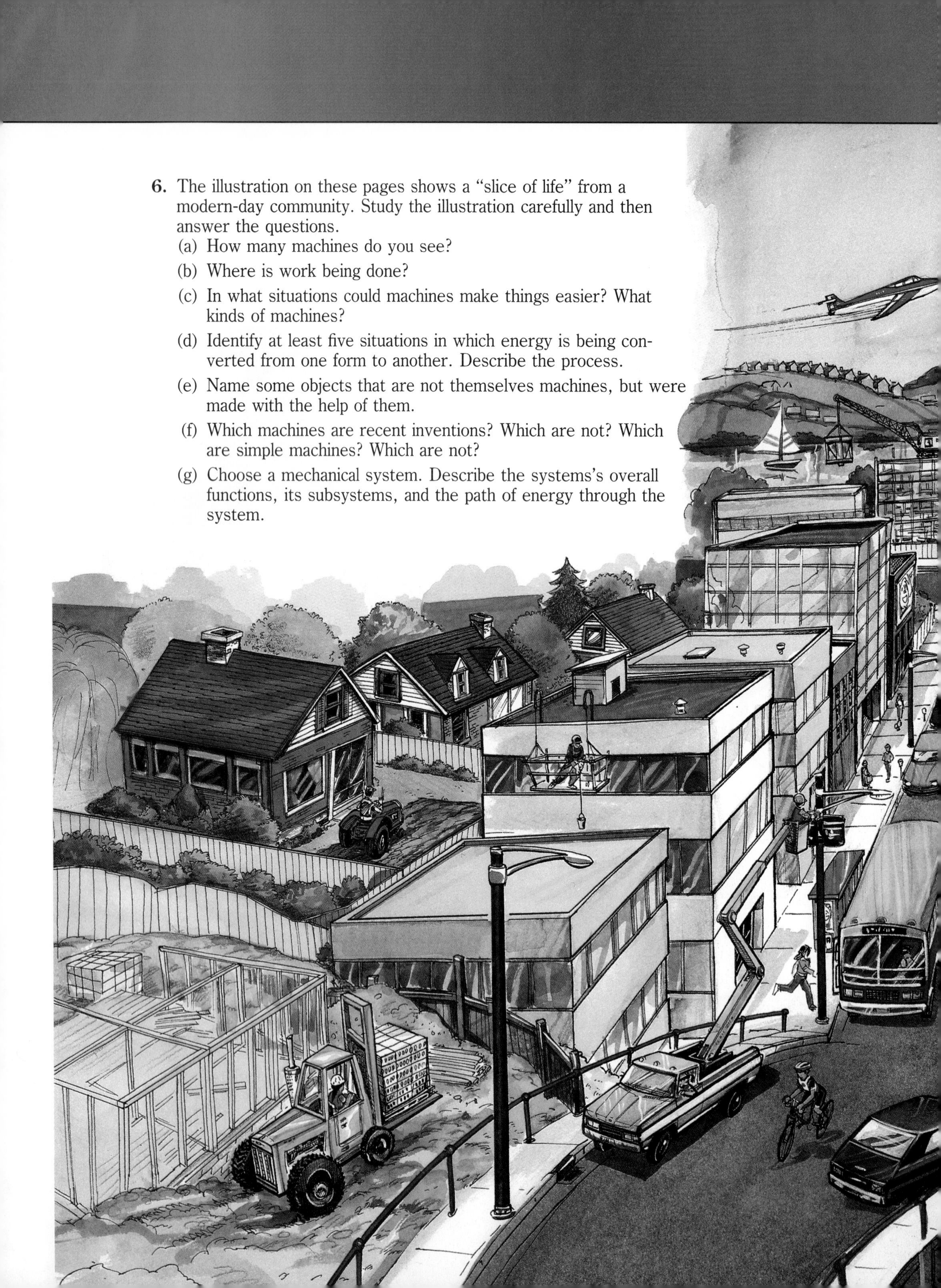

6. The illustration on these pages shows a "slice of life" from a modern-day community. Study the illustration carefully and then answer the questions.
   (a) How many machines do you see?
   (b) Where is work being done?
   (c) In what situations could machines make things easier? What kinds of machines?
   (d) Identify at least five situations in which energy is being converted from one form to another. Describe the process.
   (e) Name some objects that are not themselves machines, but were made with the help of them.
   (f) Which machines are recent inventions? Which are not? Which are simple machines? Which are not?
   (g) Choose a mechanical system. Describe the systems's overall functions, its subsystems, and the path of energy through the system.

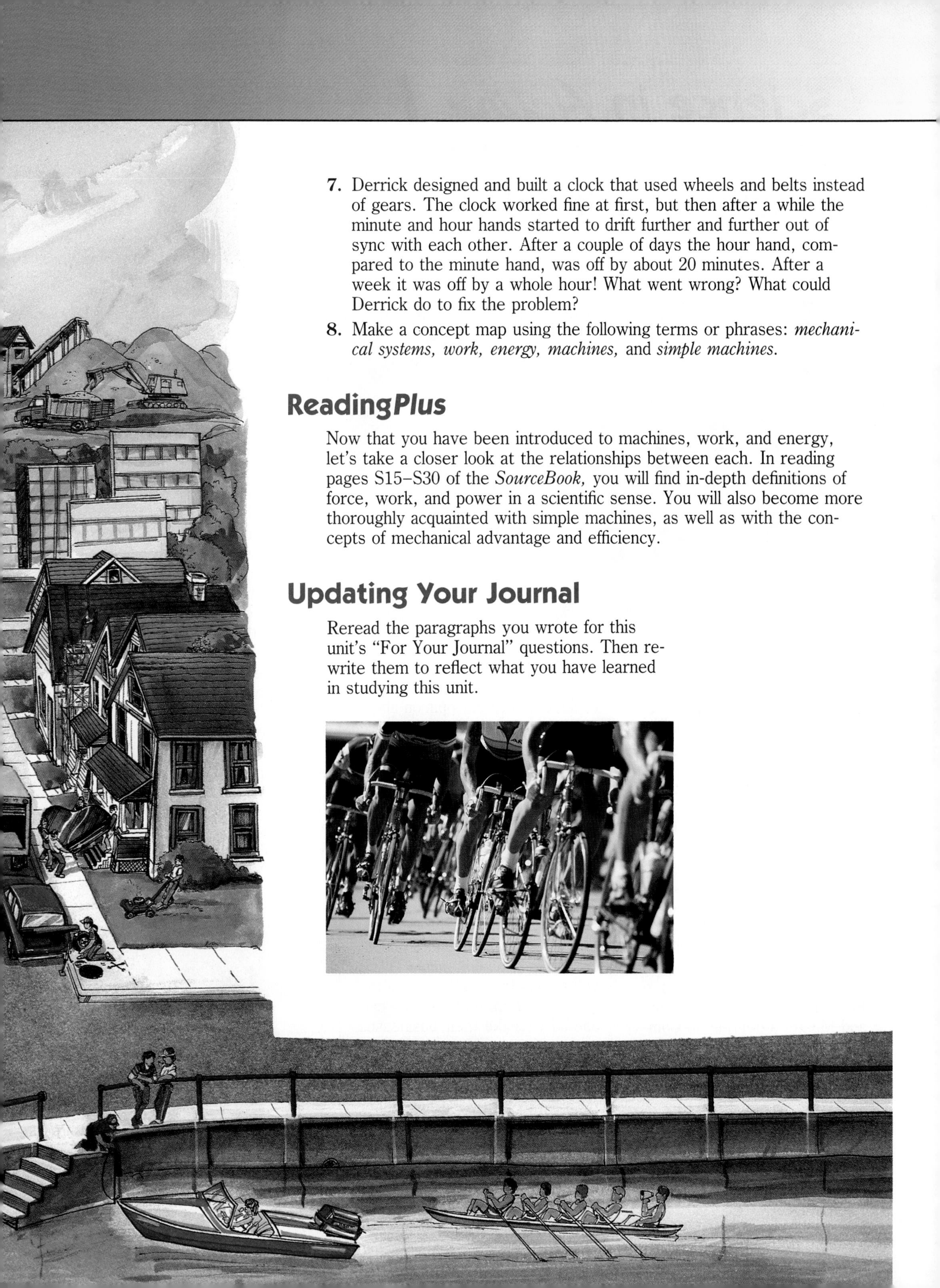

7. Derrick designed and built a clock that used wheels and belts instead of gears. The clock worked fine at first, but then after a while the minute and hour hands started to drift further and further out of sync with each other. After a couple of days the hour hand, compared to the minute hand, was off by about 20 minutes. After a week it was off by a whole hour! What went wrong? What could Derrick do to fix the problem?
8. Make a concept map using the following terms or phrases: *mechanical systems, work, energy, machines,* and *simple machines.*

Reading*Plus*

Now that you have been introduced to machines, work, and energy, let's take a closer look at the relationships between each. In reading pages S15–S30 of the *SourceBook,* you will find in-depth definitions of force, work, and power in a scientific sense. You will also become more thoroughly acquainted with simple machines, as well as with the concepts of mechanical advantage and efficiency.

Updating Your Journal

Reread the paragraphs you wrote for this unit's "For Your Journal" questions. Then rewrite them to reflect what you have learned in studying this unit.

Science in Action

Spotlight on Artificial Intelligence

Mehran Farahmand demonstrates a computer system that makes airport gate changes simple and automatic.

Is it possible for a machine to think? Could it use common sense? Could the thinking capabilities of a computer surpass those of humans? These are just a few of the many questions that surround the science of artificial intelligence, or AI.

Mehran Farahmand uses what he knows about AI to design computer systems. These systems complete tasks that ordinarily require human intelligence. Mehran's educational background includes Bachelor's degrees in French and Computer Science and Master's degrees in Computer Science, International Management, and Business Administration.

Q: What is artificial intelligence?

Mehran: AI is one of the fastest-growing areas of computer science. Everyone seems to have a different opinion about what it is, but most people agree that AI is the science of making machines intelligent.

A nice benefit of AI research is that we can apply what we find out about the computer to understanding more about human intelligence.

Q: What does your daily work involve?

Mehran: I am a senior consultant for a large company. This involves meeting with customers to advise them on how they can use AI to make their businesses run more smoothly and efficiently.

Usually, there is a definite problem to be solved. Often before I can solve a problem, I have to read a lot and meet with colleagues and experts in the field. We brainstorm ideas for new technology that will accomodate the client's needs. Once I have some ideas, I build a "thinking" program. That is, I develop computer software that can perform sophisticated skills and functions to eliminate my client's problem.

Q: Can you give us a specific example of what you do for your customers?

Mehran: Sure. Let's use the airline industry as an example. I might build a "smart" computer system that can handle airplane connections and changes, delays, cancellations, and any other problems you might encounter at the airport. The computer performs these tasks by automatically assigning new seats or making new schedules—just like a person would.

One of my specialties is making the computer into a teacher! By using different media, I can create a package that companies send over networks to various locations across the country. With this package, people can be trained by computer at their own pace.

The control center for the AT&T Worldwide Intelligent Network.

I feel this training process allows people to learn better and faster than they would if they were using traditional training methods.

Q: What are the most frustrating aspects of your job?

Mehran: AI is such a new technology that occasionally a client just can't grasp the scope of the solutions I can offer. They can't believe that computers can handle so many tasks previously reserved only for human beings!

Q: What inspired you to enter the field of AI?

Mehran: I happened upon computer science by accident! I wanted to do something different—to be in a field that would involve many other fields. To be successful in AI, it's essential to have knowledge of human psychology and linguistics, and to have skills of creativity, communication, and technical science. When I was younger, I spent a lot of time studying languages, psychology, and writing. My technical skills came along, and suddenly I had the basis for a career that combined all my interests!

Q: Does your work involve travel?

Mehran: A lot! I'm always visiting customers to study their specific needs. I like to know how they think. I find out by interviewing the people and observing the conditions of their working environment. Then I can incorporate the thinking processes I believe my client needs into modern technology. This technology will help the company become more profitable and competitive.

Q: Do you do most of your work independently or in cooperation with others?

Mehran: It's fair to say I do a lot of independent work. It's important, however, for me to have a lot of interaction with my colleagues and customers. Customers should feel as if they've been a part of every solution. If they don't, they might not see the solution as the best answer. Brainstorming with my colleagues is also essential—we're constantly helping each other.

Q: What personal qualities are the most important for your field?

Mehran: Open-mindedness and having an inter-disciplinary background are the most important qualities. AI is a new field that requires a lot of innovation. A person must be able to see a situation differently than anybody who has ever looked at it before. You have to be able to take a triangle and a circle and find similarities. Then you can't be shocked when you find out that the triangle makes a good circle!

This robot learns from its mistakes.

A Project Idea

Isaac Asimov asked, "Will computers become so complex and versatile that they will develop an intelligence approaching or surpassing that of the human being?" This fear has been the subject of much discussion surrounding the science of artificial intelligence. Research the subject. Do you feel Asimov's concern is valid? Why or why not? Prepare a report listing your findings and your opinions.

Machines in Art

In modern life machines are everywhere. Without a doubt, they affect the way we see, think, and feel about things. But we might need an artist to make us aware of this.

Artists often have unusual ways of seeing things. For example, the Swiss painter Paul Klee (1879–1940) seems to have seen the world in a way that was both poetic and humorous. In the painting on the right, Klee has shown his subject as a machine. What does the painting suggest to you? Do you think it is a view of nature? Or might it be a picture of a person who talks too much?

KLEE, Paul. Twittering Machine. *1922. Watercolor and pen and ink on oil transfer drawing on paper, mounted on cardboard, 25¼ × 19". Collection, The Museum of Modern Art, New York. Purchase.*

TINGUELY, Jean. Metamechanical Sculpture with Tripod. *1955. Tate Gallery, London.*

A Career Is Born

Jean Tinguely (1925–1991) was another Swiss artist with an unusual way of seeing things. He spent most of his life building fantastic machines.

When he was 12 years old, Tinguely began to build his first machines, which were water wheels with sound effects. He rigged up little wooden wheels and placed them at intervals in a mountain stream. As the wheels turned, they caused hammers to strike tin cans of different kinds. The overhanging pine trees, which Tinguely says were like a cathedral, enhanced the sounds. His "concerts" could be heard for a distance of 100 meters through the woods.

"Art never entered my head," he says, speaking of these early creations. But with the encouragement of one of his teachers, he continued his experiments. He made constructions out of wire, straw, wood, and paper—and even food. His artistic career was under way.

Tinguely's Anti-Machines

Many of Tinguely's machines are deliberately ungraceful. And as machines, they could hardly be said to work very well. His cog wheels jump their tracks. Or they get stuck and then start turning again, unpredictably. But this is the way Tinguely wanted it. Machine-like precision that would be valued by an engineer seemed cold and limiting to him. Unpredictability makes his machines seem more "friendly"—in fact, almost human. And for many people, the effect is quite hilarious.

Tinguely had mixed feelings about machines. On the one hand, he certainly loved machines, including Formula I racing cars. In 1955, however, he was present at the Le Mans 24-hour race in which a catastrophic accident occurred. A car traveling 260 km/h was forced to the side and thrown off the track. It exploded, throwing its engine and rear axle into the crowd. Eighty-two people were killed. The shock of this event had a profound effect on Tinguely.

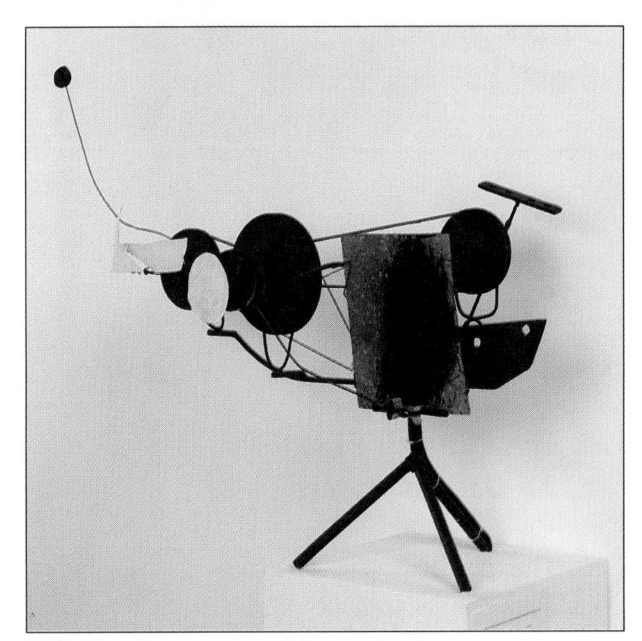

TINGUELY, Jean. Metamechanic Number 1. *1959. Musee National d'Art Moderne, Paris*

Much of Tinguely's work seems to be a social commentary on the negative effects of machines in our lives. In fact, Tinguely even produced a number of machines that were designed to destroy themselves! But in one famous instance, everything went wrong. His "Homage to New York," which was a machine that was supposed to self-destruct, failed to destroy itself properly. It did, however, start a fire that had to be put out by firemen.

CALDER, Alexander. Big Red. *1959. Sheet metal and steel wire. 74 × 114". Collection of Whitney Museum of American Art, New York. Purchase, with funds from the Friends of the Whitney Museum of American Art, and exchange 61.46. Photography by Sandak, Inc./G. K. Hall, Inc., Boston, MA*

Think About It

Tinguely's earliest water wheels were projects purely for the sake of enjoyment. However, the art world of the time was taking an interest in such original and playful creations. The way had been paved by artists like Marcel Duchamp (1887–1968), whose creations and writings challenged people to see art in new ways.

One of Duchamp's most famous works was a bicycle wheel mounted on a stool. What do you think? Can something ordinary or absurd be a work of art?

One artist whose work is often compared with Tinguely's is Alexander Calder (1898–1976). Some of his earliest works, like Tinguely's, were motor-driven. His famous "mobiles," however, have no motors. They are moved by hand or by air currents. As they turn, they trace beautiful forms in space. Study the mobile on the left. Can you visualize its motion? What mechanical principles can you find in this work of art?

Unit 3 Oceans and Climates

For Your Journal

Write a paragraph summarizing your thoughts about each of the following:

1. *When you hear the word "greenhouse," what does it mean to you?*
2. *What factors combine to make your climate what it is?*
3. *How and why do climates differ thoughout the world?*
4. *How are ocean and atmosphere interacting in this photograph?*

An Oasis in Space

You know that the Earth has many different climates, some cold, some warm, some in-between. But even if you combined the extremes of its hottest and coldest places, Earth is still a very pleasant place compared to the other planets and satellites in our solar system.

Take, for example, the moon—our nearest neighbor. The moon's surface temperature shoots to above 130 °C during the day and plummets to below −170 °C at night. Does the temperature in Eureka, California change that much from night to day? over the whole year? How about Lincoln, Nebraska? How about where you live? Do you think any place on Earth has such a wide range of temperature? What reasons would explain the very large temperature difference that the moon experiences? Why does Earth experience a more favorable climate?

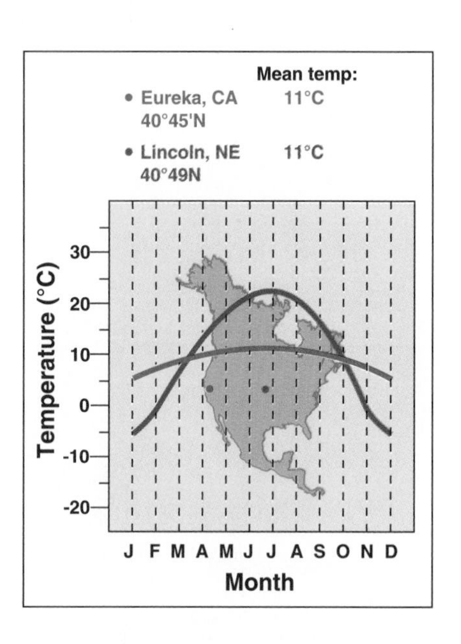

"Distance from the sun is definitely the most important factor in determining how hot or cold a planet is."
"Another factor is if the planet has an atmosphere or not."

"I think the kinds of gases that make up the atmosphere affect a planet's temperature."

"Another thing that is important is how dense or thick a planet's atmosphere is."

Earth is unique. It has an average temperature that allows living things to flourish. You might wonder: Why is that? Is Earth fortunate enough to be just the right distance from the sun, or are there other factors that contribute to its temperate climate?

Examine the data regarding Earth and its closest neighbors on the pages that follow. Compare these planets. How do their average temperatures vary? What factors seem to determine the temperature of the planets? One group of students made the suggestions shown. Can you find data that supports each suggestion?

What other factors might affect the climate of a planet? Look at the accompanying photographs for some clues.

MERCURY

Profile:

Average distance from sun: 58 million km

Average surface temperature: 450 °C, day; −350 °C, night

Atmosphere: none

Atmospheric pressure: N/A

Water: none

Life: none

VENUS

Profile:

Average distance from sun: 108 million km

Average surface temperature: 500 °C

Atmosphere: 96% carbon dioxide, 4% nitrogen; clouds of sulfuric acid surround planet

Atmospheric pressure: 9.6 million pascals*

Water: none

Life: none

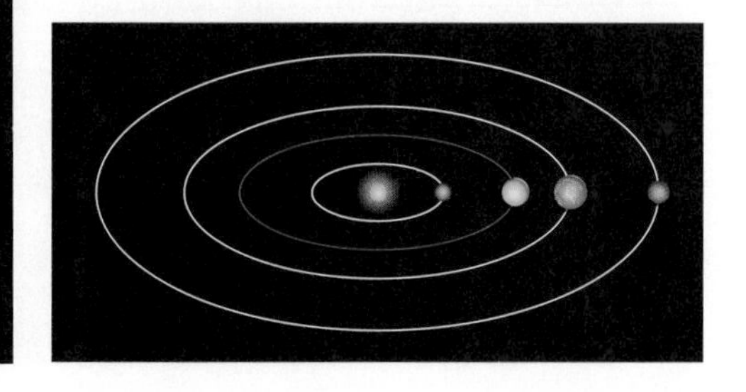

newtons per square meter, a unit of pressure

EARTH

Profile:

Average distance from sun: 150 million km

Average surface temperature: 15 °C

Atmosphere: 77% nitrogen, 21% oxygen, 1% argon, 0%–4% water vapor (variable), 0.03% carbon dioxide; traces of other gases

Atmospheric pressure: 101,000 pascals

Water: abundant; large oceans, polar icecaps

Life: extremely abundant and varied

MARS

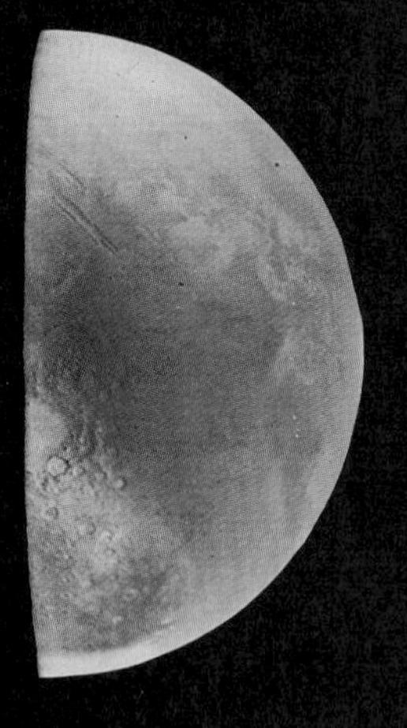

Profile:

Average distance from sun: 228 million km

Average surface temperature: −55 °C

Atmosphere: 95% carbon dioxide, 3% nitrogen

Atmospheric pressure: 1500 pascals

Water: some contained in polar icecaps and subsurface permafrost

Life: none detected

Understanding Our Planet

Did you conclude that a planet's temperature is determined by more than just its distance from the sun? What other factors play a role? What is the evidence?

In the following Exploration, you will learn about the heating of the Earth by setting up a *simulation*. A **simulation** is a type of experiment intended to model actual conditions. Scientists often use simulations to study complex problems. In Exploration 1 a glass jar will represent the Earth and a light bulb will represent the sun. If the weather is clear you can do the simulation using the real sun.

EXPLORATION 1

A Simulation

You Will Need

- 2 thermometers
- 2 large glass jars with covers
- a lamp with a 100-W or greater bulb (if not sunny)
- stiff paper strip
- tape

What to Do

Your setup should look like the diagrams shown here. One jar is your control. The second jar represents Earth and its atmosphere. Your challenge is to modify it in such a way as to cause a change in the temperature. Change anything you want, but change only one factor or variable at a time. Why is this? Now follow the steps below.

1. Get together with two or three other students. Discuss the factors or variables that could influence the temperature in the jar.
2. Choose one that your group would like to test and modify the jar to determine the effect of this variable.
3. Place both jars near the light bulb (or in the sun). Record the temperature changes in both jars.

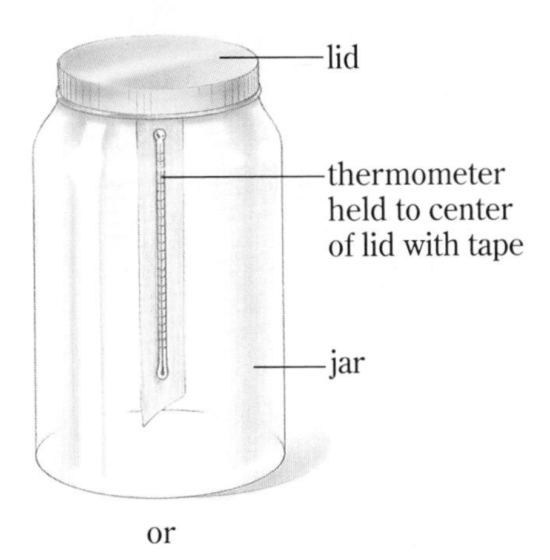

or

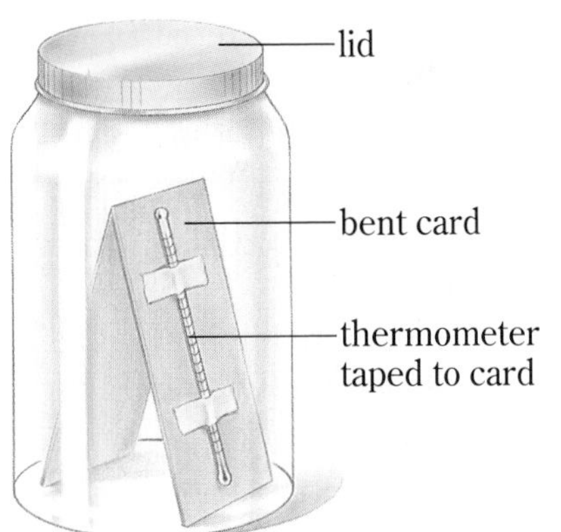

What did you do to make this a controlled experiment?

Conclusions

1. Did your modified jar reach a different temperature than the control? Was it higher or lower? What do you think caused the difference?
2. Make a list of the different modifications the class made to the simulation as well as the results obtained with each. Which modifications had the greatest effect?
3. Did the temperatures in your jars stabilize after a time? Why did this happen? How is this similar to the real Earth?
4. What are the shortcomings of this model? Compile a list.
5. Suppose you add a layer of dark soil to the bottom of the bottle. After 20 minutes the temperature in this jar is 1°C higher than in the control. What might you conclude?
6. Evidence suggests that the average temperature of the Earth has gone up by about 0.5°C during this century. What might be causing this change? Do the results of any of the simulations suggest possible answers? Explain.

2 The Greenhouse Effect

From Today's Headlines

Have you seen headlines like those above? What do they seem to imply? It seems that Earth is becoming a warmer place. What could possibly be bringing this about? Could the sun's energy output be increasing? Could Earth be moving closer to the sun? Could the warming be part of a natural cycle? (After all, for most of its history Earth had a much warmer climate than it does now.)

You have probably heard a great deal of talk about the *greenhouse effect.* A lot of people are concerned about it and blame it for the apparent *global warming* trend. What is causing this greenhouse effect? Are people somehow responsible?

With all this talk about global warming, we need to find out more about it. Only then can we understand and deal effectively with it.

Look at the graph on the right. This graph shows the average global temperature over the last century. It also shows the average global temperature through the year 2050 as predicted by some scientists. After examining the graph, think about the following questions:

- Is the prediction realistic?
- If the prediction turns out to be right, what will the consequences be?
- How do you think the greenhouse effect is connected to all of this?

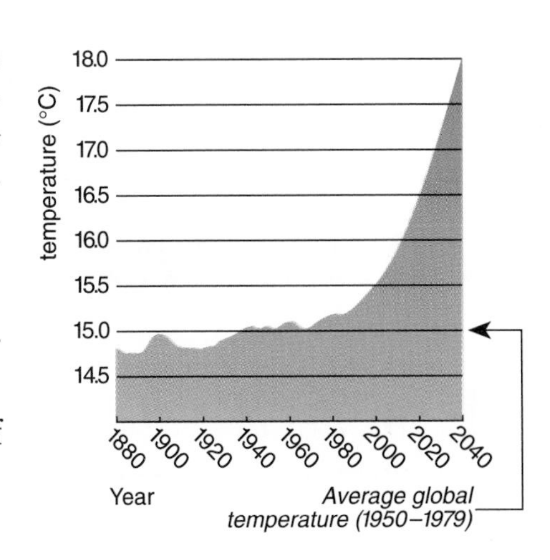

Dateline: Earth 2050

Imagine that you could be transported to the year 2050. How might the headlines read then? Make up your own headline and article dealing with the consequences of global warming. Will Earth be a very different, warmer place? Will the ice caps melt, causing the sea level to rise? Will our temperate Earth become tropical? Perhaps you are skeptical, and think that people are overreacting. If that's the case, your headline and article could reflect how all the predictions about global warming turned out to be wrong.

The Greenhouse Effect Explained

THE DAILY CHRONICLE

Without Greenhouse Effect, Earth Would be Cold, Lifeless Place

Here's one headline you probably haven't seen. It suggests an aspect of the greenhouse effect that is rarely noted in news about the dangers facing our climate. As you will see, the greenhouse effect is something we all depend on more than we realize.

The term **greenhouse effect** was coined in 1822 by Jean Fourier, a French mathematician. Fourier noted that in many ways the Earth's atmosphere behaves like a greenhouse. The glass of a greenhouse lets solar radiation pass through but traps the heat given off by the ground and plants within as they absorb the sun's energy. The glass walls and roof also keep the warm air from blowing away. As a result, the greenhouse warms up. If you have ever been in a greenhouse on a sunny but cold day, you are familiar with this effect. In Exploration 1, did the bottle in your simulation behave as a greenhouse?

Certain gases—principally carbon dioxide, water vapor, and methane—are especially good at blocking the flow of heat from the Earth back into space. These gases together make up a tiny percentage of the atmosphere, but have an enormous impact on Earth's climate.

The fact is that the greenhouse effect is essential for maintaining Earth's moderate temperature. If not for the greenhouse effect, Earth's average temperature would be much colder, at least 33 °C colder than it is—too cold for life as we know it to exist.

Venus has a large greenhouse effect. Compare the composition of its atmosphere to those of the other planets. What gases are trapping the heat there? Earth has a much smaller greenhouse effect by comparison, and Mars has an even smaller one. Can you begin to see why?

As you have seen, there is evidence to suggest that Earth is warming up. If this is the case, could human activities be responsible? Are we increasing the greenhouse effect? (Consider that for at least the last 200 years or more, humans have had a major impact on the environment.) Or could the warming trend be attributed to other, natural causes? Scientists are debating these questions.

Form some conclusions of your own. Examine the evidence in the following Exploration. Then do any additional research you may need. Work with two or three others to prepare a position paper presenting your views on the following questions.

- How does the greenhouse effect work?
- Is Earth, in fact, heating up? What is the evidence?
- If there is a warming trend, could it be caused by increased amounts of carbon dioxide and other gases in the atmosphere? What is the evidence?
- Could other causes account for the apparent warming? What evidence suggests this?

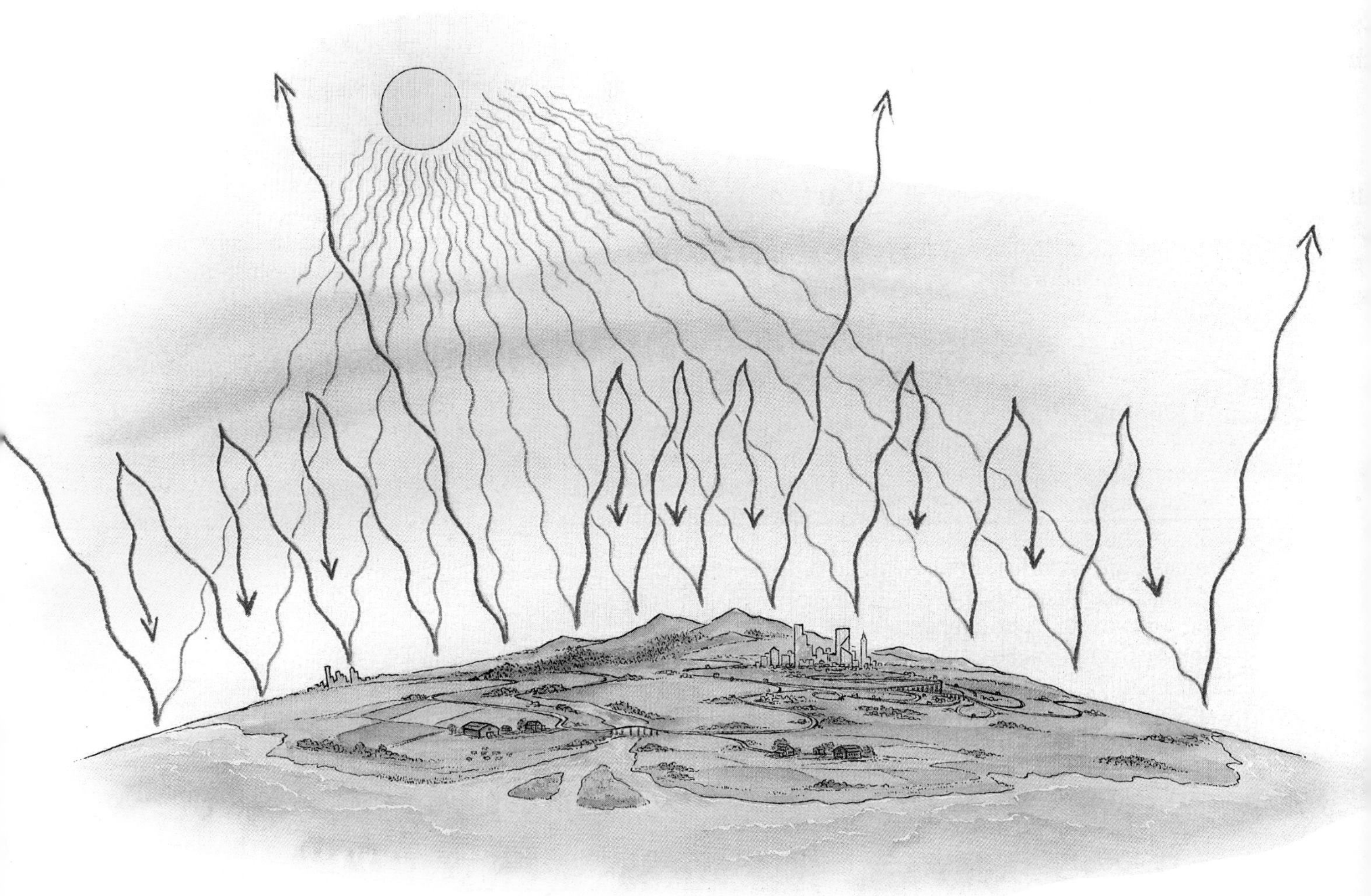

How the greenhouse effect works

Drawing Conclusions

Part 1

How Does the Greenhouse Effect Work?

Refer back to the diagram on page 137, which illustrates how greenhouse gases such as carbon dioxide, water vapor, and methane help heat the Earth. Make a rough sketch of this diagram in your Journal and place the letter of each statement in the appropriate location.

(a) Sunlight is made up of radiation of many different wavelengths.

(b) Ultraviolet and other short wavelengths of solar radiation are mostly absorbed (by ozone) high in the atmosphere.

(c) Longer wavelengths, such as visible light and infrared, pass through the atmosphere without being absorbed.

(d) This radiation is absorbed by the Earth.

(e) The absorbed energy is re-radiated at still longer wavelengths (infrared).

(f) Some of these wavelengths escape back into space.

(g) Some of the longer wavelengths are absorbed by the greenhouse gases such as carbon dioxide, water vapor and methane; thus trapping the energy.

(h) Some of this energy is re-radiated, further heating the Earth.

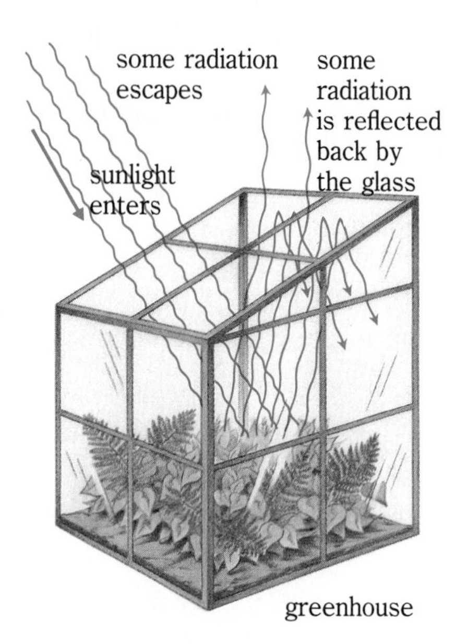

An actual greenhouse has a different kind of greenhouse effect. How does the diagram above show this?

Part 2

Temperature Versus Carbon Dioxide Concentration

The concentration of carbon dioxide in the atmosphere is about 350 ppm (parts per million). What does this mean?

Imagine having a container of one million grains of salt. If you were to replace one of these with a grain of pepper, then the concentration of the pepper in the container would be 1 ppm. Imagine replacing 350 salt grains with pepper grains. What would be the concentration of pepper in parts per million?

Using various methods, scientists have been able to estimate the Earth's average temperature, as well as the concentration of carbon dioxide in the atmosphere, over the last several hundred thousand years. This is shown in the graph below.

1. When was the carbon dioxide level greatest? the temperature?
2. Does there seem to be a correlation (connection) between temperature and carbon dioxide concentration in the atmosphere? Explain.
3. When, in the past 160,000 years, do you think there could have been an ice age?
4. Were all the variations in carbon dioxide level likely caused by human activities? Why or why not?

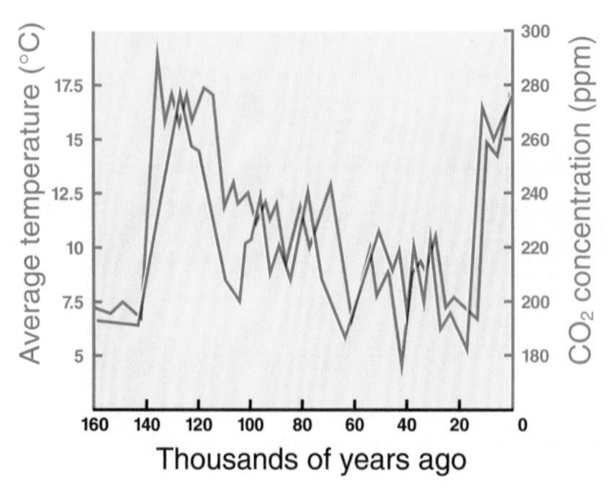

Part 3

Carbon Dioxide and Temperature

The graph below contains two sets of data. Use the data to answer the questions that follow.

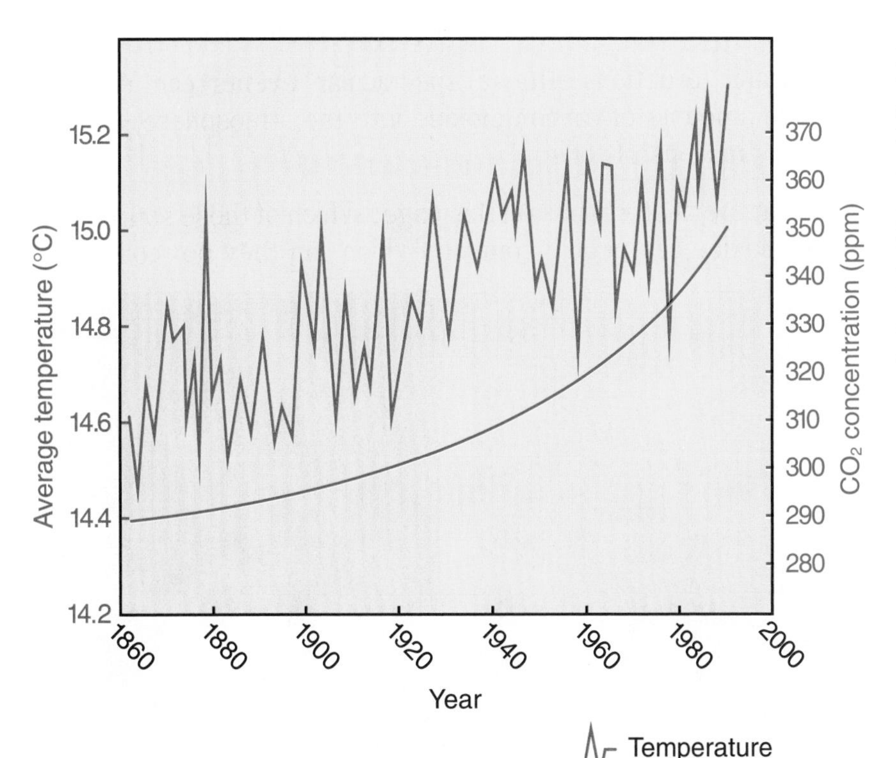

1. What was Earth's average temperature in 1990? How about in 1890?
2. Does the graph appear to show a relationship between the concentration of carbon dioxide in the atmosphere and the Earth's average temperature?
3. Does the graph *prove* that the change in the atmospheric concentration of carbon dioxide has caused the average temperature to increase?

Part 4

Methane and Human Activity

Methane is another greenhouse gas. Methane is present in the atmosphere in smaller quantities than carbon dioxide. However, a given mass of methane is much more effective as a *greenhouse gas* than is carbon dioxide. How is methane changing in the atmosphere? Why is it changing? Is there a correlation between methane levels and human populations? Prepare your own graph to help illustrate any possible connection. Then answer the questions on the following page.

Year	Methane Concentration (ppm)	Human Population (billions)
1990	1.70	5.3
1980	1.50	4.2
1970	1.40	3.5
1960	1.30	3.0
1950	1.25	2.5
1940	1.15	2.0
1900	1.00	1.5
1850	0.85	1.2
1800	0.75	1.0
1750	0.74	0.8
1700	0.72	0.7
1650	0.70	0.6
1600	0.70	0.5

EXPLORATION 2—CONT.

1. From your graph, predict world populations in the year 2000.
2. What will be the expected concentration of methane in the year 2000?
3. Here are some sources of methane. How is each related to human activities?
   - Cattle—methane is produced during digestion.
   - Rice paddies—bacteria in the flooded soil of rice paddies produce methane, which escapes into the air through the plants' hollow stems.
   - Decomposition—the decomposition of organic material in the absence of oxygen produces methane.
   - Natural gas (its main component is methane)—leaks constantly from the Earth, from pipelines, and from production and processing facilities.
   - Termites—methane is produced as termites digest wood. Termites are extremely abundant in the Earth's warm, moist regions.

Return to the questions (on page 137) that preceded this Exploration. Use them to help you prepare your position paper on the greenhouse effect and global warming. What kind of additional data would be helpful? What other questions arose in the course of your study?

Sources of Carbon Dioxide

Carbon dioxide is perhaps the most important greenhouse gas we add to the atmosphere. It is produced by a wide variety of natural processes and human activities, including:

- Burning—any time organic material (material containing carbon) is burned, carbon dioxide is produced.
- Respiration—carbon dioxide is a by-product of respiration in both plants and animals.
- Decomposition of organic matter—decomposition is like slow-motion burning, and like burning, produces carbon dioxide.
- Volcanic eruptions—these spectacular events can release huge amounts of carbon dioxide into the atmosphere, sometimes millions of tons.

Look at the photographs on this page. Which of these sources of carbon dioxide can people control? Which can they not control?

Global Warming—How Much?

Perhaps you are not yet convinced that global warming is actually occurring. If so, you are not alone. Scientists are far from unanimous on this matter. Scientists use sophisticated computer models to predict the future behavior of the Earth's climate. Different models have produced very different results. Some scientific models predict a significant rise in global temperatures in the near future. Other models predict little or no warming. The question is: which model is most accurate?

The accuracy of a scientific model depends on the reliability of the information used to formulate it. An accurate model of the causes and effects of global warming must answer the following questions. Can you think of others?

- What is the role of the oceans in absorbing and releasing heat energy? in absorbing and releasing carbon dioxide?
- How would ocean currents be affected by global warming?
- How would plants respond to a higher level of carbon dioxide in the air?
- Would a higher average temperature change the amount of cloud cover across the Earth? How would a change in cloud cover affect the heat balance of the Earth?
- What is the effect of light-reflecting dust and gases that have been pumped into the atmosphere by human activities and natural processes?

If the globe does warm up by a significant amount, what would be some of the consequences? Consider that a range of possible temperature increases (anywhere from 1.5° to 5°C) has been predicted. With two others, make a web of changes and effects that might occur. Here is a start.

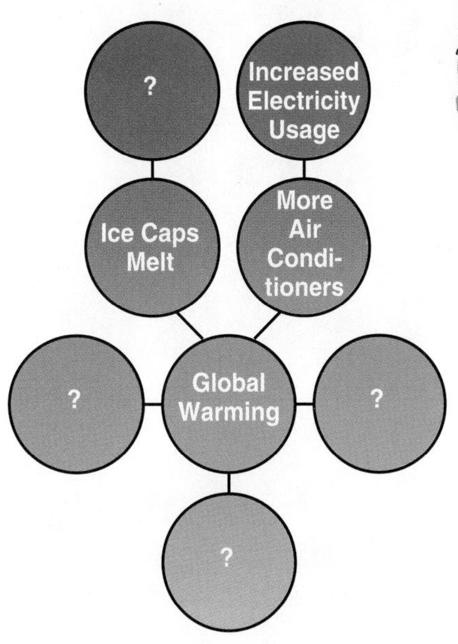

3 Land and Sea

The Earth's Thermostat

Look at the map on this page. This image, taken from space, shows surface temperatures across the Earth at the moment the image was made. What does the map tell us about the heating processes of Earth? Find North America on the map. What is the temperature at your location? Where do you think the sun's rays are most direct? What does this data suggest about the role the oceans play in distributing heat energy across the globe?

Examine the map to help you answer the following questions.

- At what time of year was the image taken?
- Where are the hottest temperatures? the coldest temperatures?
- Do all locations on land at the same latitude (degrees north or south of the equator) have more or less the same temperature?
- Do all locations on the ocean at the same latitude have more or less the same temperature?
- Does the map suggest that continents and oceans warm up and cool down differently? Explain.

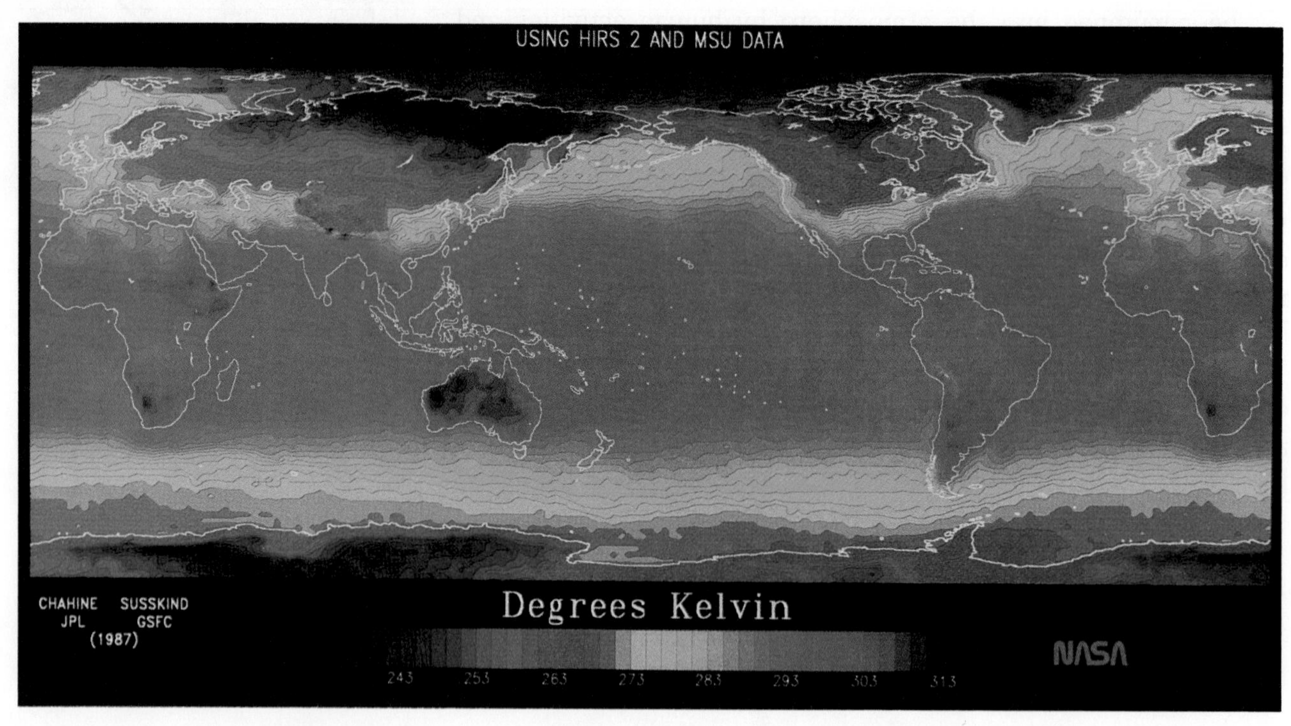

The Earth's surface temperature. To convert degrees Kelvin to degrees Celsius, subtract 273.

Heating of Land and Water

This Exploration is designed to help you answer questions raised on the previous page. It is a team effort. Each team will choose an activity, collect the materials, and do the experiment. Later you will give a report on your findings and interpretations. In doing so, make some reference to the photograph of global surface temperatures shown on the facing page. Place your data on a graph to help with your presentation.

You Will Need

- 2 test tubes
- 2 thermometers
- a beaker
- a heat source
- modeling clay
- water
- sand
- aluminum pie pans
- a lamp with 100-W or greater bulb
- charcoal powder
- graduated cylinder

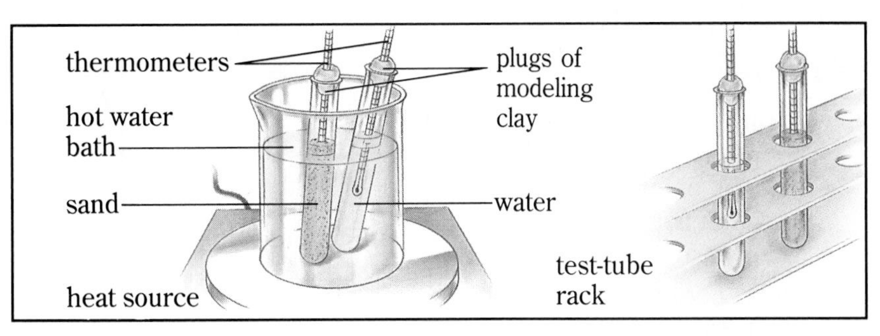

Setup for Activity 1

Activity 1

Sand Vs. Water #1

What to Do  

1. Fill one test tube halfway with water and another halfway with sand.
2. Place a thermometer in the sand and suspend another thermometer at the same depth in the water, using modeling clay to hold each thermometer in place. Place both in a beaker of hot water. Heat them to a temperature of around 70 °C.
3. Remove the test tubes and record the drop in temperature every 2 minutes for 20 minutes. Did they cool at the same rate?

Activity 2

Sand Vs. Water #2

What to Do 

1. Fill one pie pan halfway with water and another halfway with sand.
2. Place both an equal distance from the heat source.
3. Every 2 minutes for 20 minutes, record the surface temperature of each. The thermometer should be laid flat on the sand and the bulb covered with 0.5 cm of sand. The water temperature can be measured by holding a thermometer 0.5 cm below the surface. How did the temperatures change?

Setup for Activity 2

Activity 3

Heating of Different-Colored Sands

1. Measure equal quantities of sand into two separate pie pans. Mix one quantity of sand with enough charcoal powder to make it black.
2. Place a thermometer sideways in each quantity of sand. Bury the thermometer bulbs 0.5 cm below the surface of the sand.
3. Place the pans so that they are equal distances from the lamp.
4. Take readings every 2 minutes for 20 minutes. What differences did you observe?

Setup for Activity 3

Activity 4

Heating of Wet and Dry Sand

1. Measure equal quantities of sand into two different aluminum pie pans.
2. Add water to one so that the sand is quite damp. Water and sand should be the same temperature before they are added together.
3. Bury the bulb of each thermometer about 0.5 cm below the surface, one in each pan.
4. Place both pans an equal distance from the lamp.
5. Record the temperature of each pan every 2 minutes for 20 minutes. Which one warmed up faster?

Setup for Activity 4

Analyzing Activity 4

There is more to Activity 4 than first meets the eye. Certainly, water warms up more slowly than does sand. However, there is another factor involved as well. One group carried Activity 4 one step further. At the end of the 20-minute observation period, they switched off the light and watched what happened. After 15 minutes they were surprised to find that the wet sand temperature had actually dropped below that of the room. What happened? Where did the heat energy go? The next Exploration will help you solve this mystery.

Analyzing and Reporting Your Findings

What were your findings? What do these findings suggest about the role of water and land in terms of the heating processes of Earth? How does the image on page 142 relate to your findings? What further questions for investigation does this activity suggest? Summarize your findings in a brief report.

❑

The Water Cycle and Heat Exchange

You Will Need

- a thermometer
- a piece of shoelace 20 cm long
- water
- tape
- warm water
- rubbing alcohol

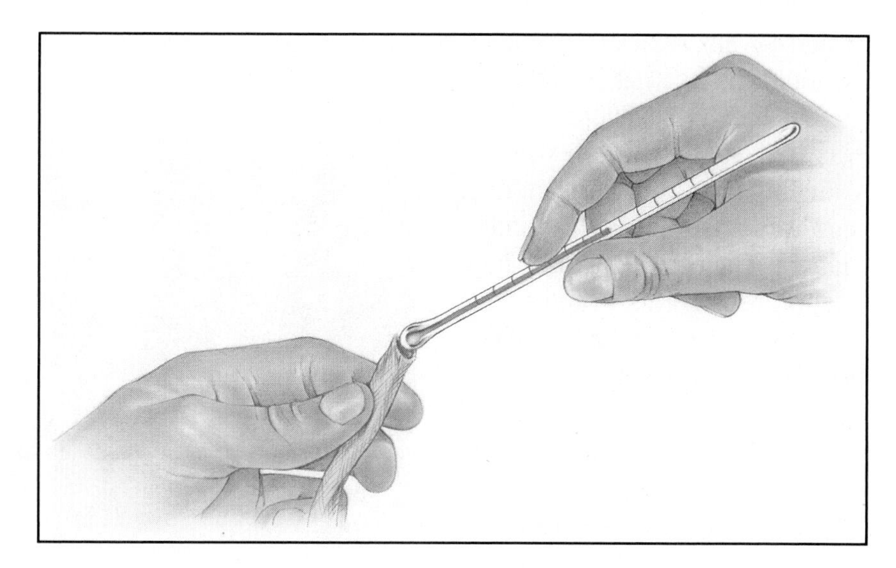

Part 1

Making a Wet-Bulb Thermometer

What to Do

A *wet-bulb thermometer* is a type of thermometer used to measure humidity. To make a wet-bulb thermometer, first cut a short (20 cm) section of shoelace. Notice that the shoelace is hollow inside. Insert the bulb of the thermometer into the shoelace as you would your foot into a sock. Use thread or a small piece of masking tape to secure the shoelace on the thermometer.

Now wet the covering and bulb with water that is at or above room temperature. Wave the thermometer vigorously in the air for a minute or so and then read the temperature again. What happened? What might have caused this to happen?

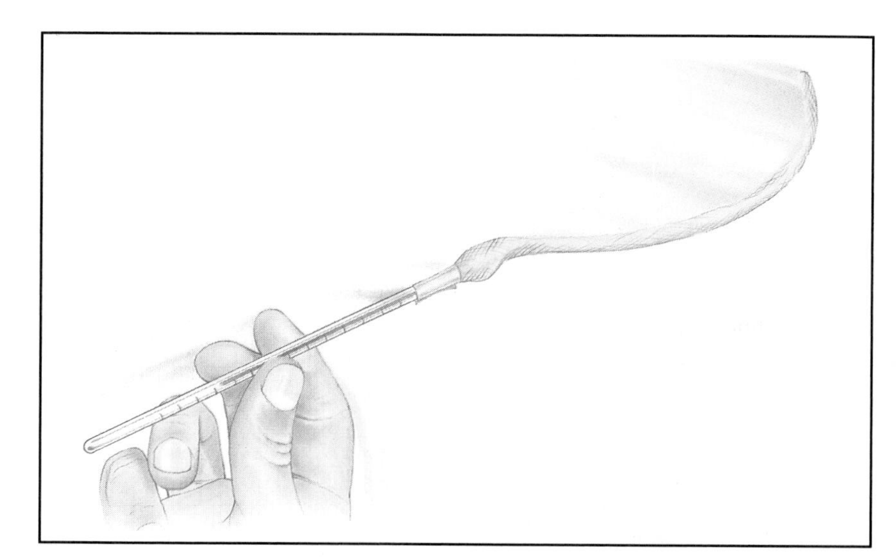

Part 2

Feeling Cool

Dip your finger in warm water and then wave it in the air. What do you feel? What do you think happened to the water on your finger? Could this help to explain what happens with the wet-bulb thermometer? If you wet your finger in alcohol, will it feel cooler than when wet in water? Why or why not?

What explanation do these activities suggest for why wet sand warmed more slowly than dry sand in the previous Exploration?

Explaining Your Findings

Your wet finger feels cool. The heat in your finger must have been absorbed by the evaporating water. What happened to this energy? Did it disappear? No, that would be impossible. What happened was that some of the heat in your finger and its surroundings was absorbed by the individual particles of water as they evaporated. If this evaporated water were to condense, that heat would be released to the surroundings once again.

Follow the exchange of heat energy in this diagram of the water cycle. Where is heat being absorbed from the surroundings? Will this make the surroundings warmer or cooler? Where is heat being released to the surroundings? Does this make the surroundings warmer or cooler?

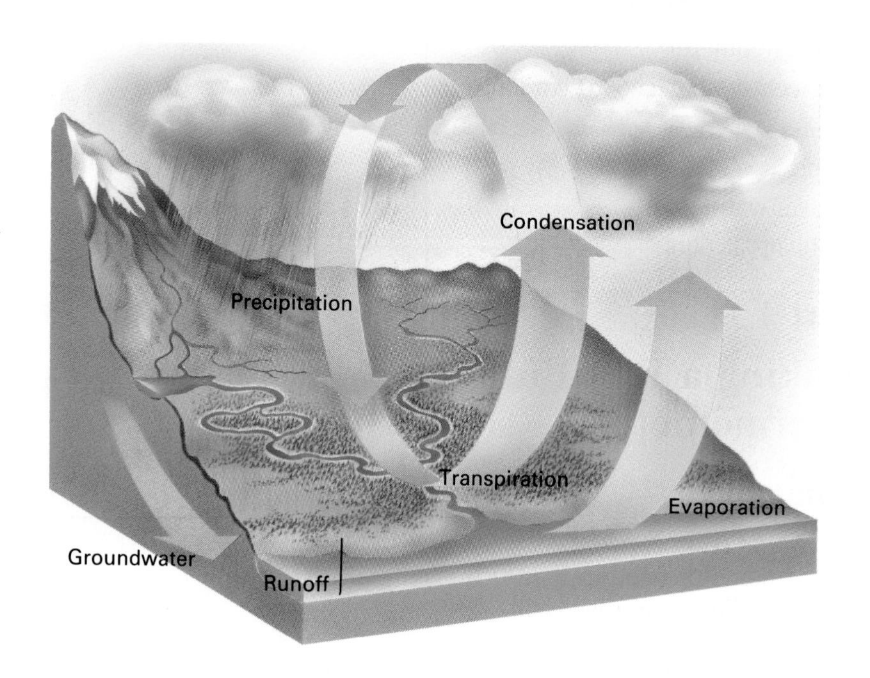

Examine again the image showing global temperatures on page 142. Where would you expect to see the greatest amount of water evaporating? When would the energy stored in the water vapor be released again?

A Mini-Activity in a Bottle

Try this at home. Add some water to a large plastic soft drink bottle. Now drop in a burning match and seal the lid. Squeeze the sides of the bottle as hard as you can and then quickly release. Do this several times. What do you observe? How would you explain your observations?

Something to Think About

Trees cool their surroundings by absorbing solar energy and providing shade. But they also provide a cooling effect in another way. How do you think this works? Why would cutting vast tracks of forest have a warming effect on the environment? How might the cutting of trees affect the world climate?

The Dew Point

You may recall from earlier studies that the **dew point** is the temperature at which the moisture starts to condense out of the air. The dew point indicates how much moisture there is in the air. The lower the dew point, the less moisture there is in the air. The amount of cooling of a wet-bulb thermometer is used to figure the dew point. Can you explain why? Think of a situation when no amount of shaking of the wet-bulb thermometer would result in a lower reading.

You can use the table provided here to figure the dew point. What happens to the dew point if the amount of moisture in the air increases? In the next Exploration you will use two methods to determine the dew point.

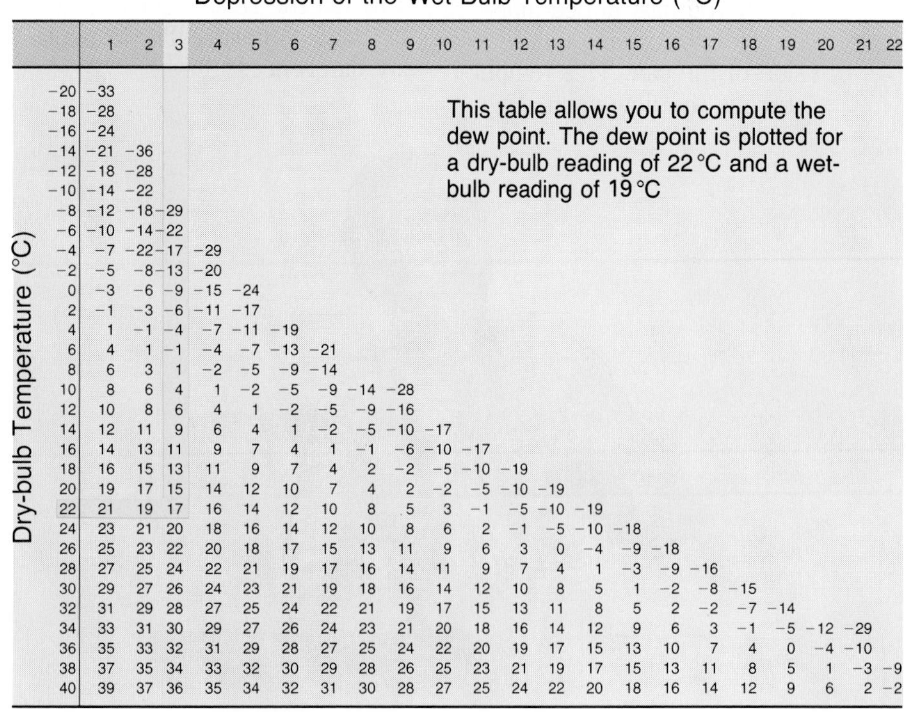

Table for Computing Dew Point

Depression of the Wet-Bulb Temperature (° C)

Dry-bulb Temperature (°C)	1	2	3	4	5	6	7	8	9	10	11	12	13	14	15	16	17	18	19	20	21	22
−20	−33																					
−18	−28																					
−16	−24																					
−14	−21	−36																				
−12	−18	−28																				
−10	−14	−22																				
−8	−12	−18	−29																			
−6	−10	−14	−22																			
−4	−7	−22	−17	−29																		
−2	−5	−8	−13	−20																		
0	−3	−6	−9	−15	−24																	
2	−1	−3	−6	−11	−17																	
4	1	−1	−4	−7	−11	−19																
6	4	1	−1	−4	−7	−13	−21															
8	6	3	1	−2	−5	−9	−14															
10	8	6	4	1	−2	−5	−9	−14	−28													
12	10	8	6	4	1	−2	−5	−9	−16													
14	12	11	9	6	4	1	−2	−5	−10	−17												
16	14	13	11	9	7	4	1	−1	−6	−10	−17											
18	16	15	13	11	9	7	4	2	−2	−5	−10	−19										
20	19	17	15	14	12	10	7	4	2	−2	−5	−10	−19									
22	21	19	17	16	14	12	10	8	5	3	−1	−5	−10	−19								
24	23	21	20	18	16	14	12	10	8	6	2	−1	−5	−10	−18							
26	25	23	22	20	18	17	15	13	11	9	6	3	0	−4	−9	−18						
28	27	25	24	22	21	19	17	16	14	11	9	7	4	1	−3	−9	−16					
30	29	27	26	24	23	21	19	18	16	14	12	10	8	5	1	−2	−8	−15				
32	31	29	28	27	25	24	22	21	19	17	15	13	11	8	5	2	−2	−7	−14			
34	33	31	30	29	27	26	24	23	21	20	18	16	14	12	9	6	3	−1	−5	−12	−29	
36	35	33	32	31	29	28	27	25	24	22	20	19	17	15	13	10	7	4	0	−4	−10	
38	37	35	34	33	32	30	29	28	26	25	23	21	19	17	15	13	11	8	5	1	−3	−9
40	39	37	36	35	34	32	31	30	28	27	25	24	22	20	18	16	14	12	9	6	2	−2

This table allows you to compute the dew point. The dew point is plotted for a dry-bulb reading of 22 °C and a wet-bulb reading of 19 °C

Measuring the Dew Point

Part 1

The Cold-Can Method

You Will Need

- a metal can (250 mL)
- ice cubes
- a thermometer
- a stirring rod

What to Do

1. Add room-temperature water to your can along with a couple of ice cubes. Fill the can one-quarter full. Do this at your work station so that you can begin to make observations immediately.
2. Place the thermometer in the water. Using a stirring rod, stir the water continuously while observing the sides of the can for any condensation. Record the temperature at which you first notice moisture (dew) forming on the side of the can. This temperature is the dew point.

Part 2

The Wet-Bulb Thermometer Method

You Will Need

- a thermometer
- a wet-bulb thermometer

What to Do

1. Measure the temperature of the room. Don't use your wet-bulb thermometer to do this, it will give you a false reading. (Do you know why?)
2. Determine the wet-bulb temperature as you did in the previous Exploration.
3. Subtract the wet-bulb temperature from the dry-bulb temperature.
4. Use the table on page 147 to determine the dew point.

How did the results for the two methods compare? Try to explain any differences.

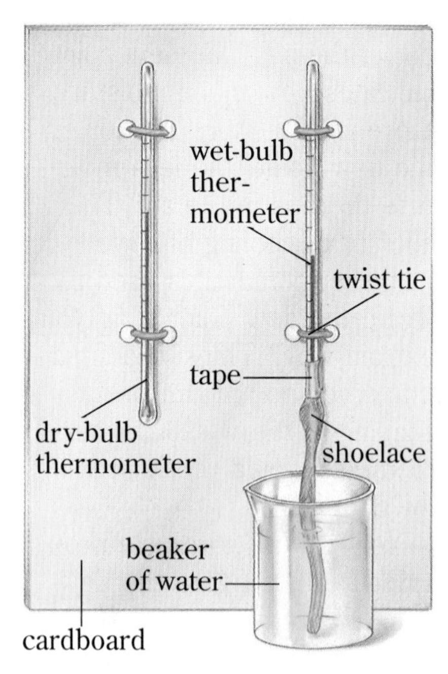

A homemade setup for measuring dewpoint

Upward Bound

Imagine taking a trip on the Aerial Tramway in Palm Springs, California. In 15 minutes the tram will travel 4 km and climb from 800 m to 2442 m above sea level. A breeze blows against the mountain and forces air to rise along its slope. As you ascend the mountain, you notice it growing cooler, and the vegetation is changing dramatically. At the base of the mountain, desert plants prevail. As you climb higher, vegetation is gradually replaced by shrubs and small trees. By the time you reach the summit you are in an evergreen forest. The pine-scented air is crisp and cool.

Why did conditions change so much from Valley Station to Mountain Station? Here is data for two trips on the tram, one on a clear day and one on a cloudy day. In both cases the temperature at Valley Station was 30 °C.

DAY 1	
Height	Temperature
1500 m	15°C
1000 m	20°C
500 m	25°C
0 m	30°C

DAY 2	
Height	Temperature
1500 m	20°C
1000 m	22°C
500 m	25°C
0 m	30°C

base elevation: 800 m

1. By how much did the temperature change on the trip up the tram on the clear day? on the cloudy day?
2. What is the dew point on the cloudy day?
3. What accounts for the change in vegetation?

As the tram ride clearly shows, the higher up you go, the cooler the air becomes. Why does this happen?

Raise the Elevation—Lower the Temperature

Let's summarize what you have just learned.

- Air cools as it rises.
- Rising air cools at a lower rate when condensation occurs.

As you might already have realized, these phenomena are very important in shaping the weather. Pause for a moment to answer the following question:

- When a mass of air sinks, what happens to its temperature?

Mountain scenes, from base (top) to summit (bottom).

As you may have already reasoned, when air sinks, it heats up. This phenomenon is important in shaping weather.

Read the following situations and think about how the phenomena just described explain each.

- On humid days when the wind blows, clouds often form around the peaks of mountains.
- The rainiest (or snowiest) places are usually in the mountains.
- Areas downwind of mountain ranges often receive very little rainfall.

Here is one additional mystery for you to ponder:

It is a bitter cold January morning in Billings, Montana. An icy stillness lays over the land. Suddenly a breath of wind stirs—a warm *wind! Within moments a stiff westerly breeze is blowing. The temperature soars 5 . . . 10 . . . 20 degrees. Snow and ice begin to melt rapidly. "Chinook!" you hear someone shout. The wind has a name! What on Earth has happened? What is this* chinook *wind?*

You already have part of the answer to this puzzle. Use the following information along with the diagram provided to help you solve the puzzle.

- Rising air cools at a rate of about 10 °C for every 1000 m of elevation. Sinking air warms at about the same rate.
- Rising air from which moisture is condensing cools at a rate of about 5 °C per 1000 m of elevation.

Can you solve the puzzle? Share your explanation of chinook winds with a classmate.

Climbing a mountain is like taking a journey northward. Climbing 1000 m is roughly equivalent to traveling north 1000 km, in terms of the change in climate. Higher elevations are not only cooler, they also tend to be wetter. How does this help explain the change in vegetation you saw in the journey on the tram?

Keeping Track

In this section you have been answering many questions and making many discoveries. In your Journal, make a list of your major findings.

Challenge Your Thinking

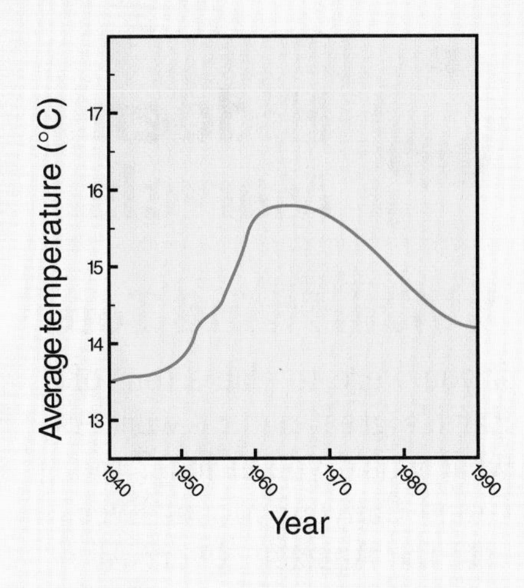

1. Towns and cities are generally warmer than the surrounding countryside.

   (a) What are some reasons for this?

   (b) What are some things that can be done to help lower the temperature within a town or city?

   (c) The chart at right shows the average yearly temperature as recorded at Green Acres Weather Station over a 50-year span. Describe the pattern shown by the graph. Explain what might have happened to result in such a pattern. (Hint: when Green Acres was first built, it was in a rural area.)

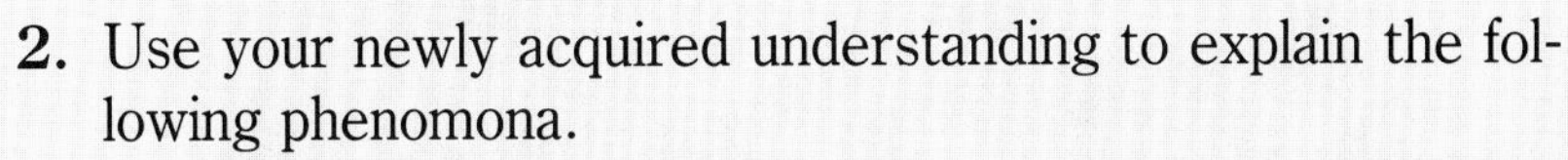

2. Use your newly acquired understanding to explain the following phenomona.

   (a) A car parked in the sun with its windows closed becomes hot very quickly.

   (b) On a stifling-hot, nearly windless day, the slightest breeze feels cool.

   (c) You are walking barefoot along the waterline at the beach on a summer day. When you walk away from the water, the sand becomes unbearably hot.

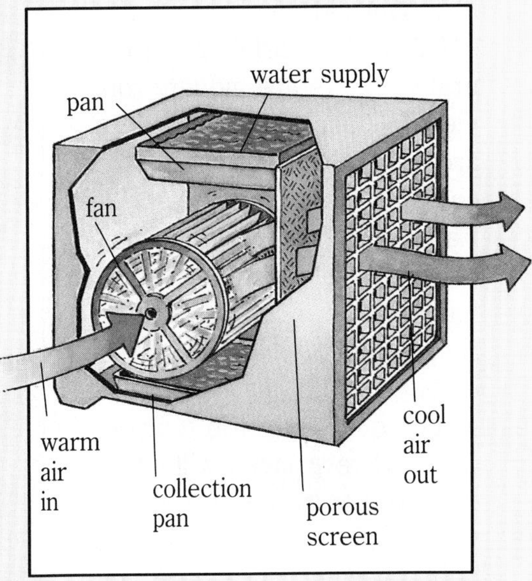

3. Before air conditioners became popular, cooling devices popularly known as "swamp coolers" were widely used. The illustration shows how a swamp cooler works.

   (a) Explain the principle of the swamp cooler's operation.

   (b) Would it work equally well in all climates? Explain.

   (c) Swamp coolers are still common in certain areas because they are as effective as air conditioners but cheaper. Where might you expect to find them in common use?

4. The graphs at right show temperatures for Montpelier, Vermont and Mount Washington, a short distance away. Explain what is happening and why.

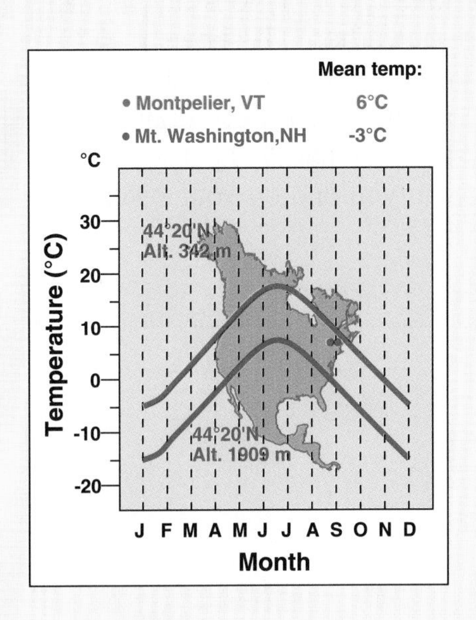

4 Hidden Marvels

A Whirlwind Tour

If you were to plan a tour of Earth's great natural wonders, what would you include? the Grand Canyon? Mount Everest? the Amazon River? All of these sights are truly spectacular. But did you know that some of the world's great natural wonders are hidden from view? These natural wonders are beneath the oceans, which cover 70 percent of the Earth's surface. The oceans conceal huge canyons, enormous peaks, and mountain ranges longer than any that exist on land.

The quick six-stop tour on the next few pages will introduce you to some of the world's great wonders you've been missing. But before you take the tour, look at the map shown here. It shows the ocean floor as it would look if all the water were drained away. What "hidden wonders" can you find on your own?

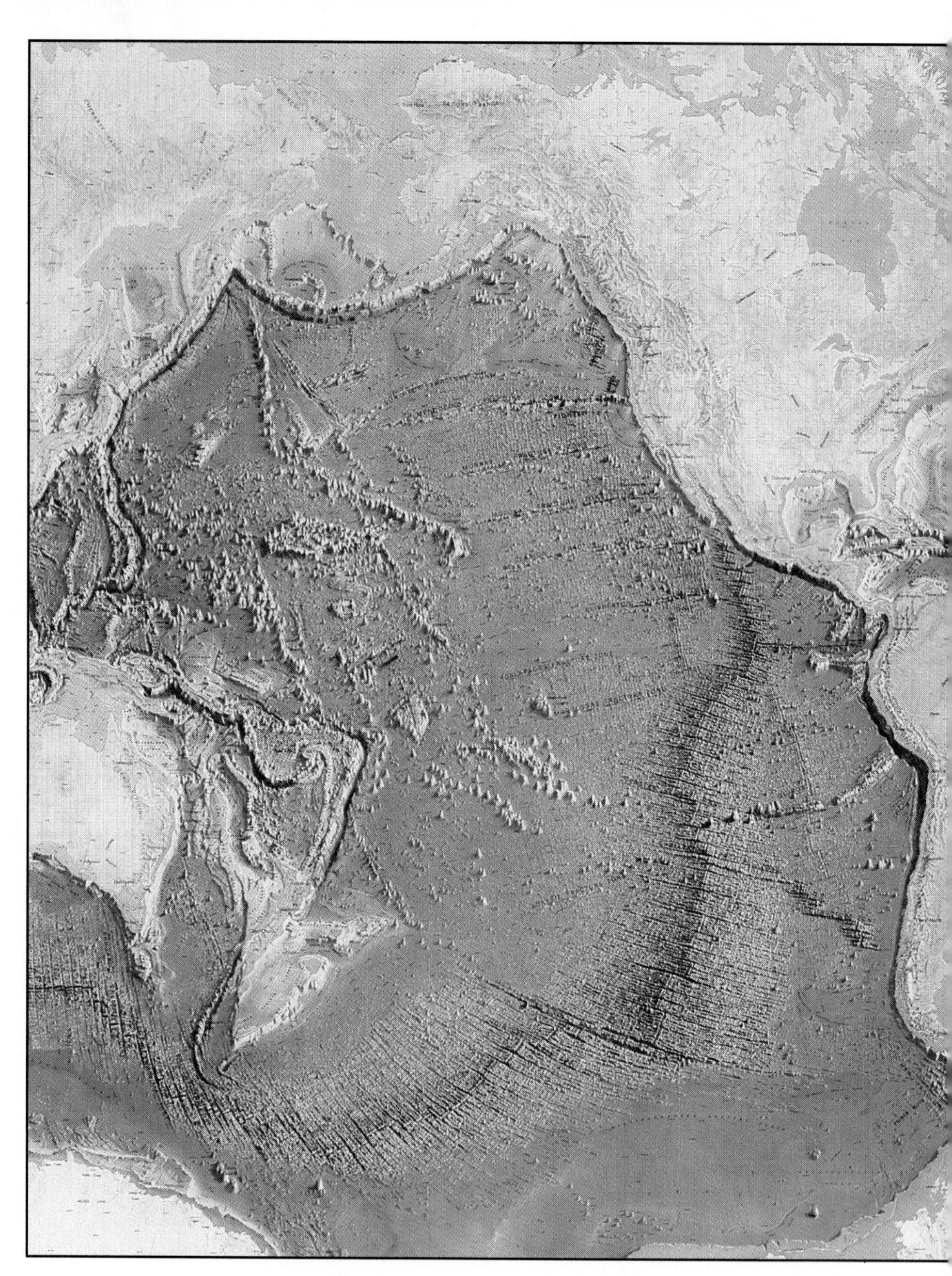

The ocean floor as it would look if all the water were drained away.

World Ocean Floor by Bruce C. Heezen and Marie Tharp, 1977.

Stop 1: Hawaii Volcanoes
The Hawaiian Islands are a familiar sight, but few people realize that only the topmost parts of these islands are visible. The rest lies beneath the ocean. Measured from the ocean bottom, the island of Hawaii (at 9850 m) would tower over Mount Everest. Volcanoes such as those that formed Hawaii play a major role in the global cycle of greenhouse gases.

The tallest Hawaiian Island is almost 10,000 m tall. Only 4000 m extend above sea level.

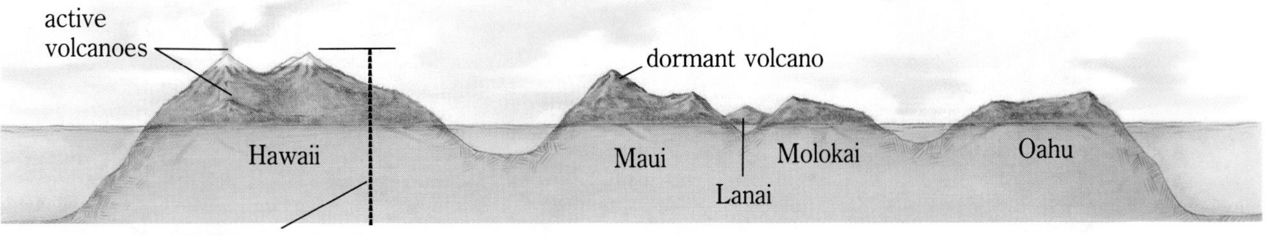

Stop 2: The Abyssal Plains

Much of the deep ocean bottom is made up of *abyssal plains,* vast stretches of featureless wasteland, flatter and more barren than any place on land. The abyssal plains are forbidding places of eternal darkness, crushing pressures, freezing temperatures, and absolute stillness, but even so, a surprising number of living things are found here. The remains of tiny dead plants and animals cover the bottom, accumulating with unimaginable slowness: 1 centimeter or less every thousand years. The slightest disturbance raises blinding clouds of mud.

Stop 3: The Mid-Ocean Ridges

Lacing the Earth's surface like seams of a baseball, the mid-ocean ridges form the world's largest mountain range. Hidden thousands of meters beneath the ocean's surface, the mid-ocean ridges are harshly rugged landscapes and deep canyons. The mid-ocean ridges form the boundaries between separating crustal plates. Here the crust of the Earth splits open, and into the gap flows molten rock from the Earth's interior. Volcanoes and *hydrothermal vents* (springs that gush hot, mineral-rich waters) are common along the ridge. Clustered about the hydrothermal vents are strange previously unknown ecosystems. The existence of these ecosystems disproves the long-held assumption that all life forms depend ultimately on the energy of the sun. Here, where not even the faintest glimmer of light from the surface can penetrate, dwell giant tube worms and clams, snow-white crabs, weird fish and crustaceans, and other bizarre creatures. This ecosystem is based on a species of bacteria that derives energy and nutrients from the mineral-rich waters of the hydrothermal vents.

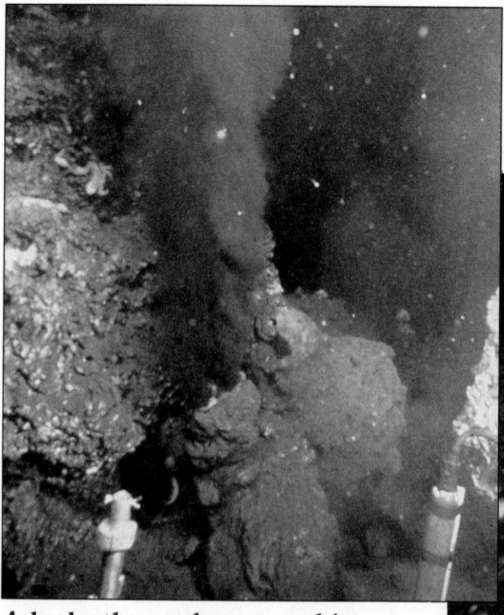

A hydrothermal vent gushing hot (350 °C) mineral-laden water.

Some of the living things that depend on the hydrothermal vents.

Stop 4: Undersea Canyons

Everyone has heard of Arizona's awesome Grand Canyon, but not many people realize that less than 300 km from San Francisco is Monterrey Canyon, every bit as impressive. Some undersea canyons, with walls 5000 m high, would even dwarf the Grand Canyon. One such canyon can be found in the equatorial Atlantic, between South America and Africa.

A computer-generated image of Monterrey Canyon

The Marianas Trench would swallow Mount Everest, which is 8925 m tall. Some parts of the trench are 10 km deep.

Stop 5: The Marianas Trench

Trenches are the sites at which the ocean crust is slowly being pushed, centimeter by centimeter, back into the interior of the Earth. Here the ocean floor dips sharply downward, forming a steep-walled valley deeper, in some cases, than Mount Everest is high. An anvil dropped from a boat floating over the deepest part of the Marianas Trench would take 1½ hours to hit bottom. Incredible pressures and an almost total lack of oxygen make life in an ocean trench a difficult proposition.

Stop 6: The Gulf Stream
The Gulf Stream is like a huge river in the ocean. In fact, if the Gulf Stream flowed over land, it would form a river 100 km wide, 1000 m deep, and move at about 7 km per hour. Fortunately, the Gulf Stream occurs in the Atlantic Ocean, where it moves north along the coast of North America. The Gulf Stream is a warm current that carries great amounts of heat away from the tropics to colder northern regions.

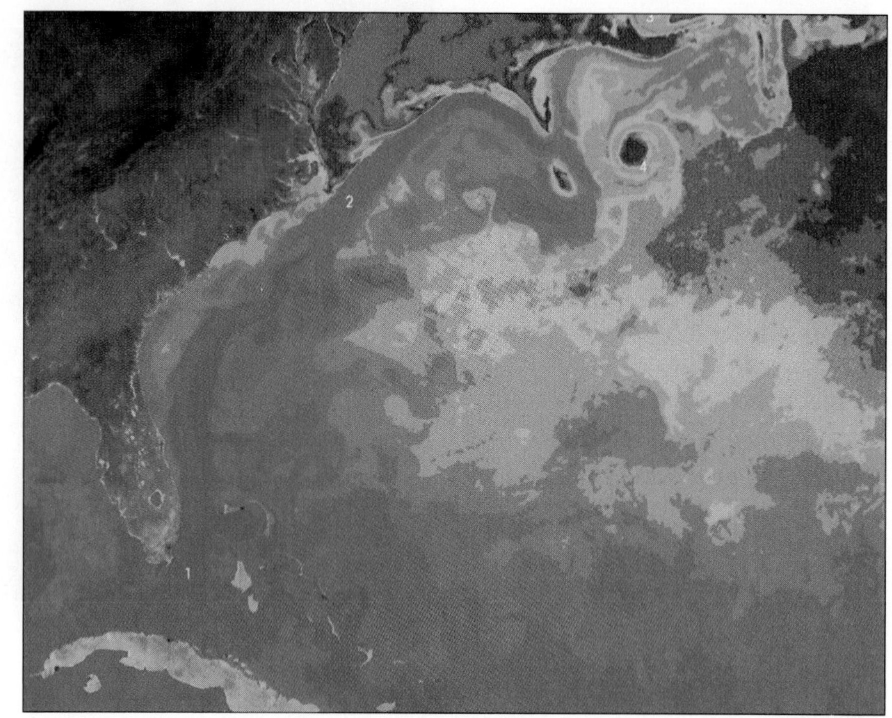

This satellite image shows the temperatures of the Gulf Stream and surrounding waters.

Points to Ponder

Refer back to each of the six stops in order to answer the following questions:

1. How could volcanoes such as those in Hawaii affect the Earth's climate?
2. Why are the abyssal plains so flat?
3. What would happen to the organisms of the mid-ocean ridges if the hydrothermal springs were to stop flowing?
4. How might the undersea canyons have formed?
5. What special adaptations would any organism living at the bottom of the Marianas Trench have to make to survive?
6. How do you think currents such as the Gulf Stream affect the global climate?

Pictures of the Deep

Now that you have had the chance to sample some of Earth's hidden natural wonders, choose one of the stops to explore further. Think of yourself as a travel agent and develop a poster, brochure, or article to inform others about your stop. You might want to read other sources to find out more about the stop of your choice.

5 The Moving Oceans

The Mediterranean Puzzle

For many thousands of years sailors have used winds and ocean currents to propel their ships. The causes of winds and ocean currents were the subject of much speculation to ancient peoples, but it was left to modern science to solve the puzzle of their origin. What causes these currents in the ocean and the atmosphere? Are they related somehow? What role do these currents play in the global climate? These are questions we will investigate in this lesson.

Let's start with an ancient current puzzle, a puzzle finally solved by Count Luigi Marsili in 1679.

Examine the map of the Mediterranean. The puzzle is this. Sailors had long known that swift currents flowed into the Mediterranean from both the Black Sea and the Atlantic Ocean. Many rivers and streams also empty into it. The Mediterranean has no apparent outlet, therefore, the water level should rise, but it does not. Many explanations were offered—for example the existence of hidden underground channels to drain the excess water. Can you solve this puzzle?

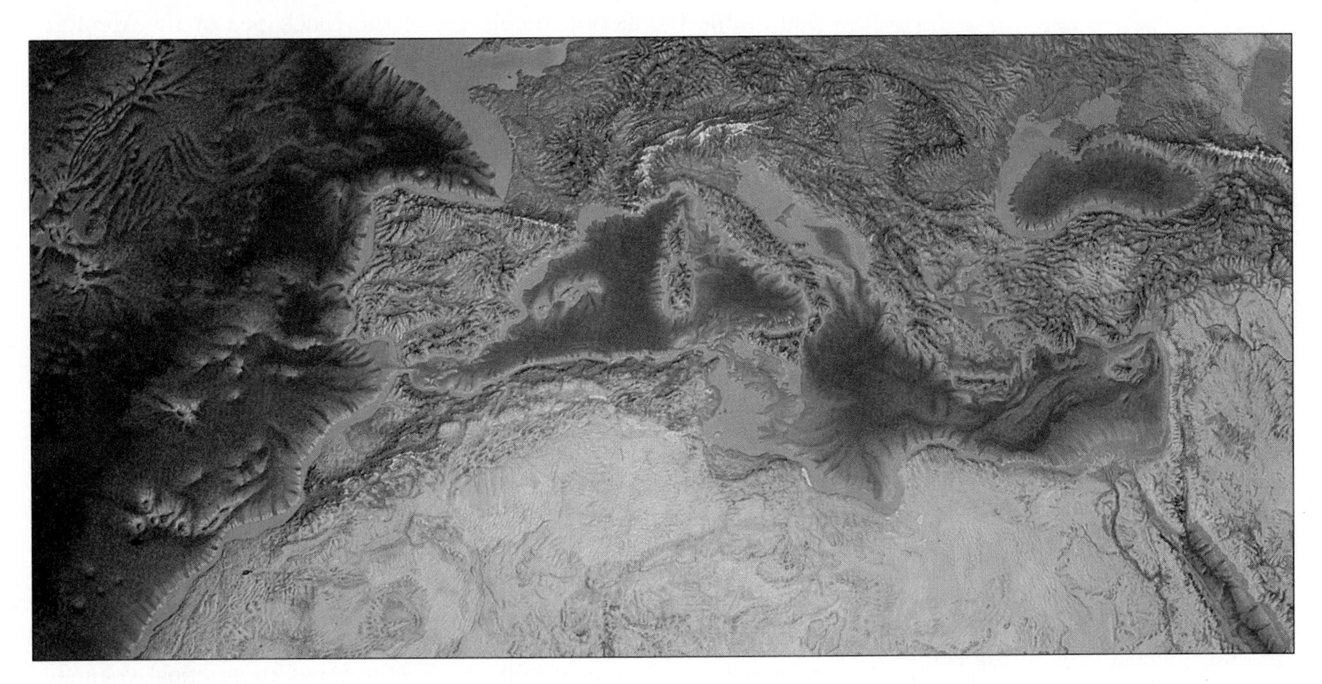

Count Marsili thought he could, and so he set up a model of the Mediterranean to test his idea. In the following Explorations you will walk in the shoes of Count Marsili as you discover the answer to the Mediterranean puzzle. At the end of Exploration 8, be prepared to describe the Mediterranean puzzle and explain its solution.

Unraveling the Puzzle

You Will Need

- salt
- an aluminum roasting pan
- aluminum foil
- masking tape
- food coloring
- pepper
- 2 containers suitable for holding at least 1 L of water
- water

What to Do

1. Cut a piece of aluminum foil slightly wider and higher than the pan. Using masking tape, make a waterproof barrier that divides the pan into two parts.

2. In one container add 1000 mL of water, 100 mL of salt, and a few drops of food coloring. Stir until the salt is completely dissolved. In the second container, dissolve 50 mL of salt in 1000 mL of water. Do not add food coloring. Save 100 mL of each solution for Exploration 7.
3. With the help of a partner, pour each solution into opposite sides of the pan at the same time (so as not to collapse the barrier). Be careful to make the water level on both sides equal. Now sprinkle some pepper into the side containing food coloring.
4. With a pencil, puncture a hole in the aluminum foil just below the surface of the water. Make another hole near the bottom of the container. Observe the system for about 10 minutes.

Interpreting Your Findings

1. Make a diagram in your Journal to record what you observed. Use arrows to show current flow. How could this demonstration serve as a model of the Mediterranean? Explain. Using what you have learned, draw a diagram of the Mediterranean showing the flow in and the flow out at both entrances.
2. From your observations, which is saltier—the Atlantic Ocean or the Mediterranean? the Black Sea or the Mediterranean? Which salt solution represents the Atlantic? Which represents the Mediterranean?
3. Why should the saltiness of the Mediterranean differ from that of the Atlantic Ocean? (Hint: Most of the Mediterranean Basin has a warm and dry climate.)
4. Count Marsili thought that he could explain how the saltiness of ocean water could set up currents. To test his idea he drew water from different depths in the Strait of Gibraltar. He then found the mass of equal volumes of each sample and compared them with the mass of an equal volume of fresh water. What do you think he found?

Test your prediction by doing the next Exploration.

Marsili's Explanation

You Will Need

- a pill bottle or similar small container
- salt solutions from Exploration 6
- tap water
- 3 test tubes
- glass tubing or clear drinking straws
- a balance
- a graduated cylinder (50 mL)
- food coloring in 3 different colors

Part 1

Comparing Masses

1. Determine the volume (in milliliters) of the pill container and its mass (in grams).
2. Fill the container with either of your salt solutions or with tap water. Determine the mass of the container and liquid combined. What is the mass of the liquid?
3. Repeat step 2, first with one of the remaining two liquids and then with the other. Be sure that you are measuring equal volumes of each liquid. How does the mass of the "Mediterranean" solution compare with the "Atlantic" solution? How do both of these compare with tap water?
4. Divide the mass of each solution by its volume. Which solution has the greatest mass for its volume? the least mass?

Part 2

Thinking About Density

In step 4 of Part 1 you actually found the *density* of each solution. What must *density* mean?

Which liquid had the largest density? the smallest density?

Density of any material can always be found using this formula:

$$\text{Density} = \frac{\text{mass}}{\text{volume}}$$

What are the densities of the following materials?

Material	Mass of 30 mL	Density
water	30.0 g	?
alcohol	21.0 g	?
ice	27.6 g	?
salt water	33.0 g	?
egg	31.5 g	?

Which would float on water? on alcohol? on salt water? Devise an experiment to test your predictions.

Part 3

Layering Liquids

Add a different color of food coloring to each sample used in Part 1. Fill a test tube with each liquid. Now use the technique shown below to try to layer the three liquids in a thin glass tube or drinking straw. Can the liquids be layered in any order?

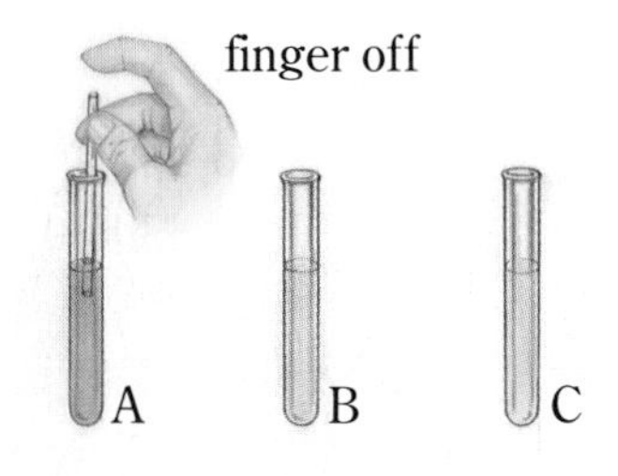

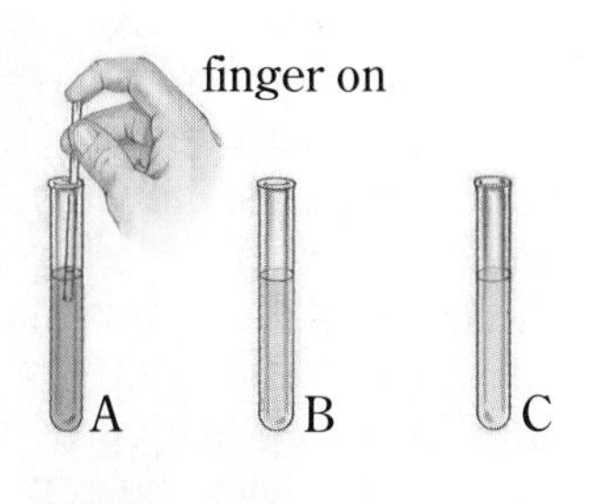

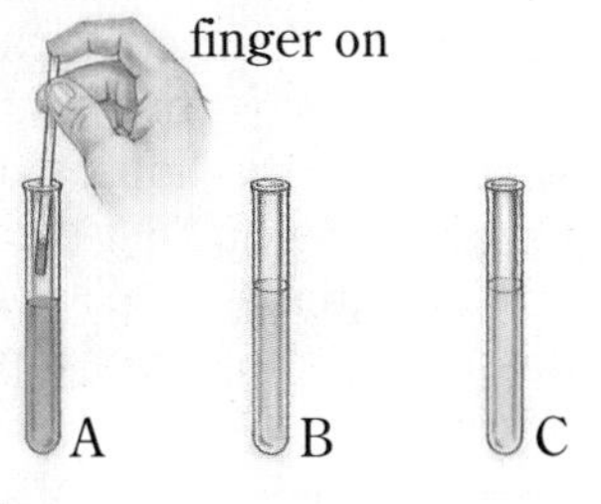

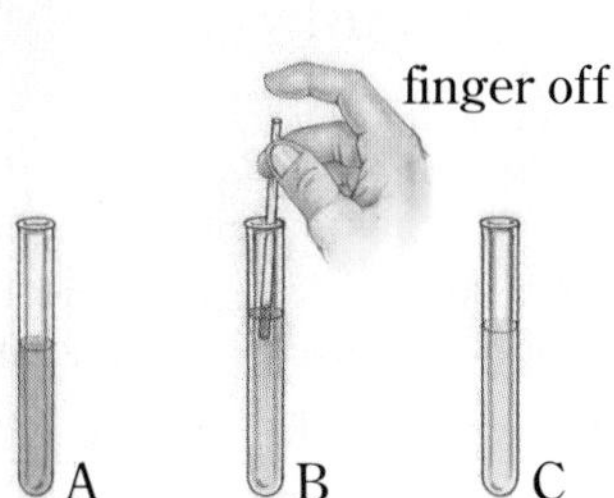

How does this relate to Count Marsili's explanation of the Mediterranean puzzle?

Predict, Observe, Explain

Could density differences in the larger oceans cause currents similar to those observed in the Mediterranean? Perhaps the following Exploration will answer this question.

You Will Need

- a 10% salt solution
- tap water (cold and hot)
- food coloring
- 2 small bottles
- a large pan or bucket
- a graduated cylinder (50 mL)

What to Do

In Activities 1–4 you will invert one water-filled bottle over another to study how solutions of different densities interact. Set up each Activity as shown in the illustration at top right. Then follow the additional instructions that accompany each Activity heading. Be sure each bottle is filled to the rim with the correct type of water. For Activities 1 and 3 only, after setting up the apparatus, turn the bottles on their sides while holding them together. To catch any spill, carry out this Exploration over a pan or bucket.

For each Activity in this Exploration, do the following:

- Predict what you expect to happen, observe what does happen, and finally give an explanation for your observations.
- Re-examine the map of global temperatures on page 142. Where in the oceans might the phenomenon you observed in this Exploration be occurring?

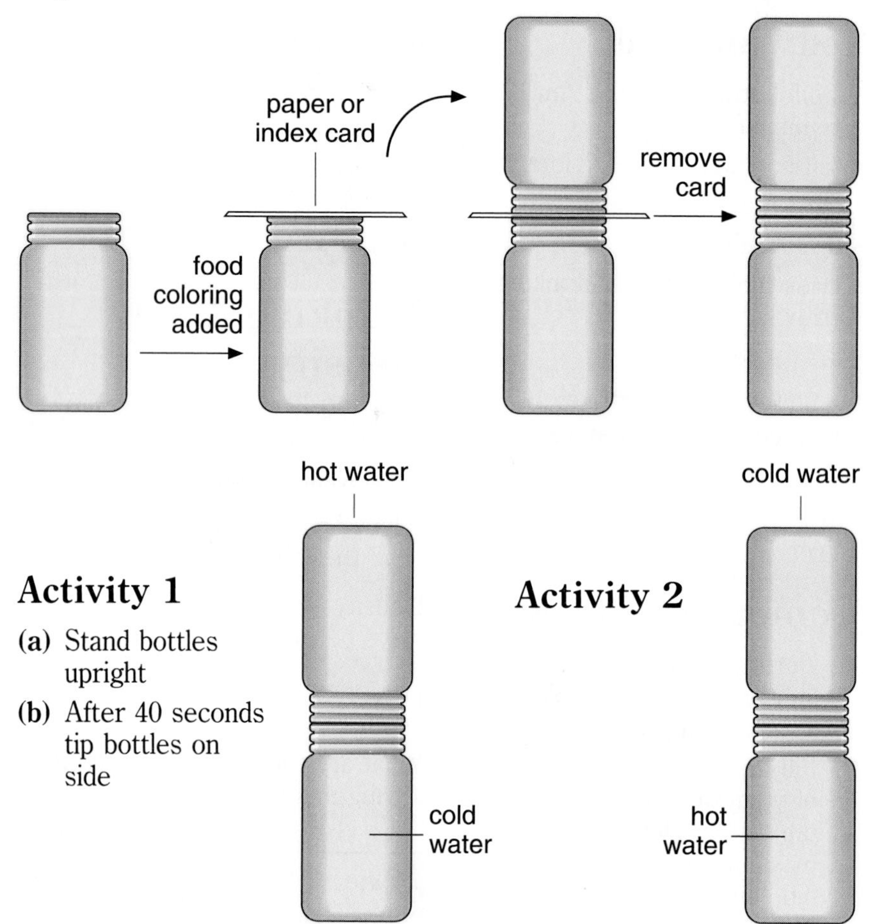

Activity 1

(a) Stand bottles upright

(b) After 40 seconds tip bottles on side

Activity 2

What conclusions would you draw from Activities 1 and 2?

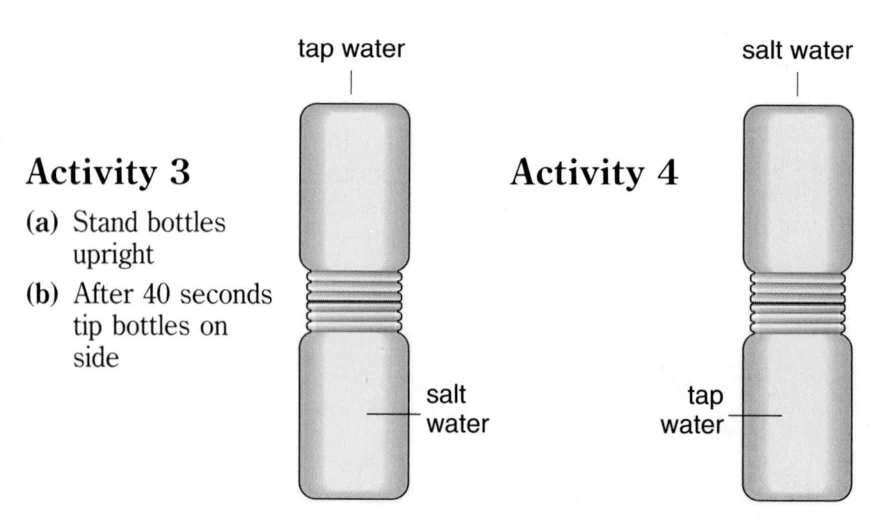

Activity 3

(a) Stand bottles upright

(b) After 40 seconds tip bottles on side

Activity 4

Count Marsili's Paper

"I have recently returned from the Mediterranean where I observed the strong surface currents at both entrances. I think I have found an explanation for them."

Finish Count Marsili's presentation. Include in your paper experiences and diagrams that would make the explanation clear to an audience less knowledgeable than you about density and currents.

Follow the Flow

How many factors have you discovered that affect the density of ocean water? The previous Exploration suggests two. The Mediterranean's water is denser than that of the Atlantic Ocean because of evaporation. When water evaporates, the salt in it is left behind. Look at the map below. Where else in the world might this be happening? Seawater in the tropics generally differs in density from the seawater of cold regions. Would tropical water be more or less dense than polar water? Why?

The Norwegian Sea bordering Greenland and the Weddell Sea bordering Antarctica are both chilled by the cold winds that blow off of glacier-covered landmasses. Does this increase or decrease the density of the surface water?

You may be surprised to find that ice formed from seawater is not itself very salty. Why? The dissolved salt does not fit well into the crystal structure of ice, so it is "squeezed" out as the water freezes. The water left behind, therefore, becomes a little more dense.

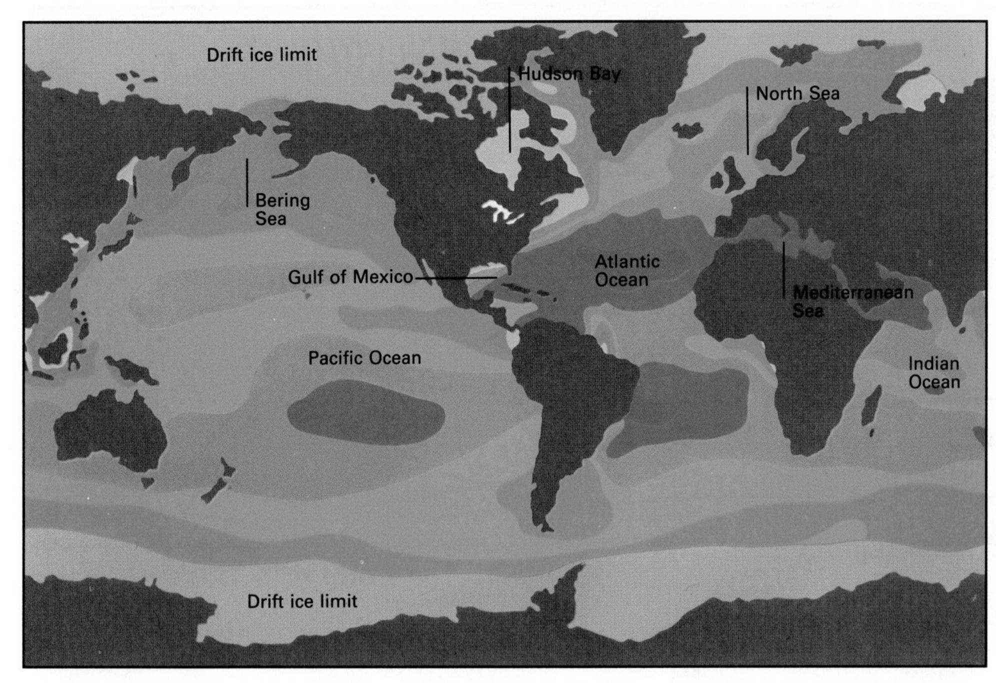

Map showing the average surface salinity of the oceans. How do you account for the differences in salinity?

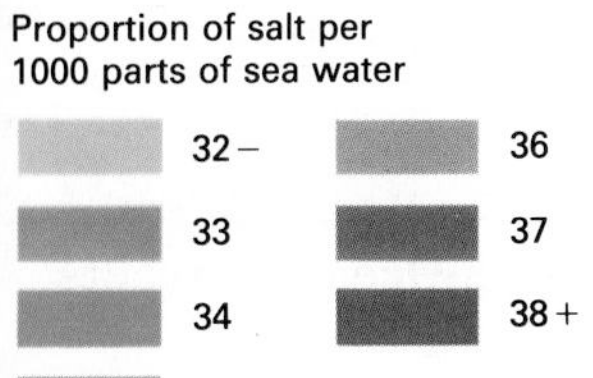

Examine the diagram of deep ocean currents. These currents are caused by density differences in ocean water. Where do they originate? Where do they go? Is there evidence in the diagram that there are density currents in the ocean at different depths and going in different directions?

These deep ocean currents play an important role in controlling Earth's temperature. Carbon dioxide dissolves well in water (think of carbonated soft drinks). The surface water of the ocean absorbs huge amounts of carbon dioxide from the atmosphere. When currents push the surface water to the bottom, the carbon dioxide becomes concentrated in the deep ocean waters. The deep ocean currents circulate very slowly. The water and carbon dioxide caught in a deep ocean current may not resurface for centuries. When deep ocean currents finally resurface, they bring with them rich nutrients collected over the years. Sites of upwelling deep currents (such as along the west coast of South America) are rich in life and make excellent fishery resources.

This penguin lives very near the equator thanks to a cold, upwelling current.

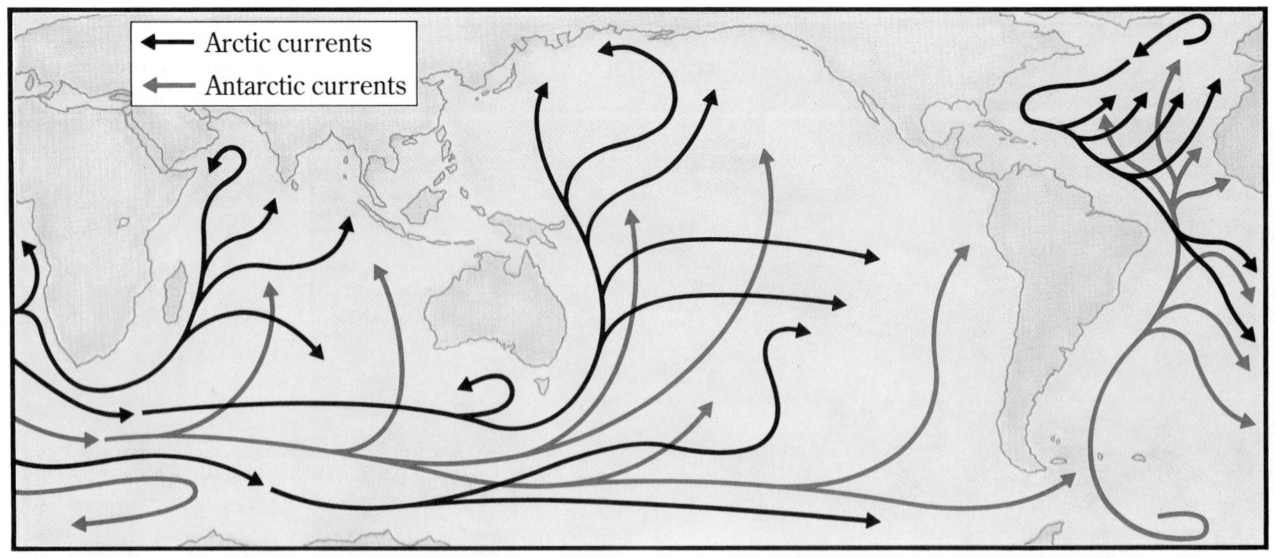

Deep-Ocean Currents

The Composition and Density of Seawater

The substances that make up seawater remain quite constant no matter where the ocean is sampled. Most seawater contains the substances shown in this table.

A Recipe for Seawater

Mix:	
Sodium chloride	23.48 g
Magnesium chloride	4.98 g
Sodium sulfate	3.92 g
Calcium chloride	1.10 g
other compounds	1.00 g

Add: enough water to form 1000 g of solution.

Here is a simpler recipe for seawater:

To 250 mL of water add 9.0 g of salt.

This will have the same salt concentration as the earlier recipe. What is the density of seawater with this salt content? How can a device such as the one in this photo be used to determine the density of a salt solution? Such a device is called a *hydrometer.* The hydrometer is calibrated in such a way that it reads the density of the liquid in which it is placed.

Believe it or not, it is easier to float in salt water than it is in fresh water. In fact, some bodies of water such as the Dead Sea and the Great Salt Lake are so salty that a person would find it almost impossible to sink. In the following Exploration you will use this idea to develop your own device for measuring the density of artificial seawater.

The Great Hydrometer Challenge

You Will Need

- an assortment of common materials such as a pencil, modeling clay, a drinking straw, cork, wire, thumbtacks and so on
- a graduated cylinder
- water
- a "standard" salt solution with a density of 1.05 g/mL (made by adding 12.5 g of salt to 250 mL of water)
- artificial seawater (using either recipe on page 163)
- a permanent marker

What to Do

1. Using the materials listed here or others of your choice, make at least three hydrometers. Test your hydrometers for their ability to float in a graduated cylinder filled with water.
2. You can *calibrate* your hydrometers by placing them in solutions of different known densities and marking the level to which they sink. Do this using plain water and the standard salt solution. Can you extend the scale above or below each mark? Which of your hydrometers appears to be most sensitive?
3. Use your best hydrometer to measure the density of the artificial seawater. How can you confirm this reading?

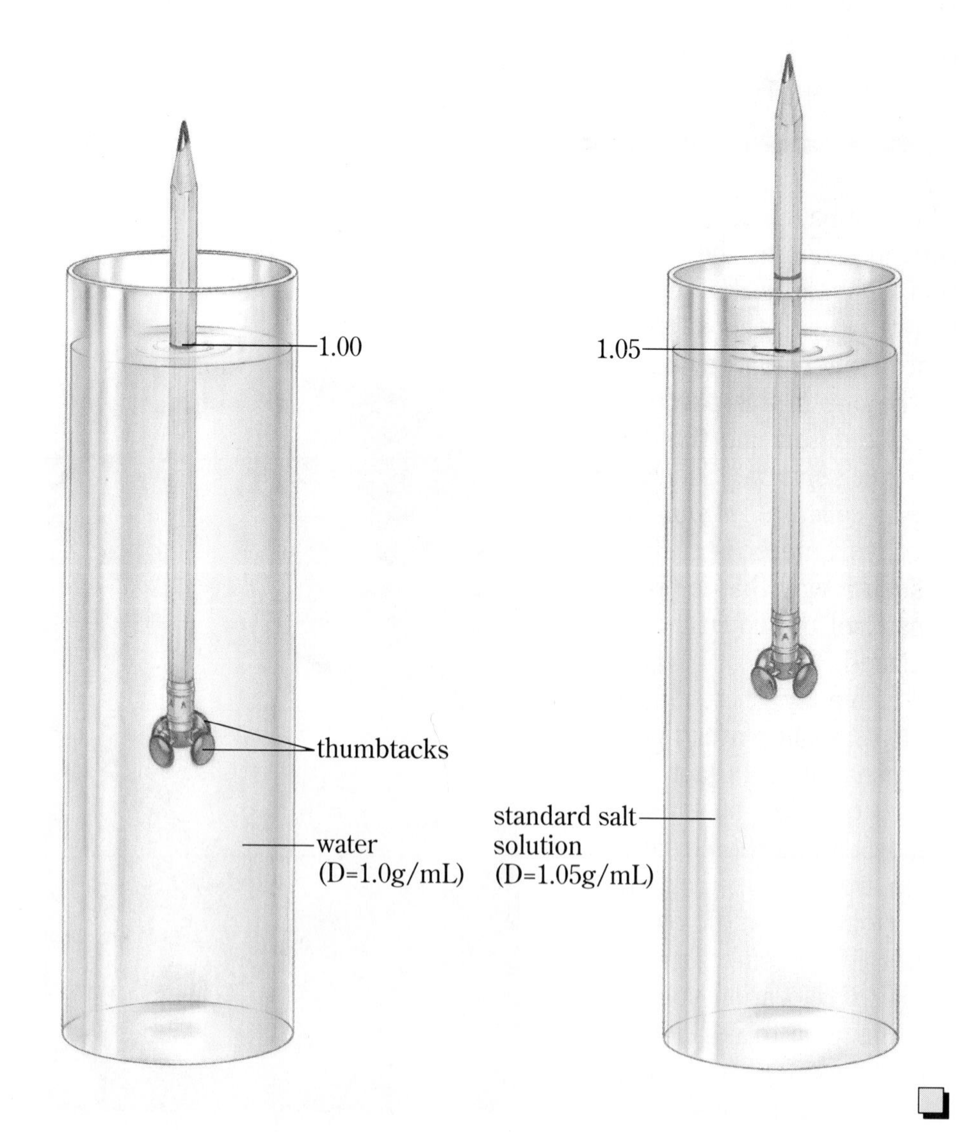

6 The Atmosphere

Density Currents in the Atmosphere

Can density differences cause air to flow as it does for water? What would cause density differences to develop in air? Examine the diagram below, which shows a breeze blowing from the ocean to the land during the daytime. Before reading on, give your explanation of what causes such breezes to blow. Then review the activities you have done in this unit for evidence that each of the following statements is true.

1. In the daytime, land heats up more quickly than does water.
2. The air over land warms and becomes less dense. This air rises (floats) on the denser surrounding air (A).
3. Cooler and denser air moves in from the ocean taking the place of the rising air (B).
4. As air rises it spreads out and cools (C), thus becoming more dense. The air then sinks (D), completing the cycle.

Draw a similar, labeled, diagram to show what happens at night, when there is no sun to warm either land or sea, and both begin to cool off.

Storm clouds over the tropics. Why do these clouds form?

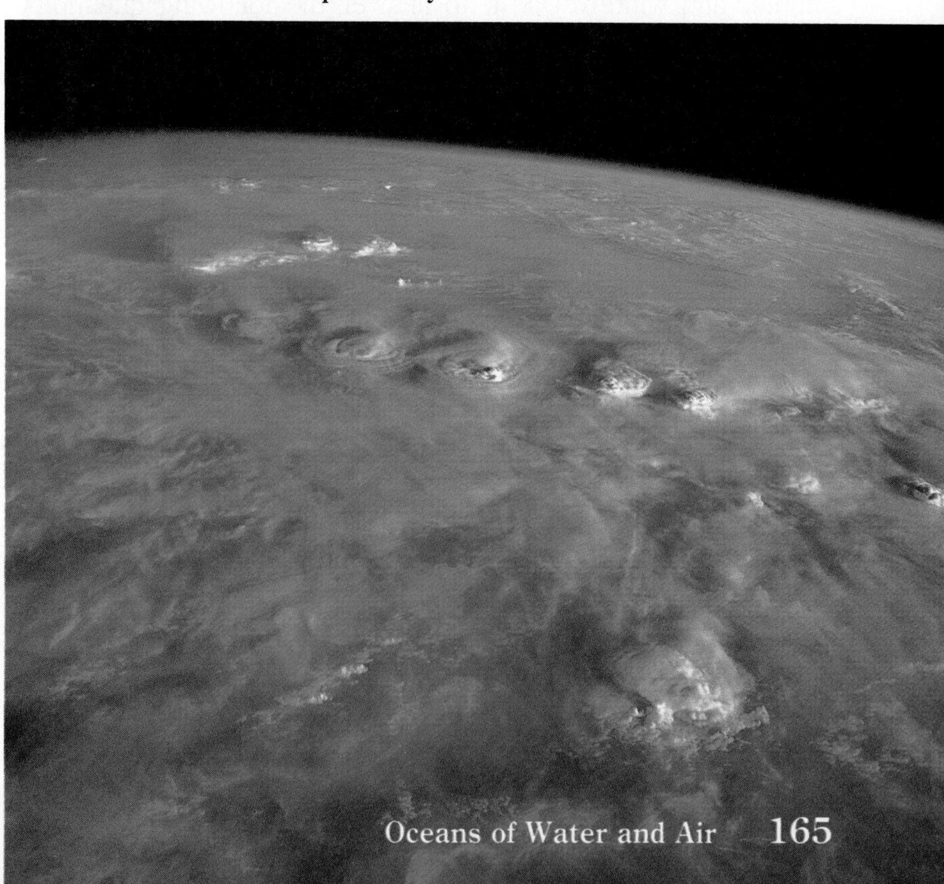

Where else in the world would you experience similar kinds of density currents in the atmosphere, only on a much larger scale? Think about the tropics. What happens to the air over the tropics as it warms? It rises, of course, and as you would expect, cooler air from farther north and south moves in to take its place. This pattern is seen throughout the tropics. The winds that result are called *trade winds.* Trade winds blow from season to season and year to year.

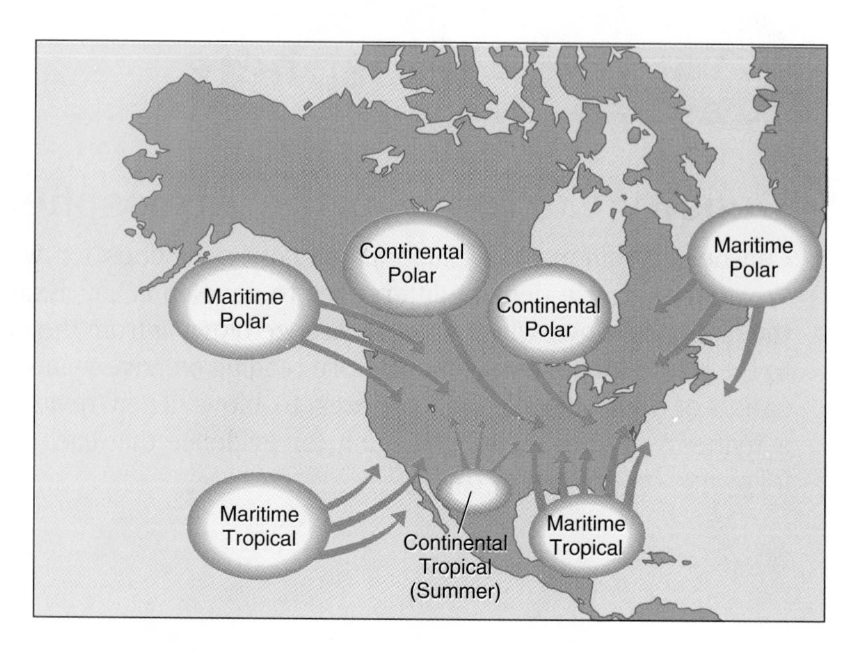

Air Masses

You have seen that bodies of water of different densities do not mix readily. Would you expect air to behave similarly?

There is a constant interaction between a body of air and the land or water it happens to lie over. Air over warm water becomes warm and moisture-laden while air over cold water is cold and moisture-laden. What kind of characteristics would air over the Sahara have? air over Siberia in winter? Would these two *air masses* have more or less the same density?

Air masses are bodies of air covering large areas and having nearly uniform temperature and humidity. The diagram above shows the air masses that affect weather in North America in summer and winter. Use it to answer the following questions.

1. Air masses are either polar or tropical, and maritime or continental. The name of an air mass is made up of two parts, for example: *polar continental.* Here is what each part indicates:

polar:	cold	continental:	dry
tropical:	warm	maritime:	moist

   Suggest how each type of air mass compares in terms of its temperature, moisture content, and density based on its location and name. Look back over the last Exploration for observations to support your answer to this question.
2. Would the different types of air masses readily mix together? What activity did you do that suggested that they do not?
3. What type of air mass is causing your weather today?
4. What type of air mass(es) will probably be causing your weather six months from now?
5. What type of air mass do you think you would find over the Atlantic Ocean near the equator?

Weather Maps

If a weather forecaster said "Look for a Canadian air mass to sweep into our area tomorrow," what kind of weather would you expect? Since much of our weather is the result of collisions between air masses, what would you see if you could see the boundary between air masses of different densities? The following diagrams provide such a picture. Answer the questions that go with each diagram.

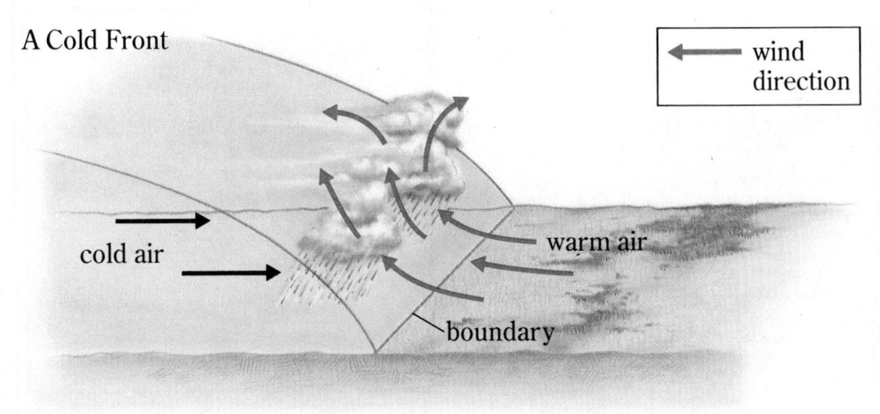

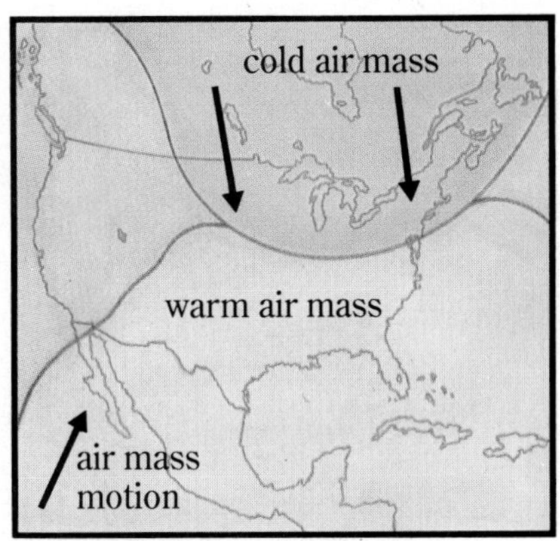

1. Which air mass is more dense?
2. What happens to the warmer air as the colder air approaches?
3. What causes the clouds to form?
4. What will happen to the temperature at location A in the near future?
5. The boundary between the warm air and the cold air is called a cold front, not a warm front. Why?

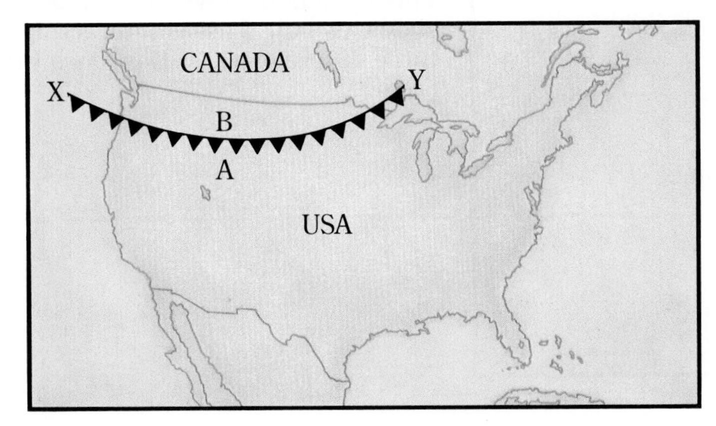

6. What does the line X Y on the weather map represent?
7. What do the triangles on the line indicate?
8. How does the temperature at A compare with that at B?
9. Compare the weather at A with that at B. How will it change shortly?

10. In the illustration below, which air mass is moving forward faster—the cold or the warm air mass?
11. What is happening to the warm air as it overtakes the cold air mass?
12. Why are clouds forming?
13. On tomorrow's weather map, where will the warm front be?

A Warm Front

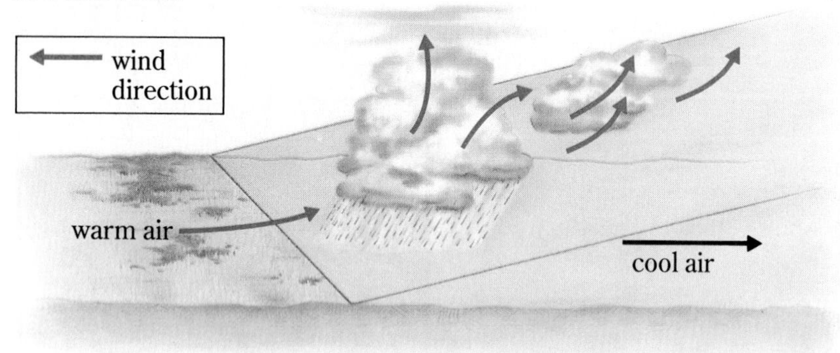

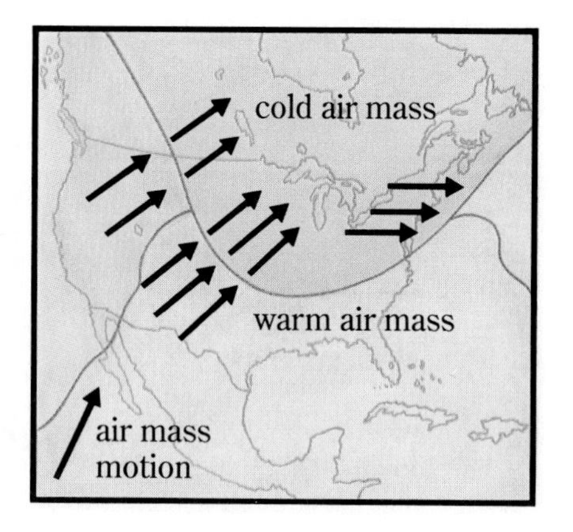

14. What does the line W X on the weather map below represent? the line X Y? the line Y Z?
15. How does the temperature at A compare with that at B? How will it change?
16. What do the "half moons" on the line indicate?
17. As the map shows, one section of the front is not moving. This is called a *stationary* front. How is it symbolized?

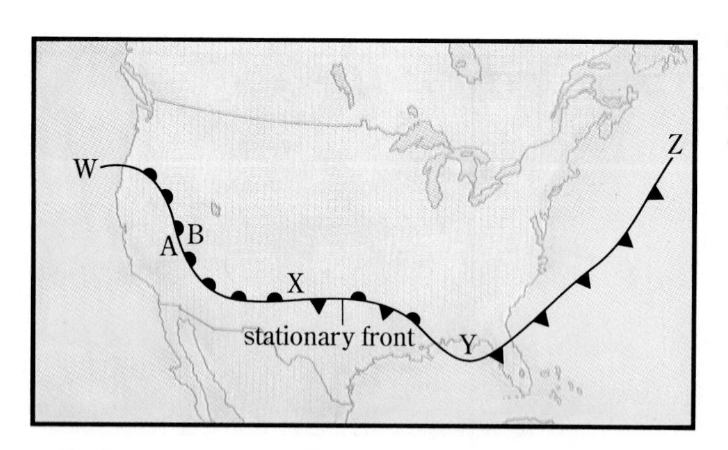

This hypothetical weather map shows two air masses interacting in three different ways. How does this occur?

An At-Home Task

Follow your local weather forecasts for a week. What elements of weather are included in the forecasts? How are forecasters (meteorologists) able to make predictions about future weather?

A Map Sequence

The weather maps below represent a span of six days. Examine the maps. Then answer the following.

1. Describe in general what happens during this sequence. How do the various weather systems move?
2. Indicate any fronts. What types of fronts are there?
3. What time of year do you think this probably is?
4. What types of air masses do you see in each map?
5. What happens at the boundaries between air masses?
6. What happens to the weather conditions of an area after one air mass overtakes another?

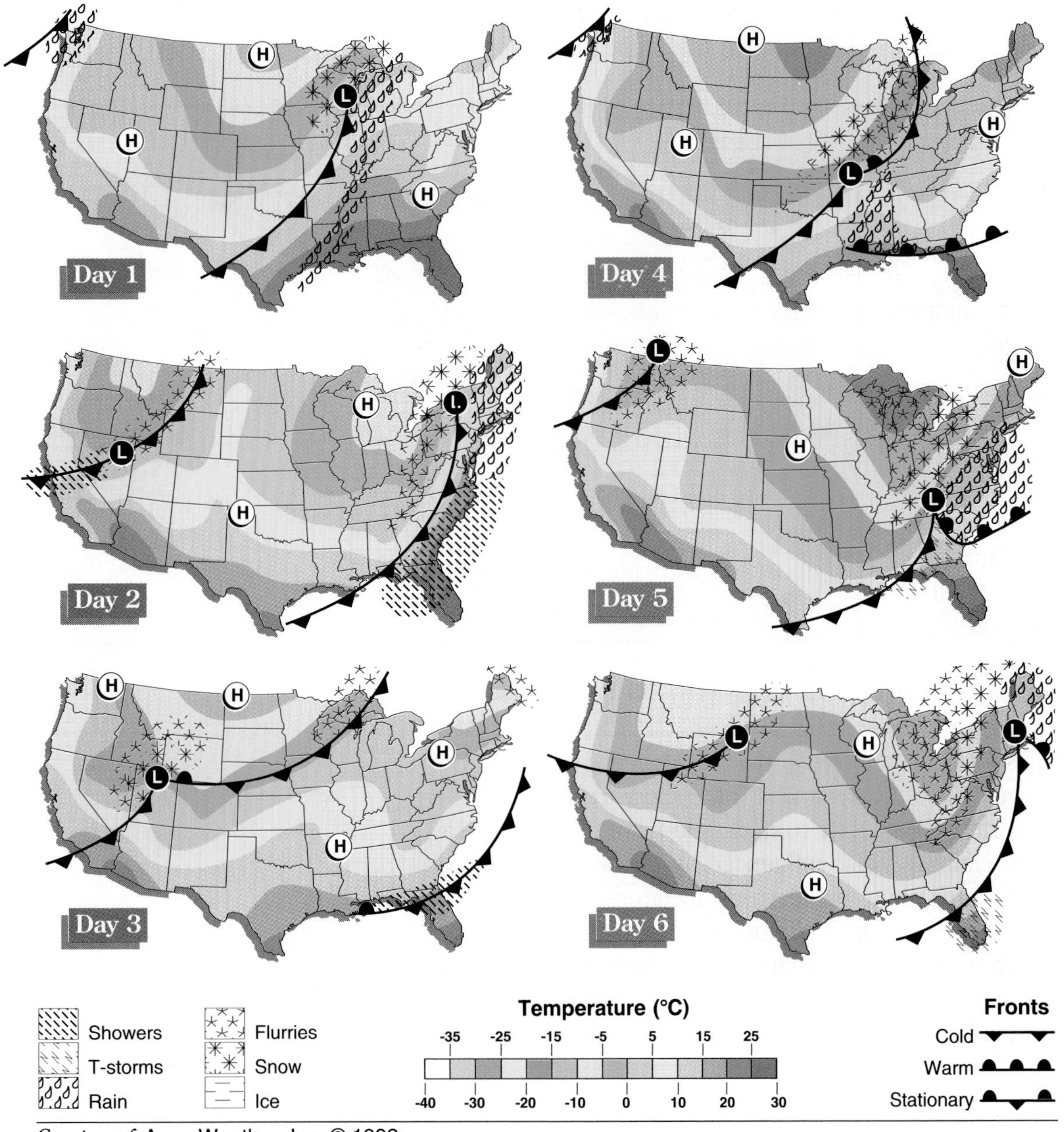

Courtesy of Accu-Weather, Inc. © 1992

A Climate Sampler

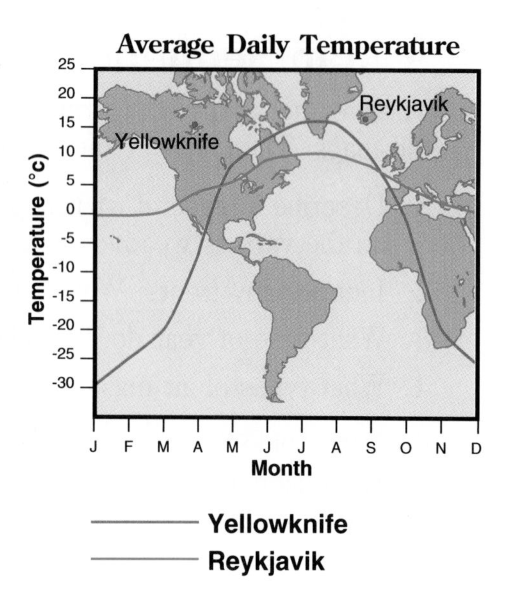

Reykjavik, Iceland and Yellowknife, Canada are both at about the same latitude, but, as the graph shows, most months these cities have very different weather. In fact both cities have very different climates. Why? What clues does the map provide?

Climate is a word we use quite often, but what exactly is a climate? Could you define it? How does it differ from weather?

Write each of the following sentences in your Journal. In each blank write either *weather* or *climate.*

1. "What—rain again! This kind of ______ really depresses me."
2. "In January we normally have clear, cold ______."
3. During the ice age, the Earth had a much colder ______.
4. The ______ of the Amazon rain forest hasn't changed in millions of years. The ______ is almost the same every day.

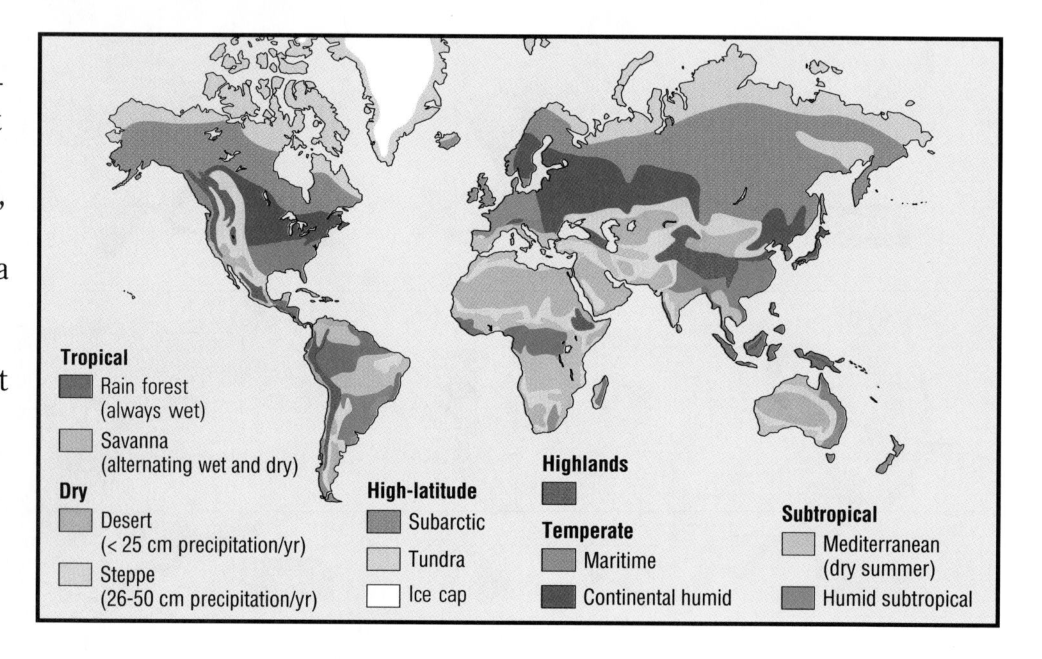

As you probably realize, weather is the condition of the atmosphere at a particular time and place. On the other hand, climate is the average weather conditions over a long period of time—years, decades or even centuries. The diagram at right shows some of the world's major climate regions. What seem to be the most important factors in determining climate? Why do you think this is so?

Back to our earlier example, though. Why do Yellowknife and Reykjavik have such different climates?

Reykjavik has a *maritime climate,* that is, a climate typical of the ocean. Maritime climates are moist and show relatively little change in temperature from day to night and from season to season. Reykjavik's climate is also influenced by the warm waters of the Gulf Stream, which you will learn about later.

Yellowknife has a *continental climate.* Continental climates are drier than maritime climates and typically have large daily and seasonal temperature ranges. Why are continental and maritime climates so different? Where in this unit have you seen evidence of this?

Keeping Track

What have been your major findings in this section? Continue the record that you started in the last section.

Challenge Your Thinking

1. Robert filled a tub with water. The water ran cold for about a minute before it finally got hot. Robert expected the water to be uniformly warm throughout, but instead the hot water floated on top and the cold water stayed underneath. He had to slosh it around before it was comfortable.
   (a) What did Robert mean when he said "This is like the Mediterranean effect."
   (b) How do you explain this occurrence?
2. The Arctic Ocean is covered year-round by a layer of ice. If this ice melted would you expect sea level to rise world-wide? Why or why not?
3. Ships loaded to the limit in cold, northern waters would be sailing into danger if they headed for the tropics without first unloading some cargo. To avoid this problem, international marine regulations require most ships to bear *Plimsoll marks*. Plimsoll marks are a kind of scale that show the maximum safe load for a given water condition. The diagram below shows a typical set of Plimsoll marks. Freshwater marks are on the left and saltwater marks are on the right.
   (a) Why are the saltwater marks lower than the freshwater marks?
   (b) Why is tropical fresh (TF) the highest mark?
   (c) Why is Winter North Atlantic (WNA) the lowest mark?
   (d) In January, a captain plans to sail his fully-loaded ship from Oslo, Norway across the Atlantic to Manaus, Brazil 1500 km up the Amazon River. How does the captain decide how fully to load the ship?

7 Pressure Differences

Structure of the Atmosphere

From the moon, space looks black. From the Earth, the sky looks blue. The difference comes from the thick blanket of air that surrounds the Earth. It extends upward 500 km until it merges with space. From a space shuttle, the atmosphere appears as blue bands against the blackness of space.

Imagine taking a trip through the atmosphere on board a space shuttle. The trip only takes about 10 minutes. You would pass through the various layers shown in the diagram at right.

The boundaries between layers are not as distinct as those shown here, nor are the elevations of each layer the same in all parts of the world. As you travel in the shuttle, at what altitude do you think half of the atmosphere is beneath you? three quarters of the atmosphere? Use the following data to make a graph to find out. This data gives the weight of the atmosphere pressing on an area of 1 m^2 at different elevations. The weight is expressed in *newtons* (the gravitational force on a 100-g mass).

Altitude (km)	Weight per square meter (N/m^2)
0 km (sea level)	100,000
5.6	50,000
16.2	10,000
31.2	1,000
48.1	100
65.1	10
79.2	1
100	0.1

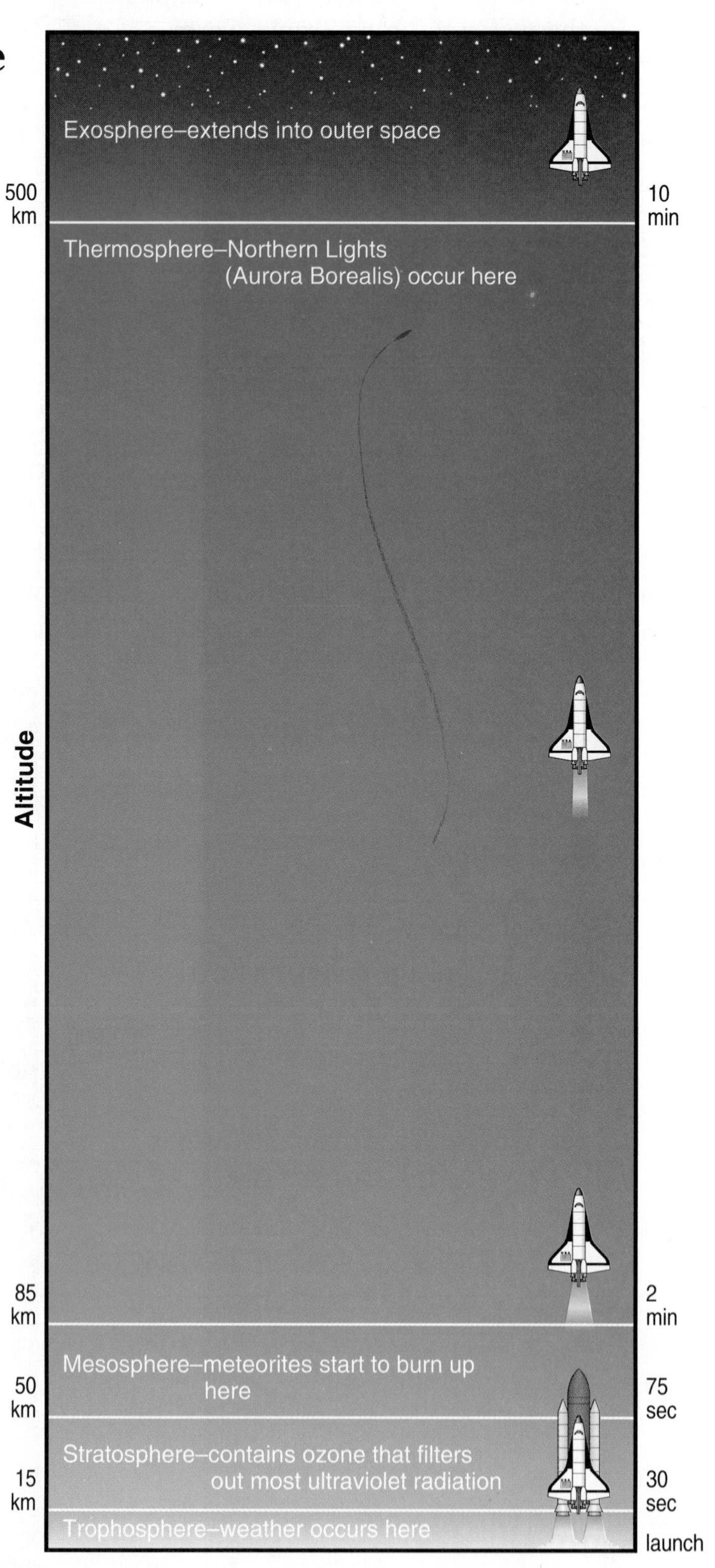

Understanding Pressure

You are beginning to work with the concept of **pressure.** The pressure of the atmosphere at sea level is about 100,000 N/m^2. This means a force of 100,000 N presses down on every square meter of the Earth at this elevation. Does this pressure increase or decrease at a level 10 km above sea level? Review the table on the previous page. What estimate can you give for atmospheric pressure at 10 km? at 50 km? How does pressure vary with altitude?

How would you define pressure now? What relationship does it have to force? After completing the next Exploration, return to these questions. Then provide a scientific definition for the meaning of pressure.

EXPLORATION 10

Pressure Situations

Part 1

A Discussion Task

Pressure is an important factor in each of the following situations. With a classmate, discuss the explanation for each one.

Why is it possible to crawl across thin ice, while walking across it will end in disaster?

Why is it that a high heel stepping on your foot hurts more than a pair of sneakers or flat sandals?

Now share with others a pressure situation of your own.

Part 2

A Brick Study

You Will Need

- a brick with a string (1 m long) tied securely around it
- a spring scale or force meter (marked in newtons)
- a metric ruler

What to Do

Predict which face the brick must rest on in order to exert the least pressure against a table. What about the most pressure? Test your predictions. Determine the pressure of the brick against a table when it is lying on each of its faces. Here is the information you will need.

> Area of a face of a brick = length × width
> Weight = reading on the spring scale in newtons
> Pressure = force (weight of brick) ÷ area

Questions

1. In what units is pressure measured?
2. Does the weight of the brick change with the face it is sitting upon?
3. Does the pressure that the brick exerts change depending on the face resting on the table? Why or why not?
4. If the brick was placed on a piece of foam, on which face would it sink to the greatest depth?

Area = length × width

What is the force in newtons?

Pressure in the Ocean

How does water pressure change as you travel deep into the ocean? Is it similar to the changes in atmospheric pressure at different altitudes? The following Exploration focuses on the differences and similarities between pressure exerted by gases and liquids.

Activity 1

Water Pressure and Depth

You Will Need

- a large can with a hole punched in it close to the bottom
- a meter stick
- a tripod stand
- water
- a stream table

What to Do

Place the can on a stand in a set-up like the one shown below. Fill the can with water and observe what happens. Record the height of the water and the distance the water squirts.

Thinking About It

1. The distance a volume of water squirts out of the hole at the bottom of a can is a measure of the pressure exerted by water. How can you use this information to compare the pressure of different depths of water?
2. Place your data in a table similar to the one shown. Prepare a graph of your data.

Height of water in can (cm)	Distance water squirts (cm)

3. Look at your graph and at the one you made earlier of atmospheric pressure at different altitudes. Do you think there is an altitude where the atmospheric pressure is equal to zero? Is there a depth of water for which the pressure in the can is equal to zero? Can you think of an important way in which the air in the atmosphere is different from the water in the can?

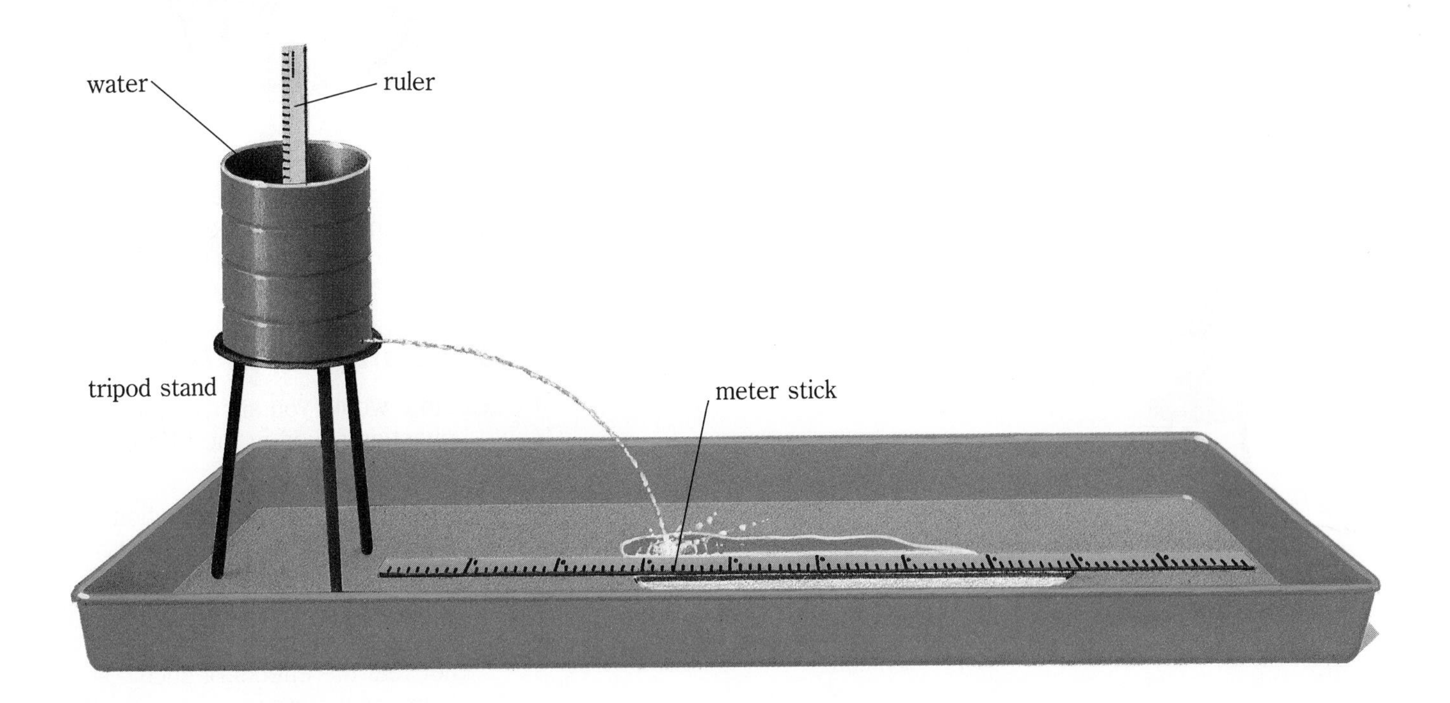

Activity 2

The Effect of Shape and Volume

You Will Need

- 2 cans of different diameters with a hole punched close to the bottom of each
- a meter stick
- a tripod stand
- a stream table
- water
- a graduated cylinder

What to Do

Arrange the materials as shown below. Measure and record a volume of water. Place one of the cans on a tripod stand and pour in the water. Measure the distance the water squirts. Replace the first can with the second can and repeat the experiment. Record and then compare your results.

Thinking About It

1. What conclusions can you make from this activity?
2. What would you observe if you filled both cans with water so that the water was at the same depth in each can?
3. Would there be any difference in water pressure if you swam 2 m below the surface of the water in a swimming pool as opposed to 2 m below the surface in a lake? Does pressure depend on the volume or the depth of water? Explain your reasoning.

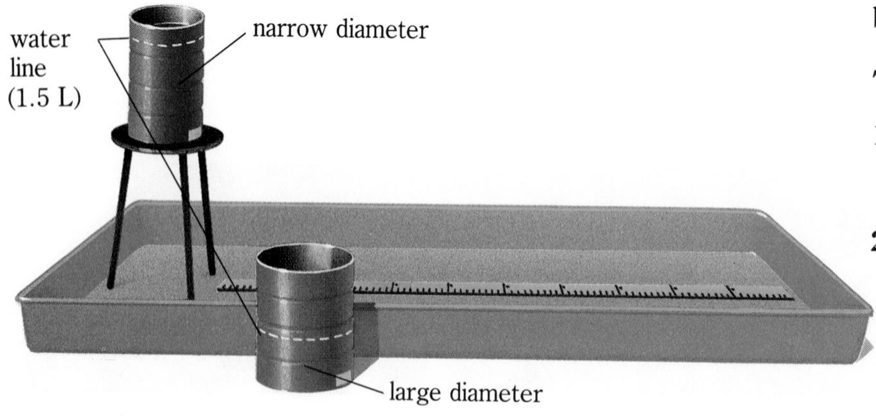

Activity 3

Is Water Pressure Exerted Equally in All Directions?

You Will Need

- a plastic bottle
- a paper match
- water
- a watch or clock

What to Do

Fill the plastic bottle with water. Drop a paper match into the container. Screw the top on tightly. Wait 10 minutes and then squeeze the bottle. Can you make the match rise and fall? Does it matter where you squeeze the bottle?

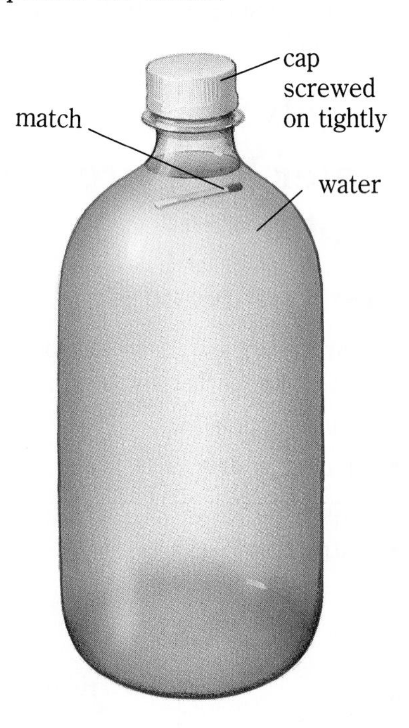

Now turn the bottle sideways. Can you still control the movement of the match? Does it matter where you squeeze the bottle?

Thinking About It

1. How would you answer the question asked in the title of this activity?
2. The atmosphere exerts about 100,000 N/m^2 of pressure at sea level. This is equivalent to two elephants sitting on a card table. Why is it you do not feel the effects of such tremendous pressure?

Historical Flashbacks

Here are three episodes from history. Each illustrates something about the pressure exerted by the atmosphere. Test your understanding by responding to the questions and completing the tasks that follow each episode.

The Magdeburg Experiment

During the 1650s, Otto von Guericke was both an inventive scientist and the mayor of the German town of Magdeburg. In 1652 Emperor Ferdinand III heard of his experiments and asked to see them. So Otto von Guericke gave him a dramatic example. He placed two copper hemispheres together so they formed a hollow sphere with a diameter of about 45 cm. He removed as much air from inside the sphere as he could by using a crude vacuum pump. After creating a vacuum inside the sphere, how did the air pressure inside the sphere compare with the air pressure outside?

Now comes the part that astonished the Emperor. Two teams of horses—one attached to each hemisphere—could not pull the two hemispheres apart! What does this tell you about the magnitude of atmospheric pressure?

Consider this: Air moves from regions of high pressure to regions of low pressure. Where have you heard of high- and low-pressure areas before? What do you think would happen in von Guericke's experiment if a hole could be punctured in the copper sphere?

Try It Yourself 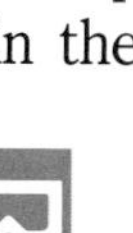 

The following simulation will help you understand the effect of air-pressure differences in the atmosphere. Add a small amount of water to an aluminum can. Boil the water until you see steam escaping from the can. Using oven mitts, turn the can upside down in a large container of cold water. What happens? How do you explain this in terms of pressure?

Blaise Pascal's Experiment

Blaise Pascal

Just before von Guericke performed his demonstration, Pascal, a French scientist, expanded on the study of atmospheric pressure. He used a new invention created by an Italian named Torricelli—the mercury **barometer.** Torricelli had filled a tube with mercury and turned it upside down in a pool of mercury. Not all of the mercury ran out. Instead, the atmosphere supported a column 76 cm high at sea level. Whenever the pressure of the atmosphere changed, so did the height of the mercury column.

In 1648, at the age of 25, Pascal had his brother-in-law carry a barometer to the top of a 1500-m mountain. At the bottom of the mountain the column of mercury in the barometer was 71.1 cm high. At the top of the mountain the mercury column was 62.6 cm high. What had Pascal proved? Why was the measurement at the bottom of the mountain different from the one at the top?

For his contribution to science, Pascal had a unit of measurement named after him. A pressure of 1 N/m^2 is equal to 1 **pascal** (Pa). Normal atmospheric pressure is about 100,000 N/m^2, or 100,000 pascals.

Try It Yourself

Below is one method for making a homemade barometer that will detect daily changes in atmospheric pressure. Build this or another barometer of your own design and keep a record of the increase and decrease in atmospheric pressure each day. Also, keep a record of the kind of weather you have each day. Is it sunny or cloudy? hot or cold? humid or dry? Can you find a relationship between pressure changes and the changes in weather?

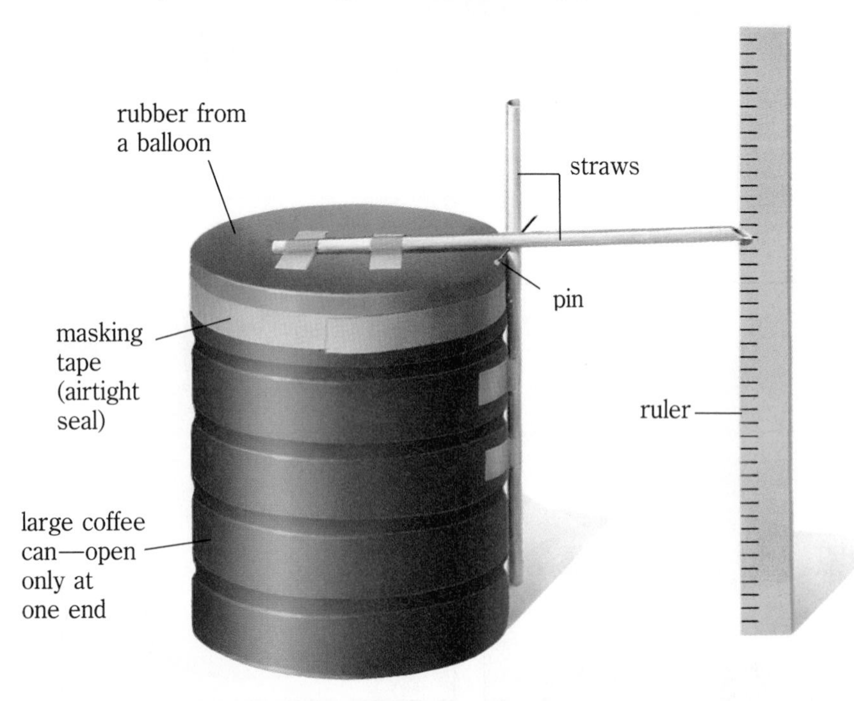

vacuum
76 cm
mercury

A model of Torricelli's barometer

Robert Boyle's Experiment

In 1667 Robert Boyle improved an air pump so that it created a nearly perfect vacuum in a container. When he pumped air out of a pipe that was placed in a container of water, he was able to raise the water a little over 10 m—but no more.

- What caused the water to move up the pipe?
- Why was he able to lift only 10 m of water this way?

Try It Yourself

Place an index card over a full glass of water. Turn the glass upside down over a large container. Why does the water stay in the glass? According to Boyle's experiment, how high can a water column be supported in an experiment of this kind?

Applying Pressure

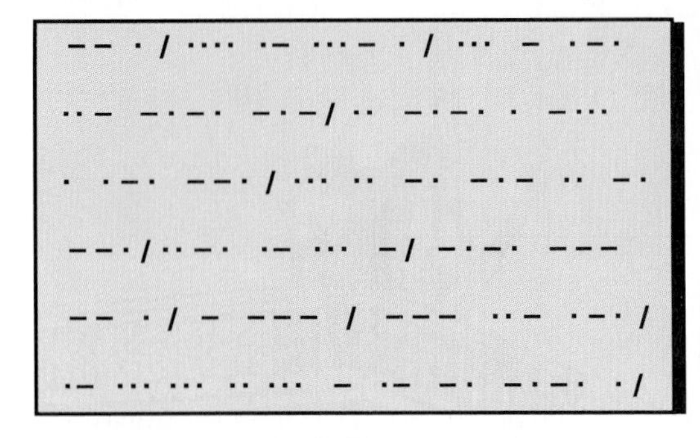

On April 14, 1912, the greatest passenger ship of its time sent out this message in Morse Code. A short time later it sank, killing 1523 people.

Decipher the Titanic's final message using Morse Code.

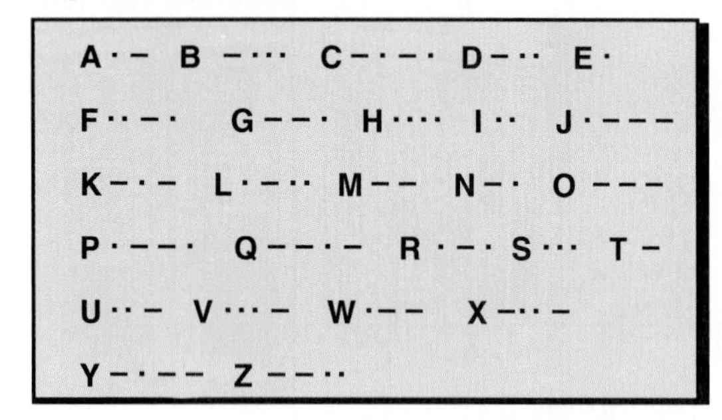

With the help of submersibles, the Titanic has been visited several times at its final resting place—nearly 4000 m below sea level. The diagram at right shows pressures at different ocean depths, including the depth of the Titanic.

1. What is the water pressure in pascals at the depth where the Titanic now lies?
2. Look at the points marked A, B, and C. What is the water pressure at each location?
3. Create a message about pressure using Morse Code. Share your message with someone else.

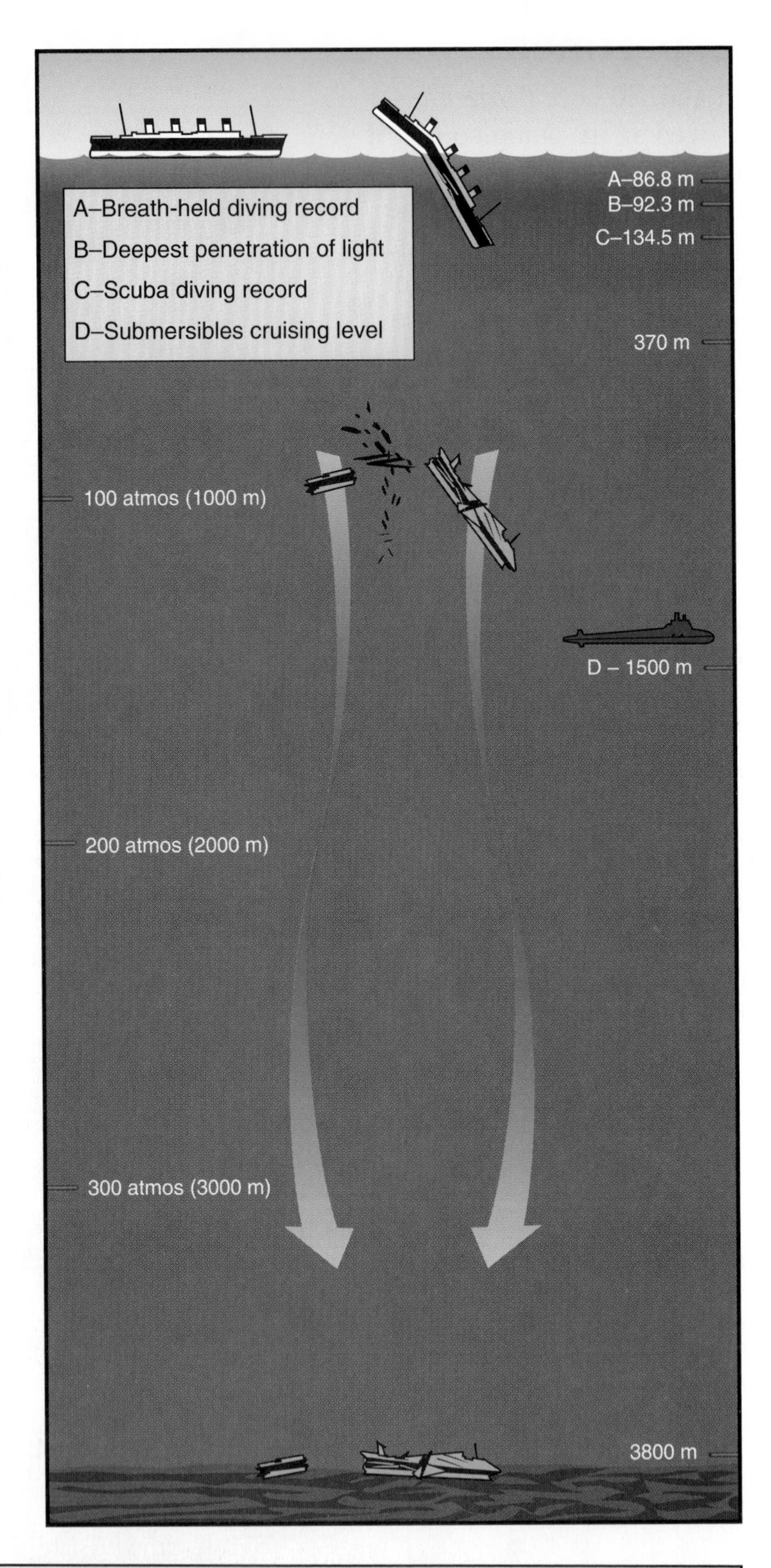

Pressure is sometimes given in *atmospheres.* One atmosphere is the amount of pressure exerted by the Earth's atmosphere at sea level. You can use the following formula to convert pascals to atmospheres.

101 kPa = 1 atm

8 The Direction of Flow

Mystery 1

On August 3, 1492, Christopher Columbus lifted anchor and set sail for the Orient. He stopped first at the Canary Islands just off the coast of Africa. There he knew he could catch steady, northeasterly winds—these later earned the name *trade winds*. It was his idea to use these winds to blow him all the way to China. Just about one month later, he reached the Caribbean and the New World.

After three months he sailed north, hoping to find suitable winds to carry him home. Luckily, he found the *westerlies,* which carried him back to Portugal. No other explorer had documented the great winds that blow in different directions across the oceans.

What was the mystery? No one knew what caused the winds or why they blew in such a consistent and predictable manner. What do you think causes the trade winds?

Mystery 2

In the mid-1700s another maritime mystery unfolded. Mail ships sailing from England to New York took weeks longer than ships sailing from England to Rhode Island. And the two destinations were only a single day's sailing apart! Benjamin Franklin solved the mystery by talking to Atlantic whalers. They told him about a strong surface current that flowed across that section of the ocean. This current was known as the *Gulf Stream* because it was believed to originate in the Gulf of Mexico.

While traveling between Europe and America, Franklin discovered that this flow of water was warmer than the water surrounding it. He used his information to draw the first map that included the Gulf Stream. That way sailors could make use of the stream—or avoid it.

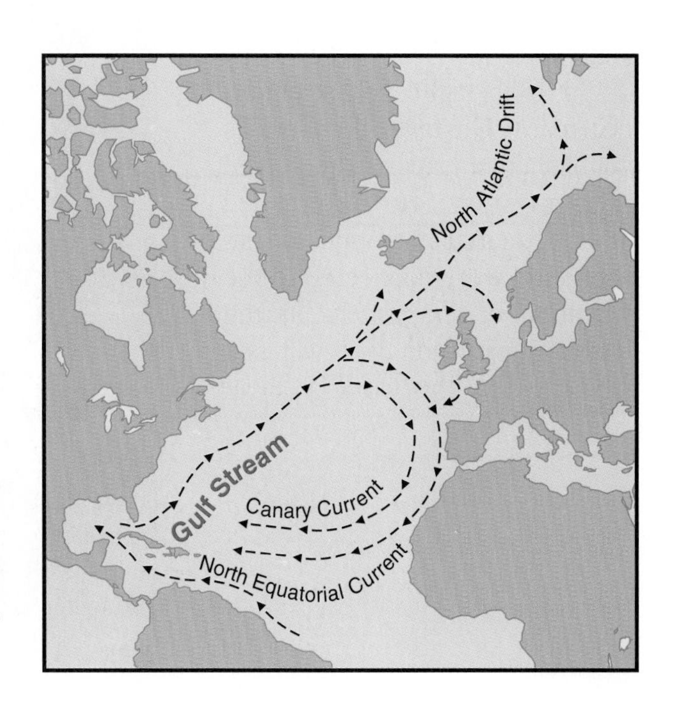

A Windy Solution

How are these two mysteries connected? The same winds that blew Columbus back to Europe are also responsible for pushing the ocean's water toward Europe. Thus, the ocean's surface currents are wind-driven.

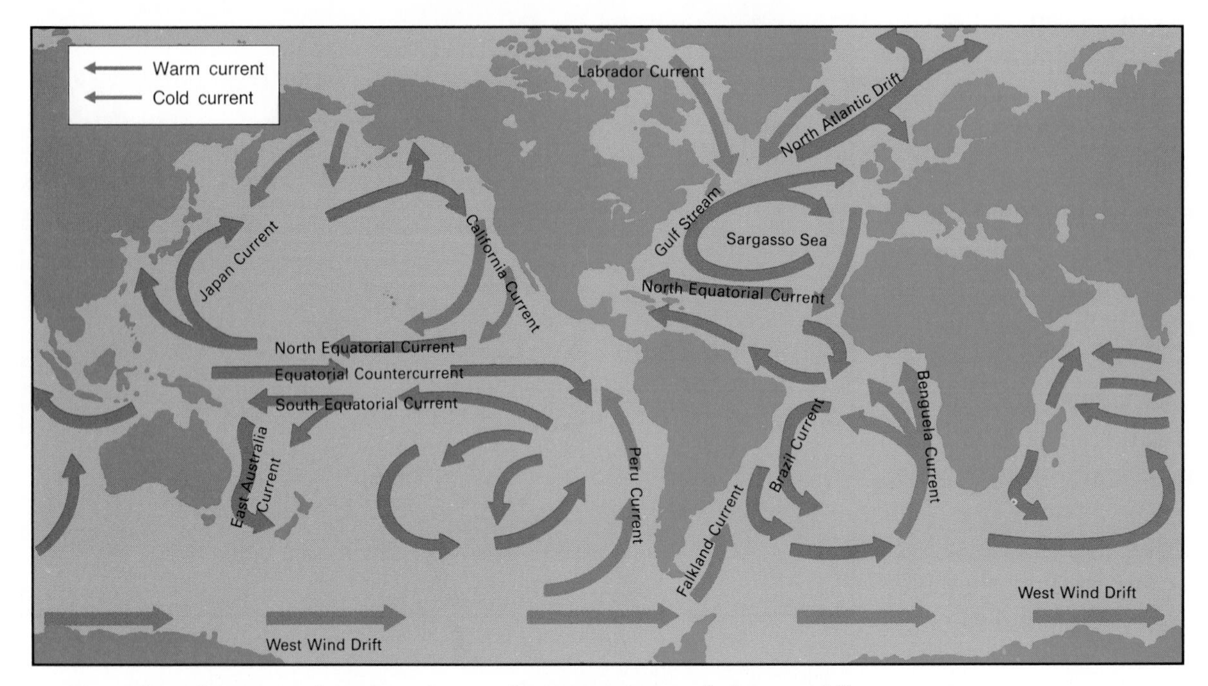

Examine the map showing the surface currents of the world's oceans. What do you notice about the direction of flow of the major currents north of the equator? south of the equator? If surface currents such as the Gulf Stream are created by winds, what can you infer about the direction of these winds? Can you think of any reason for the circular motion that you observe in the ocean currents?

In 1835, a French scientist, Gaspard Gustave de Coriolis published a paper that solved this mystery. He determined that the circular motions of the currents were connected to the rotation of the Earth. In the next Exploration you will see exactly how this works.

The Coriolis Effect

You Will Need

- ¼ sheet of poster board
- baking soda in a salt shaker
- a large ball bearing or marble

Part 1

Direction of Rotation

What to Do

1. In the center of the poster board, write the letter N as a symbol for the North Pole of Earth. On the backside of the poster board, write an S to symbolize the South Pole.
2. The Earth rotates counterclockwise when viewed from above the North Pole. Rotate your poster-board model of Earth to simulate this. Be sure to turn it very slowly.
3. Continue rotating the model in this direction, but look at the back of the poster board. You are now viewing the Earth's rotation from above the South Pole. In what direction does it rotate now?

Part 2

Coriolis' Explanation

What to Do

1. Place the poster board on a table. Lightly dust it with baking soda.
2. Without rotating the model, roll a wet marble across the poster board. Was its path straight?
3. Now simulate the Earth's rotation as observed from above the North Pole. While someone is very slowly rotating the model counterclockwise, roll the wet marble across the surface. Describe the path the marble takes.
4. Try rolling the marble from different directions across the rotating model. As you look in the direction of the marble's motion, does its path always swing in one particular direction?
5. Rotate the model Earth clockwise. Now what pole does the center of the poster board represent?
6. Roll the marble across the model Earth while it is rotating clockwise. As you look in the direction of the marble's motion, does its path swing one way or another?

Your Analysis

From your observations in this Exploration, find proof for the following conclusions. Then write a statement in your Journal that explains how you came to each conclusion.

1. When viewed from above the North Pole, the Earth rotates in a counterclockwise direction.
2. When viewed from above the South Pole, the Earth rotates in a clockwise direction.
3. A moving object in the Northern Hemisphere swings to the right as an observer looks in the direction of the object's motion.
4. A moving object in the Southern Hemisphere swings to the left as an observer looks in the direction of the object's motion.

Explaining Your Discoveries

If you found evidence to support the statements on the previous page, congratulations! You have rediscovered what Coriolis first explained in 1835—the **Coriolis effect.**

Return to the map of surface ocean currents on page 182. Apply your knowledge of the Coriolis effect to an analysis of the Gulf Stream. Can you now explain why the current in the Atlantic Ocean moves in a clockwise pattern? Why are there currents in the Southern Hemisphere that follow a counterclockwise pattern? Remember that the major factor that drives surface currents is the wind. What can we infer about the direction of the winds from the direction of the surface currents?

The Sargasso Sea

As the currents of the Atlantic Ocean slowly flow in their clockwise pattern, a calm area forms at the center of rotation. This is called the Sargasso Sea. Under clear skies the sea is warmed by the sun, which increases the evaporation of water in the area. Thus, the Sargasso Sea is saltier than the surrounding ocean waters.

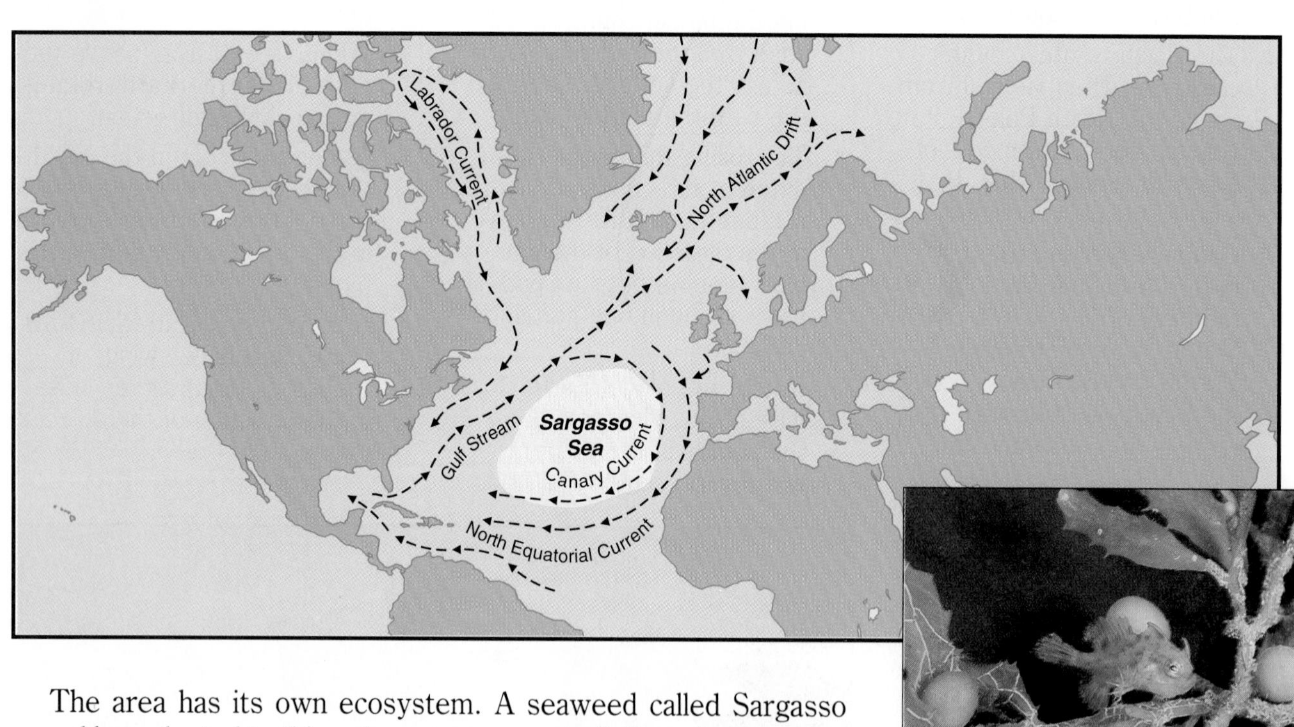

The area has its own ecosystem. A seaweed called Sargasso weed has adapted to this unique environment. It floats in the warm water, often forming wide patches that cover more than an acre of open sea. In addition to this kind of seaweed, certain animals are found only in the Sargasso Sea. One of these, the European eel, crosses huge distances each year to return to the sea to breed. Do research to discover more about the eel and other unique organisms that live in the Sargasso Sea. Then write a paragraph or two in your Journal summarizing what you've learned.

A Sargasso Sea Ecosystem

World Winds Explained

What causes the winds that helped Columbus reach America and return home to Portugal again? These same winds also drive the ocean currents in the Atlantic. What direction do they blow? Why? The answers lie in ideas that you have already investigated. First, you know that air moves from areas of higher pressure to areas of lower pressure. Secondly, the movement of air is influenced by the Coriolis effect. Keep these ideas in mind as you complete the Exploration.

What to Do

In small groups discuss the following information. Copy each diagram into your Journal and respond to the questions.

1. Permanent high- and low-pressure areas exist at certain latitudes because of temperature differences. Why is a low-pressure area indicated at the equator? Why are high-pressure areas indicated at the poles?

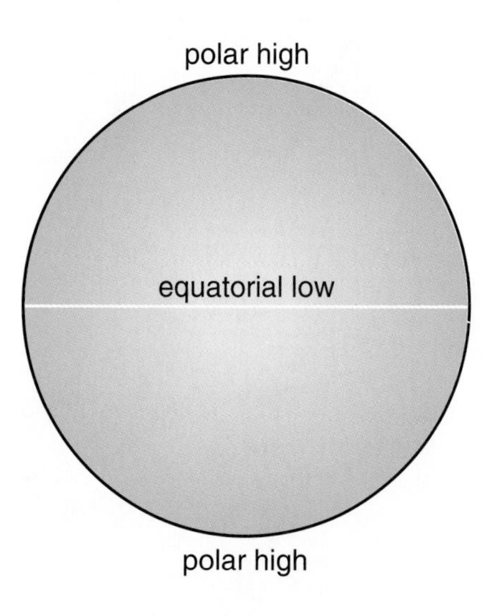

2. Other factors, especially the rising and sinking of air at certain latitudes, create other permanent high- and low-pressure areas.

What is the approximate line of latitude for the United States?

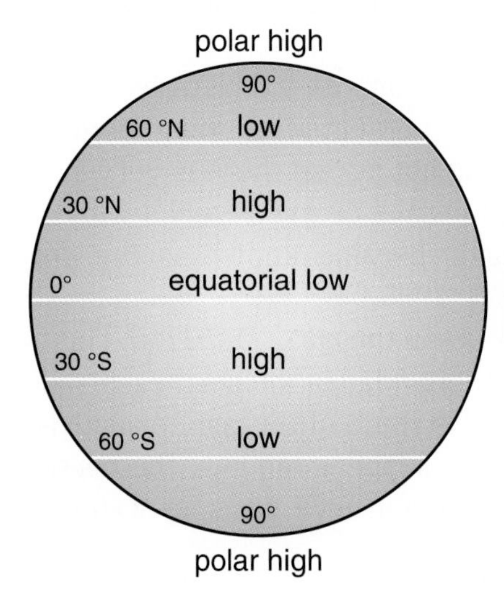

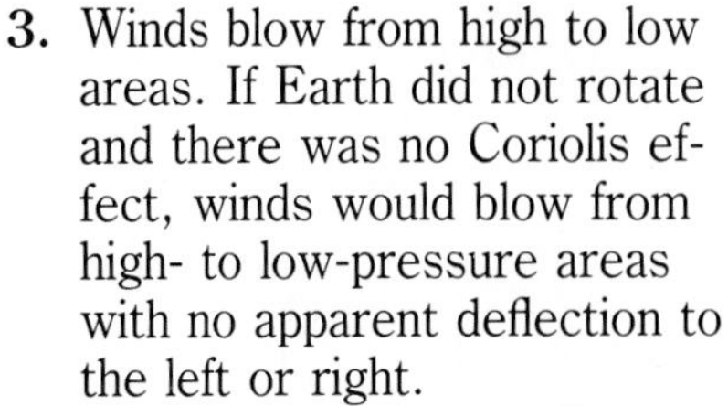

3. Winds blow from high to low areas. If Earth did not rotate and there was no Coriolis effect, winds would blow from high- to low-pressure areas with no apparent deflection to the left or right.

A few wind patterns are shown on the map below. Copy the diagram and draw in the missing arrows.

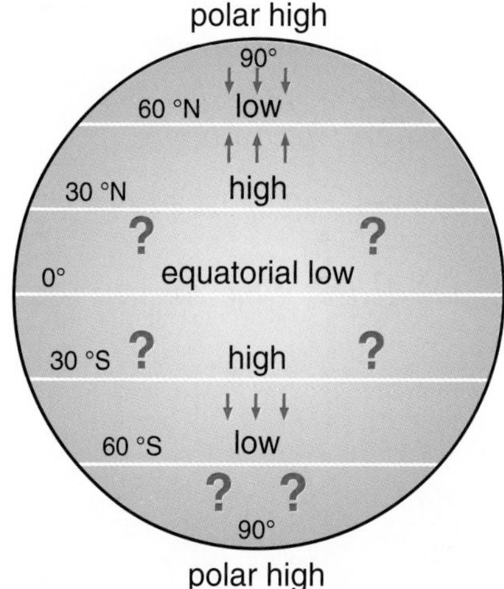

4. Since the Earth rotates, the winds are deflected to the right or left depending on the hemisphere. A few wind directions are shown here. Copy the diagram in your Journal and draw in the missing information.

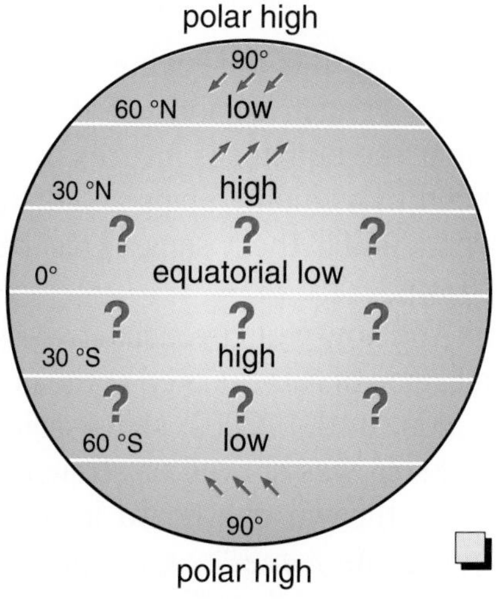

"Into that Silent Sea . . ."

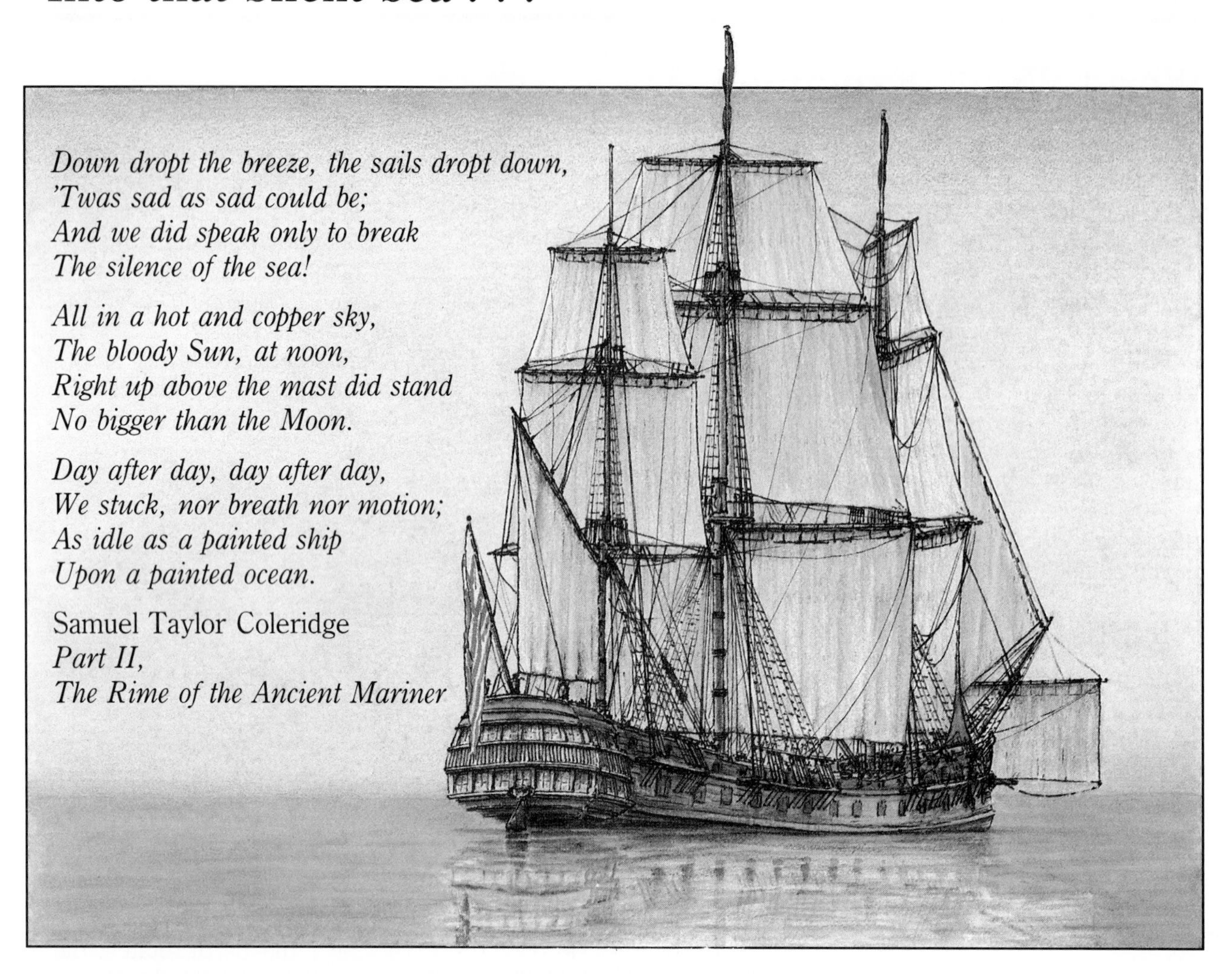

Coleridge published this poem in 1798—an era when sailing ships were the only means of travel across the vast oceans. How did he use the characteristics of winds and currents in his poetry? Where do you think the sailors are stranded? (**Hint:** See the second stanza, which describes the position and appearance of the sun.) Did sailors really have cause to worry that the wind would stop blowing?

During the time when ships relied almost exclusively on the wind, certain regions of the ocean presented serious dangers. Near the equator is an area of rising hot air, so there is very little wind movement. Ships could sit for days or even weeks with no wind to fill their sails. This region became known as the *doldrums*.

And how would you like to be stuck in the *horse latitudes?* This is a region of calm located at the northern edge of the northeast trade winds. In the 1700s, ships sailing this route often carried animals. If a ship was slowed or stranded and supplies ran low, the horses were thrown overboard in order to conserve drinking water—hence the name "horse latitudes."

El Niño

What is El Niño? Consider the following news items and then answer the questions that follow.

Experts ponder impact of El Niño phenomenon

North America could be facing dramatic weather changes due to El Niño, a variation of temperature and pressure patterns over the eastern Pacific Ocean.

This phenomenon occurs every three to five years—its severity changing from year to year. Weather patterns worldwide are affected. The strong El Niño of 1982–83 was blamed for a devastating drought in Africa and Australia. Also, severe winter storms lashed California, and parts of South America were deluged with torrential rains. The El Niño in 1986–87, however, was barely noticed.

What causes El Niño?

El Niño (Spanish for "the Christ Child") occurs in the equatorial waters off the west coast of South America around Christmas time. Normally, trade winds push surface waters away from the coastline of South America, causing cold water to well up from below. The cold surface waters suppress the formation of clouds and rain. Every few years these winds die down and then reverse, causing warm water to be pushed toward the South American coast. The warmer waters produce abundant rainfall. At the same time, normal weather patterns are disrupted over a wide area.

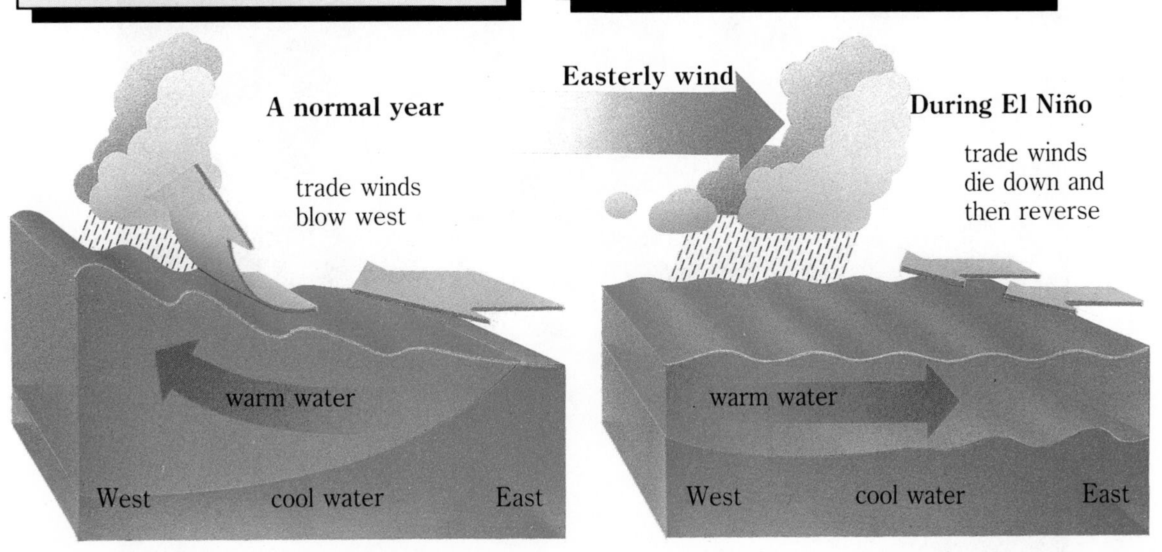

In a normal year (left), westward-blowing winds push warm waters toward the western Pacific, where they "pile up" (although only to a height of a meter or two). El Niño occurs when these winds reverse, causing warm waters to pool along the western coast of South America.

Questions

1. The El Niño effect illustrates the link between oceans and the atmosphere. Use information from the articles above to describe this link.
2. Examine the map on page 182 that shows surface ocean currents. What is the name of the surface current that flows up the coast of South America?
3. Examine the picture of global temperatures on page 142. What evidence suggests that this colder current flows up the west coast of South America?
4. What kind of damage can El Niño do? What regions suffer the most? How are plant and animal life affected? Research the El Niño phenomenon to learn more about its effects.

9 Hurricanes

Have you been following daily pressure changes on your homemade barometer? What kind of pressure change is associated with calm, sunny weather? How about with unsettled, stormy, or cloudy weather? Did you find that calm weather is usually associated with increasing atmospheric pressure and stormy weather with decreasing pressure? What is the reason for this? Through the diagrams and pictures that follow, you'll examine the relationship between low atmospheric pressure and stormy weather.

Picture Study 1

This satellite picture shows an unusual weather system—a *hurricane*. Hurricanes vary in size, but most are 150 to 450 km across. Some have winds up to 300 km/h.

1. Assuming there is a low-pressure area at the center of the hurricane, which way are the winds blowing—inward or outward?
2. If winds blow from areas of high pressure to areas of low pressure, are hurricanes examples of high- or low-pressure systems?
3. Is the wind moving clockwise or counterclockwise? Why does it blow in this direction? (In the Southern Hemisphere, the wind direction around low-pressure areas is opposite to that in the Northern Hemisphere.)

Picture Study 2

Compare the satellite picture on page 188 with the cross-sectional view of a hurricane shown below. As you read the following, locate the italicized terms in the diagram and answer the questions.

1. Most hurricanes are carried toward the U.S. coast by *westward-flowing winds.*
   - What are these winds called?
2. The *surface winds* pick up heat energy from the warm waters they blow over. In addition, a large amount of water evaporates—another mechanism of gathering energy that will later be released.
   - How is heat energy stored by the evaporation process?
3. The *warm, moisture-laden air* rises and spirals toward the center of the hurricane. Condensation creates clouds. This process releases heat energy.
   - Why should moisture condense out of the air as it rises?
4. Some of the rising air is carried away by *high-altitude winds.*
5. Some of the rising air is drawn back down into the *eye of the hurricane.* This air warms as it descends, creating the clear skies typically found at the center of the storm.
   - Why doesn't descending air form clouds like moist, rising air does?

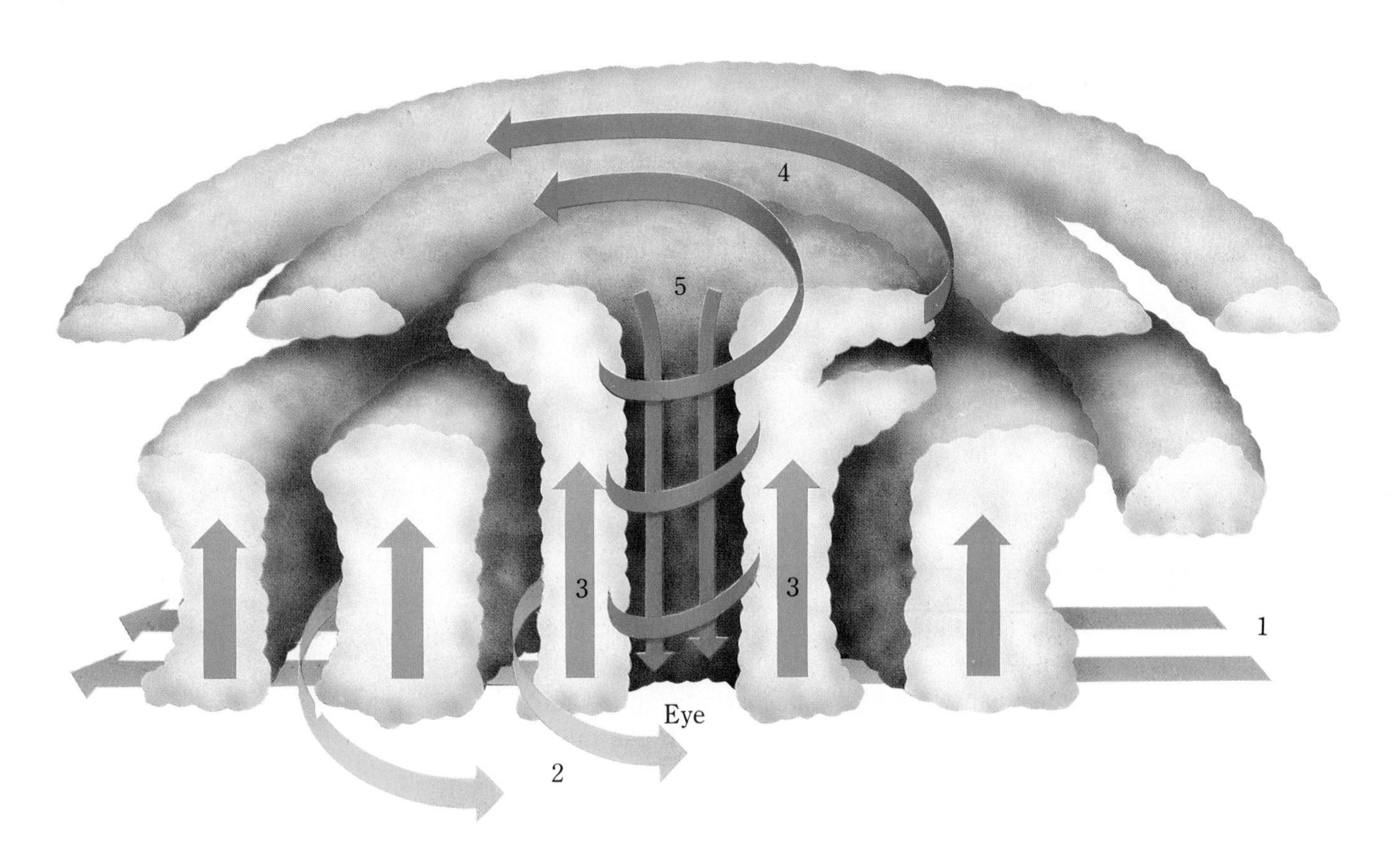

Picture Study 3

A hurricane is an extreme example of a low-pressure system. Normal atmospheric pressures are close to 100 kPa. By how much do you think a hurricane deviates from this norm?

On a weather map a hurricane looks like the diagram below. A series of solid lines called *isobars* connect points of equal pressure. The same pressure difference separates each isobar. Wind directions have been included on this map as well.

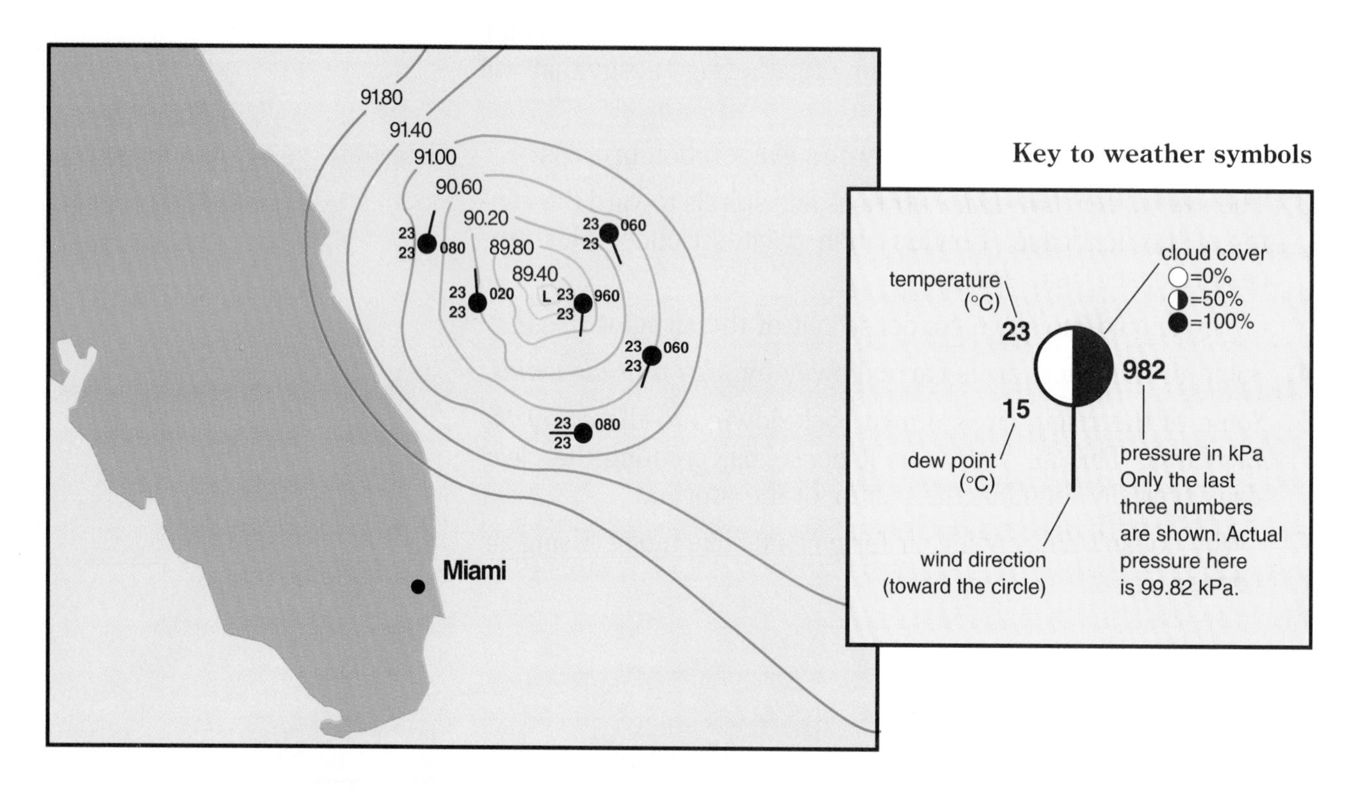

- What is the wind-direction pattern around a low-pressure system such as a hurricane?
- How does the pressure change as you progress farther from the eye of a low-pressure system?
- At each station, what number was dropped from the complete pressure reading?
- Why are the dew point and the air temperature reading the same?

Keeping Track

Make a list of the major ideas and findings you have encountered in this section. What have you discovered about pressure in the atmosphere and ocean? What causes global currents and wind? How does a hurricane form? Your list should answer these questions and more. Compare your list with one from another classmate.

Challenge Your Thinking

1. Here are some everyday experiences and observations. They all have something to do with atmospheric pressure. What is the connection?
   (a) On her last plane trip, Nora noticed that the tops of the small creamer containers bulged outward. But in restaurants on the ground they did not.
   (b) While drinking a milkshake through a straw at the ice cream shop, Joel wondered exactly how the straw worked.
   (c) Getting frozen orange juice out of the can was always a problem until Rita suggested punching a small hole in the bottom of the can.
2. What do you think would happen to the flow pattern of global winds and currents if the Earth were suddenly to start spinning in the opposite direction—from west to east instead of from east to west? How would low-pressure systems behave differently?
3. For the weather map shown here, make the following predictions:
   (a) wind direction at A
   (b) temperature differences at B and C
   (c) wind direction at D
   (d) regions of cloudy skies and rain
   (e) areas of high pressure
   (f) areas of low pressure

   Over the next day the low-pressure system will move over Boston (E). How do you think the pressure readings in Boston will change? What about the temperature? How will wind direction change?

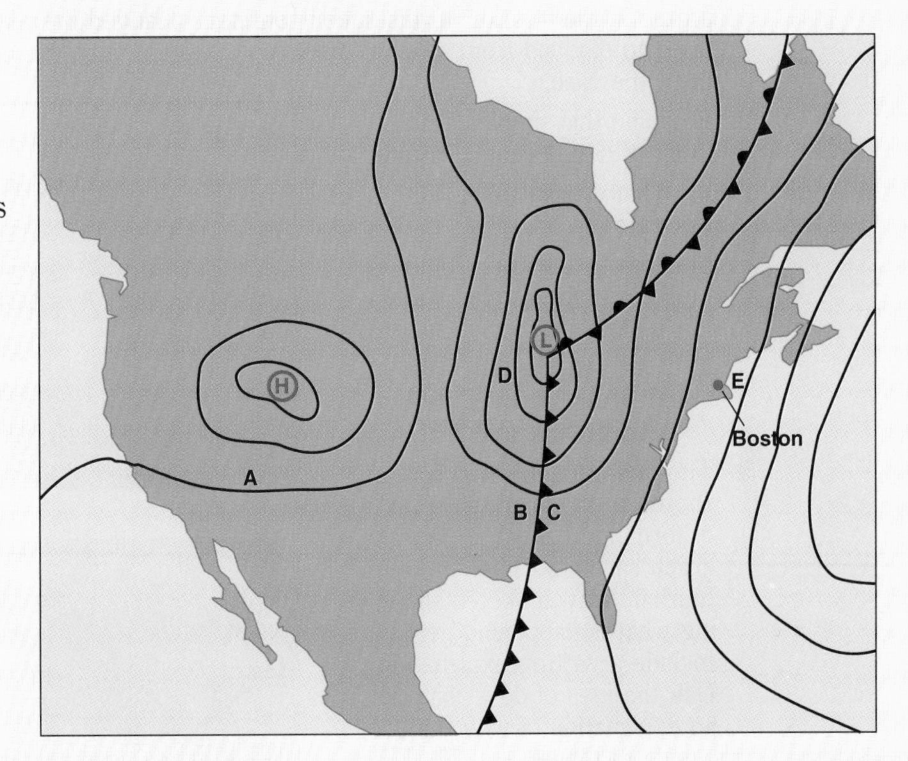

Unit 3 Making Connections

The Big Ideas

In your Journal, write a summary of this unit, using the following questions as a guide.

- What is the greenhouse effect, and what gases are responsible for it? How might it relate to global warming?
- What influence do the oceans have on the climate of the Earth?
- What is the dew point?
- What causes ocean currents? What are the causes of winds?
- What role does density play in driving winds and currents?
- What are air masses, and how do they form?
- Why do winds and surface currents move in predictable patterns?
- How does pressure affect the weather?

Checking Your Understanding

1. About 250 million years ago most of the world's continents were joined together in a single landmass. Computer simulations suggest that temperatures in the interior of this super-continent changed dramatically from night to day and from season to season. Suggest why this might have been the case.
2. Examine this simplified diagram of deep ocean currents.
   a) What causes these currents?
   b) How do these currents help to absorb the world's "excess" carbon dioxide?
   c) If glaciers melted, what effect might this have on deep ocean currents?

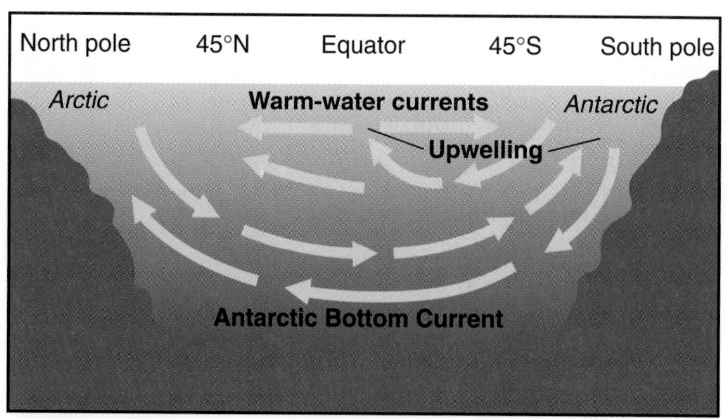

3. The photograph at right shows clouds in the process of forming. Study the photograph carefully. Then in your Journal, draw a sketch showing what is happening here. Include a written explanation. Use the idea of dew point in an explanation of why the clouds formed.

4. What is the difference between *weather* and *climate?* Illustrate the distinction by describing the climate in your area. Then describe your weather.

5. The diagram at right shows the path of a sailplane (a type of glider) as it flies along on a sunny day.
   (a) Use your understanding of air currents to explain the glider's movements.
   (b) How would the glider's path differ on a cloudy day? at night? if it passed over a city?
   (c) How do the effects shown here relate to weather and climate?

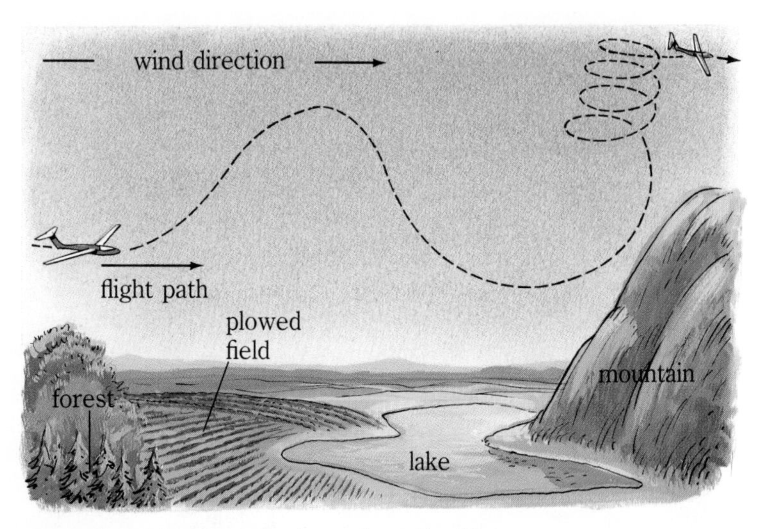

6. Every summer, the average concentration of atmospheric carbon dioxide drops significantly in the Northern Hemisphere. What reasons can you suggest for this? Would the carbon dioxide levels be more or less the same everywhere? Why or why not?
7. Make a concept map using the following terms or phrases: *Mediterranean Sea, atmosphere, trade winds, density differences, currents,* and *oceans.*

Reading *Plus*

Now that you have been introduced to Earth's atmosphere and oceans, let's take a closer look at the structure and history of each. In reading pages S31–S50 in the *SourceBook,* you will learn about the origin and evolution of the atmosphere and oceans, the structure and composition of the atmosphere, the water cycle, and the interaction of the Earth's surface features with the atmosphere to cause local climatic effects. You will also learn about the ocean's mineral and biological resources as well as some of the methods used by humans to harvest those resources.

Updating Your Journal

Reread the paragraphs you wrote for this unit's "For Your Journal" questions. Then rewrite them to reflect what you have learned in studying this unit.

Science in Action

Cristy Mitchell uses computer technology to analyze the weather.

Spotlight on Meteorology

Meteorology is the study of the Earth's atmosphere and everything that happens in it. As a meteorologist for the National Weather Service, Cristy Mitchell uses new technology to forecast weather conditions. Cristy's education includes a B.A. in Mathematics with a minor in Physics, and specialized graduate training in Meteorology.

Q: What does your daily work involve?

Cristy: To put it in a nutshell, my daily work involves asking myself several questions including: What is the weather doing right now? Why is it doing that? What is going to happen to change it? I then use a variety of computer-generated models to help me find the answers to these questions. With my results, I issue complete weather forecasts including, if needed, severe-weather, flash-flood, and marine warnings.

Q: What kinds of instruments do you use?

Cristy: The computer is an invaluable tool for me. Through it, I receive maps and detailed information including temperature, wind speed, pressure, and general sky conditions for a specific region. The information is in the form of codes, which I interpret. In addition to the computer, radar and satellite imagery are also important for showing the "whole picture" of a region's (and the nation's) weather.

Q: What school subjects were the most important for your career?

Cristy: Definitely physics, and the math to work the physics! The atmosphere behaves according to physical properties, so it is essential to have a good base in that aspect of science.

Mathematics has always been fascinating to me. Since I was quite young I've seen real-world applications for it in my life. When I was growing up, I helped my grandmother in the grocery store she owned. I think it was that experience that led to my appreciation for math. I was

After taking relative humidity readings, these meteorologists analyze the readings with the computer.

always using it—weighing and calculating the price of meat, and adding up groceries. I learned a lot about geometry as my mother taught me to sew and make my own clothes. As a teenager, I used geometry to help my father develop floor plans and build houses.

Q: What are the most interesting or exciting aspects of your work?

Cristy: There's nothing like the adrenalin rush you get when you see a tornado coming! I would say witnessing powerful forces of nature like the tornado are what really make my job interesting. One time I saw the end of a rainbow. It was only a quarter of a mile away from me, and believe it or not, it was in a bank parking lot!

This job is also very exciting to me because I love the outdoors so much. Since I always have to be aware of what's going on outside, I'm really involved in the outdoor environment.

This student-intern meteorologist is reading outdoor weather gauges for the National Weather Service.

Q: What are the most frustrating aspects of your career?

Cristy: Ironically, I would again say the powerful forces of nature. It's frustrating to know I can only do so much. For example, one time I came to work on the midnight shift when rain was falling over the tributaries of a major Texas river. Within an hour, rainfall had spread to a much larger area and it had intensified. Conditions looked so serious, a flash-flood warning was immediately dispatched.

Right away I made some calls to the affected area. I kept hoping no one would try to cross the river. Unfortunately, several buses carrying children were attempting to leave a camp that was further downstream from the storm. The first bus made it to higher ground. The second bus, however, was washed off the road when the river suddenly surged over its banks. Ten people drowned in that unfortunate accident. It's moments like that which give me a frustrating sense of powerlessness.

A Project Idea

Forecast with the forecasters! Using the barometer you made in Lesson 7 of this unit, graph barometric pressure for seven days. At the same time, graph the barometric pressure according to the reports of a local meteorologist. How do the charts compare? If you were to make a forecast from the results of your barometer, what would it be? How does your prediction compare with that of the meteorologist?

Science and Technology

Spying on the Planet

We are being watched. NOAA-9. LANDSAT 5. GOES-E. These are not code names of alien spacecraft. They are names of artificial satellites orbiting the Earth. From viewpoints high above the Earth, these satellites keep an eye on things below.

But that's not all. Radar installations all over the United States are continuously monitoring the sky. And twice a day, every day, about 200 balloons carrying instruments are released into the atmosphere.

The GOES satellite before launch, with photo taken from orbit

The Big Picture

The reason for all these detection devices is, of course, the weather. Together, they provide a very complete weather picture. However, an "overview" of the weather can be given by just a single instrument—a weather satellite.

The weather maps you probably know best are the satellite pictures you see on weather reports. They are produced by a satellite that stays above the same point on the Earth. Its orbit takes one day, so it keeps pace exactly with the Earth's rotation. Satellites with this type of orbit are known as *geostationary* satellites. Their orbits are quite high—about 36,000 km above the Earth. This gives them a view of an entire hemisphere.

The geostationary weather satellite produces a single picture about once every 20 minutes. Weather station computers usually have software that puts satellite pictures together into sequences or "loops," creating brief moving pictures. Such sequences show the way clouds are moving, and how fast. From this visual information, meteorologists make predictions about the weather.

Scanning the Sky

Meanwhile, back on the ground, radar is on the job. Radar can detect rain, snow, hail, and extremely moist clouds. This caused problems in World War II, when radar was first developed for detecting airplanes. But for weather purposes, it is what we want. Weather radar gives us up-to-the-minute, detailed information about the location of storms and storm clouds in a given area.

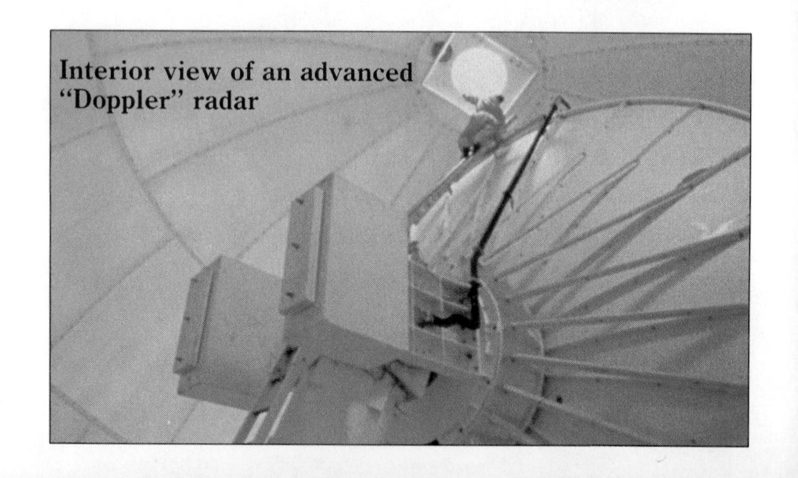

Interior view of an advanced "Doppler" radar

Upper-Air Analysis

Twice a day, at midnight and noon, Greenwich Mean Time, weather stations release instrument-carrying balloons into the atmosphere. These balloons transmit data back to computers in the weather stations. The computers quickly relay the information to a single supercomputer in Washington, D.C.

This supercomputer is one of the fastest computers in the world. Even so, it takes several minutes to process the information it receives to create weather forecasts. When local offices receive the forecasts for their particular area, the weather balloon data will be about two hours old.

As the balloons rise, the air around them gets thinner. The balloons expand, and eventually they burst. Then their instrument packages are lowered to the ground by parachutes. If you ever find one of these packages, you should take it back to the weather station so it can be used again. There will be instructions in the package that tell you where to return it.

Most of our information about the upper atmosphere still comes from weather balloons.

Closer to Home

Television stations in your area may employ their own meteorologists. It is the job of the meteorologists to use different sources of information to provide a weather forecast for their areas. In doing this, they draw upon experience and judgement. Satellite and radar images are most useful for immediate and short-range conditions ("nowcasting"). Longer-range forecasts are more tricky. They require analyzing a wide variety of data.

Find Out for Yourself

Geostationary orbits are only one type of orbit. Another type of orbit that is important in artificial-satellite technology is the near-polar orbit. How does the near-polar orbit differ from the geostationary one? What are some advantages and disadvantages of each? And what, besides weather monitoring, are some other uses of artificial satellites?

Technology is constantly changing. For example, a new type of radar called Doppler radar is being developed. In addition to detecting precipitation, it is also able to measure the wind speed and direction of clear air. What advantages would this have? Do some research on Doppler radar and other new weather technology.

Unit 4 Electromagnetic Systems

For Your Journal

Write a paragraph summarizing your thoughts about each of the following:

1. What is electricity?
2. How are electrical currents created?
3. How are electricity and magnetism related?
4. What clues suggest that the device in this photograph uses electricity?

Indispensable Energy

Your school is planning an exhibition to be called the *Museum of Today*. The idea behind the Museum is a simple one: to introduce the technology of today to imaginary visitors from the past. It is your job to plan the exhibits for the Museum. Include as many examples as you can of the technology and gadgetry that make life what it is today.

Where do you start?

You could start with simple things like electric lights. We take them completely for granted, yet we could hardly even imagine life without them. Electric lights were a revolutionary development. With their invention, the 24-hour-a-day city suddenly became a possibility.

Then, of course, there's the telephone. Who could imagine life without the telephone? A hundred and fifty years ago, the idea of being able to punch a few buttons and talk to someone far away would have been unimaginable.

And don't forget radio. How wonderful it is to be able to flip a switch and hear music, talk shows, or news—all of it originating many kilometers away.

Television! Now there's an invention! Life wouldn't be the same at all without television. Who would have imagined that you could send pictures through the air and have them appear on a screen? That would seem like magic to someone from the past.

Here are a few examples to get you started. What other examples of technology would you include in your exhibit?

What do all the devices just mentioned have in common? They all use electricity. Most use magnetism as well. Obviously, electricity and magnetism serve us in many ways. Can you identify some of these ways?

Clearly, life without electricity would be very different. Imagine going an entire day without it. How would you get by?

But what exactly is electricity? How would you explain it to your visitors from the past? Although you can see what electricity does, you can't really see *it*. In this unit you will be introduced to this mysterious, indispensable form of energy and its close relative, magnetism.

Electricity and You: Case Studies

It takes only a little knowledge and skill to make use of many helpful electrical devices. It takes more knowledge to know how they work—and still more to design them. With one or two classmates, discuss how electricity works in one of the following situations. Then make up a case study of your own.

Case Study A

Jenny arrives at the hospital to visit her friend Kim, who is on the fourth floor. Jenny uses the elevator. What information can Jenny get from the elevator lights? What control does Jenny have over the elevator's operation? Are there other controls aside from those that Jenny uses? For example, are there safety controls to prevent people from walking into an empty shaft? What features of the electrical design are specifically for the passengers' benefit? for their safety? Sometimes a system provides you with information as it operates. This is called feedback. Can you identify any controls that might be thought of as feedback mechanisms? Where does the electrical energy used by the elevator come from? The elevator is a complex electrical system. Although you cannot see the subsystems, you know that they exist because of the functions they perform.

Case Study B

"Put the laundry in the washing machine, Bill!" calls Bill's mother from upstairs. "Watch how you set it—you know how!" Bill has many options in setting the automatic washer. What are some of them? How does he control them? What is the nature of this control? How many different kinds of functions does the machine perform or control? For example, if you only had a small load of laundry to do, could the electrical system control the amount of water needed? Does Bill need to stay beside the washer throughout its cycle? Why or why not? What takes place during the washing cycle? How do you think these events occur? Are there any safety features? feedback mechanisms? What is the source of electricity for the washer?

Case Study C

"A plane is traveling at 1000 km/h. Has it broken the sound barrier, if sound travels at 335 m/s?" Fred looked at the problem, pulled out his calculator, and in a few seconds responded: "Yes, the plane is going faster than the speed of sound. It says so right here," he explained, pointing to his calculator.

Electricity certainly works for you in a calculator. What are some of the functions it performs for you? The calculator is quite complex. However, you do not need to know how it works in order to operate it. How do you control the calculator's various operations? For example, in Fred's problem, what might you do? (By the way, was Fred's response correct?) What might be the source of electricity in the calculator? in the plane?

Case Study D

"Oh no, not again!" Angela cried as she turned the car key only to hear the now-familiar "clunk" followed by silence that signals a dead battery. "That's the third time this month! The battery is brand new, so it has to be OK. I guess I'm going to have to get the electrical system checked out."

What could be wrong with Angela's car? What components make up a car's electrical system? Where does the power needed to start a car come from? What kind of electrically-operated feedback or control systems do cars have? How are different forms of energy converted to electricity and vice versa in a car?

Your Own Case Study

In your Journal, describe a situation in which electricity performs some sort of function for you. Describe any devices or processes involved, what takes place, how the device or process is controlled, any energy conversions that take place, and so on.

Electricity at Work

You have probably seen warnings like the one below on appliances or power tools. As helpful as it is, electricity in large amounts is deadly, and must be carefully controlled.

CAUTION

RISK OF ELECTRIC SHOCK
DO NOT OPEN

The familiar situations that you have been analyzing in the Case Studies involve complex electrical parts and arrangements. Large quantities of electricity are used in Case Studies A and B. This is true for the operation of most appliances in homes, stores, or industry. A moderate amount is used in Case Study D. Care must be exercised in using these amounts of electricity.

The device used in Case Study C requires only a small amount of electricity—like most of the Explorations in this unit. They are quite safe to do. Here are four experiments in which a small amount of electricity works for you. You will notice that it does this by producing other forms of energy. Watch for these energy forms.

In these experiments the electricity is supplied by dry-cell batteries such as those used in flashlights.

Activity 1

Shedding a Little Light

You Will Need

- a length of thin copper wire
- a flashlight bulb
- a D-cell

What to Do

1. Using only these three items, find as many different arrangements that will light the bulb.
2. Sketch each arrangement.
3. What form or forms of energy does the electricity produce?
4. What are other examples of electricity used in this way?
5. You have constructed an electric *circuit.* What parts make up this circuit? Check the dictionary to find out the origin of the word circuit. How is this significant? What would you say an electric circuit is?

Have you ever looked closely at a flashlight bulb?

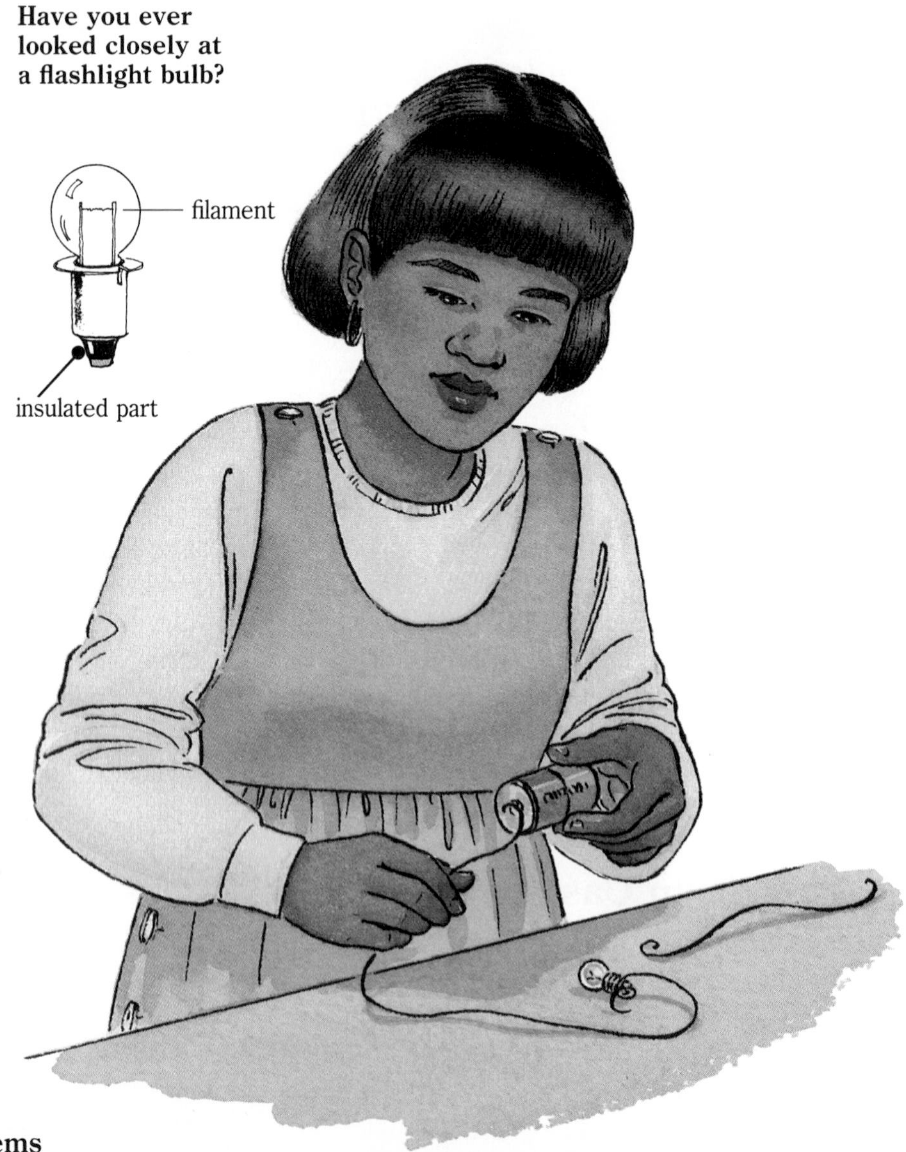

Activity 2

The Heat Is On

You Will Need

- 2 30-cm lengths of thin, uninsulated copper wire
- modeling clay
- a clothespin
- a D-cell
- a wide rubber band
- a steel-wool strand
- aluminum foil
- a thin nichrome wire (10 cm long)

What to Do

1. Make a small loop at one end of each of the two lengths of thin copper wire.
2. Bend the wires and support them with modeling clay as shown.
3. Attach the ends of the wires to the D-cell, securing them with a wide rubber band.
4. Place a strand of steel wool in the loops. What do you observe? Repeat with a narrow piece of aluminum foil, and then with a length of nichrome wire. Bring your hand close to each but do not touch them. What do you feel?
5. What form of energy does the electricity produce?
6. What are some examples in which this type of electrical energy is used?
7. Do Activities 1 and 2 demonstrate the same principle? Explain.

NOTE: Do not leave any of the wires in contact with the loops for very long since this will quickly drain the cell of its electrical energy.

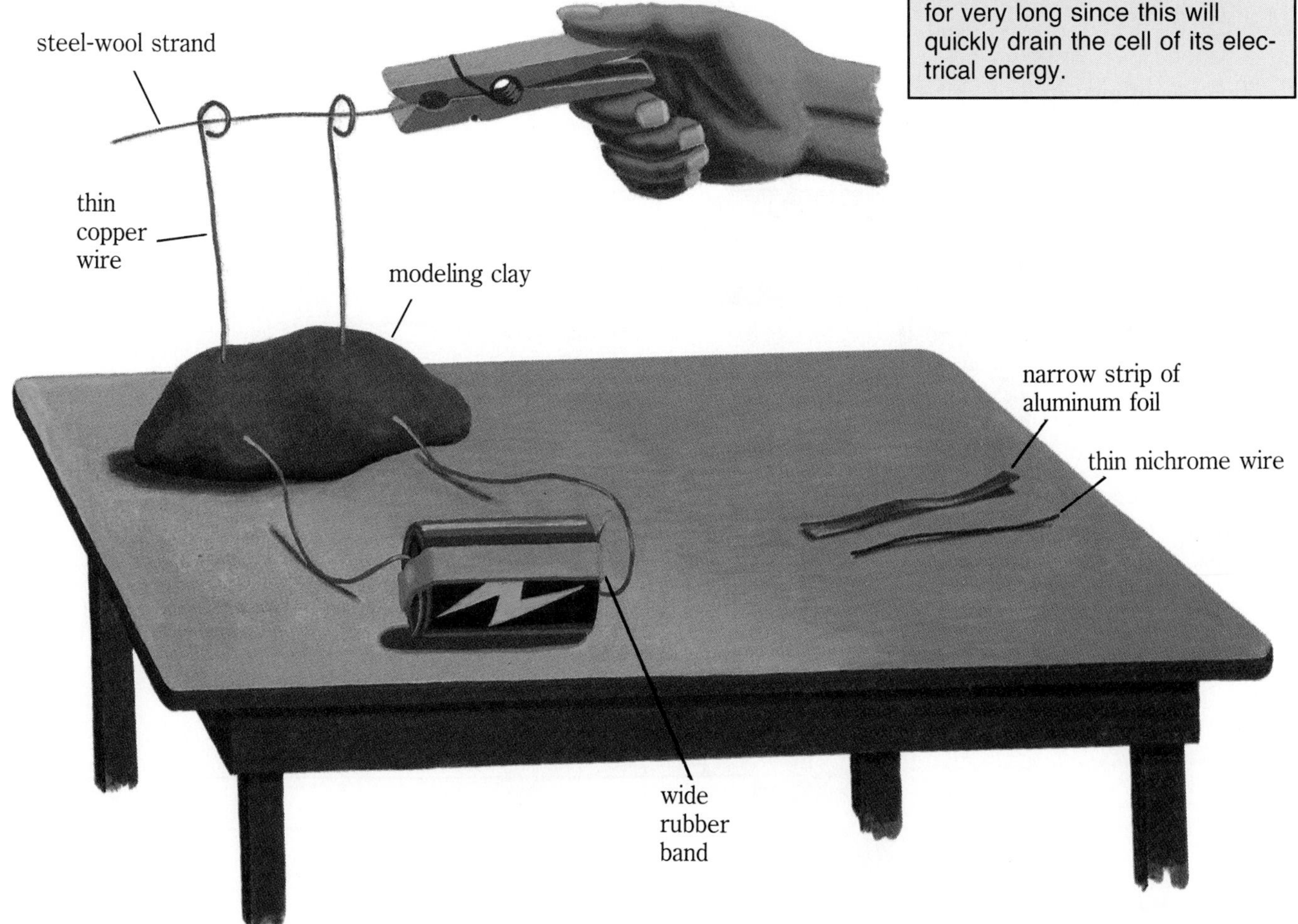

As you may have guessed from the title of this unit, there is a connection between electricity and magnetism. The following Activities will help illustrate that connection.

Activity 3

The Electricity–Magnetism Connection

You Will Need

- a D-cell
- a compass
- 2 thumbtacks or screws
- a wood block
- a paper clip
- 2 pieces of insulated copper wire (one 15 cm long, one 25 cm long)
- a rubber band

What to Do

1. Set up the apparatus as shown. Align the compass needle and place the wire over the compass in a North–South direction (so the wire lines up with the compass needle).
2. Close the electric circuit by pressing down on the contact switch. What happens? **Caution:** *Don't keep the switch closed for very long.*
3. What kind of energy does the electricity produce in this experiment?
4. Can you think of any applications in your daily life that may make use of electricity in this way?

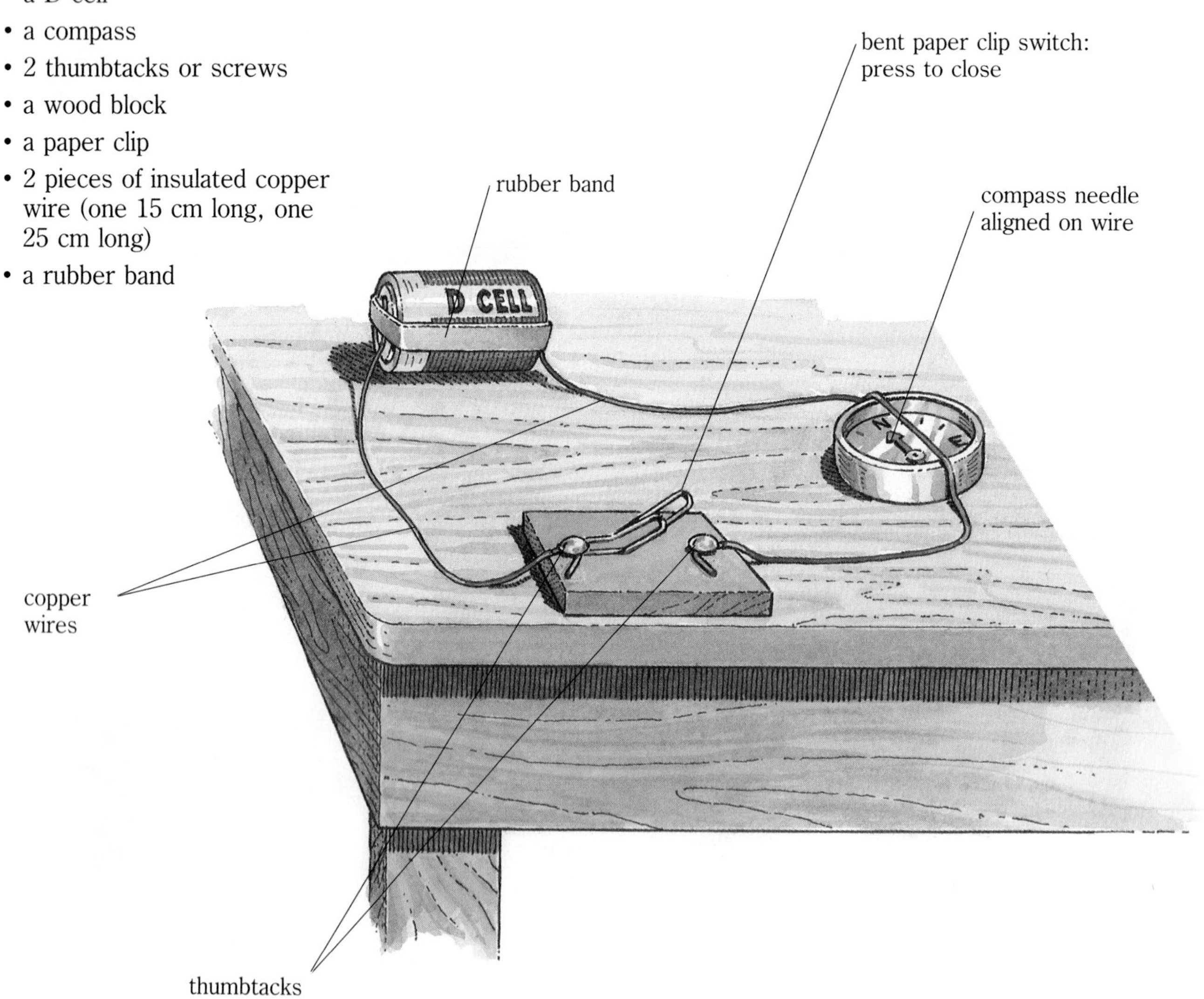

Activity 4
Let's Get Moving
You Will Need
- 2 D-cells
- 3 paper clips
- thin insulated copper wire
- a bar magnet
- a support stand
- a cork
- tape
- cardboard
- 2 thumbtacks or screws

What to Do
1. Assemble the circuit as shown in the illustration.
2. Have one person hold a strong magnet near the cork while the other person presses on the contact switch to complete the circuit. Observe what happens. Open the switch. What happens?
3. Turn the magnet around and repeat step 2. What happens?
4. What kind of energy is being produced by the electricity?
5. What are some practical examples in which energy works for you in this way?

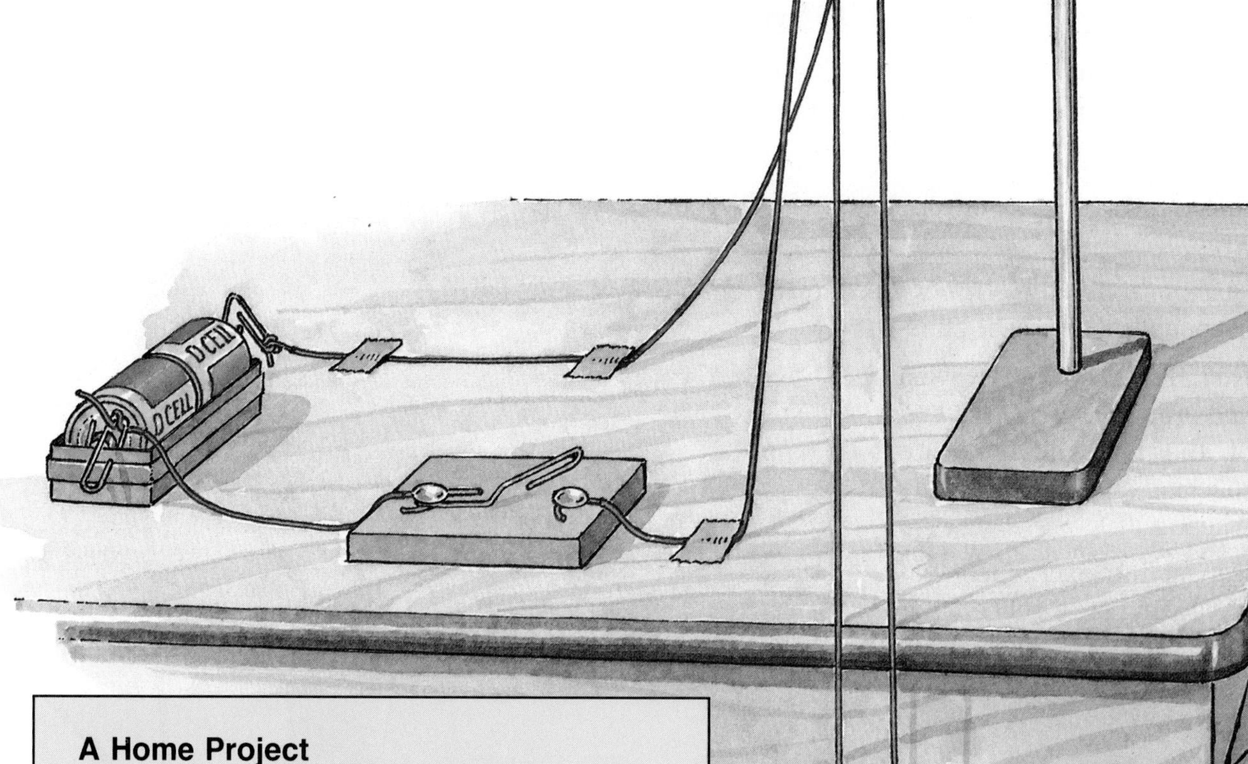

A Home Project
Make a "battery tester." Use one of the arrangements you discovered in Activity 1 of Exploration 1. Your task is to devise a tester that consists of a bulb with two wires connected to it. Touch the ends of the wire to the battery. The brightness of the light bulb will be an indication of the strength of the battery.

What Is Electricity?

In Exploration 1 you used electricity to produce other forms of energy. What were they? Later in the unit you will discover that each of these energy forms can, in turn, produce electricity. Remembering that energy can be transformed from one form to another, does it seem that electricity itself must be a form of energy just as heat and light are?

Electricity accomplishes these changes in energy when an electric **current** is flowing in a circuit. But what is a current? What is it that flows or moves in the circuit?

Have you ever gotten a shock when you touched someone or something after walking across a carpet? If so, you had become *electrically charged.* Have you ever rubbed an inflated balloon on your hair or clothes and then stuck it to the wall? The sticking of the balloon is a sign that it, too, is electrically charged. What might these charges be? In the demonstration that follows you will see these charges in action.

A Classical Current Demonstration

Here is a demonstration similar to one done hundreds of years ago. Make a setup like that shown and try it yourself. Follow the steps closely.

1. Rub a plastic strip or plastic ruler vigorously with plastic wrap.
2. Quickly, touch the strip or ruler to the end of the nail. What happens to the wheat puff?
3. Rub the plastic strip again and bring it close to the end of the nail without quite touching it. Observe the wheat puff.
4. Repeat the experiment using a vinyl strip rubbed vigorously with flannel cloth. What happens to the wheat puff now?

How do you explain the events in this demonstration?

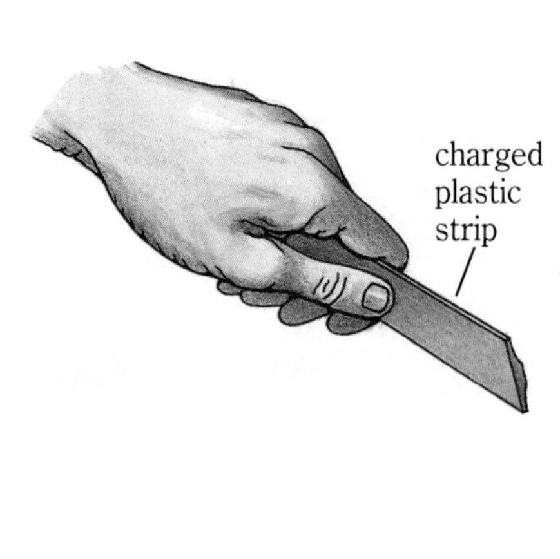

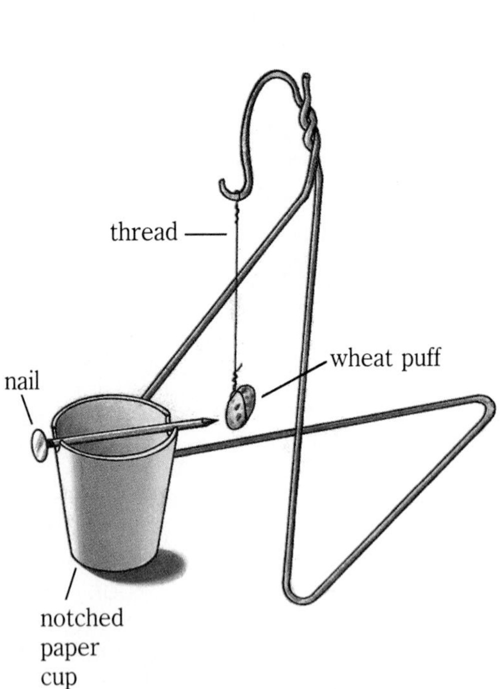

A Theory of Charged Particles

Scientists explain events such as those seen in the demonstration in this way:

1. When one material is rubbed against another, charged particles move from one material to the other. The accumulated electric charge is indicated by the ability to attract lightweight or finely powdered substances. Where did you observe this happening in the previous demonstration?
2. There are two kinds of charged particles—positively charged and negatively charged. How might the effects of each type of charge differ?
3. Objects that have the same kind of charge tend to repel one another. Differently charged objects tend to attract one another. Where did you observe these effects in the demonstration?
4. Charged particles pass easily through certain materials, called **conductors,** but cannot pass easily through other materials, called **insulators** (non-conductors). What evidence is there that iron is a conductor of electricity? You might experiment by replacing the nail in the demonstration with a glass rod, a wooden stick, a copper wire, or objects made of other materials. Which are conductors? insulators?

The drawings below apply the above theory to the classical demonstration. Express in your own words what is taking place in each drawing.

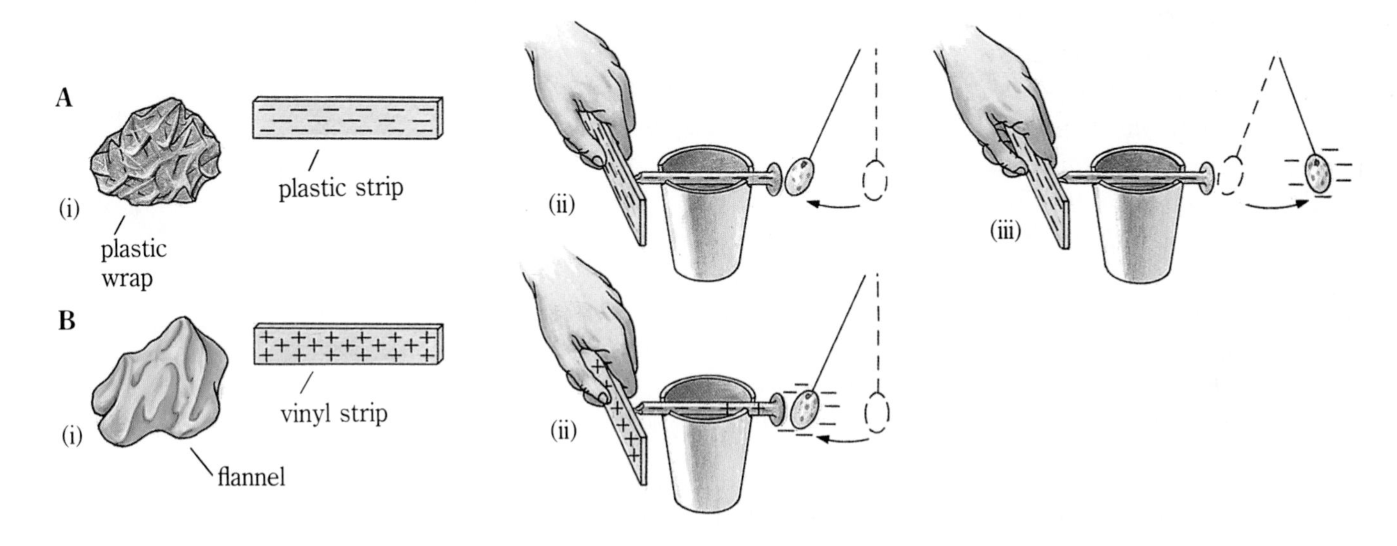

The theory we have been considering is an interesting one. It helps to explain many of your observations about electricity. But how do scientists know about these charged particles? Read the following account from the pages of history.

The Discovery of Charged Particles

It is the year 1900 at Cambridge University in England. J.J. Thomson is talking to a small group of students at the famous Cavendish Laboratory. Listen to what he might be saying:

"We believe that we have discovered the smallest particle of negative electricity. We have obtained a stream of identical negative particles from many metals. In fact, we believe that every bit of matter contains these same particles. We call them **electrons**.

Obviously, most substances are uncharged. Therefore, most substances must have some kind of particle that neutralizes the charge of the electron. We have discovered just that kind of particle in other experiments, and we call it a **proton**. *Each proton has a positive charge that exactly counteracts an electron's negative charge.*

It appears that the fundamental particles of matter are composed of electrons and protons in equal amounts, so that no overall charge occurs. Electrons are relatively light and move freely in conductors. Protons, being much heavier, remain fixed in their positions."

J.J. Thomson and his colleagues expanded the theory of charged particles. Did you follow what he was saying to the students? Can you relate it to what an electric current is—such as the current you got when you correctly connected a dry cell, wires, and a flashlight bulb in Exploration 1?

Charged or Uncharged?

In terms of the theory of charged particles, what are positively charged, negatively charged, or uncharged objects? What makes them that way? Note that in the illustrations on page 209, there were no charges shown on the plastic wrap or the flannel. Should there have been? The illustrations at right give a more complete picture of what happens when a plastic ruler is rubbed with plastic wrap. Count the number of positive and negative charges in the before-and-after situations.

before

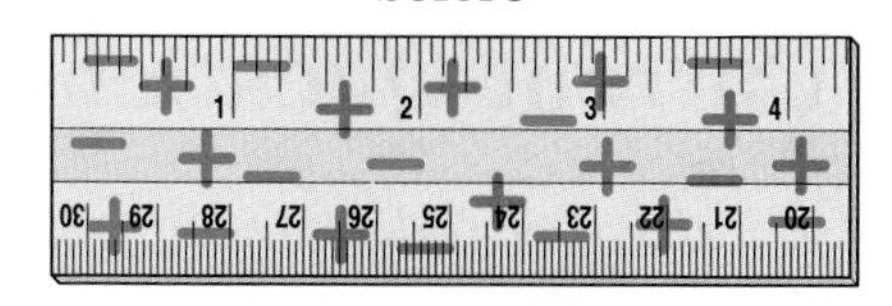

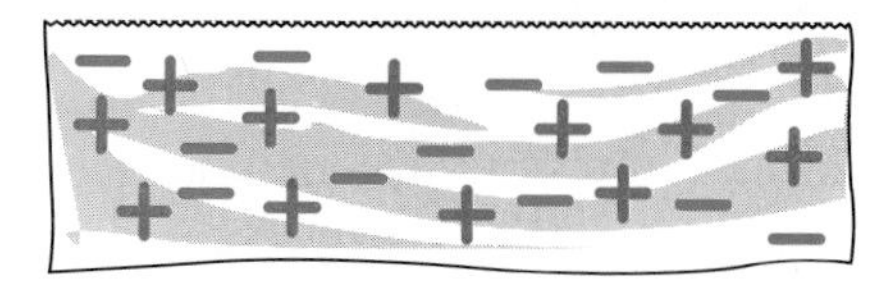

after

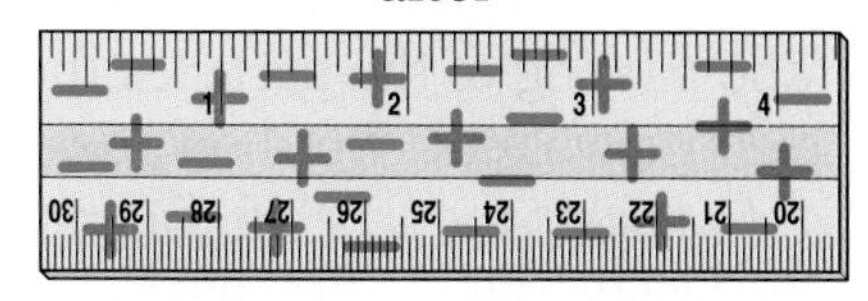

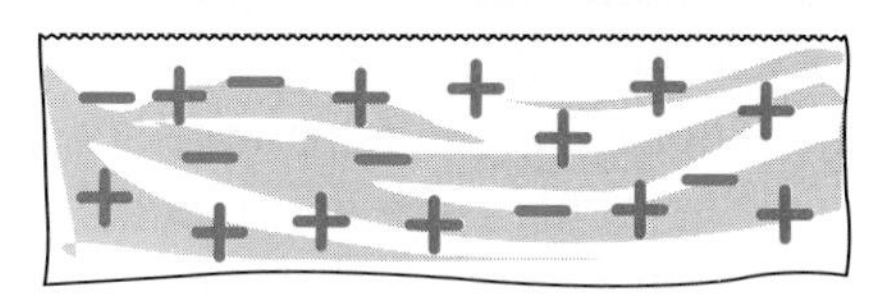

(a) Explain the movement of the electrically charged particles.

(b) Now make "before" and "after" drawings for rubbing a vinyl strip with flannel.

(c) The same balloon is charged differently in each of the following illustrations. In which figure is the balloon slightly negative, strongly negative, uncharged, slightly positive, or strongly positive? What remains the same in all diagrams? Why?

These diagrams are really unreal! Actually, a balloon would contain trillions and trillions of positive charges and an equal number of negative charges.

A Novel Use for Charges

The buildup of electric charges on surfaces is called *static electricity*. You have seen several examples of this phenomenon on the previous few pages. Static electricity was put to work in an ingenious way by Granville Woods, a self-taught American scientist and inventor. In 1887 Woods devised a system that utilized static electricity to send telegraphic messages to and from moving trains. To send a message, the operator on the train typed the message on a special system, which was wired to the frame of the train. The signal created a static charge around the train. Sensors on the train were able to pick up the static charges of messages sent from the train station or from other trains. The result of Woods' invention was a dramatic improvement in railway safety, timeliness, and efficiency. Granville Woods went on to invent many other electrical devices, including the telephone transmitter.

Granville Woods

Current Questions

What causes electrons to move in a conductor to produce a current?

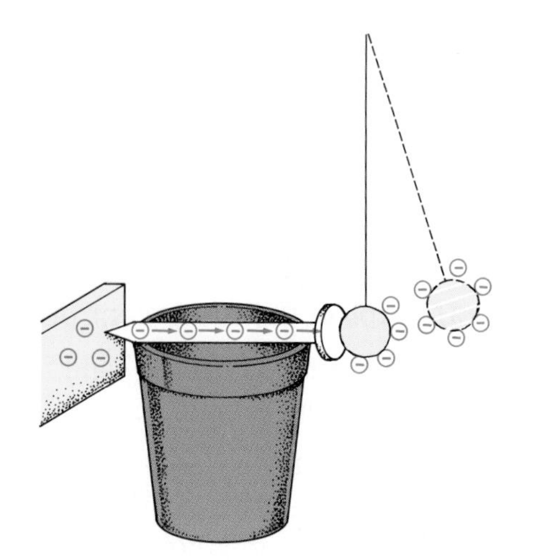

Do you recall charging the wheat puff in the classical demonstration? You brought a plastic ruler which you had rubbed with plastic wrap near the end of a nail. The wheat puff at the other end of the nail moved away. The puff moved because charged particles passed through the nail to the puff, causing the puff to jump away. The passing of the particles through the nail created a temporary electric current. But how, exactly, did it happen?

Consider the labeled diagram shown below. First, electrons in the plastic ruler repel (push away) electrons in the point of the nail A to the right.

These electrons repel electrons at B. The electrons at B move to the right, repelling electrons at C . . . and so on all the way through the nail, to the other end Z.

The electrons at Z are repelled onto the wheat puff. When the puff is sufficiently charged, the repelling force becomes strong enough to push it away from the nail. The current then stops flowing, since there is nowhere for the electrons to go.

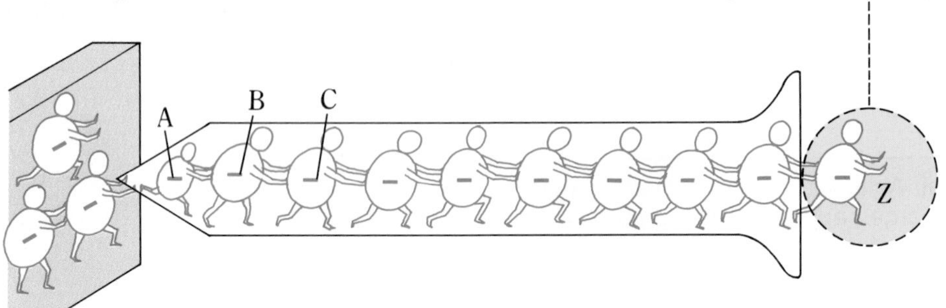

You can see that electrons behave somewhat like falling dominoes. Each electron pushes on the one next to it. If there is a gap in the "chain" of electrons, no electric current can flow.

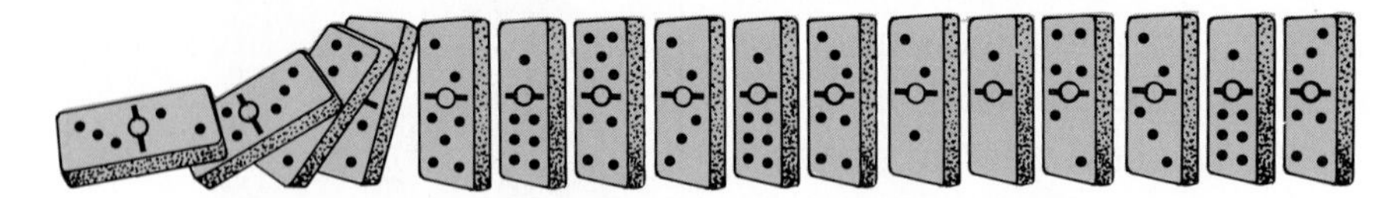

What causes a continuous flow of current?

Let's summarize. We need two things for a continuous current: a continuous supply of charges and an uninterrupted pathway—a conductor—to carry the charges. Compare current flow to the flow of water to your home. For water to get to your home, you need a source—a reservoir—and a conducting path—pipes.

The conducting path usually includes wire made of a metal such as copper, and at least one device that makes use of the electricity. Trace the conducting path in the diagram, starting at the cell, and coming back to the cell. A complete circuit consists of:

(a) a source of electrical charges

(b) a conducting path

(c) a device that uses the electrical energy

Match (a), (b), and (c) with the numbered labels in the diagram. What would you add to the circuit so that you could control the continuous flow of charges, that is, to start and stop the flow? What does it mean to make, or close, a circuit? What does it mean to break, or open, a circuit?

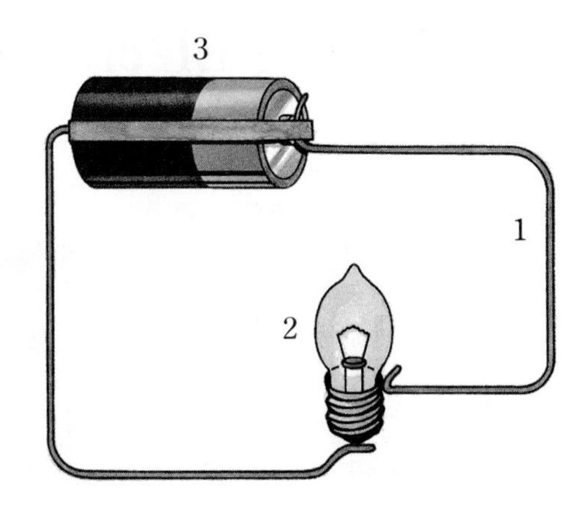

A Continuous Supply

One way to obtain a continuous supply of charges is to use a chemical **cell**. People often call a cell a "battery." However, a battery is actually a group of cells. A car battery, for example, is made up of six separate cells.

As you have observed, a cell always has positive and negative components. These are the *electrodes* of the cell. They are also conductors. The two electrodes of a cell are made of different materials and extend into a solution of chemicals called an **electrolyte**.

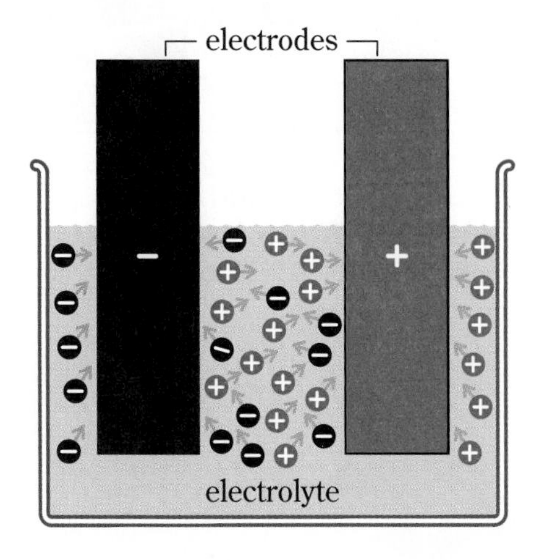

The electrodes and the electrolyte interact with each other. The result is that some electrons move from one electrode into the solution while other electrons move from the solution to the other electrode. The electrode with fewer electrons becomes positively charged. Why? The electrode with more electrons becomes negatively charged. Why? In a typical dry cell, the zinc sides and bottom constitute the negative electrode. The positive electrode is a rod of carbon (graphite) at the center of the cell.

Beginning at the negative electrode of the cell, electrons repel one another throughout the length of the conducting path to the positive electrode. At the same time, the positive electrode of the cell attracts the electrons. Thus, a current flows. The chemicals in the cell keep taking electrons from the positive electrode and giving electrons to the negative electrode. The cell, therefore, causes a continuous current to flow.

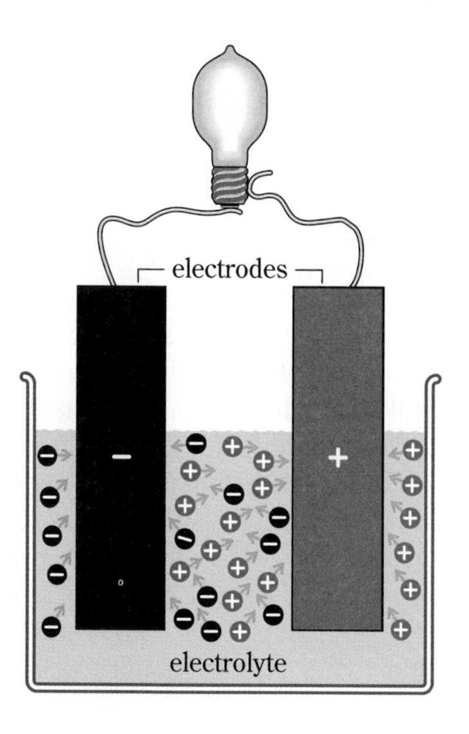

Detecting Electric Current

What these students seem to be saying is:

You can't see electricity. But you know that it's there by its effects. You can feel the heat produced in an electric stove and you know that an electric current is flowing. Likewise, when you see the light given off by a flashlight bulb, you know that an electric current is flowing. In this case, a small amount of current is enough to produce an effect.

Later, you will be making even smaller amounts of electricity. How will you detect them? Look back at Activity 3 of Exploration 1 on page 206. What effect did a current of electricity produce in that activity? This effect can be increased many times if you wrap the wire (it must be insulated) around the compass. This is the basis of sensitive current detectors called **galvanometers**.

EXPLORATION 2

Constructing a Current Detector

Your task is to design and construct a durable galvanometer consisting of a small magnetic compass, insulated wire, and anything else you need to hold the parts in place.

Hints:

1. Remember the results of Exploration 1, Activity 3.
2. Try coiling the wire around the compass. Use different numbers of turns and observe the effect.
3. Leave two ends of the wire free to attach to the source of the small current.
4. Position the galvanometer so that the compass needle and the wire are parallel to one another.
5. Test your galvanometer using a small current. A small current can be obtained from a lemon cell. To make a lemon cell, insert a straightened paper clip and a piece of sanded copper wire deeply into the lemon. The paper clip and the wire are the electrodes and the juice of the lemon is the electrolyte. Hook the free ends of the galvanometer wire to the electrodes. What happens?
6. Will your galvanometer be able to give you any information about the size of the current? Explain.

Challenge Your Thinking

1. Explanations, please!

One by one, a negatively charged plastic ruler is brought near three light, metal-covered spheres suspended by nylon threads. The ruler repels A and attracts B and C. A attracts B, and C attracts B. What are the charges on each of the spheres?

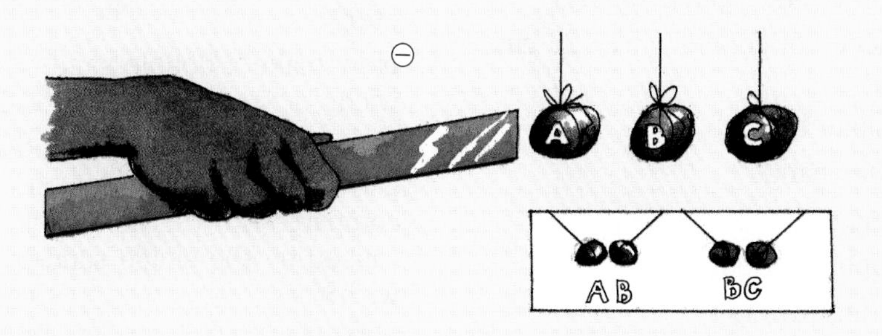

2. What is happening in this sequence? (**Hint:** Study the whole sequence of steps, then reason out your explanation, working back from (c), (d), and (e).)

3 Electricity from Chemicals

It is surprising how many of the things we use operate on cells or batteries. Most chemical cells convert the energy stored in chemicals into a form we can use—electricity. What are the advantages of chemical cells? the disadvantages? Are all chemical cells alike or are some better than others?

In the Explorations that follow, you will construct several chemical cells, which you can test with your home-made galvanometer.

EXPLORATION 3

Chemical Cells

You Will Need

- a home-made galvanometer
- 2 pieces of copper wire (20 cm long)
- masking tape
- rubber bands
- 2 zinc strips
- 2 copper strips
- blotting paper
- ammonium chloride solution
- salt solution
- 2 clothespins

Experiment 1

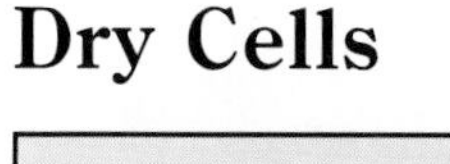

Dry Cells

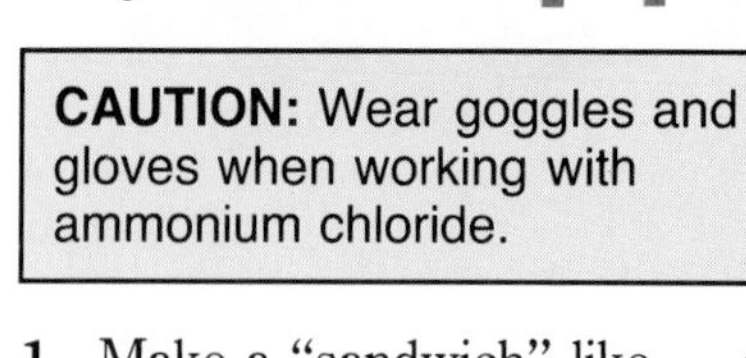

CAUTION: Wear goggles and gloves when working with ammonium chloride.

1. Make a "sandwich" like that shown below.
2. Measure the amount of deflection of your galvanometer.
3. Now make a double sandwich.
4. Measure the deflection of the galvanometer. How does it compare with the deflection caused by the single sandwich?

Because the double sandwich consists of more than a single cell, it is known as a **battery**.

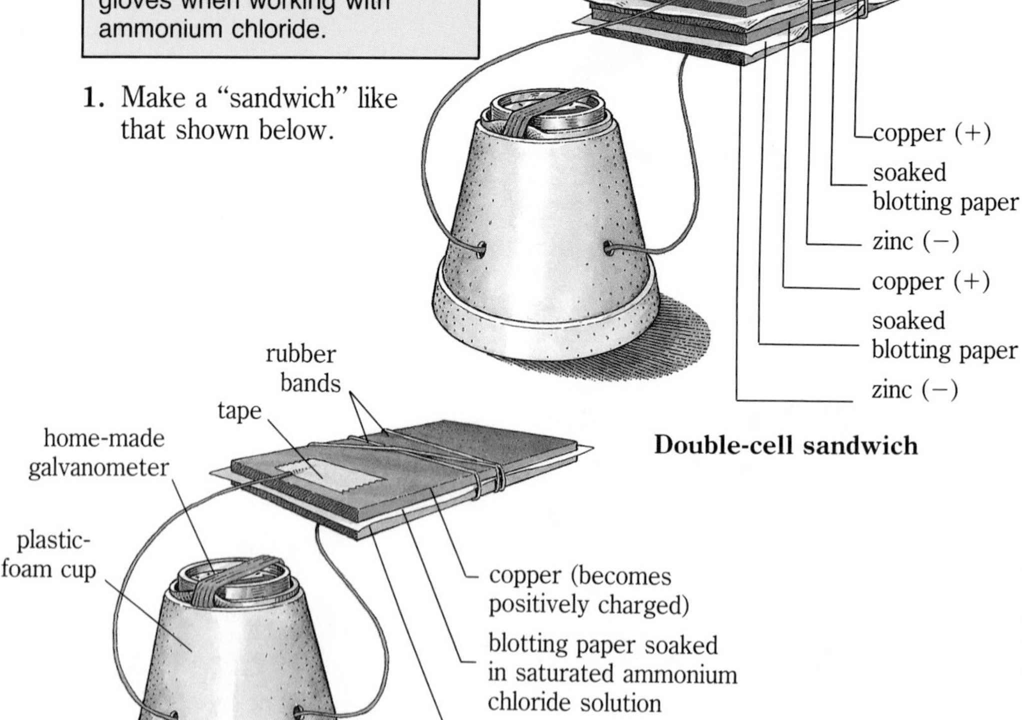

Experiment 2

Wet Cells

1. Make a setup like that shown below.
2. Add enough salt solution to cover one half of the metal strips.
3. Observe the galvanometer. Record the highest reading reached. What happens to the reading? Observe each electrode carefully. What happens to each electrode?
4. Add enough salt solution to fill the beaker.
5. Record the galvanometer reading.
6. Try other metals and solutions.

Questions

1. What accounts for the difference in the galvanometer readings for the single and double sandwiches? What would be the effect of adding more layers to the sandwich?
2. Which electrodes were linked together in converting a single sandwich dry cell to a double sandwich cell?
3. How would you connect two wet cells to get more current? Show with a sketch.
4. What is one way of increasing the current in a wet cell? How would you explain this?
5. What other factors might be altered in a wet cell to increase the current?
6. During the operation of your wet cell, did the galvanometer reading fall off quickly? If so, what might explain this?
7. What energy changes take place in the operation of dry and wet chemical cells?

Chemical Cell Technology

You have made some simple chemical cells. The current produced was very small—enough to be detected by a sensitive galvanometer, but not enough to light a bulb. Many different types of chemical cells have been invented—some tiny and some large and powerful. Chemical cells provide the small amounts of current needed to run calculators, radios, flashlights, heart pacemakers, hearing aids, and portable telephones, as well as larger amounts for mobile objects such as cars and spaceships. You will find out about modern chemical cell construction and usage in the next Exploration.

Commercial Electric Cells

In this Exploration you will examine some common chemical cells. Most you have probably seen. You may have even wondered how they worked. Here's your chance to find out.

Several types of chemical cells and batteries

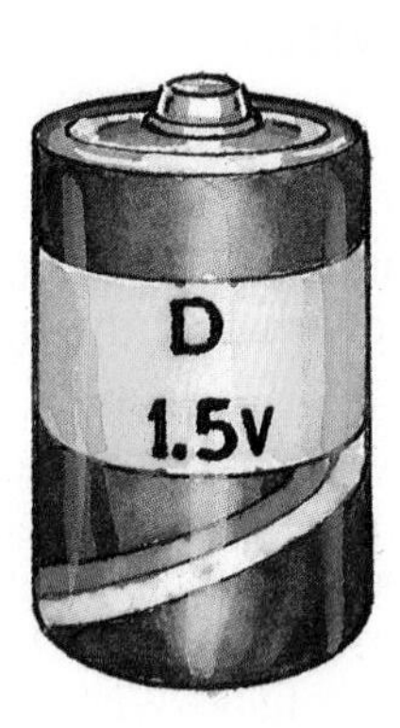

Part 1

More Dry Cells

Dry cells were invented to overcome the disadvantages of wet cells. However, dry cells are not really dry. Rather, the electrolyte is blended with other substances to make it thick and pasty.

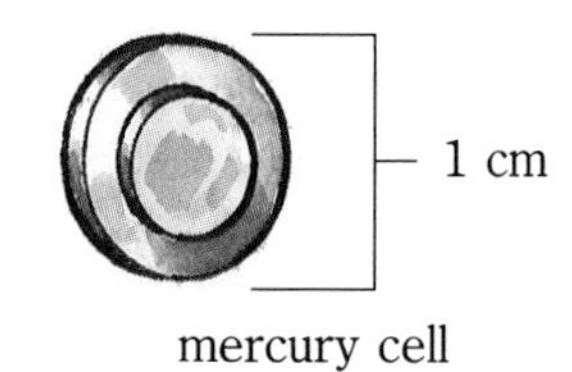

mercury cell

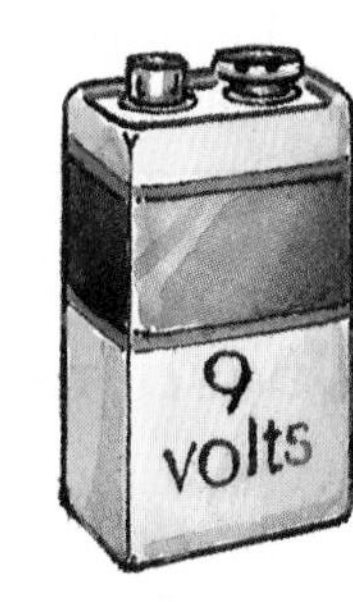

1. Look at the cells pictured at right. Which would you use in a standard-size flashlight? a penlight? Which would you use in a watch?
2. The voltages of each cell are marked. Notice how several cells of different size have the same voltage. How can that be? What does *voltage* signify to you?

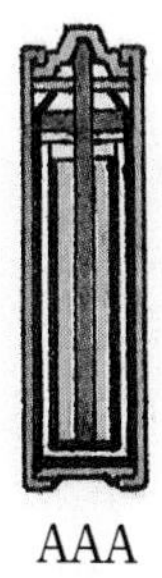

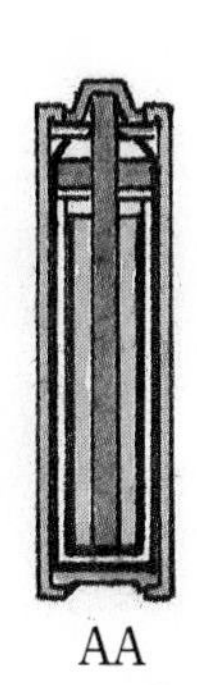

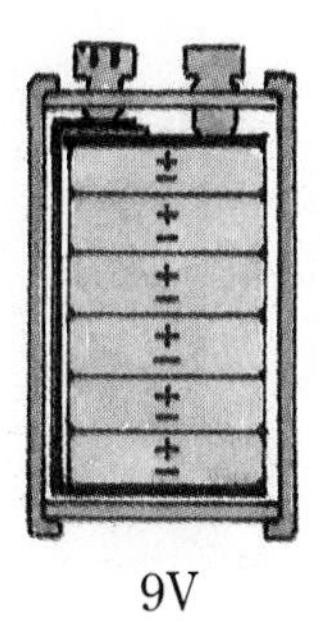

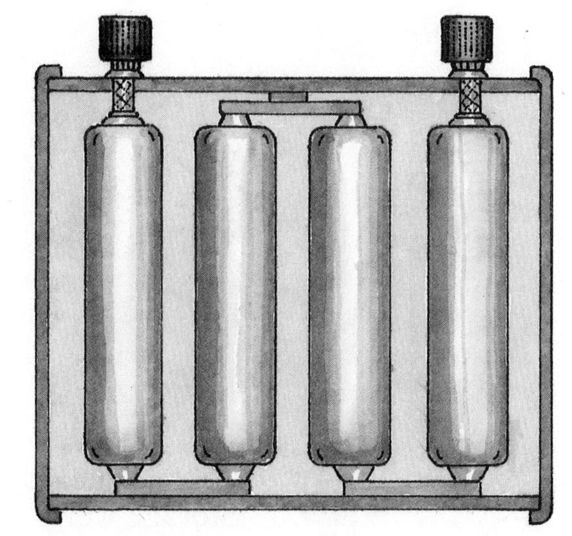

An inside view of cells and batteries. Compare the AAA, AA, and the D-cells with the 9 V and 6 V cells. What differences do you see?

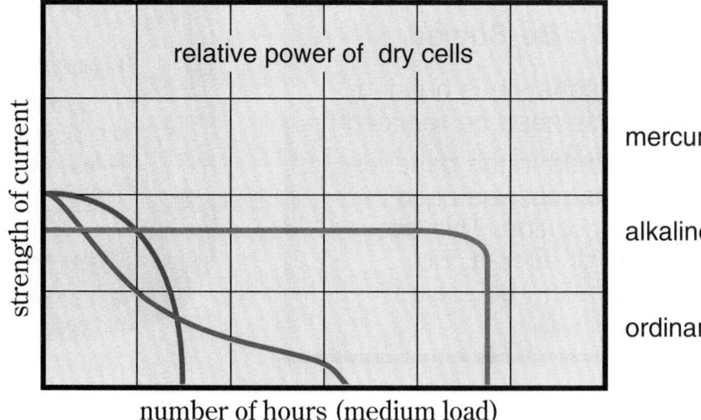

3. Look at the graph comparing electrical outputs of the three types of cells. Which type of cell gradually "winds down"? Which type loses power quickly? Which type would you probably use in devices requiring a steady current?
4. In your Journal, sketch one of the cells or batteries shown on this page. Then use the descriptions below to help you label its parts.

Like all chemical cells, the ordinary dry cell has two electrodes, or conductors, and an electrolyte solution. The *positive electrode* has two parts—a *graphite rod* in the center of the cell and a mixture of *manganese oxide* and *powdered carbon* surrounding the carbon rod. The *negative electrode* is zinc; it makes up the sides and bottom of the cell. The *electrolyte* fills the space between the electrodes. It consists of ammonium chloride paste. At the top of the cell is an *insulator. Batteries* consist of at least two *individual cells* joined together by conducting strips.

In the mercury cell, the *positive electrode* consists of a small block of zinc. The *negative electrode* is a layer of mercury oxide. The *electrolyte* is potassium hydroxide.

5. The *alkaline cell* is very similar to the ordinary dry cell, but differs from it in two major ways. One, the negative electrode is made of a spongier form of zinc. Two, the electrolyte is potassium hydroxide, a strong alkali (base). What effect do these differences have on the power output of the alkaline cell?

Part 2

Other Cells

1. A surge of electric current is needed to start an automobile engine. It is provided by a group of cells joined together in a battery. Study the drawings at right to discover or infer the answers to these questions:
   (a) What substances make up the two electrodes and the electrolyte in a car battery?
   (b) How are these cells connected to one another?
   (c) Why is such a large battery needed for a car?
   (d) How do you know when the battery needs charging?
2. Automobile batteries have a limited life span, and not all batteries last the same amount of time. Why do batteries wear out? Why do some wear out sooner than others?
3. Research "maintenance-free" batteries. How do they work? How are they different from standard batteries?
4. In many ways, the *nickel-cadmium* storage cell is replacing both lead-acid and dry cells. Find out how these cells work.
5. Unlike dry cells, lead-acid batteries can be *recharged.* What does this mean? How is this ability useful?

Electricity from Magnetism

You have discovered that magnetic effects can be caused by a current flowing in a wire or coil. Could the reverse be true? Could a magnet produce electrical effects in a wire? Try the following Exploration to find out. There are three parts—an activity that you can do and two completed experiments that you can analyze.

EXPLORATION 5

Moving Magnets and Wire Coils

Part 1

Building Your Own

You Will Need

- a commercial galvanometer
- a coil of insulated wire
- a strong bar magnet

What to Do

1. Attach a coil of insulated wire to a commercial galvanometer.
2. While watching the galvanometer, move a bar magnet into the coil, hold it there for a moment, and then remove it. Is the galvanometer needle affected?
3. Repeat step 2 several times, moving the magnet at different speeds. What do you observe? Does moving the magnet into the coil have a different effect on the galvanometer than moving the magnet out of the coil? What might this suggest about the direction of current flow through the coil?
4. Disconnect the galvanometer and move the magnet to see whether the magnet itself is affecting the galvanometer.
5. Hook the galvanometer up to the coil again. This time, hold the magnet still, but move the coil over the magnet and back. What do you observe?
6. Use your observations to suggest answers for the following questions.
   (a) How can a magnet help produce electricity?
   (b) How is the direction of the current affected by the motion of the magnet?
   (c) How does the speed of the magnet's motion affect the amount of electricity provided?
   (d) What change in energy forms occur in this investigation?
   (e) Can a stationary magnet ever produce electricity?
   (f) Would current still be generated in the wire if it were broken at some point? Why or why not?

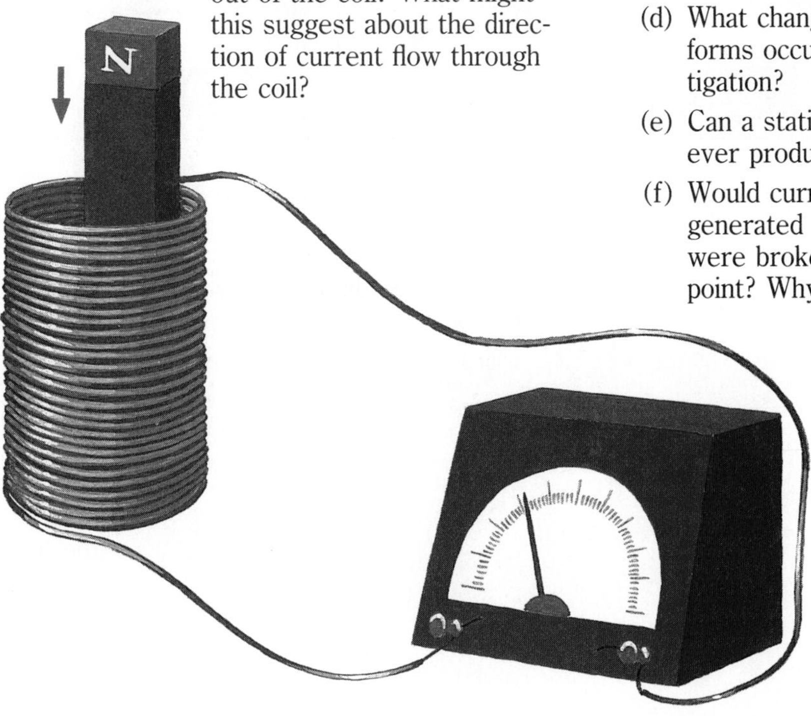

Part 2
Francesca's Experiment

Francesca devised an experiment to answer questions raised by Part 1 of this Exploration. She started with a wire coil of thirteen turns. Illustrations (a) to (c) show the average galvanometer readings she recorded. Then she used a coil with twice as many turns. Illustrations (d) and (e) show her average readings. Francesca tried each part of this experiment three times and obtained similar results each time. What conclusions do you think she drew for each part? Below the illustrations are some hints.

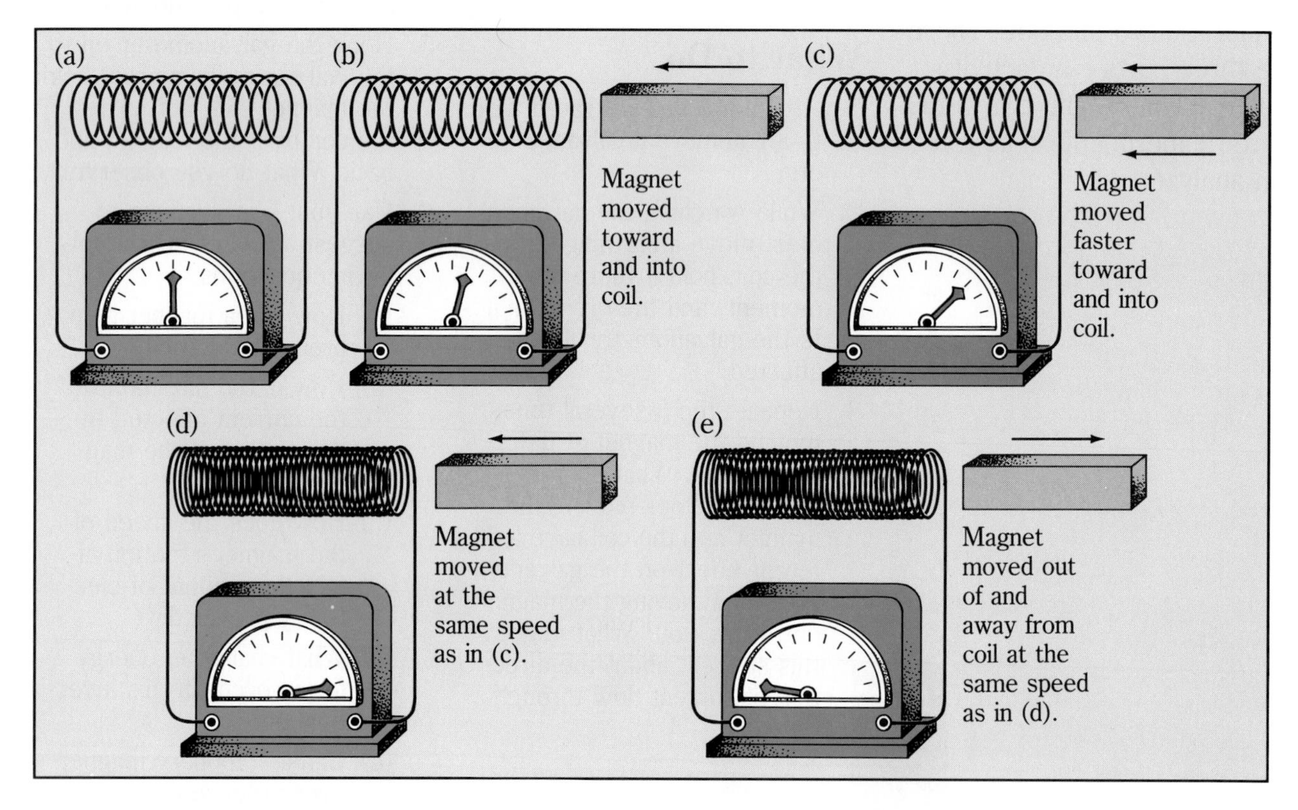

1. When a magnet moves toward a coil of wire, ___?___ is detected in the wire. It ___?___ its direction from the coil.
2. A larger current is produced if ___?___ or if ___?___.
3. Suppose Francesca moved the magnet in and out of the coil 15 times per minute. What would happen to the current? How many times per minute would the current go in one direction, then in the opposite direction?
4. What do you think would happen if Francesca held the magnet stationary and moved the wire instead? Why?

Part 3

A Related Experiment

Francesca made an important discovery: When a magnet is moved through a coil of conducting wire, electricity is generated. Both the number of coils and the speed of movement of the magnet affect the amount of current produced. Francesca also discovered that the direction in which the magnet was moved made a difference. If the magnet was moved in one direction, current flowed one way. If the magnet was moved in the other direction, current flowed the other way. Let's examine the findings of a related experiment.

But before you begin, think a little bit about how a magnet exerts its influence. Does the magnet have to touch something to have an effect, or does its force act through space? Look at the diagram at the upper right. It shows a magnet on which iron filings have been sprinkled. Do you see evidence that (a) the iron filings have been attracted, and (b) that *magnetic force* is exerted through space along curved paths? We call these paths *magnetic lines of force.*

Look at the series of illustrations on the right, representing the results of the experiment. The wire is being moved while the magnet is held stationary. The arrows between the north and south poles of the magnet represent the lines of magnetic force.

1. How do the results compare to those of Francesca's experiment?
2. What happens when the wire is momentarily motionless as it changes direction as in (a)?
3. What happens when the wire is moved parallel to the magnetic lines of force as in (d)?
4. What role do the magnetic lines of force appear to play in the generation of electricity?

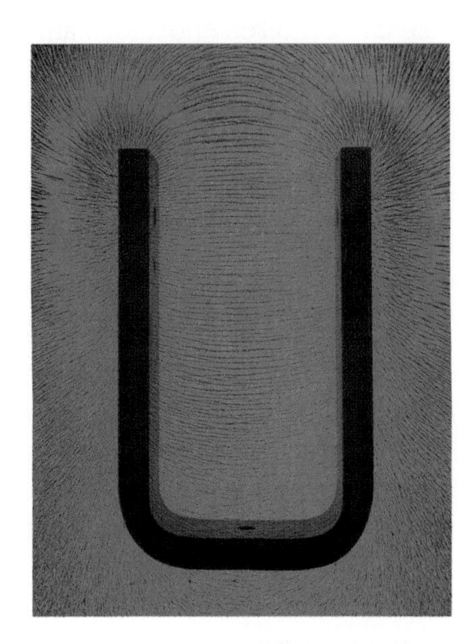

a)

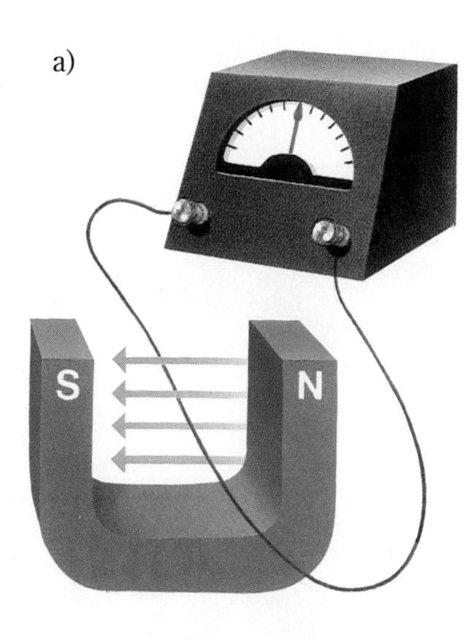

b)

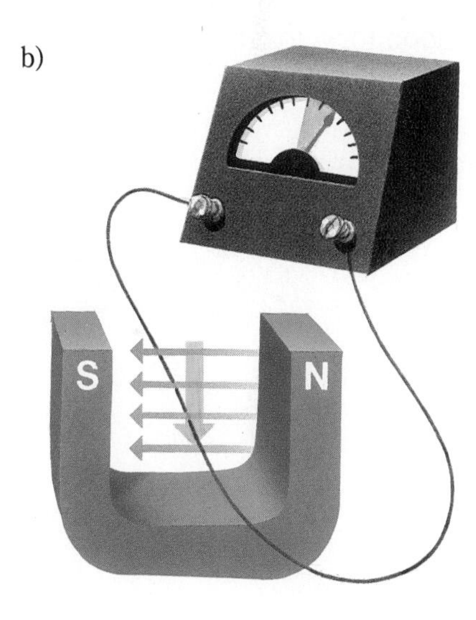

c)

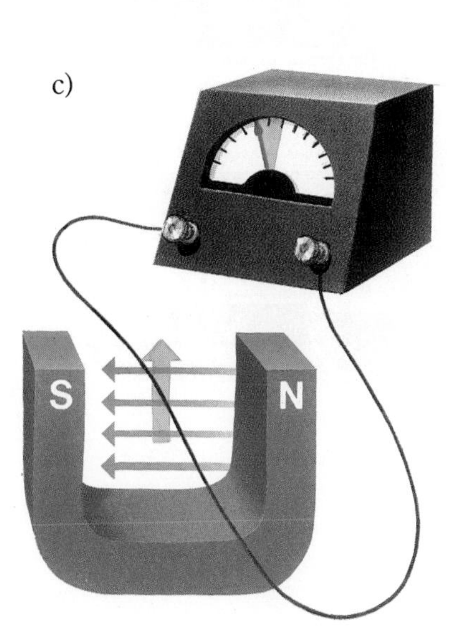

d)

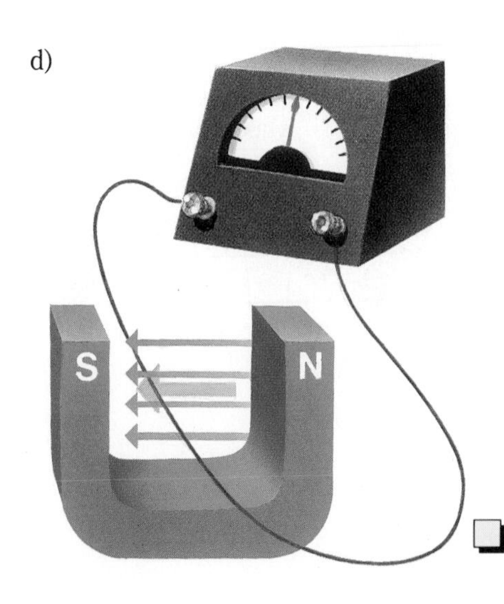

Generators—Small and Large

Our way of life requires large amounts of electricity. A medium-size city requires enormous amounts of electricity to operate normally. Do you think that the electricity-generation systems you have seen so far in this unit could meet such demands? Could they be adapted to do so?

Examine the systems on these pages. Systems such as these meet the needs of our energy-hungry society.

A Bicycle Dynamo

A bicycle dynamo is a practical application of the *electromagnetic principle* you discovered in Exploration 5. Whenever a magnet's lines of force sweep across a wire that is part of a closed circuit, an electric current is generated. It does not matter whether the magnet or the wire moves to cause this to happen; the effect is the same. In Exploration 5 you saw electricity generated by back and forth motion. However, it is easier to generate electricity by rotating either the magnet or the wire coil. In the dynamo shown here, look for a magnet that rotates near a coil of wire (B). The dynamo is shown with its parts separated to help you see how it works.

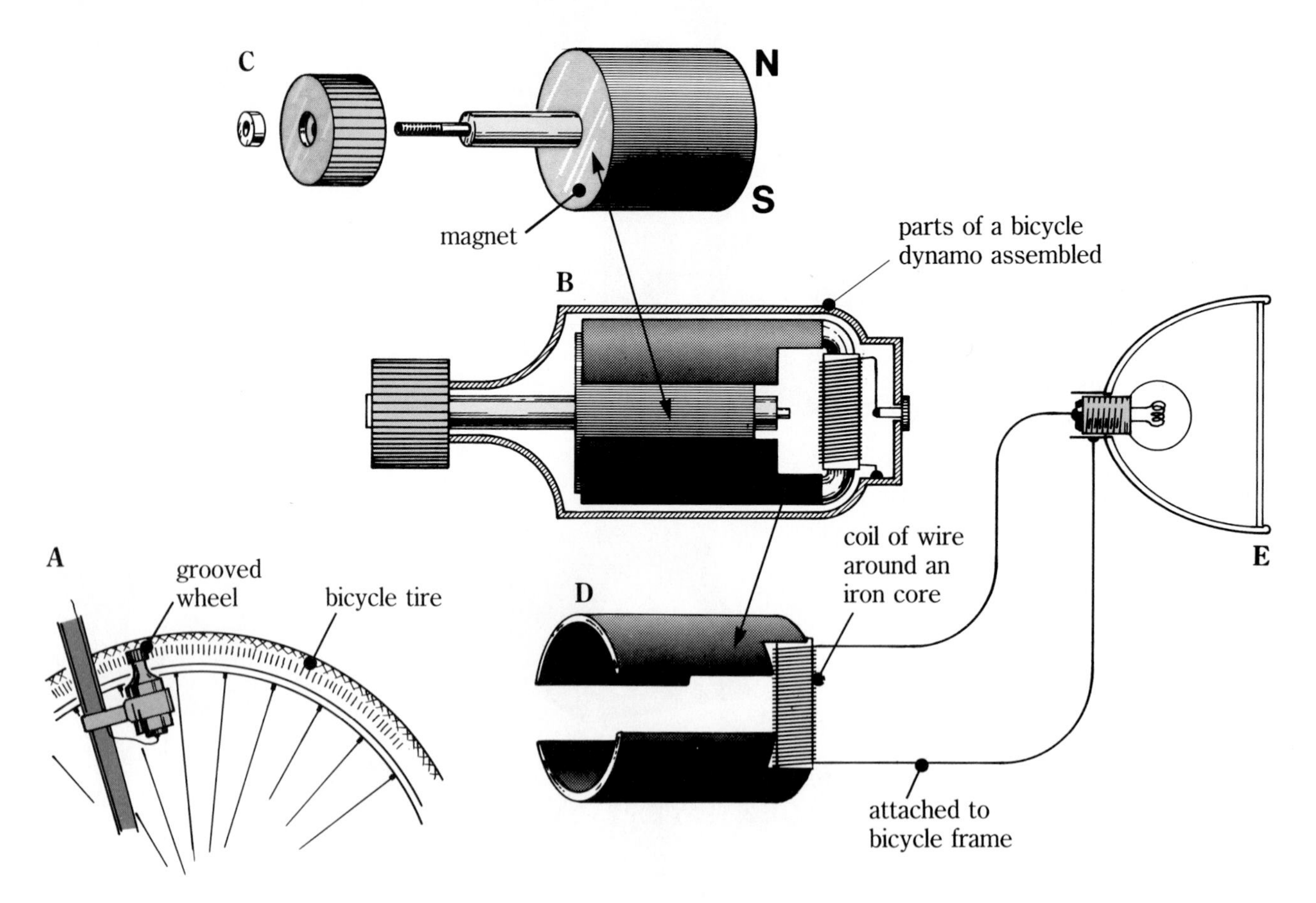

Analyze the dynamo's construction and operation:

1. Locate the magnet in B and C.
2. How does the magnet move?
3. Locate the conducting wire that is wound on an iron core attached to a metal casing. (The casing helps to transmit the effect of the magnet to the coil; it produces the same effect as if the magnet were actually moving in and out of the coil.)
4. Trace the complete electric circuit.
5. The current Francesca got was very small—it could be detected only by a sensitive galvanometer. The current developed in the dynamo is hundreds of times greater. What factors in the dynamo design could account for the larger current? Which of Francesca's experiments support your answers?

Large Generators

Generators like the dynamo use moving magnets to generate electricity in coils of wire. Study the illustration, which shows a large generator. What features of it account for the great amount of electrical energy it can produce? How do Francesca's results support your answer?

What is the energy story suggested by this diagram?

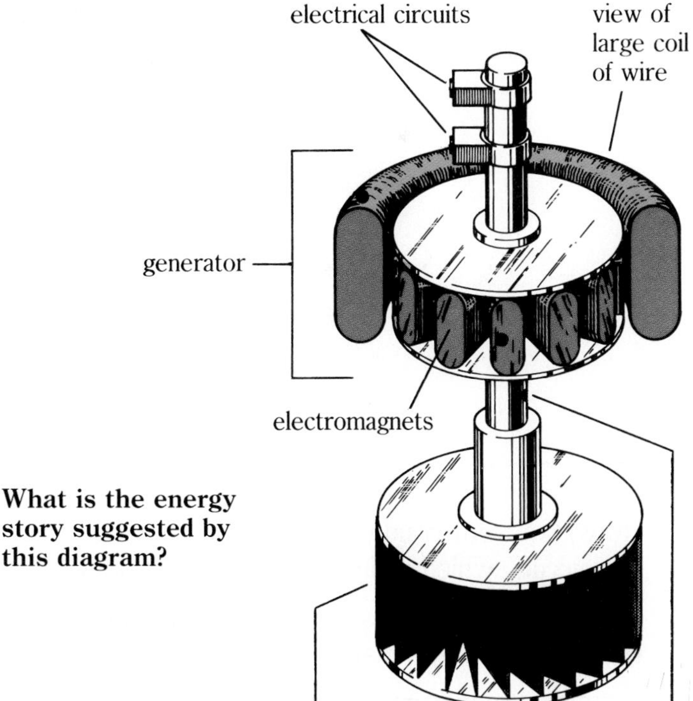

Tracing the Flow of Energy

Water in the reservoir has __(1)__ energy. As it flows down, this energy changes into __(2)__ energy of the moving water. The moving water forces the __(3)__ to turn, giving it __(4)__ to rotate. The attached __(5)__ turn inside a stationary __(6)__ in which __(7)__ is produced. The resulting energy of the generator operation is __(8)__ energy.

Alternating Current

The principles shown by the experiment on the previous page can be used to generate a special kind of electrical current called **alternating current** (A.C. for short). The device below is an example of a simple machine designed to do just that. The handle sets the device in motion. As the coil moves through the magnetic field, electrical current is generated.

Carefully study the diagram below to figure out how the device works. In your Journal, write a description of what is happening in the diagram. Then answer the questions that follow.

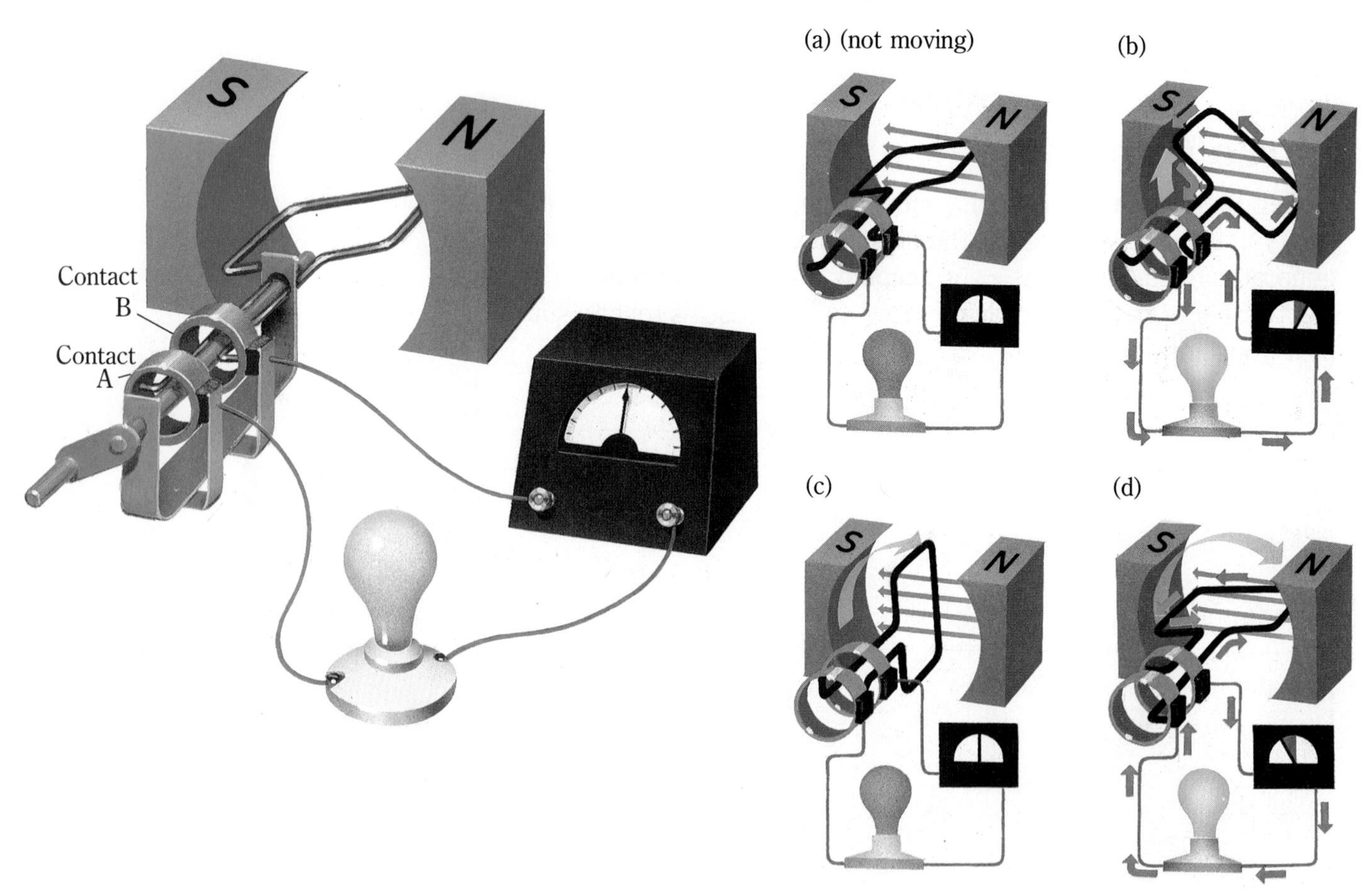

1. How is current conducted from the moving coil to the rest of the circuit, which includes the lamp and galvanometer?
2. As the device rotates, the current changes direction. Why? How do we know this?
3. In what direction is the current flowing in illustration (b)? in illustration (d)? Why the difference?
4. How often does the current reverse direction with each complete turn?
5. Why is the light bulb unaffected by the change in current direction?

6. What's happening in illustration C? Why does this happen? (Remember the second experiment in Exploration 5.)
7. Suppose that you rotated the handle 5 times a second for 1 second. How many times does the current go first in one direction and then in the other? This is a current of 5 cycles per second. What do you think a cycle is?
8. What happens to the current output if you turn the handle faster and faster?
9. How is this device different from the two electrical generation systems shown in Exploration 5? How is it similar to each?
10. Why do they call this type of current *alternating current*?

Normal house current makes 60 complete cycles every second. Have you ever noticed "60 Hz" marked on tools or appliances? Hz stands for *Hertz*, a metric unit meaning "one cycle per second." This mark on a device means that the device is designed to run on 60-cycles-per-second alternating current.

To generate alternating current, either the magnet or the coil can rotate. Both types of generators are used.

Every time alternating current switches direction, for an instant no electrical current flows. If this is so, why don't we notice it? The reason is that it happens so quickly that we can't detect it. There are ways to detect this change indirectly, though. Here is one. Wave a meter stick back and forth in the light from a single fluorescent (tube-type) or neon light. The room should be dark except for the single light. An ordinary incandescent light will not work. As you sweep quickly back and forth you should see several repeated images of the meter stick. Each image represents the time during which the light is on as the current flows in one direction or another. Every time the current drops to zero as it changes direction, the light actually goes off. When this happens you see no image.

Many electrical devices will work with either alternating current or *direct current* (non-alternating current). Why? What kind of systems do you think produce direct current?

Direct Current

Previously, you were introduced to chemical cells. Why does the current produced by a chemical cell go in just one direction? Study the diagram at the right to help you answer this question.

Chemical cells produce **direct current** (D.C.). Direct current is needed instead of A.C. for many circuits, for example, those in a car. Why can't you charge a battery with alternating current?

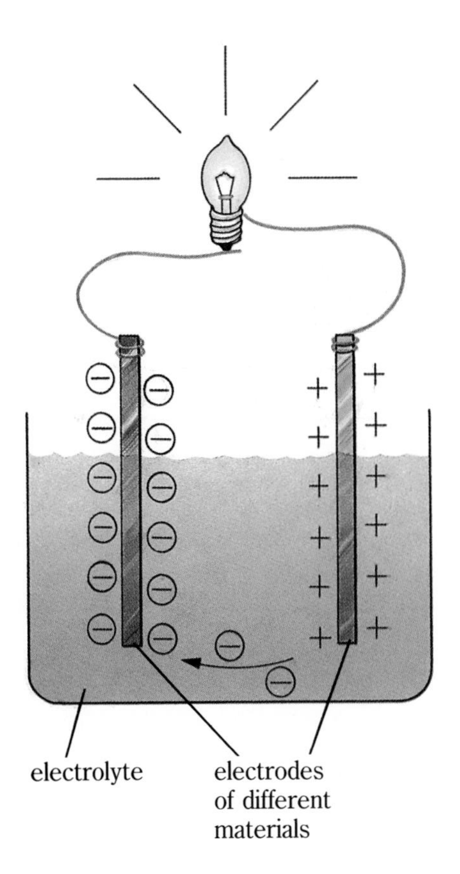

5 Other Sources of Electricity

You have produced electricity on a small scale from (a) kinetic energy alone, (b) chemical energy, and (c) a combination of kinetic energy and magnetism. You have also investigated applications of (b) and (c). In this lesson you will find that electricity can also be produced from light energy, heat energy, and mechanical-pressure energy.

You may not have heard of these last two methods of generating electricity. In fact, these methods can only generate tiny amounts of electrical current. Nevertheless, they have specialized applications, as you will see.

EXPLORATION 6

Research Projects

Project 1

Solar Cells

Solar panels generate electricity from light energy. You may be familiar with solar-generated electricity. Many calculators are powered by light energy alone, for example. Try some experiments with a solar cell using different levels of light at various distances from the solar cells. Check on the amount of electricity being produced in each case.

Generating large amounts of electricity using solar energy alone is difficult for many reasons. For one, solar panels take up a great deal of space. Advances in technology will not shrink them beyond a certain limit because there is only so much energy available in a given amount of sunlight. Solar panels are also expensive. Find out how solar panels work. Do you think that solar energy is the answer to some of our energy needs? If so, which energy needs and why do you think so? Why might the sun be a good source of energy for a space station?

Project 2 

Thermocouples

A *thermocouple* converts heat energy into electricity. It consists of wires of two different metals joined together. When heat is applied to the point where the metals are joined, an electric current flows. Make a simple thermocouple like that shown at right. Apply heat to it and check the current output with a galvanometer.

Thermocouples have only limited use as sources of usable electrical energy. They are most commonly used as high-temperature thermometers or thermostats (devices for maintaining a set temperature). How do you think they work? How do you think industries might make use of these devices?

A home-made thermocouple

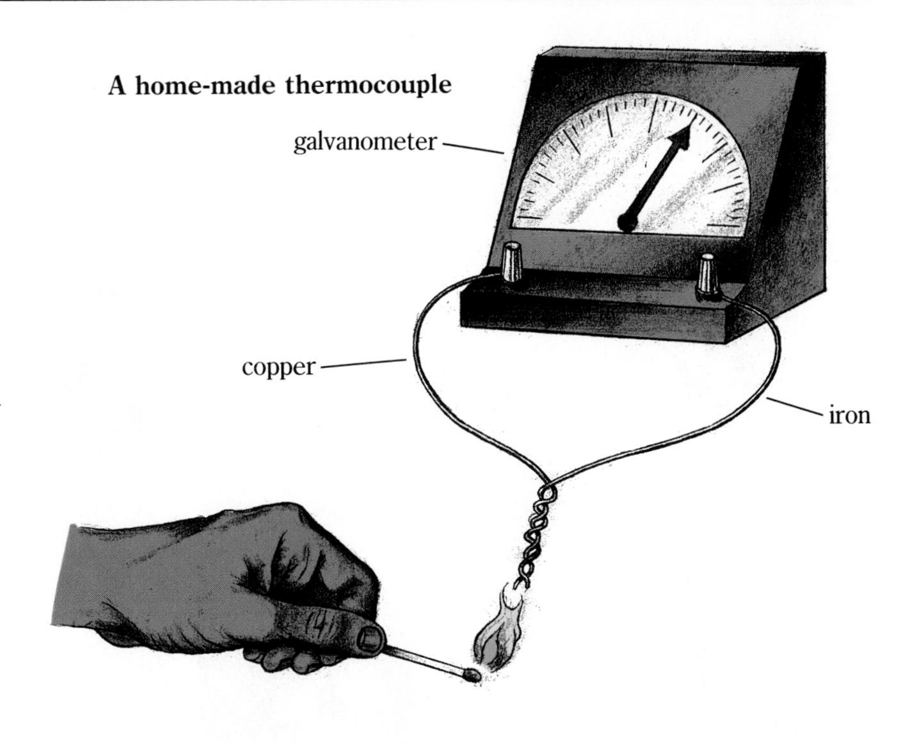

Project 3

Piezoelectricity

Certain types of crystals produce electrical currents when squeezed or stretched. This is called the *piezoelectric effect.* Piezoelectric crystals are useful for turning vibrations into electrical signals. For example, the needle of a record turntable uses a tiny quartz crystal. This crystal converts the vibrations engraved onto a record into an electrical signal suitable for amplification. How does it do this? As the needle rides along in the groove of a record, the tiny bumps in the groove, which represent the recorded sound, cause the crystal to be squeezed and stretched. Thus, the crystal produces an electrical current—a current that mirrors the original recorded sound. Piezoelectric crystals also have the property of vibrating when an electric current is passed through them.

Computers, radios, watches, and many other devices could not operate without piezoelectric crystals. Research how piezoelectric crystals are used in these devices.

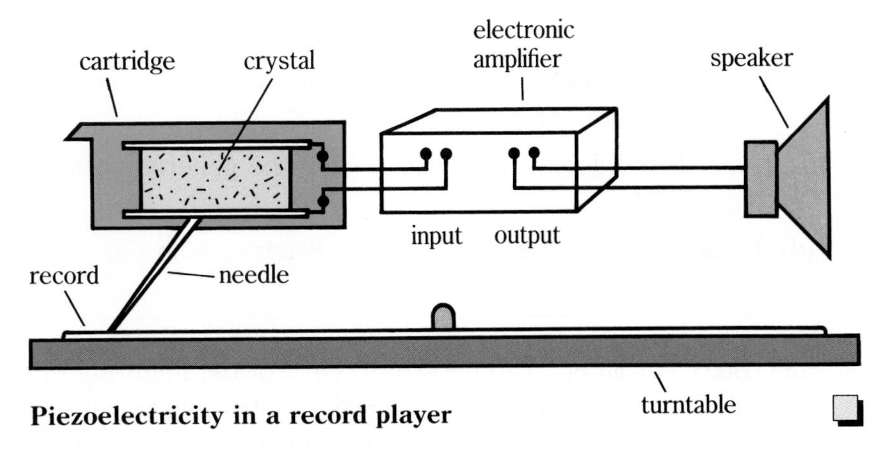

Piezoelectricity in a record player

Challenge Your Thinking

1. The sequence of pictures below shows a type of generator in action. Use the pictures to help you answer the questions that follow.

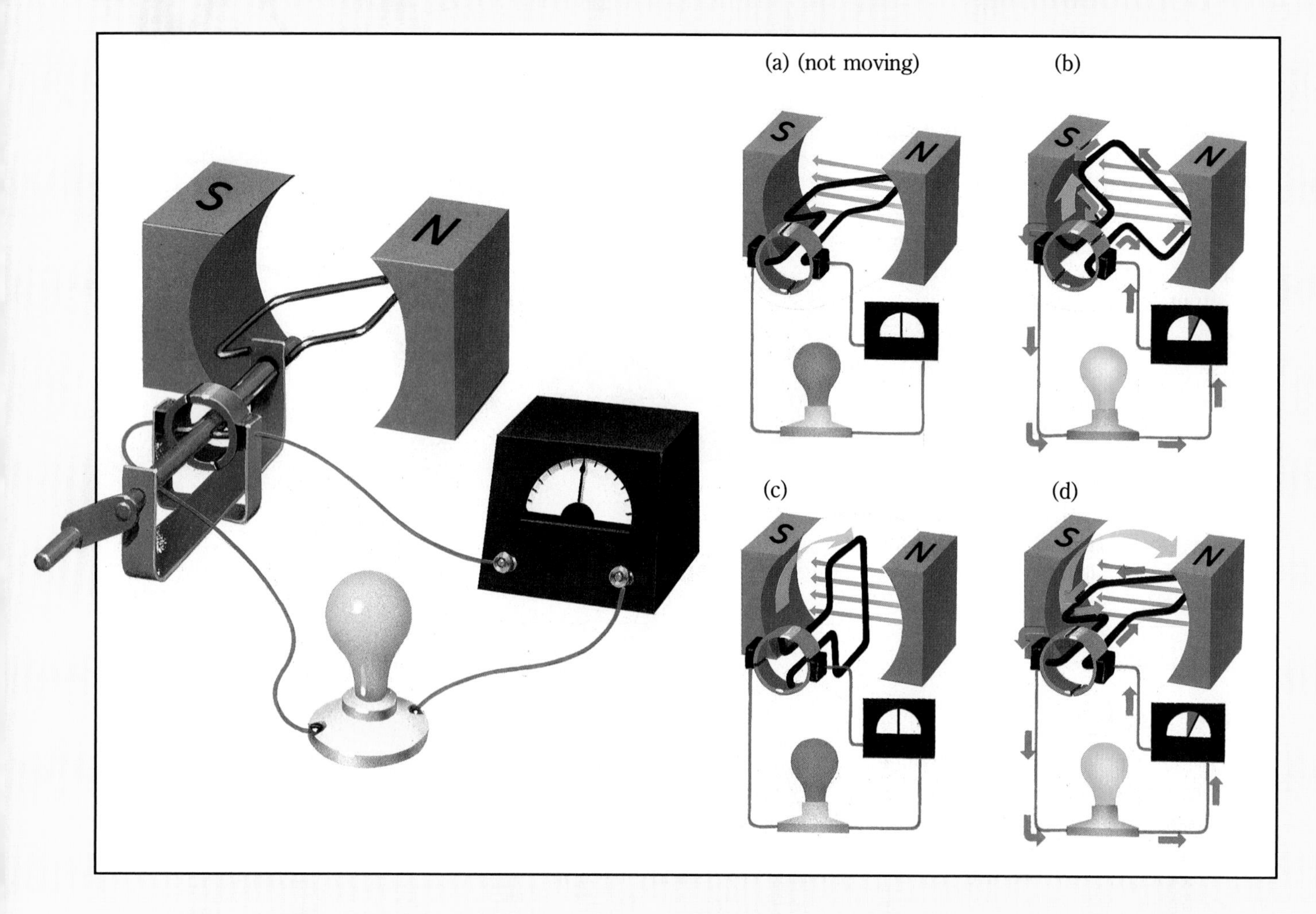

(a) Examine the construction of this generator. How does this generator work?

(b) Study the galvanometer readings. What kind of current does this generator produce?

(c) How does this generator differ from the generator shown on page 226?

(d) Explain what is happening in each illustration in the sequence.

2. Electricity is related in some way to each of the following energy forms: light, heat, magnetic, chemical, kinetic, and pressure. Identify this relationship in each of the following convertors: dry cell, solar cell, wet cell, light bulb, generator, crystals, storage cell, and thermocouple.

6 Circuits: Channeling the Flow

In examining many of the circuits that follow you will see bulbs, cells, and switches. If you have not already designed bulb and cell holders, switches, and connecting clamps, it would be good to do so now. All of these are available commercially but you may wish to build your own.

The illustrations show some circuit components designed and built by students.

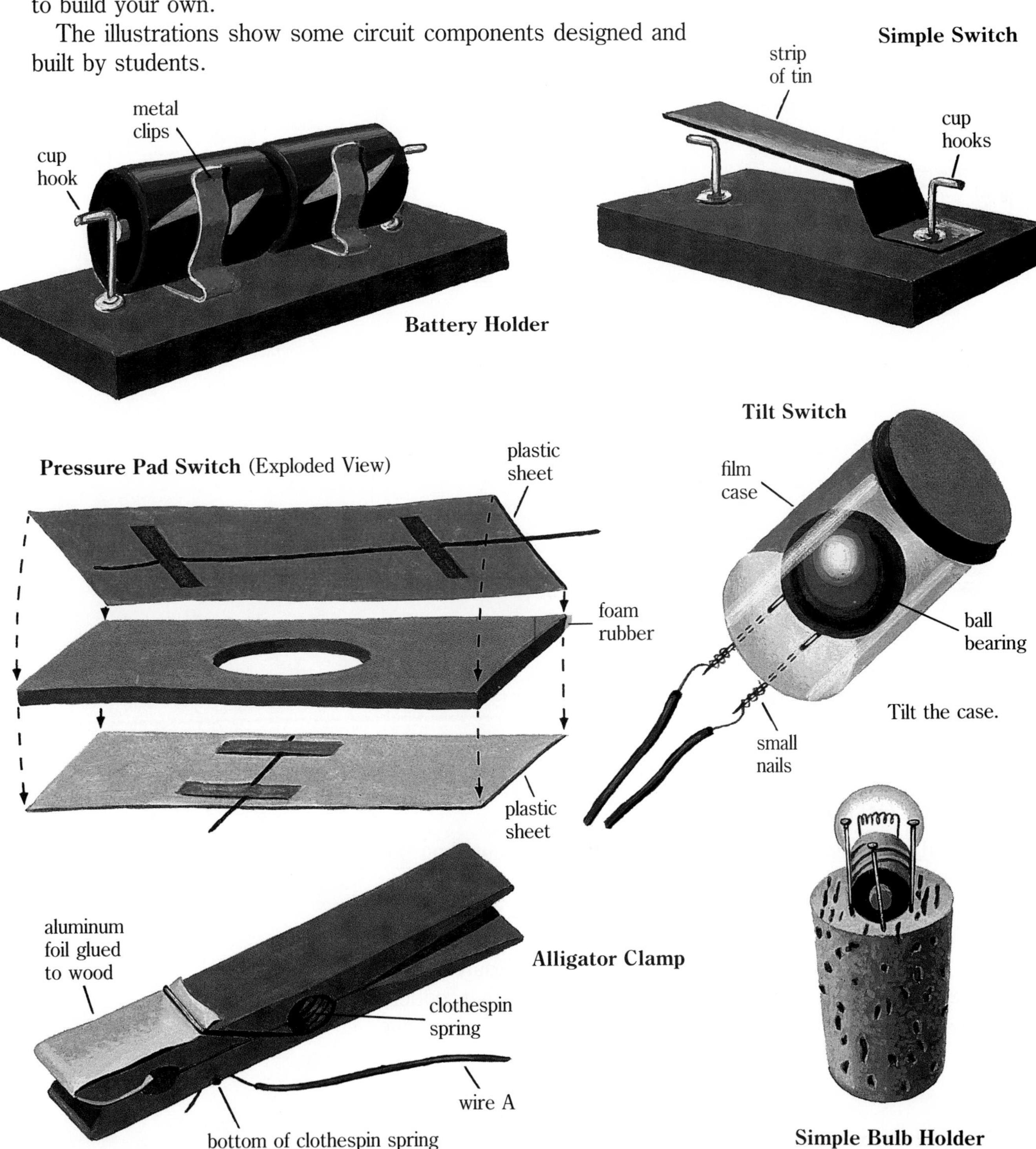

Circuit Symbols

Imagine trying to describe the circuitry of a radio using words alone. It would be almost impossible! Electricians, engineers, and circuit designers use symbols instead of words in their diagrams of circuits. Familiarize yourself with the following symbols.

= cell (The long line represents the positive end.)

= battery of three cells

= conducting wire

= lamp

= coil of wire

= switch

= galvanometer

See how they are used to represent this circuit.

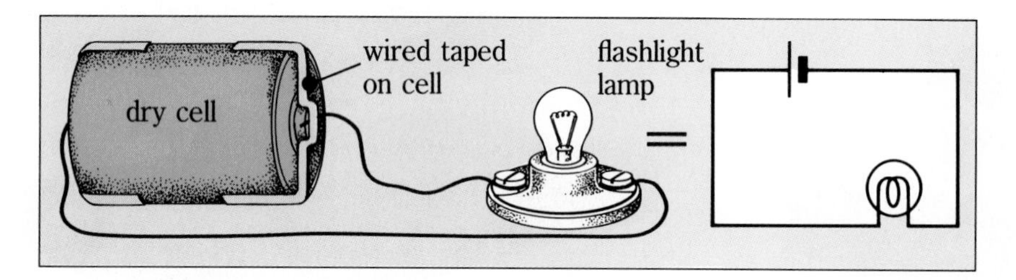

How would you draw a circuit diagram for the circuit below?

Try drawing circuit diagrams for each of the circuits in Explorations 1 and 3.

An Incredible Journey

The world at the scale of the electron is far different than our everyday world. We would find little familiar there. Imagine that you could ride along with an electron as it made its way through each of the circuits in the above illustrations. Choose one or more circuits to write about. Be sure to incorporate any ideas you have about electrons, charges, and circuits. Here's another option: Imagine taking a journey through the circuit shown on page 226.

A Conduction Problem to Investigate

The Problem: Do all wires conduct an electric current equally well? Here are some questions to investigate as you seek an answer to this problem.

A. Does the kind of metal affect the transmission of current?

B. What is the effect of having different thicknesses of wire?

C. Does the length of wire influence the current?

D. What happens if a wire resists the flow of current?

You Will Need

- thin copper wire
- thin and thick nichrome wire
- a wooden dowel
- D-cells
- a flashlight bulb
- newspaper
- a coin
- a rubber band

Part 1

Investigating Questions A and B

1. Prepare equal lengths of the three wires that you will test.
2. Set up the circuit as shown in the illustration, using one of the three wires. Observe the brightness of the light.
3. Do the same for each of the other wires. Observe the bulb in each case. Does the intensity of the light vary? Double-check your results.

Conclusions

(a) Is it easier for a current to flow through thin nichrome wire or thin copper wire?

(b) Is it easier for a current to flow through thin nichrome wire or thick nichrome wire?

(c) How might you explain your observations?

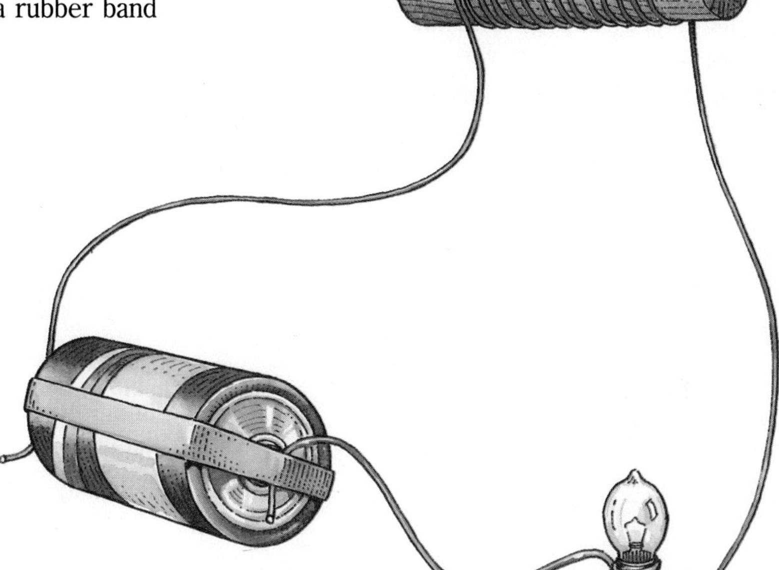

Part 2

Investigating Question C

Vary the length of the thin nichrome wire in the circuit by placing the contact wires at different points along the wire as shown below. Observe the bulb.

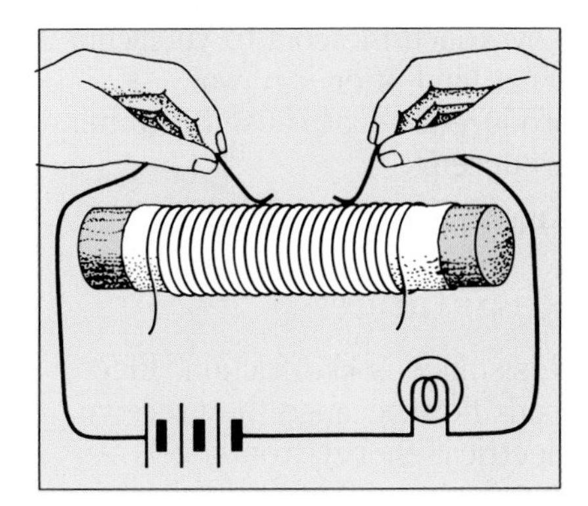

Conclusions

(a) Is it easier for a current to pass through a long or a short piece of nichrome wire?

(b) How might you explain this observation?

Do your answers for (a) and (b) match the information below?

Interpreting Parts 1 and 2

Some wires do not allow electric charges to move through them as readily as others. In other words, these wires offer more *resistance* to the flow of the charges. Which offers more resistance: nichrome or copper wire? thin or thick nichrome wire? long or short nichrome wire?

Resistance can be compared to friction. In what ways do you think they are similar?

Friction

Flick a coin with your finger. It moves a little, slows down, and stops. What causes this slowing down? As the coin slows down and stops, where does the kinetic energy of the moving coin go? Here's how you can find out. Place your finger firmly on the coin and rub it back and forth a dozen times or so on a table top. Now touch the coin to your chin. What kind of energy was produced? What caused it to be produced?

Kinetic Energy → Heat Energy

Resistance

Resistance is like friction. Electrons flow because they receive electrical energy from a cell. As the electrons flow through a piece of nichrome wire, the wire resists the flow of the electrons—in much the same way that the nails resist the rolling marbles in the illustration.

If the nails were a little closer together, what effect would they have on the rolling ball? Would this situation represent a wire with more, or less, resistance?

What form of energy do you predict will be produced from the electrical (kinetic) energy of the electrons as they are slowed down? You will check your prediction in Part 3.

Part 3

Investigating Question D

Connect the circuit as shown. Note that no bulb is included. Wrap the wire with newspaper, then watch it for one minute. Now disconnect the circuit and unwrap the newspaper. Touch the wire carefully.

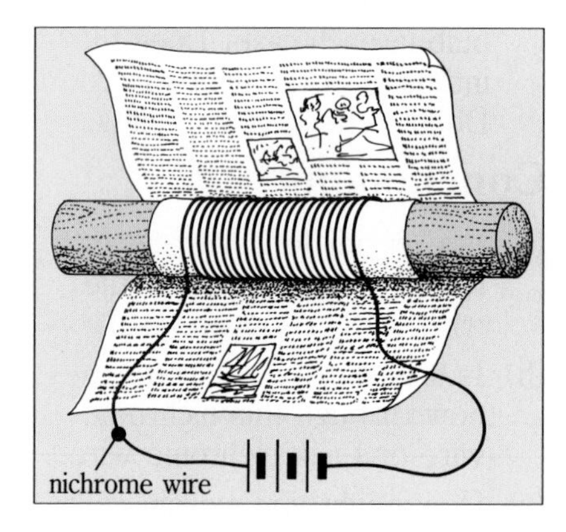

Conclusions

(a) When a wire resists the flow of current, what happens?

(b) What energy change is taking place in the nichrome wire?

Any conductor that offers considerable resistance is called a **resistor** and is represented by the symbol -WW- . Using circuit symbols, draw a circuit containing two D-cells, a coil of nichrome wire, a switch, and a bulb.

Applications of Resistance

You have found that resistors produce heat, and you have inferred that they reduce the flow of current. Each of these characteristics is useful.

- Resistors are used to produce heat in appliances like toasters or irons. Appliances that produce heat when an electric current flows through them are called *thermoelectric* devices.

Why is this a good name for them? Make an on-site survey of the thermoelectric devices in use in your home. How much power do they use? (This will be the number, in *watts,* marked on the appliance.)

- Resistors also reduce current flow. A *variable resistor,* such as the one used in Part 2 of Exploration 7, varies current. Such resistors are in common use—the volume control on a radio is one example.

Here is a project for you to try. Convert the arrangement used in Part 2 of this Exploration into a usable rheostat (device for varying the current). What might you use it for?

7 Making Circuits Work For You

How have you made use of electric circuits in the last twenty-four hours? How have they worked for you? In the following Exploration, you will build and test a number of simple yet functional circuits.

Dry cells quickly drain if left in a closed circuit without something to provide resistance (for example, a bulb). Placing a switch in the circuit and keeping it open until you check the circuit's operation will help to conserve the dry cells.

EXPLORATION 8

Constructing Circuits

You Will Need

- 6 copper wires (each 10 cm long)
- masking tape
- 3 D-cells
- 3 flashlight bulbs in holders
- 2 contact switches

Part 1

Exploring

You can devise any circuits you wish. Use some or all of the equipment listed above to construct your circuits. Make any arrangements desired. If you want to use more than one dry cell and do not have a holder, you can use masking tape to hold them together. If the bulbs light up, you have complete circuits. After constructing your circuits, draw them in your Journal using circuit symbols. Suppose that a certain number of electrons flow out of the cell(s) in a given time. Describe the path(s) taken by these electrons.

Part 2

Solving Circuit Problems

The following are a series of circuit problems. What arrangement of circuit components would you make to accomplish the functions described in each problem?

First make a circuit diagram of your proposed solution. Then construct:

1. A circuit that will light two bulbs, A and B, at the same time when the switch is closed. If either bulb burns out or is unscrewed, the other bulb goes out too. This type of circuit is called a **series circuit.**
2. A circuit that will light two bulbs, A and B, at the same time when the switch is closed. If either bulb burns out or is unscrewed, the other bulb stays lit. This type of circuit is called a **parallel circuit.**
3. A circuit that contains three bulbs, A, B, and C. If A is unscrewed, then B and C go out. If B is unscrewed, A and C stay lit. If C is unscrewed, A and B stay lit.
4. A circuit with two bulbs and two switches, P and Q. When P and Q are closed, both bulbs come on. If either switch is opened, no bulbs light.
5. A circuit that contains two switches and two bulbs. If both switches are open, no light shines. If either one of the switches is closed, both bulbs light.
6. Analyze your findings.
   (a) Is there any parallel circuitry in the room where you are now? How could you find out?
   Caution: *Don't expose any wiring.*
   (b) Identify the series and parallel circuits in each of your designs.

Part 3

Current Questions to Investigate

Remember: The brightness of the light bulb is a measure of the amount of current flowing.

You Will Need

- 6 light bulbs
- 3 contact switches
- 3 D-cells
- copper wire

What to Do

Construct each of the circuits shown in the table, and record the results in your Journal in a similar table. Make certain that switches are included in the circuits you construct.

Question	Experimental Design	Result of Experiment/Conclusion (Select from list on the next page)
1. How is the amount of current affected by the number of cells in a circuit?	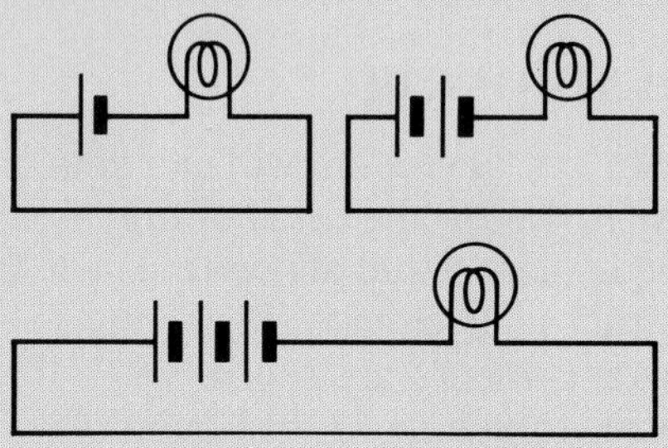	A
2. How is the amount of current affected by the number of bulbs connected *in series,* that is, one right after the other?	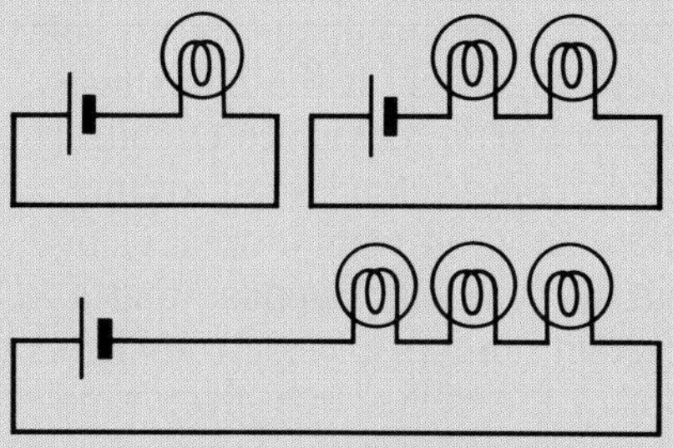	B
3. How does the current through each bulb in a *parallel*, or branched, circuit compare with the current through each bulb in a series circuit?	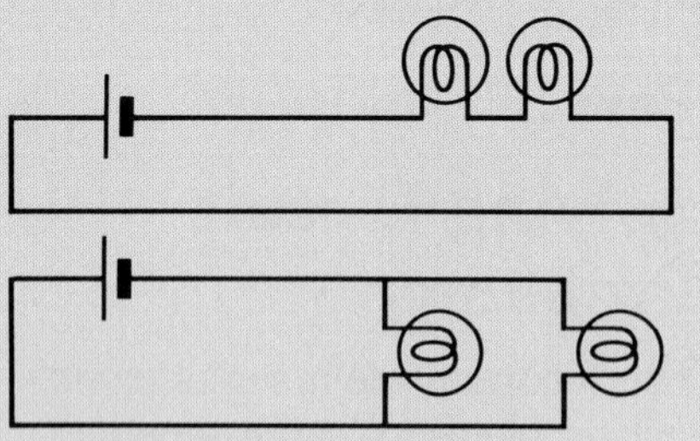	C
4. What difference is there between the current through the battery whe:. two bulbs are connected in series, and the current through the battery when two bulbs are connected in parallel?	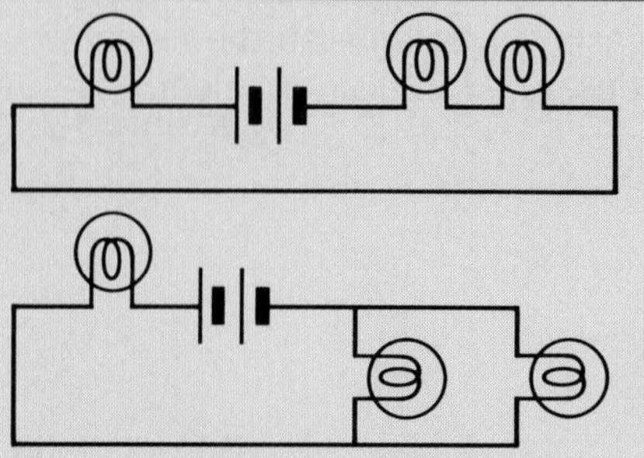	D

How Bright Are You?—A Quiz

How well can you apply what you discovered in Exploration 8? In the circuits shown on the next page, choose the correct brightness (S for standard brightness; L for less than standard; M for more than standard) for each of the 20 numbered bulbs. Assume that all bulbs are identical.

First note the three illustrated degress of brightness L, S, and M. The standard brightness, S, to which L and M are compared, is the brightness of a single bulb connected to a single dry cell.

In your Journal make a table similar to the one below. Record your choices in it. Be prepared to defend your choices. The first situation is done for you.

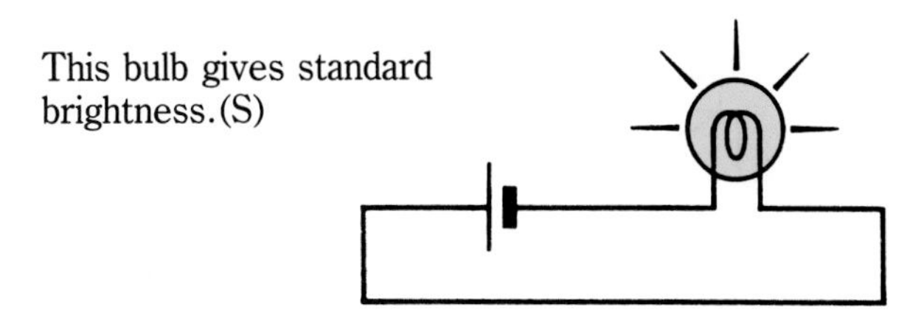

less than standard (L)

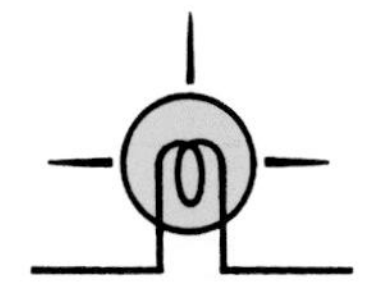

more than standard (M)

Situation	L, S, or M
1	S
2	
3	
etc.	

Scorecard

Over 15	You dazzle me!
10–15	You are bright!
Under 10	You need enlightenment!

Below are a number of statements with options. Each statement, with the correct option, is a valid conclusion for one of the four experiments you just did. Choose appropriate conclusions for each experiment.

(a) Connecting bulbs one after another in a circuit (decreases, increases) the amount of current flowing in the circuit.

(b) If the number of cells are increased in the circuit, the amount of current is (decreased, increased).

(c) If more bulbs are connected in series in a circuit, the resistance of a circuit is (decreased, increased).

(d) If two bulbs are placed in a branched circuit rather than in an unbranched circuit, the current through the battery is (decreased, increased).

(e) The current through each bulb in a parallel circuit is (greater than, less than) the current through each bulb in a series circuit.

(f) The resistance of a circuit is decreased when the bulbs are placed in (series, parallel).

(g) The resistance of a circuit is (more, less) with three bulbs in series than with two bulbs in parallel connected to a third in series.

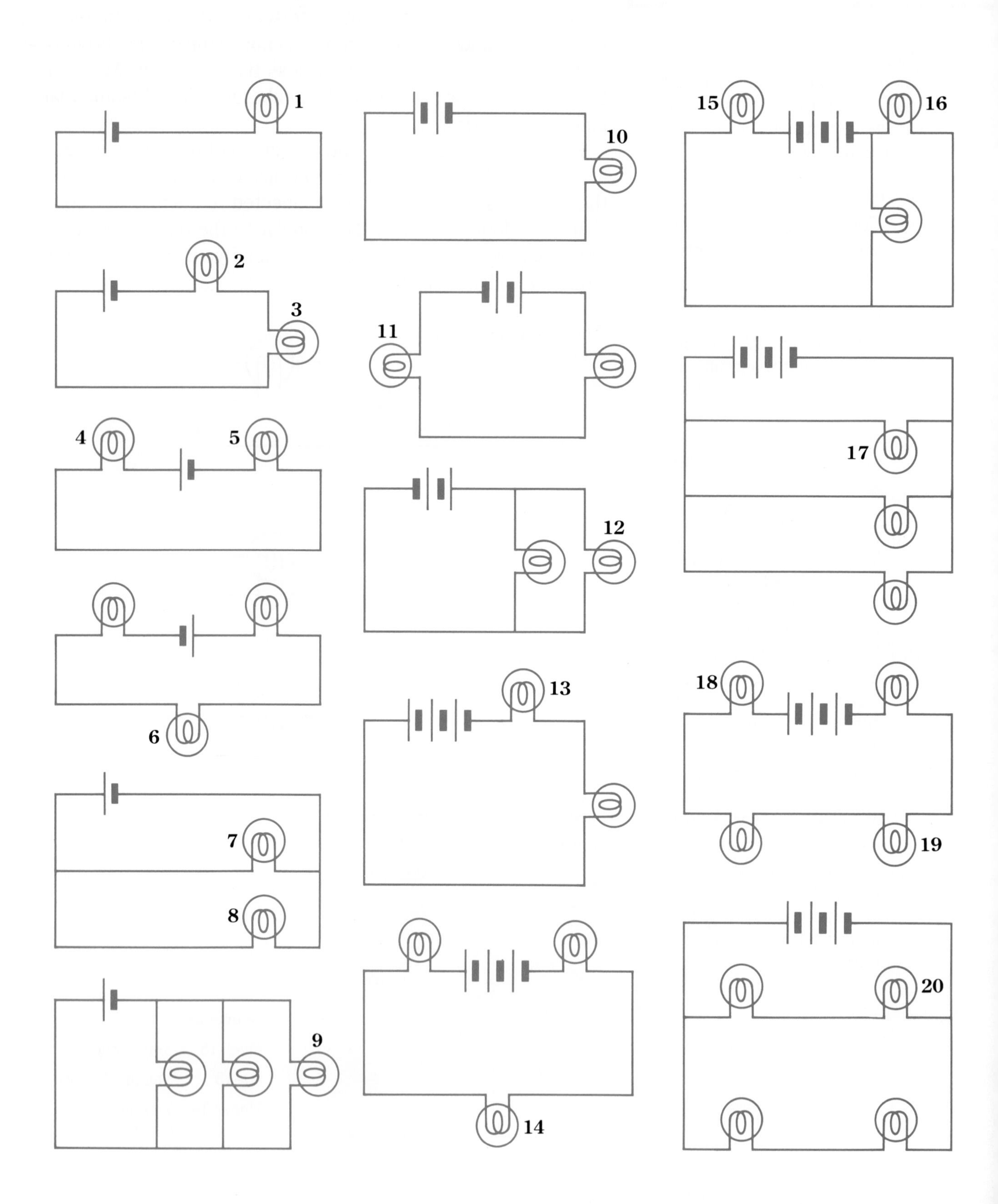
1
2
3
4
5
6
7
8
9
10
11
12
13
14
15
16
17
18
19
20

8 Controlling the Current

The radio blared suddenly to life, jolting Nick from a sound sleep. He sat up, looked at the time on the clock radio that had just come on, and jumped out of bed. In the bathroom he turned on the fan-timer switch for ten minutes. Now at breakfast, he pushed down the bread in the toaster. Shortly, golden brown toast popped up. Knowing he would be getting home late that night, Nick set the oven to come on at 7:00 P.M. to cook the casserole for one hour. Already late, Nick jumped into his car and backed out of the driveway to the sound of the beeper telling him to buckle up. When Nick returned home that night, his apartment was already lit up. A light-sensitive switch triggered the lights to come on at sunset. Best of all, there was a delectable smell in the air. He dimmed the kitchen lights and sat down to a great meal.

Automation! People depend on various types of circuits, switches, and controls to make life easier and more convenient. Although you may not know how they work, you are still able to make use of them.

1. How many automatic devices are mentioned in this story?
2. Which of these devices have switches that just turn electricity on or off?
3. What are some of the kinds of controls that affect what is being done by the appliances and electrical devices?
4. Which of these switches or controls operate by mechanical means? by some other means?

Simple Current Controls

Here are two circuits that control the operation of a light bulb. Which contains two on/off controls? What does the other control do? How?

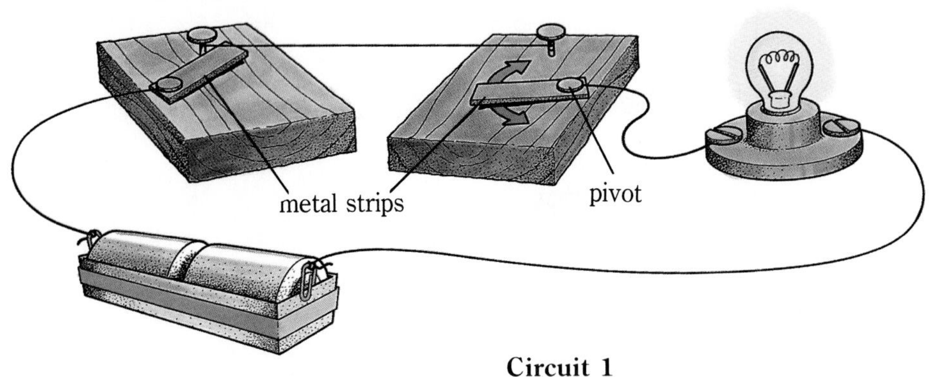

Circuit 1

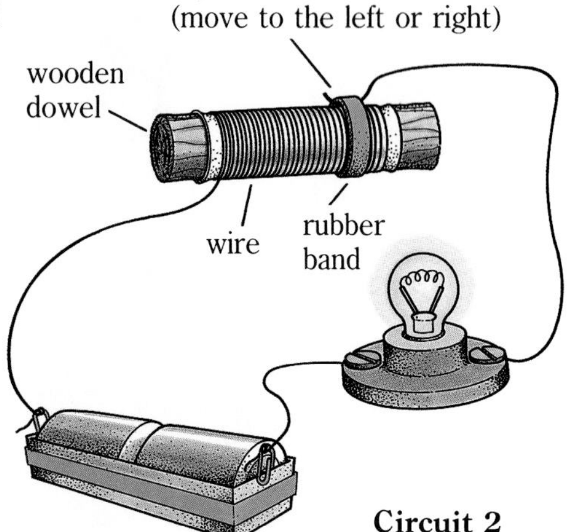

Circuit 2

Draw a circuit diagram for Circuits 1 and 2.

Anne converted the switch in Circuit 1 into another kind of switch by adding more nails to the wooden blocks as shown in the diagram.

The metal strips on each wooden block can pivot easily between the two nails A and B, and between C and D. She then connected wires between the two blocks so that the following things happened:

1. When the strip in the first block moved from A to B, the light came on.
2. Then she moved the strip on the second block from D to C and the light went out.
3. Next she moved the first strip from B to A; the light came on.
4. Finally she moved the second strip from C to D; the light went out.

Where did Anne connect the two wires between the two blocks? Where have you seen switches that operate in a similar way to those Anne constructed?

Circuit 2 causes a variation in the amount of electrical energy supplied to the light bulb. Circuit 2 has an application in the common dimmer switch, which you sometimes find in place of the normal flip switch controlling the lights in a room.

Can you imagine what the inside of a dimmer switch looks like? As you turn the knob counter-clockwise, the lights in the circuit become dimmer. Sketch what you think might be inside the dimmer switch.

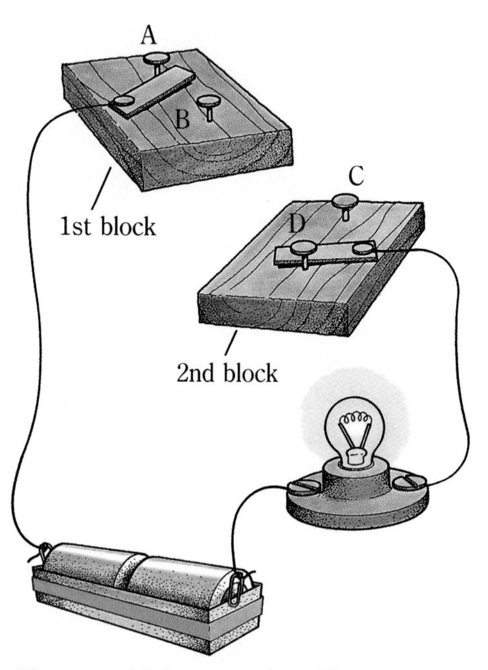

How could Anne make this circuit work?

> **A Design Challenge**
> Use the lead in a pencil to construct a dimmer switch for a flashlight bulb powered by a dry cell.

A Switch in Time!

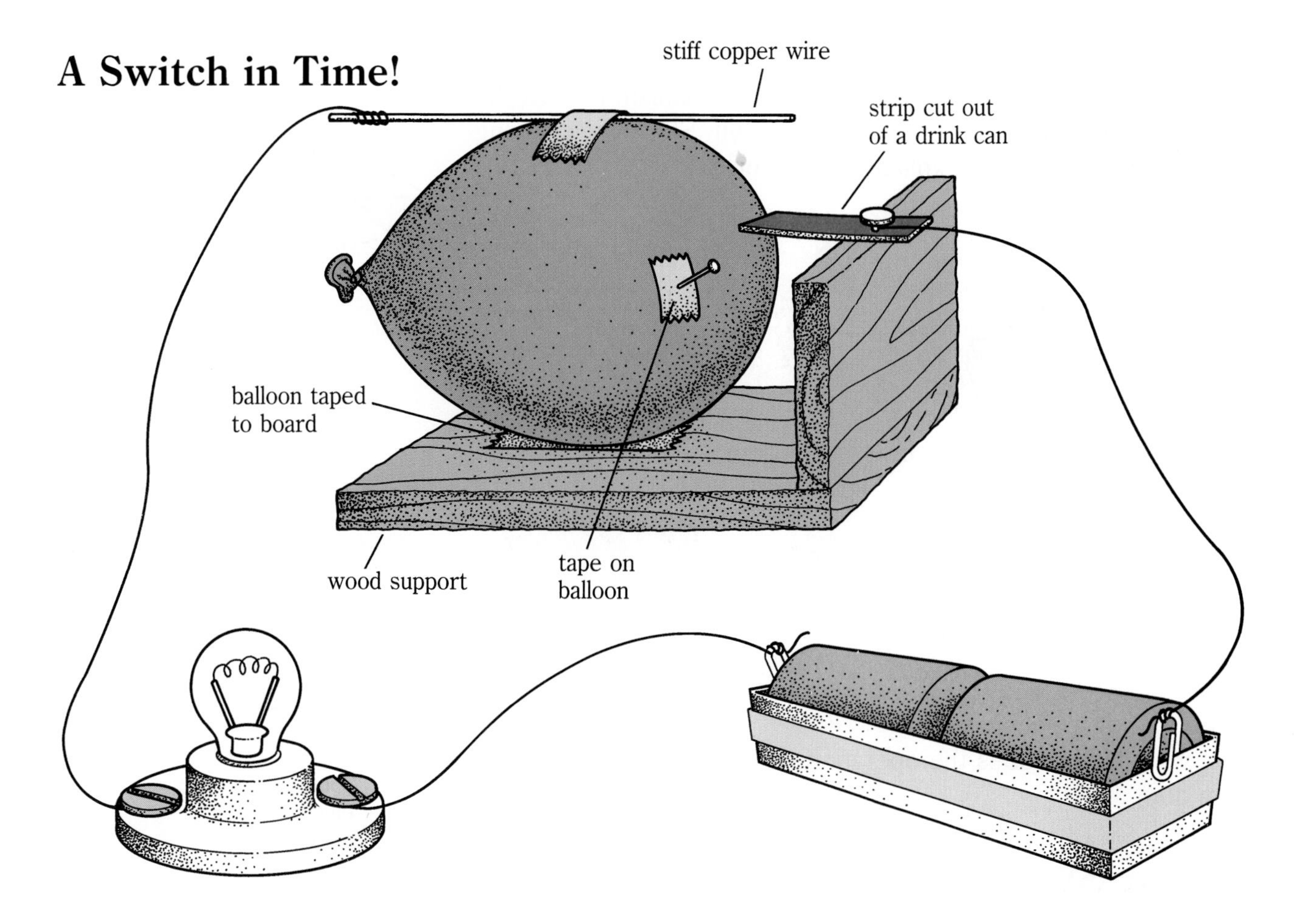

Here is a switch for you to construct that illustrates the principle of time-delayed operation. To operate it, puncture the tape on the end of the balloon with a small, sharp pin. (Don't worry—the balloon won't pop.) This allows the air to escape slowly from the balloon. Record the exact time that you puncture the balloon.

1. How does this switch operate?
2. How long was the delay from the puncturing of the balloon to the lighting of the bulb?
3. What is one disadvantage of this switch?
4. Can you think of some switches that have a time delay?
5. Timer switches can turn a circuit on immediately, or after a time delay, and then turn the circuit off after another delay. Identify some possible uses for timer switches.

For example, here is a common type of timer switch. It could be used to operate a fan for a given period of time. For how many minutes is the timer switch set? Which direction is the switch moving? How might the circuit be broken when zero is reached?

What are some common applications of "timed" switches? You will find some in Nick's story. How many of these seem to operate on a mechanical clock mechanism?

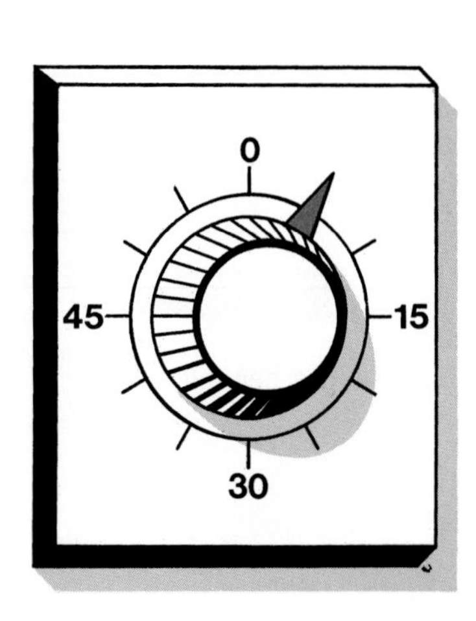

You remember that Nick woke up when the clock radio clicked on. Some radio alarm clocks have a motor mechanism to turn the electrical circuit containing the radio on and off. Study the drawing of the mechanism. Then explain how it works.

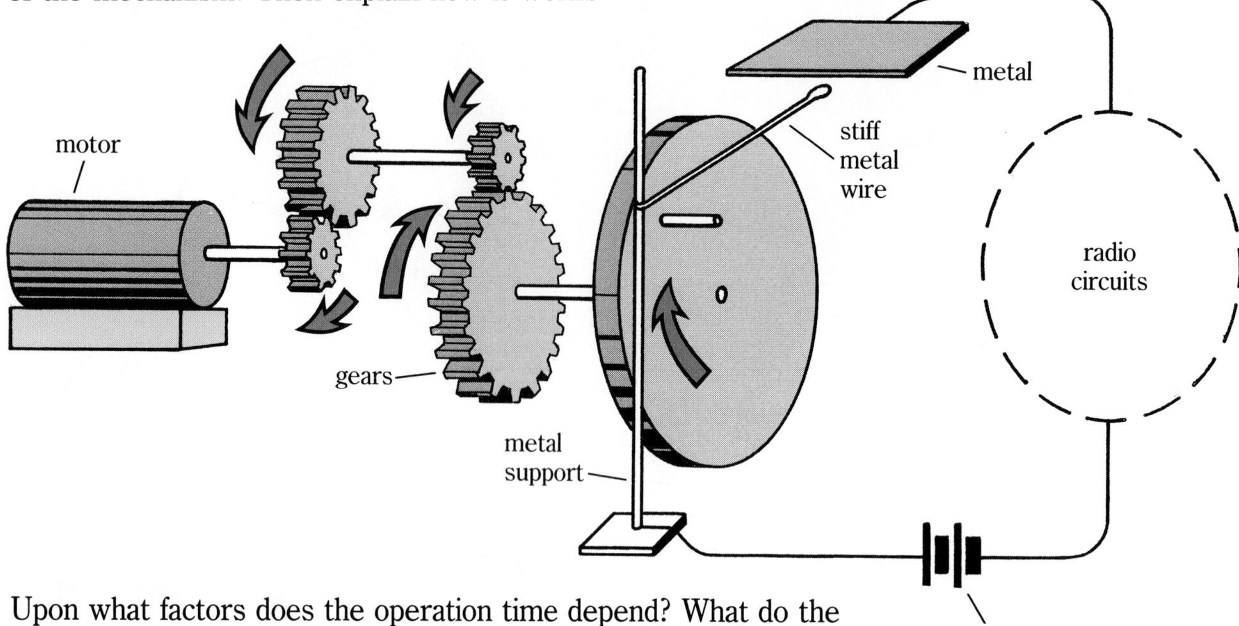

Upon what factors does the operation time depend? What do the gears do?

A Heat Switch

In the pictured circuit the bimetallic strip is the switch. (The term *bimetallic* means that there are two metals that are bonded together.) Note what happens when a bimetallic strip is heated. Why does this occur? How does this characteristic break, or close, a circuit? What type of energy is operating this switch?

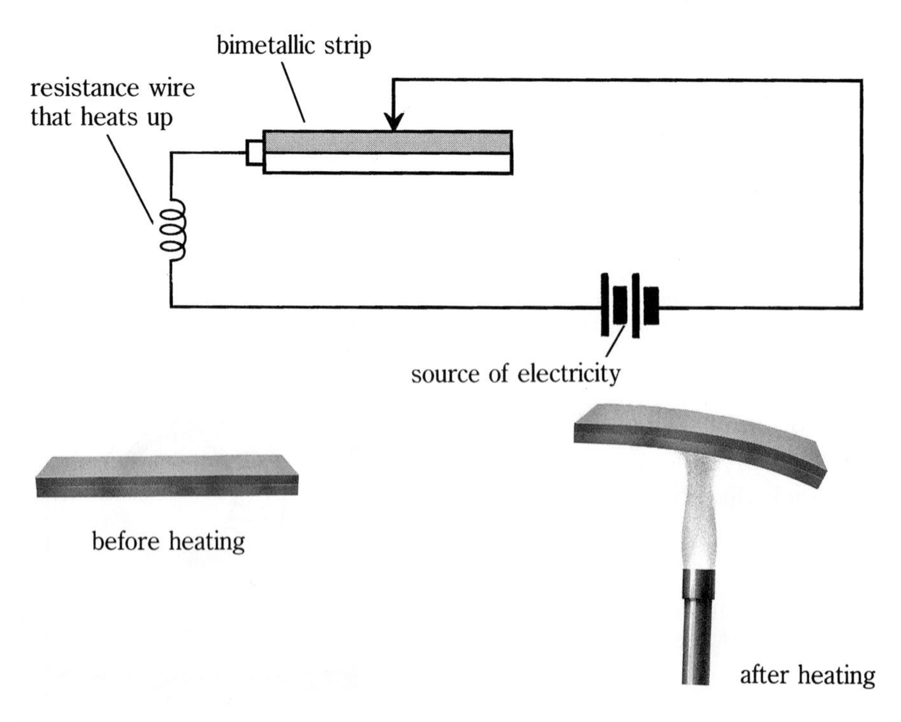

Comfort-Maintaining Circuitry

Sometimes, as in thermostats, bimetallic strips are found in the form of spring-like coils. Changes in temperature cause them to coil or uncoil. Study the illustrations below. How does the bimetallic strip break the circuit? How would the circuit be closed again? Where might this method of control be used?

One application of the bimetallic strip is in thermostats or temperature controls. Here is a sketch of a thermostat that controls the operation of a furnace. Study it closely. How does this switch work? Can you visualize the whole electric circuit and the conducting path through the thermostat?

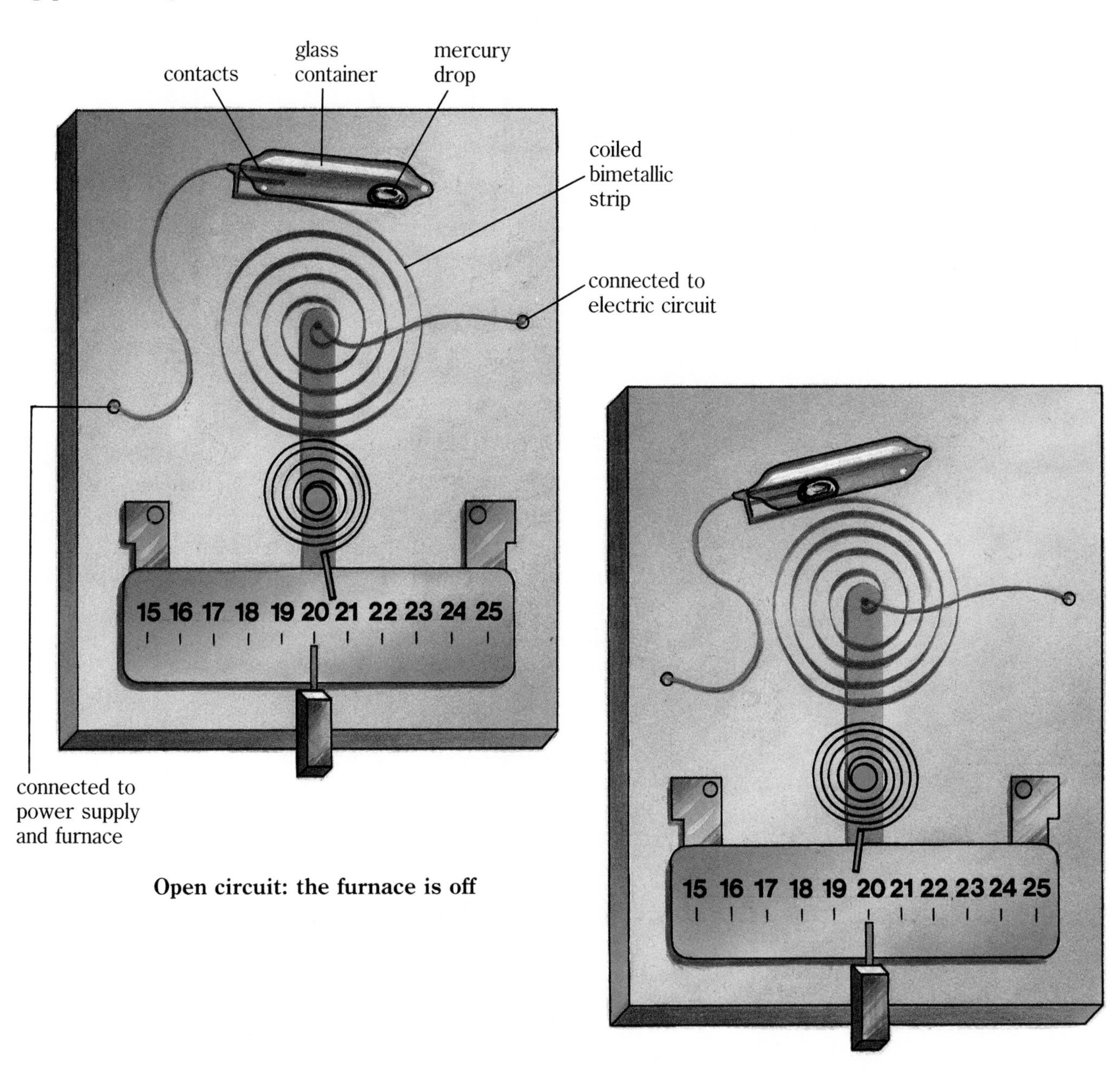

Open circuit: the furnace is off

Closed circuit: the furnace is on

A Magnetic Switch

Observe the two illustrations of the magnetic switch. How does it operate? What must be a characteristic of the metals making contact in the switch if the switch is to open when the magnet is moved away? Could just any metal be used, for example, copper? (Try bringing a magnet close to copper wire.) What kind of energy operates this switch? Why would this magnetic switch be appropriate for poor environmental situations such as water, mud, or snow?

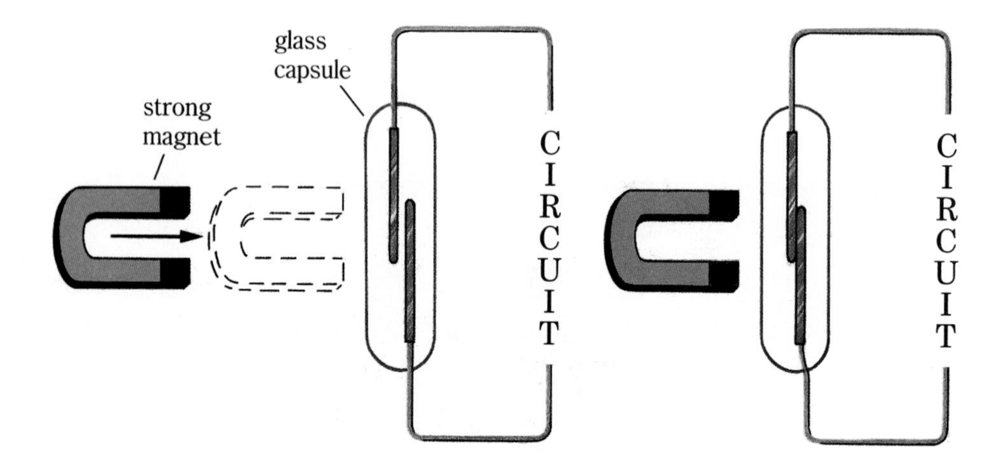

A Lever Switch

The switch shown below is an example of a common lever switch. Note in the illustration that if the lever is pushed upwards lightly, Circuit 1 is opened and Circuit 2 is closed. Try to picture the arrangement inside the switch. Draw it. Invent a situation where a lever switch might be of value.

If you visit an electrical shop, you will find many other kinds of switches, components, and circuit controls.

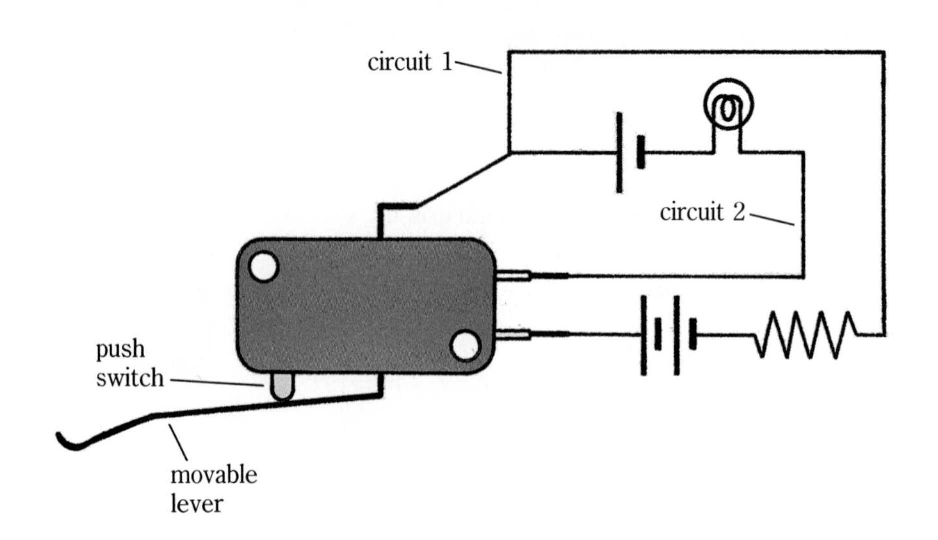

Electromagnets

Electromagnets are magnets created by flowing electric currents. You made a kind of electromagnet when you constructed your galvanometer. The current-bearing coil of wire became magnetized and deflected the magnetic compass needle. Look at the circuit in the diagram below. This arrangement of a spike and a coil of wire is a simple electromagnet.

A typical electromagnet consists of a core of soft iron or soft steel, surrounded by a coil of insulated wire. (Why must the wire be insulated?) When the current flows through the wire, the core quickly becomes a temporary magnet. When the current is interrupted, though, the core loses its magnetism. Note that there is a switch in the circuit to allow the circuit to be disconnected easily.

EXPLORATION 9

Constructing an Electromagnet

You Will Need

- paper clip
- washers
- some light insulated wire
- 2 D-cells
- an iron spike
- a switch

What to Do

Get together with one or two classmates. Your task will be to make a functioning electromagnet, and then to discover how its strength can be increased.

Making the Electromagnet

Use the materials illustrated to make an electromagnet capable of supporting a paper clip from which several washers are hanging. Identify the parts of the circuit and trace the path of the current. Are the spike, paper clip, and washers part of the circuit? How many washers can your first design hold?

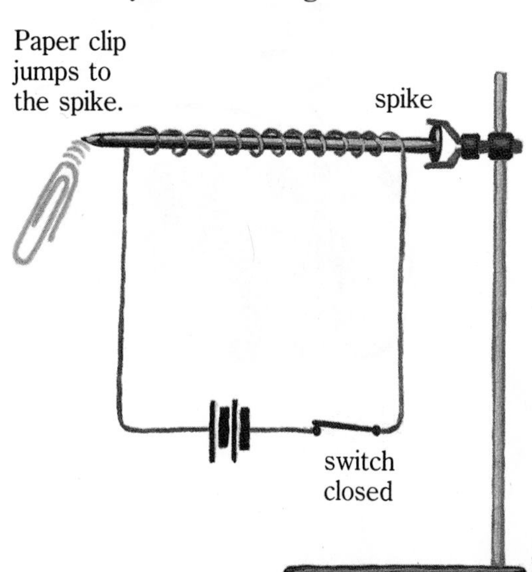

Increasing the Electromagnet's Strength

How can you make your electromagnet hold more washers? Take some time to discuss the following:

- the apparatus
- factors or variables that might be altered
- ways to measure the magnet's strength
- safety precautions
- recording your results

Get your design approved by your teacher. Then assemble the necessary apparatus, do the experiment, record the results, and finally, draw conclusions based on your results.

How does this electromagnet work?

Figuring out Electromagnetic Circuits

Have you ever used a diagram to put something together or figure out how something works? A diagram can show how something operates, or how parts should be assembled. The following are diagrams of circuits containing electromagnets. In groups of three or four, determine how either of the circuits works. A few hints are provided in some of the drawings.

A good plan is to begin with the source of the electricity, then follow the path of the complete circuit back to its source. For example, in the doorbell circuit, when the button is pushed down, the circuit is completed. Electrons flow along the parts of the circuit: battery, wire, contact screw, springy metal strip attached to the armature, wire, coil, and back to the battery. Now trace the path and think about what is happening in each part of the circuit, especially the electromagnet.

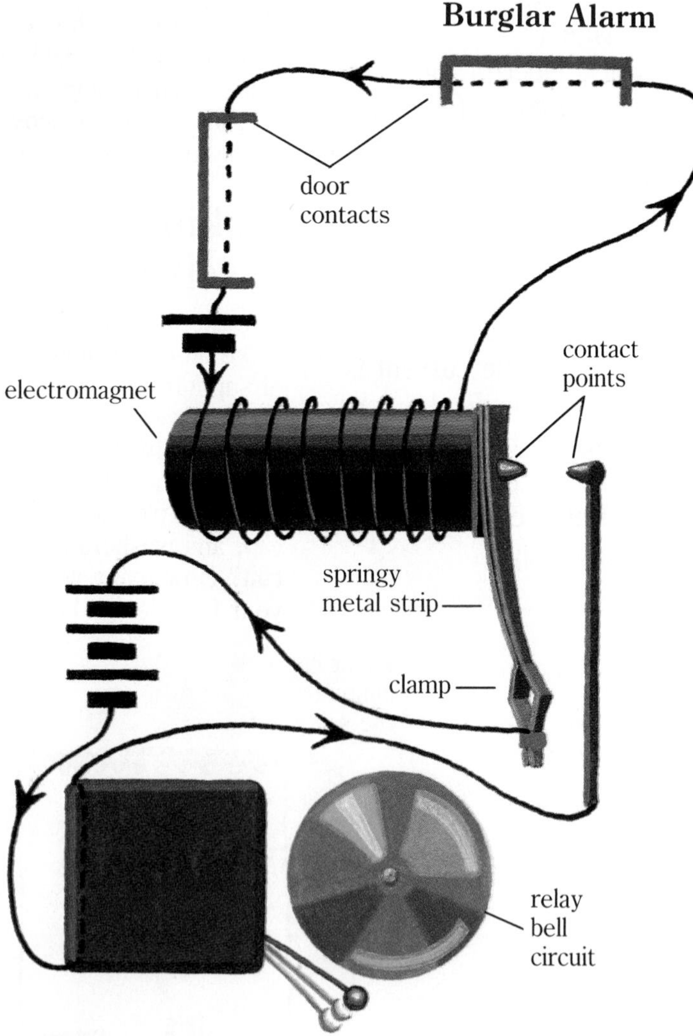

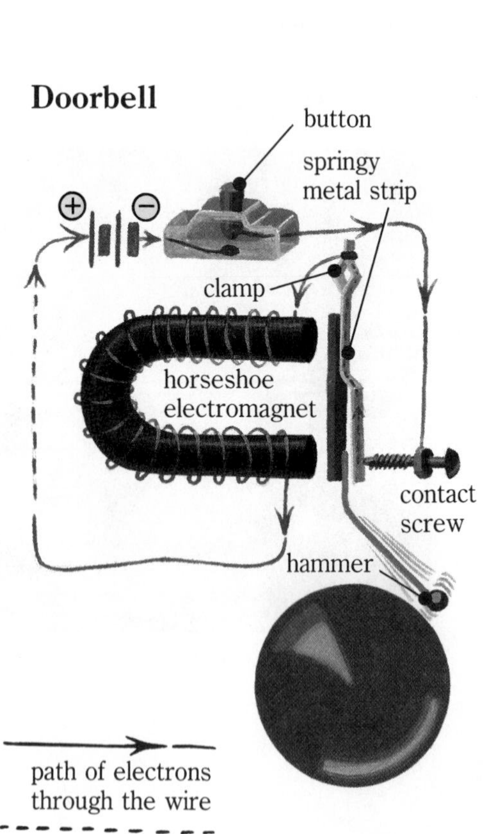

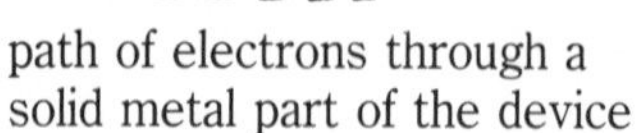

Moving Coils—Revolutionary!

One of the most common applications of electromagnets is the motor. Can you imagine life without motors? The invention of motors was revolutionary in more than one way!

Take a look at a small motor, such as the one in a washing machine or furnace, and notice the metal tag attached to it. You may see information similar to that shown here. Some of these words and symbols may be unfamiliar to you. What do you think "RPM" means?

"RPM 1500" means that the motor makes 1500 revolutions per minute. A *revolution* is one complete turn about a central axis.

What part does an electromagnet play in a motor? What part of a motor revolves? To find out, assemble the device below and close the circuit. What happens?

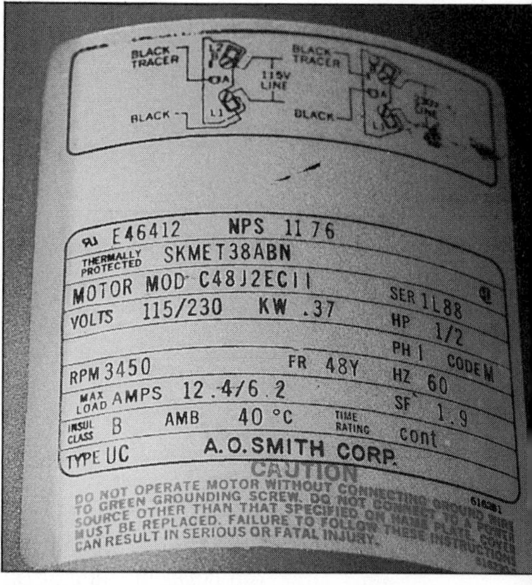

How many complete turns will this motor make in a minute?

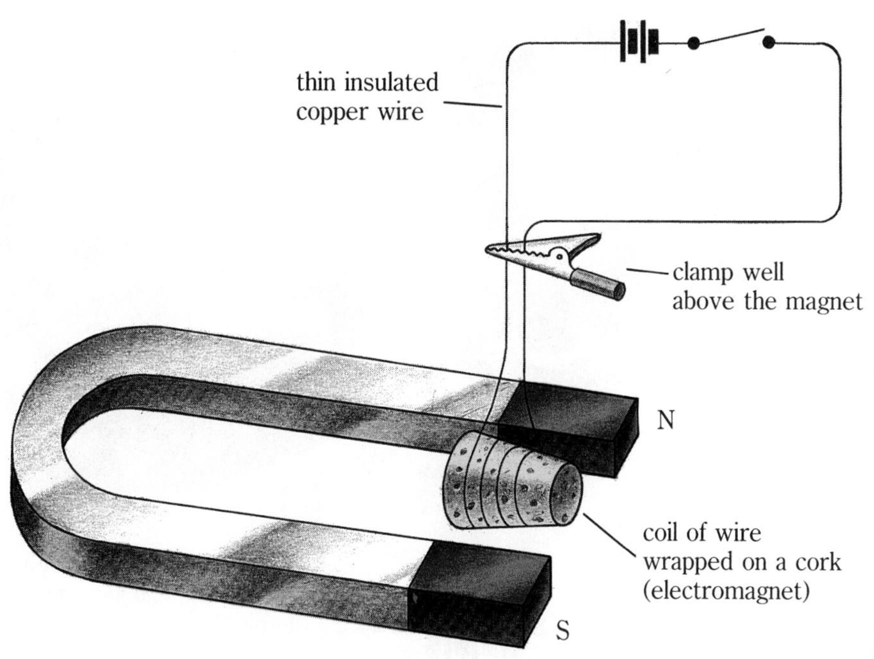

The coil of wire acts like a magnet. When a current flows, the coil has North and South poles. If, as in the illustration, you grasp the coil with your left hand so that your fingers go in the direction of the flow of electrons, the position of your thumb gives you the North pole. Locate the N and S poles of the coil. What will be the interaction between the N pole of the magnet and the N pole of the coil? between the S pole of the magnet and the N pole of the coil? What other interactions are there? You should realize that the coil will rotate in the presence of the horseshoe magnet when a current is flowing. Will it turn clockwise or counter-clockwise? Can the coil make a complete turn? If you reverse the coil's connections with the battery after it has turned halfway, what happens? What is it that causes a continuous rotary motion in a motor? Constructing a motor will help you see the importance of coil connections with the wires of the circuit.

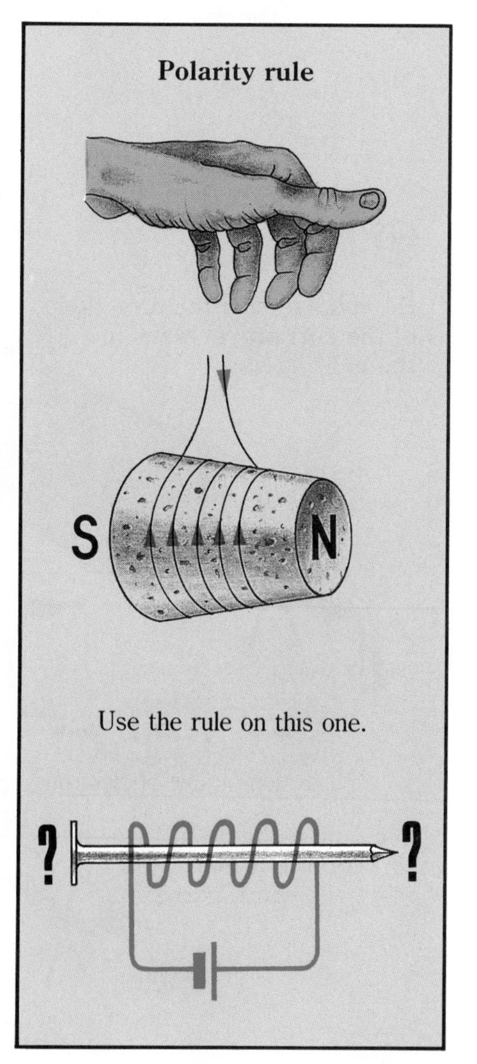

A Model Motor

Before building this motor, read the labels and comments in each step.

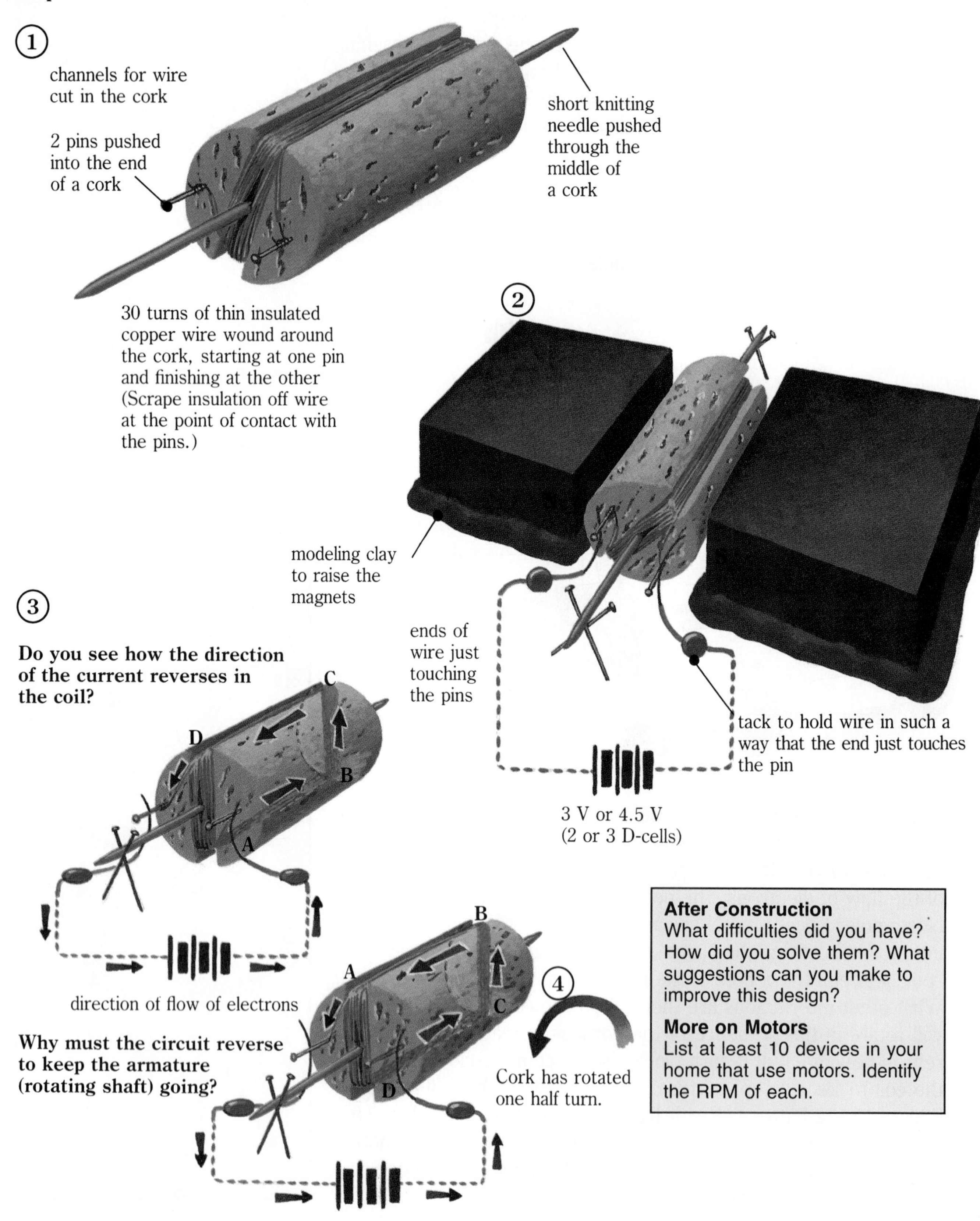

After Construction
What difficulties did you have? How did you solve them? What suggestions can you make to improve this design?

More on Motors
List at least 10 devices in your home that use motors. Identify the RPM of each.

How Much Electricity?

Look again at the motor tag pictured on page 247. This tag bears information about the motor. We already know what *RPM* and *Hz* mean, but what do the terms *amps* and *volts* mean? These terms relate to certain characteristics of electricity. In this case, these terms and the numbers that accompany them indicate how much current the motor needs to run properly.

How would you go about measuring electrical current? Electrical currents are often compared to water flowing through a hose. We can use this as a model.

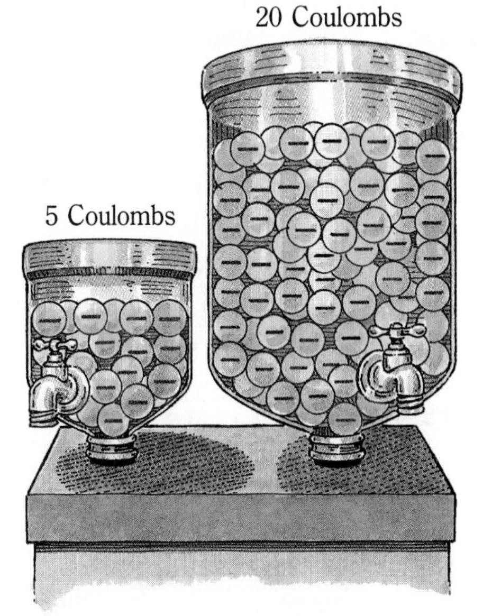

Charge is like capacity.

How Much Charge?

If you were to talk about amounts of water, you would say "liters" or "cubic meters" or "milliliters." In describing amounts of electricity, you might say "How many electrons?" But electrons are far too small to be easily used as a unit of measurement. Charge is measured in *coulombs* (C, pronounced KOO lahm). One coulomb is the charge carried by 6.24 quintillion (billion billion) electrons.

The more charges you have, the more current can flow. How would you measure the amount of electricity available from a given source? Let's use the water analogy. Which container do you think contains the greater "charge"? How did you arrive at this decision?

How Much Current?

Look at the diagram to the right. Which hose has a greater rate of flow? That is, which hose would carry a greater amount of water in a given amount of time? What is your clue?

The electrical equivalent of flow rate is **current.** Current is measured in *amperes* (A, *amps* for short). A current of 1 ampere is the rate of flow in which 1 coulomb of charge passes a given point in 1 second. The more coulombs passing a given point in 1 second, the greater the electrical flow rate, or amperage. Which hose has the greater "amperage"? What is your clue?

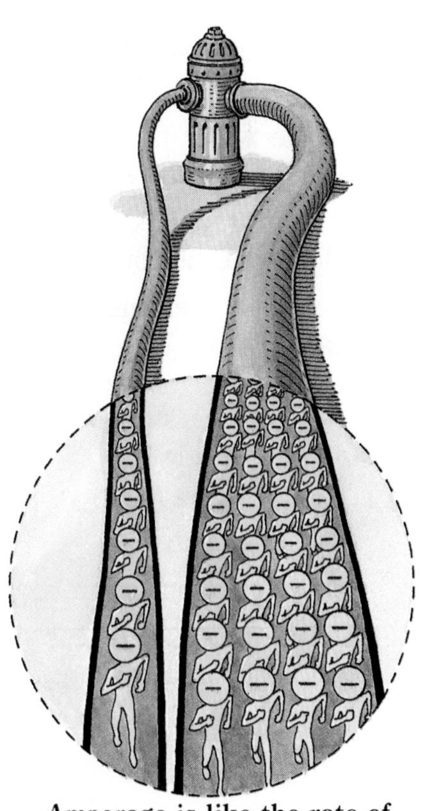

Amperage is like the rate of flow through a hose.

How Much Energy?

Have you ever tried to hold your thumb over the mouth of a garden hose at full blast? Most likely, the water was able to push your thumb out of the way and you got wet in the process. Why was it so hard to close off the hose with your thumb? The answer, of course, is pressure, which the water has due to the energy it possesses. This energy, which drives the water forward through the hose, is given to the water by a pump or by a reservoir in an elevated position. The greater the energy given to the water, the greater the pressure. The electrical equivalent of water pressure is **voltage.** Voltage is measured in *volts* (V). Voltage is a measure of the energy given to the charge flowing in a circuit. The greater the voltage, the greater the force, or "pressure," that drives the charge through a circuit. How do you think that voltage and current are related?

Which hose in the illustration on the right has the higher "voltage"? What is your clue?

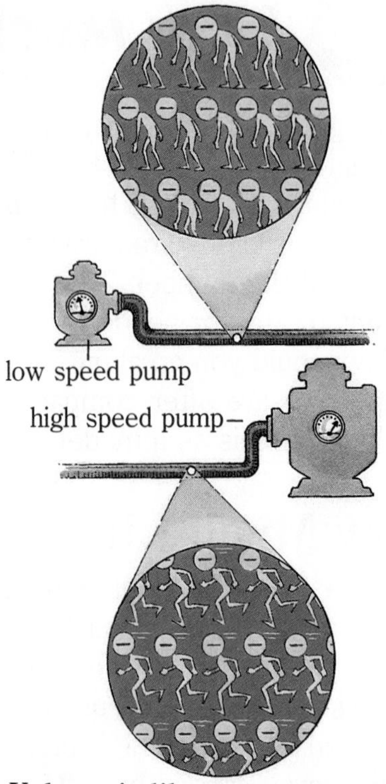

Voltage is like water pressure.

Resistance Revisited

We can also use our water model to explain the principle of resistance. What happens to the flow of water when you crimp a garden hose? The flow rate decreases, but what about the pressure, does it also decrease? What do you think would be the effect of a resistor on the voltage and the current in a circuit?

Putting It All Together

Charge, current, voltage—how do they all work together? So far our water model has worked well. The flow of electricity behaves remarkably like that of water. However, we have to complete the analogy by considering that electricity flows in circuits. Look at the diagram on the right, which shows a "water circuit." Let's analyze how it works.

1. The pump raises the water to the top. In doing so it boosts its energy level.
2. As the water flows downward, its potential energy is converted to kinetic energy in the form of moving water.
3. At the water wheel, work is done as some of the energy in the falling water is converted to mechanical energy.
4. The water returns to the pump, where the cycle begins again.

Let's compare our water circuit to an electrical circuit.

1. A cell or generator boosts the potential energy of the electrons to a high level.
2. Closing the circuit allows the potential energy to be converted to kinetic energy.
3. As the electrons do work, they lose kinetic energy.
4. The electrons, their energy mostly expended, return to the electrical source, where the cycle begins again.

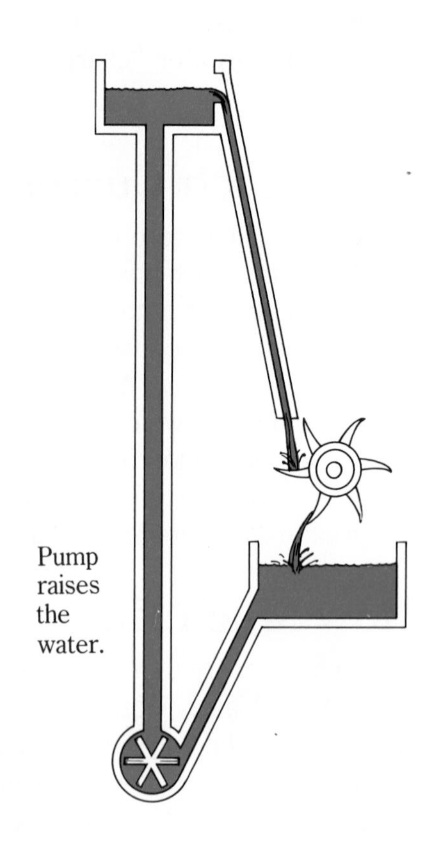

Using Electrical Units

Here is a summary of all you need to know to make practical use of electrical units.

- Quantity of electrical charge is measured in coulombs.
- Electrical flow rate, or current, is measured in amperes. *One ampere of current is the same as 1 coulomb of charge passing a given point in 1 second.*
- Voltage is the electrical energy given to a unit of charge flowing in a circuit. *One volt is the same as 1 joule of electrical energy given to 1 coulomb of charge.*
- Electrical energy, like other forms of energy, is measured in joules.
- Power, measured in watts, is the rate at which electrical energy is used. *One watt is 1 joule of electrical energy used in 1 second.*

You will draw on these definitions to complete the activities that follow.

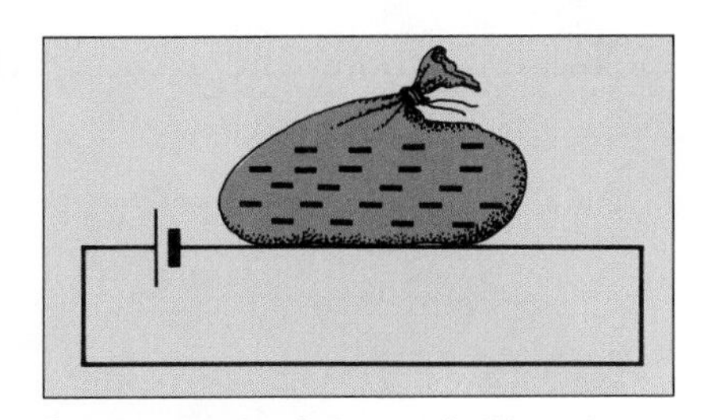

One coulomb of charge is like a bag of 6.24 billion billion electrons.

10 C of electrons passing a point in 1 s = 10 A of current.

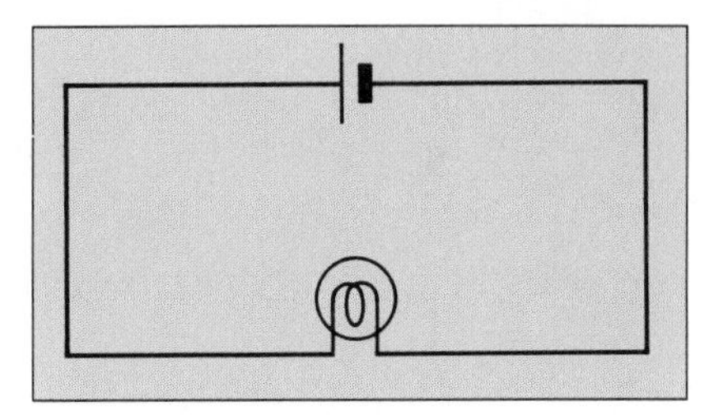

1.5 J of electrical potential energy given to 1 C of charge by the cell

Understanding Amperes, Volts, and Coulombs

1. Imagine using a toaster. Suppose the toaster draws 5 A and that it takes 3 minutes to toast a slice of bread.
   (a) How many coulombs flow through the toaster in 1 second? in the length of time it takes to toast the bread?
   (b) To figure the number of coulombs, multiply the number of ___?___ by the number of ___?___.
2. (a) A 1-V electrical source supplies ___?___ J of electrical potential energy to 1 C of charge; ___?___ J to 500 C of charge.
   (b) To figure the number of joules of energy used, multiply the number of ___?___ by the number of coulombs, or to put it mathematically:
   joules = ___?___ × ___?___.

Understanding Watts

The rate at which appliances use energy is measured in watts. This is also known as the power of the appliance.

3. If something uses 1 W, it uses ___?___ J of electrical energy every second.
4. An 800-W iron uses ___?___ J of electrical energy every second. If the iron is used for 15 minutes, it uses ___?___ J of electrical energy.

5. To figure the number of joules of energy used, multiply the number of watts by the number of ____?____. To put it mathematically:
   joules = watts × ____?____.
6. In questions 2 (b) and 5, you found that electrical energy, measured in joules, is equal to two different products of electrical units. By setting these two different products equal to one another, see if you can show:
   watts = volts × amps.
7. A 1320-W electric heater is plugged into a 110-V household circuit. A current of ____?____ A flows through the heater. Explain in your own words what each of the numbered quantities in this question means.

Power by the Hour

As you saw earlier, multiplying power by time gives you the total amount of energy used. For example, watts multiplied by seconds equals joules. Utility companies use another unit to measure the energy used by consumers: the kilowatt-hour (kWh). It is obtained by multiplying the power in kilowatts by the time of energy usage in hours.

Imagine that you have left a 2000-W electrical heater on for five hours.

8. A kilowatt is equal to 1000 W. The power of the heater is ____?____ kW.
9. In five hours an electric heater uses ____?____ kWh.
10. The heater used ____?____ J of energy. (**Hint:** first figure out how many seconds are in five hours.)
11. In questions 9 and 10 you found the energy used by the heater in both J and kWh. Therefore: 1 J = ____?____ kWh.
12. Why do you think kilowatt-hours are commonly used instead of joules to measure energy usage?

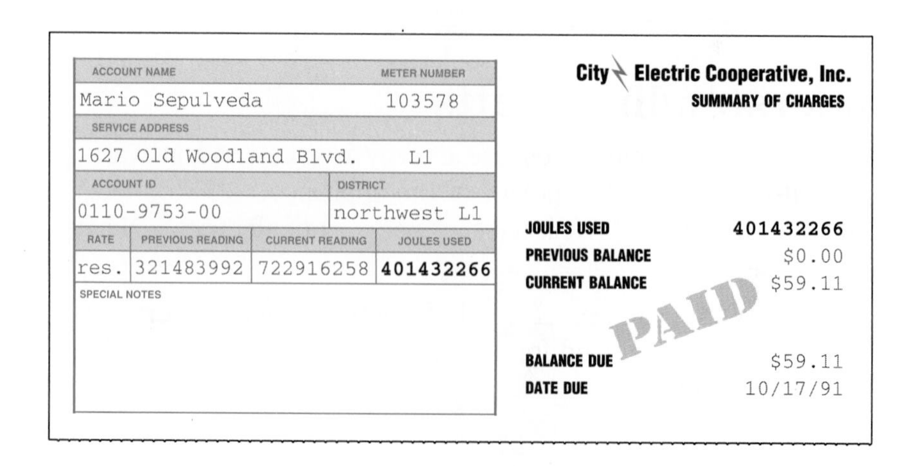

ACCOUNT NAME			METER NUMBER
Mario Sepulveda			103578
SERVICE ADDRESS			
1627 Old Woodland Blvd.			L1
ACCOUNT ID		DISTRICT	
0110-9753-00		northwest L1	
RATE	PREVIOUS READING	CURRENT READING	JOULES USED
res.	321483992	722916258	**401432266**
SPECIAL NOTES			

City Electric Cooperative, Inc.
SUMMARY OF CHARGES

JOULES USED	401432266
PREVIOUS BALANCE	$0.00
CURRENT BALANCE	$59.11
BALANCE DUE	$59.11
DATE DUE	10/17/91

PAID

Challenge Your Thinking

1. Many apartment buildings have security doors that can be opened from any apartment to admit visitors. The diagram below shows the circuitry of one such security door. Explain how it works.

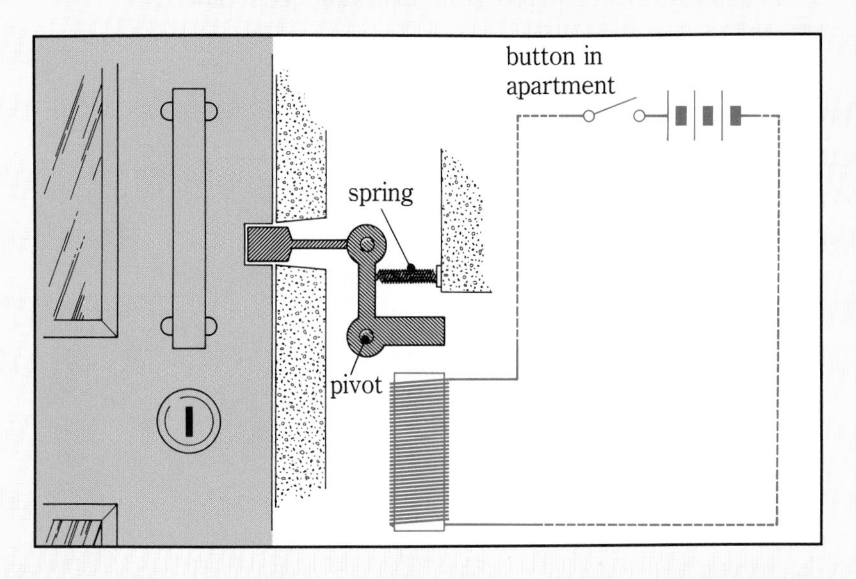

2. Ms. Alvarado asked her class to design a circuit containing two switches (S_1 and S_2), two light bulbs (L_1 and L_2), and one dry cell. She asked them to arrange the circuit in the following ways:

   (a) If only S_1 is closed, L_1 will light.

   (b) If only S_2 is closed, nothing will happen.

   (c) If S_1 and S_2 are closed, both lamps will light.

   Their work is shown below.

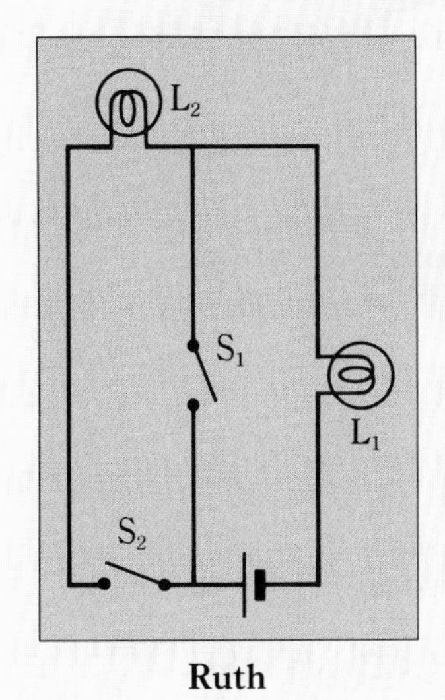

Ruth

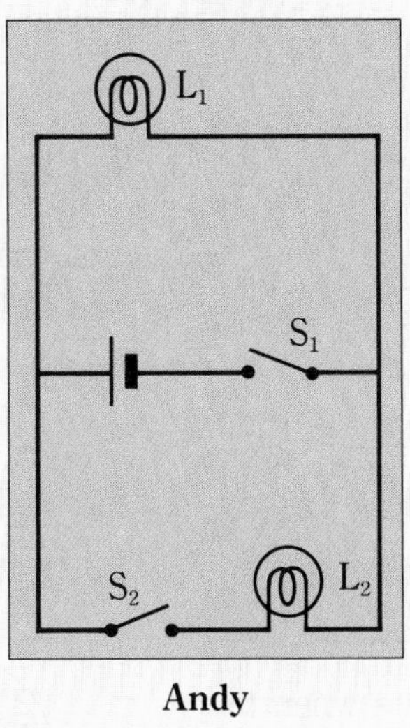

Andy

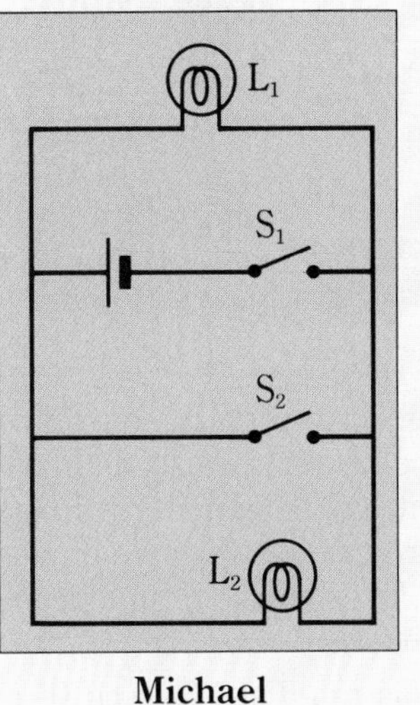

Michael

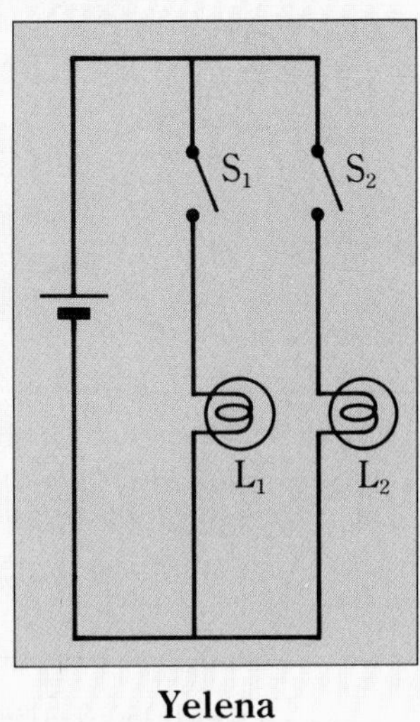

Yelena

Take the role of Ms. Alvarado and grade your students' work. Indicate how you know whether each design is right or wrong.

3. Redraw the circuit shown here so that the upper switch moves from contact A to contact C and the lower switch moves from contact B to contact D. Then suggest a function for it.

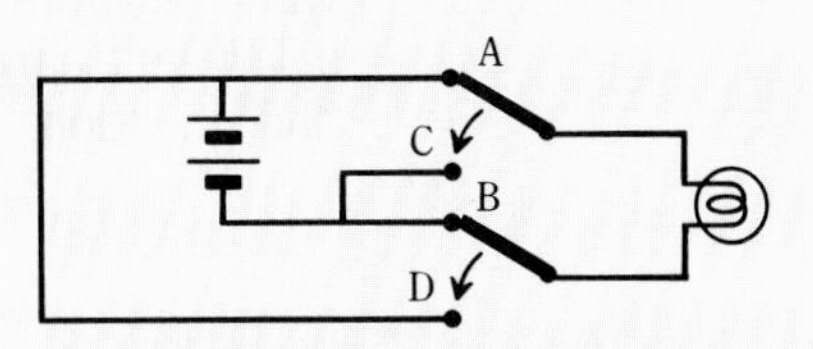

Unit 4 Making Connections

The Big Ideas

In your Journal, write a summary of this unit, using the following questions as a guide.

- What is electricity?
- How is electricity produced?
- What is a cell? a battery?
- What are the two types of current? How do they differ?
- What factors affect the size of currents in circuits?
- What are the different types of circuits? Give examples of each.
- What are electromagnets? How do they work?
- How is the flow of water like that of electricity?
- By what units is electricity measured? How are these units related?

Checking Your Understanding

1. A typical carpet shock can deliver a jolt of thousands of volts. Why doesn't it seriously injure you?
2. The diagrams below show two different types of microphones. Use the diagrams and your knowledge of electrical principles to explain how each of these microphones works.

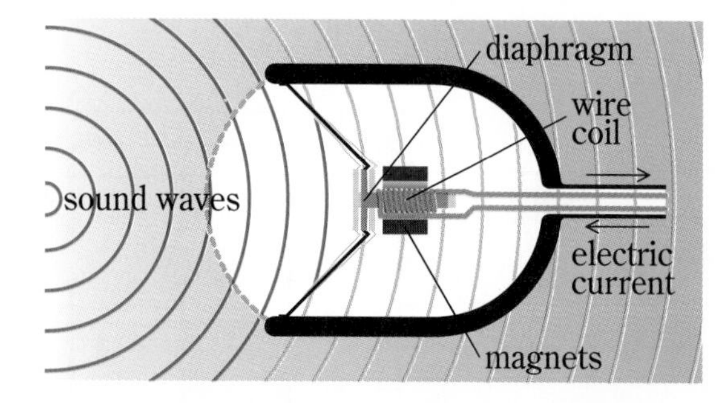

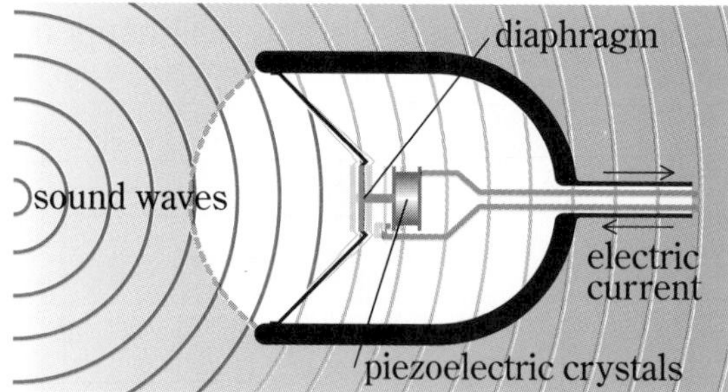

3. *Fuses* are designed to protect household circuits against damage due to electrical overload. Fuses are made of metal alloys that have low melting points. When a circuit carries more current than is safe, the fuse strip melts. How do fuses work? (Hint: Think about the effects of resistance.)
4. Suppose a fuse is rated at 15 amps (it melts when the circuit carries more than 15 amps). How many 100-watt light bulbs would it take to blow the fuse? Assume a 110-volt current. (Don't try this yourself!) You will find questions 6 and 7 on page 252 helpful.

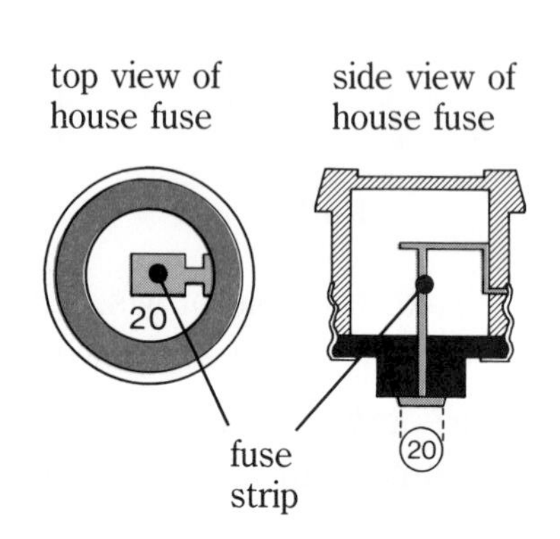

5. What's wrong with these pictures? Redraw them to make them correct.

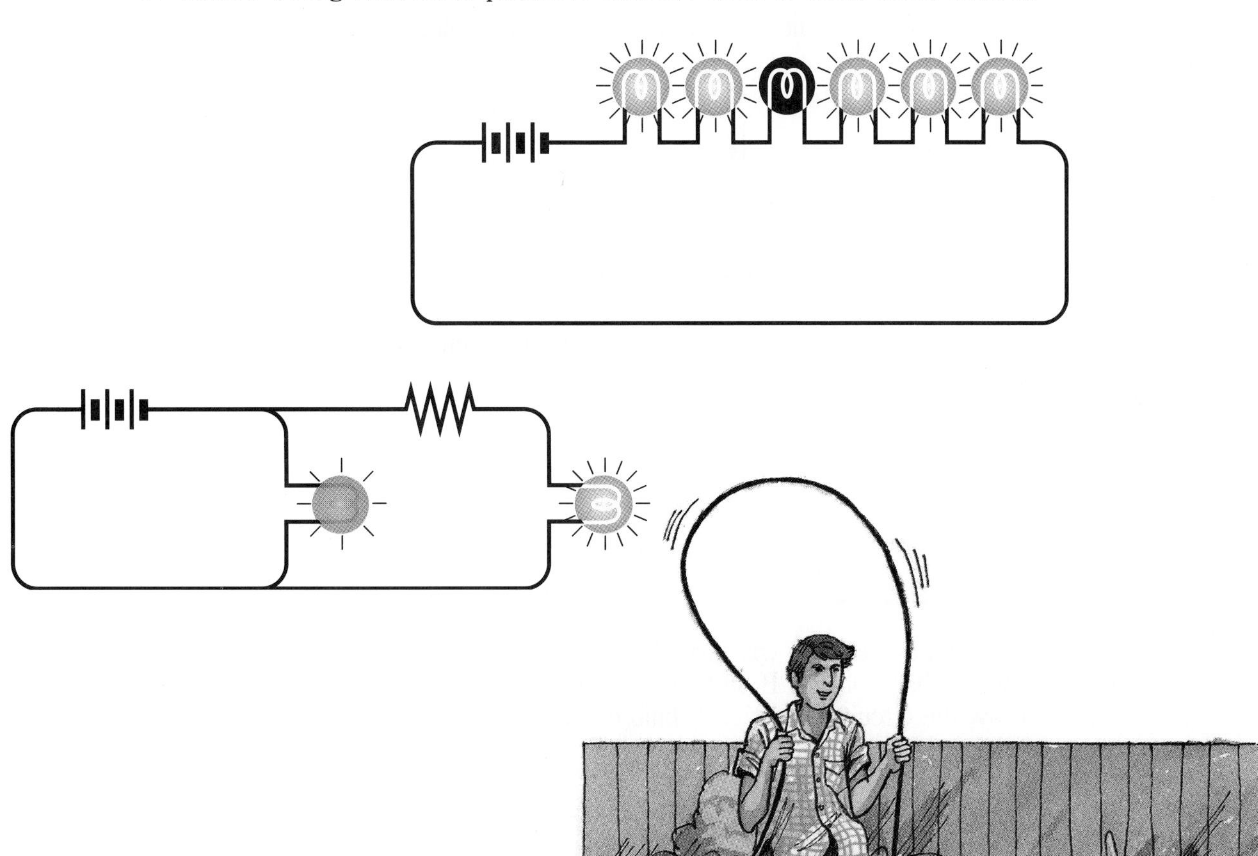

6. Stan made a jump-rope out of a loop of wire and then performed the activity pictured at right. The galvanometer showed that a current was being generated.
   (a) Explain what happened. **(Hint:** What makes a compass work?)
   (b) Did Stan generate direct current or alternating current? Explain.

7. Electricity is often compared to flowing water. Read the following examples and decide whether each suggests high or low amperage, high or low voltage, or any combination of these.
   (a) the Mississippi River
   (b) Niagara Falls
   (c) the blast from the spray-gun nozzle at a do-it-yourself carwash
   (d) a dripping faucet

8. Test your circuit savvy! Solve each of the problems that follow.
   (a) Add something to the circuit to prevent the cell from being drained of its energy too quickly.

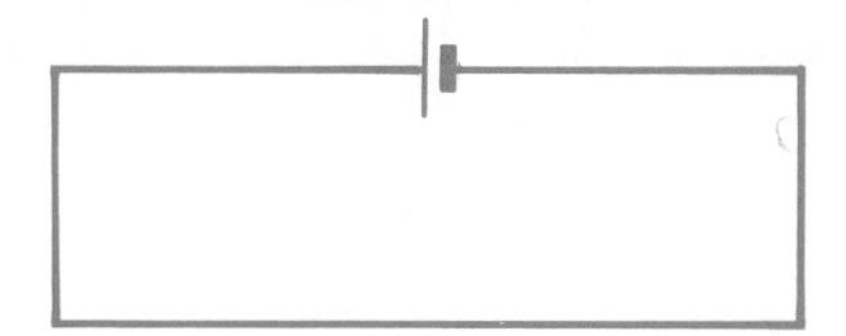

   (b) Redraw the circuit so that each light can be turned on and off independently.

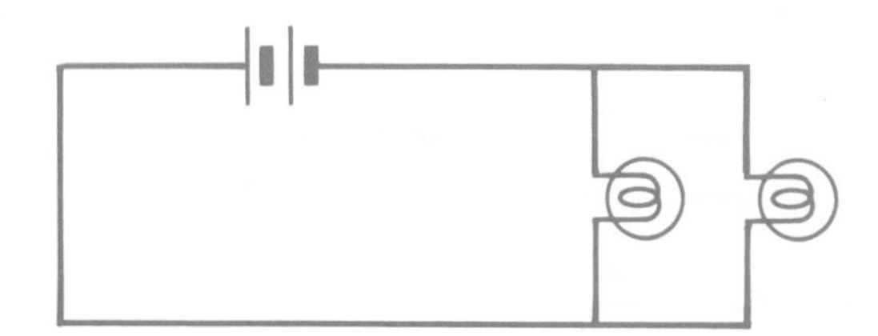

   (c) Redraw to add a switch that will turn all the bulbs on and off. Add another switch that will turn B off when the first switch is closed. Then redraw the circuit so that each bulb can be controlled with its own switch.

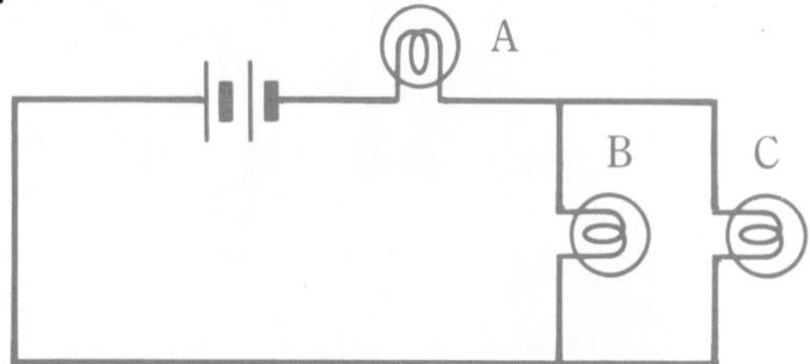

9. You have been asked to write the "How It Works" feature of a science magazine for kids aged 10–13. You have a two-page spread for each issue. Here are some topics.

   - How a flashlight works
   - How a telephone works
   - How a generator works
   - How a light bulb works
   - How an electric motor works

   Choose one topic. Write in a clear style, explaining any new or difficult terms you use. Design an attractive layout that includes photos, drawings, and cartoons. Be creative. When finished, share your feature with your classmates. Use their suggestions to make revisions to your work where necessary.

10. Make a concept map using the following terms: *volts, current, electricity, amperes, voltage, coulombs,* and *charge.*

11. At right is a diagram of a *circuit breaker,* a device that mechanically performs the same task as a fuse, breaking the circuit when too much current flows through it.

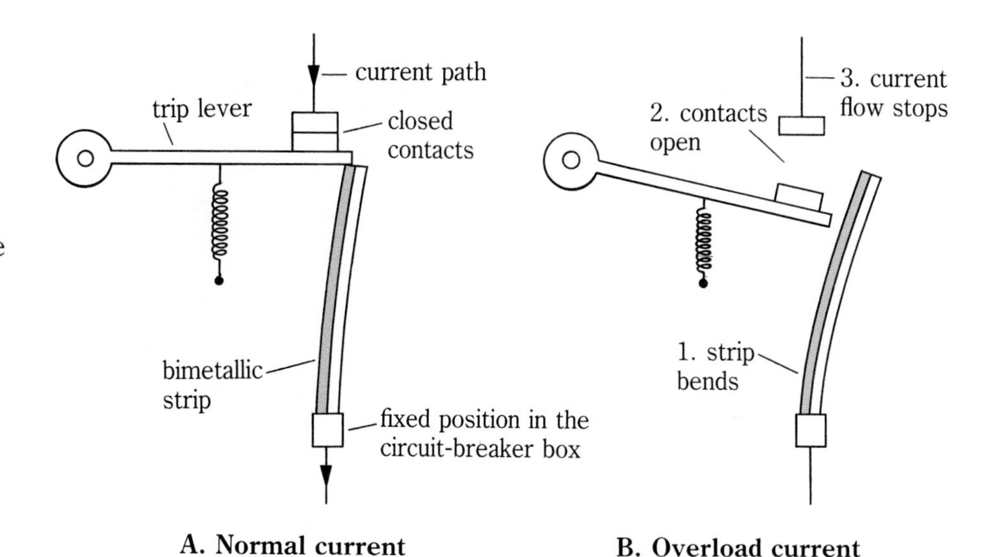

How does this device work?

12. How is a water faucet like a variable resistor?

Reading *Plus*

Now that you have been introduced to the basics of electromagnetic systems, let's take a closer look at electricity and magnetism. By reading pages S51–S68 in the *SourceBook,* you will learn more about the basic types of electricity, and about magnetism and its relationship to electricity. You will also learn about some of the ways in which electricity is harnessed by humans and some inventions that make this possible.

Updating Your Journal

Reread the paragraphs you wrote for this unit's "For Your Journal" questions. Then rewrite them to reflect what you have learned in studying this unit.

Science In Action

Spotlight on Magnetic Resonance Imaging

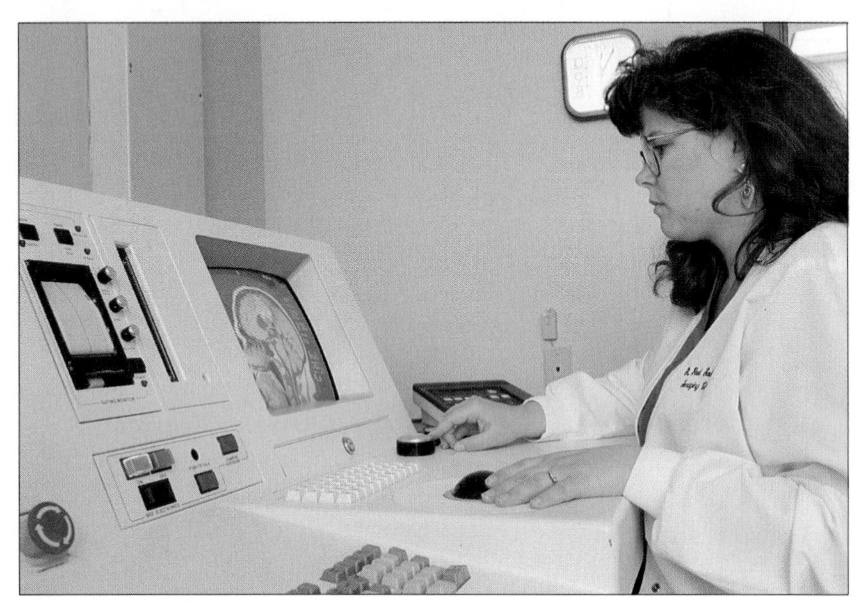

MRI technologist Kathy Jacobsen

Magnetic resonance imaging (MRI) is a revolutionary new medical technology. It allows doctors the opportunity to view the internal structure of the human body at work. Unlike X rays, MRI does not expose the body to potentially hazardous radiation. And unlike exploratory surgery, MRI is quick, painless, and often safer in helping doctors diagnose human ailments.

Kathy Jacobsen is an MRI technologist. Kathy's education includes a four-year bachelor's degree in Radiographic Science, although a two-year associate's degree fulfills the basic requirement for an MRI technologist.

Q: How does MRI work?

Kathy: Using a large machine that is basically a tunnel-shaped magnet, MRI sets up a powerful magnetic field around the patient. Since the body is made up mostly of water, there are lots of hydrogen atoms in the body. The magnetic field in the machine lines up those hydrogen atoms. Meanwhile, a radio frequency is transmitted in a brief burst to knock the atoms out of alignment. After the brief burst, the computer measures the activity of the atoms as they return to their original lineup. By tracing that process, the computer can produce an image.

Q: How is MRI used in medicine?

Kathy: MRI helps us to see what the internal organs and skeleton of a person's body look like. It can give us clear pictures of brain and spinal tumors, fatty tissues, muscles, tendons, arteries, vertebrae, blood flow, and a lot more.

MRI also allows us to observe the function of specific body parts. It does this by taking a series of individual images, combining them, and then speeding them up. The process is similar to that of making cartoons, which are a series of drawings linked together to create the effect of movement. Similarly, we can take numerous MRI scans of the jaw, for example, and use the computer to rapidly link them together. The result allows us to see if the jaw is working properly.

Q: What do you do as an MRI technologist?

Kathy: The first thing I do is review a patient's file to see what the doctor is looking for.

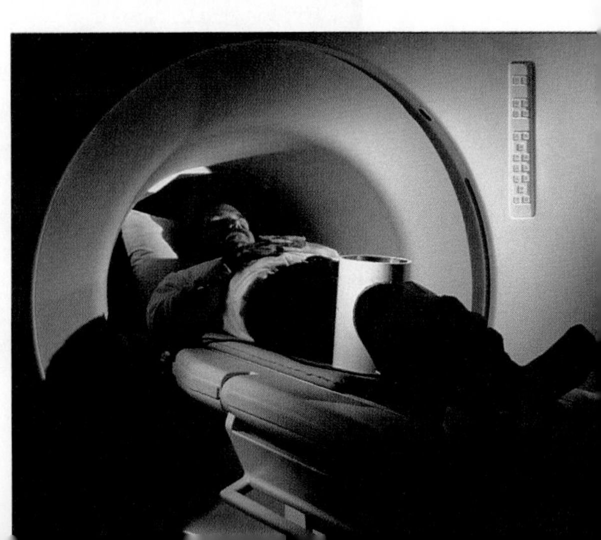

This patient, entering an MRI tunnel, is wearing a cylinder on his knee for greater detail of the bone and soft tissues in that area.

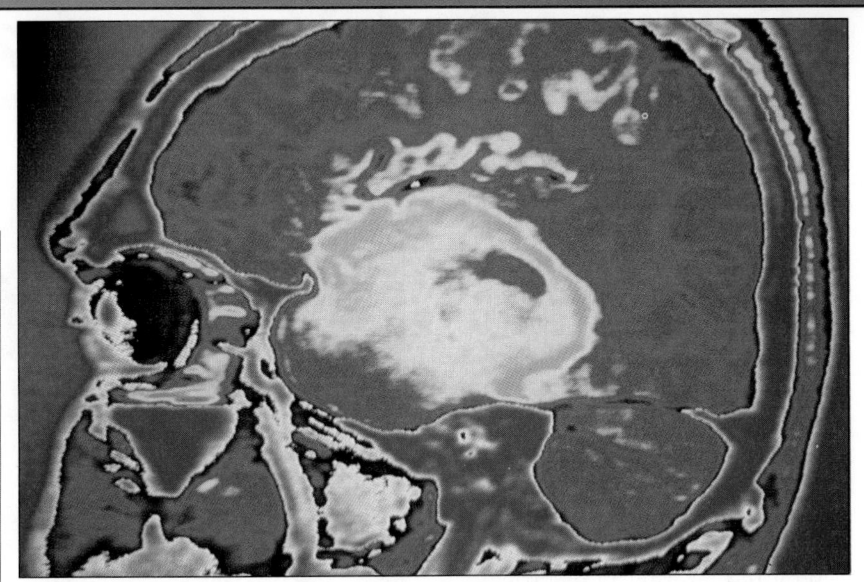

This color-enhanced MRI image of a brain shows a tumor (tinted yellow). The tumor was removed and the patient resumed a healthy life.

Then I meet with the patient and ask him or her a series of questions. The questions are designed to make sure the magnet will not harm the patient. For example, I need to be certain that the patient is not wearing a pacemaker and that no metal implants or metal scraps are in their body. (Some types of metal are okay, it depends on the location of the metal and how long it has been in the body.) After that, I ask the patient to lie on a table and I slide him or her into the large machine. From my computer console, I then program the computer to scan specific regions of the patient's body. The result is a set of images that allows the doctor to analyze the patient's problem.

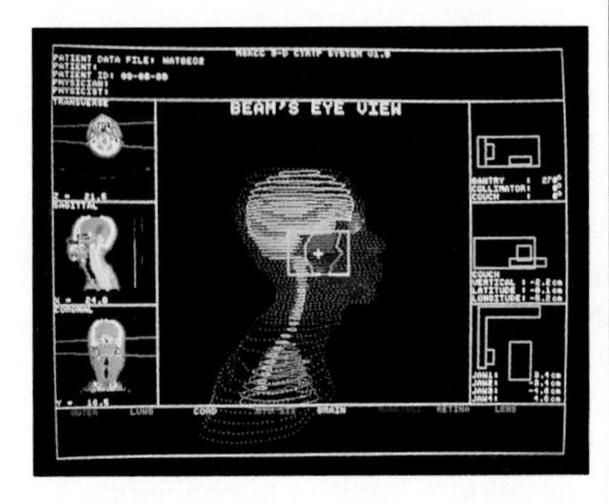

Q: What is the most exciting aspect of your job?

Kathy: I really enjoy the patient contact. There are some patients I see over and over again because they have serious problems. For instance, a patient may have a brain tumor. As their doctor treats the tumor, a series of MRI scans are usually ordered to monitor the progress of the treatment. It's incredibly rewarding to see the patient's tumor shrink and disappear—especially when I've had the time to get to know the patient.

Q: Are there any unforgettable people or situations that you've encountered using MRI?

Kathy: One time a nurse in the hospital didn't realize the magnet is kept on all day, even when it's not being used on a patient. She walked into the room with a portable oxygen tank. The tank flew half way across the room and became attached to the magnet! We were able to dislodge the tank only after momentarily turning off the machine. Ever since then, we've had to guard the room so that no serious accidents occur!

Q: What personal qualities do you feel are most important for an MRI technologist?

Kathy: Patients are often nervous about the test and the results. Compassion, communication, and patience are qualities the technologist must have in order to make the patient feel comfortable. For example, it's never my position to coax an unwilling patient into the machine. But because the exam is important, I try to be calm and reassuring so that the patient will successfully complete the exam.

A Project Idea

Prepare a poster or report illustrating the use of MRI to treat one specific condition. For example, you may ask one of the following: How does MRI help a doctor to locate a spinal tumor? to analyze heart beats or blood flow? Include in your report a summary of why a doctor might use the MRI process rather than another technique.

Consult your library for books and magazines on magnetic resonance imaging. For additional assistance, contact a radiologist at a hospital or a medical clinic. If you have an MRI facility in your hometown, make an appointment for a tour!

Science and Technology

A Cleaner Ride

If you think electric cars have to be slow and plain-looking, like overgrown golf carts, it might be time to have another look. The electric car at the right can support the "creature comforts" you might want, like air conditioning and stereo. And it can actually outrun some of today's top sports cars!

Why the interest in electric cars? For one thing, electric cars do not pollute the air. Since they don't burn fuel, they don't produce toxic emissions. For another, they make better use of energy resources than gasoline-powered cars do.

Are electric cars the way of the future?

The *Gossamer Albatross* taking off using human power

Trimming Off the Fat

In electric car design, *efficiency* is everything. Batteries, at present, provide very little energy for their weight. Electric-car makers must use that limited energy wisely. Paul MacCready, whose company helped a car manufacturer engineer the car shown above, has a passion for efficiency.

You may already be familiar with some of Paul MacCready's work. For example, he designed the *Gossamer Albatross,* which was the first (and only) human-powered aircraft to be flown across the English Channel. Such projects prepared him well for electric-car design. When power has to be provided by human muscles alone, you develop a healthy appreciation for efficiency—but fast. No excess mass can be tolerated. The *Gossamer Albatross,* whose wingspan is an astonishing 29 meters, has a mass of just 32 kilograms.

Recycling Energy

In a conventional car, the brakes stop the car by means of friction. This heats up the brakes. The mechanical energy of the car is converted to useless heat energy in the brakes. When you think about it, this is a serious waste of energy.

MacCready's car, on the other hand, uses a **regenerative braking system.** When the driver's foot is lifted from the accelerator, the car's motor becomes a generator. The turning of the car's wheels drives the generator, which puts a drag on the wheels and slows the car down. The electrical energy produced by the generator is recycled back to the batteries.

A test run of the *Sunraycer*

Plugging In

Like all batteries, electric car batteries go dead. To keep the car going, the batteries must be recharged from time to time. MacCready's car can travel a little over 190 km before its batteries need recharging. Then you have to plug them into an energy source. With standard house current, a complete recharge takes about 8 hours. The parking lots of the future might have electrical power sources for recharging electric-car batteries.

Solar Energy

Solar energy is another source of power that may prove useful. Paul MacCready's team has been a leader in this field. They worked with the car manufacturer in developing the solar-powered *Sunraycer*. In 1987 this experimental vehicle won the Australian 3150-km cross-country competition. Capable of a top speed of 80 kilometers per hour, it came in a full two and a half days ahead of its closest competitor!

The Real Answer?

In the near future, however, **hybrid cars** may very well prove to be the most satisfactory. These would be cars that use electric power in combination with some other source of energy, such as gasoline or alcohol. Such cars, for example, might carry their own gasoline-powered generators to charge their batteries.

Think About It

Fossil fuels are most often used to generate the electricity needed by electric cars. But the use of fossil fuels pollutes the environment. Are there alternative ways of providing electricity that would be pollution-free? Could inexhaustible resources be used? Do you think they would be practical?

Many people doubt that a completely solar-powered car will ever be satisfactory. Would this mean that solar power has no place in automobile design? Explain your answer.

Unit 5 Sound

For Your Journal

Write a paragraph summarizing your thoughts about each of the following:

1. If a tree falls in a forest and no one is there to hear it, will there be a sound?
2. How does sound travel from one place to another? How fast does it travel?
3. How can a guitar string be made to produce a higher sound? a louder sound?
4. How might the unusual objects hanging from the ceiling affect sounds heard in this auditorium?

WHAT IS SOUND?

1 A World of Sounds

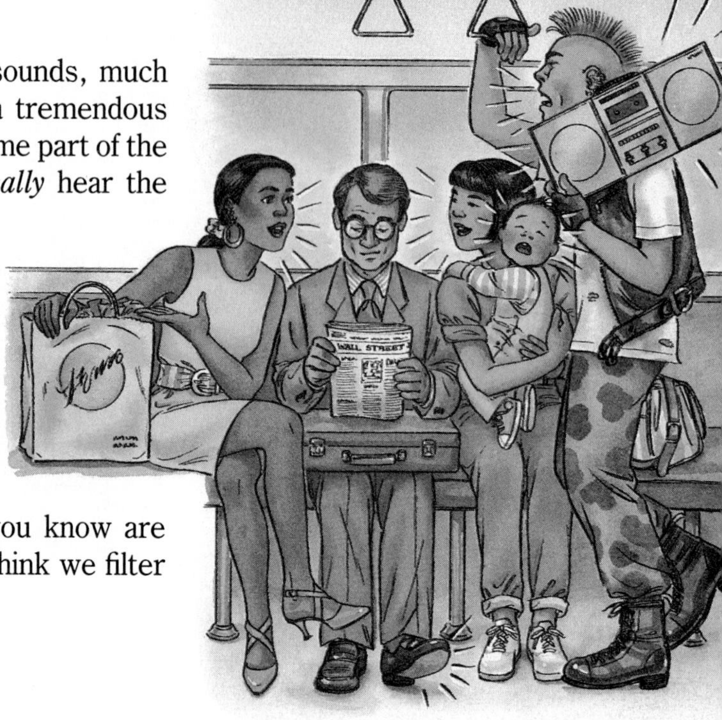

Some Thoughts on Sounds

You are surrounded by sound—an atmosphere of sounds, much like the atmosphere of air you breathe. There is a tremendous variety of such sounds. At times, some of them become part of the background. For example, this morning did you *really* hear the following sounds?

- the water running in the sink
- the orange juice splashing into the glass
- the rattle of dishes
- the slam of the door
- the rumble of the school bus engine

How many other sounds can you think of that you know are there but that you don't really notice? Why do you think we filter out certain sounds?

Investigating Sound

Studying sound raises many interesting questions to investigate. Think about these for starters . . .

- What is the loudest sound ever recorded?
- How do the blind use sounds to "see"?
- Why can a person imitate a bird by whistling, but not by using his or her own voice?
- How does a trombone player make the instrument produce a sound?
- Are there sounds you can't hear?
- How can a faint whisper sometimes be heard 60 km away?
- What is the lowest sound you can make? hear?
- Why is it so silent after a snowfall?

Can you add to these questions? With your friends, generate a list of other questions about sound that you would like to investigate. At the end of your study of sound, see how many you can answer.

The Sound of Silence

Is there such a thing as total silence? For instance, if you were put in a completely soundproof room, would you hear anything? Actually, you would. People placed in soundproof rooms hear two sounds—one high and one low. Sound engineers have discovered that the high one is the person's nervous system, and the low sound is the person's blood circulating. Can you hear *your* body's sounds?

The tiles in this room eliminate almost all sound. This specially designed room is being used to check the static transmitted over a telephone system.

Noise

In many cities today, the continuous background of sound is increasing to an alarming level. What are some sources of offending sounds? Studies show that many people are suffering physically and mentally from excessive noise. What are some possible harmful effects?

At the other extreme, people can become nervous and irritable when there is too little sound. Sound engineers have constructed modern buildings that are so quiet that even ordinary noises can come as shocks. To overcome this problem, sound is piped back into the rooms! It sounds like the hiss of a steam engine and is called "white noise." Why do you think it might be given this name? Sometimes quiet background music is used for the same purpose.

Airport workers wear earphones to protect their ears from noise.

Describing Sounds

Make or obtain an audiotape of various sounds. Have others guess what is making the sounds. Then brainstorm to think of words that describe each sound, such as *harsh, loud,* or *grating.* Afterward, think of all the ways you might classify these word descriptions.

Classifying Sounds

Sounds in 4-D

As you read the following article, write a descriptive heading for each paragraph, using no more than three or four words.

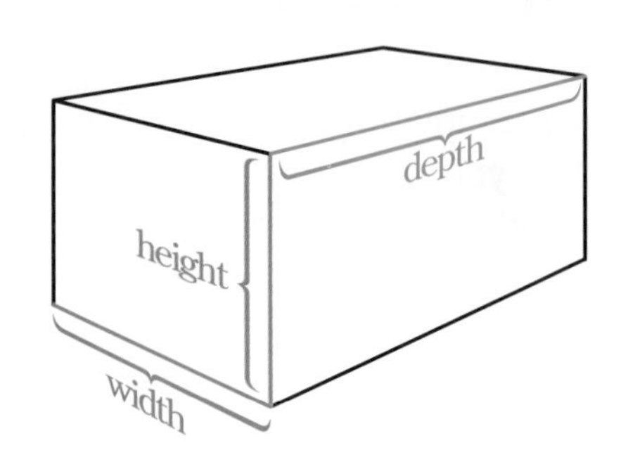

(a) ______________
(paragraph heading)

Sounds have certain characteristics that can be thought of as "dimensions." When measuring objects, we often speak of dimensions, such as height or length. Buildings, trees, and people can be described as tall, short, or in-between. Can you think of some dimensions for sound?

(b) ______________
(paragraph heading)

In a way, sounds also have height. A sound may be high like the chirp of a cricket, lower like the croak of a frog, or still lower like the grunt of a pig. The "highness" or "lowness" of a sound is called its **pitch**. *We speak of the cricket's chirp as high-pitched, and the pig's grunt as low-pitched.*

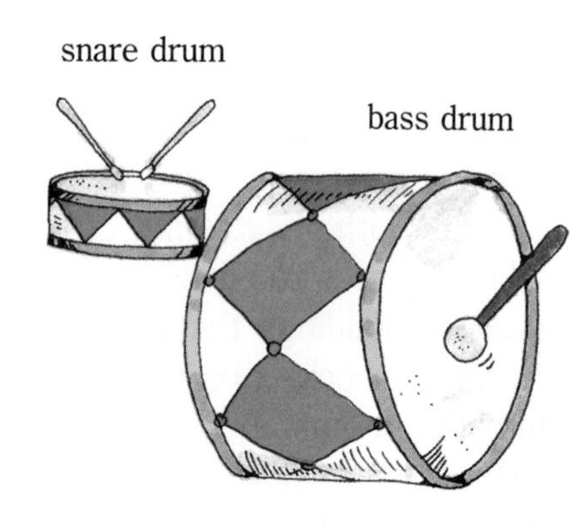

Which sound has a higher pitch?

(c) ______________
(paragraph heading)

Sounds like rain falling may be quite soft. Or, like a book falling on the floor, they may be louder. A speeding diesel truck on the highway makes a still louder noise. Loudness *is another characteristic, or dimension, of sound.*

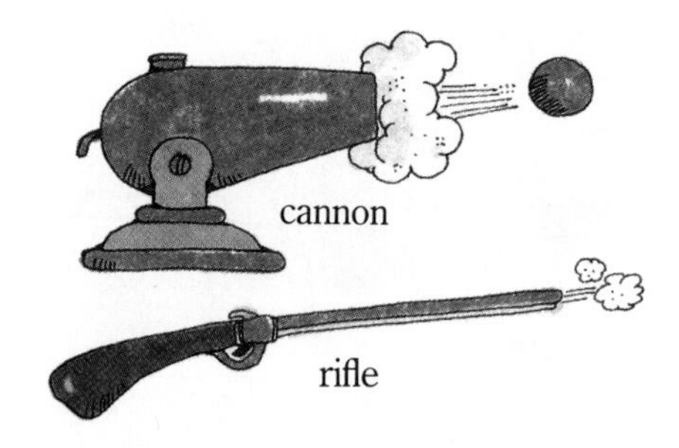

Which sound is louder?

(d) ______________
(paragraph heading)

The rat-a-tat-tat of an air compressor is a sequence of very short sounds; each lasts for only a fraction of a second. Hit your pencil on the table. How long did the sound last? Other sounds seem to linger, such as the ring of a large bell, a sustained note made by a trumpet, or the continuous background sound of the air conditioning in some buildings. Are musical sounds generally long or short? The length of time, or duration, *for which a sound can be heard is another dimension of sound.*

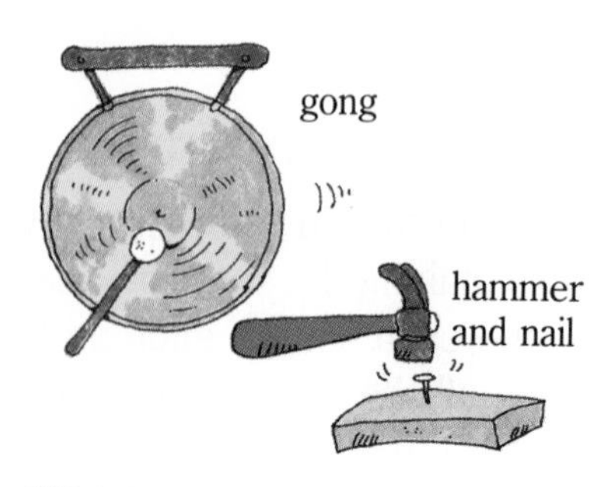

Which sound lasts longer?

(e)
(paragraph heading)

Which sound is "sweeter"?

When we try to classify sounds further, we may think of sweet sounds, dull sounds, nasal sounds, bright sounds, blaring sounds, or harsh sounds. We may also classify sounds by means of such descriptive words as cracking, buzzing, clanging, and tinkling. Each sound source generally has its own characteristic type of sound, such as the blaring tones of the trumpet; the pure, bird-like notes of the flute; the nasal sounds of the bagpipes; and the thud of heavy steps on the staircase. We use these descriptive words in an attempt to describe another dimension of sound—its quality. *A bell makes a sound of a different quality from that of a car horn, even if the dimensions of loudness, pitch, and duration are the same for each.*

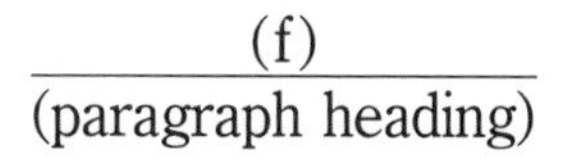
(f)
(paragraph heading)

Many sounds at one time

In music, quality is also called timbre, *or* tone. *Why is there this variation in quality? When you mix paints, the shade you get depends on the colors you start with and on how much of each color you use. This is also true of sounds. The quality you hear depends on the "mix" of sounds being produced by an object. In other words, the quality depends on the number of different sounds and on the varying loudness of each of these sounds.*

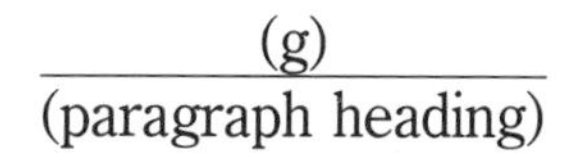
(g)
(paragraph heading)

What is music, and what is noise? Think of the pure musical tones of the flute, the less musical sound of clanging cymbals, and the unmusical noise of the electric lawnmower. Music and noise are aspects of the quality *of sound. Musical sounds have fewer component sounds. We might call them "purer" sounds. In fact,* purity *of sounds can be thought of as a sub-dimension of quality. More on this later.*

Music and noise with a saw

(h)
(paragraph heading)

Thus, the dimensions of sound are loudness, pitch, duration, and quality (also called timbre or tone). Changing one or more of these dimensions is what creates the incredible variety of sounds we hear around us.

3 Making Sounds

EXPLORATION 1

A Symphony of Sound

It's not difficult to make sounds, but it is sometimes difficult to see what is happening when sounds are made. Do several of these experiments. For each one, record answers to the following questions:

- How do I make the object produce the sound?
- What is the object doing as it produces the sound?
- How long does the sound last?
- How can I stop the sound?
- Can I change any characteristics (dimensions) of the sound, such as loudness and pitch? If so, how?

You Will Need

- tuning forks—different sizes
- a rubber stopper
- a glass container of water
- 2 rulers—one wooden, one metal
- a table
- a drinking glass
- a cardboard box
- 2 rubber bands
- a pencil
- a plastic drinking straw
- scissors
- a balloon
- a cardboard tube (12 cm long)
- rubbing alcohol
- cotton balls
- paper
- miscellaneous objects

What to Do

Tuning Fork Sounds

1. Strike a tuning fork on the edge of a rubber stopper.

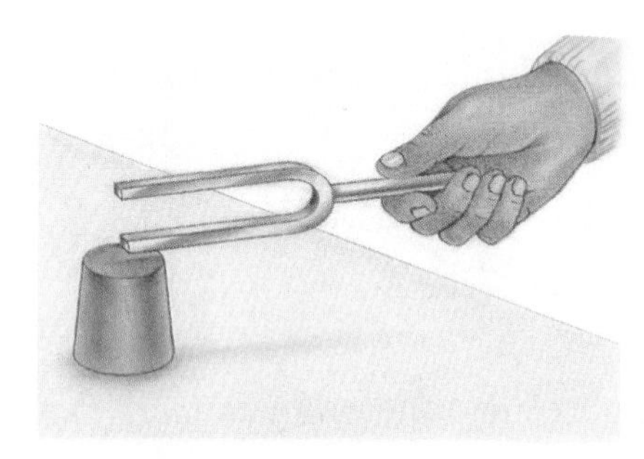

2. Hold it close to your ear. What do you observe?
3. Touch the prongs lightly to various parts of your body such as your ear lobes and nose. **Caution:** *Do not touch the prongs to your eyes, eye glasses, or teeth. Clean the tuning fork with rubbing alcohol and a cotton ball before another student uses it.* After striking the fork again, touch the prongs to the surface of a glass container of water and to a loosely held piece of paper. What do you observe?
4. Next, while the tuning fork is sounding, touch its base to your teeth, to the table, to a cup held over your ear, and to other objects.
5. Try some of these same experiments with a tuning fork of a different size.

Ruler Sounds

1. Hold one end of a wooden ruler firmly on the edge of a table. Push down on the other end. Let it go. Try this several times, with various lengths of the ruler extending over the end of the table. What do you observe?
2. Substitute a metal ruler for the wooden one, and repeat the activity. What do you observe?

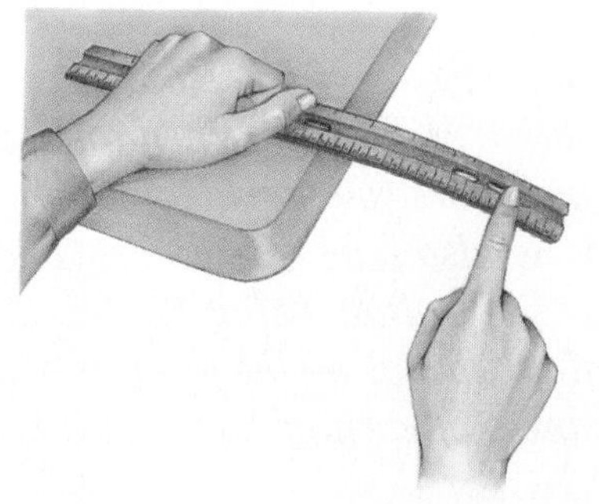

Rubber Band Sounds

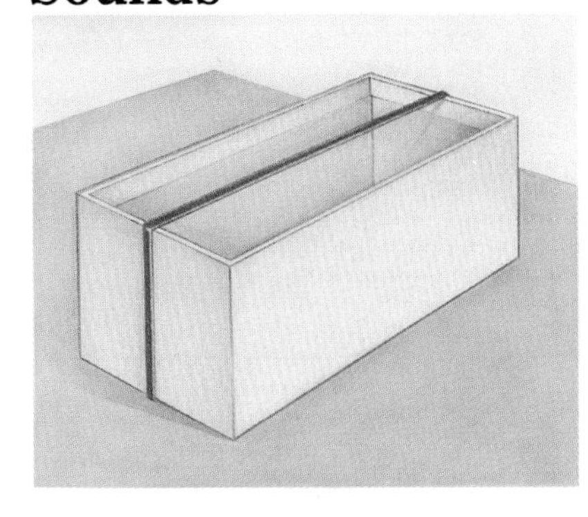

1. Pluck a rubber band that has been stretched across a cardboard box, as shown. Listen carefully and observe what happens.
2. Tighten the part of the rubber band that is on the top of the box. Pluck it again. Is there any difference in sound? Why or why not?
3. Put a pencil across the top of the box (the short way), under the rubber band, and pluck again. Do you hear or see any differences? Why or why not?

Straw Sounds

1. Make a "straw saxophone," as shown in the drawing to the right.
2. Adjust the position of the "sax" in your mouth until a steady sound is produced. Try producing different sounds by blowing in different ways.
3. While blowing a steady sound, use scissors to cut the straw shorter and shorter. What happens?

Simulated Voice Sounds

1. Blow up a balloon. Make it squeal as air is slowly released. Try for variations of sound. What do you have to do to get a higher sound? a lower sound?
2. Place your fingers on your neck, near your vocal cords. Say "Ah" loudly. What do you feel? Try some loud sounds and some low sounds. (What is happening in your throat is similar to what happened to the opening of the balloon.)
3. Make a working model of human vocal cords: Stretch a piece of balloon over the end of a cardboard tube, but not too tightly. Cut a narrow slit in the piece of balloon and blow into the tube from the opposite end. Now tighten the balloon (this makes the slit wider), and blow again. What do you observe?

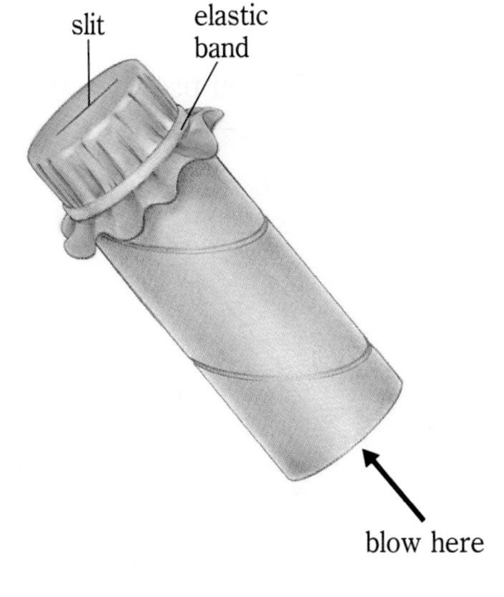

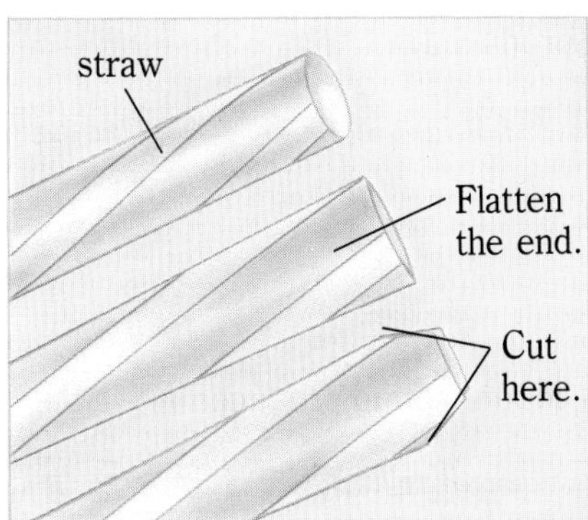

Making Sense of Your Observations

After doing Exploration 1, you now know the following ideas about sound:

1. *Sound is caused by vibrating objects.* What vibrates in each of the experiments in Exploration 1?
2. A force is exerted on a part of an object to move it a certain distance and to start it vibrating. This means that work is done on the object. Whenever work is done on an object, energy is given to it. In the case of sound sources, this energy causes a vibration, which results in sound—one of the forms of energy produced by the vibrating object. *Energy must always be added to an object for sound to be produced.*

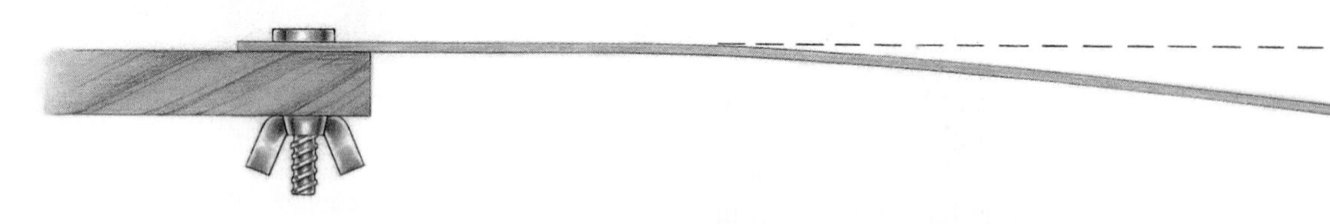

How did you provide energy in each of the experiments in Exploration 1—particularly the last two?

3. *If more energy is added to an object, it passes through a greater distance as it vibrates, and the sound is louder.*

ruler still—no sound

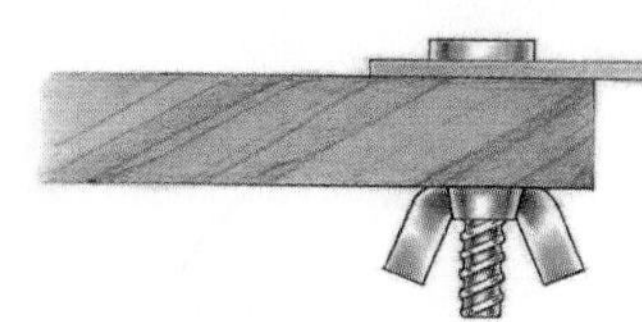

ruler pushed down and released—sound produced

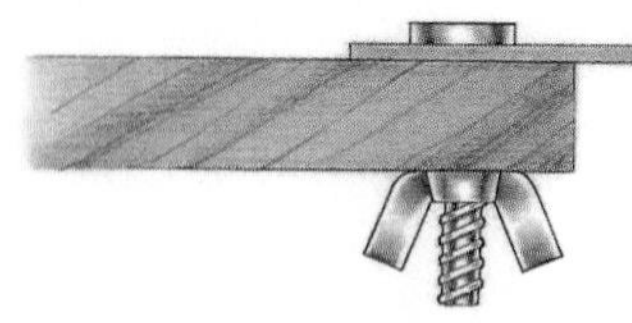

ruler pushed farther down and released—louder sound produced

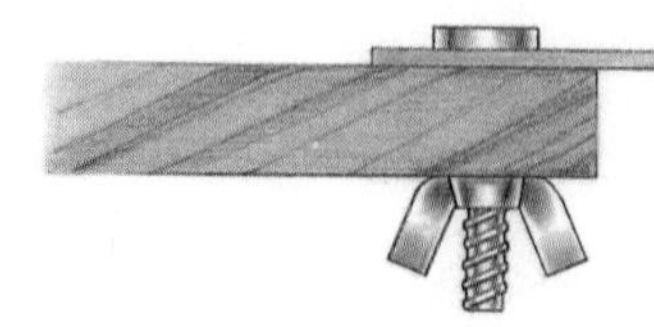

How did you produce a louder sound in each experiment in Exploration 1—particularly the last two?

4. *The more often an object vibrates within a given period of time, the higher the sound's pitch.*

vibrates slowly—low sound

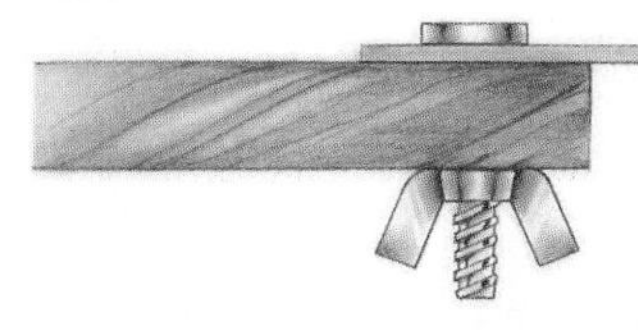

vibrates faster—higher sound

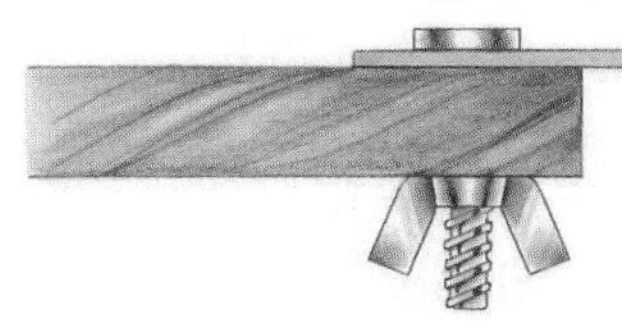

vibrates still faster—still higher sound

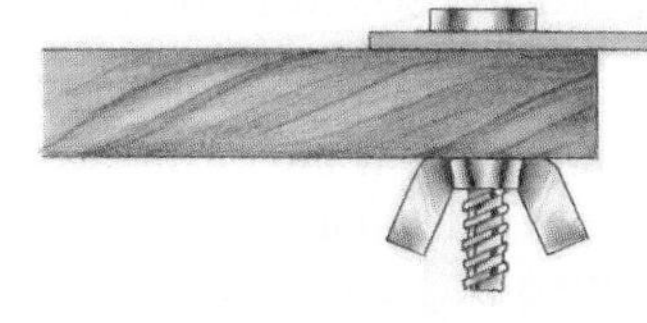

How do our vocal cords produce higher-pitched sounds? How did you get higher-pitched sounds with the rubber band? How did you get higher sounds with the straw saxophone?

5. *Many vibrating objects have a natural rate of vibration, which depends on their length or size.* Can you use this idea to relate the sound of an object to its length or size? How did the length and size of the tuning fork affect the sound it produced? How did the length of the straw affect the sound? If you dropped a long pencil and a short one, what different kinds of sounds would you expect to hear? Try it. Then try dropping a large book and a small one. What would you do if you wanted to keep the loudness approximately the same?

Some Puzzlers

1. What is vibrating when you perform each of the following actions?
   - blow over the mouth of a soft drink bottle
   - tear a piece of paper
   - boil a kettle of water
   - open a soft drink bottle
   - play a record
   - tap-dance on an uncarpeted floor

2. What is vibrating when you play each of the following musical instruments?
   - a piano
   - a harmonica
   - a trumpet or bugle
   - the drums
   - a violin
   - an oboe

EXPLORATION 2

Sound Questions to Investigate

Part 1

Design an experiment with a meter stick to answer these questions.

- Must vibrations occur at a certain rate to be audible (capable of being heard)?
- Must vibrations move a certain distance to be audible?
- Is a minimum amount of energy needed for sound to be audible?

Write your procedure in your Journal and then perform the experiment. The diagrams in "Making Sense of Your Observations" (pages 270 and 271) may give you some ideas.

Sounds Heard by Animals

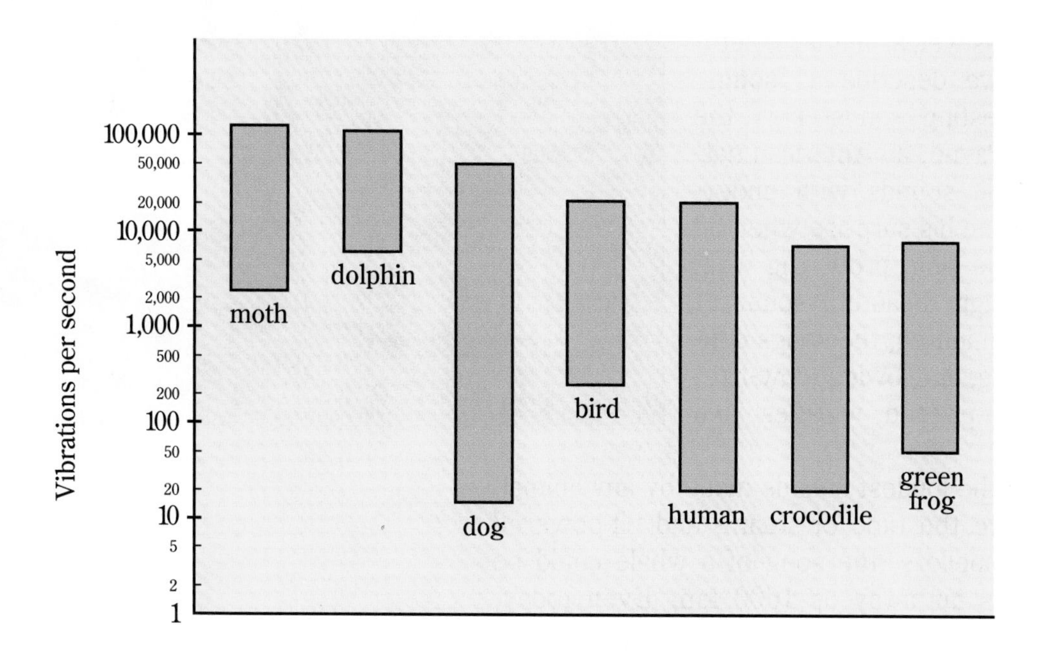

Part 2

The graph above shows the approximate vibration rates required to produce sounds that various animals can hear. From the graph, find out the information below. Hint: First write down the approximate range of vibrations for each animal.

1. Note that a vibrating object must move back and forth about 20 times per second to produce a sound that humans can hear. Above what rate of vibration are there sounds that humans cannot hear?
2. Does any animal hear sounds that people can't hear? Which animal(s) can hear higher sounds than are audible to people? lower sounds?
3. Tawanda blows a whistle. She hears nothing, but her dog runs to her. Why?
4. Which animal can hear the largest range of vibration rates?
5. What sounds can be heard by a green frog but not by a bird?
6. Which animal misses most of the sounds that humans can hear? What range of vibration rates do humans and that animal share?

Animal Sounds

Animals produce sounds of many kinds and make them in many ways. A dog, for instance, whines, whimpers, barks, growls, and howls. To make these sounds, it uses vocal cords just as humans do. In fact, this is true of most mammals, as well as other animals, including frogs. Many kinds of frogs also have balloon-like air sacs, which amplify the sound from the vocal cords.

Some other mammals (as well as humans) actually make sounds that we describe as songs. Perhaps the most interesting songster is the humpback whale. Roger Payne, an American naturalist, investigated whale sounds with underwater microphones. Imagine his surprise when he discovered that humpbacks sing, producing what he described as "hauntingly beautiful sounds." Lasting from six to thirty minutes, the songs consist of low moaning, wailing, and lowing (cow-like) sounds, along with high-pitched whistles and screeches.

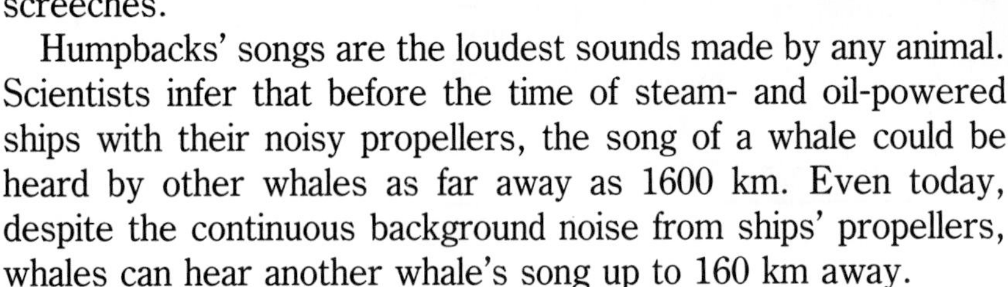

Humpbacks' songs are the loudest sounds made by any animal. Scientists infer that before the time of steam- and oil-powered ships with their noisy propellers, the song of a whale could be heard by other whales as far away as 1600 km. Even today, despite the continuous background noise from ships' propellers, whales can hear another whale's song up to 160 km away.

Humpback whales have no vocal cords. Their sounds or songs are probably produced by forcing air back and forth through the **larynx,** or "voice box."

The most familiar animal songs are those produced by birds. Birds' songs are made up of chirps, trills, whistles, and sometimes quick mechanical sounds, made by clicking the beak. The nonmechanical sounds are produced by an organ called the **syrinx,** or "song box." The syrinx is located at the bottom of the windpipe (the tube connecting the mouth to the lungs). It contains a pair of membranes that vibrate when air passes over them, producing sound. Muscles attached to the membranes change the pitch of the sound.

Human vocal cords cannot produce the high sounds that birds are capable of making. That's because the membranes of our vocal cords cannot vibrate fast enough. However, we can imitate birds in a crude way by whistling. When we whistle, we cause air to vibrate at a very high speed, which produces sounds as high-pitched as those of birds.

Picturing Bird Songs

(a) musical notation (meadowlark)

(b) audiospectrograms of songbird songs

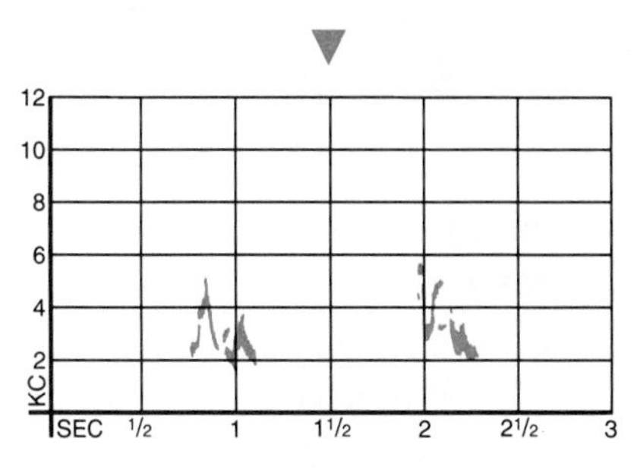

Red-eyed Vireo

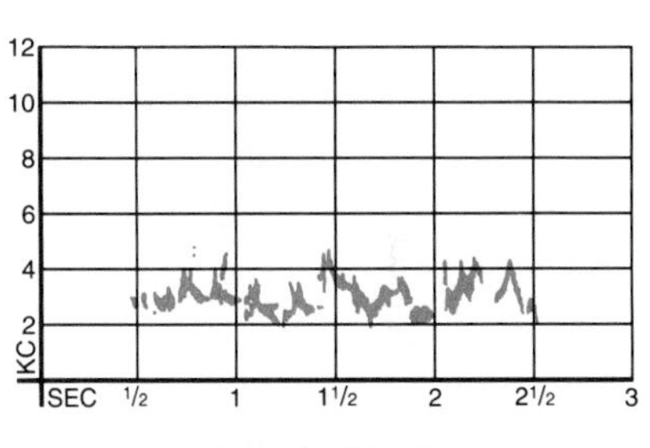

Purple Finch

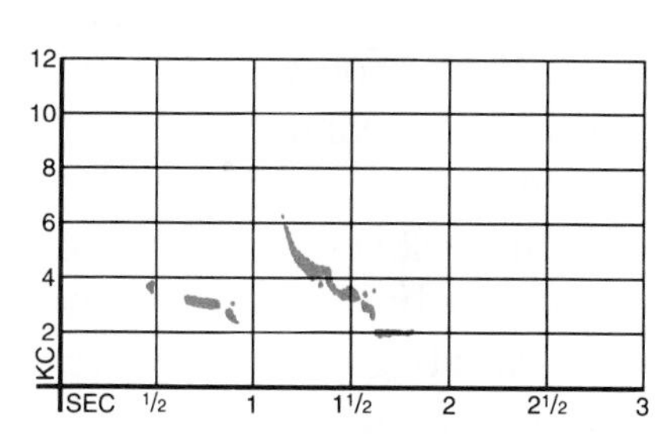

Eastern Meadowlark

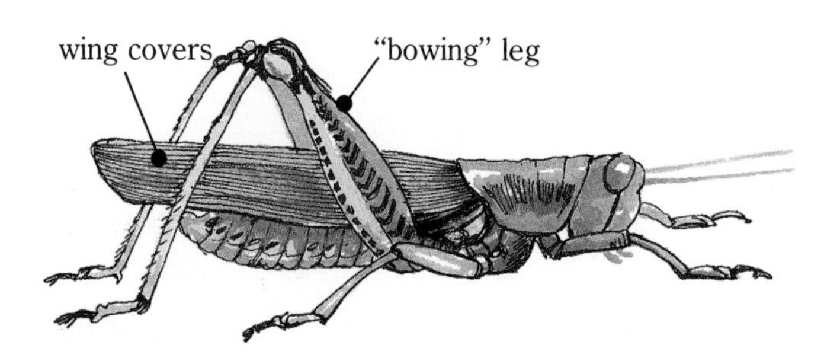

Grasshoppers produce their mechanical sounds by rubbing or "bowing" their long, hard, rough legs against their hardened wing covers. Mosquitoes produce their high-pitched whine by flapping their wings in the air many hundreds of times a second. All male crickets and females of certain species chirp by moving a body part that looks like a toothed file, fastened under one wing, against a scratcher located under the other wing. Each of the approximately 4,000 kinds of crickets produces its own distinct pattern of chirps. This is how members of the same species recognize each other. There may be 50 kinds of crickets singing in a meadow, but they do not get confused. Female crickets can recognize the call of the male crickets of their species.

Remember that sound is energy—and that it takes energy to make sound. For each chirp of a cricket, 19 pairs of muscles are doing work!

Other types of animals besides insects use parts of their bodies to produce sounds. For example, lobsters on the ocean floor "clack" their messages with parts of their external skeleton. Underwater sounds such as these, unknown until recently, are actually abundant. Satin fish produce purring sounds. And by grunting, the male cod signals to a female that "this is a good place to lay eggs." Fish may produce these sounds by the vibration of gases in their swim bladders.

4 Vibrations: How Fast and How Far?

When a robin sings, the membranes of its syrinx vibrate as many as 20,000 times in one second. When you pluck middle G on a guitar, it vibrates 400 times in one second. The lowest moaning of a humpback whale comes from forcing air back and forth in its larynx 20 times in one second.

Since the vibrating objects that produce sound move so fast, how can you see them? The answer is to take pictures of them with a high-speed motion picture camera and then project the pictures at a very slow speed. You would then see the vibrations in slow motion.

In the following Exploration, you will analyze a slow, "soundless" vibrating situation. This will help you understand the much faster vibrations that produce sound.

EXPLORATION 3

Investigating an Object Vibrating Very Slowly

You Will Need

- a heavy button, washer, or large paper clip
- a metric ruler
- a watch or clock with second hand
- fine thread
- scissors

What to Do

1. Make a pendulum by hanging a button on a thread that is 25 cm long. Pull the button 10 cm to one side and let it go.

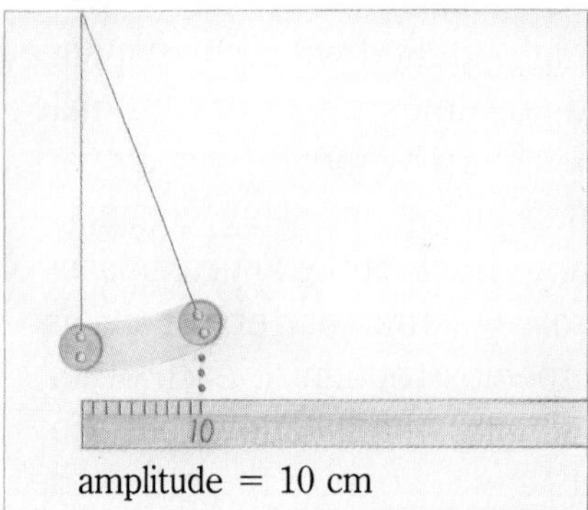

2. A **vibration** is defined as one complete back-and-forth movement of an object from one side to the other and back again. Approximately how long does one vibration take? Does the time seem to change from one swing to the next? Devise a way to find the time for one vibration. About how many vibrations are there in one second?

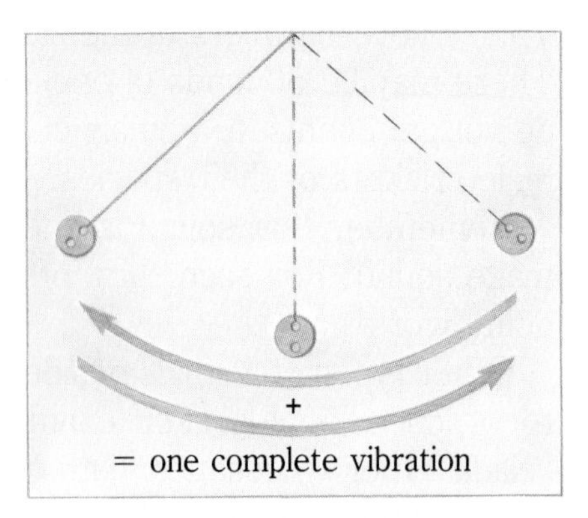

3. Repeat Step 2, but pull the button 20 cm to one side. The size of the swing—the horizontal distance from the side position to the central position, as shown below—is called the **amplitude.** Does the time for one vibration appear to change when the button swings twice as far?

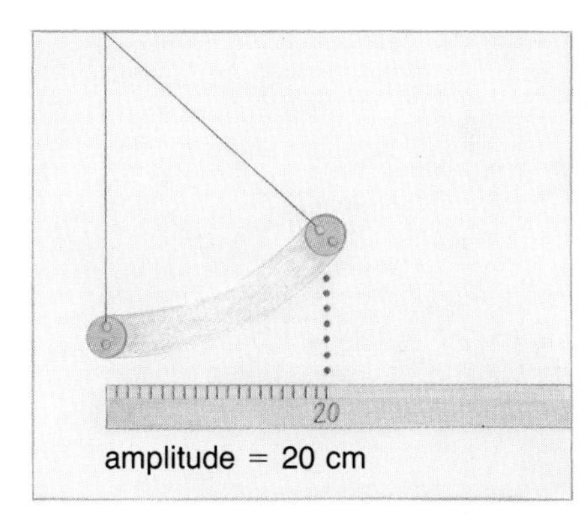

amplitude = 20 cm

4. Shorten the thread to 6 cm. Pull the button 3 cm to one side and let it go. Find the time for one vibration. About how many vibrations are there in one second? The number of vibrations in one second is called the **frequency** of vibration.

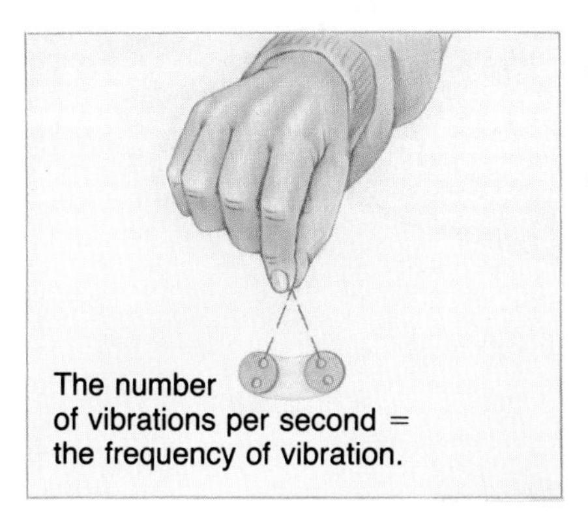
The number of vibrations per second = the frequency of vibration.

Suppose you had to explain the meaning of *vibration, amplitude,* and *frequency* to Mason, a fourth-grader. How would you do that, using a metal ruler as your teaching tool? In your Journal, write down what you would say.

Check It Out

Review what you discovered in doing the previous Exploration. From the statements below, choose those you agree with, those you disagree with, and those you are unsure about.

1. Changing the amplitude of the swing did not change the time needed for one swing very much. If I could do the experiment accurately enough, it probably wouldn't change the time at all.
2. If I double the distance the button swings, I can make a big change in the time it takes for one vibration.
3. If I were to start the swing only 5 cm away, there would be fewer vibrations in one second.
4. If I shorten the thread, the time needed for one vibration will be reduced.
5. If I shorten the thread, the frequency will be greater (the button will complete more swings in one second).
6. If I use about one-quarter of the length of thread, the time it takes for one vibration is about half that required for using the whole length of thread.
7. Changing the length of the thread does not change the number of vibrations that occur in one second.
8. If I were to make the thread 100 cm long, the time of vibration might be 2 s.

Jennifer's Sound Experiment

Jennifer wanted to actually count the vibrations of a sound-producing metal ruler. She took pictures of it with her high-speed movie camera and then projected the film at a much slower speed—about 1/200 of the original speed. Suppose one back-and-forth motion (one vibration) took one second on the screen. Satisfy yourself that the actual frequency of vibration of the ruler would be 200.

Jennifer counted the number of vibrations in a given time from her screen projections, as observed in slow motion.

Case 1: 20 vibrations in 10 s.

Case 2: 31 vibrations in 15 s.

Case 3: 30 vibrations in 10 s.

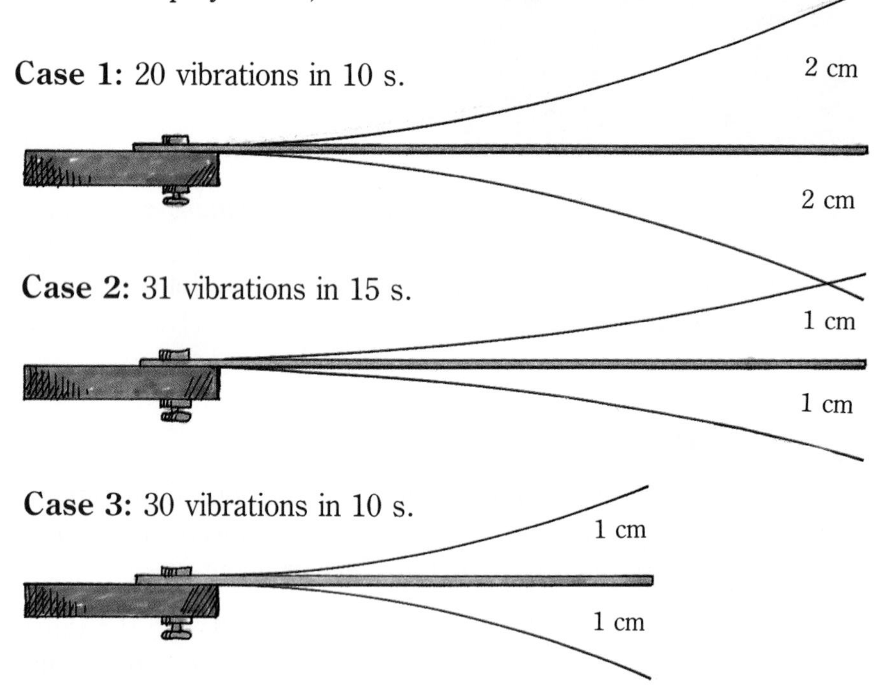

For Case 1:
What is the frequency in slow motion? Try calculating the actual frequency.

For Case 2:

(a) What is the frequency in slow motion? in real motion?

(b) How do the amplitudes in Cases 1 and 2 compare?

(c) Does the amplitude affect the frequency of vibration?

For Case 3:

(a) What is the frequency in slow motion? in real motion?

(b) How do the lengths of the two blades in Cases 1 and 3 compare?

(c) Do their lengths affect the frequency of vibration?

Conclusions:

(a) In which case is the loudest sound produced?

(b) In which case is the highest-pitched sound produced?

Challenge Your Thinking

1. Sara is totally deaf and has never heard a sound. You are able to communicate with her only by writing. What would you write to help Sara understand the ideas of high- and low-pitched sounds, soft and loud sounds, and how we distinguish one sound from another?

2. Ryan attached a piece of nylon fishing line to a nail at one end of a board. He stretched it along the board and hung a pail of sand from the other end. When he plucked the string, he got noise. When he slipped a wooden dowel between the string and the board, though, he got a pleasant musical sound. Why? What different ways could he devise to make the musical sound higher? louder? of a different quality?

3. How would you demonstrate or explain convincingly each of the following to someone who has never studied science before?
   (a) Energy is needed to produce sounds.
   (b) Sound is produced from vibrating objects.
   (c) Not all vibrating objects produce sounds that you can hear.
   (d) Bats often communicate in sounds that humans can't hear.
   (e) The vibrations that produce noise and music are different.

Sound Travel

How does sound energy travel from the sound source to your ear? Do you have a theory that might explain this process? What evidence do you have for your theory? The six experiments in the next Exploration will help you build evidence for a theory of sound travel.

EXPLORATION 4

Transmitting Sounds

What to Do

Experiment 1

How are other forms of energy transmitted?

1. Place five pennies flat on the table.
2. Flick a sixth penny so that it hits penny E.
3. What happens to penny A? What kind of energy was transmitted from E to A? What was the energy transmitted through?

A

B

C

D

E

Scientists tell us that air is made of tiny particles. Is sound energy transmitted through these particles just as kinetic energy (energy of motion) is transmitted through the pennies?

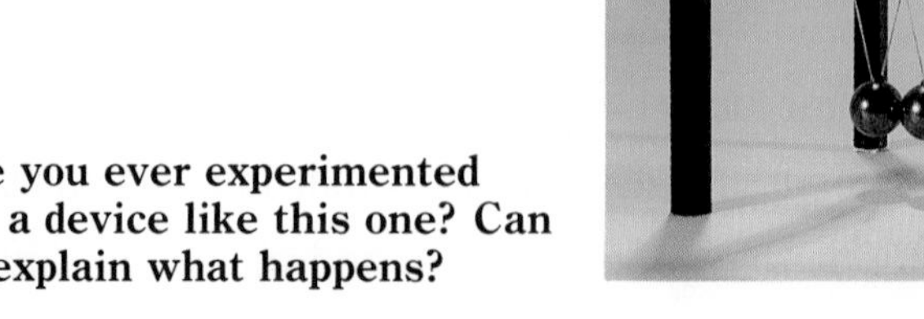

Have you ever experimented with a device like this one? Can you explain what happens?

Experiment 2 

If there was no air (or anything else) between a sound source and your ear, would sound still be transmitted to you?

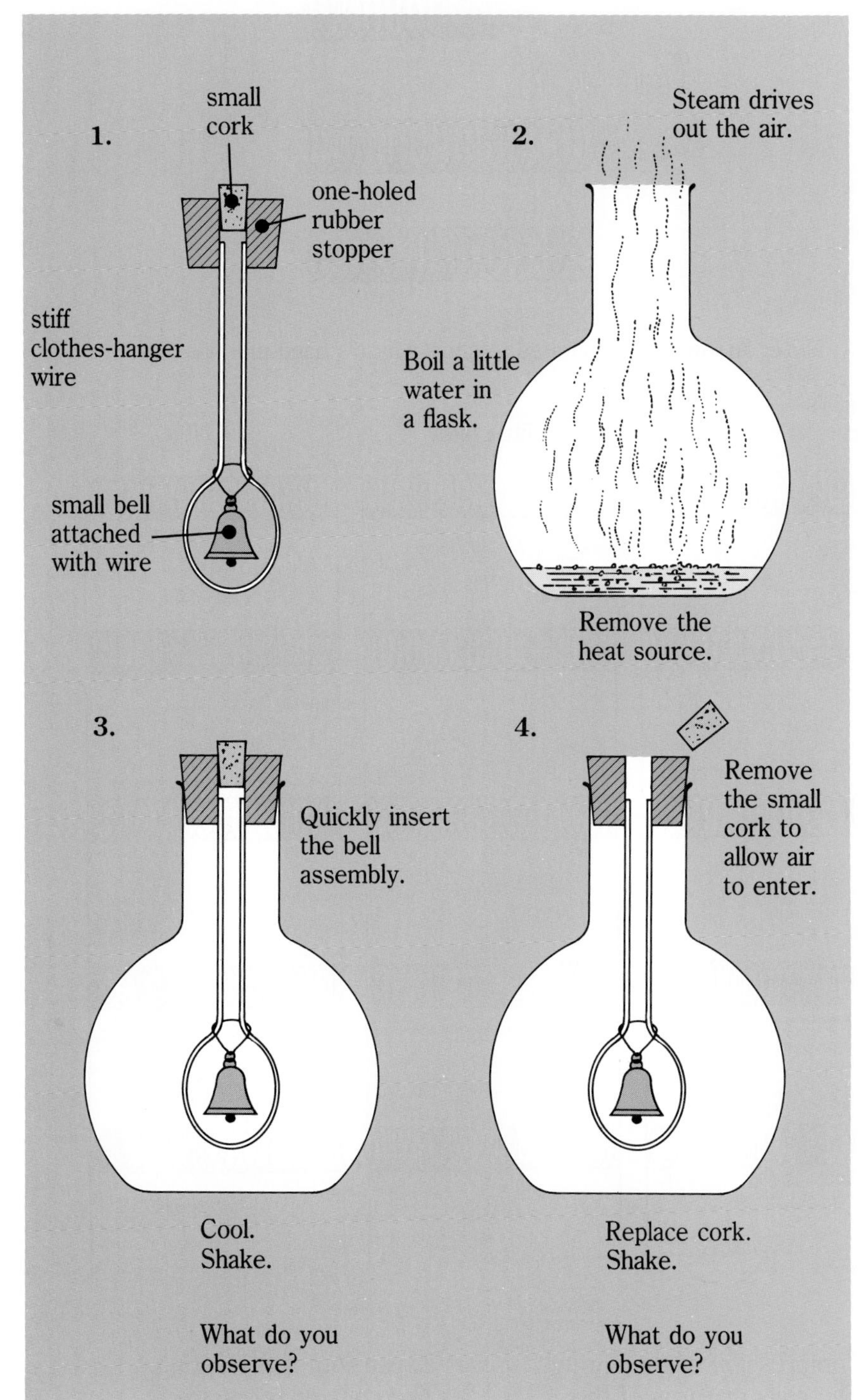

Experiment 3  

Does the air between a sound source and your ear move?

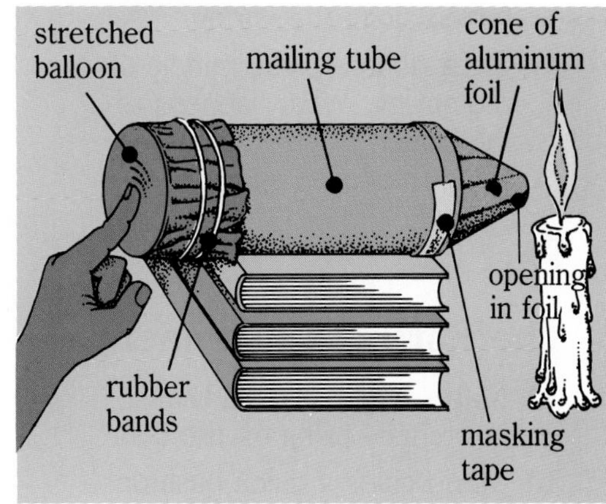

1. Touch the stretched balloon lightly with your finger. What happens? Is air moving in the tube?
2. Make a loud sound by hitting the bottom of a wastebasket with your hand near the stretched balloon. Observe the flame.
3. Try counting "1-2-3-4" loudly, close to the balloon. What happens?

When sound energy moves from a vibrating object through the air to the ear, what do you think happens to the air?

Experiment 4

Part 1

A Model of Air

How is energy transmitted through a coiled spring?

1. Hold one end of a spring and have a classmate hold the other end. Send pushes and pulls along the spring. Do you see energy being transmitted?

2. What do you feel when the "push" or "pull" reaches your hand?
3. How fast is the energy transferred along the spring?
4. Tie a ribbon to the middle of the spring. What happens to the ribbon as energy passes along the coiled spring?

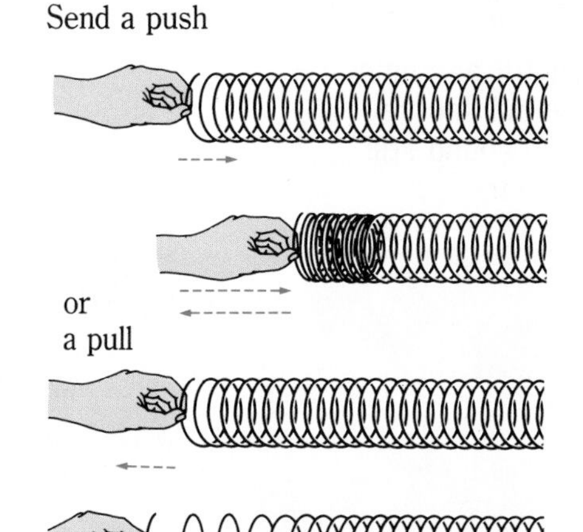

Note: Arrows are separated to show Mario's hand motions.

Part 2

The Spring Experiment

(a) Mario and Kate hold the stretched spring on the floor.

(b) Mario gives a quick push on the coiled spring, toward Kate. He then returns it quickly to its original position. Do you see the energy being transferred?

(c) When the coiled spring is still again, Mario gives a quick pull on the spring toward himself. He returns it quickly to its original position. Do you observe the energy being tranferred?

(d) Mario pushes his end of the spring quickly and then pulls it back quickly. Next, he pushes the end of the spring back to its starting position. Mario has produced one complete vibration of the end of the spring. *One compression and expansion together create a sound wave.*

(e) Mario produces two complete vibrations of the end, one right after the other.

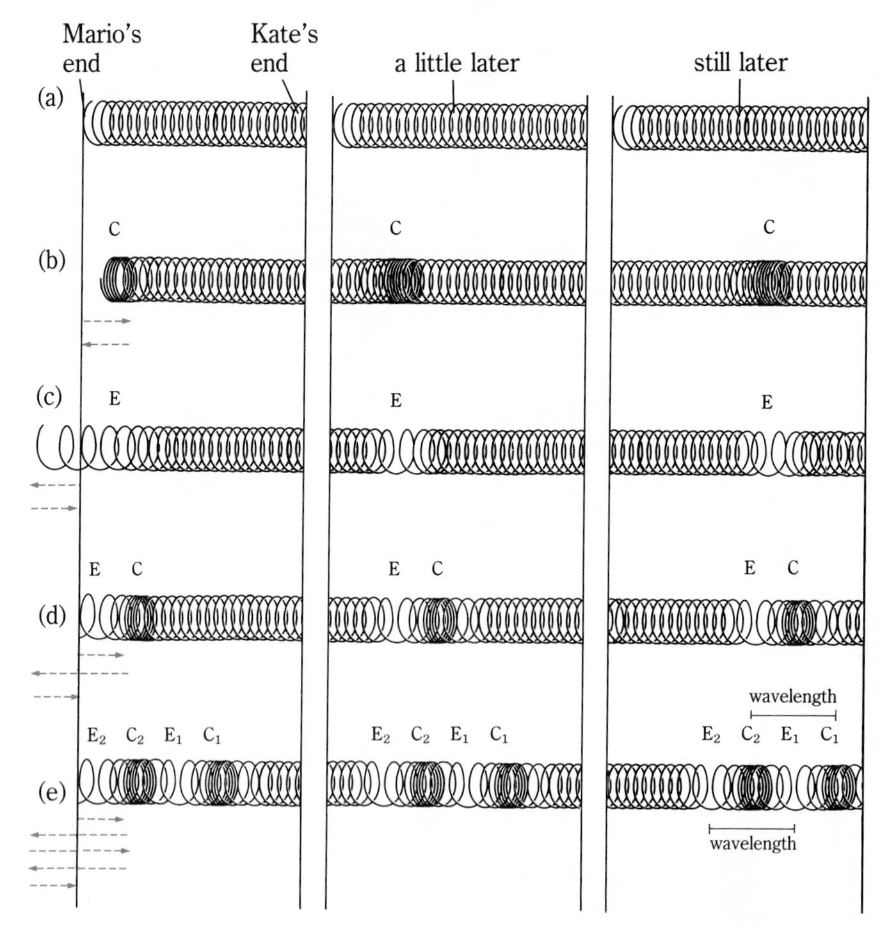

Key:

C = Compression of the spring. The coils or turns of the spring are closer together than normal.

E = Expansion of the spring. The coils or turns are spread farther apart than normal.

Use what Kate and Mario observed on page 282 to help you answer the following questions.

Questions

1. When Mario pushes in (b), what happens to the turns (coils) of wire?
2. What happens to the spring compression after Mario stops pushing?
3. When Mario pushes on the end of the spring in (b), he gives energy to the spring. What happens to Kate's hand when she receives the compression?
4. When Mario gives a pull as in (c), what happens to the turns of wire?
5. What happens to the spring expansion after he stops pulling?
6. Every time Mario makes his end of the wire go through one vibration, as in (d), what is given to the coiled spring?
7. How is the energy that Mario gives to his end of the spring passed along to Kate?
8. What is happening to each turn of wire as Mario sends a series of "waves" along the spring? (Review your observations of the ribbon.)
9. Suppose that air particles act like the turns of wire in a coiled spring. How would sound energy from a vibrating tuning fork pass through the air to the ear?

A Problem to Ponder

What could you do to the spring to make the compressions and expansions move faster? Try it.

Experiment 5

Elastic Air

Are air particles elastic? That is, do they act like the turns of wire in a coiled spring?

1. Fill a syringe half-full of air, and plug its tip with modeling clay.

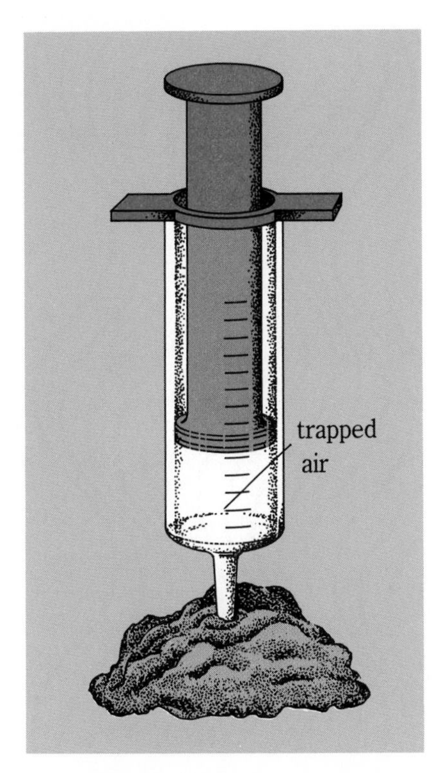

2. Push down on the piston, compressing the trapped air. Let the piston go. What happens?
3. Now pull up on the piston, letting the trapped air expand. Let the piston go. What happens?
4. Try compressing and extending a ball of clay. Which behaves more like a spring, clay or air?

Experiment 6

Viewing Moving Waves

Here's another way to view what may be happening in air.

Cut a narrow slit 10 cm long and 1–2 mm wide in a piece of cardboard. Place the slit over the diagram on the next page and move it down at a constant speed.

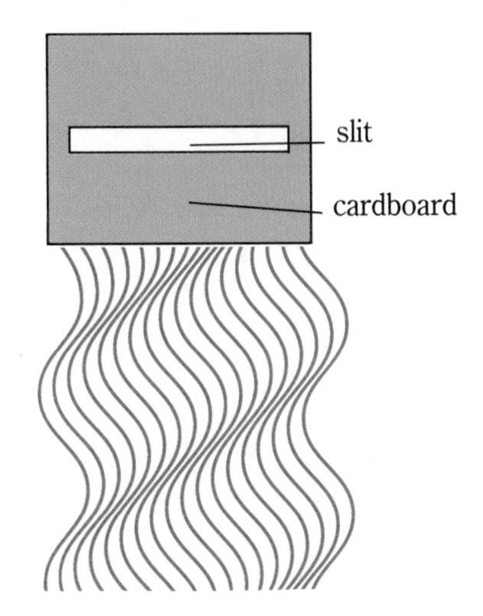

Questions

1. In which direction does the compression move? (It's difficult to see expansions move.)
2. How can you make compressions move in the opposite direction?
3. How can you make compressions move faster?
4. How many compressions can you see at one time?
5. What do you think is the **wavelength?** Measure it.
6. As compressions move along, what is happening to each line as the card goes down the page?
7. As a compression travels along a spring, what is each turn of wire doing?

Drawing Conclusions

In which experiment did you draw the following conclusions?

1. Sound energy is not transmitted through a vacuum. In other words, sound cannot be transmitted where there is no air.
2. Air is elastic like a spring.
3. Energy often uses a **medium** (consisting of particles) for its transmission.
4. When sound moves through air, the air moves.
5. One way in which energy may be transmitted is in the form of waves of compression and expansion.

Using this evidence, construct a theory of how sound energy is transmitted through air from a sound source to your ear.

On the next page is a series of diagrams that will help you visualize what is happening in the invisible air around a vibrating object. Later, you will be able to compare the theory that these diagrams suggest with the theory you constructed here.

Sound Thinking

Figure 1

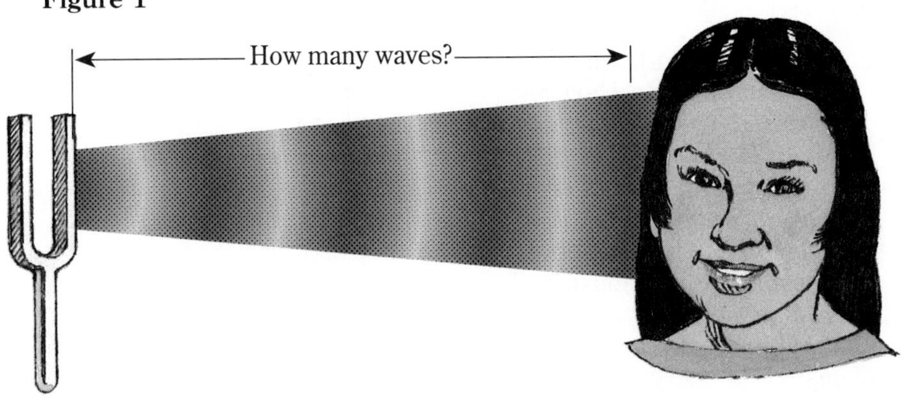

Figure 2

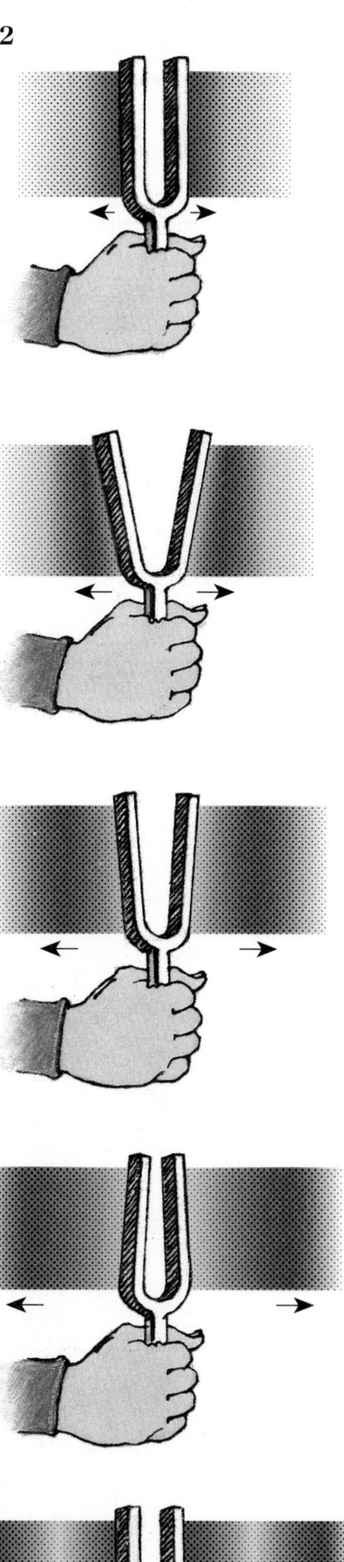

1. Here are illustrations of what you might see if the air around a vibrating tuning fork were visible. Locate where the particles are pushed together (compression) and where they are much farther apart (expansion). In Figure 1, how many waves are shown in the bracketed region? How many times has the fork vibrated in that same region?
2. Measure the wavelength in Figure 1 with a ruler. For waves of the length shown, how many centimeters would a compression move to the right when the tuning fork vibrates once? when it vibrates five times?
3. Figure 1 is a scale drawing of what would happen in air with a tuning fork of a certain frequency. In the figure, 1 cm along the wave (in red) represents approximately 40 centimeters in reality. How far would the sound travel during four complete vibrations of the tuning fork? during one vibration?
4. The tuning fork in the illustration vibrates 440 times in one second. How far (in centimeters) would a wave travel through the air in one second? How many meters would this be?
5. In question 4 you calculated the distance the wave traveled through the air in 1 second. In so doing, you estimated the speed of the sound wave. Complete the following equation:

   wavelength × __?__ = speed of sound waves
6. Suppose a tuning fork vibrates at the rate of 220 times in one second. What would be its wavelength, given the value for the speed of sound that you obtained in question 5? How would an illustration of the sound waves produced by this tuning fork differ from Figure 1 above?
7. Figure 2 shows what happens during one complete vibration of the fork. Add four more drawings to show what happens during the second vibration of the fork. Include the motion of the first vibration in your new drawings.

EXPLORATION 5

Investigations of Sound Transmission

Is sound transmitted through wood, metal, cardboard, plastic, and other materials? Can you hear a scratch through wood that you cannot hear through air? Find out by doing the following activities.

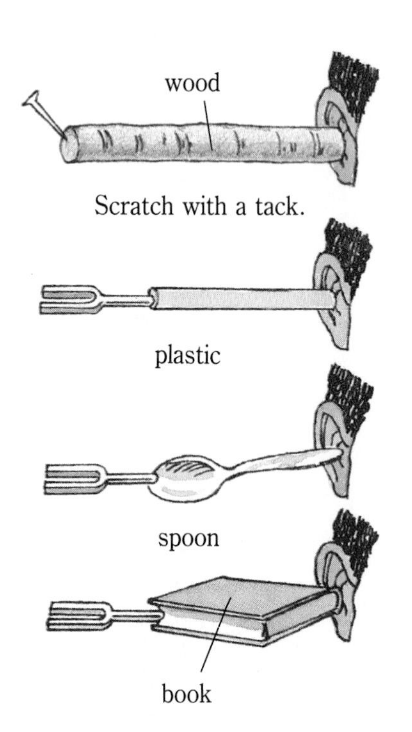

1. Use a tuning fork with different materials. **Be Careful:** *Rest each material carefully so that it touches the outside of your ear canal. Clean each material with rubbing alcohol before use.* Which material transmits sounds most loudly? In other words, which material is most efficient in conducting sound energy? Is there a material that will not transmit sound?
2. Which transmits sounds more efficiently—air or water? Try the following activities: Hit two spoons together in air and then in water, listening carefully to the sounds in each case.
3. Try the following combinations at home in the bath tub. **Caution:** *Keep electrically operated devices away from water.*
   (a) Sound source in water, ears in air, and ears in water;
   (b) Sound source in air, ears in air, and ears in water.
4. You've probably seen a movie in which someone put his or her ear to the ground to hear the sound of horse hooves before the sound could be heard through the air. Design an experiment to test this idea, using the floor. You will need a very quiet room.
5. Can you hear better through your bones? Design an experiment that uses bones in the transmission of sound. **Caution:** *Take care not to cause undo stress on any part of the body. Each student should use a fresh rubber band for the experiment shown below.*

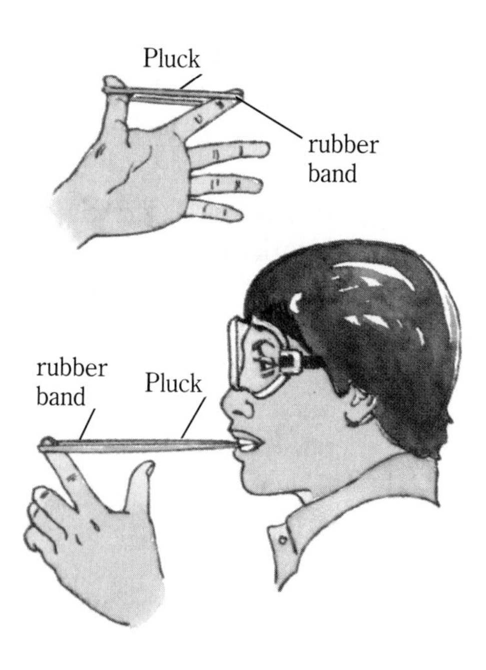

6. Do some humming. While you hum, plug your ears. What difference(s) do you notice in the sound? Why does this occur?

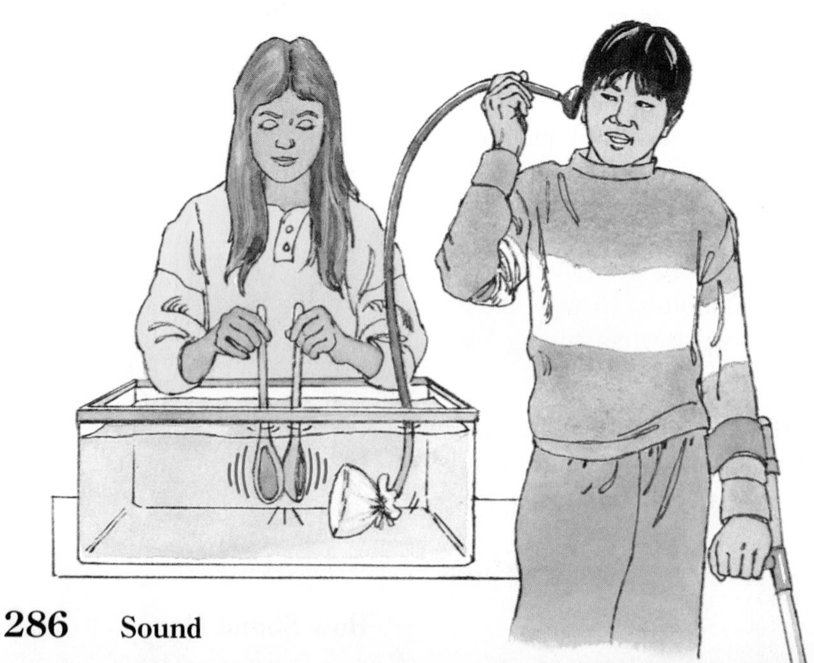

6 The Speed of Sound

What Is Speed?

How did Travis calculate the speed? What facts did he need to know? Travis divided the distance in meters by the time in seconds and found the speed in meters per second. Lena's motorcycle speed tells us that she traveled 50 m in one second. (To change this unit to kilometers per hour, the speed unit used in racing, multiply by 3.6.) How would you define *speed?*

An Unusual Race!

Was Lena's speed really that good? Suppose Lena raced for 5 seconds with each of the following:

(a) a cheetah (the world's fastest animal sprinter)
(b) a peregrine falcon (the world's fastest bird)
(c) the fastest human runner
(d) the *Bluebird* (a record-holding racecar)
(e) the sound from a pistol shot
(f) the *Concorde* (a supersonic jet airplane)
(g) *Apollo 11* (a spaceship)
(h) the flash of light from a pistol shot
(i) the bullet from a pistol

Scale: 1 cm = 50 m
Drawn to scale, the distances shown represent how far each competitor went in 5 seconds.

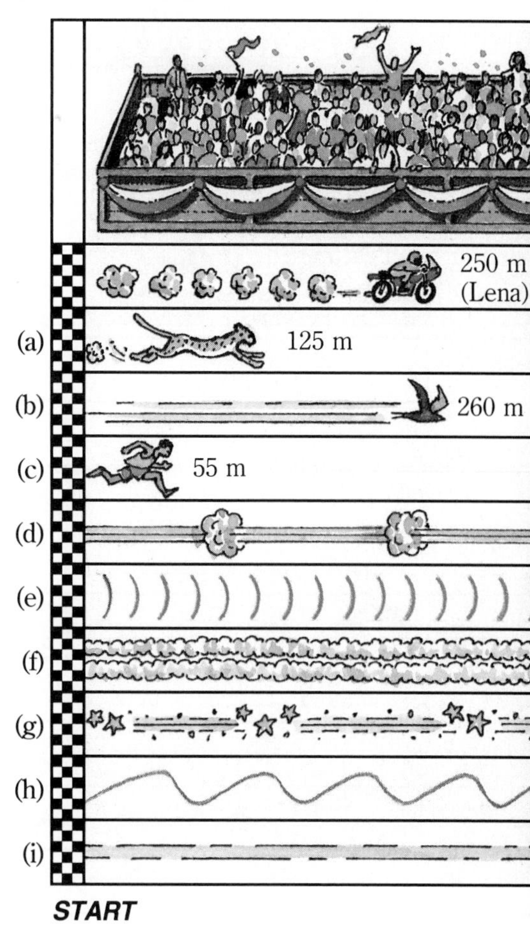

1. Which of these would Lena outrace?
2. Which would outrace Lena?
3. Which of all these has the fastest speed?
4. A pistol is fired 1 km away from a target. Which reaches the target first—the bullet, the flash of light, or the sound of the pistol shot? Which arrives last?
5. What is the speed record (in meters per second) for a racecar?
6. What is the speed of the fastest human runner in meters per second?
7. What is the speed of sound?
8. Objects that approach the speed of sound compress the air ahead of themselves until the air is almost like a solid wall. At the speed of sound, a shock wave is created that makes a thunderous bang. Such a bang is known as a **sonic boom.** Which of the things shown might cause a sonic boom?

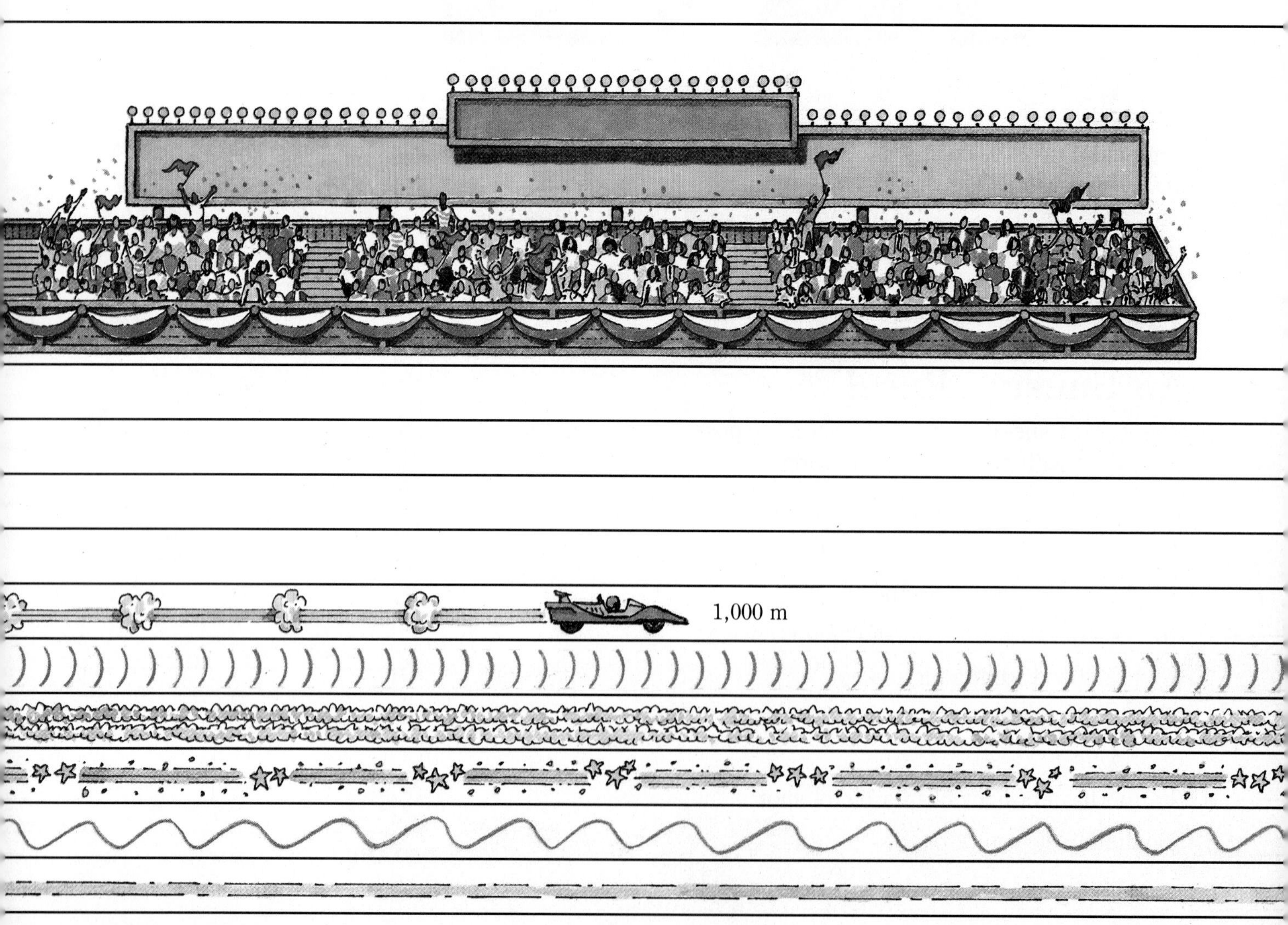

EXPLORATION 6

An Echo Experiment to Measure the Speed of Sound

You Will Need

- a meter stick
- a wall
- a watch or clock with second hand
- Information: the time for a sound to travel a certain distance
- an Equation: speed = $\frac{\text{distance}}{\text{time}}$
- a Sound: clapping
- a Distance: to the wall and back to your ears
- a Time: the time between the clap and its echo

What to Do

1. Stand about 50 m from the wall. Clap your hands. Listen for the echo. Clap your hands twice. Listen for two echoes.

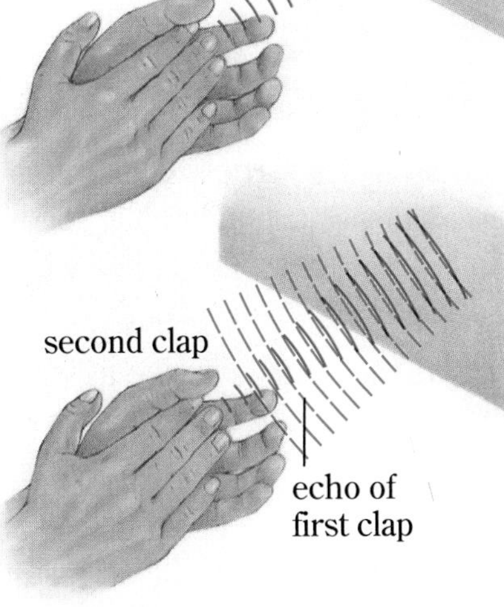

Try clapping your hands at such a speed that you begin a new clap just when you hear the echo from the previous clap. You won't hear the echoes any more.

What does the time between claps tell you?

2. Have someone time how long it takes to clap 20 times in such a way that you cannot hear the echo. Try this several times.

What is the average time between claps? Where has sound traveled in this time?

3. Measure the distance to the wall in meters by pacing the distance (once you've found the length of your pace). Have several people check the distance. What is the total distance traveled by the sound of the clapping—the distance from your hand to the wall and back to your ear?

4. From the distance and time of travel, calculate the sound's speed.

Blake's Data: How did he calculate the speed?

20 claps in 8 s
115 paces to wall
10 paces are 6 m long

Distance sound traveled	= 138 m
Time per clap	= 0.4 s
Speed of sound	= 345 m/s

Did you notice that the speed of sound Blake got was considerably higher than the speed you calculated from the race on page 287? Blake performed his experiment at 25 °C. The speed from the race is what it would have been at 0 °C. What effect does temperature have on the speed of sound? What is the average increase in the speed of sound per °C?

1,650 m

Some Sound Puzzlers

1. What causes thunder to roll, sometimes for quite a long while?
2. The speed of sound at 25 °C = 345 m/s, and the speed of light = 300,000,000 m/s. Can you explain the different observations below? How far from the lightning bolt are the two people in the middle of the drawing? Why doesn't the last person hear the thunder?

3. Suppose you are riding in a car down a city street. Close your eyes. Can you tell when you are passing a building, a space where there are no buildings, or an intersection? How is the sound of the car helping you?
4. Your ears separate sounds as long as the sounds arrive at least 0.1 s apart. How often must the striker hit the bell in order to keep the bell sounding with a steady ring?
5. The Colorado River winds its way through the deep gorge of Colorado Canyon. In this canyon, an outboard motor on a raft can be heard 15 minutes before it actually arrives. Why?
6. Why can't we hear the tremendous nuclear explosions that occur on the surface of the sun?
7. Why is it so silent after a snowfall?
8. Why does your singing in the shower sound so great?
9. A Challenge! How far away from a wall should you be if you want to hear a complete, distinct echo of your voice? (Hint: To solve this, you'll need to estimate how long it takes to say "hello.") How would this distance differ for the two people on the far right of the illustration?

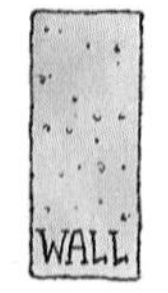

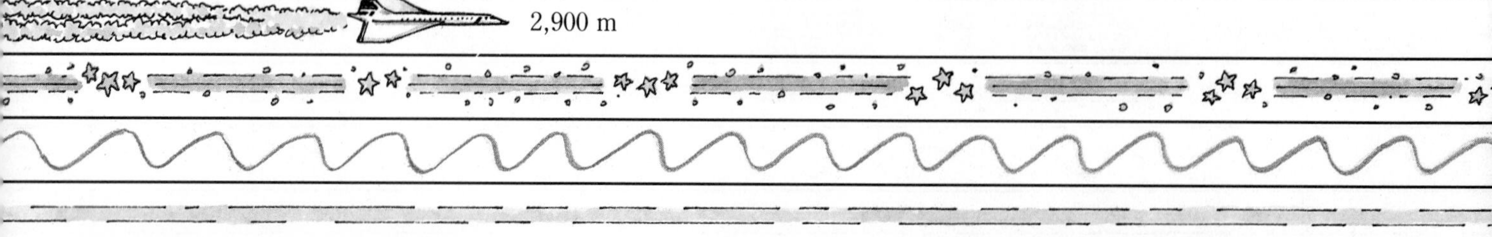

7 More About Echoes

How Do the Blind "See"?

Ask a blind person how he or she is able to approach an obstacle, pause, and go around it. You may get answers like these:

"I feel the presence of the object."
"There are pressures on my skin that tell me an object is there."
"I sense danger."

There is a story about a six-year-old blind boy who learned to ride his tricycle on the sidewalks near his home. He never had an injury or an accident. He could veer around people, and he knew when to turn corners without going into the street. How did he "see"?

For years, there was a blind cyclist who rode his bike in downtown Toronto, Canada. With less than 10 percent vision, he could not see traffic lights, jaywalkers, or road repair crews. How did he "see"?

Blind "beep baseball" player listens for the beeps the ball makes to help him find the ball. Notice the speaker holes on the ball.

A Research Project

A Cornell University professor and two of his students, one of whom was blind, wondered how blind people can avoid obstacles even though they can't see them. The research team designed a project to find out.

Suppose you are one of the research scientists. As a researcher, consider the steps you might follow. The questions below will assist you.

Cover the story to the right of the questions with a card. Gradually move the card down the page to see what the Cornell researchers actually did.

(a) What theories (hypotheses) about how the blind "see" would you want to test?

The researchers were particularly interested in testing the following:

- *Skin-Pressure Theory:* Obstacles send out signals that produce pressure on the skin, enabling the blind to be aware of the obstacles.
- *Sound Theory:* Sounds hit obstacles and bounce back, telling the blind person of their presence.

(b) What kind of people would you use for the experiment—blind, blindfolded, or both?

The Cornell team decided to use both blind and blindfolded subjects. They gave special training to the sighted persons. After a little practice, they, too, could avoid obstacles placed in their path.

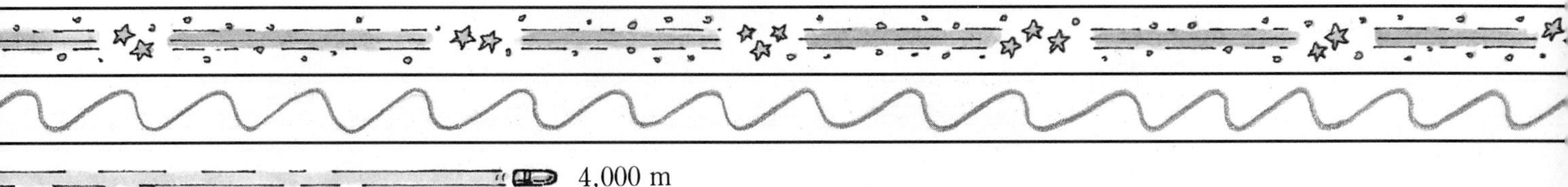

(c) What kind of experimental situation would you set up to test the avoidance of obstacles?

The experimenters walked down a long hallway toward a fiberboard screen. The experimenters changed the position of the screen from trial to trial. Before long, each subject could walk down the hallway, detect the presence of the screen, and stop. On average, subjects detected the screen when they were about 2 m away from it.

(d) Which theory would you test first?

The researchers tested the theory that blind people mention most often: the skin-pressure theory.

(e) What would you do to the subjects to prevent their skin from being affected by the screen?

The subjects had to wear an "armor" of thick felt over their heads and shoulders and heavy leather gloves on their hands. While clothed this way, they couldn't even feel the air from an electric fan, but they could still hear sound.

(f) If they bumped into the obstacle when wearing the armor but didn't when they were without it, what would you conclude? If they still avoided the obstacle when wearing the armor, what would you conclude?

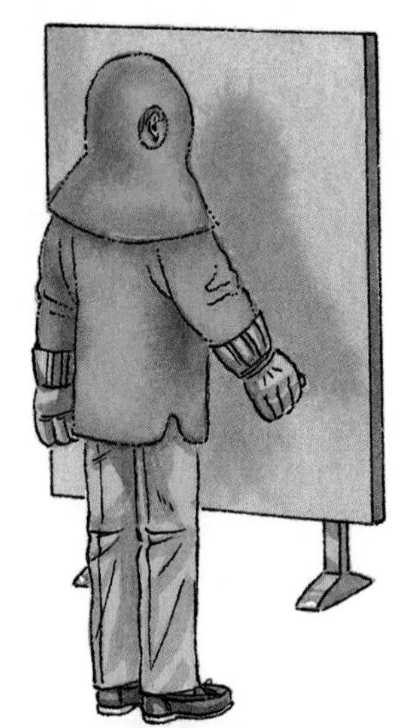

The subjects wearing the armor still avoided the obstacle. However, they tended to approach it more closely, on average, than they did before. They stopped 1.6 m away, as compared with 2.0 m without the armor. The researchers concluded that the skin-pressure theory was not correct. Blind people do not "feel" the pressure of obstacles with their hands and face.

(g) What would you do to test the sound theory?

Earplugs of wax, cotton, earmuffs, and padding were worn by the subjects. They could not even hear their own footsteps. Their faces and hands were left uncovered.

(h) If they bumped into the screen while their ears were plugged, what would you conclude? If they stopped short of the obstacles, what would you conclude?

Spectacular results! Both blind and blindfolded bumped into the obstacle. The blind said all sensation of "feeling" had gone. The sound theory explained the researchers' observations the best.

(i) Some scientists objected to this conclusion. Perhaps the ear coverings changed the pressure on the ear and ear canal—pressure that would have resulted from the obstacle's presence. What could you do now to check whether avoidance of the obstacle was due to sound rather than to some sort of pressure on the ears?

The subjects were placed in a soundproof room some distance away from a hallway where an experimenter walked with a microphone. The sounds of the experimenter's footsteps were picked up by the microphone and carried by a telephone line to the subjects. The subjects were asked to locate the screen by listening to the sounds picked up by the microphone.

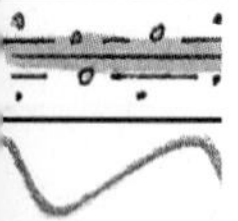

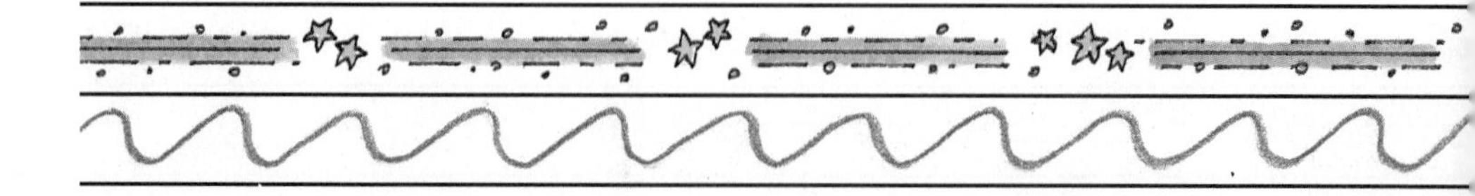

(j) What would you expect to happen if the sound theory were true?

The subjects were able to detect the screen from the sounds picked up by the microphone. The closest approach was 1.9 m. That's more proof for the sound theory! The ability of blind people to avoid obstacles is apparently due to sound—not to pressure on the ear.

(k) Some scientists saw weaknesses in this last experiment as well. Can you think of any?

The experimenter's pace and breathing might have given the subjects hints about the obstacle's presence.

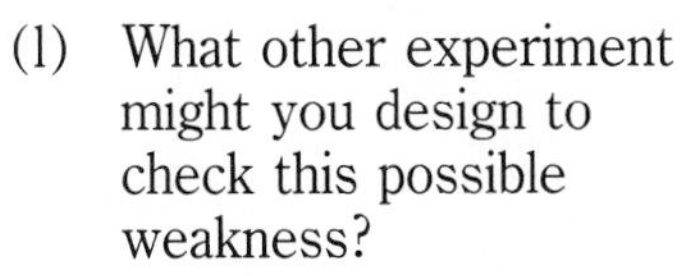

(l) What other experiment might you design to check this possible weakness?

The experimenters decided to use a motor-driven cart to carry the microphone toward the screen, and a loudspeaker to make sounds. The movements of the cart could be controlled remotely by the subjects in the soundproof room.

(m) What conclusions would you draw if the subjects detected the screen?

Although the average closest approach was a little closer than in previous experiments, the subjects always detected the screen. Therefore, signals from the breathing or pace of the walker could not explain why the subjects were able to detect the screen. The experimenters made three general conclusions.

1. Blind people locate obstacles by sound and by the reflection of this sound—that is, by echoes from the obstacles.
2. With practice, sighted people can also learn to detect obstacles by this means.
3. We are often not aware of how our senses and brains are working. The blind sense that they "feel" the presence of objects around them. But these and other scientific experiments show that they are, instead, hearing changes in sound as they approach obstacles.

(n) What further experiments might you now do?

The researchers are still interested in sending different kinds of sounds through the loudspeaker. Through experimentation, they may discover what sounds are best suited for the **echolocation** of objects.

Reporting Your Findings

Often, the final step of a research project is a written report in a scientific journal. In addition, scientists, like many people, enjoy sharing their interests with friends by writing letters. (Galileo, for instance, wrote many fascinating letters about his work to his friends.) Write a letter to a friend, telling him or her of your exciting discoveries in your role as a researcher into how the blind "see."

Echolocation at Its Best!

Whales' and dolphins' sounds include clicks and rasping noises that reflect off of objects in the form of echoes. River dolphins can thread their way among logs and fallen trees in large, muddy rivers. In the open sea, dolphins accurately echolocate the fish they eat.

Bats, too, use echolocation to find food. They catch flying insects in the dark by sending out pulses of high-frequency sound waves (about 45,000–100,000 vibrations per second—a frequency that is inaudible to humans). The waves bounce off the insects, and the bats find their prey by listening for the echoes. In one experiment, a North American brown bat was able to echolocate and catch 1 g of mosquitoes (175 insects) in 15 minutes. If a bat was unable to echolocate, it might have to fly all night with its mouth open before catching one mosquito.

A bat usually chirps about 20 times a second. This number increases to 200 shorter chirps a second as it approaches its prey. How far away from its prey could a bat be and still hear a separate echo?

Many ships have a navigation system that uses echoes to find the depth of the water, the shape of sunken vessels, the presence of submarines or schools of fish, the location of icebergs, and even the structure of the rock on the ocean bottom. This system is called *sonar,* which stands for **so***und* **na***vigation* **r***anging.* Sonar works in very much the same way that a bat uses echolocation. The sonar device sends short pulses of sound waves through the water. When the sound waves hit the sea floor, some of the waves are reflected back as an echo. The echo is then detected by a receiver. The information gathered can then be displayed on video monitors.

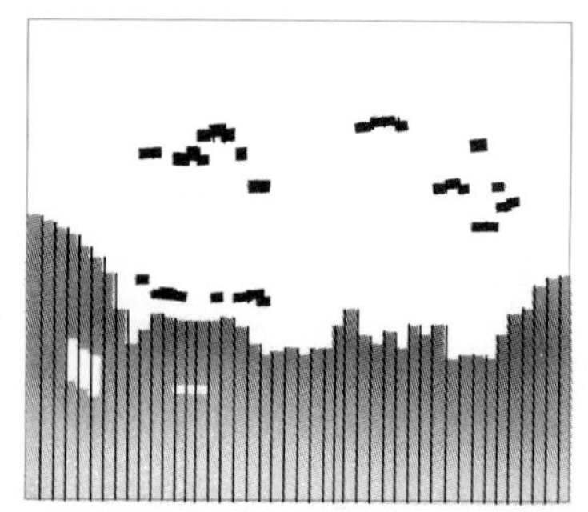

A reading from a ship's echolocation device might look something like this. The dark area at the bottom represents the ocean floor. The dark squares at the top of the screen represent fish.

Completing the Sound Story

You know that all vibrating objects can start sound waves by pushing and pulling on particles of a medium (such as water, wood, or air). These particles then alternately crowd together and spread apart, just as the coils of a spring compress and expand.

But the sound story isn't complete until sound waves activate a receiver, such as your ear. Sound is really the effect that sound waves have on our sense of hearing. So let's complete the sound story by looking at what happens between the time an object *vibrates* and the time we actually *hear* the sound produced by the vibrations.

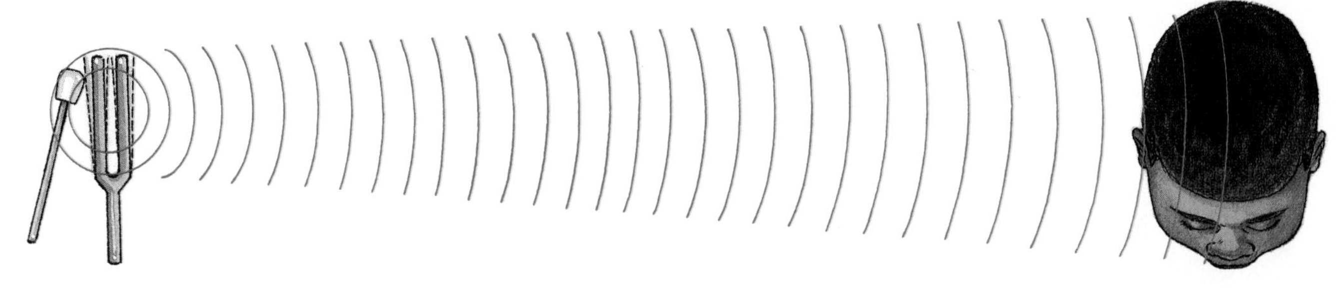

Designing an Instrument for Hearing

An instrument for hearing must be able to:

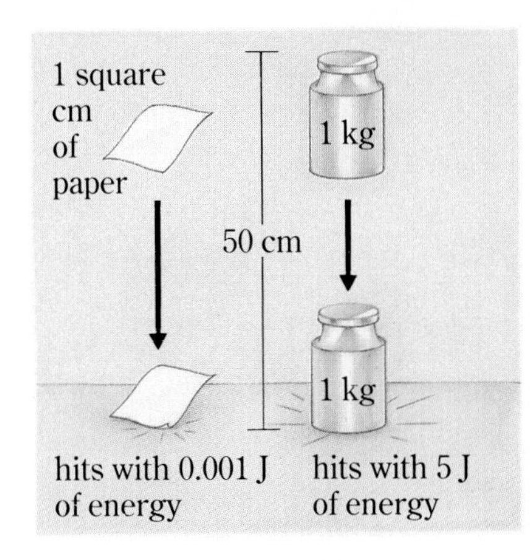

- detect air particles moving back and forth in a sound wave
- detect particles hitting the instrument with only 0.000001 J of energy. (To give you an idea of how subtle this is, a square centimeter of paper dropped from a height of 50 cm hits with 0.001 J of energy!)
- function when subjected to very great air pressure (very loud sounds)—up to 5 J of energy
- detect air particles vibrating as slowly as 20 times per second and as quickly as 20,000 times per second
- distinguish between air particles vibrating 500 times per second and those vibrating 505 times per second. Compare this with the difference in vibrations per second between notes B and C′ on the piano at the right.
- convert waves in air into waves of greater pressure in a liquid
- change the information it has detected into electrical impulses and carry this information to a central recording place
- detect its own position in space—whether it is level, on a slant, upside down, or about to fall over

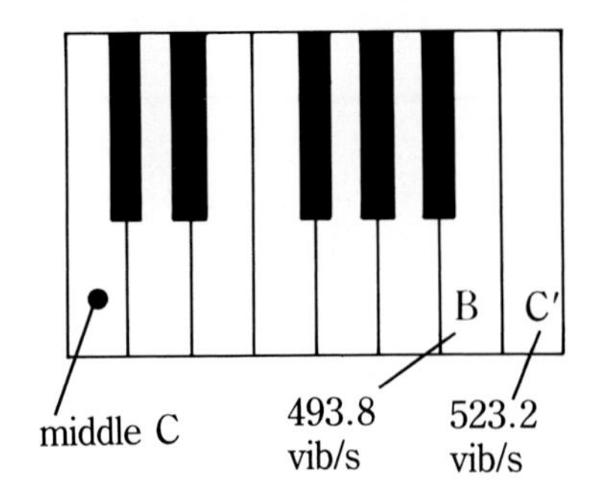

If your device can do all this, congratulations! You have designed the human ear!

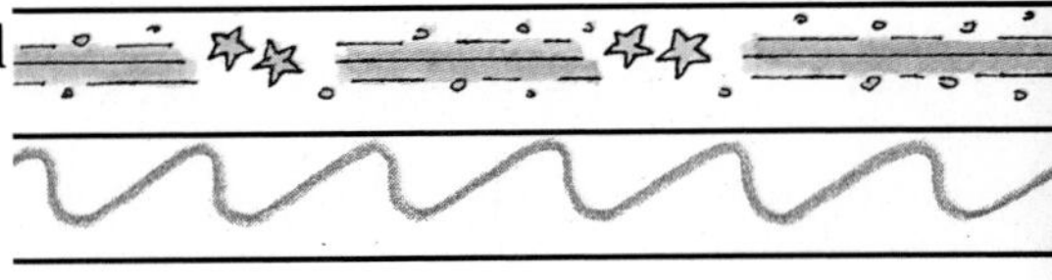

The Human Ear

Here are three unlabeled diagrams of that amazing instrument, the human ear. Figure 1 shows the whole ear, while the two smaller figures show sections of the ear, enlarged for easier viewing. You'll understand how the ear works after reading the story of hearing on the next two pages.

On the next page, the left-hand column describes the three figures shown below. Read it carefully and use the information to label a copy of the diagrams in your Journal. (The labels are in boldface type in the descriptions on the next page and can be matched to the numbered parts in the diagrams.) Then read the story of hearing, located in the right-hand column. It describes how each part of the ear works.

When you have finished, you'll put your understanding of how we hear to use in answering some questions.

Figure 2
Close-up of middle ear and part of inner ear

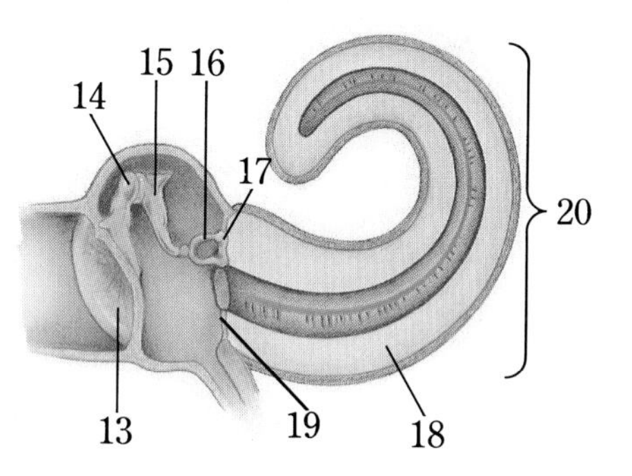

Figure 1

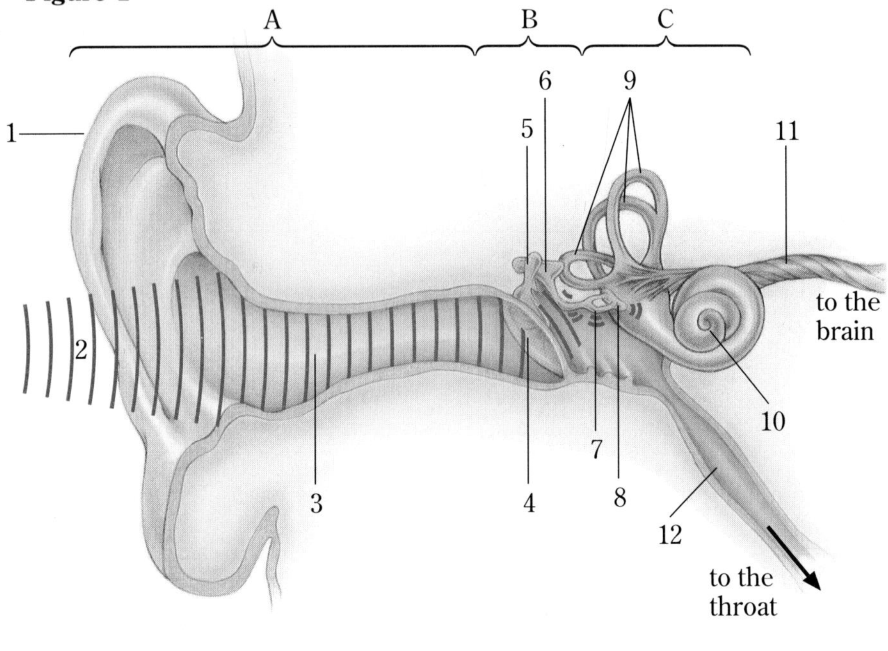

A
B
C
1
2
3
4
5
6
7
8
9
10
11
12
to the brain
to the throat

Figure 3
Cross-section of the cochlea

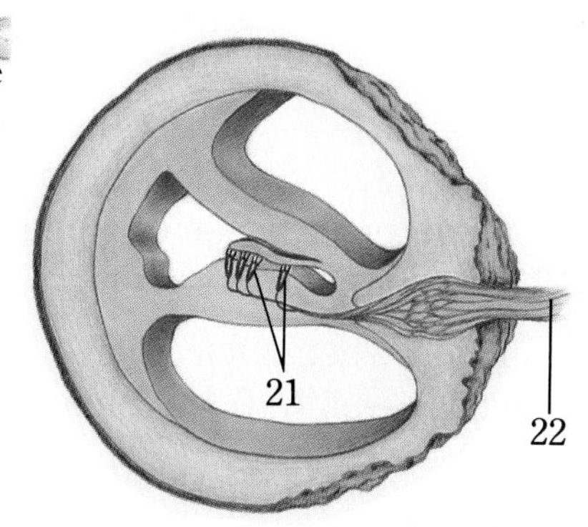

Figures 1, 2, and 3

In Figure 1, the braces show the three regions of the ear: the **outer ear,** the **middle ear,** and the **inner ear.**

The outer ear includes the large external flap of cartilage and skin called the **pinna,** which receives the **sound waves.** It leads into a narrow tube called the **canal.**

At the end of the canal is the **eardrum.** This is the beginning of the middle ear.

Next to the eardrum are the three bones of the middle ear: the **hammer,** the **anvil,** and the **stirrup.** They fit neatly into each other. You can recognize them in Figure 2 by their shapes.

The Story of Hearing

The story of hearing tells how the vibrations of air particles in sound waves are passed, step by step, from one region of the ear to the next.

First, the pinna catches the sound waves and directs them into the canal.

The vibration of the air particles causes the eardrum to vibrate. The energy of the sound waves determines how far and how fast the eardrum vibrates.

The eardrum passes the vibrations on to the three bones of the middle ear, which in turn vibrate. The bones act like simple machines (levers). Remember that simple machines sometimes multiply force at the expense of distance. By reducing the distance of the vibration, the bones increase the force of the vibration. The stirrup then passes the vibration on to the next part of the ear.

vibrating
source of sound

vibrating
air particles

vibrating
eardrum

vibrating
middle ear bones

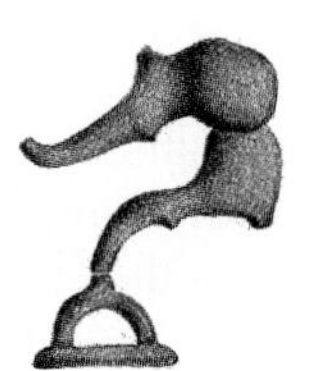

vibrating
liquid particles
of inner ear

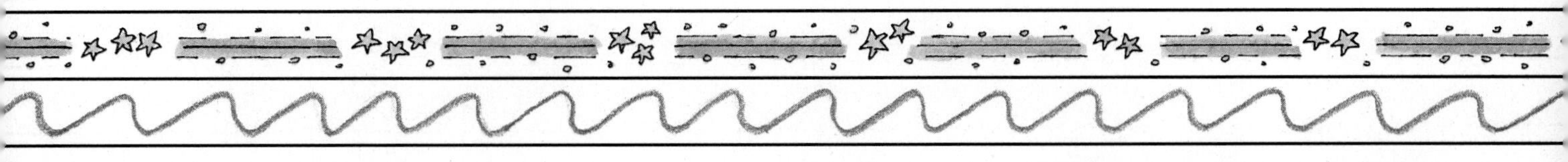

Figures 1, 2, and 3

The middle ear is connected to the throat by a long, narrow canal called the **Eustachian tube.**

The stirrup fits into the small **oval window** in the inner ear (see Figures 1 and 2). The inner ear is a cavity in the bone of the skull that is filled with **watery liquid** (shaded area in Figure 2). The cavity in the bone is made up of a snail-shaped part called the **cochlea** and three **semicircular canals.** The **round window,** located below the oval window, is another membrane-covered opening in the inner ear. A close-up of the cochlea is shown in Figure 2.

The cochlea contains a membrane that runs the length of the tube, down its middle. Located on this membrane are a large number of little **hair cells.** Figure 3 shows a cross-section of the cochlea.

Located at the base of the hairs in the cochlea are nerve cells that join together to form the **auditory nerve,** which is shown in Figure 1. The auditory nerve connects the ear to the **brain.**

The Story of Hearing

Loud sounds push the eardrum far into the middle ear. When this happens, the Eustachian tube can be opened to prevent an excess of air pressure from building up in the middle ear. The excess pressure is released into the throat.

The stirrup is connected to a membrane of the inner ear called the oval window. The inner ear is filled with fluid. When the stirrup vibrates, the energy passes through the oval window into the cochlea as waves in the fluid. The round window vibrates at the same time, maintaining a constant pressure in the fluid. The semicircular canals have nothing to do with hearing. Instead, they help us keep our balance.

Movement of the liquid in the cochlea causes the little hairs there to vibrate, stimulating nerve endings within the inner ear. The hairs at the beginning of the cochlea are quite short. They move when affected by sound waves of high frequency. High-frequency waves are caused by high-pitched sounds. Farther down the cochlea, the hairs are longer and respond to sound waves of lower frequency.

The nerve cells change the movements of the hair cells into electrical impulses (which are much like vibrations). The nerve impulses travel through nerve fibers to the auditory nerve, which carries the impulses to the brain. The brain then interprets the impulses as various kinds of sound.

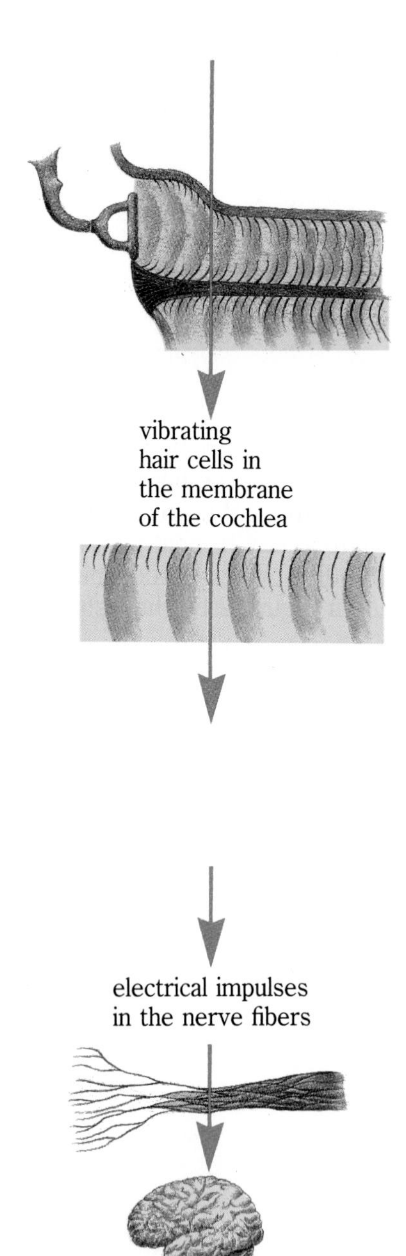

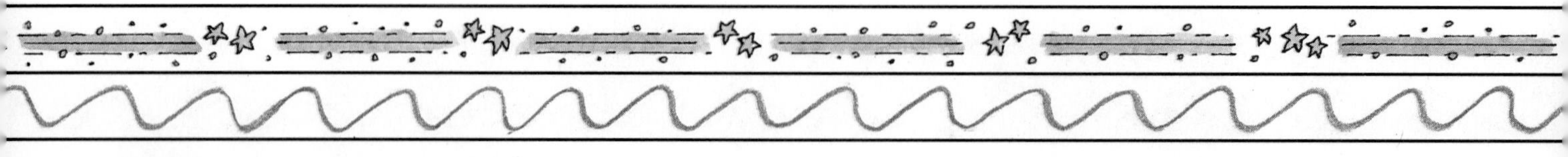

Questions to Think About

1. When a compression in a sound wave in the air hits the eardrum, in which direction does the eardrum move? In which direction does the eardrum move when hit by an expansion of a sound wave?
2. If you hear a bird sing a note with a frequency of 2000 vibrations per second, how many times per second does each air particle vibrate? How many compressions (and expansions) reach the eardrum per second? How many times does the eardrum vibrate per second?
3. How does the ear distinguish between a loud sound and a soft sound? a high sound and a low sound?
4. How does the ear strengthen the sound waves in air so that they will be strong enough to affect the liquid of the inner ear?
5. If your Eustachian tube is blocked because you have a cold, how might your ears feel? How would this affect your hearing?
6. You turn around fast a few times and find you can hardly stand up. Suggest what might be happening in your ear to "upset" you. (A clue: Swirl water in a glass and then set it down. What happens?)
7. Fernando's model of the inner ear is shown in the illustration below. Which part of the inner ear is represented by each of these items?

    (a) the tubes of the stethoscope

    (b) the water in the container

    (c) the stretched rubber sheet

    (d) the tuning fork

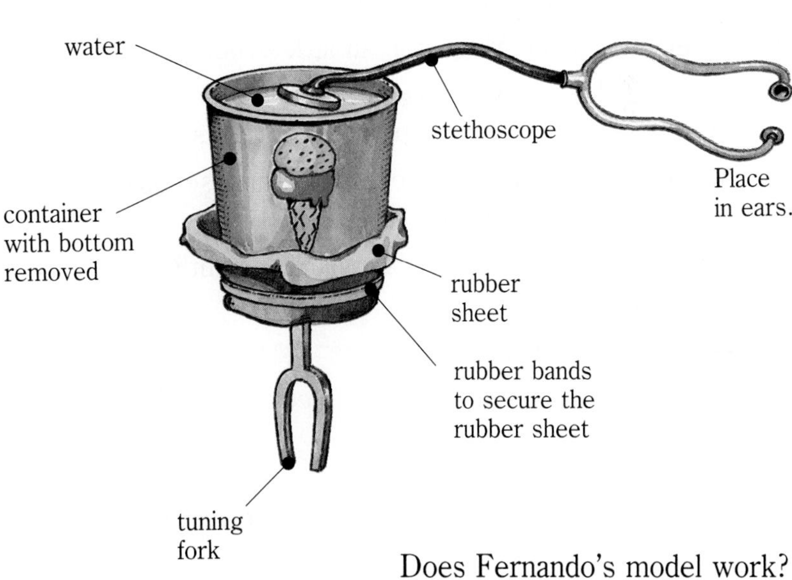

Does Fernando's model work? Try it.

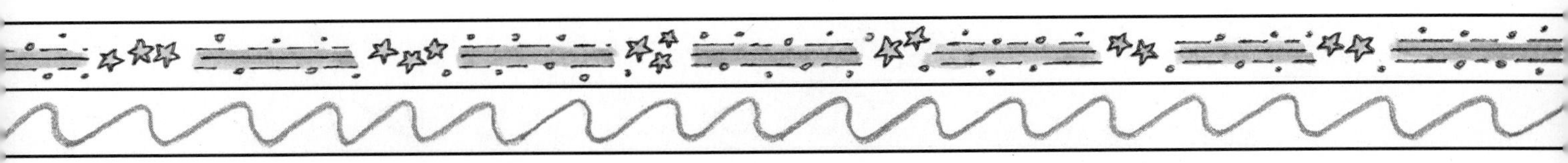

A Centuries-Long Debate!

If a tree falls in the forest when no one is around to hear it, is any sound produced? Write your opinion, and back it up with some principles and facts that you have learned in this unit, such as:

- Sounds are produced by vibrating objects.
- Sounds are transferred from place to place by vibrating particles.
- Sounds are heard as a result of vibrations transmitted through parts of the ear and the nerves to the brain.
- Dogs hear sounds that people cannot hear.
- Bats make sounds that people cannot hear but that can be heard by the bats.

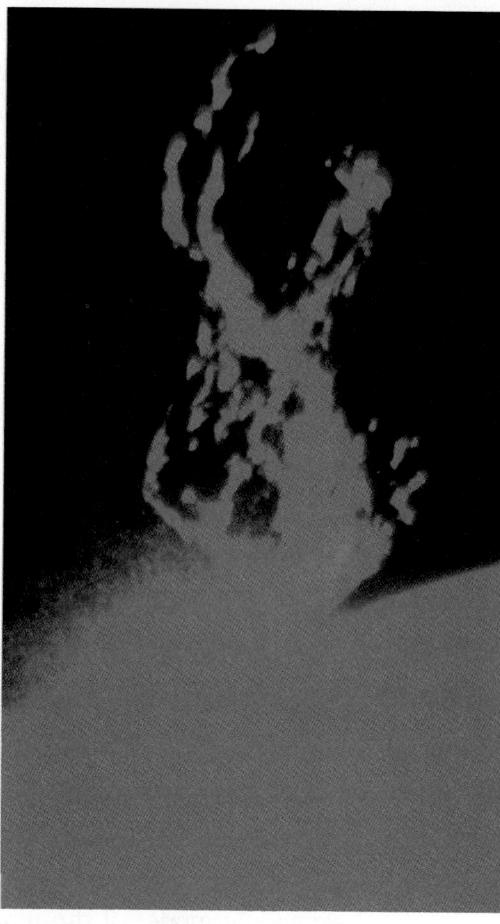

A nuclear explosion on the sun's surface

More Items for Debate

Is sound produced in each of these situations? Give reasons for your answers.

1. Giant nuclear explosions frequently occur on the sun's surface.
2. In the middle of a quiet night, you are in a house all alone.
3. A tiny square of facial tissue falls to the rug.
4. Your heart beats.
5. There is a running brook in the middle of the woods.
6. An electric bell is ringing in an airless bell jar.
7. Moths make clicks to confuse the echolocation sounds of bats. These clicks are inaudible to humans.
8. An electronic audio oscillator vibrates and sends out waves at 25,000 vibrations per second (25,000 Hz).
9. Particles of the gases in air are constantly hitting all the objects around us.
10. You are in outer space.
11. A plane moving faster than the speed of sound has just passed you.

You already know that the *frequency* of vibration is the number of vibrations produced by an object in one second. Frequency is measured in units called **hertz.** One hertz (Hz) is the frequency of one complete vibration, or sound wave, per second.

The *Apollo* spaceship would travel 50,000 m in 5 s. That's another 53 book pages. But what about the flash of light? This book would need another 20,000,000 pages!

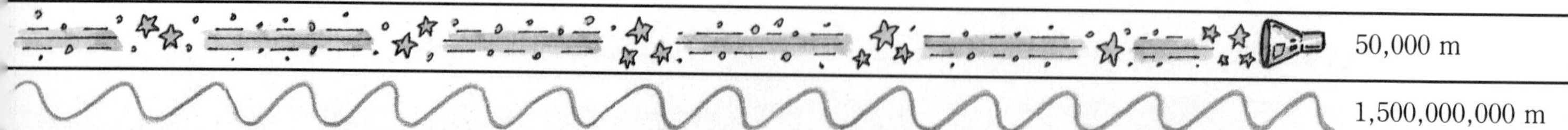

Challenge Your Thinking

1. How might you explain each of the following?
   (a) When you strike tuning fork A, the paper ball jumps off its resting spot on an identical tuning fork B. Why?
   (b) A grasshopper produces sound by rubbing its rough legs against its hard wing covers. This sound can be heard almost 100 m away. There are over 1000 tons of air in one cubic hectometer of air (1 hectometer = 100 m). How does a grasshopper have the strength to move that much air as it chirps?

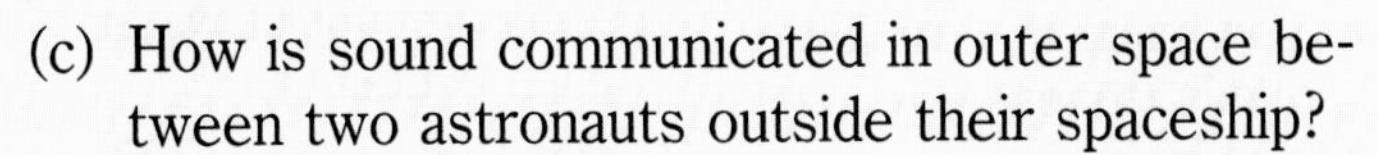

   (c) How is sound communicated in outer space between two astronauts outside their spaceship?
   (d) The sound of a dentist's drill is louder to the patient than to the dentist. Why?
   (e) In the Alps, why does the sound of a yodel last so much longer than the original yodel?
   (f) Marching soldiers always break step to walk across a bridge. Why?

2. Yale was watching Janelle play her bass drum on the playing field. She kept a steady rhythm going—about 20 beats in 10 seconds. When Yale backed away from Janelle a distance of 85 m, he noticed that the *Boom!* came precisely when the drumstick was farthest away from the drum.

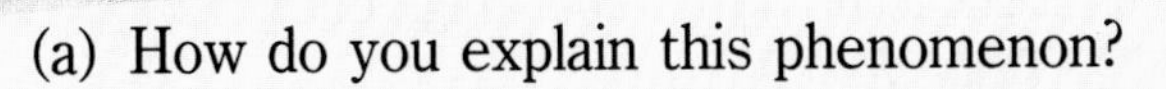

   (a) How do you explain this phenomenon?
   (b) From the mathematical information given, find: (i) the time between drumbeats; (ii) the time for the drumstick to get to the farthest point from the drum; and (iii) the speed of sound in air.

A Closer Look at Pitch

EXPLORATION 7

An Investigation: Strings and Pitch

Before you do this Exploration, think about these questions:

- How many musical instruments can you name that use strings to produce their sound?
- How can you make high or low musical sounds (tones) with these instruments?
- What characteristics of strings can be changed to make musical tones higher or lower?

You Will Need

The diagram shows one setup that will enable you to do this Exploration. You may, however, be able to find other materials that work just as well. Try to design your own apparatus if you can.

Be Careful. *Do not let the plywood extend over the table top.*

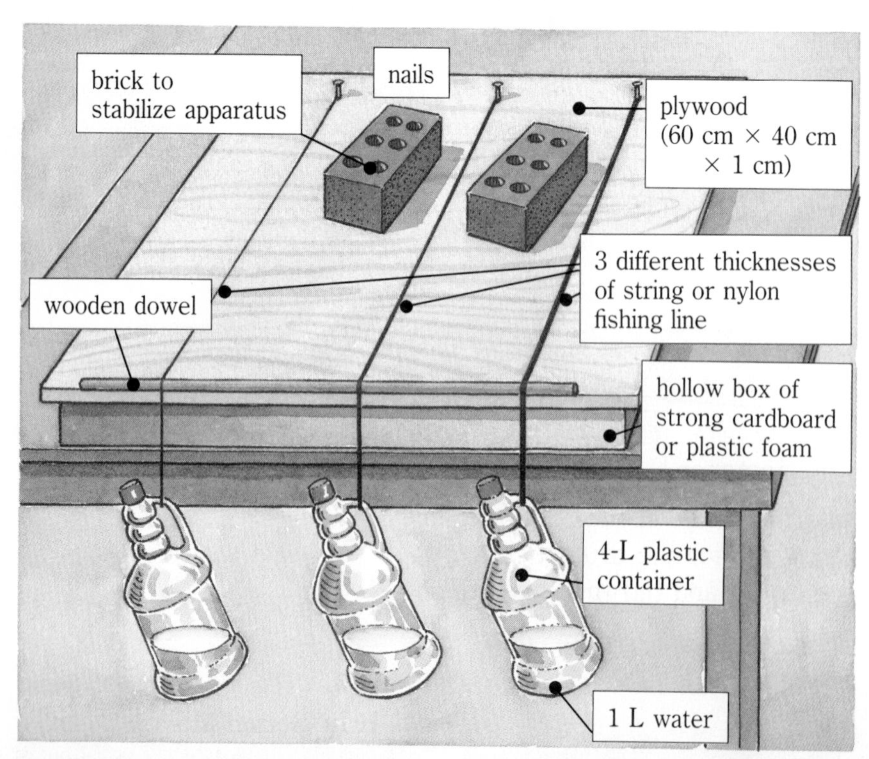

What to Do 

Consider the following:

Does the thickness of a string or fishing line affect pitch?

1. Stretch the three different sizes of string or fishing line by the same amount. (Why?) You can do this by adding the same amount of water to each container.
2. Pluck the strings. What do you observe? By singing the musical scale *(doh, re, mi . . .)* you may be able to locate the tones produced by these strings on the scale. What conclusions can you draw?

Does a string's tension (the amount by which it is stretched) affect pitch?

3. The strongest of the three nylon strings is best (a 9-kg fishing line works well). While one person plucks the string, another person can gradually add water to the jug.
4. What do you observe? What conclusions can you draw from your observations?

   Try this:

   Add enough water to a jug so that the jug and water have a total mass of 1 kg. Hang the jug on the strongest string, and pluck the string. Sing this tone as *"doh."* Then sing the scale *"doh-re-mi-fa-sol-la-ti-doh'."* (The scale from *doh* to *doh'* forms an **octave**.) Try to keep in mind the sound of *doh'* (high doh).

Now add 3 L of water to the jug (that is, 3 kg of water. What is the total mass of the jug and water now?) Again, pluck the string. Does the tone sound like the high *doh'* you sang earlier? If not, can you tell how far away it is on the scale?

You may have discovered an important idea. If the tension of a string is increased by four times, the frequency is doubled; in other words, *doh* becomes high *doh,* or *doh'*.

Does the length of a string affect pitch?

5. You can make a vibrating string shorter by using a pencil as shown. How does the musical tone change as you shorten the vibrating part of the string?

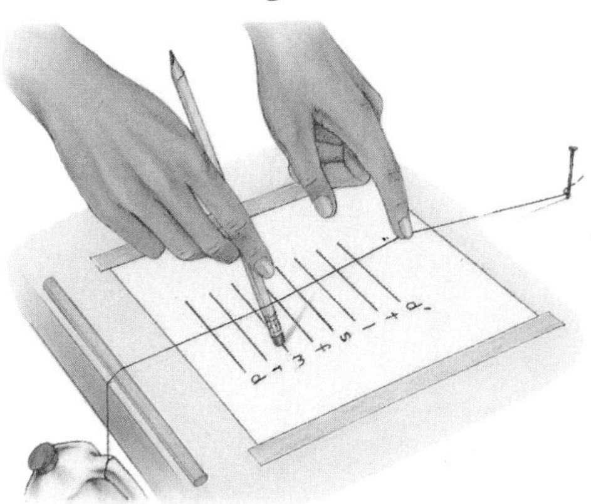

6. Suppose the tone produced by the whole string vibrating is *doh.* Find the length of string that gives *doh'* (twice the frequency). How does this length compare with the whole length?
7. By moving the pencil along one string, locate the tones of the entire octave. Mark them as *d, r, m, f, s, l, t, d'* on a paper taped under the string.
8. Now play the mystery tune below. See if you can discover what it is.

mi re doh—mi re doh—sol fa mi—sol fa mi—
sol doh' doh' doh' ti ti ti
doh' doh' doh' sol (repeat these two lines twice more)—
fa mi re doh

Questions

1. What effect does the size of a string have on the pitch of a sound?
2. What effect does the tension of a string have on the pitch of a sound?
3. What change in tension causes the frequency to be doubled?
4. What effect does the length of a string have on pitch?
5. What change in length causes the frequency to be doubled?
6. What change in length might cause the frequency to be halved?

What factors affect the pitch of this cello? Does it have a low or high pitch?

Puzzling Pairs

Pick out the pairs whose elements belong together. When unscrambled, the letters that identify the incorrect pairs spell a five-letter word. What is it?

(a) short string/large frequency
(b) long string/low sound
(c) low sound/large frequency
(d) high tension/large frequency
(e) thick string/large frequency
(f) increase the tension/decrease the pitch
(g) thin string/high sound
(h) half the length/half the frequency
(i) four times the length/sound one octave lower
(j) half the length/sound one octave higher
(k) small tension/low frequency
(l) four times the tension/two times the frequency

An Orchestra

Do research on the main groups of instruments in an orchestra—strings, pipes, and tubes.

- How is sound made in each musical instrument?
- How are sounds made higher? lower? louder?
- How is the size of the instrument related to the pitch of its musical tones?

10 Loudness

Making Sounds Louder: A Problem

You are given a tuning fork. When you strike it, you can scarcely hear it. How can you make the sound of the fork louder? Before you read any further, write down some suggestions in your Journal.

Some Solutions to the Problem

Darcel came up with a simple solution. He asked himself a question: How is the loudness of a sound related to the energy put into the object that produces the sound? His answer can be seen in the illustrations.

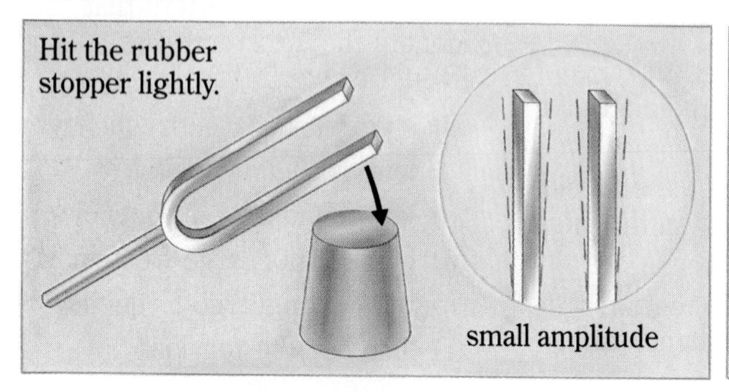

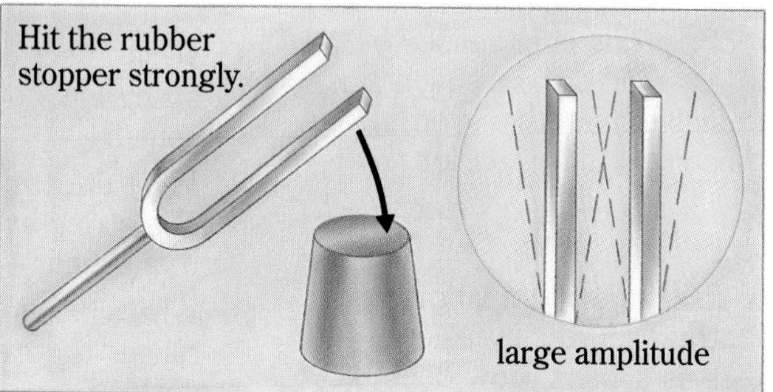

Nicole also found a simple solution. Her question was, "What is the relationship between loudness and the distance between the ear and the source of the sound?"

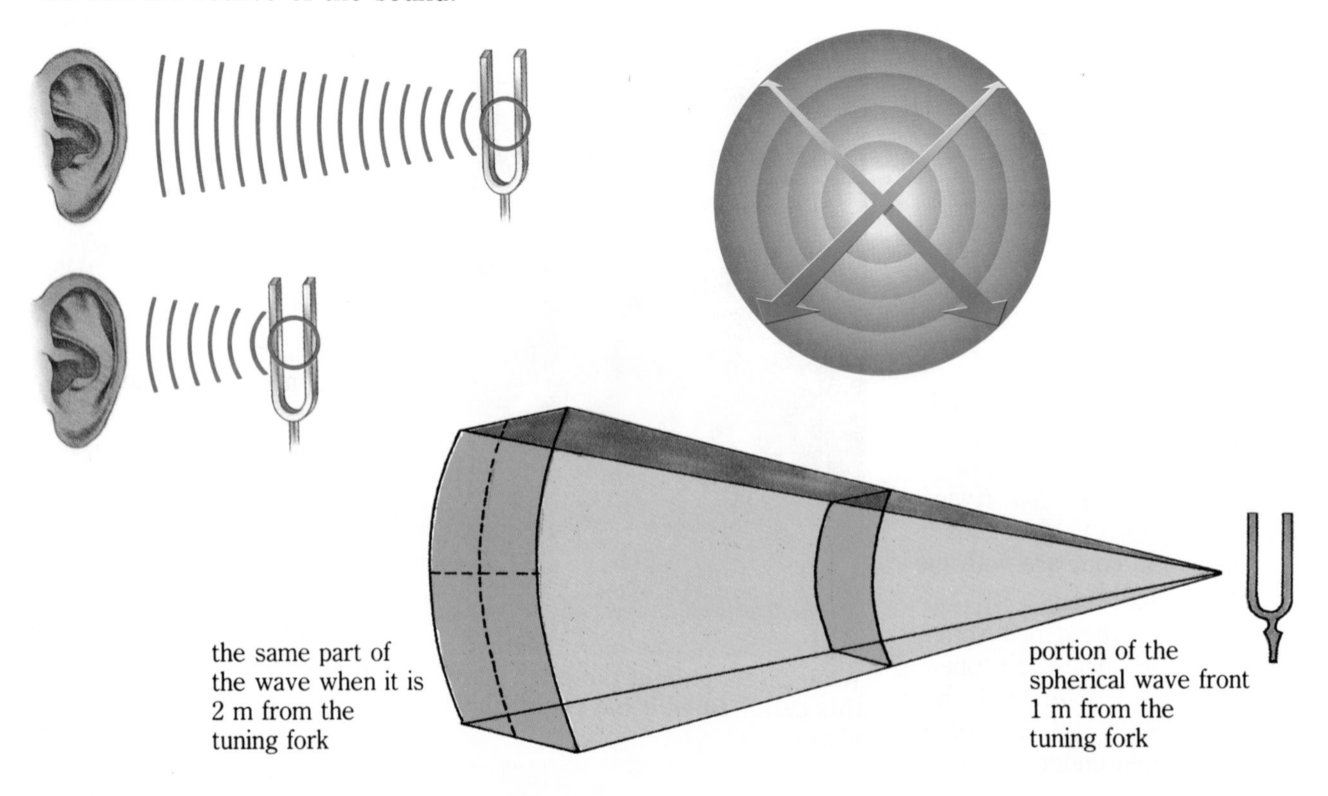

Look at Nicole's drawings and answer these questions. Over how much more area is the same sound energy distributed when the wave surface is twice as far from the source of the sound? How loud will the sound be to a listener located twice as far away? three times as far away?

Akira discovered that placing the base of the tuning fork on a table-top or a hollow wooden box makes the sound louder.

Akira's solution makes use of **forced vibration**—getting something to vibrate by touching it with an already vibrating object. What is the loudest sound you can get from a tuning fork by means of forced vibration? Try objects made of various materials, such as glass, metal, and plastic. What do you discover about the effects of different sizes of objects and different materials?

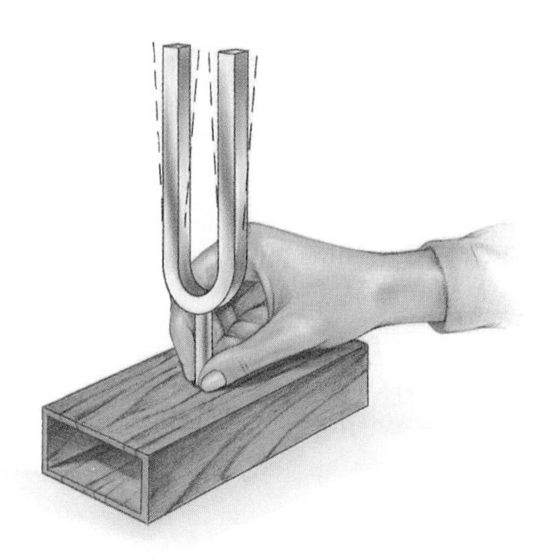

Yolanda found that the apparatus shown below helped make the sound of the tuning fork louder. Why did it work? She concluded that the cone shape prevented some of the sound waves from spreading out in all directions. The cone sends the sound energy in a specific direction. Therefore, the sound appears louder to an observer in the path of the sound energy.

Have you ever used a **megaphone?** What are some situations in which a megaphone is used? Design an experiment to show how much farther a sound can travel when a megaphone is used.

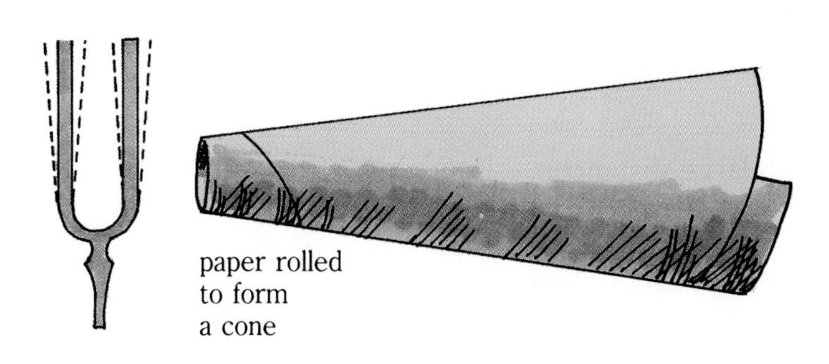

Leonor discovered that she could make the sound louder by holding the tuning fork over an air column. If the air column is just the right length—such that it has the same natural rate of vibration as the fork—it will start to vibrate along with the fork. As a result, the sound is louder. The air column is vibrating in **resonance** with the tuning fork. What do you think causes this to happen?

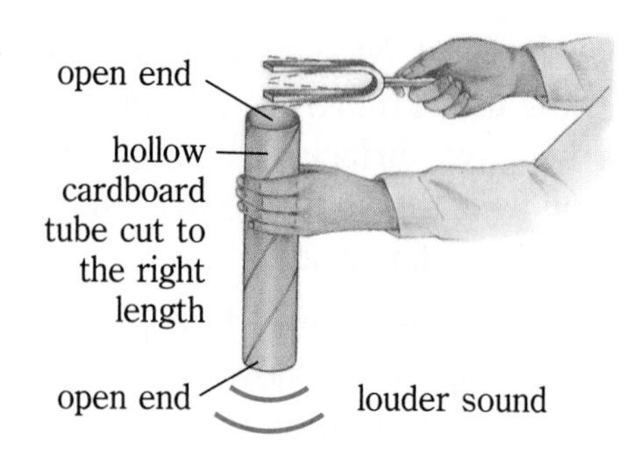

Resonance occurs when you strike a certain key on the piano and something in the room begins to vibrate. What would you say about the frequency of the piano string and the natural frequency of the object in the room? What other experiences have you had that involved resonance? Try writing the meaning of *resonance* in your own words.

How is *resonance* different from *forced vibration?*

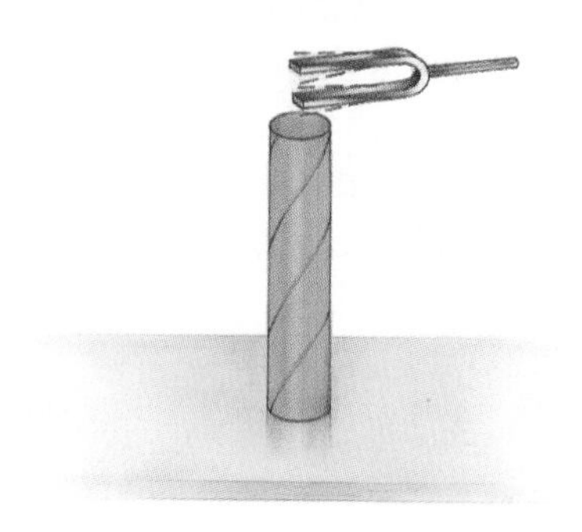

Leonor's design for making the tuning fork louder is quite simple. To try it, cut a cardboard mailing tube to the right length using this approximate rule:

$$\text{length of tube in cm} = \frac{17{,}200}{\text{frequency of fork}}$$

For a 512-Hz tuning fork, what length of tube would you cut? To get it just right, you may have to shorten or lengthen the tube a little by using masking tape.

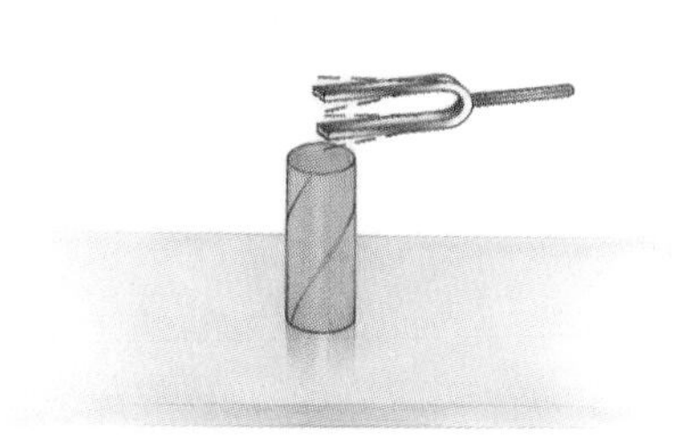

After you get this to work, place the tube vertically on the table so that one end is closed. Next, hold the vibrating fork above the tube. Does the air column now vibrate in resonance with the fork?

Now cut the tube in half and try again, first with the tube open at both ends and then with it closed at one end. What do you observe? Why do you think this happens?

Who Said It?

Which student might have made each of the following comments?

"As more and more air particles are affected, their movement back and forth decreases."

"If two objects are vibrating at the same frequency (their natural frequency), the sound will be louder."

"If the disturbance of the particles is directed, their energy lasts longer."

"If the air particles vibrate vigorously through large distances, the sound will be loud."

"The more air particles that are set into vibration, the more that are likely to hit the eardrum."

Measuring Loudness

The loudness of a sound can be measured with a *sound meter.* The scale on such a device is marked in units called **decibels (dB).** On the meter shown below, what would happen to the needle if the meter were moved closer to the source of sound? farther from the source?

Estimating Loudness

How sensitive are your ears? Can you tell whether one sound is louder than another? On the next page, List A contains a number of sounds in alphabetical order. List B gives the approximate loudness of these sounds, as measured on a sound meter, in increasing order. Match each sound with its loudness and arrange the sounds in order of increasing loudness by their item number. You will find the answers on the bottom of the next page.

List A

1. Air drill breaking cement (nearby)
2. Breathing (ordinary)
3. Conversation (ordinary)
4. Diesel truck traveling at 65 km/h (15 m away)
5. Drag strip (near the starting line)
6. Farm tractor (the sound heard while sitting on it)
7. Hard-rock band (live)
8. Jet takeoff (nearby)
9. Noise in a town on a weeknight
10. Private office
11. Rifle blast
12. Rustle of a newspaper
13. Sound that is just audible to you
14. Sports car moving at 80 km/h (the sound heard inside the car)
15. Whisper

List B

0 dB, 10 dB, 20 dB
30 dB, 40 dB, 50 dB,
60 dB, 70 dB, 80 dB,
90 dB, 100 dB, 110 dB,
120 dB, 140 dB, 150 dB

The Decibel—An Unusual Unit

The sound of a person breathing registers about 10 dB, while the rustle of a newspaper registers about 30 dB. Yet the sound energy provided by the newspaper is about 100 times that of breathing. The table below shows the relationship between sound energy and decibels.

Sound Energy and Decibels—Some Simple Rules

Change in Sound Energy	Increase in Loudness
2 × sound energy	3 dB more
4 × sound energy	6 dB more
8 × sound energy	9 dB more
10 × sound energy	10 dB more
100 × sound energy	20 dB more
1,000 × sound energy	30 dB more
? × sound energy	40 dB more
1,000,000 × sound energy	? dB more

A Puzzler

One sound has a loudness of 30 dB. Another sound has 400 times as much energy. What is the loudness of the second sound, in decibels?

Here is the correct order for the "Estimating Loudness" sensitivity test above: 13, 2, 15, 12, 9, 10, 3, 14, 4, 1, 6, 7, 8, 5, 11.

Noise Pollution—The Effect of Sound on Health

As you read, *Hey! Hear This!* find clues that will help you answer these questions:

1. What is crammed into the small space of your right ear? (You know this from material you studied earlier in the unit.)
2. What does the author think are the most damaging sounds to a person's ear?
3. As you grow older, what kind of change in your hearing range will occur?
4. Higher tones lie in the upper part of the hearing range—about 20,000 cycles per second. Using this information, what kinds of sounds will be easier for you to hear as you get older?
5. What causes ringing in the ears?
6. What causes middle-ear infections?
7. What can you do to protect your hearing now?
8. What are four types of noise pollution that can really hurt your ears?

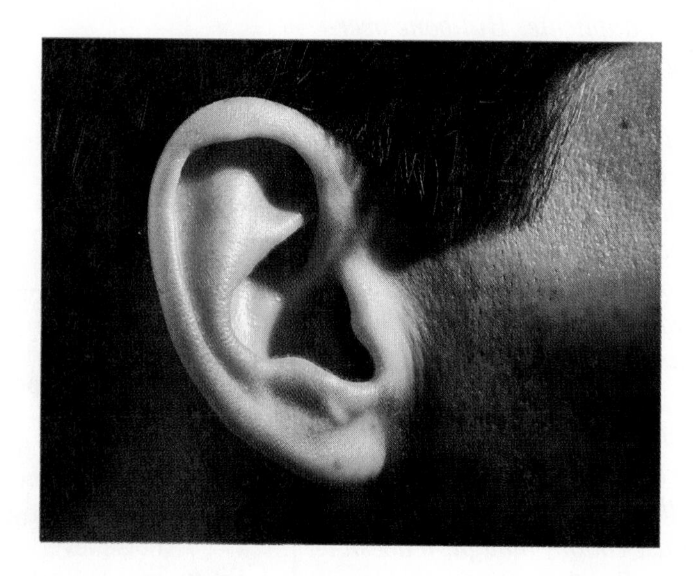

Hey! Hear This!

Got your headphones on? Going to the concert and gonna stand right in front of the speakers? Got the TV, the radio, the stereo, or the CD player turned all the way up? Well you'd better hear this—that's right, I'm talking to YOU! And just who am I? I am the high-powered, turbocharged, anatomically amazing right ear. I could be the right ear on your head.

I think you and I need to have a little talk. I mean, we've hung out together all our lives, but how much do you really know about me? Did you know that there's enough electrical circuitry packed into me to light a city? Or at least to give it phone-service. And that's no joke!

Without me, you'd be missing out on the world of sound—everything from the sounds of horns honking, babies crying, music playing, and friends talking. And I do all of this in a space so small that I rival computer microchips. But our lines of communication seem to be breaking down. Those tiny parts and delicate pieces that make me so incredible, well, they're beginning to deteriorate. So you gotta take care of me, if you want to keep me working for you.

I was at my best the day you were born. From that time on, my abilities have been on the decline. What causes this? Loud sound, mainly. Sounds too loud, too often, and too long. As we get older, my tissues begin to lose their flexibility. We may be pretty young, but just ten or fifteen years down the road, we're going to feel the effects of any abuses now. Besides the loss of elasticity, hair cells begin to fall apart, and deposits of calcium build up. I'm sure you didn't know it, but that's all going on right now, right here, inside your right ear.

When you were born, I had a hearing range of 16 to 30,000 cycles (vibrations) per second. But by now, the upper limits of our hearing are just about 20,000 cycles per second. By the time we're as old as our grandparents, I could be down to as low as 4,000 cycles per second. Think of all the sounds—good and bad—you'd be missing out on.

I tell you I'm cool, but I'm fragile. Drum punctures are frequent. But these usually heal themselves. Ringing in here is another problem—the doctors call it tinnitus, and it can come from almost anything: drugs, fever, circulation changes, or tumors on my acoustic nerve. Sometimes the ringing can be stopped—but not always. I can also get infected, most often inside the middle ear. The Eustachian tube exposes me to those nasty infections. The Eustachian tube goes from the middle ear to the throat. In the throat there are a lot of microbes. Those microbes can make their way into me, and there ya go—a middle-ear infection.

Another way I get damaged is by an overgrowth of bone from my middle ear. This freezes the motion of the bones, causing conduction deafness. If you have conduction deafness, a hearing aide might help. Surgery helps, too, about 80% of the time. A surgeon can go in and replace the stirrup bone with a metal duplicate. But bone overgrowth is not the biggest cause of hearing loss. I think you know what it is. You got it—LOUDNESS!

When you crank up the volume, my tiny muscles tighten up. I'm better at standing up to sounds like thunder or a bass guitar. But high-pitched, screaming sounds from things like jet airplanes, factory machines, and lead guitarists in rock bands destroy my delicate hairs and wreck my tiny muscles. I mean, when I am at my best, I am a finely-tuned organ. But when you make me listen to high-pitched, glass-shattering, screeching sounds over and over again—well, then I'm done for.

So that's why I'm asking you to listen to me. We've got to do something to protect your hearing. Hearing is too precious for you to lose—take care of your ears. Keep them away from damaging sounds—plug 'em if you have to. Don't use cotton, though; it doesn't work. Use special ear plugs instead. Hey, I'm the only right ear you got. And I'll keep hearing for you if you look out for me. Deal?

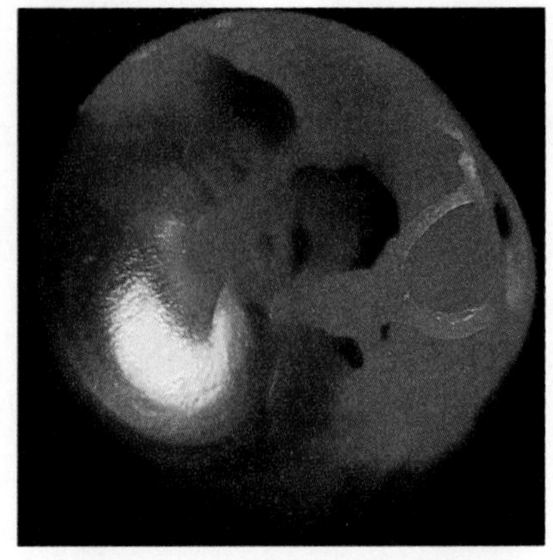

See if you can locate which part of the middle ear shown here would be replaced by stainless steel during surgery.

Levels of Noise

Listed below are the decibel levels at which you'd experience the following effects:

- 70 dB: Difficulty in hearing conversation
- 80 dB: Annoyance at the noise level
- 90 dB: Hearing damage if noise is continuous
- 120 dB: Permanent hearing loss
- 130 dB: Beginning of pain
- 140 dB: Sharp pain in the ear
- 170 dB: Total deafness

The Quality of Sounds

"Seeing" Sounds

You can often recognize things simply by looking at their shapes. For example, you probably can distinguish the leaves of maple, oak, poplar, and other types of trees by their shapes.

How can you distinguish between the different types of sounds made by a trumpet, a piano, a violin, and a guitar, each playing middle C at about the same loudness? It's easy! All you have to do is listen. Your ear and brain can tell the difference. But what makes this difference? If you could see sounds the way you can see a leaf, you could understand more easily what gives each of these instruments a unique sound.

There is, in fact, a way to change sound waves into wave shapes that can be seen. Study the method by examining these illustrations.

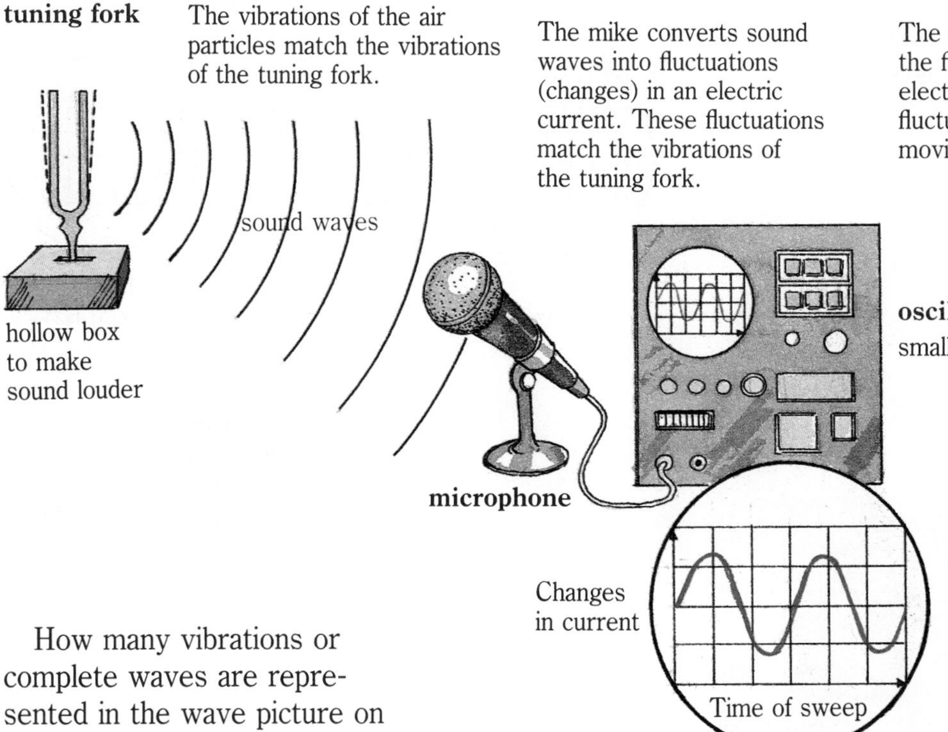

How many vibrations or complete waves are represented in the wave picture on the oscilloscope?

As the tuning fork sounds, the light path sweeps from left to right. Suppose the oscilloscope was set in such a way that this happened in one-hundredth of a second. You wouldn't see the path as it was being traced out, since it would happen too fast. Instead, you would see the complete wave picture at once; this picture would continue as long as the tuning fork sounds.

If two complete waves appear on the screen in 1/100 s, how many vibrations does the fork make in one second? Did you figure out that the frequency would be 200 vibrations per second?

As you can see, you can determine the frequency of a sound by knowing how long it takes the light path to sweep across the oscilloscope screen and by knowing the number of waves appearing on the screen. How many complete waves would appear on the screen if the frequency of the sound source is 300 vibrations per second (300 Hz)? if it is 550 Hz?

If the time of sweep on an oscilloscope is set at 1/50 s and the frequency of a sound source is 200 Hz, how many complete waves are pictured on the screen?

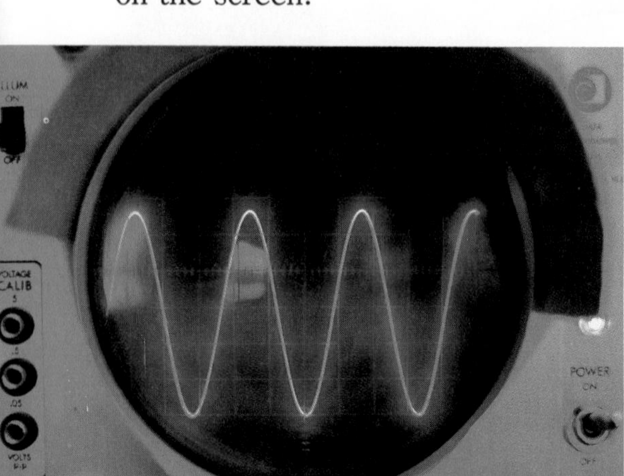

The wavy line on this oscilloscope shows how air pressure changes when a sound wave strikes a microphone.

EXPLORATION 8

Analyzing Wave Pictures on an Oscilloscope Screen

What to Do

Discuss the following four questions with a partner. Be sure to write your conclusions in your Journal.

1. How does the wave picture on an oscilloscope show the loudness of sounds? Study the picture and form your own conclusions.

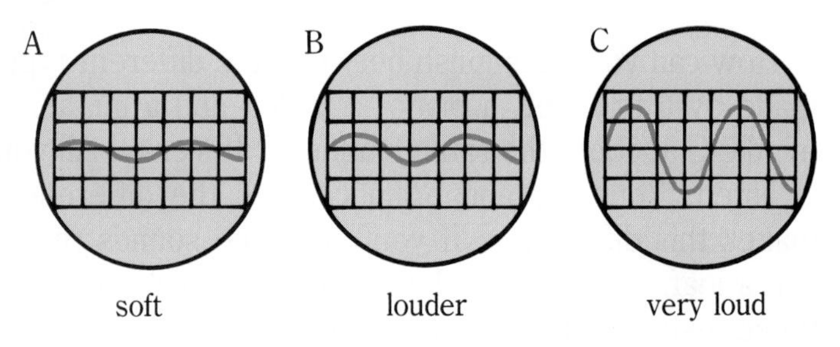

2. How does the wave picture show the pitch of sounds?

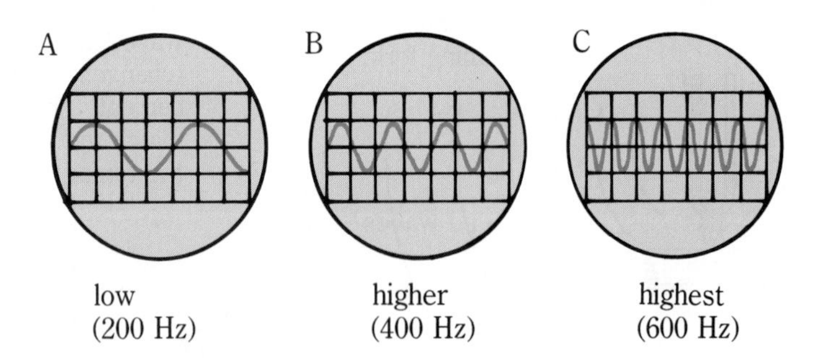

3. Check your conclusions to questions 1 and 2 by studying the five screen displays on the next page and by answering the questions that follow.

A 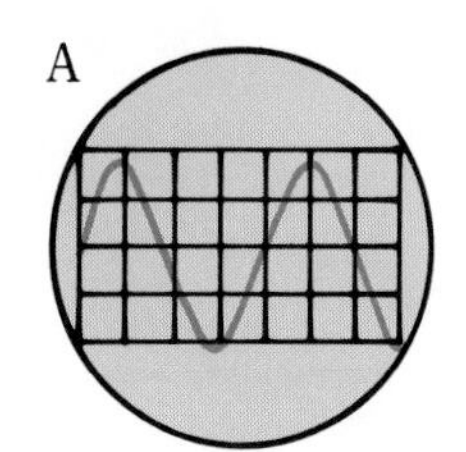 B 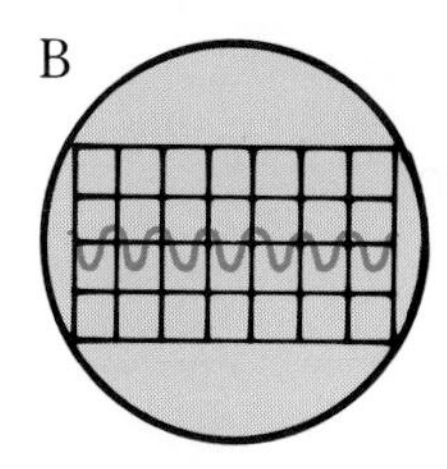 C 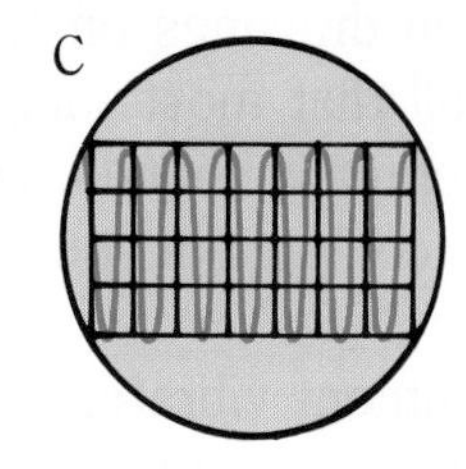 D 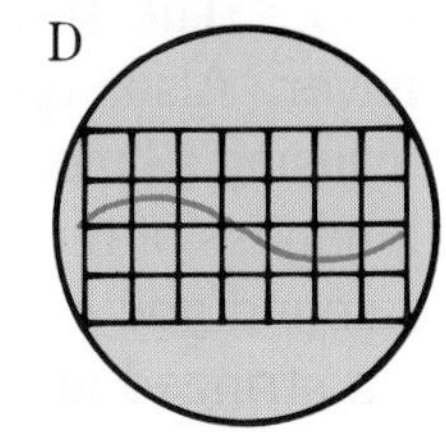 E 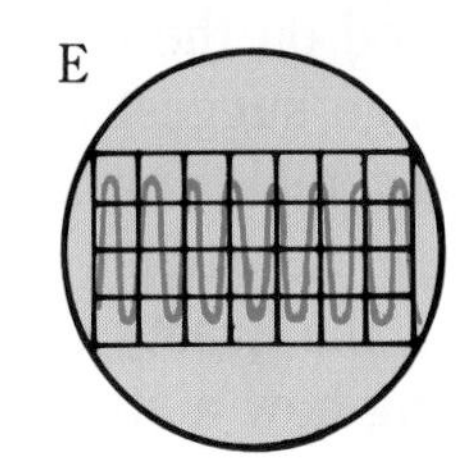

(a) Which screen display above is made by each of the following?
   - (i) A sound that is both soft (quiet) and high-pitched
   - (ii) a sound that is both loud and low-pitched
   - (iii) a sound that is both loud and high-pitched
   - (iv) a sound that is both soft and low-pitched

(b) Which two screen displays above show sounds of the same loudness? of the same pitch?

(c) Which screen display above is made by the lowest-pitched sound? by the softest sound?

(d) What are the frequencies of the sounds in (D) and (E) above if the time of sweep on the oscilloscope is 0.01 s?

4. Here are the screen displays of five different sounds that you would probably recognize. The sounds differ primarily in quality.

A 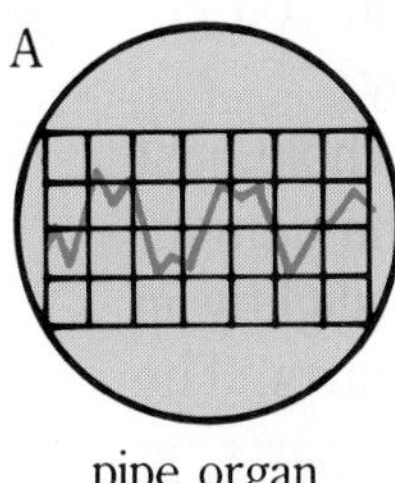
pipe organ

B 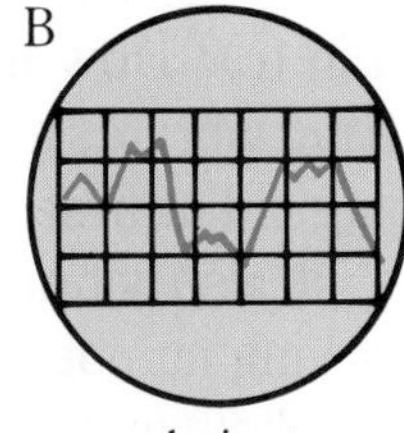
clarinet

C 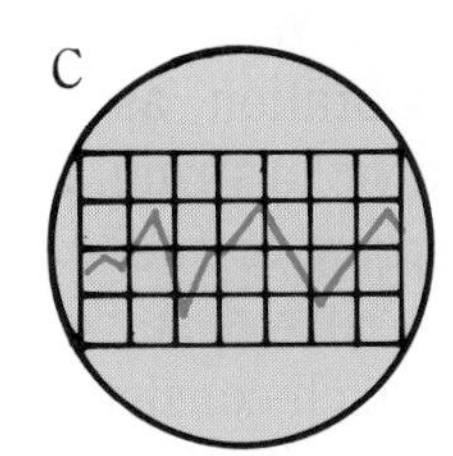
violin

D 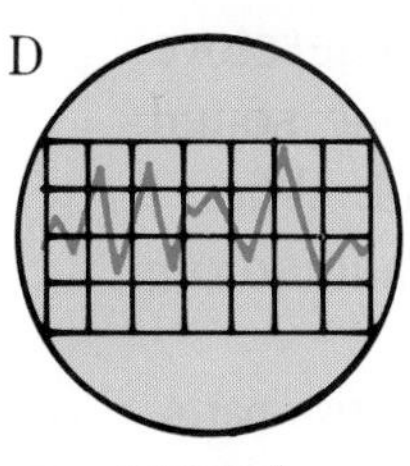
trumpet

E 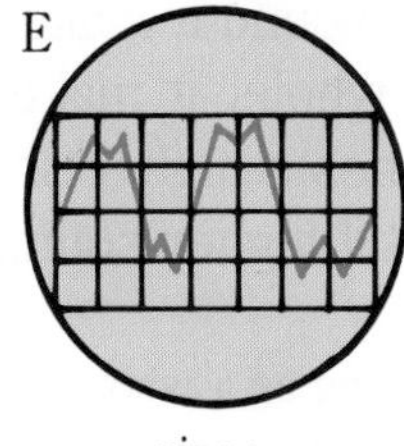
piano

(a) Do you recognize any similarities in these drawings? What are they?

(b) What differences are there among the wave patterns? Suggest some reasons for these differences.

How a String Vibrates

What accounts for the different qualities of sounds that you have seen pictured on the oscilloscope? The screen drawings on the previous page suggest an answer. Also, recall that most things vibrate in a complex manner, causing a "mix" of many sounds. Do you see how the pitches and loudness of these component sounds could affect the quality of the sound you hear?

Examining how a string on a stringed instrument vibrates will make this clearer. First, though, do the following activity with a partner.

One of you will hold one end of a coiled spring or tubing steady. The other will slowly move the other end back and forth rather widely until the spring or tubing vibrates as a whole, as shown in (a) at the right.

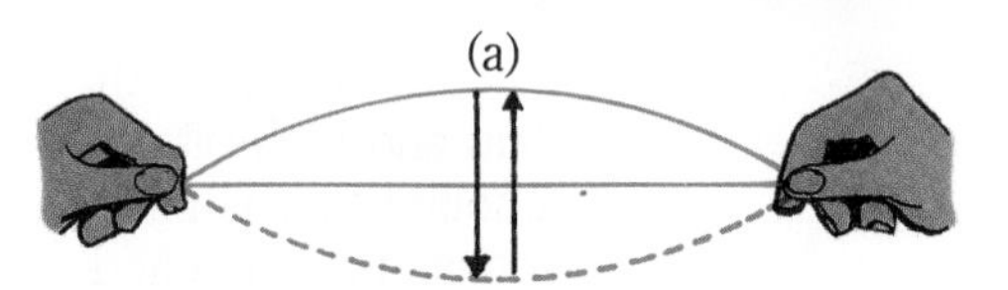

Now, gradually put in more energy; that is, make the spring or tubing move faster and faster until it assumes a more stable pattern of vibration. What happens? Were you able to produce the pattern shown in (b)? With even more energy input, can you get other patterns of vibration, as shown in (c)? How many?

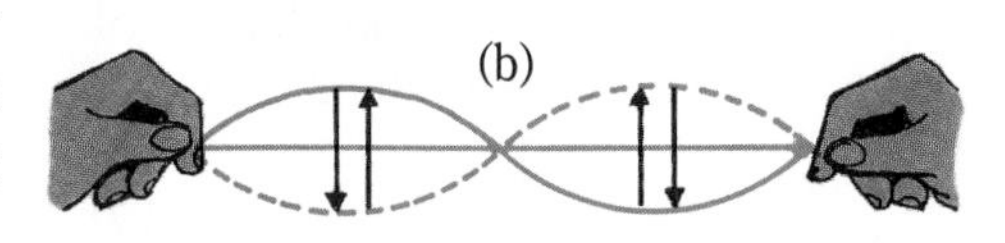

When you bow a violin string, the string vibrates as a whole, just as the stretched spring or rubber tubing did at first in the demonstration. However, at the same time, it also vibrates slightly in various patterns, which consist of a varying number of parts. The second drawing shows it vibrating as a whole and in two parts, at the same time. You know that the length of a vibrating string affects the frequency and pitch of the sound it makes. Therefore, when the violin string vibrates as a whole, it makes one sound. When it vibrates in two parts, it makes a higher sound. When it vibrates in three parts, it makes a still higher sound, and so on.

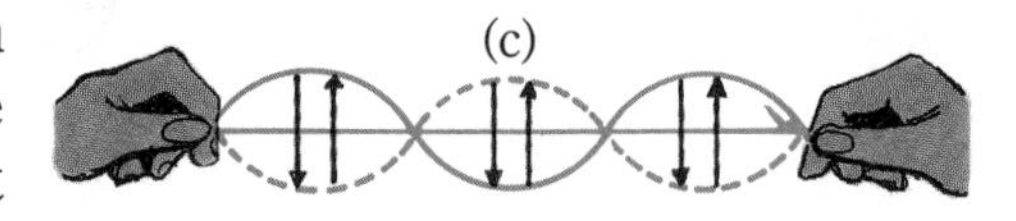

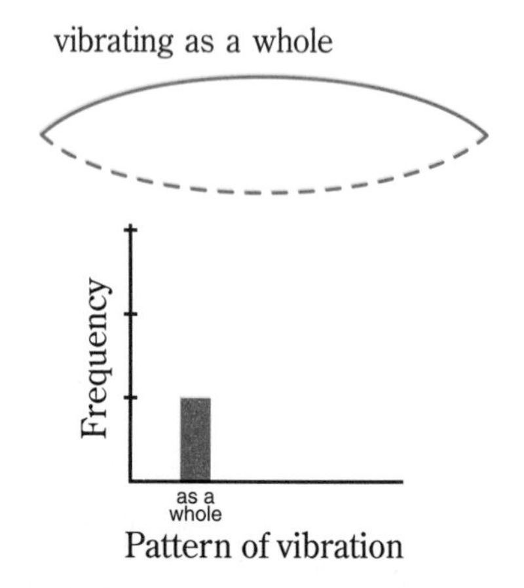

vibrating as a whole and in two parts at the same time

Frequency

as a whole | in 2 parts

Pattern of vibration

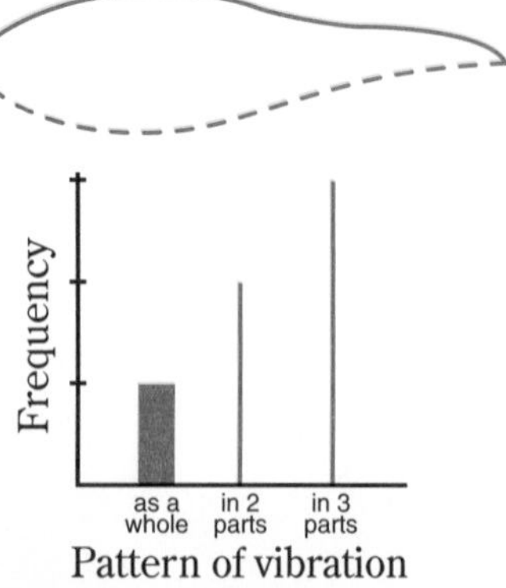

width of bar = loudness

A wave picture of a vibrating string

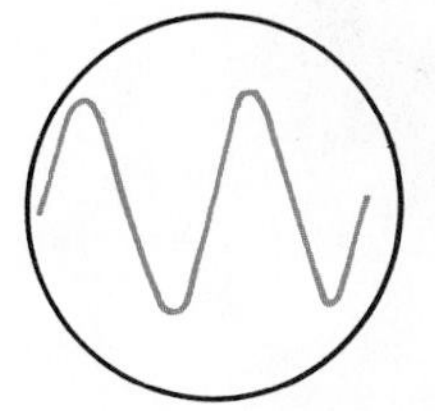

string vibrating as a whole

string vibrating as a whole and in two parts

string vibrating as a whole, in two parts, and in three parts

In reality, a violin string vibrates in many more parts than three. Generally, the lowest vibration as a whole is greatest, so this sound is loudest. The higher sounds are softer. These higher sounds add the "wiggles" to the waves you see on the second and third oscilloscope screens above. Put all the sounds made by the string's different vibrations together, and you have the quality of the sound produced by the violin string.

Locating the Higher Sounds

1. Pluck the highest-sounding string of a guitar (or violin) one-third of the way along its length (a).

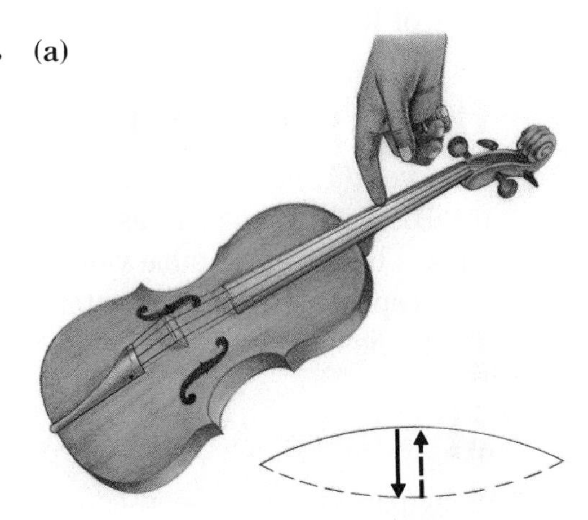

(a)

2. Then touch the string exactly at its middle to stop it from vibrating as a whole (b). Listen for a higher note. (This may take a little skill.) The note is the sound from the string when it is vibrating in two parts. Does it sound one octave higher?

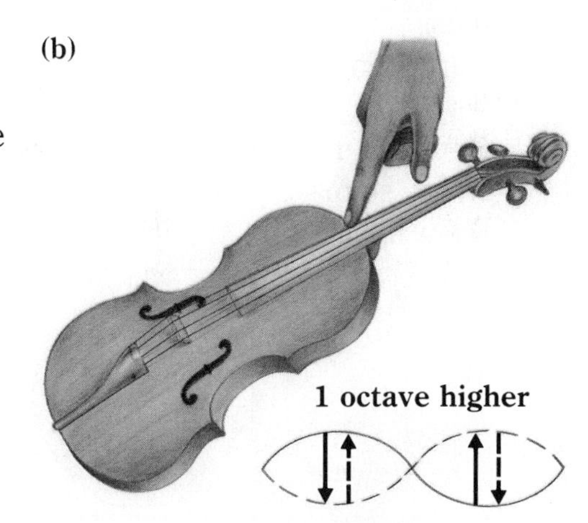

(b)

3. Pluck the string again. Now touch it one-third of the way along its length. With a little skill, you can get the next highest note (c). This is the sound made by the string as it vibrates in thirds.

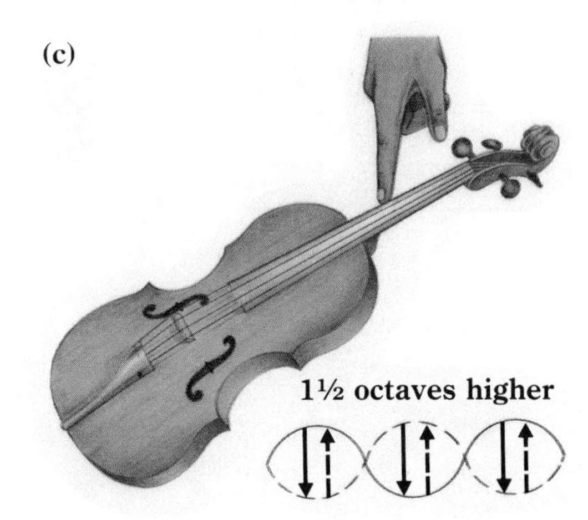

(c)

Other Sound Qualities

Brasses and Woodwinds

Air columns in such instruments as the clarinet and trumpet also vibrate as a whole, in parts, and in varying numbers of parts simultaneously. The dominant low sound mixed with the different higher sounds gives each instrument its own distinctive quality. Which higher sounds are produced depends on the shape of the instrument, the material it's made of, and the way it's blown. The wave shown on the oscilloscope, with its unique set of "wiggles," is a sort of fingerprint of the instrument producing the sound.

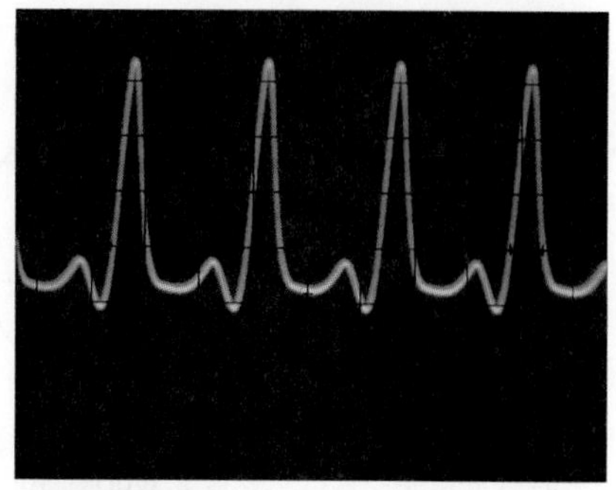
Saxophone

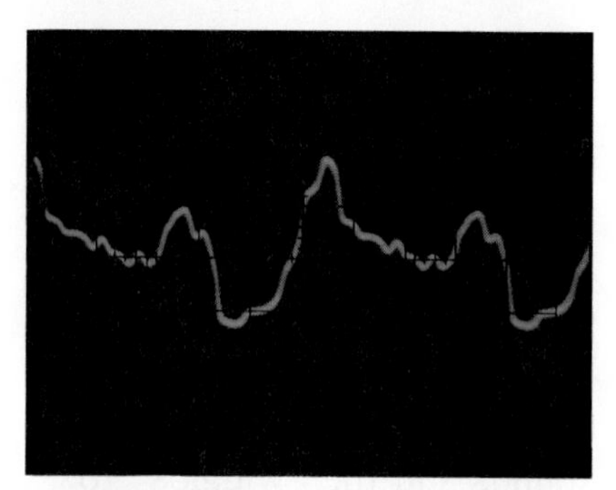
Trumpet

How Noise Is Produced

Slap your hand on the table. This sound is not musical. On an oscilloscope, it appears as a jagged, haphazard, irregular display. This is because the table vibrates in a very irregular way. No part of it repeats itself in a constant way, like a vibrating string. The sound dies out very quickly.

Music or Noise?

Your parents have been complaining lately, "That *noise* you call music is driving us crazy! Turn it off!" Drawing on what you've learned in this unit about sound, music, and noise, explain to them why your favorite group really *does* make music—not noise. Remember that your parents probably haven't studied science in a while, so make your explanation clear and instructive. If you help them understand *scientifically* the difference between music and noise, you just might win them over to your kind of music!

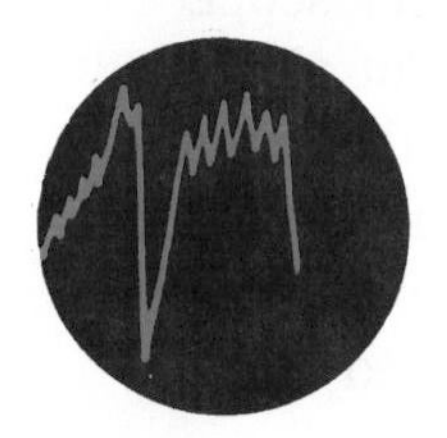
A sharp slap

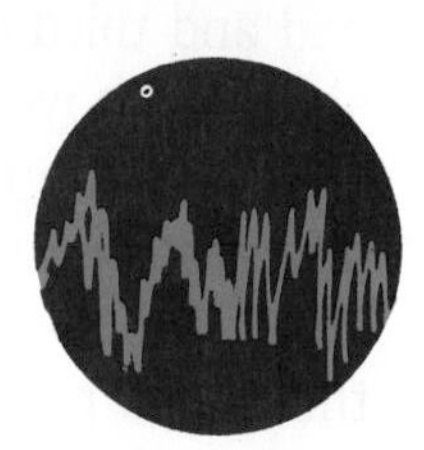
office noise

The Final Word Is Yours

In scientific terms, explain to someone how you can recognize a person just by hearing his or her voice.

Challenge Your Thinking

1. We are able to detect the position of a sound because both of our ears hear it. In order to appreciate how sensitive the brain can be in interpreting sound signals, calculate the difference in the time taken by each ear to hear a sound, if the sound source is on your left.

2. Paka selected six different pieces of steel wire. She hung different numbers of identical masses from the wires, as shown at right. Then she plucked the strings and arranged them in order of pitch, from low to high. Predict this order (and learn something about yourself)!

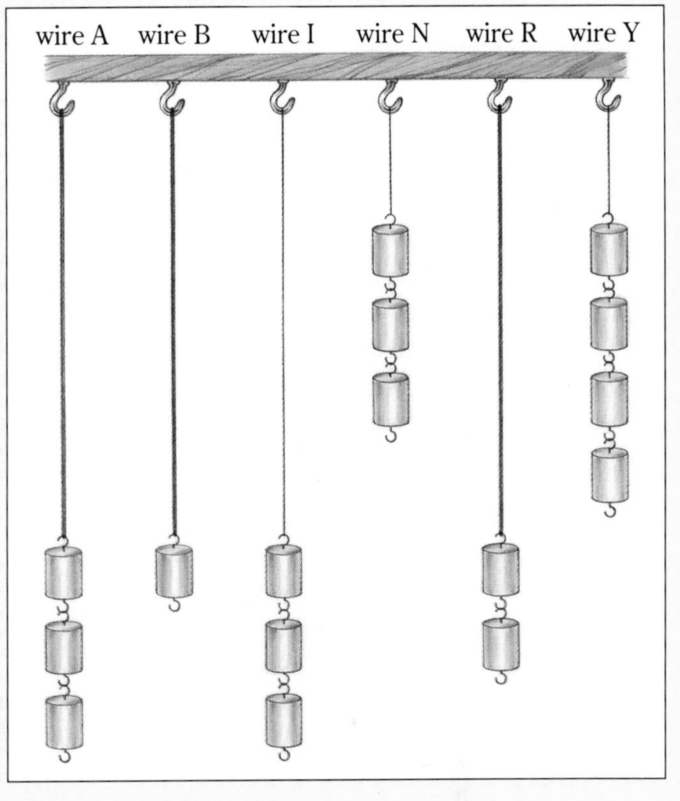

3. Alejandro connected a microphone to an oscilloscope to see what sounds looked like. Here are some of the pictures he saw.

   (a) Which screen shows the noisiest sound?

   (b) Which shows the purest musical sound?

   (c) Which shows the quietest sound?

   (d) Which shows the highest sound?

   (e) Which shows the loudest sound?

   (f) Which shows the lowest sound?

   (g) Which shows the sound that is an octave above the sound on screen G?

   (h) Which shows the sound that is an octave lower than that in screen F?

   (i) Which shows the sound made by the same instrument as is shown in screen C?

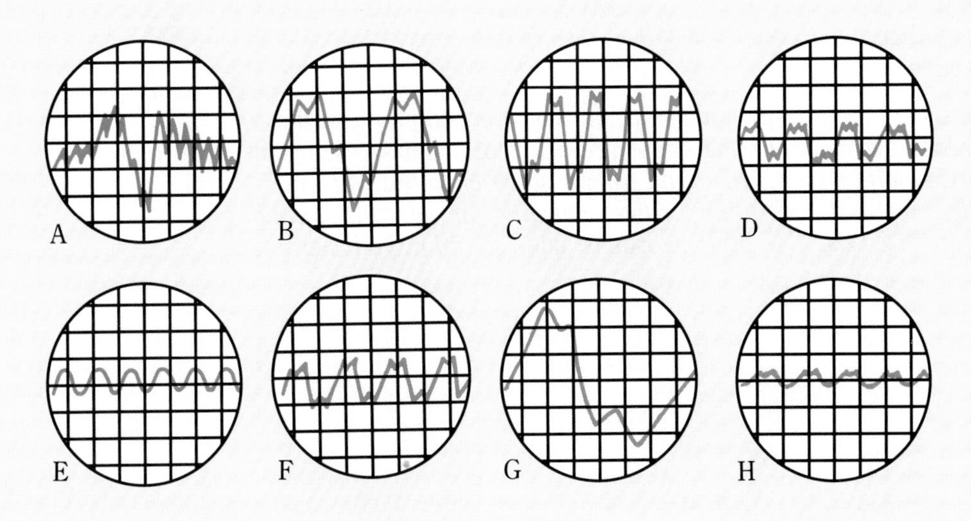

Unit 5 Making Connections

The Big Ideas

In your Journal, write a summary of this unit, using the following questions as a guide.

- What are some characteristics of sound?
- What causes sound?
- How is sound transported from a sound source to a sound receiver?
- Why can some animals hear sounds that people can't hear?
- How do you hear sound?
- How can you calculate the speed of sound in air?
- In what ways can you increase the loudness of a sound? the pitch of a sound?
- How are the terms *frequency* and *amplitude* of vibration related to pitch and loudness of sound?
- How do microphones and oscilloscopes "hear" sounds?
- How does an oscilloscope represent the pitch, loudness, and quality of musical sounds and noise?
- How could you use a vibrating guitar string to illustrate what gives rise to the distinctive quality of the sound it makes?

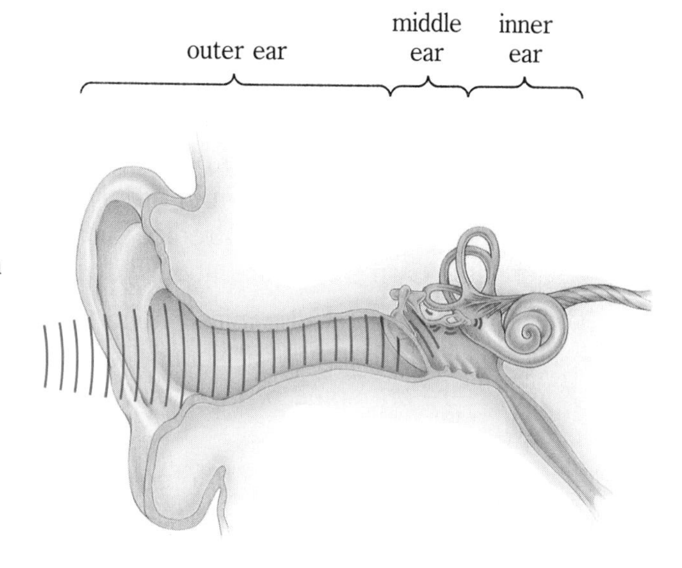

Checking Your Understanding

1. Word Puzzles!

Soft
s**O**unds
Use less
e**N**ergy than
lou**D** sounds

That's an acrostic of just one sound principle you learned about in this unit. Try making acrostics that illustrate principles of sound by using the following terms: *loudness, pitch, sound waves, speed of sound, music, noise, quality, amplitude, vibrating strings, vibrating air columns, swinging pendulums,* and *decibels.* Use other words besides *sound* in your acrostics.

2. Make an acrostic puzzle for someone else to solve, such as the following:

What's the Mystery Word?

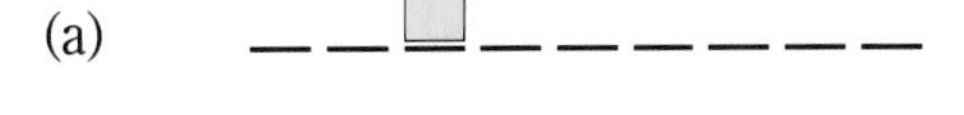

(a) The distance a vibrating object moves from its resting position

(b) A unit for measuring loudness

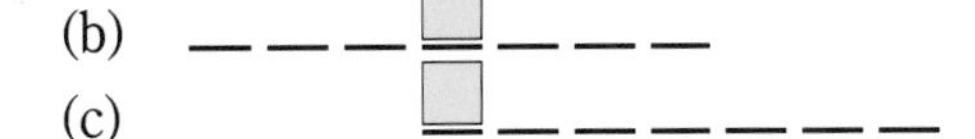

(c) Increasing this raises the pitch of vibrating strings

(d) Blind people "see" by this

(e) Short pipes produce __?__-pitched notes than long pipes

3. Try It and Explain!

(a) If you blow over the top of the bottles, what will happen to the pitch of the sound made by the bottles as you go from left to right? What do you think will happen if you clink the bottles with a spoon in turn from left to right?

(b) Predict what will happen to the sound of a plucked rubber band when you increase the number of masses hooked to it.

(c) Are the sounds of filling a bottle the same as those of emptying a bottle? As it fills, are there changes in the sound? Are there any changes when it empties?

(d) Make a straw whistle. Cut six holes in it. Cover the first hole from the top with your finger and blow. Next try covering the first two holes, and so on. Is there any change in sound? What do you observe?

(e) Tape a circular piece of paper onto a record. (It might be wise to use an old, scratched record.) Place the record on the turntable and turn on the record player. As the record turns, hold a fine felt-tipped pen on the paper at the center, and move it outward slowly and steadily along a straight path. Note the design that forms. View the turning design through a narrow slit in a card. What do you see? Is this a good model of sound waves? As a model, where is its weakness?

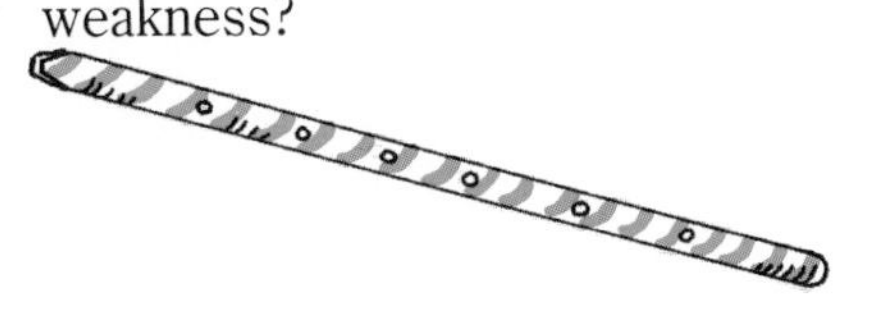

4. What's wrong?

5. If you placed your alarm clock inside a bell jar and used a vacuum pump to remove the air in the jar (as shown), would the sound of the alarm wake you up in the morning? Explain your reasoning.

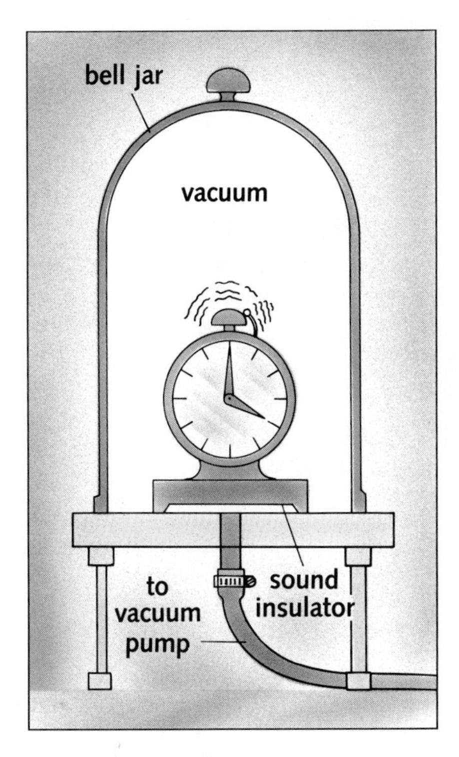

6. Draw a concept map that shows how the following ideas are related to each other: *sound, energy, vibrations, frequency, amplitude, pitch, loudness, waves,* and *hearing.*

Reading *Plus*

You have been introduced to the nature and qualities of sound. By reading pages S69–S86 in the *SourceBook,* you will take a closer look at the physical basis for sound—waves. You will also learn about ways in which sound waves can be used—for purposes that range from locating earthquakes, to measuring the depth of the ocean, to examining a fetus inside its mother.

Updating Your Journal

Reread the paragraphs you wrote for this unit's "For Your Journal" questions. Then rewrite them to reflect what you have learned in studying this unit.

Science in Action

Spotlight on Aeronautical Engineering

Richard Linn reviews an aerial photo of Dallas-Fort Worth Airport.

Aeronautical engineers specialize in the design, construction, or testing of aircraft. Also called aerospace engineers, these people are responsible for anything from the design of new airplanes to the design of missiles and space capsules.

Richard Linn received a B.S. in Aeronautical Engineering and an M.S. in Industrial Engineering. He has applied these studies to 30 years' worth of airplane noise control and reduction.

If you've ever had a jet airplane pass close overhead, you know how loud jet engines can be. As Senior Coordinator of Environmental Planning for a major U.S. airline, Richard analyzes and offers solutions for airplane noise problems.

Q: What inspired you to enter aeronautical engineering and how did you get involved in the noise-control issue?

Richard: Two kids to feed! Actually, I've loved airplanes ever since I was a kid. I built model planes and flew them all the time. It seemed natural that I would spend my life working around them.

My specific job was born by accident. Back in the '60s the noise of airplanes really began bothering people. It became a social problem that needed attention.

Q: Could you give us some background on what was done about the problem in the '60s?

Richard: Initially, all kinds of research was done on how to stop the escalation of noise. At the same time, homeowners and political groups began putting pressure on Congress to do something about the problem. As a result, laws were passed that forced airplane manufacturers to make quieter jet engines.

Soon we had the technology to produce new airplanes that were quieter. As the years passed, however, air traffic increased. The result has been that people find the constant noise just as bothersome as the louder jets had been.

Q: What do you do on a daily basis?

Richard: I work with engineers and citizens' groups to develop plans for manufacturing airplanes that will keep everybody happy. I also spend time lobbying for more money to cover the high cost of noise-reduction research.

Q: What is the most memorable event of your career?

Richard: The first flight of a brand new airplane soaring over the neighborhood is always a memorable event to me.

It's great knowing that all your research and experimentation has paid off and you can demonstrate to the community that you have new, quieter technology.

Q: Do you use any modern technological tools in your research?

Richard: We have new computer technology that allows us to use what we call "noise footprints." These are airplane simulation programs that help us determine the effects of engine noise on a community.

Q: What do you feel are the most interesting aspects of your career?

Richard: Traveling and working with lots of different people makes my job really interesting. I travel across the world to visit airports to see their unique flight-path situations. It's great to have the opportunity to see so much of the world!

Some aeronautical engineers help design jet engines.

Q: What are the most frustrating aspects of your career?

Richard: I think the noise-control problem is one that should be tackled by both airlines and communities. It's frustrating to realize that in spite of the well-known noise problems, cities often zone or rezone the land around airports for homes, churches, and schools. That makes my job really difficult because no matter how quiet we make an engine, the airplane will always produce some noise! The airlines can do only so much by themselves—without the communities' help in zoning, we'll never completely solve the problem.

Q: What personal qualities do you consider essential for a person to be successful in your field?

Richard: A sense of humor is invaluable. You also have to be able to keep your cool—you can't let the chaos get to you! It takes someone who can really commit himself or herself to improving the environment.

Some Project Ideas

Visit your local (or the nearest) airport. Determine the kind of community that immediately surrounds it. Check your local library for some statistics on the airport, such as the number of people that travel through it, the importance of the airport to the community, and the services and businesses that surround it. Think about who would be affected if the airport were suddenly closed.

Find a local map and chart the flight paths over your city. Which neighborhoods are most seriously affected?

One particularly loud aircraft is the Concorde. This aircraft flies at supersonic speeds. Research the Concorde to find out all you can about its noise problems.

Science and the Arts

Synthesized Sound

"All possible sounds are available on this equipment," Charles Ditto says enthusiastically. "I used to go to gigs with my spinet piano. Three guys would load it into a truck. Back then, all I could be was a piano player. Now I'm a totally mobile one-man orchestra."

Charles Ditto is an independent musician living in Austin, Texas. He creates and performs his music using state-of-the-art synthesizer equipment.

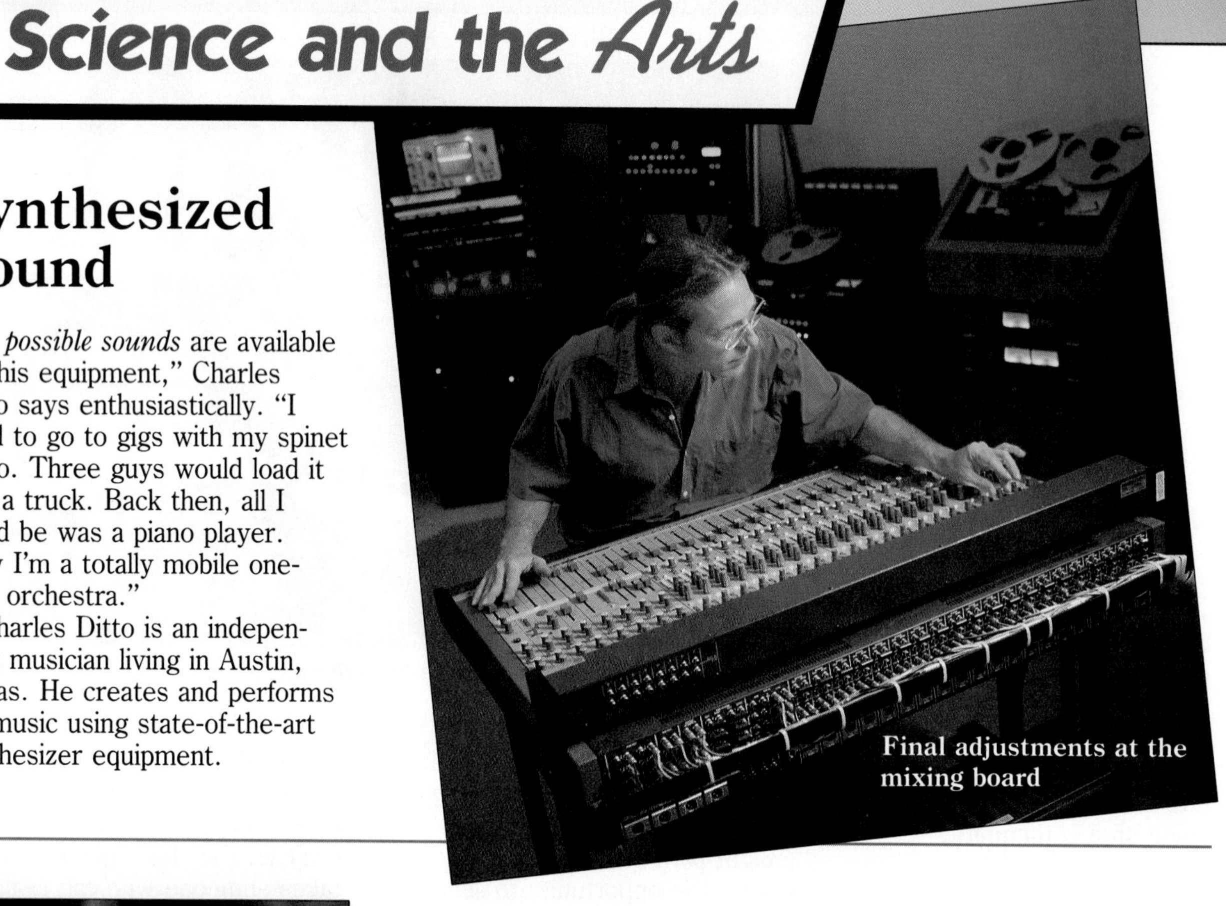

Final adjustments at the mixing board

Charles Ditto in performance

Going It Alone

With synthesizer equipment, you can create an entire band by playing all the parts yourself. For each musical instrument or sound you want, you play a separate part on your keyboard. A computer known as a *sequencer* stores the necessary information, such as the notes you played, how hard you struck the keys, and how fast you played. When you play back the performance, the sequencer selects the appropriate sounds from the synthesizer.

During actual performances, Ditto often uses musical backgrounds he created earlier and stored on computer disks. He plays the solo parts himself on the keyboard. Sometimes a friend will join him on a saxophone or other instrument.

Sounds Made to Order

A synthesizer can produce any sound you want. It might be natural, like the sound of wind and waves, or of seagulls. Or it might be industrial, like the sound of machinery. Another important possibility is *white noise,* which sounds like static on a radio. "White noise can be very beautiful when it's filtered and processed," Ditto explains. "It's used in a lot of new music."

Synthesizers come with a large assortment of sounds. But some of the most interesting sounds may be the ones you create yourself. Just throw in a little of this and a little of that until you've got something you like. And experiment with different ways of changing sounds—like *filtering,* which can screen out high or low frequencies.

Complex sound waves

Violinist Sha-Zhu producing a wave sample

Using the Sampler

Some sounds are difficult to create artificially. The sounds of human voices, for example, or of acoustic instruments like a piano or violin, are very complex. "When you look at their wave patterns on an oscilloscope, you see all sorts of things going on," Ditto says, "and the more you magnify them, the more you find."

For this reason, most artists use a lot of *sampled sounds.* These are the recorded sounds of actual instruments or voices—or anything else—that are stored so they can be played from the keyboard.

Promoting Your Product

Suppose you're just starting out, and you've got some music you want to get on the market. What do you do? Ditto says, "Well, first you'll need a demo tape of some of your stuff. With a synthesizer this is easy. You just hook it up directly to a tape recorder. I can make a top-quality tape without ever leaving the house.

"I like to send tapes to music magazine editors and reviewers. Sometimes they will listen to your music and write about it in their magazines. Radio stations are important, too. One college music journal puts out a list of stations that play new music. They're in the business of promoting new artists.

"Of course, you've got to send actual product—not just demos—to *distributors.* These are the people who will get your product into the stores. Some distributors will buy it from you wholesale. But a lot of them will just take it on *consignment.* This means they'll pay you if and when they sell it. There's risk involved. These businesses can pop up one day and be gone the next. You may never get your money or your product back. It's a lot of work, and there are a lot of headaches. But it's a lot of fun, too."

Find Out for Yourself

Find some tapes or CDs of synthesizer music. Which music do you like best? Describe the different kinds of effects you hear.

Try to find a musician in your area who has some synthesizer equipment. Ask him or her to demonstrate it for you. Write a report about what you learned and the sounds you heard.

Unit 6 Light

For Your Journal

Write a paragraph summarizing your thoughts about each of the following:

1. How can you produce light of various colors?
2. What kind of mirror makes you look bigger? smaller? upside down?
3. Why do blue jeans look blue?
4. How do faraway objects look in a magnifying glass?
5. Why is there variation in the color of light in this photo?

1 What Is Light?

Can you think of some popular songs in which the word "light" is used? Have you heard of any societies that worshipped the sun? Have you noticed the use of torches and candles in many religious ceremonies and sports events?

Questions like these remind us of how important light is. While you know about the significance of light, how much do you know about *what* light is? Here are some questions that may help direct your thinking about the nature of light.

1. When a flashlight is turned on, it gives off light. Does the flashlight lose mass?
2. Why is a room with light-colored walls brighter than a room with dark walls?
3. Why is the sky blue when seen from the Earth, yet black when seen by an astronaut on the moon?
4. How can we go backward in time by looking through a telescope?
5. How do we see? Why can we see an object when the light is on, but not when the light is off?

Make a list of other puzzling questions about light that you would like answered. By the time you have finished this unit, you'll be able to shed some light on your questions!

Light Brigade

Each of the activities in this Exploration will give you a better understanding of what light is.

You Will Need

- an eyedropper
- water
- a clock or stopwatch
- 2 saucers or watch glasses
- a magnifying glass
- a houseplant
- aluminum foil
- a newspaper
- a light bulb or light source
- 2 thermometers
- white paper
- black paper
- transparent tape
- scissors
- a radiometer
- blueprint paper
- a spring scale or balance
- copper wire
- a flashlight

What to Do

Form groups of three. Perform at least four of the activities, dividing them among the groups. While you're waiting for some of the light effects to occur in the longer activities, you can start on a different activity. Keep a record in your Journal of what you find out about light. Later, you can report what you discovered to the other groups.

Light and the Vanishing Water Drop

Use an eyedropper to place a drop of water on each of the two saucers or watch glasses. With a magnifying glass, focus light from the sun or a strong light source on one of the drops. Which drop disappears first? How long did it take to disappear? How long did it take the other drop to vanish? What does this demonstrate about light?

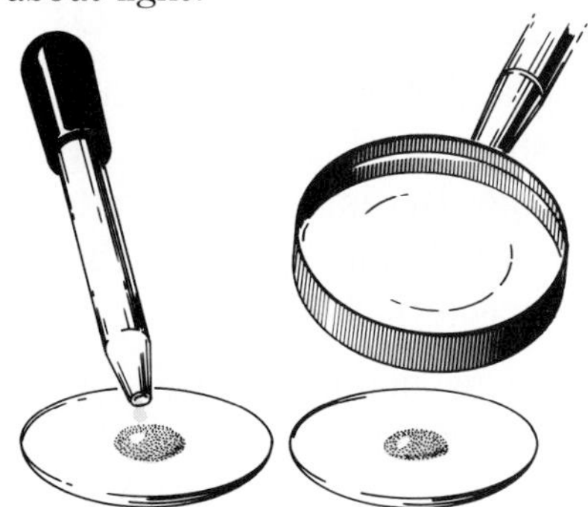

Light and Leaves

Carefully wrap a leaf of the houseplant in aluminum foil. Be careful not to damage the plant. Leave it for several days. Then remove the foil and compare the leaf with the rest of the plant. Explain what effect the foil had on the leaf. What does this indicate about light?

Light and Paper

Take a piece of newspaper 10 cm square or larger. Put it in a window where the sun will shine on it. Leave it for half a day. Compare its appearance with a piece of newspaper that has not been exposed to sunlight. Will a bright light bulb produce the same effect? Try it. What does this activity tell you about light?

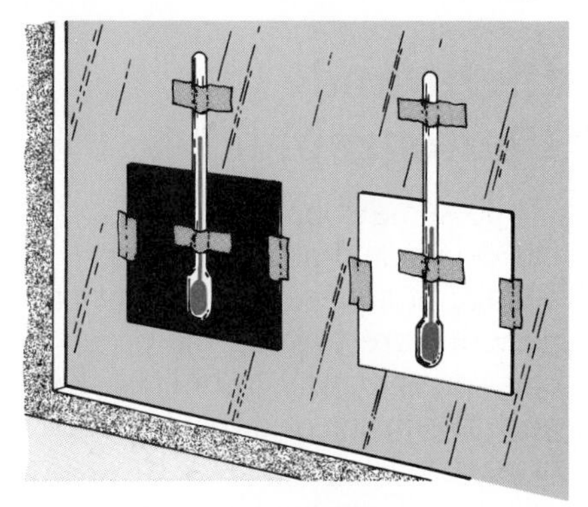

Light and Paper Color

Tape a piece of white paper and a piece of black paper to the inside of a window that faces the sun. Tape a thermometer to both pieces, as shown above. Record the temperature shown on both thermometers every minute for five minutes. What do your findings tell you about light and white paper? about light and black paper? What effect do you think light has on the temperature of other colors of paper?

EXPLORATION 1—CONT.

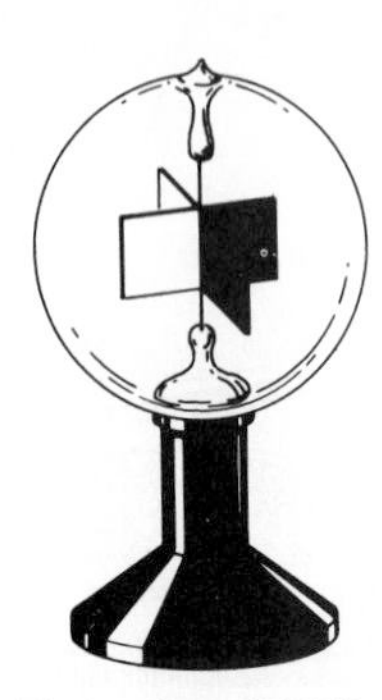

The Light Windmill

Examine the light windmill. The scientific name for this device is *radiometer*. Put the radiometer in direct sunlight for several minutes. What happens? Place it in the dark. What happens? Place it at different distances from a bright light bulb. What do you notice? What direction does the radiometer rotate? What does the radiometer show you about light?

Light and Photographs

Shape some copper wire into a flat design and place it on the surface of a piece of blueprint paper. Be sure to keep the paper covered and away from light. Expose the paper to direct sunlight or to a sunlamp for 5 to 10 minutes. Then immerse the paper in water. Allow it to dry. What does your "photograph" look like? How was it made? What did light have to do with it?

Flashlights and Mass

Find the mass of a flashlight. Then turn it on for at least 10 minutes. Now check its mass again. Is there any change? What does this show you about light? Do you think light occupies space?

What Do You Think?

What did you learn from the Light Brigade? Working in small groups, discuss the statements below. Do you *agree* with, *disagree* with, or are you *uncertain* about the truth of each statement? List your conclusions in your Journal.

1. Light speeds up evaporation.
2. Lack of light can make a leaf change color.
3. Light can make some things move.
4. Light can slowly change the color of newsprint.
5. Light can quickly change the color of blueprint paper.
6. Light can heat up materials.
7. Light has no mass.
8. Light does not take up space occupied by something else.

Light—Matter or Energy?

The activities in the Light Brigade certainly reveal a great deal about light. Light can do things: it can make things move or change. But is it matter, like air or water? Consider the following.

- A fully inflated basketball has a greater mass than a deflated basketball.
- A flashlight has the same mass after you've used it as it did before it was used.
- A room has the same mass whether it is well-lit or dark.
- When you wave your arm through water, you can feel a force on it.
- You can wave your arm through a light beam without feeling anything but air.

Clearly, light does not consist of matter. It does not seem to have mass the way matter does. It does not fill up space as matter does. Nor does light seem to push against things that move through it the way matter does.

What, then, is light, if it is not a form of matter? Is it energy? You know that energy can make things move or change things. Because of light, radiometers turn, some materials change color, the temperature rises, and water evaporates. Light can also produce electricity in a solar cell. Where do you find solar cells? Like other forms of energy, such as heat and electricity, light does not have mass or occupy space.

Photovoltaic cells like these convert sunlight into electricity.

Light—A Quick-Change Artist!

Energy is able to change from one form to another. In a standard light bulb, electrical energy is changed into heat energy and light. In the Light Brigade, the black paper became warmer in the sunlight; in other words, light changed into heat energy. What are some other examples of energy changes that involve light? Since light is so often a part of changes in energy, what do you think light must be?

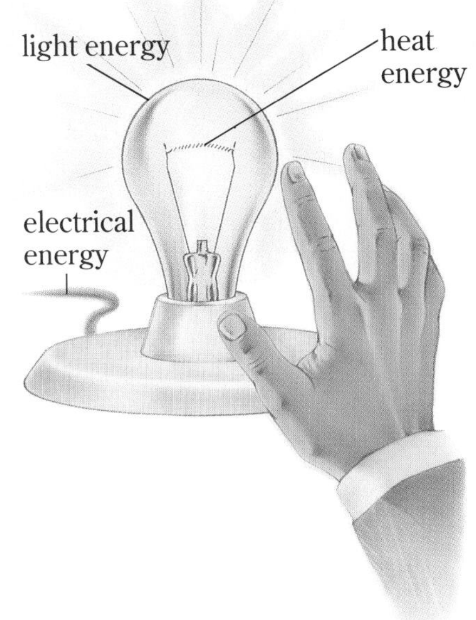

2 Light, Heat, and Color

Think of several things that produce light. Are all of them hot? Do things that produce light have to be hot? Fluorescent lights and neon signs for example, are relatively cool—even when lit for a long time. But these are exceptions to the general rule that light sources also produce heat. Complete Exploration 2 to discover more about the relationship between heat and light.

EXPLORATION 2

Observing Hot Solids

You Will Need

- 3 D-cells—1 weak and 2 strong
- a copper wire (uninsulated, approximately 20 cm long)
- a flashlight bulb
- masking tape
- a clothespin
- a Bunsen burner
- a hotplate or stove burner

What to Do  

Part 1

Set up the D-cells as shown below. Feel the wire and bulb and observe the color in each case. **Be Careful:** *The flashlight bulb may be hot.*

(a) A weak cell—just strong enough to make the filament start to glow

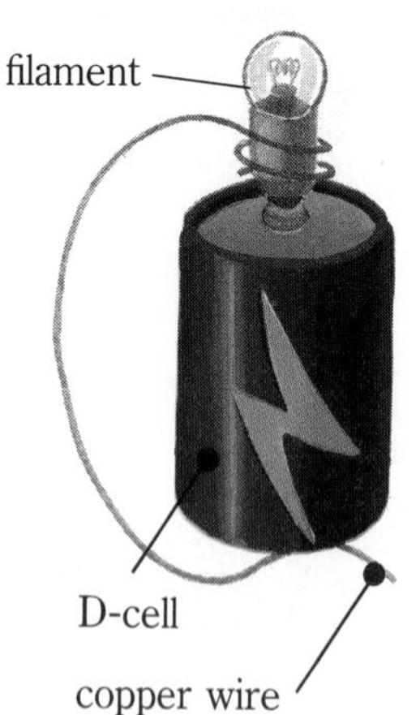

(b) A strong cell

(c) Two strong D-cells held together with masking tape

Part 2

Using a clothespin, hold a piece of copper wire in a flame until it glows (gives off light). What color do you see? Place your hand about 5 cm from the wire. What do you feel? **Be Careful:** *Do not touch the wire!*

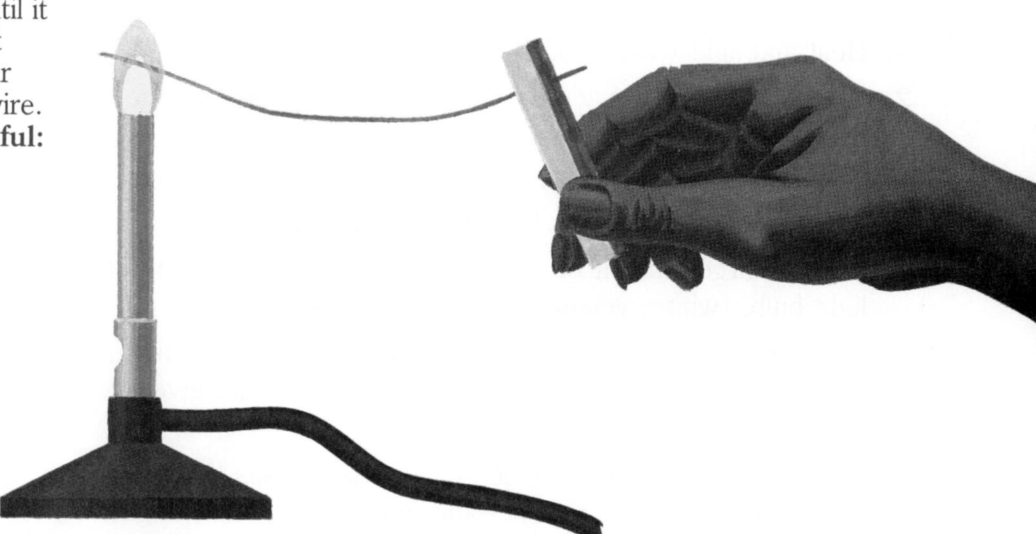

Part 3

Turn a hotplate or stove burner to low and gradually increase the heat to high. Observe the color of the burner and the heat it gives off in each case. **Be Careful:** *Keep your hands away from the heated surface!*

Part 4

What is the color of the molten steel in the photo below? What do you think its temperature is?

Light Check

Consider all the evidence you obtained from Exploration 2. Then complete the following statements by selecting the correct word or phrase from the choices provided. Record your choices in your Journal.

1. Heat and light (rarely, often) occur together.
2. When electrical energy passes through a wire, the wire (gets cold, stays the same temperature, gets hot).
3. When a small amount of electrical energy passes through a light bulb, (white, yellowish-orange) light is produced.
4. When a large amount of electrical energy passes through a light bulb, (white, yellowish-orange) light is produced.
5. When a small amount of electrical energy passes through a coiled wire, the wire gets hot; it (gives off, does not give off) light.
6. When a larger amount of electrical energy passes through a coiled wire, the wire gets hot and (gives off a reddish light, gives off a white light, does not give off light).
7. As a coiled wire becomes hotter it gives off light that gets (brighter—going from red to white; darker—going from white to red).
8. Heating a copper wire in a hot flame (does not affect its appearance, causes it to glow).
9. When metal is heated to a very high temperature, it gives off (red, white) light.
10. The color of light often indicates (how large, how hot, what shape) a substance is.

You might summarize your findings from Exploration 2 like this: When solids are heated enough, they start to glow red. With more heat, the light they emit becomes yellowish and then white.

How Hot Is Red-Hot?

Place the items listed below in order of their increasing temperatures.

- the surface of the sun
- a human body
- boiling water
- the filament in a 100-W bulb
- red-hot iron
- solid carbon dioxide (dry ice)
- melting ice
- white-hot iron

Now place the corresponding temperature next to each item on the list: −20 °C, 0 °C, 37 °C, 100 °C, 500−650 °C, 1500 °C, 2200−2700 °C, 5000−6000 °C. Compare your lists to those of other students. Are you surprised at the results?

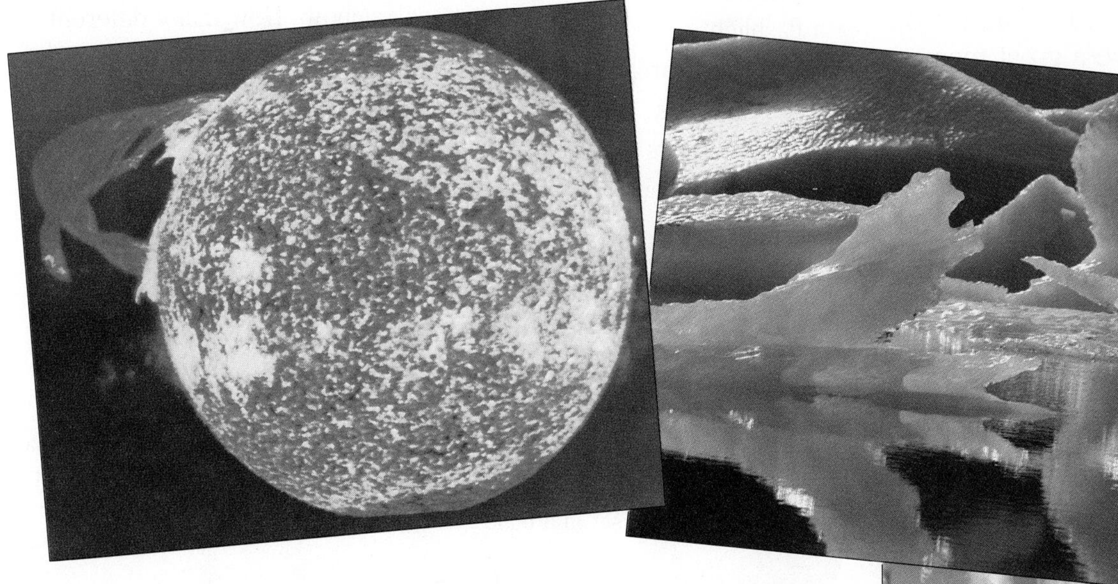

Light and Temperature

You have seen that light can indicate the temperature of objects. Recall the color change in copper wire held over a flame. When the wire is red-hot, it is hotter than when it is brown. When the wire is white-hot, it is hotter than when it is red-hot.

As an object becomes hot, it usually gives off colored light. The color of emitted light changes as the temperature of the object rises. Certain colors generally represent hotter temperatures. To understand the relationship between temperature and the colors of light you need to consider the following:

- Where does color come from?
- Is white a color?
- Is black a color?
- What is the relationship between white light and the other colors of light?

You may be surprised by how much you already know about color!

3 A Colorful Theory

In 1666 Isaac Newton, age 24, was experimenting with prisms. People had known about and had used prisms since the time of Aristotle (about 350 B.C.). But Newton made some remarkable new discoveries. In Exploration 3, you can relive a great moment in science—one of Newton's first discoveries.

EXPLORATION 3

Lights, Prisms, and Filters

This series of experiments works best in a darkened room. The activities should be done in the order given. Use the questions provided to help you think about and develop your own "theory of color."

You Will Need

- 2 prisms
- a flashlight
- white cardboard
- 3–6 red filters
- 3–6 blue filters
- 3–6 green filters
- white paper
- aluminum foil
- tape

What to Do

Experiment 1—Light and Prisms

Shine a narrow beam of light on a prism in such a way that a rainbow of colors (called a **spectrum**) forms on a white cardboard screen. Notice how the colors run into one another in a rainbow. How many different colors do you see on your cardboard screen? What are the colors at each edge of the spectrum?

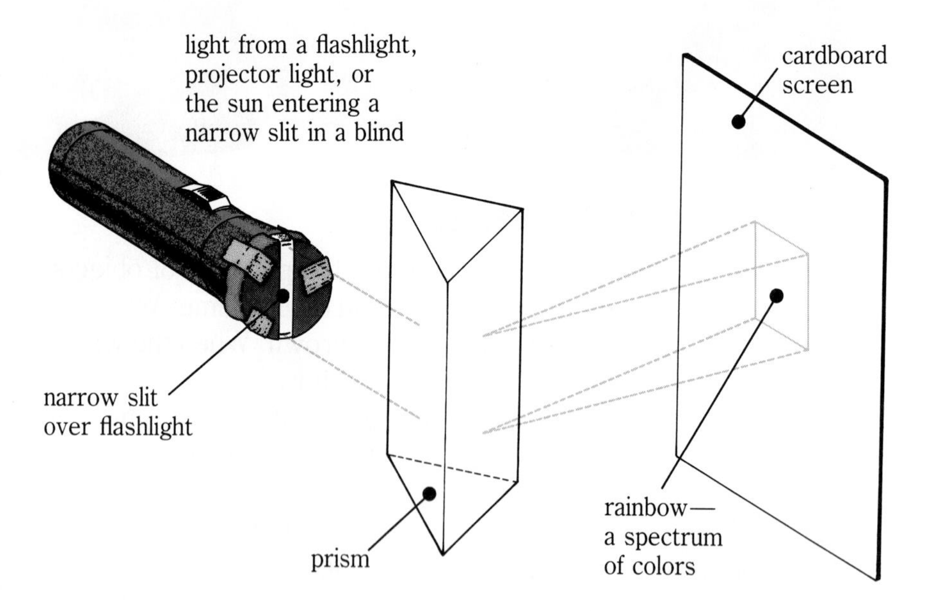

Where is the prism the thickest? Which color seems to come through the thickest part of the prism?

The people who lived before Isaac Newton believed that the production of colors in the spectrum was caused by the thickness of the prism. According to this theory, white light passing through the thinnest part of the prism changes to red; white light passing through the thickest part of the prism changes to blue; and white light passing through a medium thickness of the prism changes to green. Do you agree with this explanation? How else could the rainbow of colors be produced from white light?

Isaac Newton didn't believe the common explanation of his day. Complete the following experiments and you will understand Newton's reasoning.

Experiment 2—Light and Filters

Place a red filter between the flashlight and the prism. Observe the screen. How is the light on the screen different from before? Slip a piece of white paper between the filter and the prism. What is the color of the light between the filter and the prism? Did the prism change this color of light? Did the various thicknesses of the glass prism alter the red color in any way?

Repeat this experiment replacing the red filter with a blue one. Again, observe the color of the light. What is the color of the light between the blue filter and the prism? Did the prism cause any color change?

Now summarize the information you have acquired about light and color. In your Journal, state what you think filters do, what white light may be made of, and what prisms do to light.

Note: another way to do the following experiments (3, 4, and 5) is to lay the filters on an overhead projector and look at the projection screen.

Experiment 3—More About Filters

Prepare the setup shown in the figure below. Does the red filter add something to white light, or does it take something away from it?

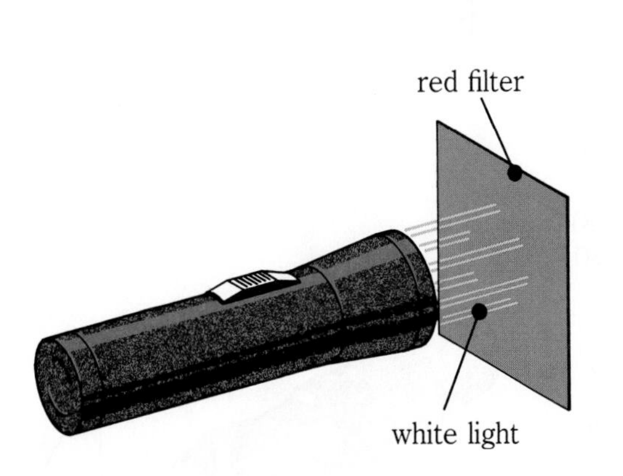

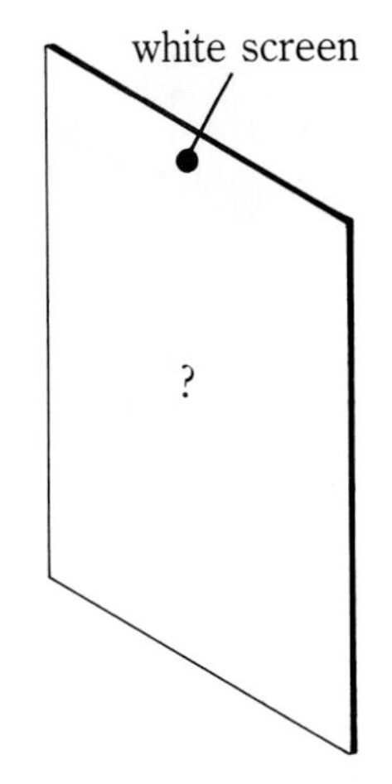

What color comes through the filter?

Repeat this experiment using a green filter in place of the red filter. Does the green filter add something to white light or take something away from it?

Consider This

A good analogy for light passing through a filter involves a rubber ball. When a ball moving along the floor hits a puddle of water, does the ball gain or lose energy? Remember that light is energy. When light passes through the red filter, does it lose energy, gain energy, or retain its original energy?

Experiment 4— A Combination of Filters

Return to the setup from Experiment 3.

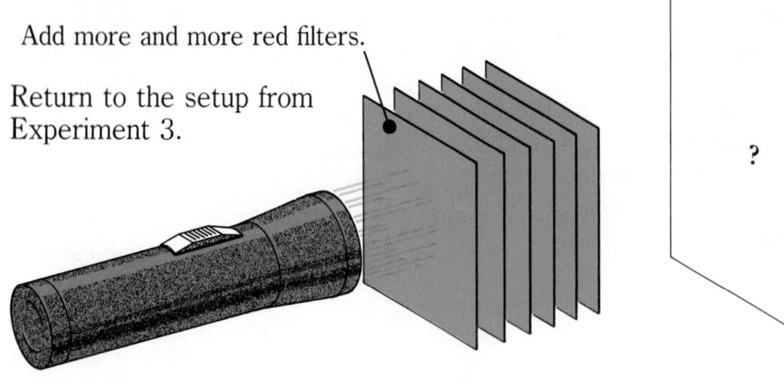

What happens to the white light as you add more red filters? Are the filters removing light energy? Could you add enough filters so that no light energy gets through? If no energy got through, what color would you see on the screen? In other words, what color is "no light"?

Repeat this activity using green filters and consider the same questions.

Now, shine the light through a single red filter, and then add either a dark blue or a dark green one. Was the energy loss similar? Recall how many red filters you had to use so that no light energy reached the screen. With a single red filter, how many blue or green filters did you have to use to get the same effect? Why do you think this occurred?

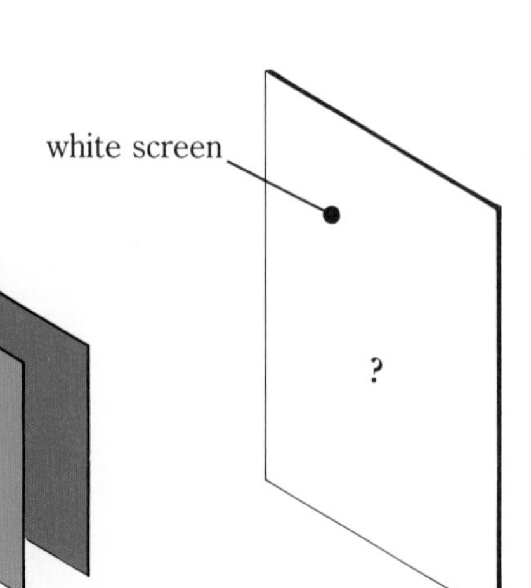

After considering the results of Experiments 3 and 4, would you agree that filters absorb some of the light energy in white light? In the following Experiment, you will discover the nature of the energy loss caused by filters.

Experiment 5— Disappearing Colors

Shine white light through the prism and obtain a spectrum on the white screen. What do you observe?

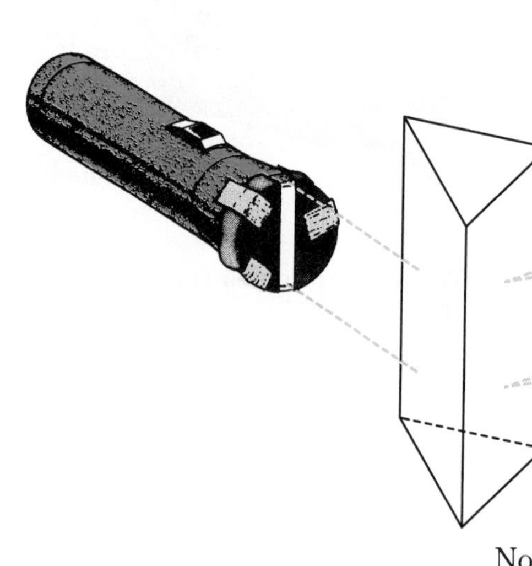

Now, place a red filter between the prism and the screen. What colors seem to have disappeared? In other words, what colors are absorbed by the filter? What color(s) pass(es) through the red filter? If just red light was shone on this filter, what would happen to the light? What effect would this filter have if green light fell on it? If blue light fell on it?

Repeat this experiment, this time using a dark blue filter between the prism and the screen. Consider the same questions.

Experiment 6—Double Prisms

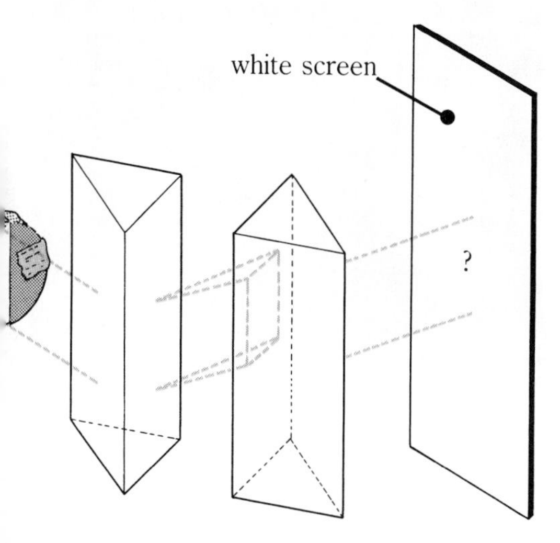

Use the same setup as in Experiment 5, but remove all filters. Place a second prism facing in the opposite direction from the first one. Make sure the prisms are close together. The spectrum produced by the first prism should shine on the slanted side of the second prism.

What color of light comes from the second prism? What did the first prism do to the white light? What did the second prism do to the colors of light?

Your Theory

The goal of the six Experiments in this Exploration was to allow you to develop your own theory about color—what color is, where the different colors of light come from, what white light is, and what is meant by "black." Think through all you have discovered, and write down your theory. Then you can see how it compares with Isaac Newton's explanation, which is described next.

Newton and You!

Consider Newton's theory about light and color. Newton said that light is not changed when it goes through a prism. Instead, it is physically separated. Here is Newton's theory, in his own words. (Some words are changed from the original to make the reading a little easier.)

> *I procured a Triangular glass-Prism, to try . . . the celebrated 'Phenomena of Colors.' And in order to do this, having darkened my chamber, and made a small hole in my windowshuts, to let in a convenient quantity of Sunlight, I placed my Prism at this entrance, that it might be thereby directed to the opposite wall. It was at first a very pleasing entertainment, to view the vivid and intense colors produced . . . but after a while, applying myself to consider them more carefully, I became surprised to see them in an (oblong) form which . . . I expected would be circular.*

> *And I saw . . . that the light, at one end of the Image was bent considerably more than the light at the other end. And the true cause of the (oblong) length of the Image was reasoned to be that Light consists of Rays, bent by different amounts to be transmitted towards different parts of the wall.*

The underlined words are Newton's conclusion about why the spectrum appears as it does. Newton also reasoned that, if the old theory was true—that is, if the thickness of a prism did change the color of white light—then another prism should change the resulting colors again. This did not happen when he tried it. What he found was that once the colors of white light were separated, they could not be changed any further. Newton called this the "critical experiment."

Newton also noticed that the spectral colors always occurred in the same order. Thinking back to your investigations, can you recall what the order is? Some people remember the order of the colors of the spectrum by thinking of "Mr. Color"—ROY G. BIV. What does each letter stand for?

On the next page are a few more excerpts from Newton's findings on light and color.

But the most surprising, and wonderful composition was that of Whiteness. There is no one sort of Ray which alone can show this. 'Tis ever compounded, and to its composition there are required all the previously mentioned primary Colors, mixed in the right proportion. I have often with Admiration observed, that all the Colors of the Prism being made to come together again, and thereby to be again mixed, reproduced light, entirely and perfectly white.

Hence therefore it comes to pass that Whiteness is the usual color of Light, for Light is a confused collection of Rays possessing all sorts of Colors, as they are randomly darted from the various parts of shining bodies.

Isaac Newton

Newton's theory of light and color was a totally new one. Many scientists of his day disputed it and attacked his ideas. Newton eventually wrote a letter saying that he was so harrassed by arguments about his theory that he regretted losing his peace of mind in order to "*run after a shadow*" of an idea.

Things to Think About

1. Translate a few of Newton's sentences into contemporary English. Do you find his writing to be scientific? Explain.
2. According to Newton, what is white light? red light? blue light?
3. How do the colors red, green, and blue differ from each other as they pass through a prism?
4. How did Newton prove that any color that is produced when white light passes through a prism cannot be changed to any other color?
5. Why do you think other scientists disputed Newton's theory so strongly?
6. Using the illustration to the right and your own theory of light and color, predict the color of the light that enters the blue filter. Why do you think the screen would appear black?

Performed in a darkened room

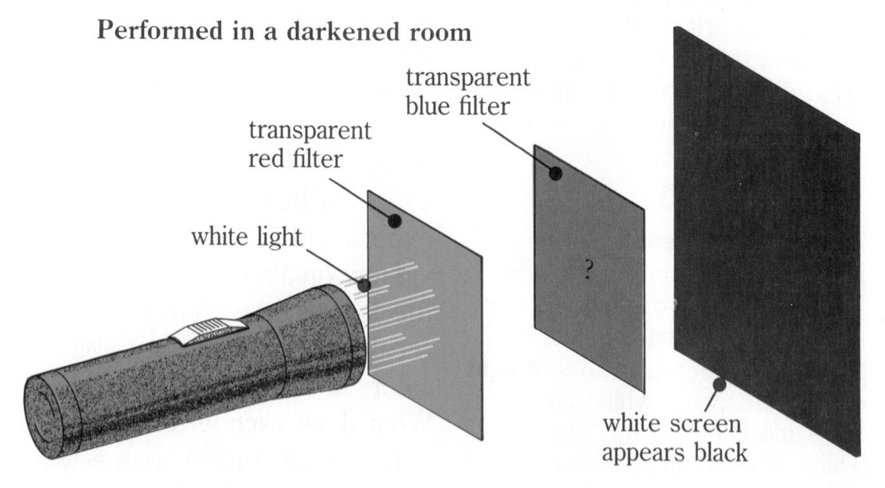

7. Predict what you will see in the situation shown to the right, and explain your prediction.
8. Write a letter to a friend, relating your personal discoveries about light, in much the same way that Isaac Newton wrote about his discoveries. (Your letter can be written in a more modern style!)

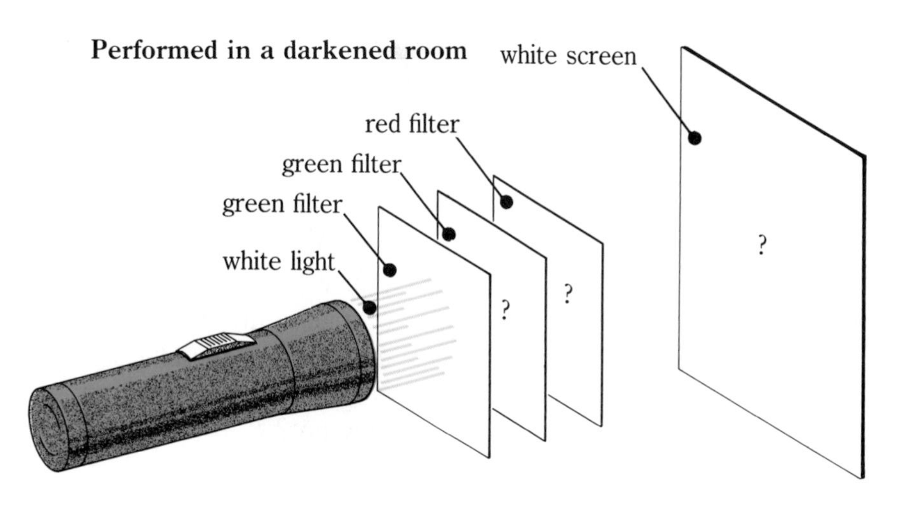

Activities to Try at Home

Each activity below shows a way of breaking white light into its colors.

Judging Temperature by Color Change

White light is composed of, and can be separated into, light of seven different colors. The opposite of this is also true. Light of the seven different colors can be recombined to make white light. With this theory, you can explain how to measure the temperature of cool, warm, and hot objects. First recall the observations you made of the flashlight bulb in Exploration 2.

- A flashlight bulb attached to a nearly dead D-cell emits (gives off) light that appears reddish-orange.
- A flashlight bulb attached to a new D-cell emits a white light.
- An extremely hot object will emit all colors of the spectrum, with a greater proportion of violet, indigo, and blue colors.

Consider the illustrations on the right.

1. The temperature of the filament is just hot enough that it gives off reddish-orange light. Would you expect that *any* blue light is emitted, too? Why or why not?

2. In this case, the filament is hotter than in the first illustration. As a result, it gives off all of the colors of light, which combine to make white light.

3. If a light source is hotter than even white light, the amounts of blue, indigo, and violet increase. This light appears bluish-white. How would you draw this light source? Make a sketch in your Journal similar to the ones shown here. In your drawing, illustrate the colors given off from a bluish-white light source.

Fingerprints of Matter

How do you think the illustration at right was made? This colorful, *continuous spectrum* was formed when white light passed through a prism and separated into its different colors. Why do you think this is called a continuous spectrum?

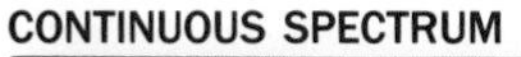

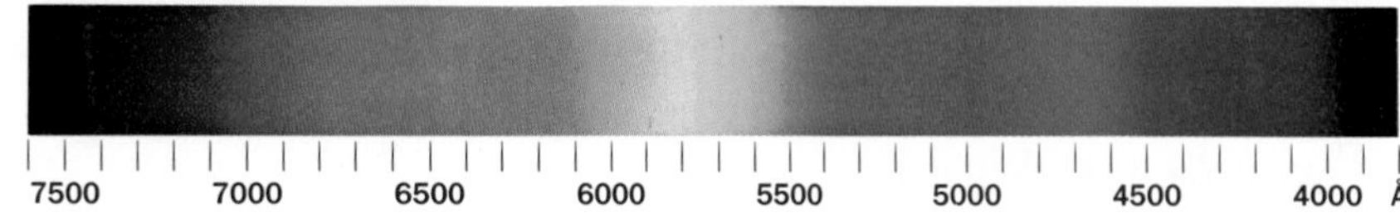

Spectra (the plural of spectrum) are usually studied with an instrument called a **spectrometer,** which contains a prism and a small telescope that together produce a magnified spectrum. The spectrum is displayed against a numbered scale to measure the distribution of different colors of light. In the drawing below, you can see how a spectrometer is designed. Can you trace the path that light takes when it passes through the device? Where do you think the prism is located?

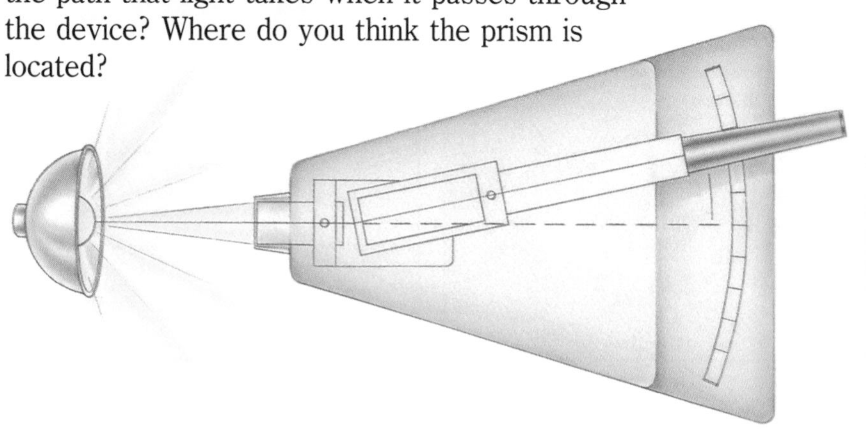

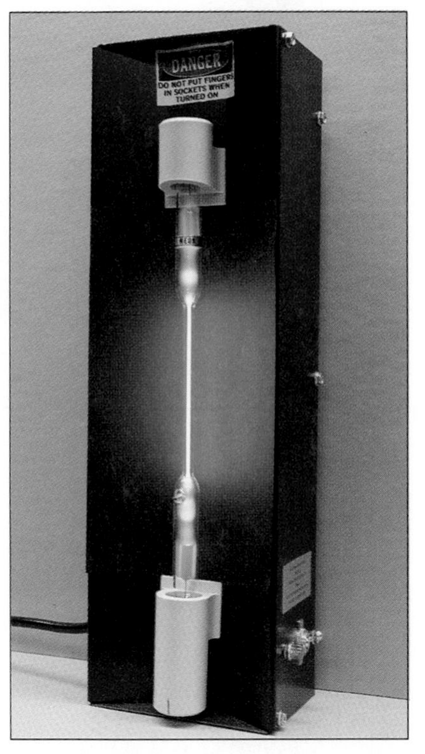

Spectrometers aren't just used to study white light. Consider the facts, photos, and questions below.

- Every element can be converted into a glowing gas. The photo at the upper right shows the red light produced by glowing neon gas. Where have you seen red neon light before?
- When the light from the neon passes through a prism, you get what is called a *bright line spectrum.* How is it different from a continuous spectrum?
- Sodium gas gives off a yellow light, while hydrogen gas gives off a lavender light. What can you infer about the bright line spectrum for each of these elements? What about all of the other elements?

If each element has its own spectrum, what special uses do you think a spectrometer has? Studying spectra can tell you many things—it even led to the amazing discovery of the element helium. Do some research to discover how an *absorption spectra* helped identify helium on the sun before it was discovered here on Earth!

Lines from the spectra of helium (yellow), neon (red), and argon (lavender)

Challenge Your Thinking

1. A device that can change one form of energy into another is called an *energy converter.* Identify the energy conversions that occur in each picture. What role does light play in each?

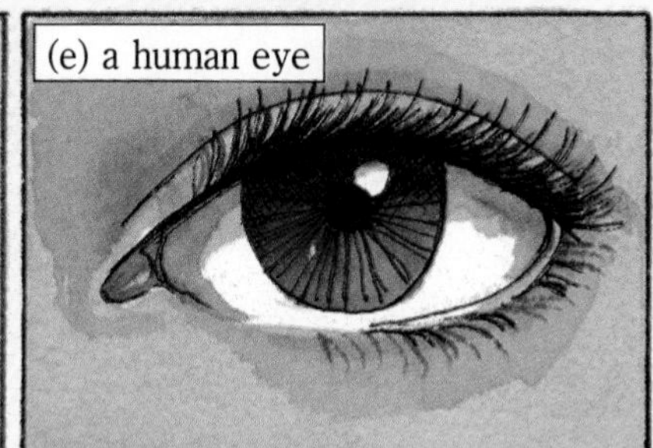

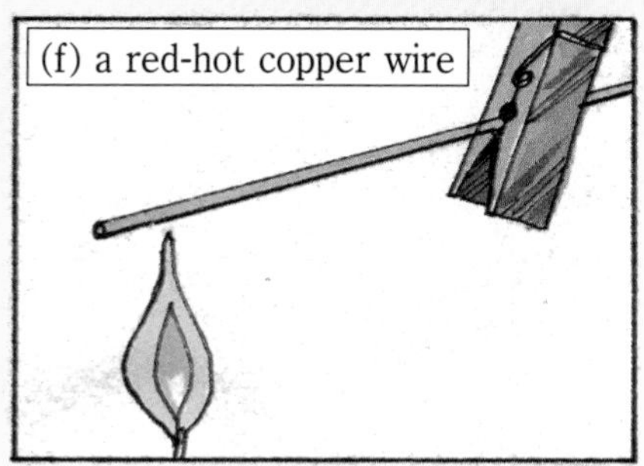

2. Brent made a color spinner like the one shown at the right. He wound the cord tightly and started the disc spinning by pulling back and forth on the cord. What did he see? What would he see if the colors on the wheel were mainly red and orange? blue, indigo, and violet?

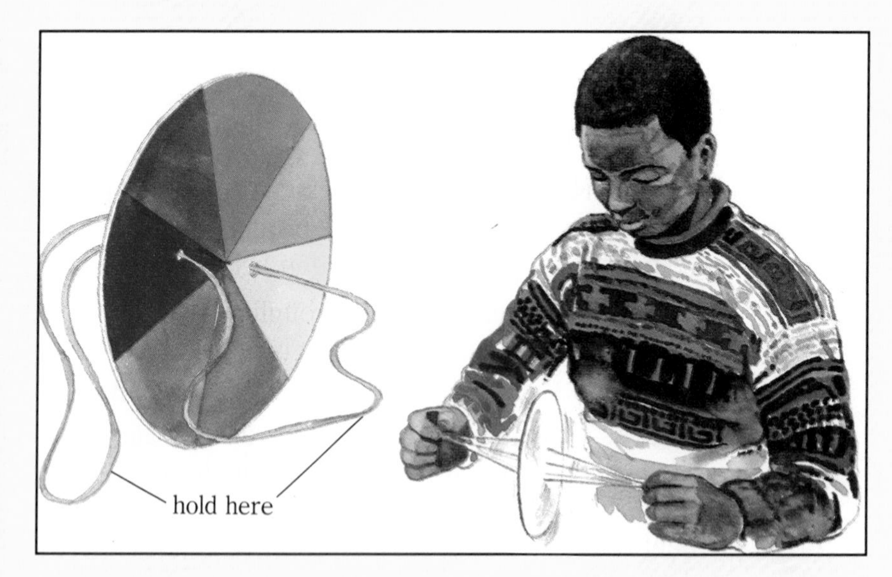

3. Photographers observe that when pictures are taken using the light from ordinary light bulbs, the developed photos have a reddish tint. Why? If a flash is used, the color is better. Why? Another way to improve the photo is to place a blue filter over the camera lens. How does the filter help?

Scattering, Transmission, and Absorption

Scattering, transmission, absorption. What do you think these words mean? Read the four statements below, which use forms of these words, to increase your understanding.

- When a light beam shines on a mirror, it bounces off in one particular direction. However, when a light beam hits a white piece of paper or the white wall of a room, the light bounces off in all directions, or **scatters.**
- When a white light shines on a blue piece of paper, the blue color in the white light is scattered by the paper, while all the other colors are **absorbed.**
- When white light shines on a red filter, mainly the red color is **transmitted** through the filter; that is, it passes through. All the other colors are absorbed by the filter.
- When a beam of white light shines through smoky air, part of the light is transmitted through the smoke. The rest of the light is scattered by the smoke particles.

A Light Box

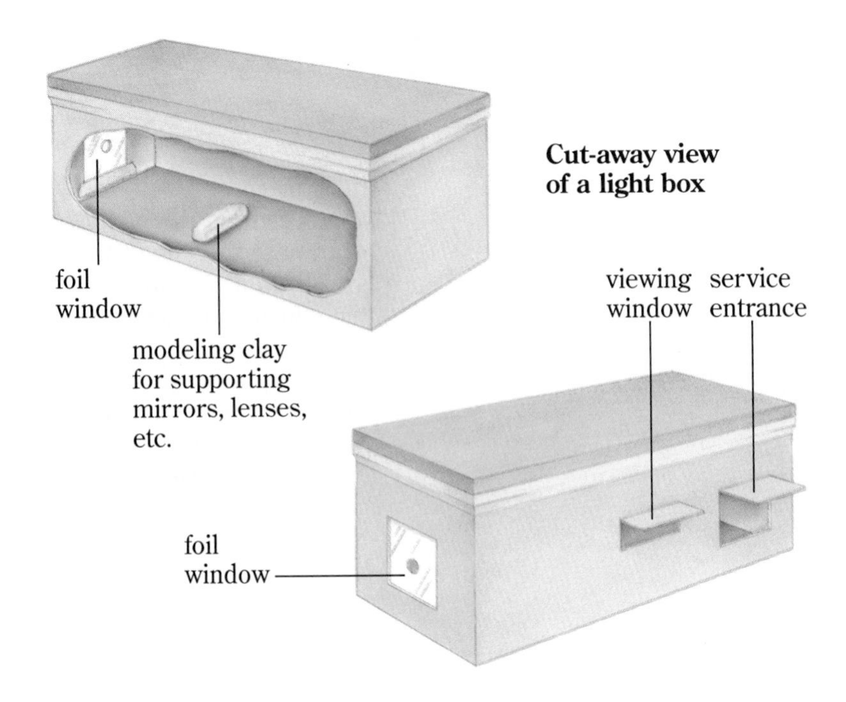

Cut-away view of a light box

To get a better understanding of scattering, transmission, and absorption, you can perform a series of experiments using a light box. A light box shuts out most of the unwanted light in a room, making it easier for you to see the light used in the experiments. Follow the written instructions on page 346 to build your own light box.

How to Build a Light Box

Here is how to make a light box from simple, readily available materials. You will use the light box for many experiments in this unit.

First, get an empty cardboard box about 40 to 50 cm long and 25 to 30 cm wide and high. A large shoe box will work.

On one end, near the bottom, cut out a square window about 6 cm on each edge. Then cut a square of aluminum foil about 8 cm on each edge. In the center of the foil, cut a round hole about 1.5 to 2.0 cm in diameter. Tape the square of foil over the window at the end of the box.

Cut two flaps in one side of the box—one near the center, the other one lower and to the right. Each flap should be about 10 cm wide and 8 cm high. The center flap is for looking into the box, while the other flap is for putting things into or removing things from the box. Finally, tape the box closed with masking tape.

EXPLORATION 4

Scattering, Transmission, and Absorption Experiences

You Will Need

- a light box
- a small flashlight
- rough black paper or cloth
- modeling clay
- white paper
- colored paper
- a wooden splint
- matches
- a piece of window glass
- paraffin wax
- a small beaker of water
- skim-milk powder or milk
- a stirring rod

Part 1

A Black-Surface Experiment

Set up the light box that you made earlier. Position a strong flashlight so that light shines into the foil-covered hole. Look through the observation flap.

What to Do  

1. There is now some light in the box. Where does most of the light come from? Is it from the beam as it passes through the air, or from light scattered when the beam hits the end of the box?

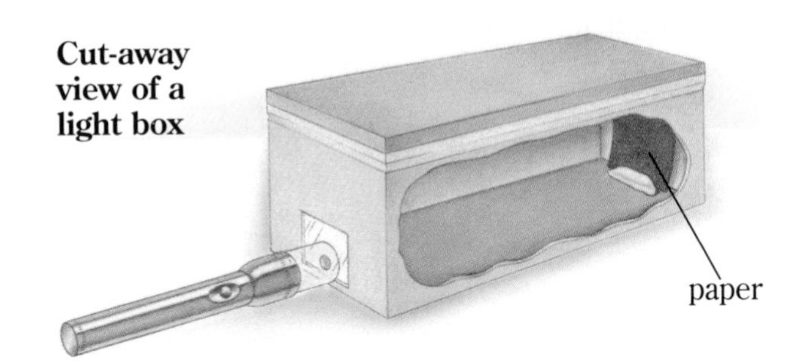

Cut-away view of a light box

2. Put a piece of rough black paper or cloth in the path of the beam near the end of the box. Prop it up with something such as modeling clay. What happens to the brightness of light in the box when the black surface is added? What happens to most of the light when it strikes the black surface?

3. Replace the black paper with a sheet of white paper. What does this do to the amount of light in the box? Which color of paper absorbs more light—black or white? Which color reflects more light? Try other colors of paper as well.
4. Add some smoke from a wooden splint. **Caution:** *Beware of fire. Be careful not to use too much smoke; try to keep it confined to the light box. People may be adversely affected by too much smoke.* Blow gently into the observation hole to spread the smoke around. Can you see through the smoky air? The particles of smoke scatter some of the light in all directions. The smoke lets you see the light more clearly.
5. You should now be able to complete the following statements. Write them in your Journal, and fill in the answers.
   (a) As a beam of light passes through the air, it (lights up, does not light up) the box.
   (b) As the beam of light hits the end of the box, it (lights up, does not light up) the box.
   (c) As the beam of light falls on the black screen, the box is lit up (less than, more than, to the same extent as) before.
   (d) White light falling on a black surface is absorbed (less than, more than, to the same extent as) when it falls on a white surface.
   (e) White light falling on a black surface is scattered (less than, more than, to the same extent as) it is scattered by a white surface.
   (f) A green surface scatters (more light than, less light than, the same amount of light as) a white surface, and (more light than, less light than, the same amount of light as) a black surface.
   (g) A green surface scatters (what color?) light from its surface. Therefore, it must absorb (what colors?) of light.
   (h) A black surface absorbs (what colors?) of light.

Part 2

A Window-Glass Experiment

Replace the black paper or cloth that is in the path of the light beam with a rectangle of window glass. Place it near the center of the box, just behind the viewing window. Add smoke so you can see the light beam. Move the glass at various angles and observe the beam.

What to Do

1. Place the glass at a 45° angle to the light beam. Does the beam of light scatter as it hits the glass, or does it bounce off *(reflect)* in a certain direction? Can you see light pass through the glass? In other words, can you see any transmitted light? Which seems brighter—the reflected beam or the transmitted beam? Why?

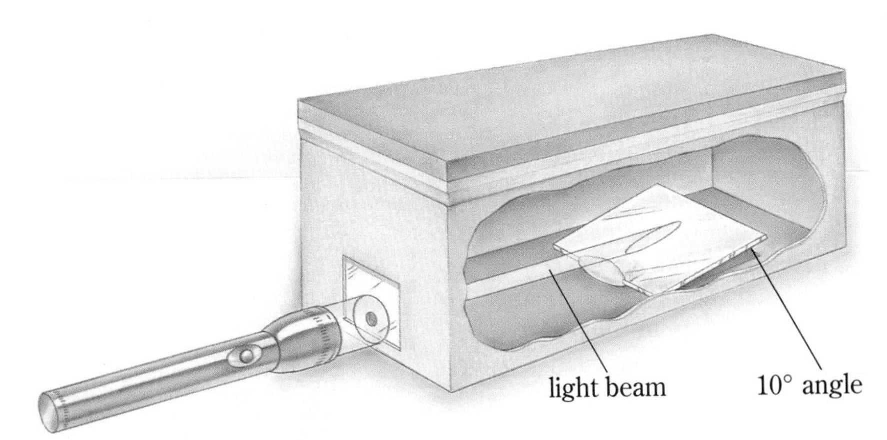

2. Repeat step 1 again, first placing the glass at a steep angle such as 80°, then at an angle such as 10°. For each angle, answer the questions in step 1.
3. Now, write several statements like those you completed after Part 1 of this Exploration. They should describe what you have found out by doing this experiment.

Part 3

A Paraffin and Light Experiment

What to Do

1. Replace the piece of glass with a piece of paraffin wax. Hold the wax at various angles in the light beam. Use smoke to help you see the beam.
2. How does the appearance of the paraffin compare with the glass when each is in the beam? Is the box brighter when the paraffin is in place or when the glass is present?
3. Record your findings in a series of simple statements.

Part 4

A Water and Light Experiment

What to Do

1. Put a small beaker nearly filled with water in the light box. Turn on the flashlight to make a light beam shine through the water. Can you see the beam in the water?
2. Add a few crystals of skim-milk powder or a drop or two of milk to the water and stir. Is the beam now visible in the water? What are the milk particles doing to the light? Is there evidence that light is transmitted through the slightly cloudy water? Is there evidence that light is scattered by the cloudy water?
3. Write a conclusion for this experiment using the words "transmitted" and "scattered."

Part 5

Another Water and Light Experiment

What to Do

1. The light box is not needed for this experiment. Take a large beaker or jar of clean water, and hold a small flashlight against one side of it. (If only a large flashlight is available, cover its face with a piece of foil that has a hole 1 cm in diameter in it.) Look at the light from the opposite side of the container as well as at right angles to the light beam. You can obtain the best results in a darkened room. What color is the light? What color, if any, is the water when viewed at right angles to the beam?
2. Add some granules of skim-milk powder or a few drops of milk to the water and stir. What color does the light seem to be? What color, if any, is the water?
3. Repeat step 2, adding more milk or powder until faint color effects are observed.
4. Describe the findings you have discovered in this experiment.

Interpreting the Absorption, Transmission, and Scattering Experiments

Recall the experiments you did in Exploration 4. Compare the concluding statements you made for each part of the Exploration with the corresponding information given here.

Part 1

The inside of the light box was darker when black paper was used than when white paper was used. Why? The reason is that the black surface absorbed most of the light, while the white surface reflected most of the light. Light energy is absorbed by black objects and changed into heat energy. Less light is absorbed by white objects; instead, much of it is reflected. Consider the clothing you might wear on a hot summer day. Would you be cooler wearing light or dark colors? Why?

Part 2

Glass is **transparent**. Light can pass through it, and you can see objects on the other side of it. However, when light meets glass, not all of the light passes through. Instead, a small amount of light is reflected at the surface of the glass. The amount of light that is reflected depends on the angle at which the light meets the glass. When light skims along the glass at an angle such as 10°, more is reflected than if the angle is increased to 70° or 80°, for example.

Now consider this situation. When Jacob looks at the front window of his building, he can see the building across the street. At the same time, he can see the stairs behind him from inside his own building. Why? At what angle should Jacob stand in relation to the front window so that he sees mainly the building across the street—a large or small angle? Sketch this situation in your Journal.

How does the surface of a still pond or puddle resemble the window glass used in Part 2?

Part 3

A block of paraffin is different from either glass or white paper. When you shine a light at glass, most of the light beam passes through. When you use white paper, the light bounces off in all directions. But when you use paraffin, something different happens. Some of the light is scattered in all directions *within* the block itself, while a little of the light passes through. Paraffin is **translucent**. It allows some light to pass through it, but much of the light is scattered within it. You cannot see things clearly through a translucent material. What other materials can you think of that are translucent?

Glass and paraffin wax are positioned between you and a light source. See how only part of the light passes through the paraffin wax.

Parts 4 and 5

The experiments with the milk and water showed something else about the way light behaves. In Part 4, the presence of a few milk particles in water caused light to scatter. The light beam then became visible to you as you viewed it from a position at right angles to the beam.

When you added a little more milk powder in Part 5, you observed something different. Looking directly at the beam of light through the slightly milky water, you saw faint colors. The beam itself looked faintly orange-red. The surrounding water looked faintly blue. Why?

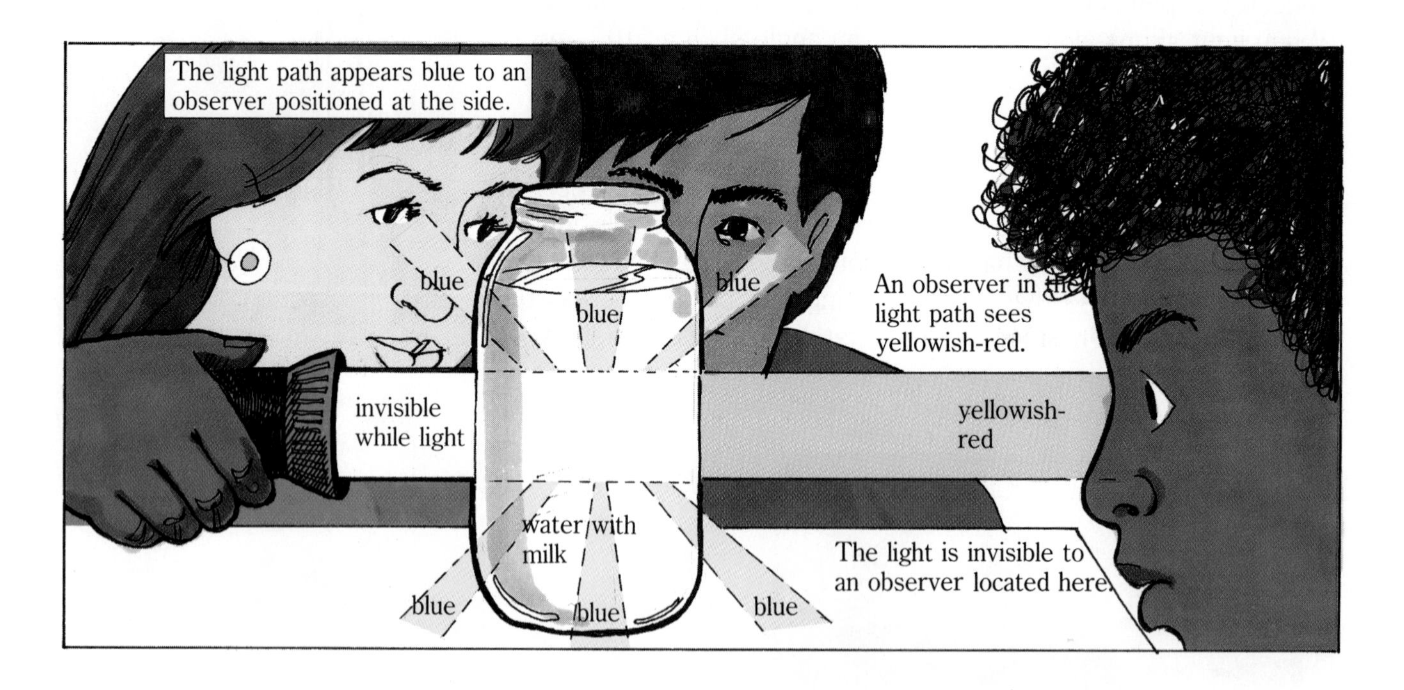

The reason is that particles scatter more blue, violet, and indigo light than any other colors. This light makes the water around the beam look bluish. The light that passes through the jar of milky water is somewhat lacking in blue, indigo, and violet while it still has most of its red, orange, and yellow. Therefore, it looks orange-red.

The results of these experiments will help explain some interesting questions: Why is the sky blue? What is the color of space beyond the atmosphere? Why is a sunset red? Why does the ocean look blue? Think of some possible answers to these questions before reading further.

Light Beams, Black Space, and Red Sunsets

Whenever light hits an object, some of the light scatters. Before you put smoke into your light box, the light beam is not visible; the air particles are too small to scatter much light. The smoke in the air makes the beam visible. Smoke also consists of extremely small particles—you would need a microscope to see one. But the tiny carbon particles that make up smoke are not nearly as small as air particles. Some light bounces off each carbon particle. Because there are millions of these particles in smoky air, the total amount of light bouncing off all of them together makes the beam visible. The carbon particles tumble through the air in all directions, so they reflect light in all directions. The light is therefore scattered, and some of it reaches your eyes.

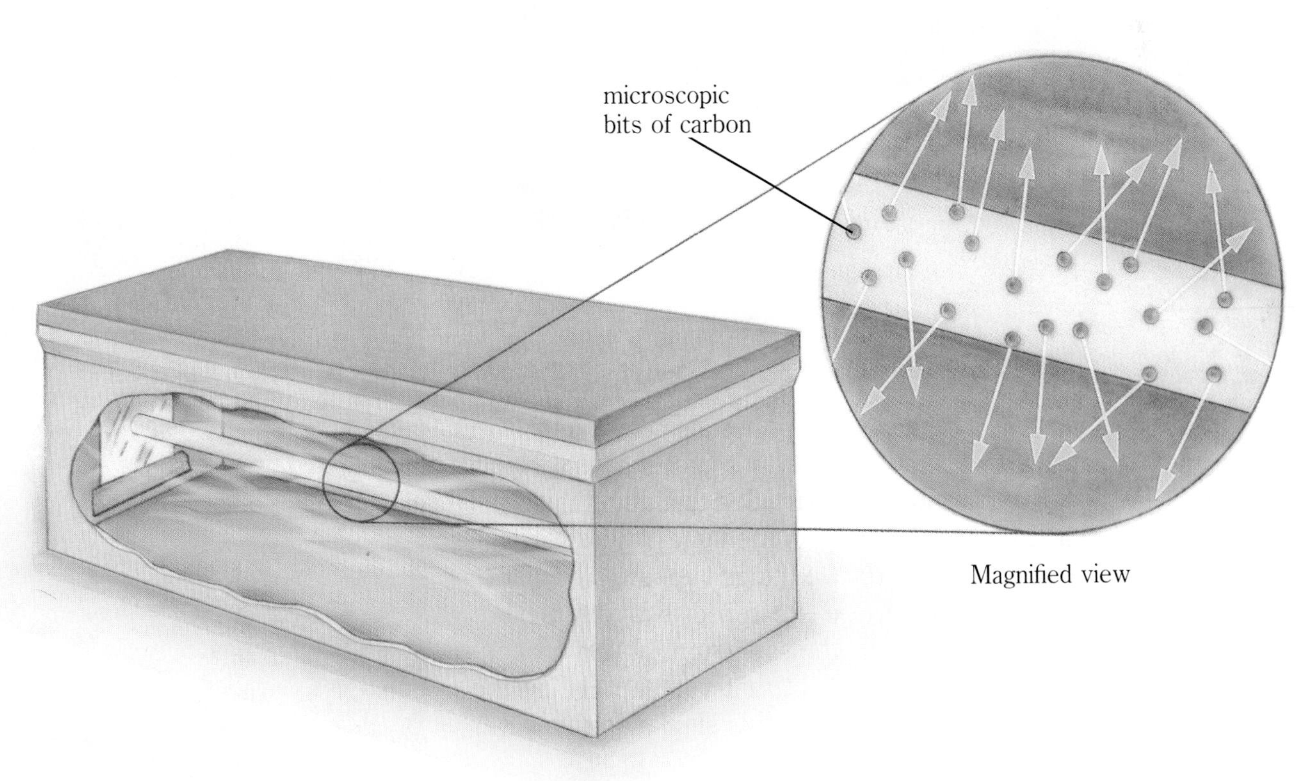

Like milk particles in water, the particles forming the air of the atmosphere scatter light—especially the light at the blue end of the spectrum. It is for this reason that the sky has a blue color. If you were to go up 10,000 m in a jet aircraft and look upward in the sky, it would appear darker blue than it does from the surface of the Earth. Why? There is less air at such a high elevation, so less light is scattered and the blackness of empty space shows through. Astronauts who travel out of the Earth's atmosphere see no scattered light in the sky at all—the sky is black. Of course, the sun and other stars still shine in this blackness. To the astronauts, the stars appear like the lights you see from a highway at night—white lights shining out of the darkness.

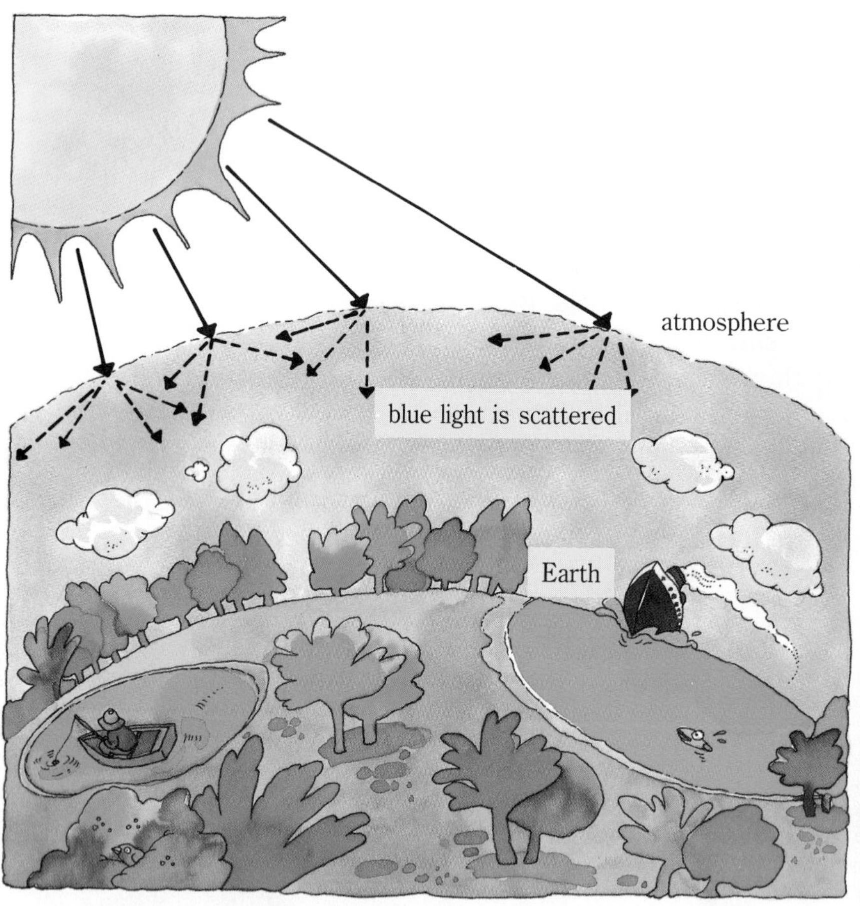

The reddish appearance of the setting sun is also caused by light scattering. When the sun is low in the sky, its light shines through a much thicker layer of air and other particles (including dust and smog) than when it is high in the sky. As the light travels through this added thickness of atmosphere two things happen. The colors at the blue end of the spectrum are scattered away from your eyes. The colors at the other end of the spectrum, however, continue toward Earth. It is this part of light that you see. Thus, the setting sun's light seems reddish. How is this similar to the water and milk experiment from Exploration 4?

Why does the ocean appear blue? Use the illustrations below to help you explain this phenomenon.

Other Color Phenomena to Wonder About

Here are some challenging questions. If you're not sure of the answers, go ahead and speculate about them. Form small discussion groups and share your ideas. As you continue to study light, you'll be able to come up with better answers.

1. You read earlier in this unit that rainbows are very large spectra. Can you recall what colors are on either edge of a rainbow? What is a double rainbow? Where might it appear? What are the colors on either edge of a double rainbow?
2. During the day, why do the windows of a house look darker than the outside walls of the house—even if the walls of the house are painted a dark color?
3. Why are many freshwater lakes and streams light bluish-green in color?

5 Reflection

No, these symbols are not strange markings or hieroglyphics carved on an ancient stone. Try to guess what they are. Can you draw the symbols that come next in the series? First, you need to place a mirror through the center of each symbol. Check the reflection. If you're still having trouble, try covering up the left half of the symbol. Use an index card and your mirror to add to the series.

What you have done so far will help you solve the next puzzle. Can you crack the code? You know that mirrors are good reflectors of light. But do you realize that *every* object you see reflects light? If it didn't, you would not be able to see the object. The activities in Exploration 5 will help you understand more about the reflection of light.

EXPLORATION 5

Reflection Investigations

You Will Need

- a flashlight
- a light box
- modeling clay
- a flat mirror
- a protractor
- matches
- a wooden splint
- white cardboard 15 cm × 10 cm
- green cardboard 15 cm × 10 cm
- cardboard of various colors

Part 1

The Reflection of Light from a Flat Mirror

What to Do  

1. Shine the flashlight into the light box. Using a piece of modeling clay to support the mirror, place it in the path of the light beam. Use a smoldering splint to add smoke to the box, and then observe the light beam. On its way to the mirror, the light beam is called the **incident beam.** After it bounces off the mirror, it is called the **reflected beam.**

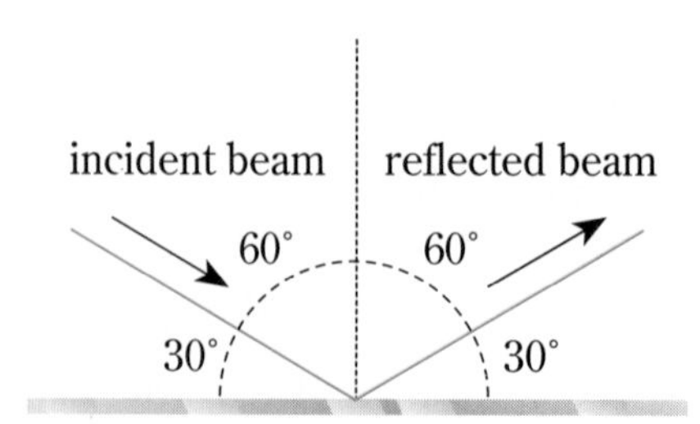

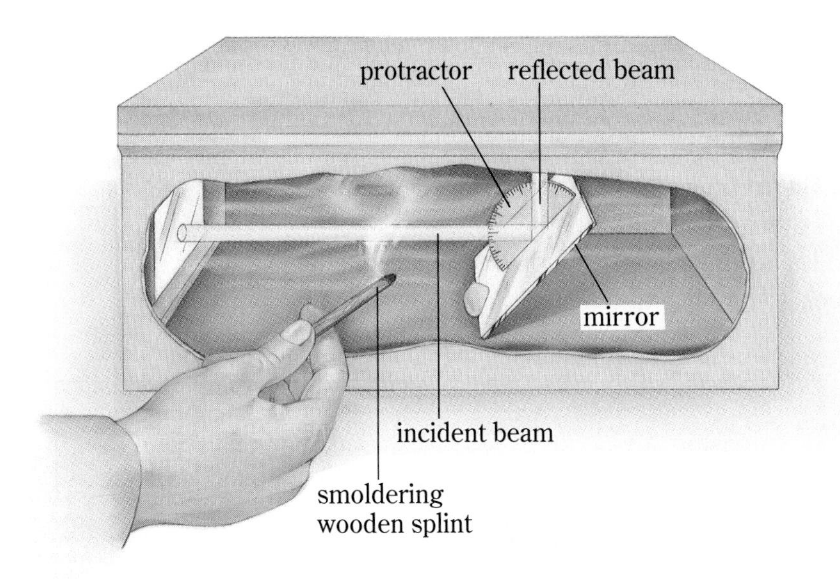

2. Hold a protractor with its base against the mirror in such a way that it lies in both the incident and the reflected beam at the same time. The center of the two beams, where the light strikes the mirror, should be at the midpoint of the base of the protractor. What is the size of the angle between the incident beam and the mirror? between the reflected beam and the mirror?
3. Rotate the mirror to a different position. Again measure the angles between the incident beam and the mirror, and between the reflected beam and the mirror.
4. What is the rule about the direction a reflected beam takes after it strikes a plane mirror?
5. Suppose a beam of light were to hit a mirror in a light box. The mirror is rotated backward by 20 degrees. How many degrees will the reflected beam rotate? If necessary, use the light box and the figure below to help you answer the question.

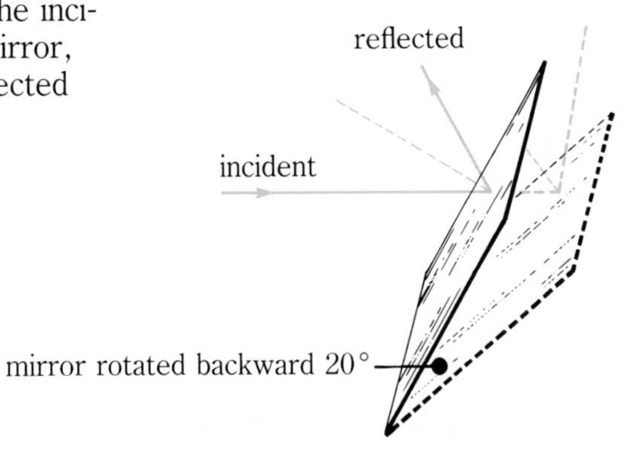

Part 2

The Reflection of Light from Cardboard

What to Do

1. In this activity, you will use white cardboard and green cardboard instead of a mirror. Use a smoldering splint to add some smoke to your light box. Place the white cardboard in the incident beam. Describe the appearance of the beam reflected by the white cardboard. How does this beam differ from the beam reflected by the mirror?
2. Rotate the white cardboard in the light beam. In which position does the cardboard reflect the brightest beam? the weakest beam?
3. Describe the appearance of the reflected beam at the point where it hits the side wall of the light box.
4. Repeat steps 1 through 3 using the green cardboard in place of the white cardboard. Answer the same questions as you go along.
5. How does the appearance of the beam reflected from the green cardboard differ from that of the beam reflected from the white cardboard?
6. Repeat the experiment, using different colors of cardboard.

Thinking About Your Findings

From the activities you have just done, you have begun to develop an understanding of reflection. Now, review each investigation, check for new ideas, and answer some of these interesting thought questions.

Part 1

What type of reflection occurred when the mirror was used? The incident light beam was reflected from the mirror without being scattered. This type of reflection is called **specular** reflection.

Here is an illustrated review to check what you discovered in this part of Exploration 5.

(a) How large is angle X?

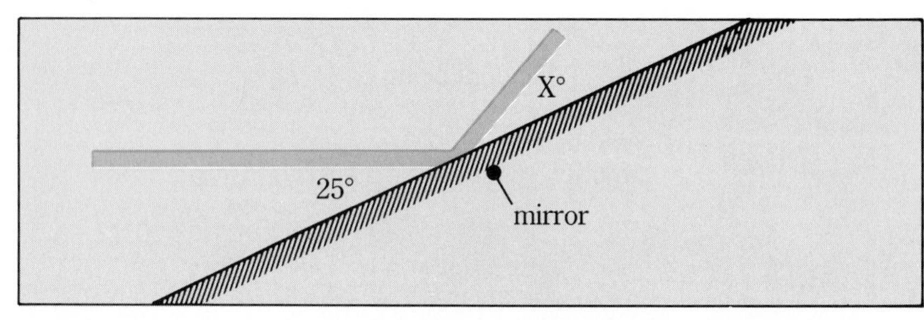

(b) Which is the correct reflected beam, P, Q, R, or S?

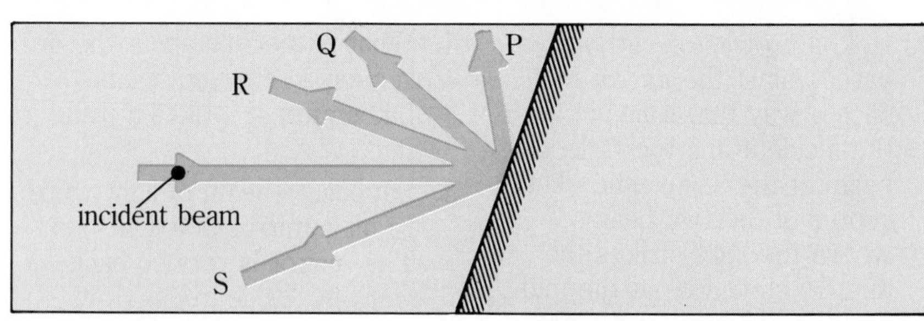

(c) Now that the incident beam has changed its position, which is the correct reflected beam?

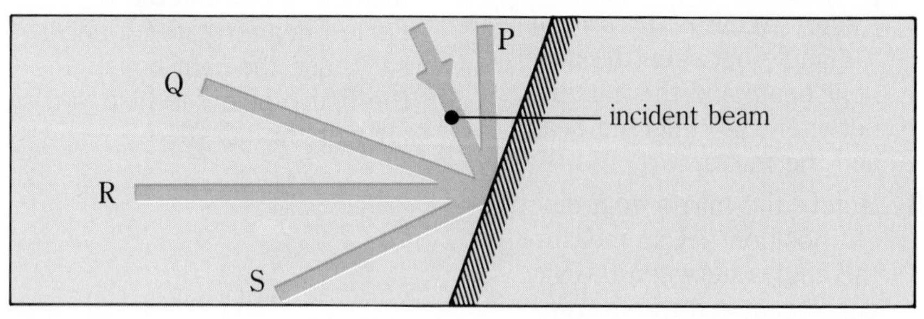

(d) Here the mirror is placed in a different position. In this case, which is the incident beam and which is the reflected beam?

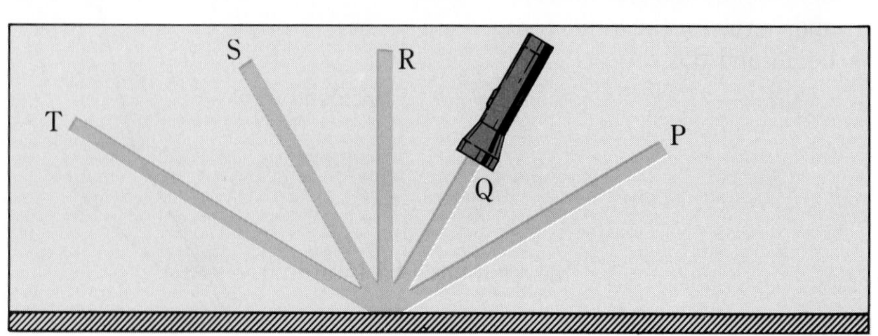

(e) There doesn't appear to be any reflected beam. Explain.

(f) What angle does the reflected beam make with the mirror?

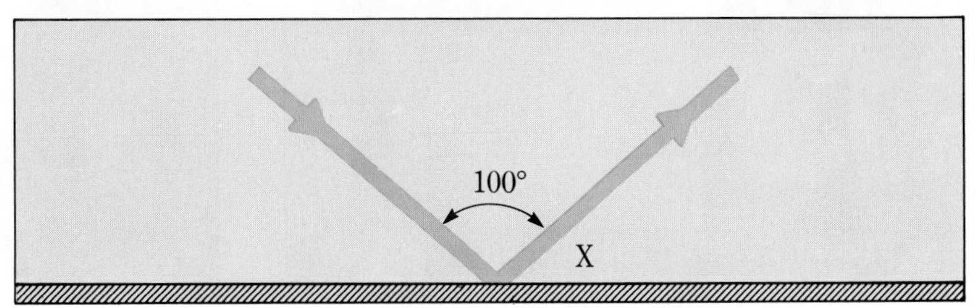

A mirror reflects a single beam of light in only one direction. Therefore, once you know the angle of the incident beam, you can predict the angle of the reflected beam. Let's discuss why a mirror reflects light in this way and does not scatter it.

You will recall that smoke particles scatter light in different directions. The tiny particles of soot and smoke tumble about in the air facing every direction, so they reflect light in every direction. However, the metal particles that make up the reflecting surface on a mirror all face the same direction. Therefore, they all reflect the light in the same direction.

Now that you know this, here are two more questions to consider.

(a) Devise a rule for the reflection of light in a flat mirror. Here is a start: In specular reflection, the angle between a flat mirror and the incident beam is (equal to, greater than, less than) the angle between ___?___.

(b) List some other surfaces in which specular reflection occurs. (Some surfaces that are smooth to the touch, such as paper, are not smooth when examined with a microscope. These surfaces are not good specular reflectors.)

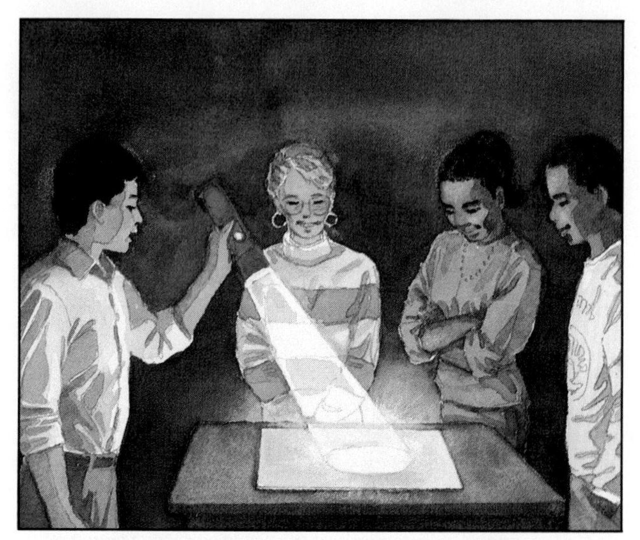

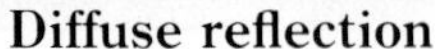

Diffuse reflection

Specular reflection

Part 2

How did the white cardboard reflect the beam of light? Which of these two descriptions best fits your observations? "Specular reflection occurred just as it did with the flat mirror," or "The reflected beam generally followed the expected direction, but it scattered more than the reflection from the mirror."

The scientific term for "scattered reflection" is **diffuse** reflection. Diffuse reflection is more common than specular reflection. Why do you think this is so?

Was the reflection from the green cardboard specular or diffuse? What color was the scattered light from the green cardboard? What color was the smoke? What color were the walls of the light box? When the light entered the light box, it was white. What colors found in white light were not visible? The following illustration suggests what happened.

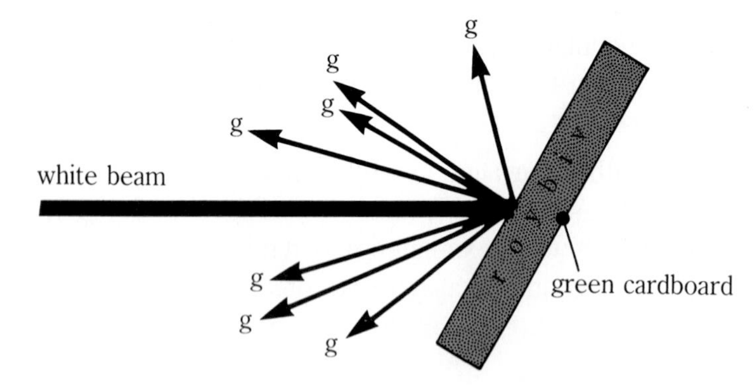

The green cardboard reflected the green part of the white light onto the smoke, the walls of the light box, and your eyes. It absorbed all the other colors. Green paper looks green when light falls on it because the green is scattered in all directions—including toward your eyes.

Challenge Your Thinking

1. Study the illustration below. Where is white light scattered? reflected? What about blue light? What are the transmitters of light? Where can you identify light absorption occurring?

2. How do you think a carnival's House of Mirrors takes advantage of the way light behaves? Using the words *specular, incident beam, reflected beam,* and *angle,* describe how funhouse illusions can confuse you.

3. Are you a color expert? Consider the illustrations shown here. They all involve light of various colors. It's up to you to identify the color you would expect to see at points A–F.

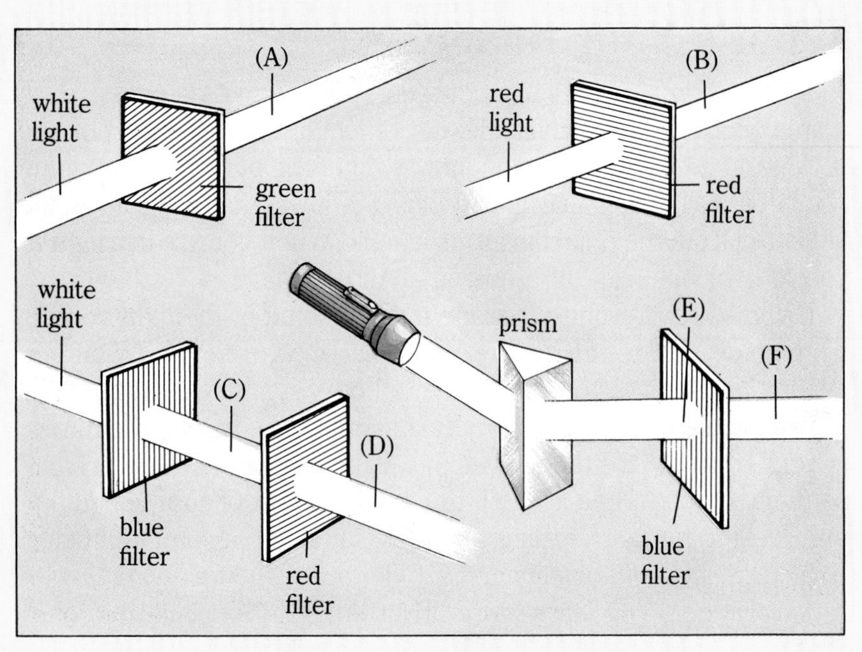

6 Images in Plane Mirrors

Considering the Plain Plane Mirror

Flat mirrors are also known as **plane mirrors**. They are the kind of mirrors used in bathrooms, clothing stores, and make-up cases. Plane mirrors show reflections that are the same size and shape as the original object.

What's Your Image?

In a lighted room, every object reflects light. You read earlier that, if an object (such as a chair) did not reflect light, you could not see it. The fact that people standing at different places in the same room can all see a particular object shows that light is being reflected from the object in all directions. When you are in a lighted room, light bounces off of *you* in all directions.

If you stand in front of a plane mirror, some of the light coming from you is reflected from the mirror back to your eyes. What you see in the mirror is your image. Your body, standing in front of the mirror, is the **object**; the body that seems to be behind the mirror is your **image**. Are the images in a plane mirror exactly the same as their objects? Hold a watch or a book in front of a mirror to find out. Telling time or reading the book suddenly becomes difficult! In fact, images and objects are not identical. The following Exploration will help you learn more about images in a plane mirror.

EXPLORATION 6

Plane-Mirror Insights

You Will Need

- graph paper
- a plane mirror
- a pencil
- a flat piece of colored glass
- modeling clay
- metric ruler
- index card

What to Do

1. Mark over one of the horizontal lines on a piece of graph paper. Make sure the line is very dark. Stand the plane mirror straight up along the line you made.
2. Put your pencil in front of the mirror. Where is the image of the pencil in the mirror?
3. Now place the pencil six squares in front of the mirror. Where is the image now? How many squares behind the mirror? Place the pencil in another position, and check the location of the image again.
4. Replace the plane mirror with a piece of colored glass. The glass reflects light just as a plane mirror does. However, you can also see through the glass. Both of these properties will help you make your plane-mirror insights.
5. Draw a straight line across the center of a sheet of graph paper. Place the glass along the line. Use modeling clay to hold the glass upright. Draw a triangle in front of the glass, near the middle of one half of the page.

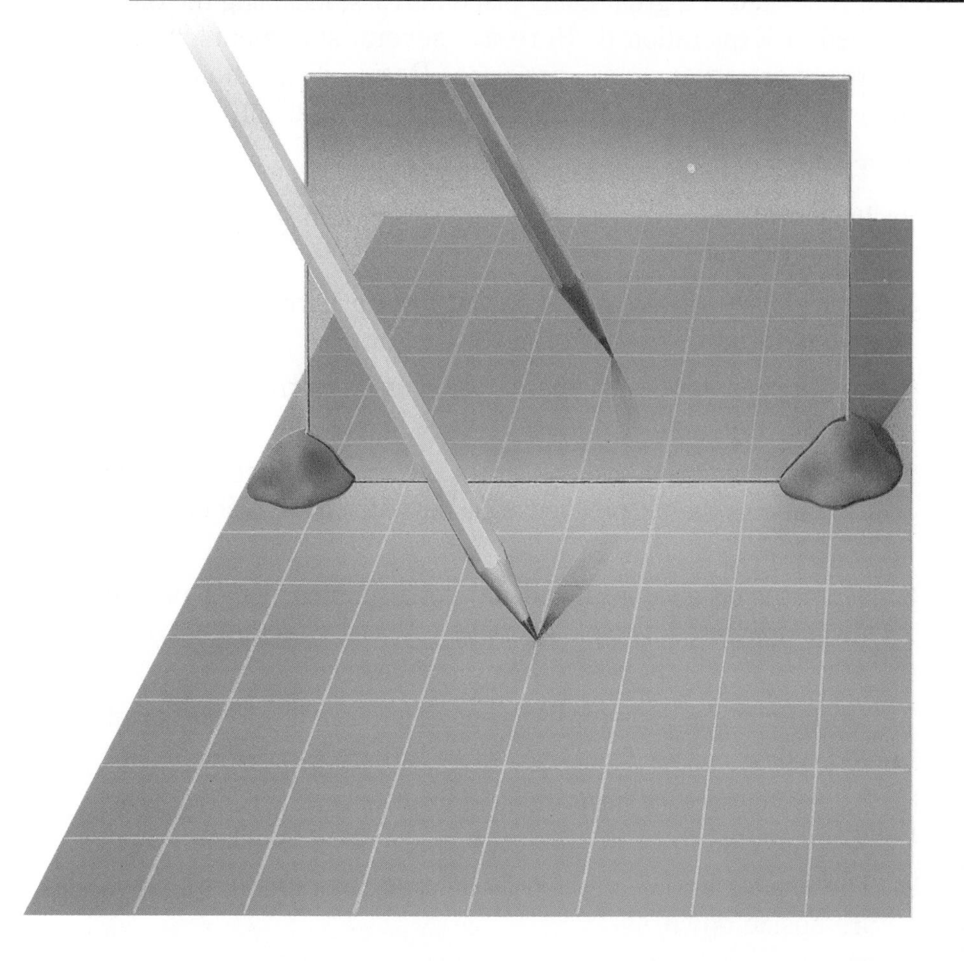

Look at the image of the triangle. While looking through the glass, trace the lines of the image on the paper behind the glass. Now remove the glass.

6. Draw a line from one point on the triangle to the corresponding point on its image. Measure the distance along this line

(a) from the glass to the image (called the *image distance*) and

(b) from the glass to the object (*object distance*).

How do the two distances compare? What is the angle between the line the glass was sitting on and the line joining the point on the image? What conclusion can you make about where you will see the image of an object that is placed in front of a plane mirror?

7. Measure the sides of the object triangle and those of the image triangle. How do the sizes of the object and the image compare?

8. Draw a center line on a new piece of graph paper. Then set the glass vertically on the line. Imagine that your pencil point is a car on a road driving toward the glass. Put your pencil on the paper and slowly move it toward the glass while drawing a line on the paper. Observe the image of the pencil and line. Now, gradually curve your pencil line—your car—to the left. What direction does the image turn? Now trace over the image line that you see on the paper behind the glass. As you follow the line—the car's image—toward the glass and around the curve, in which direction is the image car turning?

9. Make a card with the words KID, OXO, POP, and WOW printed on it. Predict what will happen to the words when you lay the card down in front of the glass. Be sure the words are facing you right-side up. What are the images of the words like?

10. Next, predict what will happen when you hold the card upright in front of the glass. How are the images of the words different? Turn the words upside down. What does the image in the glass look like now? Where have you seen that image before?

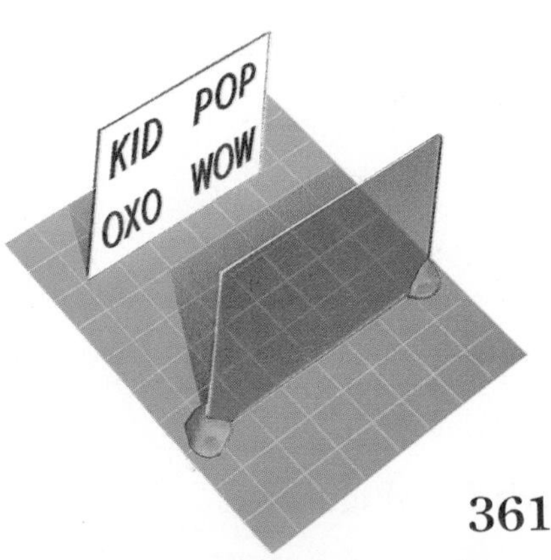

EXPLORATION 6–CONT.

11. Write your name on a card in such a way that it can be read correctly in a mirror or a piece of glass. This is how Leonardo da Vinci wrote his notes—so that other people couldn't read them.

Checking the Facts

You have looked at a great many plane mirrors, including the ones you used in Exploration 6. Here are several statements. Some statements are true; others are false. Decide which statements you agree with, disagree with, or feel uncertain about.

1. The image in a plane mirror is the same size as the object.
2. Images of objects in plane mirrors appear to be on the surface of the mirror.
3. If an object moves farther away from a mirror, the image seems to move backward in the mirror.
4. If an object moves farther away from a mirror, its image appears to be smaller.
5. The perpendicular distance from the image to the mirror is the same as the perpendicular distance from the object to the mirror.
6. If a point on an object and the corresponding point on the image are joined by a straight line, the line cuts the mirror at 90°.
7. You need a plane mirror half your size to see all of yourself at one time.
8. If a moving object turns right in front of a plane mirror, the image seems to turn left.
9. The images of objects held at right angles (90°) to a mirror are upside down.
10. The images of objects held in front of a plane mirror are reversed left to right.
11. The images of letters "H," "O," and "X" look the same as the objects in a plane mirror, no matter how the letters are held in front of it.
12. The images in a plane mirror aren't where they seem to be.

Real or Virtual?

Images are representations of objects. Your bathroom mirror *reflects* light to produce an image of your face. A movie screen *projects* light to produce an image from a frame of film onto a screen. You can see that both the movie projector and the bathroom mirror give you images. But one of these images is **virtual** while the other image is **real**. What is the difference between a real image and a virtual one?

First, consider your bathroom mirror. When you look into it, what you are seeing is the result of light bouncing off a flat plane. This is a virtual image. The image appears to be the same distance *behind* the mirror as you are *in front of* the mirror. But the light is not really passing through the mirror—it is bouncing off of it! Study the diagram below.

Now think about a movie projector. Instead of bouncing off a flat surface, the light begins at a light source and *passes through* the film in the projector. After the light leaves the projector, it is projected onto a screen. All real images can be projected onto a surface—this distinguishes real from virtual.

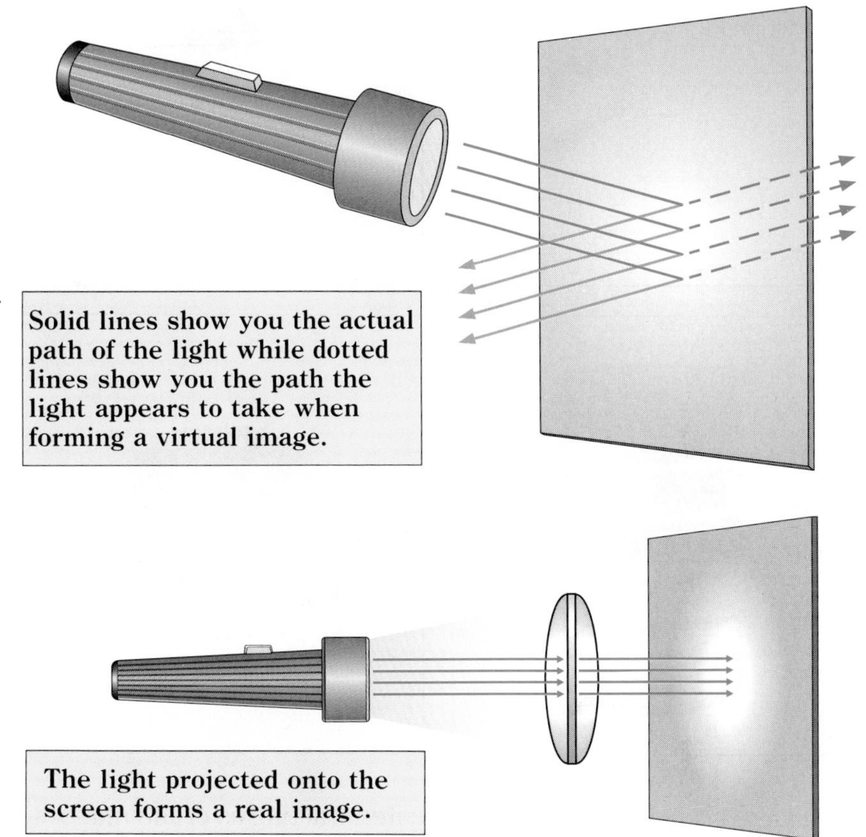

Solid lines show you the actual path of the light while dotted lines show you the path the light appears to take when forming a virtual image.

The light projected onto the screen forms a real image.

Now think about what happens when someone stands up in the middle of a movie and interrupts the light's path to the screen. The image from the projector can be seen *on the person's body*. Does the same thing happen when someone steps in front of you as you are looking into a mirror? Of course not! The difference between these two situations is an example of the difference between real and virtual images. Compare the illustrations below, and then write your own definitions for real and virtual images.

7 Convex Mirrors

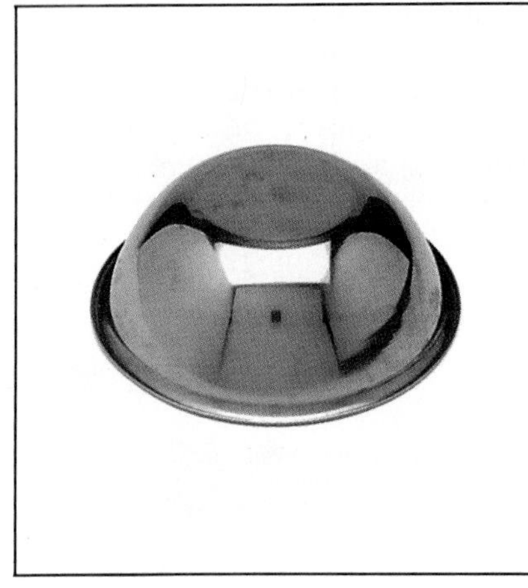

You have probably seen all of these items acting as crude mirrors. What do the images reflected in these mirrors look like? What is similar about each of these mirrors? They are all irregular **convex mirrors**. Convex mirrors have a curved surface with a center that bulges outward toward the viewer.

How do convex mirrors differ from plane mirrors? What are objects and images like in convex mirrors? First, recall what the images in plane mirrors are like.

- How do they compare in size with their objects?
- Are the images *erect* (right-side up) or *inverted* (upside down)?
- Are they behind or in front of the mirror?
- Are they reversed from left to right?
- Are they real or virtual?

How would you answer these same questions with respect to convex mirrors? The best way to find out is to do the following Exploration.

EXPLORATION 7

Experiments with Convex Mirrors

You Will Need

- a large spoon
- a light box
- aluminum foil
- a convex and a plane mirror
- glue
- a wooden splint
- matches
- a flashlight
- a pencil
- a candle or light bulb
- modeling clay
- a sheet of white paper

Part 1

A Spoon Mirror

Examine the images on the back of a spoon where the curve is greatest.

(a) How do the sizes of the images and objects compare?

(b) Is the image right-side up or upside down?

(c) Is the image reversed?

(d) How does the field of view in this convex mirror compare with that in a plane mirror?

Part 2

Using a Light Box with a Convex Mirror

Make a circular hole about 2 cm in diameter in a piece of aluminum foil. Glue the foil over the window in a light box. Use modeling clay to support a convex mirror in the box. Insert a smoking splint and shine a flashlight or

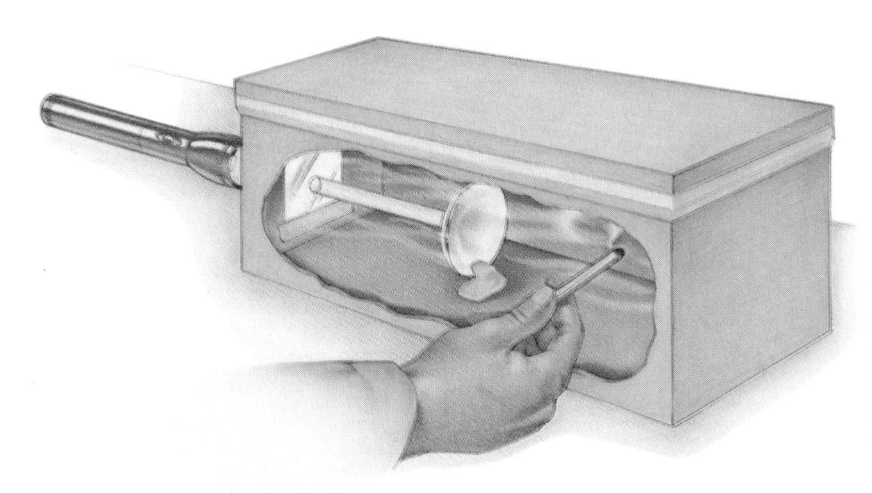

projector beam through the window. Draw a diagram to show what happens to the light that strikes the convex mirror.

Part 3 

Looking for Real Images with a Convex Mirror

You will recall that plane mirrors produce virtual images. Which type of images do convex mirrors produce—real or virtual? The following activity will help you answer this question.

In a darkened room, place a lit candle near one end of a table. Place a convex mirror supported with modeling clay at the other end of the table, facing the candle. Hold a piece of paper between the candle and the mirror, as shown in the illustration. Move the paper slowly from side to side and back and forth to see whether you can focus an image of the candle on it. Are you able to get a real image—one that can be located on a screen?

Part 4

Locating the Image

Hold a pencil between your eyes and a convex mirror. Now move the pencil toward the mirror. How does the image seem to move? Move the pencil away from the mirror. How does the image seem to move?

On the basis of the observations you made in the last two parts of this Exploration, state whether images appear behind or in front of a convex mirror.

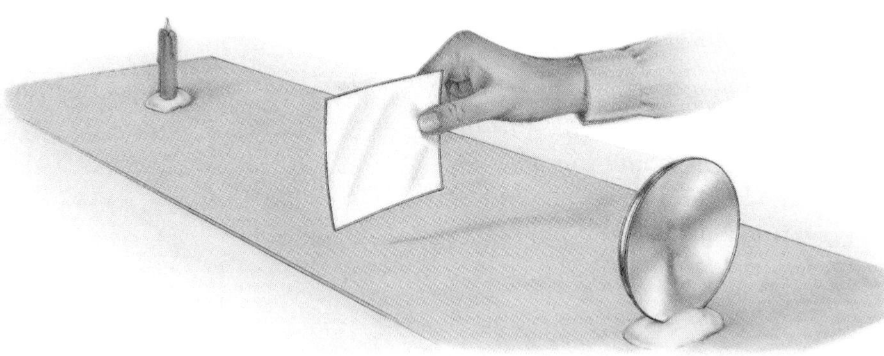

Uses of Convex Reflecting Surfaces

Consider the following applications of convex mirrors and think of the characteristics that make them so useful.

- Large convex mirrors can be found in many stores. Why?

- Not all of the mirrors in cars are convex, but the wide-angle side-view mirrors always are. Why is this so? Some side-view mirrors have a warning attached: "Objects may be closer than they appear." Why?

How many more uses for convex mirrors can you find?

Converging Lenses and Real Images

What do all of the following items have in common?

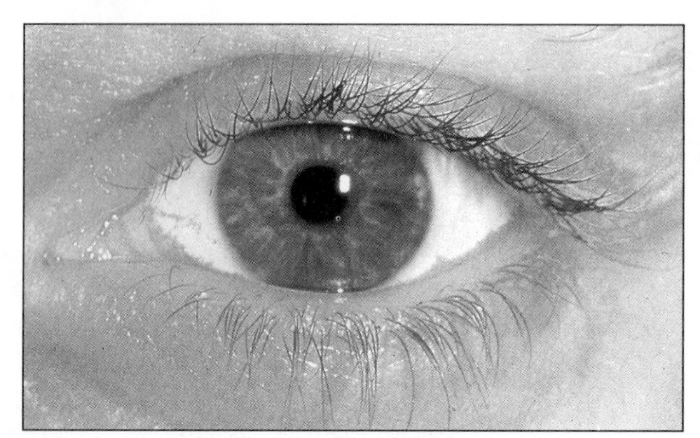

You are correct if you said "lenses." A **lens** is a transparent solid with curved sides. Lenses that are thicker in the center than at the edges are called **converging lenses**. What do you think a converging lens does to light? Check the dictionary for the definition of *converging* to get you started.

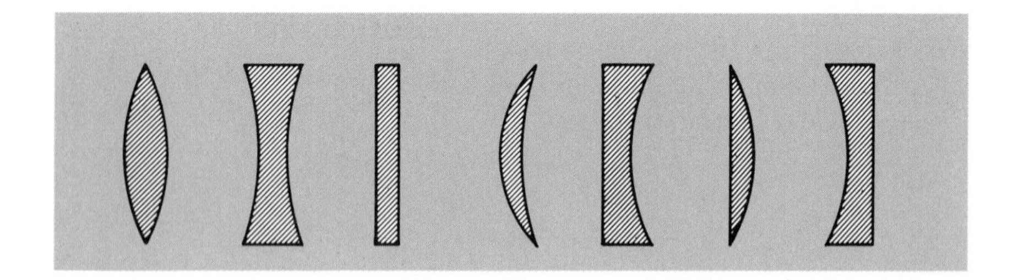

Examine the cross sections of lenses shown here. Which would you classify as converging lenses? One of these converging lenses is used in eyeglasses for farsighted vision. Which do you think it is and why? The lens shown at the far left is called a double-convex lens. What do you think it could be used for?

Converging Lens Experiences

You Will Need

- a book or piece of paper with writing on it
- a double-convex lens
- a light box
- modeling clay
- a candle
- a screen
- a wooden splint
- matches
- a projector or flashlight
- white paper

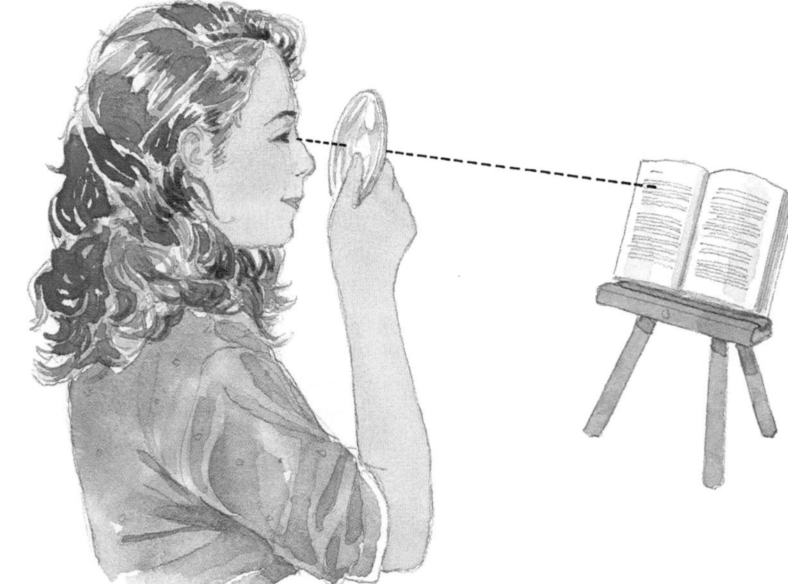

Part 1 

Strange Sights

1. Place a book or a piece of paper with writing on it 1 m from your face. Hold a double-convex lens about 50 cm from your eyes, and look through it at the book. Slowly move both the lens and your eyes closer to the book. Continue until the lens touches the book.
2. Describe the appearance of the print when you first looked at it through the lens. Was the print erect or inverted? Decide whether the image of the print was larger than, smaller than, or the same as its actual size.
3. Describe the appearance of the print when the lens is only a few centimeters away from it.
4. Repeat the procedure, only this time observe the image *on* the lens instead of looking *through* the lens. Notice how the image of the print changes as you bring the lens closer and closer to it.
5. Describe the changes in the appearance of the print, as you did in steps 2 and 3.
6. On the basis of your observations, choose the word or phrase that best fits in each of the following blanks. You may have to double check your observations.
   (a) When the converging lens is far away from the object, the image of the object formed by the lens is (erect, inverted).
   (b) When the converging lens is moved slowly toward the object, at first the image of the object gets (smaller, larger).
   (c) As the converging lens moves closer to the object, at one point the image becomes (clear, distorted).
   (d) When the converging lens gets even closer to the object, the image is changed. It is now (inverted, erect). It is also (smaller, larger) than the object.
   (e) As the converging lens moves still closer to the object, the image (increases, decreases) in size.

Part 2 

Converging Lenses and the Light Box

1. Set up your light box, and insert a smoking wooden splint. Mount a double-convex lens on some modeling clay so that the lens lies in the beam of light from a projector or flashlight.

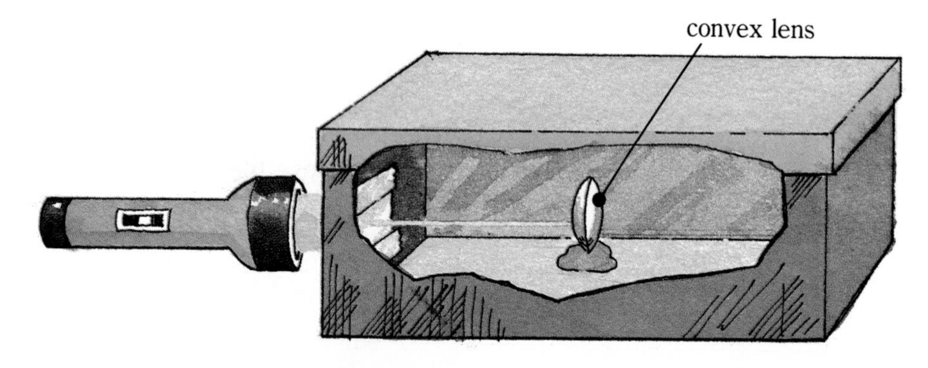

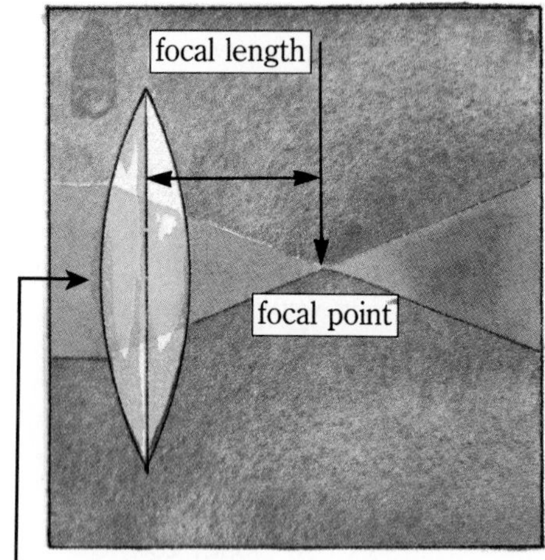

2. Draw a diagram of the beam of light before and after it reaches the lens. Is there any difference in the behavior of a beam of light passing through a double-convex lens and a beam bouncing off a convex mirror? If so, describe the difference.

3. When the beam of light passes through a converging lens, it converges to a point at a position known as the *focal point* of the lens. Label the focal point in your diagram.

   Estimate how far, in centimeters, the focal point is from the lens. This distance is called the *focal length* of the lens.

Part 3  

Looking for Real Images with Converging Lenses

1. Using a double-convex lens, find the image of a lighted candle (or light bulb) on a piece of white paper or cardboard. The light source should be on the side of the lens opposite the paper. In what position should the lens, candle (or bulb), and screen be located in order to obtain
   (a) an image of the object larger than the object itself?
   (b) an image the same size as the object?
   (c) a smaller image?

   It is sometimes easier to leave the lens in one position. Start with the object fairly far from the lens. Try to locate its image on the screen. Then systematically change the position of the object.
2. For each situation in step 1, draw a diagram in your Journal.
3. You have been looking at real images made by a double-convex lens. Before you answer the next question, think back to the first activity in this Exploration. Do converging lenses ever form virtual, erect images? Where do objects have to be placed for this to happen? Position the candle (or light bulb) and lens in such a way that you get this type of image. Where does the image seem to be? How does it compare in size to the object?

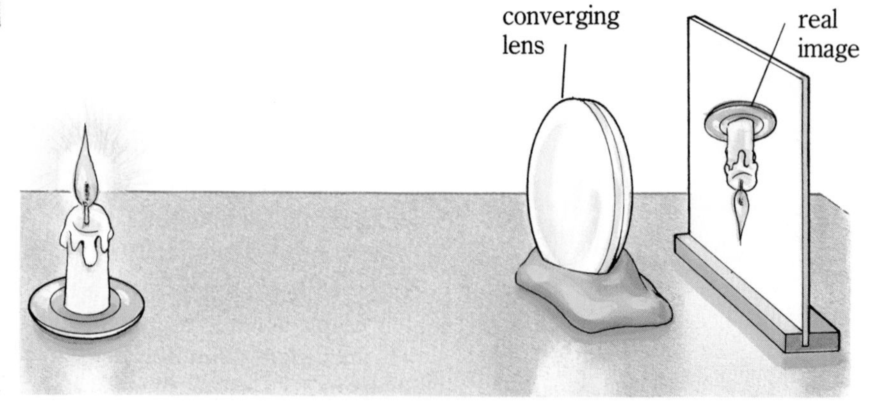

Uses of Converging Lenses

Two important applications of converging lenses are found in a slide or film projector and a camera. Study the illustrations below to see how each of these devices works. Drawings of two setups for Part 3 of the last activity in Exploration 8 are provided for your assistance.

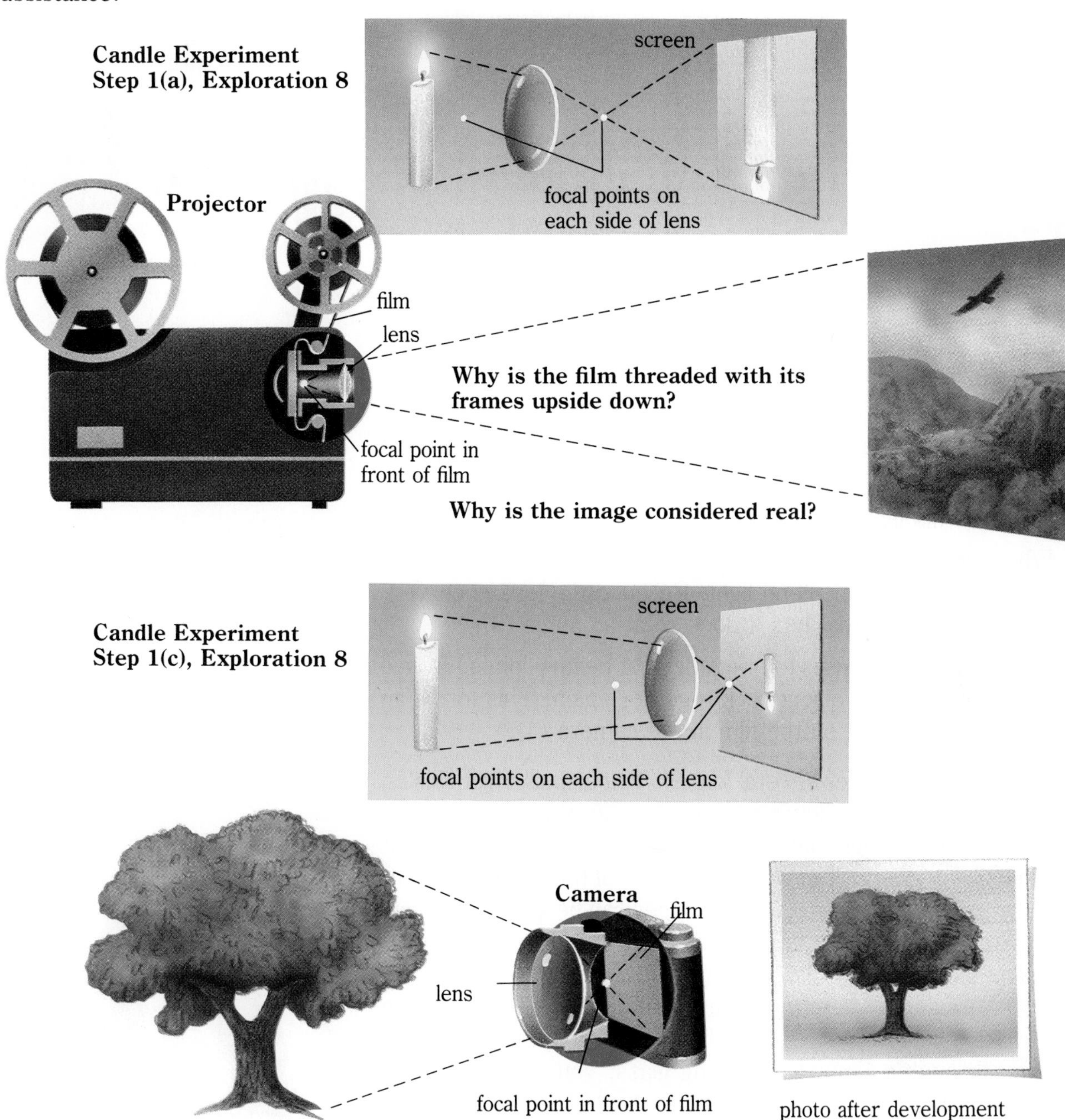

In what ways are a projector and a camera similar? How do they differ? Imagine you have been asked to tell a fifth-grade class how a camera and a projector produce the images they do. Make a lesson plan, as if you were their teacher. Note what you would say or do to help the class understand how the two devices work.

More Information for the Photographer

The diagram shows how an image or picture of an object is formed on the film in a camera.

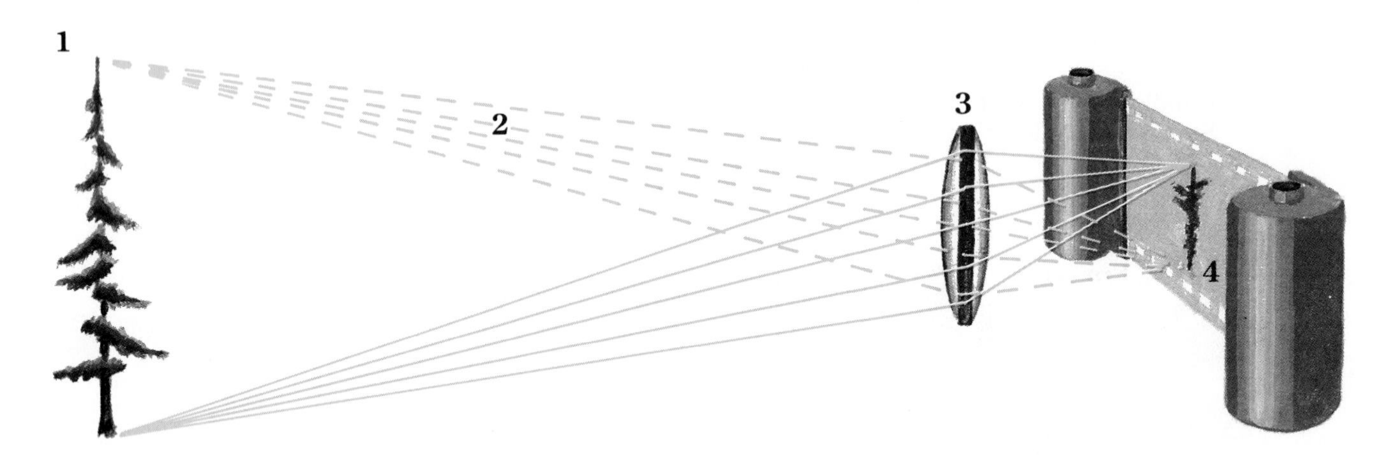

- Light goes out in all directions from each point on an object (1).
- Some of this light (2) reaches the lens of the camera.
- When light hits the lens at an angle, the light bends. Recall, for instance, that prisms bend light. Because the lens is curved, most of the light that hits it does so at an angle (3).
- The effect of the lens is to bring the light beams coming from a given point on the object back together again. This forms an image of that point on the film in the camera (4).

A camera actually uses several lenses. You can discover this on your own by doing the following demonstration. With your back to a bright window or a light bulb, hold a camera in front of you so that you can look into the lens and find reflections of the window or light bulb. You will find that several small images can be seen in the lens of most cameras. By counting the number of images, you can tell how many lenses are in the camera.

A simple lens distorts light, because some colors in the light are *refracted* (that is, bent) more than others. If you look at the print on this page through a simple magnifying glass, you may be able to see the distortion. Try to magnify the letter "A" as much as possible. Do you notice that only the part of the letter near the center of the lens is in focus, while the rest of it is distorted? Can you see a slight fringe of colors at the edges of the letter? Combinations of lens shapes and different types of glass can correct these distortions. The need to make such combination lenses is one reason why a high-quality lens for a camera or projector is so expensive.

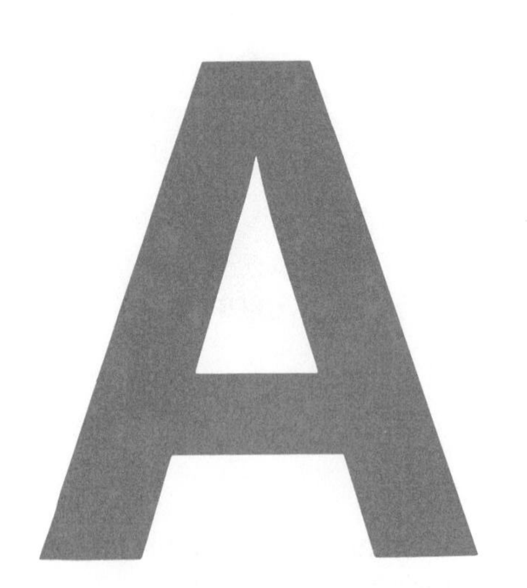

Real Images and Concave Mirrors

Here are two candles and their images. Which image was produced by a plane mirror and which by a convex mirror?

Now look at the illustration below. What kind of mirror would produce images of a candle that look like these?

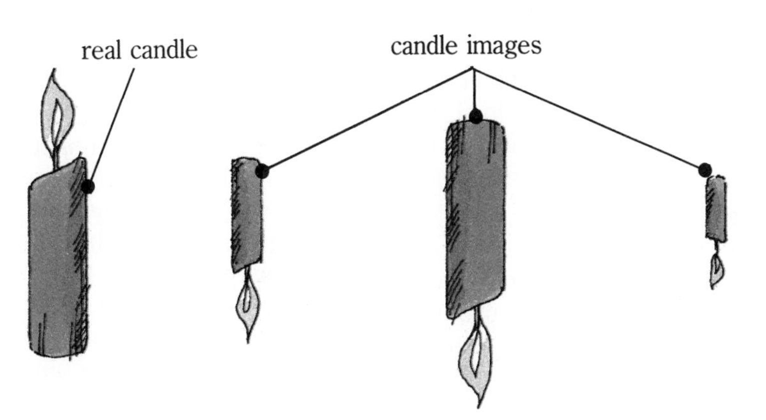

You know that convex mirrors bulge outward. Now you are going to study mirrors that curve inward—like a cave. Such mirrors are called **concave mirrors**. A spoon is a simple example of both a concave mirror and a convex mirror. Which side is which?

EXPLORATION 9

Experiments with Concave Mirrors

You Will Need

- a large, shiny spoon
- a concave mirror
- a metric ruler
- a light box
- a wooden splint
- matches
- modeling clay
- a flashlight or projector
- a candle
- paper

Part 1

A Concave-Spoon Mirror

Look at your image in the concave side of the spoon, holding it at arm's length from your body. (The concave side is the side where the food goes.) What do you see? Now hold the spoon very close to you. How has your image changed?

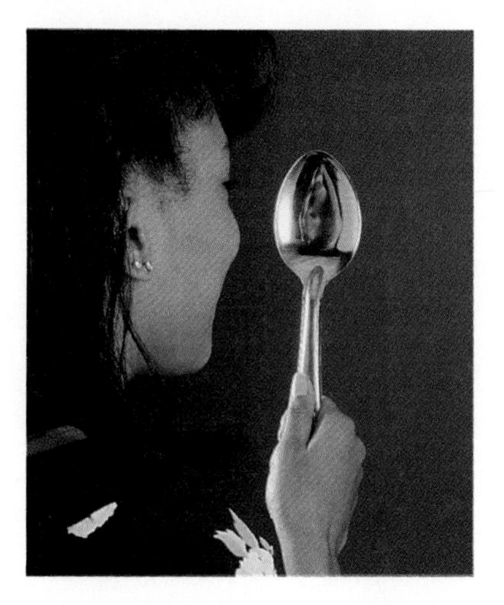

Now hold the spoon far away from your body. Gradually move it closer and closer. Record all the changes in the images you see.

To understand why concave mirrors form the images they do, you are going to perform more experiments.

Part 2

Using a Light Box with a Concave Mirror

Set up the light box as before, but this time include a concave mirror supported by a piece of modeling clay. Use a smoldering wooden splint to put smoke into the light box. Aim the flashlight or projector through the hole and onto the concave mirror.

(a) Draw a diagram in your Journal to show what happens to light when it strikes the concave mirror.

(b) Label the focal point on your diagram. As you know, the distance from the focal point to the mirror is the focal length. Label the focal length on your diagram.

Part 3

Looking for Images with a Concave Mirror

Hold a sheet of paper between a concave mirror and a candle. The paper should be slightly off to one side, so light from the flame can reach the mirror. Move the mirror slowly back and forth toward the paper, as shown in the illustration. Look for an image on the paper.

(a) Was a clear image of the candle formed? If so, draw a diagram to show the locations of the candle, the paper, and the concave mirror when the image was clearest.

(b) Are the images formed by concave mirrors real or virtual? How can you tell?

(c) Do the images appear to be behind the mirror or in front of it?

(d) Gradually move the candle toward the mirror. Keep the paper between the candle and the mirror, but adjust the paper to get a sharp image. What change takes place in the location of the image? Is the size of the image always the same?

(e) Continue the experiment until you can construct three diagrams. The first should show where the candle, paper, and mirror are placed if a very small image is desired. The second should show the setup that produces an image the same size as the object. The third diagram should show how the candle, paper, and mirror need to be arranged to produce an image larger than the object.

(f) Now place the candle very close to the mirror. Look into the mirror. What do you see? What kind of image is it—virtual or real? How can you tell?

Lenses and Mirrors: Time to Compare

1. Using a candle as an object, you were able to get images with a concave mirror that looked like those at the right. Which images were real? Which were virtual? What other optical device gave you the same kinds of images?

2. When an image is real, what other characteristic do you expect the image to have? When an image is virtual, what other characteristic do you expect it to have?
3. If a real image is smaller than the object, which is located closer to the device forming the image—the object or the image?
4. All of these mirrors form virtual images:

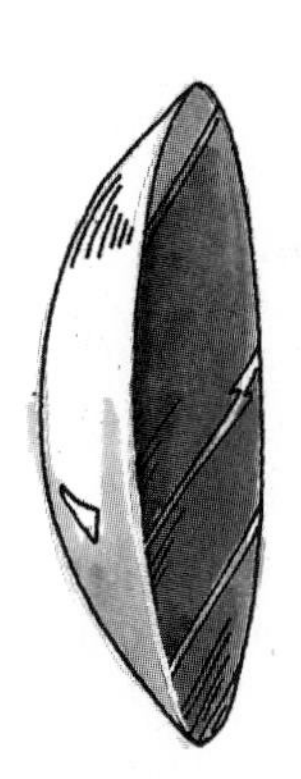

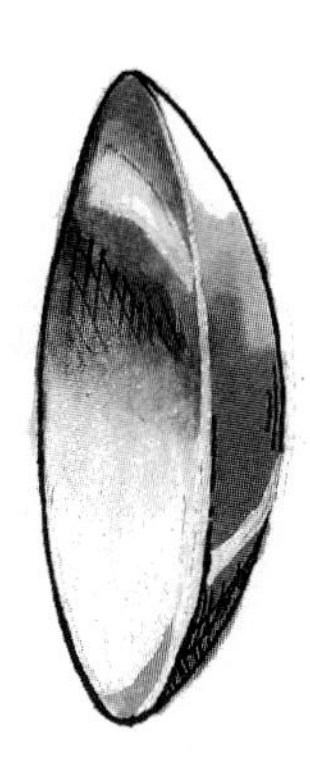

Which mirror(s) give(s) only virtual images? Which mirror(s) can give(s) both virtual and real images? Do virtual images appear in front of the mirror, on the mirror, or behind the mirror? Do real images appear in front of the mirror, on the mirror, or behind the mirror?

5. Which mirror or lens would you use to do the following:
   (a) magnify your finger?
   (b) examine a freckle on your nose?
   (c) see yourself as you are—life-size and right-side up?
   (d) see yourself smaller than you are, but right-side up?
   (e) see yourself smaller than you are and upside down?

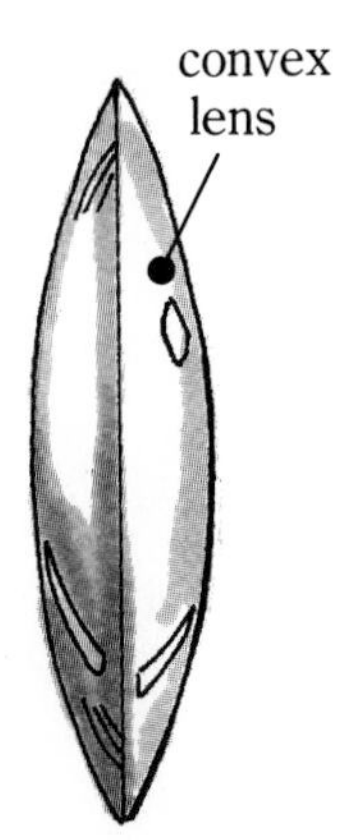

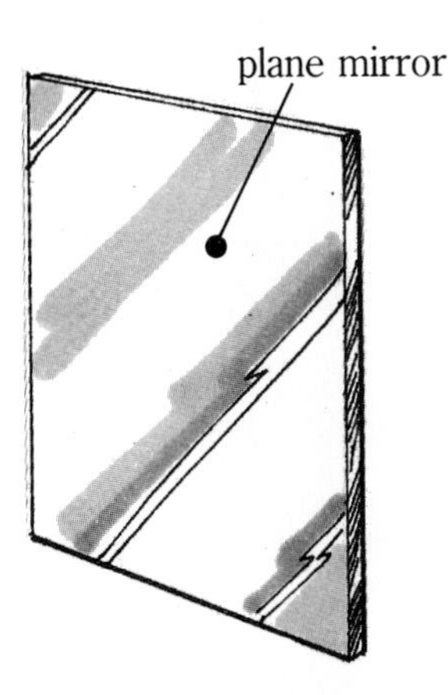

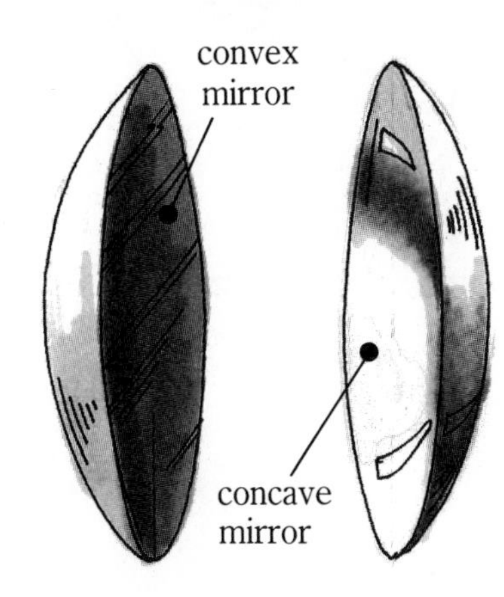

Uses of Concave Mirrors

As you look around, most of the curved reflecting surfaces you notice are convex. But there are also a surprisingly large number of uses for concave mirrors. For example, what is the purpose of the concave mirror in the device at the right?

Recall what the focal point of a converging lens is. Similarly, when a beam of light is aimed at a concave mirror, almost all of the light goes to the focal point.

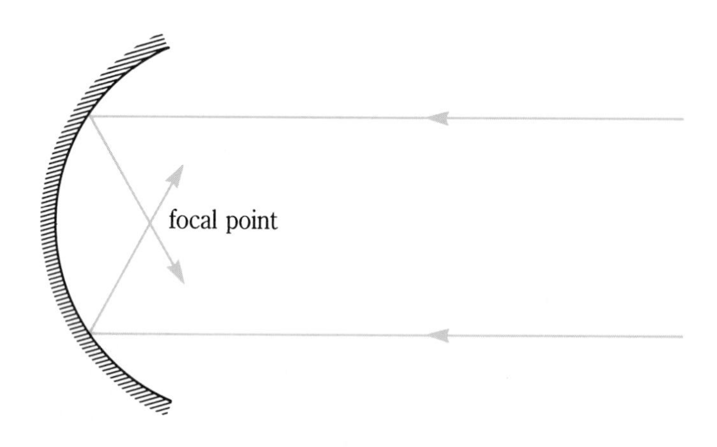

The opposite is also true; if you place the filament of a light bulb at the focal point of a concave mirror, the light reflects away from the mirror in a beam. Suggest other examples of concave mirrors that work on the same principle as a flashlight reflector. One type of concave mirror is shown below. How does it work?

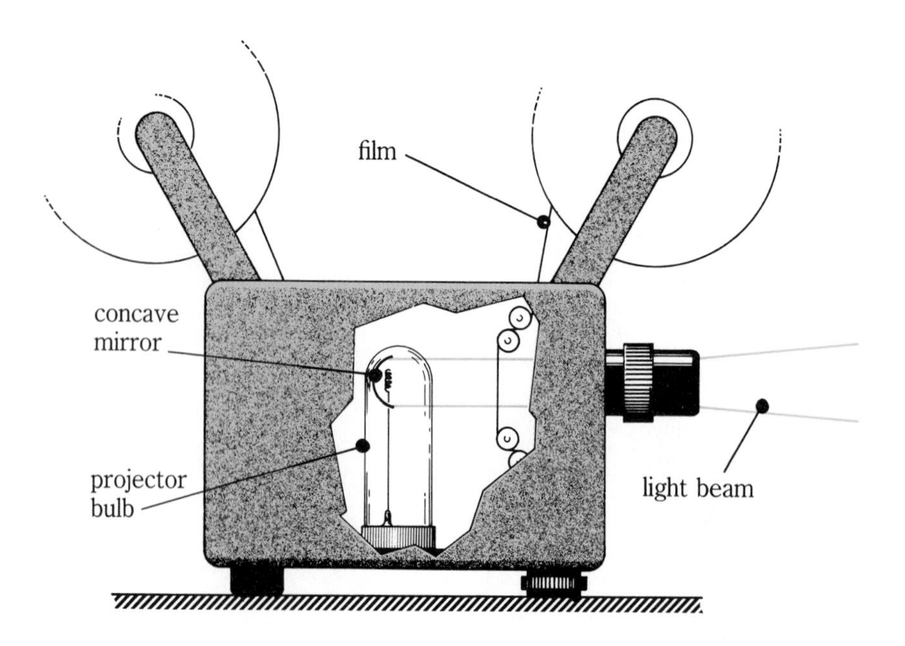

Diagrams of a solar cooker and searchlights are shown below. How do they differ from each other?

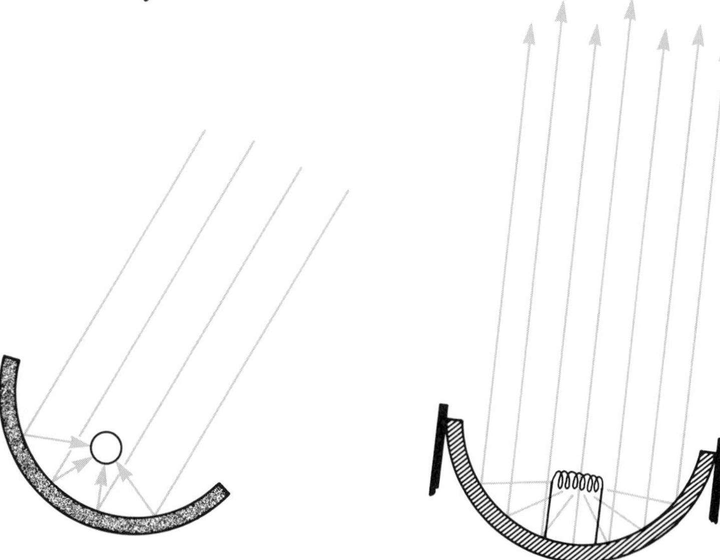

This diagram shows a solar cooker. Where is the focal point?

A searchlight

Giant reflecting telescopes have concave mirrors that measure 5 or 6 m in diameter. Why are the mirrors so large? The bigger a telescope is, the more light it can collect from very faint stars. Study the photograph and the diagram shown to determine how a telescope works.

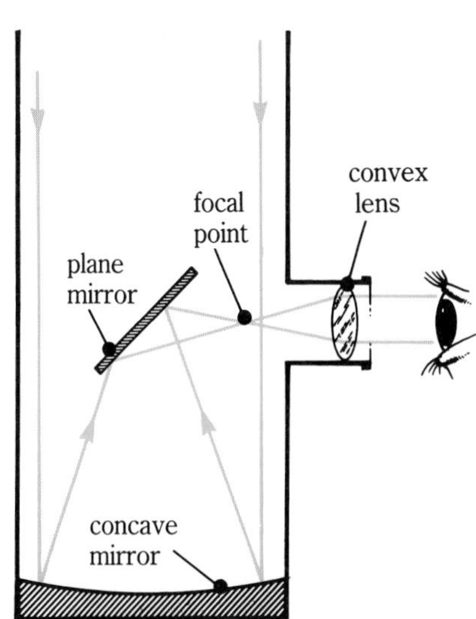

An observer views the image of the star at the focal point of the concave mirror by looking through a convex lens.

Refraction

EXPLORATION 10

The Bending of Light Beams

Light travels through empty space at a speed of about 300,000 km/s. Moving at this speed, someone could travel around the world about 7.5 times in one second! In air, light travels slightly slower than it does through empty space. In water, light travels even slower—at about 225,000 km/s. In glass, light's speed is about 200,000 km/s. That's still pretty fast!

When light passes through one material to another, its speed changes and some unusual things happen. Complete the following experiments to find out more about this phenomenon.

Refraction Experiences

You Will Need

- 2 pennies
- water
- 2 clear plastic cups
- a pencil
- a felt-tip marker
- rubbing alcohol and vegetable oil
- a 500-mL or 1-L beaker
- skim-milk powder
- a flashlight
- masking tape
- chalk dust

Part 1

Which Penny Is Closer?

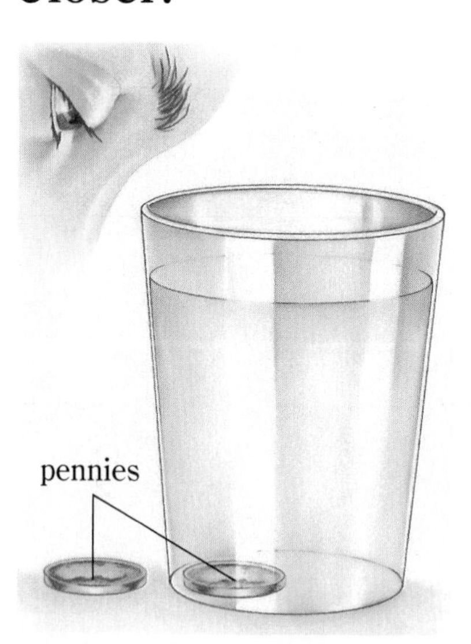

Fill a clear plastic cup nearly full of water. Drop a penny down the side of the container. Put another penny on the table outside the container, but as close as possible to the penny that is inside. Look directly down at the pennies.

(a) Describe any differences in the appearance of the two pennies.

(b) Which penny seems closer to you than the other?

(c) Move your head slowly side to side while observing the two pennies from above. Does either one appear to move across the background of the table? If so, which one?

Remember, reflected light from the outside penny comes from the penny, through the air, and to your eye. Reflected light from the penny in the water passes through the water first and then through the air to reach your eye. What effect occurs when light must pass through two materials to reach your eye?

Part 2

Measuring Apparent Depths

Place a penny in a clear plastic cup filled with water. With a partner, take turns looking down at the penny. Indicate where the penny appears to be by pointing to the spot on the side of the cup that is closest to the penny's position. Make a permanent mark there.

Then substitute alcohol and vegetable oil for the water in the cup. Compare the depths at which you see the pennies.

Part 3

Bent Objects

Fill a large container, such as a 500-mL or 1-L beaker, two-thirds full of water. Stir in a few grains of skim-milk powder—just enough to give the water a faint milky appearance. Put a long pencil into the container. Observe the pencil from all sides and from above.

(a) Describe the appearance of the pencil from different angles.

(b) When seen from above, is the pencil bent upward or downward where it is below the surface of the water?

Part 4

Bent-Light Paths

Remove the pencil. Take a flashlight and cover the top of the container with a ring of masking tape so that only a hole about 0.5 cm in diameter remains at the center. Shine the light through the hole and at an angle into the top of the water. Observe the light from the side. Shake a small amount of chalk dust into the air above the water so that the beam of light is visible above the water, as well as in the water.

(a) How does the beam appear to bend when it goes into the water? Look at the beam through the side of the container. Draw a diagram showing how the beam bends.

(b) Is the beam of light going into or coming out of the water? When you observed the underwater part of the pencil in Part 3, did you see it as a result of light going *from* your eyes to the pencil or light coming *to* your eyes from the pencil?

(c) Why did the part of the pencil that was underwater appear to bend upward, while the light beam seems to bend downward where it enters the water?

Part 5

Total Internal Reflection

Move the container near the edge of the table and direct the flashlight beam upward from below the water level through the side of the container. Again shake a bit of chalk dust over the water to make the beam visible as it comes out of the water.

(a) Do you observe bending? Make a sketch showing how the beam bends.

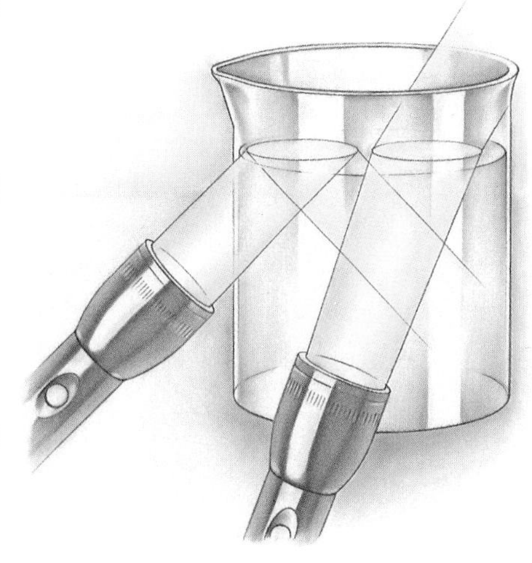

Slowly raise the flashlight so that the angle between its beam and the surface becomes smaller. What happens to the beam in the air? At one particular angle the beam stops coming out of the surface. Then for smaller angles the beam is reflected back into the water as though there were a mirror at the surface. This phenomenon is called *total internal reflection.* Total internal reflection occurs only when light reaches a boundary between one material and a second material that light can travel through at a faster speed. Thus, light traveling in water can be totally internally reflected when it comes to a boundary between water and air—provided the light hits the boundary at a sufficiently small angle.

(b) Estimate the angle between the light beam and the water surface when total internal reflection is first observed.

Interpreting Your Findings

Check your understanding of refraction by answering these questions in your Journal.

1. Does viewing a penny under water make it appear deeper, shallower, or at the same depth as it really is?
2. In this case, is light going by the path: eye → air → water → penny; or by the path: penny → water → air → eye?
3. In the illustration below, which path will the light beam most likely follow—1, 2, or 3?

4. In the situation below, light is traveling from the water to the air to the eye. Which beam of light—1 or 2—will most likely reach your eye?

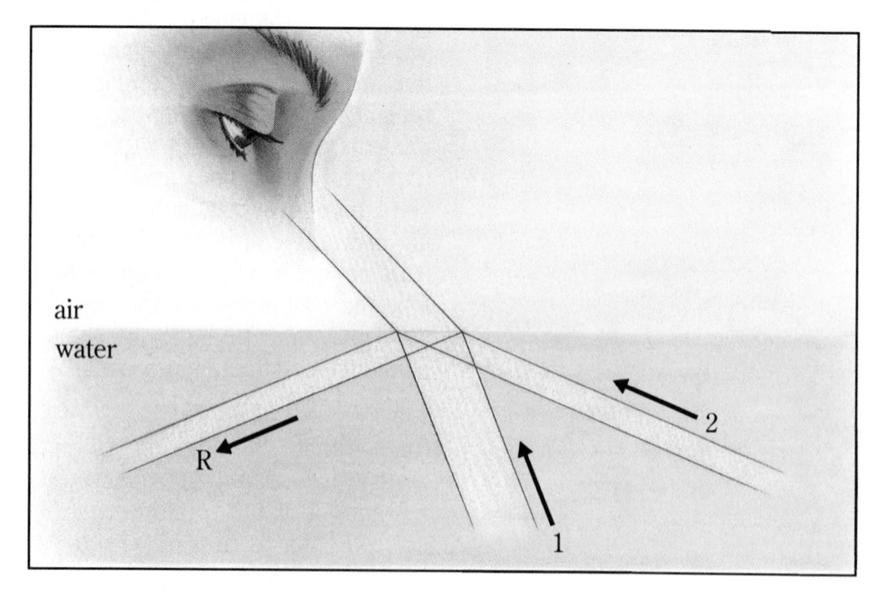

5. Which beam of light—1 or 2—will probably not get out of the water, but will instead be reflected as R?

Challenge Your Thinking

1. Explain what's wrong in each of these three pictures. Then redraw each one so that it's correct. The lens has a focal length of 10 cm.

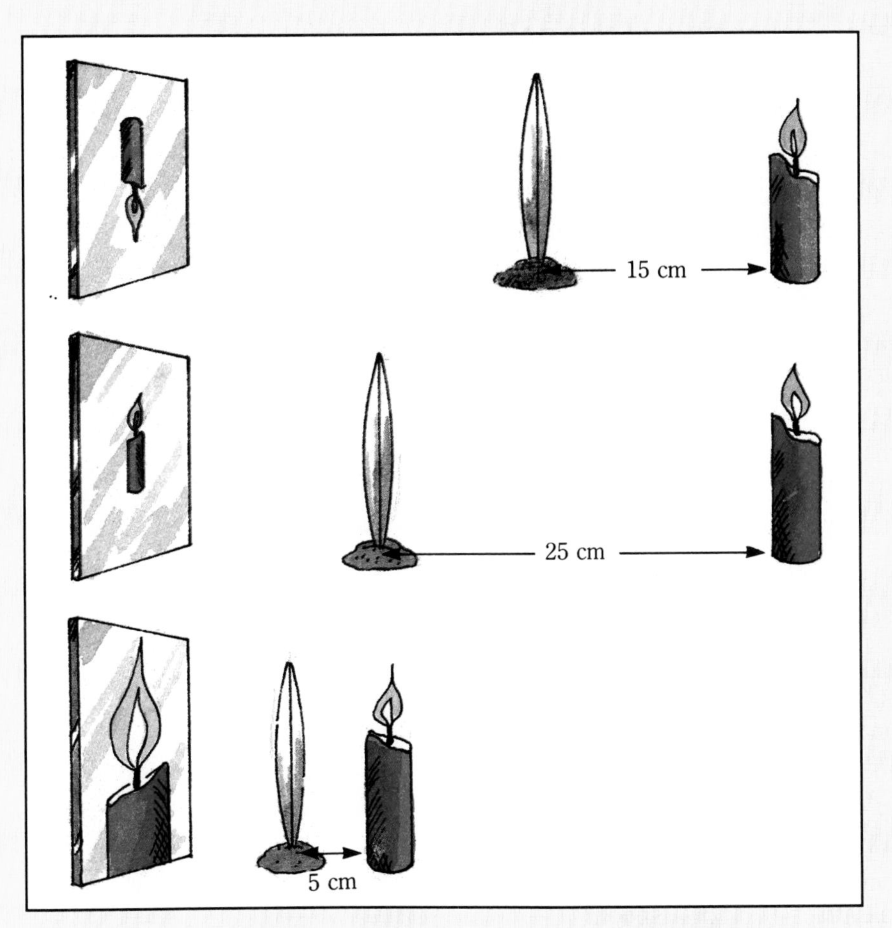

2. Dr. Hughes uses a small concave mirror when she examines her patients' teeth. Lorenz uses a convex mirror on his motorcycle to see the traffic behind him. Why do these two people use different types of curved mirrors?
3. Your eye always views an object as if its light path never changed direction. This explains why you saw a pencil bent upward. Write a brief explanation of the illustration to the right based on what you know about light paths.

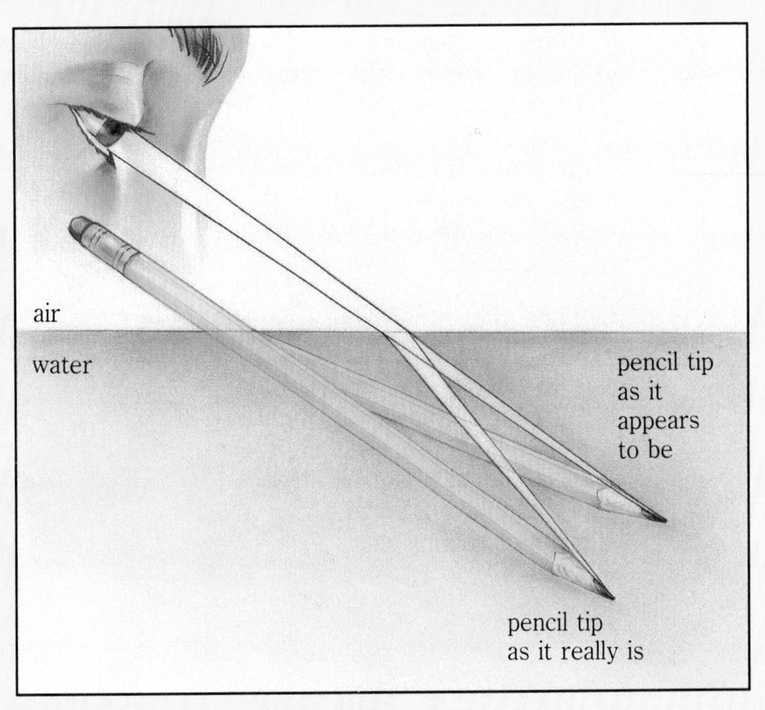

Unit 6 Making Connections

The Big Ideas

In your Journal, write a summary of this unit, using the following questions as a guide.

- How can you tell whether light is matter or energy?
- How can colored light be produced from white light?
- What happens to light energy when light hits a surface?
- What is the difference between scattering, transmission, and absorption of light?
- How can you distinguish between transparent and translucent objects? between diffuse and specular reflection?
- What are the characteristics of the image of an object when it is placed in front of a plane mirror? a convex mirror? a concave mirror?
- How do you distinguish between a virtual and a real image?
- What are some practical applications of a converging lens? a concave mirror? a convex mirror?
- What is refraction? How does it distort our observations?

Checking Your Understanding

1. You are in charge of creating a reference booklet with the title: "Questions About Light That People Often Ask." You must include at least eight questions and brief answers. Use the three questions listed below to get started, then add five questions and answers of your own.
   (a) What does a flat mirror do to handwriting?
   (b) How can you make rainbows?
   (c) Why does your image appear distorted when you look at the back of a spoon?
2. Con-vex, con-cave, con-verging . . . CON-FUSING! Sometimes it's difficult to keep track of what each kind of lens and mirror does. Construct a chart listing each lens and mirror discussed in the unit. For each, record whether it produces a real or virtual image (or both), if it makes images that are smaller or larger than their objects, and two of its practical uses.
3. Trace the path of light from the sun, through the Earth's atmosphere, and onto various surfaces on the Earth. Light should be shown reaching the ocean; a dark, asphalt street; and the roof of a green car. What happens to sunlight at each location?

4. What Black Is

What light is not . . . .

What light that is not reflected or transmitted is . . . .

What the size of an image is in comparison to the size of an object in a convex mirror . . . .

What light is . . . .

What a mirror that produces both an upright and inverted image is . . .

What a material that transmits only scattered light is . .

What the color transmitted by a yellow filter is . . . .

What a mirror bulged out in the center is . . . .

What scattered reflection is . . . .

What an image that can be located on a screen is . . . .

What the color of light with the greatest amount of energy is . . . .

What an object makes after light from it is reflected or transmitted . . . .

What converts into light energy when a lamp is turned on . . . .

What an image behind a flat mirror is . . . .

5. Which of the illustrations below shows total internal reflection?

(a) (b) (c)

6. You are in charge of teaching a class of fourth graders about light and color. One student insists that light ALWAYS has to be white and that filters only add their color to the light that passes through. You are equipped with two prisms, several filters of varying colors, a flashlight with a strong battery, and a white screen. What will you say and do to convince the student that she is mistaken?

7. **The Mysterious Black Box**

What's inside the box? The only clues you have are the light beams that enter and leave each mystery box.

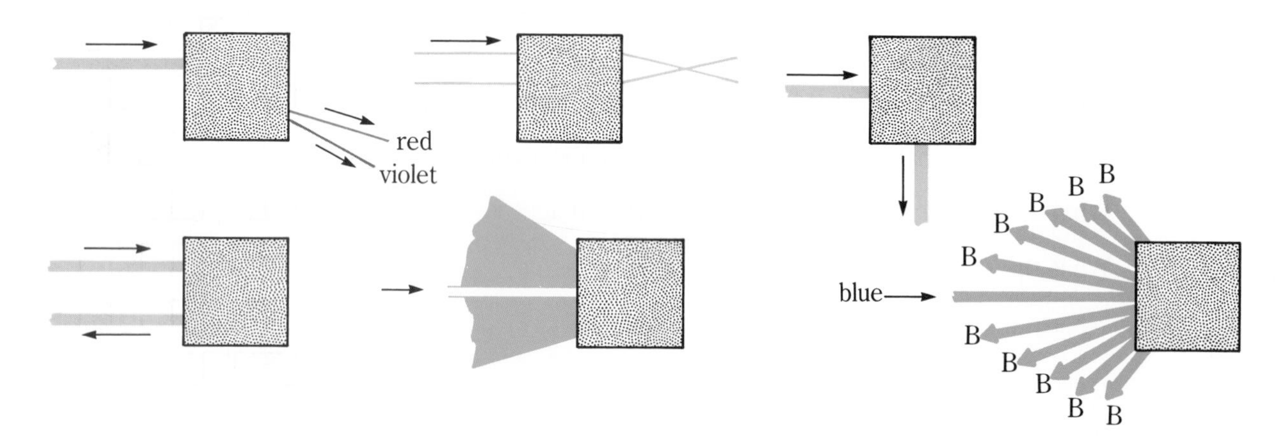

8. Use the principles of light refraction to explain why it is hard to judge the depth of a still, clear pond.

9. Red giants and white dwarfs are different types of stars found in the universe. A red giant is a large star that emits red light. A white dwarf is a smaller, denser star that emits white or bluish-white light. Which star do you think is the hottest? Why?

10. Write a letter to a blind person that explains what you now know about light and color. How will you communicate the concepts of light and color so that the person will understand what they are?

11. **Asking Lightly**
    (a) If you hold up a pencil in class, students on all sides of the room can see it. Why? Is the pencil a source of light?
    (b) Why do sports players often smear a dull, black substance below their eyes on bright, sunny days?
    (c) How could you use two mirrors to see the top of your head?
    (d) How could a small candle be used to produce a light that would warn ships of the presence of dangerous rock?

12. Draw a concept map showing the relationship between these words: *converging lens, reflected, real image, refracted, virtual image, light,* and *plane mirror.*

Reading *Plus*

In this unit you have learned much about the behavior and properties of light. You will be able to continue making discoveries about light by reading pages S87–S106 in the *SourceBook*. There you will find a discussion on theories of light's composition, as well as an introduction to the electromagnetic spectrum and various technologies involving the use of light.

Updating Your Journal

Reread the paragraphs you wrote for this unit's "For Your Journal." Then rewrite them to reflect what you have learned in studying this unit.

Science in Action

Spotlight on Laser Research

Luis Elias has been researching laser technology for over 20 years. With bachelor's and doctoral degrees in Physics, Luis is a professor of Physics and the director of the Free-Electron Program at the University of Central Florida. This program is designed to set up a facility that will provide scientists and researchers access to a special type of laser called the free-electron laser. Luis played a major role in the initial development of this unique laser.

Luis Elias working on the free-electron laser

Q: What is a free-electron laser?

Luis: This is a new type of laser that uses particle accelerators to generate beams of electrons that move close to the speed of light. The interaction of the electron beam with magnetic fields causes the emission of an electromagnetic wave in the form of laser light.

The difference between the free-electron laser and most other lasers is that it provides more control and flexibility. Most lasers only emit one frequency, or color, of light. This laser, on the other hand, can emit radiation throughout a large portion of the electromagnetic spectrum. The free-electron laser is unique because we can control the frequencies by "tuning" the electron speed to a proper value.

Q: What are some of the uses of the free-electron laser?

Luis: There are lots of uses! For one thing, physicists can use it to better understand the physical properties of matter. Matter responds to the frequencies of lasers. Therefore, the response of matter to different frequencies can be observed much more easily with a free-electron laser.

There are biological and medical applications as well. For example, a doctor performing eye surgery can use the precision of the fine-tuned free-electron laser to operate directly on the retina in the back of the eye. This is accomplished without burning the cornea in the front of the eye, as most other lasers would. Naturally, this eliminates a step and provides a more efficient form of surgery.

Q: What does your daily work involve?

Luis: All of my activities basically center around the Free-Electron Laser Program. I supervise researchers, teach several classes, prepare a lot of proposals, and attend meetings!

Lasers are used in medicine to locate precisely where radiation will be delivered.

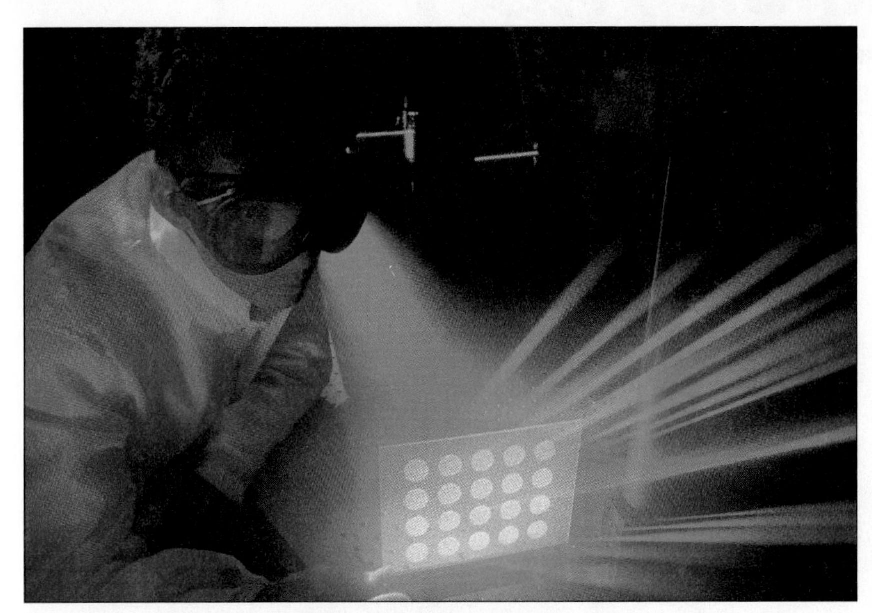

Technician viewing beams of an argon laser

Q: What are the most interesting or exciting aspects of your work?

Luis: I think the most exciting part of my research activities are the innovations it makes possible in other fields. For example, with a system I set up in California, scientists are currently learning more about DNA. They are researching the way in which DNA replicates. Using the free-electron laser, they can illuminate a DNA strand with a laser light and then watch the way it vibrates as a result of the energy it absorbs from the laser light. By fine-tuning the laser, the scientists can then affect the replication of the DNA by making it vibrate in a different manner.

An argon laser, pretested in a beaker of water, can be used to clear blocked arteries.

Q: How would you describe the importance of your work to high school students?

Luis: Sometimes people think that research is not important because it is difficult to assign a monetary value to it. I believe, however, that pure research is crucial because from it the potential for real technological advancements are limitless! Many exciting technologies have come about by accident, because somebody was curious about something and dedicated themselves to finding out about it.

The laser research I perform, like any other scientific endeavor, is important because it helps us better understand the things around us. When I began thinking about the possibility of a laser that could be fine-tuned, I had no idea that one day surgeons would be able to use it for eye surgery. But it's been a remarkable tool for surgeons. And that's just one example of the exciting progress I feel I have somehow played a part in! I think a scientist's greatest achievement is progress for the common good.

Circular holograms at the Museum of La Villette in Paris

A Project Idea

Since the 1950s laser light and its possiblities have fascinated many scientists. Why? Think about how the laser works. What does the word "laser" stand for? How does the light from a laser differ from light emitted by an ordinary light bulb? What do the words "metastable" and "stimulated emission" have to do with lasers? Research the subject and use your findings to determine how a hologram or a compact disc (CD) player works.

Science and the *Arts*

Neon Art

In 1898 two chemists made an exciting discovery. They passed an electric current through a gas they had collected in their experiments with liquid air. The gas began to glow with a fiery, reddish-orange light. It was a light that no one had ever seen before. The chemists, whose names were Sir William Ramsay and Morris Travers, realized that they had discovered a new element. They named the element *neon,* from the Greek word meaning "new."

The discovery of neon led to new art forms.

LIVINGSTON, Ben, Where the Roses Get Red, *1991. Neon art, 6 × 2 m. Austin, Texas.*

A New Art Form

It was not long after the discovery of neon light that people began to think of artistic and decorative uses for it. By the 1920s, neon signs were in wide use. Today they are so common that we are likely to take them for granted. But have you ever wondered how neon signs are built, or how they work?

Neon art requires skill and patience.

Handle with Care

The first step in creating a neon sign or work of art is to fashion the desired shape. To do this, glass tubing is heated to make it flexible and then is gently bent into shape. Next, electrodes are installed in each end. Finally, a vacuum pump is used to pump the air out of the tubing. The tubing is now ready for the neon gas.

Neon gas comes from suppliers in 1-L containers that look like large light bulbs. The gas is invisible, so the containers look empty. Actually, there is enough gas in a single container for about 75 m of tubing. The gas is put into the tubing through a valve on the pump hookup. Then the tube is closed off by heating it at a point close to the pump connection.

Color Magic

The completed neon light produces a reddish-orange light. But if a few drops of mercury are added to the tube of glowing gas, something almost magical happens. The color of the light changes to a brilliant blue!

The color change occurs because different gases give off different colored light when they are energized. The atoms of neon gas give off reddish-orange light. Mercury vapor atoms, on the other hand, give off blue light.

Neon artist Jim Austin performing "color magic" with neon and mercury

Fluorescent rock under normal light

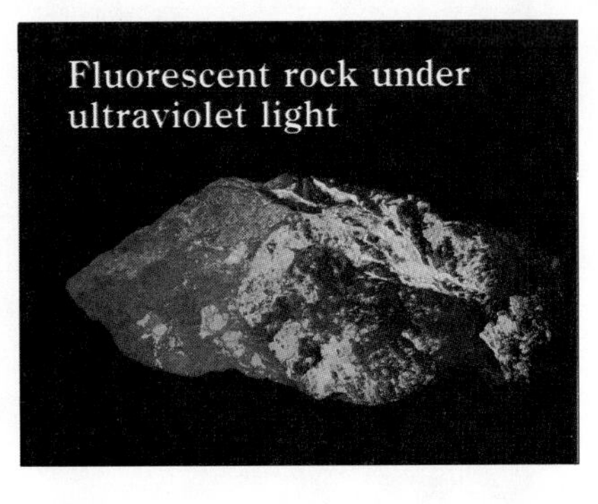

Fluorescent rock under ultraviolet light

More Tricks of the Trade

You may have seen how certain substances *fluoresce,* or glow, when they are exposed to ultraviolet light. Such substances are known as **phosphors.** The colors of glowing phosphors can be both beautiful and fascinating. Green, yellow, and orange are some of the colors you might see.

There are a lot of ultraviolet rays in the blue light produced by mercury vapor. This suggests some interesting possibilities. Taking a clue from natural fluorescence, neon light designers coat the inside of light tubes with powdered phosphors. When ultraviolet rays from the mercury vapor strikes them, the phosphors give off vivid light of various colors.

There are still other ways of creating new colors with light. One way is to use gases other than neon, such as argon. Another way is to use colored glass for the tubes. With all the techniques at their disposal, neon artists have a complete palette of color for their work!

Find Out for Yourself

Look around you. Neon signs and art are everywhere. Make a list of some of your favorite neon light works. Compare your list with those of your classmates. Discuss what you like most about the works you chose. What do you think the future of neon art will be like? What kind of neon art would you like to create?

Do some research to find out why passing a current through a gas produces colored light. Then write a brief report in your Journal.

Unit 7 Particles

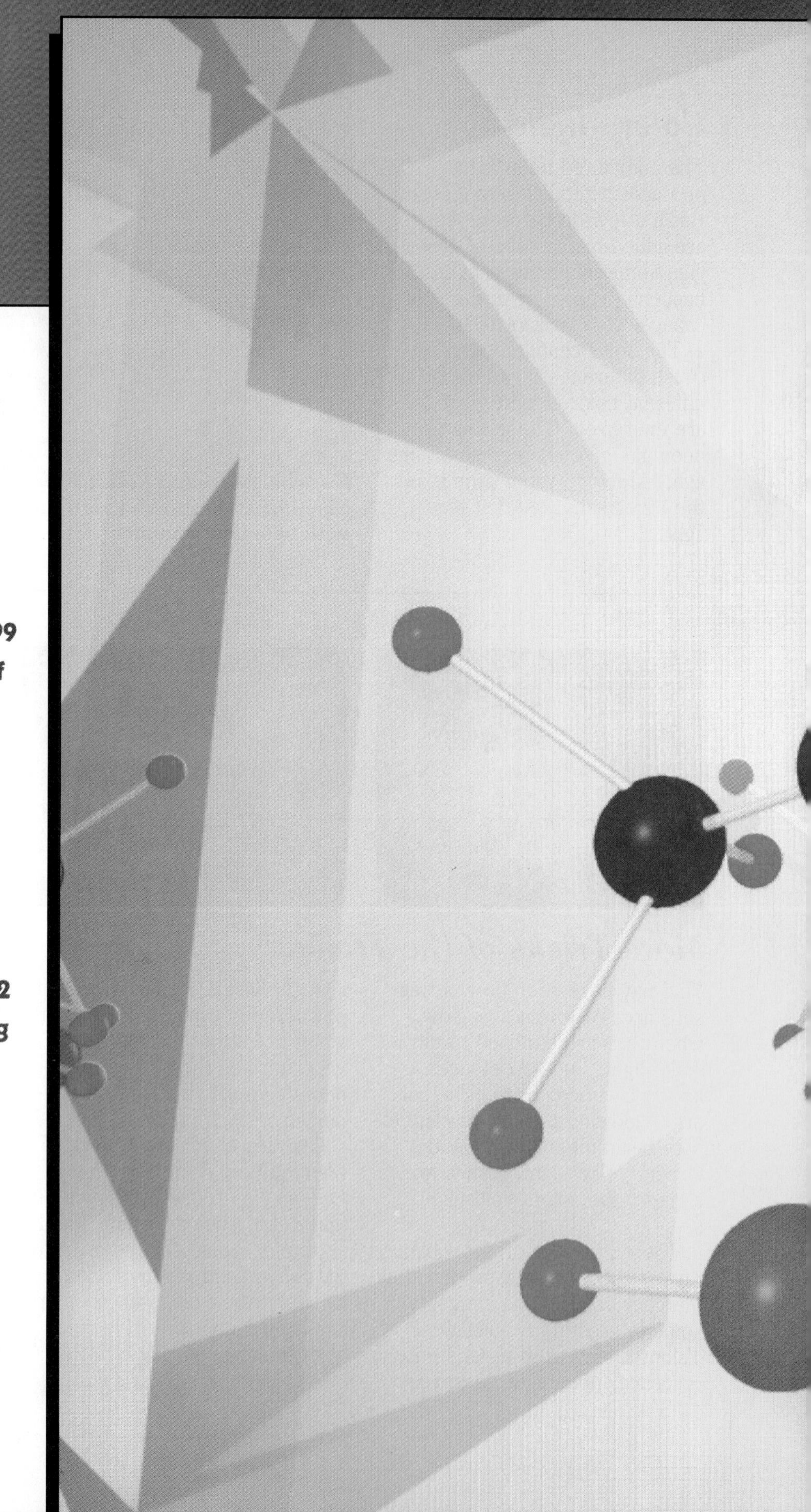

For Your Journal

Write a paragraph summarizing your thoughts about each of the following:

1. How small can a particle of matter get and still behave like the kind of matter it is?
2. If you could see what air is made of, what might it look like? Would warm air look different from cold air?
3. Why does an inflated balloon get larger when it is heated?
4. What does the model of matter in this picture show?

1 A Search for Explanations

What Is the Smallest Object You Can See?

Angelina's class went to the planetarium. As they sat back in their seats and looked up at the domed ceiling, the lights went out. Stars appeared on the dark ceiling. Angelina felt as if she were in outer space. Then a voice began to speak:

> *Imagine you have journeyed far, far away to another galaxy such as Andromeda (the closest galaxy to our own, the Milky Way). The Milky Way appears as a mere point of light in space because of the vast distance separating the two galaxies. As you come closer, the Milky Way begins to resolve into millions of stars, of which our sun is only one. Move even closer and the planets can be seen shining in the reflected light of the sun. Come closer still and Earth is seen as a sphere of land, ice, and water. When you land on Earth, you are confronted with a myriad of objects in dazzling colors: hundreds and hundreds of shades, and every shape and size. With a sharp eye, you can detect tiny particles of dust in the air of a sunlit room. And, with the aid of optical instruments, you can view even smaller particles.*

What is the limit for detecting smaller and smaller bits of matter? Is there no limit to the size of objects that can be detected? The answers to these questions are important steps in the search for explanations about why matter behaves the way it does. That is the purpose of this unit—to lead you on a search for these explanations.

The Sizes of Things

Some objects and distances are very large. The diameter of the solar system, for example, is about 10,000,000,000,000 m. Other objects and distances are very small. The size of one cell in your body is about 0.000001 m. But it can be very awkward to work with such large or small numbers. One useful way to refer to the sizes of very large and very small things is to use **exponents.** Exponents are a kind of shorthand. Count the number of zeroes in the figure that shows the diameter of the solar system. When it is expressed using an exponent, the figure becomes 10^{13} m. The size of a body cell expressed with an exponent is 10^{-6} m.

Now examine the chart below. Explain the rule for using exponents to express both very large and very small numbers.

Exponents for measurements that are 1 m or greater
10^0 m = 1 m
10^1 m = 10 m
10^2 m = 100 m
10^3 m = 1000 m
10^4 m = 10,000 m
10^{20} m = ?
Exponents for measurements that are 1 m or less
10^0 m = 1 m
10^{-1} m = 0.1 m
10^{-2} m = 0.01 m
10^{-3} m = 0.001 m
10^{-4} m = 0.0001 m
10^{-20} m = ?

The list to the right gives more examples of the sizes of different objects, using exponents. The measurements are approximate, not exact.

Where would you place the following objects in the list?

- width of a pencil
- diameter of a hair
- distance from your home to school
- diameter of Venus' orbit

Suggest other objects with sizes that could be added to the scale.

meters	
10^{25}	
10^{21}	length of the Milky Way
10^{13}	diameter of the solar system
10^{12}	diameter of Jupiter's orbit
10^{11}	diameter of Earth's orbit
10^{10}	
10^9	diameter of the moon's orbit
10^8	
10^7	diameter of the Earth
10^6	distance across the continental United States
10^4	deepest part of the Pacific Ocean
10^2	height of a 25-story building
10^1	length of your classroom
10^0	height of a young child
10^{-1}	width of a hand
10^{-2}	width of a finger
10^{-3}	diameter of a thread
10^{-4}	diameter of a fine sand particle
10^{-5}	
10^{-6}	diameter of a body cell
10^{-10}	

2 Give Me the Facts

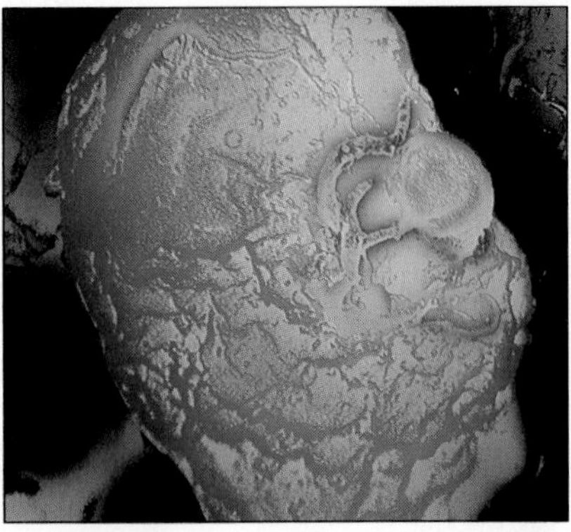

The AIDS virus is approximately 1.2×10^{-7} m across.

A Question of Proof

You have probably heard the saying, "I've got to see it to believe it!" With average vision you can see objects as small as 10^{-4} m. With a microscope, objects as small as 10^{-7} can be detected. What are the facts about even smaller objects? You cannot see them, so how do you know they exist?

You must rely on *circumstantial evidence.* In scientific terms, that means you must make observations and use those observations (facts) to make inferences.

The Daily Fact

How does an inference differ from an observation? This article, excerpted from the newspaper, *The Daily Fact,* offers some clues. While reading the story, try to find at least three inferences made by the defense lawyer and prosecuting attorney. What facts (observational evidence) support each inference? Are all the inferences true? In your Journal, record your answers in a table similar to the one shown at the right.

Inference	Supporting Observation (Evidence)
1.	
2.	
3.	

Jury Still Out

Phoenix, AZ

The jury retired at 11:00 A.M. today to consider the evidence and arrive at a verdict in the trial of Phineas Swipe, who is accused of robbing the corner store. Earlier in the day, Judge Pedrosa summed up the case. She reminded the jury of the facts brought out by the prosecution:

1. The accused party was seen in the area of the robbery.
2. His blood type was found on the doorknob. He had a cut on his right hand.
3. His fingerprint was found on the counter top.
4. He was observed spending more than the usual amount of money at the horse races the next day.
5. The defendant has a past record of robbery.

Prosecuting attorney J.D. Mortimer stated that since Swipe was in the area of the crime and had a past criminal record, he committed the crime. The blood type found on the doorknob also matches Swipe's. So, Swipe cut himself while breaking the store's window. Finally, the prosecutor declared that the money the defendant spent was the money taken during the robbery.

On the other hand, defense lawyer Leo Kostas claimed that many people were seen in the area of the robbery and Swipe is no more guilty than any of them. The cut was the result of a dishwashing accident and had nothing to do with the broken window at the store. The fingerprint was probably left on the counter top when the defendant bought a paper at the store a few hours earlier. The judge went on to discuss how the case hinges on circumstantial evidence. For a verdict of guilty to be reached, the jury must find evidence that shows—beyond a reasonable doubt—that the accused is guilty.

Excerpt From Misha's Journal

"Jury Still Out"—From what was said in court, I think the jury should give a verdict of "not guilty." No one saw the accused at the scene of the crime, and there is no real *proof that he did it. I just don't think the circumstantial evidence is good enough to let anyone make an inference that will send the man to prison.*

Do you agree or disagree with Misha's opinion? Explain why. Do all inferences come from circumstantial evidence? Describe the similarities and differences between the inferences that scientists make and those that the lawyers and Misha made.

Observation or Inference?

You live in an ocean of air, but you cannot see it. Why not? The answer is based on circumstantial evidence. Such an explanation depends on knowledge about the unseen structure of matter. For instance, it is easy for a scientist to observe how liquids or solids behave, but it is more difficult to explain *why* they behave as they do. A scientist must draw inferences based on observations. In this sense, a scientist does exactly what a jury does!

Copy the chart below into your Journal. Working in small groups, examine the following statements about air. Do you agree with each one? Place a check mark next to the statements you agree with. Are your decisions based on observations, or are you making inferences? What evidence supports each decision? Discuss your decisions with your classmates. As a group, do you agree or disagree?

Statements About Air	Observation	Inference	Evidence Supporting Your Decision
1. Air can be squeezed into a smaller space.	✓		I can pump air into my bicycle tire.
2. Air is invisible.			
3. Air has volume.			
4. Air has mass.			
5. Air moves.			
6. If we could see air, we would see many particles.			
7. Air behaves like a sponge.			
8. There is water in air.			
9. Create your own statement about air.			

Using Models

Based on your observations and inferences, you can come to some conclusions about air. Do you think air is made of particles, as suggested in statement 6 on page 395? Or, is air more like a sponge that can be compressed and expanded, as suggested in statement 7?

These kinds of questions help you develop a mental picture or idea about the structure of air. They help you form a *model* for the concept you are thinking about. So far, you have considered two types of models. In statement 6, the model is an idea: air is made up of particles. You cannot actually see air—you can only observe how it behaves. Therefore, you form an idea about what air is really like, based on your observations of its behavior. In statement 7, you compare air to an actual object—a sponge. In what ways is it like a sponge? In what ways is it different? Do you think the particle model is better or worse than the sponge model? Why? How can the models help you understand air and its behavior?

More About Models

In Unit 5, you explored the workings of the inner ear and saw how sounds are transmitted across the eardrum. Recall Fernando's model on page 299, which is shown again below. What does each feature of the model represent? (Check your Journal—you probably have that information recorded already.) Do you find that Fernando's model helps explain how the ear works? Do you think it is a good model? Why or why not?

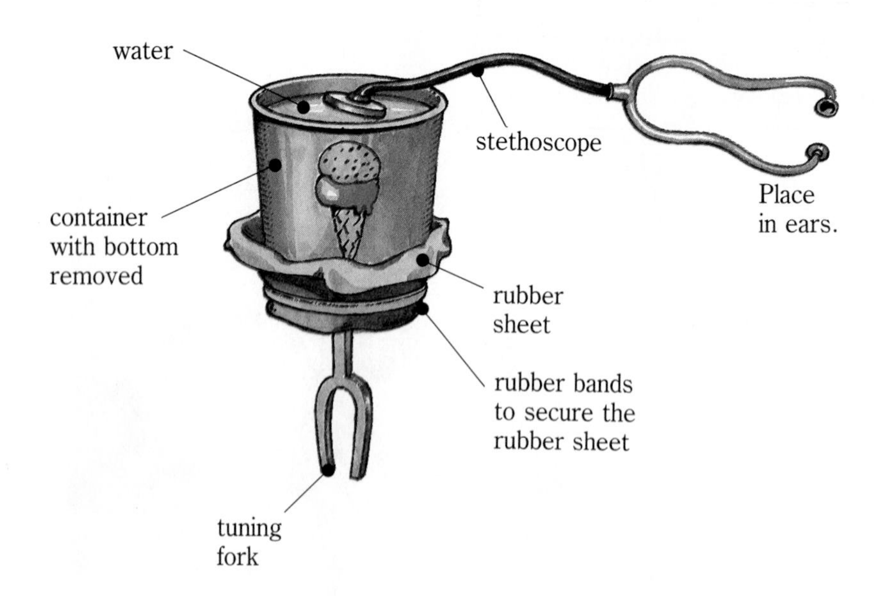

Consider the model of the Mediterranean Sea, which was used in Unit 3 on page 158. This model is a representation of a phenomenon that occurs on Earth. On a reduced scale, the model demonstrates how currents actually behave in the Mediterranean. What is the food coloring used for? Why are different concentrations of salt solutions used? How helpful was this model in explaining how the currents behave?

You have considered two examples of models. Were they useful? Consider these previous examples of models. Then, in your Journal, describe how models help us do the following things:

- Visualize a complex idea or structure
- Explain observations and make more inferences
- Make predictions that can be tested through further observations and experiments

In the following Exploration, you will study one student's model of air. Do you think the model is both accurate and useful? Explain.

EXPLORATION 1

Making Kevin's Model

Kevin agreed with statement 6, "If we could see air, we would see many particles," so he devised the following model. Now it's your turn to build Kevin's model.

You Will Need

• a balloon • 5 small ball bearings

What to Do

Place the ball bearings inside the balloon. Then blow up the balloon—but not too much! Tie the balloon closed.

The balloon now contains not only real air but also the ball bearings, which represent air particles. Gently shake the balloon until the bearings rattle around inside.

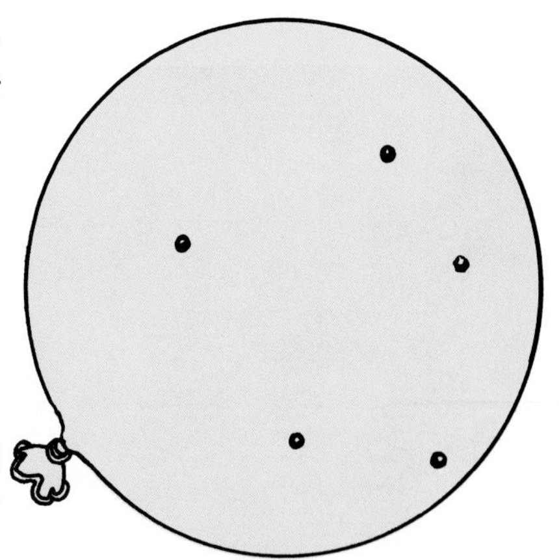

Can you see the model air particles in motion? Can you feel them in motion? What observation about air does this model explain?

A Final Word

What single word or phrase best describes a model? Suggest another example of a model.

Challenge Your Thinking

1. *Nobody noticed me until I bought my Shred Jeans. Now that I've got 'em, I'm the head cheerleader, an A+ student, and I look goooood. Get your Shred Jeans today . . . they'll change your life. They sure changed mine.*

   Obviously the advertisement is making some pretty ridiculous claims. It gives several examples of drawing an incorrect inference from available observations. With another student, make a list of five ridiculous claims that you can support without actually lying. Identify the circumstantial evidence you used to make your claim. Then explain how you can make better inferences based on your observations.

2. Design a model for water that illustrates the observations you have made about water. What happens when liquid water turns into ice? into steam? Can your model represent these changes? What observations does your model have trouble explaining? After you create your design, summarize its strengths and weaknesses as a model.

3 Building the Case

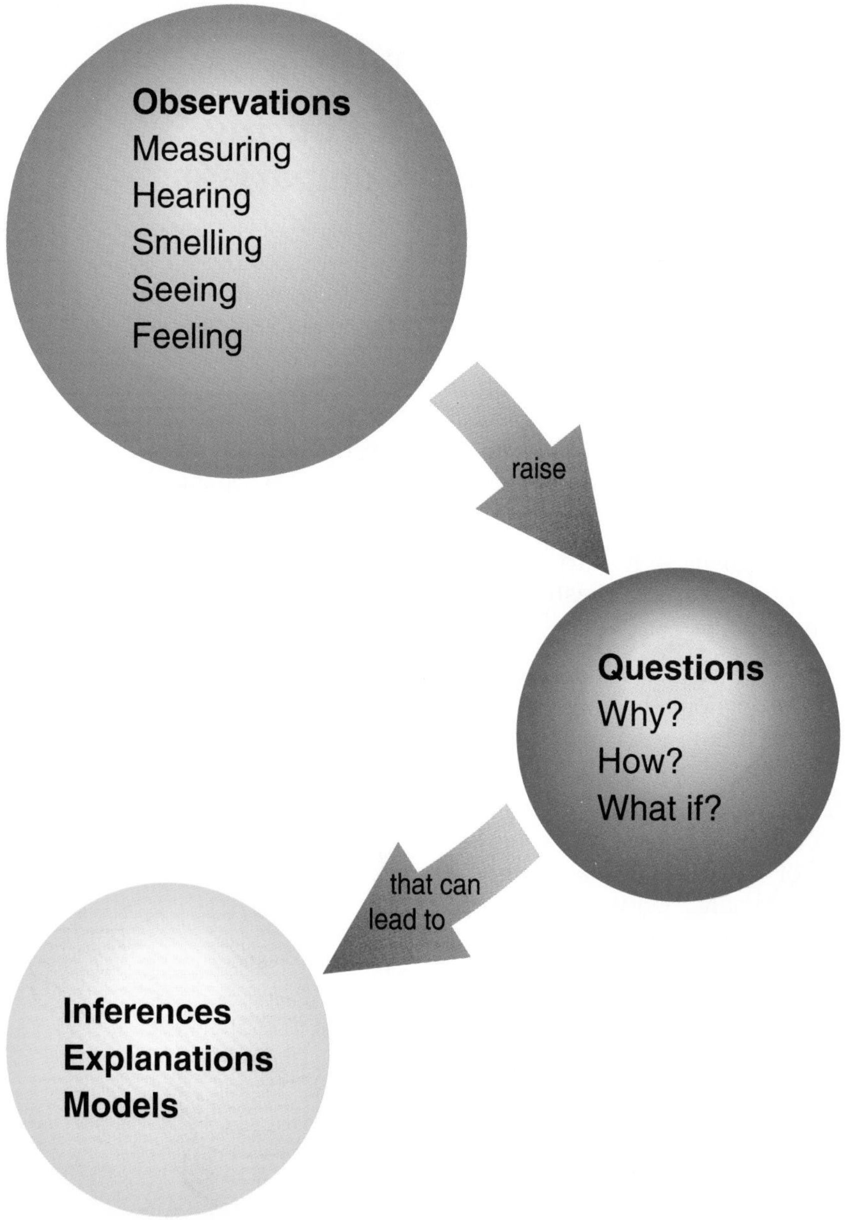

Observations
Measuring
Hearing
Smelling
Seeing
Feeling

raise

Questions
Why?
How?
What if?

that can lead to

Inferences
Explanations
Models

You are the judge, jury, and attorney in a landmark case—a case to determine whether all matter is composed of particles. This case may raise as many questions about matter and its behavior as it answers. To build the case for or against particles, the following experiments will provide you with the observations and information needed to make some all-important inferences. These will help you prepare your case in favor of a particle theory of matter—or against it.

EXPLORATION 2

Making the Case

You Will Need

- red food coloring
- an eyedropper
- a stirring rod
- a 100-mL beaker
- a half-liter of sand
- water
- a half-liter of dried peas or beans
- 40 mL rubbing alcohol
- 4 large containers
- 25 mL salt
- 2 graduated cylinders
- a stopwatch or clock

Part 1

What to Do

A Thought Experiment

Read the observations and inferences about liquid and frozen water. Then answer the questions that follow.

Observation

In the freezer, ice cubes become smaller over time.

Questions

- Where does the ice go?
- How does it disappear?
- Can ice be prevented from disappearing?

Inferences, Explanations, and Models

- Perhaps ice (water) is made up of particles.
- Maybe some of these particles escaped from the solid state to form a gas—a process called *sublimation.*

Follow Up

1. Name another substance that changes directly from a solid into a gas.
2. Could a gas change directly into a solid? If so, think of some examples.
3. Do these observations and explanations support the idea that water is made up of particles? Why or why not?

Part 2

What to Do

Coloring Water Red

How many times can you divide a drop of food coloring in a beaker of water and still detect its red color? Here is a way to find out.

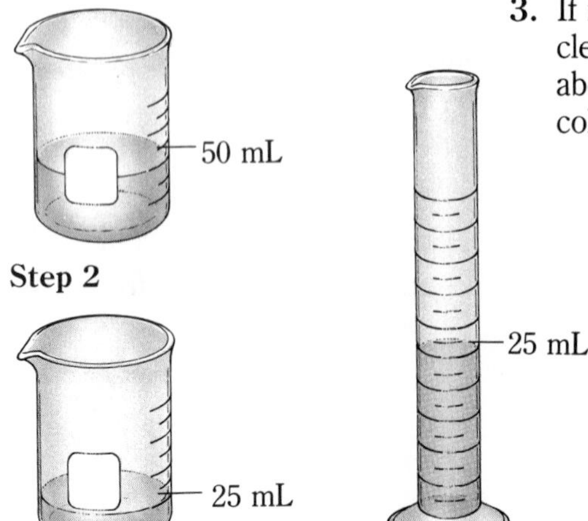

Thoroughly dissolve a drop of food coloring in 50 mL of water. Now divide this solution into two equal parts. Wash 25 mL down the sink, and add 25 mL of water to what remains. Once again the total volume of the solution is 50 mL. Is the solution colored red?

The concentration of the food coloring has been diluted to one-half the original amount. Repeat the dilution process once more. Can you still see the red coloring in the water?

Your beaker now contains one-quarter of the original drop of food coloring. Repeat the procedure—keeping accurate records—until you can no longer see the red color.

Before going on to Part 3, discuss the following questions with a partner.

1. Is the color spread evenly throughout the solution, or are bits of food coloring clumped together?
2. How much of the food coloring do you have in the beaker of water at the end of the experiment? Do you think there may be some food coloring left in the solution at the end, even though you cannot see any?
3. If matter is made up of particles, what can you infer about the size of the food-coloring particles?
4. Does the experiment support a particle theory of matter? Why or why not?

Part 3

What to Do

A Pouring Comparison

Fill three large containers using each of the different substances listed below. Keep the substances in separate containers.

- dried peas or beans
- sand
- water

Now pour each substance into an empty container. Did either of the first two resemble water in the way they poured? What can you infer about matter from this experiment?

Part 4

Where $1 + 1 \neq 2$

Carry out the following three activities. After making careful observations, use them to generate inferences about the unseen structure of matter.

1. Pour 50 mL of sand into a 100-mL graduated cylinder. Then pour 50 mL of water into another graduated cylinder. Now carefully pour the water into the sand. Record the volume of the mixture. Suggest why the combined volume is not 100 mL.

2. Put 25 mL of salt into a graduated cylinder. Add enough water to bring the combined volume of salt and water to 100 mL. Without spilling the contents, gently shake the cylinder for two or three minutes. Record the volume after shaking. How do you explain your value for the final volume of salt and water?

3. Pour 50 mL of water into a graduated cylinder. Then pour 40 mL of alcohol into a second cylinder. Pour the alcohol into the water and stir. Does the combination read 90 mL?

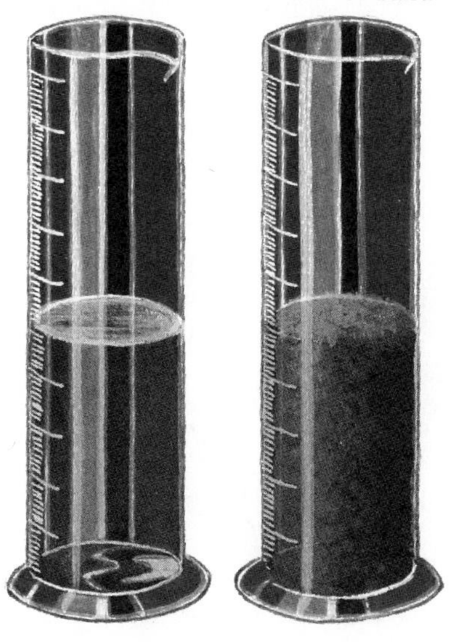

25 mL of salt

water

50 mL of water 40 mL of alcohol

Drawing Conclusions

How does the particle model of matter help to explain the observations you made in these experiments? Do your observations support the idea that matter consists of particles? Why or why not? In your Journal, summarize your case for or against a particle theory of matter.

Particles on Trial

A ninth-grade class was asked to build a case for or against the particle model. Here are a few excerpts from their replies. Imagine you are the teacher—what comments would you write on each report? Has each student built a good case for the existence of particles? How could they improve their cases?

Nikki's Case

I think the particle model is correct. In Part 4 of the Exploration, water filled the spaces between the sand particles. This is also what probably happened when the water and alcohol were mixed.

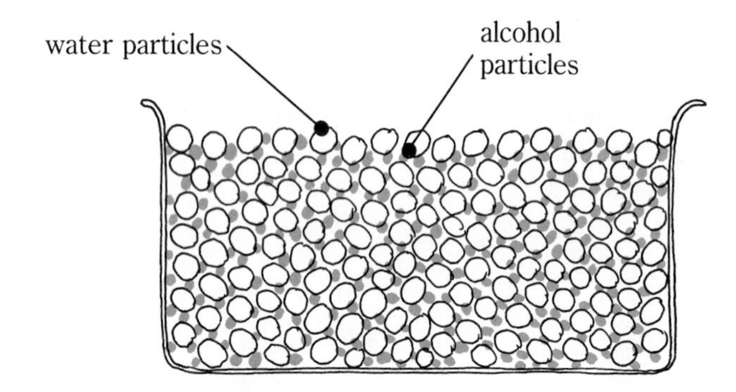

Marco's Case

In sixth grade we did this experiment where a jar full of sand was turned upside down in a container of water. Then we took the top off the container. We thought that the water would fill the jar when the sand came out, but the water only went halfway up. This showed that the sand had only half-filled the jar. Air filled the other half, by fitting into the spaces between the sand particles. Otherwise, the water would have filled the jar. When salt dissolves in water, I think the same thing happens. The salt fills in the spaces between the water particles.

Bob's Case

Matter must be made of particles. It says so right in the book.

Your Case

In a few paragraphs, revise your own case for or against particles. Share it with a few classmates and ask for their opinions. You should use your summary and evidence from the Exploration as well as any other evidence you have observed. Now test the strength of your case: does it explain why you cannot see air? Can it explain why you see sugar when it is in the sugar bowl, but not when it is dissolved in water? How does your case for or against particles answer these questions?

The Hidden Structure of Matter

Democritus (460–370 B.C.) said:

> *"According to convention, there is a sweet and a bitter, a hot and a cold, and according to convention there is color. In truth there are atoms and a void."*

John Dalton (1766–1844) said:

> *"I should apprehend there are a considerable number of what may properly be called elementary particles, which can never be metamorphosed [changed] one into another."*

These two men were born more than 2000 years apart in two different countries, Greece and England. But while thinking about the nature of matter, they both arrived at the same conclusion. Matter is made up of particles—what Democritus called **"atoms."** During the two millenia between the existence of these two men, this idea was largely ignored. How did Dalton arrive at a conclusion that had been neglected for 2000 years? Read "Constructing a Particle Theory" to find out for yourself.

Constructing a Particle Theory

Recall your study of chemicals from earlier science courses. You found out that all pure substances on Earth can be classified as either *elements* or *compounds*. All matter is made of elements. During chemical changes, elements can combine to form compounds or compounds can decompose into elements. Compounds can also be changed into other compounds.

By Dalton's time, much had been discovered about chemical change and the elements involved in these changes. Antoine Lavoisier had discovered the role oxygen plays in combustion. Water had been decomposed into the elements hydrogen and oxygen when an electric current was passed through it. Metals could be obtained from their ores by chemical changes, and at least 40 different elements had been identified.

Time Out for Fact-Gathering

1. About 2000 years ago, people had already discovered the elements gold, silver, iron, lead, tin, mercury, sulfur, and carbon.
2. By 1735, the alchemists had added zinc, arsenic, and phosphorus to the list of elements.
3. Hydrogen, oxygen, and nitrogen were discovered between 1765 and 1775.

John Dalton began his scientific investigations at a very early age. When he was twelve, he organized a school in an old barn and became a schoolmaster.

As an observant and curious scholar, Dalton explored many scientific questions. One of these explorations led to some rewarding conclusions. Trace Dalton's path of investigation by completing the flowchart to the right. You'll find the information you need in the reading that follows.

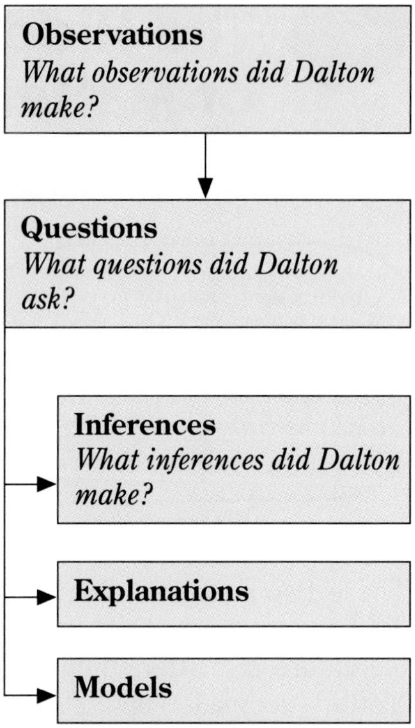

By passing an electric current through water, Dalton observed that water could be separated into the elements hydrogen and oxygen. He also observed that, by identifying the amount of one element in a quantity of water, he could predict the amount of the other element. For example, if 16 g of oxygen were formed by decomposing water, then 2 g of hydrogen were always formed as well. In other words, the mass of the oxygen was always eight times greater than the mass of the hydrogen. Dalton asked himself, "From this information, what can I infer about the structure of matter?"

Perhaps this activity will help you understand the answer Dalton found. Measure 16 g of modeling clay and form it into a ball. This ball represents an oxygen atom. Next form two 1-g balls of clay. These represent two hydrogen atoms.

Attach the model atoms so that they look like the illustration at right. You have just simulated a chemical change, between oxygen and hydrogen atoms, that forms a new substance—water.

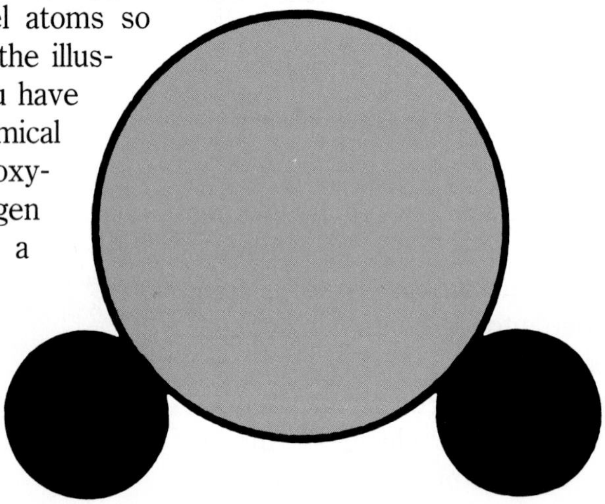

Observations
What observations did Dalton make?

Questions
What questions did Dalton ask?

Inferences
What inferences did Dalton make?

Explanations

Models

How would you complete the flowchart?

Time Out for Discovery

1. How many times heavier is your model oxygen atom than your model hydrogen atom? Does your answer support Dalton's observation?
2. Add your water-particle model to those formed by your classmates. How does the total mass of model oxygen atoms compare to the total mass of model hydrogen atoms?
3. Consider a glass of water. If the water consists of oxygen and hydrogen atoms, how will the total mass of oxygen atoms compare with the total mass of hydrogen atoms?

In 1803 Dalton wrote in his journal: *"An enquiry into the relative weights of the ultimate particles is, as far as I know, entirely new. I have lately been prosecuting this enquiry with remarkable success."*

Of course the "ultimate particles" Dalton spoke of were the same as Democritus' atoms. Although he could not see atoms, Dalton inferred the following:

1. An atom of one element has a different mass than an atom of any other element. (How much heavier was your model oxygen atom than your model hydrogen atom?)
2. An atom of one element is identical to any other atom of the same element. (Did you make your two model hydrogen atoms the same or different?)
3. An atom of one element cannot be changed into an atom of another element. (Did you alter or destroy your model oxygen and hydrogen atoms by making your model water particle?)

Because Dalton made careful observations about the masses of different elements and asked the right questions, evidence emerged to support a new model for the structure of matter: *matter is composed of atoms.*

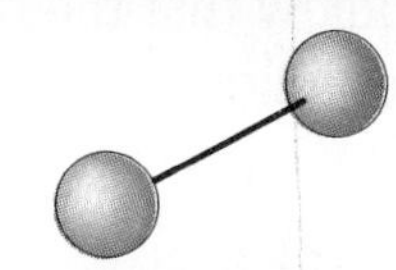

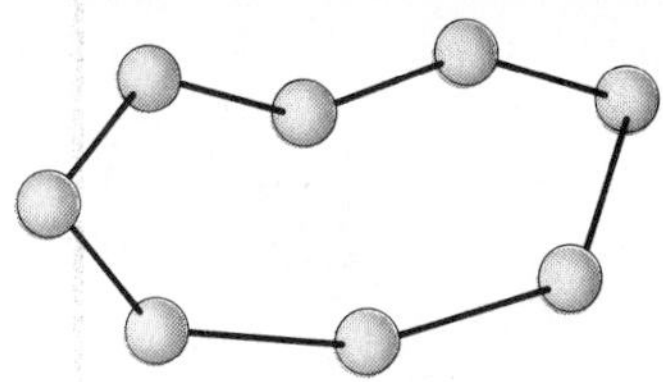

Iodine (top) and sulfur (bottom) are examples of elements. What do you think the drawings show? Can you see how each element is made up of only one kind of atom?

Time Out for Analysis

1. What inferences were made in designing the model? Consider the examples below. Not all these inferences are necessarily true:

   (a) Atoms are round.

   (b) The two atoms of hydrogen are identical.

   (c) Atoms have different colors.

   (d) In forming water, two hydrogen atoms combine with one oxygen atom.

   (e) An atom of oxygen is sixteen times heavier than an atom of hydrogen.

   Which of these inferences are similar to Dalton's? Which of these inferences can you support? Which can't you support?

2. Here are the symbols Dalton used to represent some atoms:

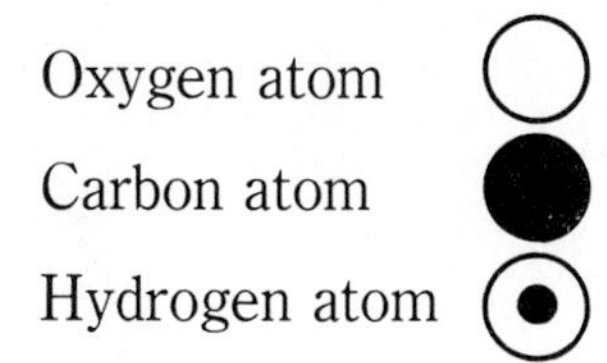

3. He represented water this way: 

4. Another substance, called marsh gas or methane, he represented this way:

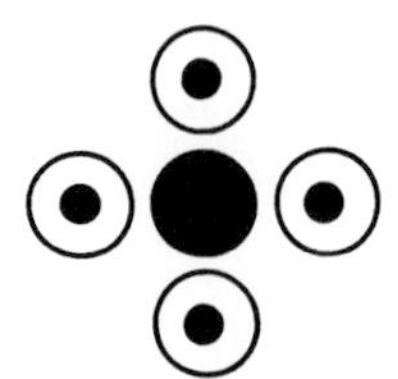

5. Dalton inferred that there were four hydrogen atoms for each carbon atom.

Dalton knew that the total mass of the carbon atoms in methane was three times greater than the total mass of the hydrogen atoms. How much heavier would each carbon atom be than each hydrogen atom?

Try this—measure out a 12-g ball of clay and four 1-g balls of clay. Construct a model of a methane molecule. How does the mass of the model carbon atom compare with the total mass of the four model hydrogen atoms?

This lesson started with quotations from Democritus and Dalton. Rewrite what each said in your own words.

Elements and Compounds

Both Dalton and Democritus concluded that all matter is composed of particles called atoms. Gold, silver, and sulfur are pure substances; each consists of one kind of atom. These substances are called **elements.** Ninety-one elements, and therefore ninety-one kinds of atoms, are known to exist naturally on Earth.

When you simulated a chemical change between oxygen and hydrogen atoms, you formed a new substance—water. Water particles are called molecules. **Molecules** are particles that are a combination of two or more atoms. Water, sugar, and carbon dioxide are substances that consist of molecules made of two or more different kinds of atoms. These substances are called **compounds**. The number of existing compounds is practically limitless.

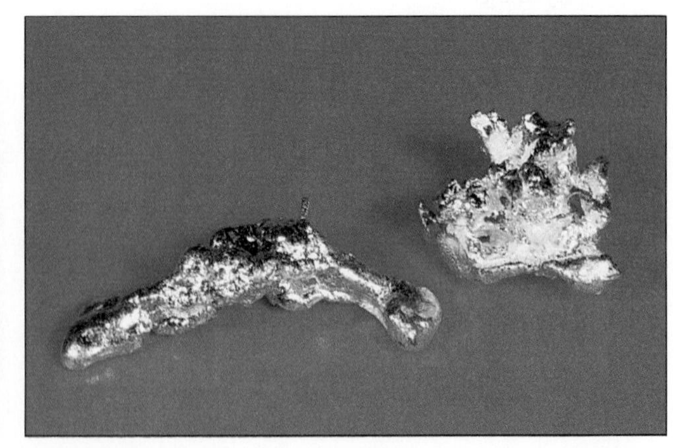

Gold and silver are two naturally occurring elements.

Summing It Up

- Elements consist of atoms; each element is made of one kind of atom.
- Elements cannot be divided into simpler substances by chemical changes.
- There are 91 naturally occurring elements on Earth.
- Elements combine with each other to form compounds.
- A molecule of an element consists of two or more of the same kind of atoms. A molecule of a compound consists of two or more different kinds of atoms. A molecule is the smallest particle of a compound that still has the characteristics of the compound.
- Compounds can be broken down into elements by chemical changes.
- The number of compounds that exist or that can be made through chemical changes is essentially unlimited.

Testing Your Understanding

1. Use the ideas in the summary to draw a concept map showing the relationship between atoms, elements, molecules, and compounds.
2. Both carbon dioxide and carbon monoxide consist of carbon atoms and oxygen atoms. Yet the two have different properties. Carbon monoxide will burn and is poisonous, while carbon dioxide will not burn and is not poisonous. Here is how Dalton represented carbon dioxide.

   (a) How would he have represented carbon monoxide?

   (b) How are the masses of these two compounds different?

3. Earlier you made models of water molecules, using clay balls to represent oxygen and hydrogen atoms. The ball representing the oxygen atom had a mass of 16 g. Each model hydrogen atom had a mass of 1 g. Why do you think these masses were chosen?

Even More Models

Here are more models of molecules you can make. Use different-colored clay balls to represent each element. In order to make your representation fit Dalton's findings, your model atoms should have the masses listed below. Or, you can use a fraction of the suggested masses. (For example, you can divide all of the masses by two and your models will still show how much heavier one type of atom is than another.)

For a model carbon (C) atom, use 12 g of clay.
For a model hydrogen (H) atom, use 1 g of clay.
For a model nitrogen (N) atom, use 14 g of clay.
For a model oxygen (O) atom, use 16 g of clay.
For a model sulfur (S) atom, use 32 g of clay.

Compound	Elements Forming Each Molecule	Atoms in the Molecule
water	hydrogen (H), oxygen (O)	H—O—H
hydrogen sulfide	hydrogen (H), sulfur (S)	H—S—H
carbon dioxide	carbon (C), oxygen (O)	O=C=O
methane	carbon (C), hydrogen (H)	H | H—C—H | H
butane	carbon (C), hydrogen (H)	H H H H | | | | H—C—C—C—C—H | | | | H H H H
ammonia	nitrogen (N), hydrogen (H)	H—N—H | H
glucose (sugar in honey)	carbon (C), hydrogen (H), oxygen (O)	H H H | | | O H O H O O | | | | | ‖ H—C—C—C—C—C—C—H | | | | | H O H O H | | H H
alcohol (from fermentation)	carbon (C), hydrogen (H), oxygen (O)	H H | | H—C—C—O—H | | H H

In Three Dimensions

Below are representations of some fairly common molecules. These are three-dimensional models of molecules that are so tiny even a powerful microscope has difficulty capturing their image. Carefully study each and answer the questions that follow.

Methane

Carbon dioxide

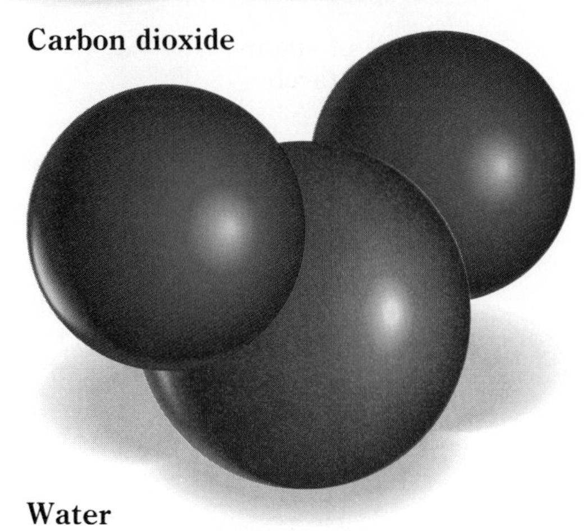

Water

Questions

1. How are these molecules the same? different?
2. How many atoms are there in each molecule? How many kinds of elements are shown?
3. In forming a molecule of a compound, were the individual atoms of the elements destroyed?

EXPLORATION 3

Sugar and Starch Molecules

Starch and sugar are two compounds that consist of the same elements (just arranged differently). The molecules of starch and sugar are made up of carbon atoms, hydrogen atoms, and oxygen atoms.

After doing the following experiment, name three differences in the properties of sugar and starch.

You Will Need

- a graduated cylinder
- 5 mL cornstarch
- 5 mL dextrose
- a jar with a lid
- a stirring rod
- a large beaker
- 100 mL hot water
- an egg
- a straight pin
- iodine
- Benedict's solution
- a watch or clock
- a hot-water bath
- 2 test tubes
- an oven mitt or test-tube tongs
- an eyedropper

What to Do

1. Shake 5 mL of cornstarch with 5 mL of dextrose. (Dextrose is a form of glucose—a sugar. Table sugar will not work in this Exploration.) Add this mixture to 100 mL of hot water in a beaker. Stir.
2. Crack an egg in half and save the larger end, which contains the air sac.
3. Using a straight pin, carefully break into the large end of the shell to expose the air sac. Be careful not to puncture the air-sac membrane. (You *will,* however, be breaking the membrane that lies flush with the egg shell.)
4. Pour 5–10 mL of water into the shell, and float the shell in the sugar-and-cornstarch mixture.

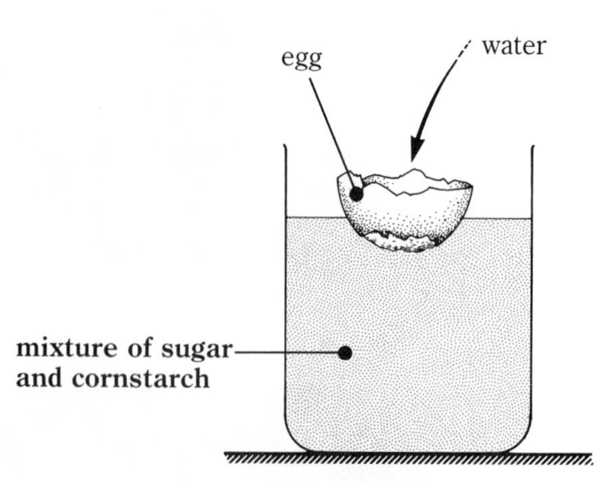

5. After 15 minutes, divide the liquid in the egg into two halves. Pour each half into its own test tube.

6. Test one half of the liquid with a few drops of iodine solution. A blue color indicates the presence of starch. Did starch molecules move through the air-sac membrane into the liquid in the egg?
7. Test the other part of the liquid by adding eight drops of Benedict's solution. **Be Careful:** *Benedict's solution is corrosive.* Heat the liquid *gently* in a hot-water bath to avoid splattering. A red or yellow color indicates the presence of sugar. Did sugar molecules pass through the air-sac membrane?

Something to Think About

1. As a conclusion to the experiment, which of the following statements do you think is most appropriate? Which statement is an inference?
   (a) Sugar molecules have properties that are different from those of starch molecules.
   (b) The experiment showed that some molecules pass through an egg membrane.
   (c) Since sugar molecules passed through the membranes and starch molecules did not, sugar molecules may be smaller in size.
   (d) Iodine solution is a test for starch, while Benedict's solution is a test for glucose.
2. What is the purpose of the air-sac membrane? What molecules do you think pass through the membrane during the development of the egg into a chicken, and why?

Flashback! Osmosis in Vegetables

You observed the passing of water through a membrane in "Life Processes"—Unit 1. You may recall that this process is called *osmosis.* Osmosis is an important process in every living organism. By osmosis, water is transferred from cell to cell. Review and explain the observations you made in the "Osmosis in Vegetables" Exploration (Unit 1), using the particle model of matter in your explanation. Can you define "osmosis" using the particle model?

EXPLORATION 4

A Mini-Experiment—Locking in the Smell

Wrap a mothball in a piece of plastic wrap. Can you still smell the mothball? How many layers of plastic wrap are needed to keep in the smell? What can you conclude from doing this mini-experiment?

Huge Numbers of Tiny Particles

If some particles are so tiny that they can pass through openings invisible to the human eye, then atoms and molecules must be extremely small. Because of the extreme smallness of particles, a great number are needed to make up even a tiny bit of matter. The following examples illustrate just how small atoms and molecules are.

- If you magnified a flea to the size of a 20-story building, you would still not be able to see the atoms that make up the flea.

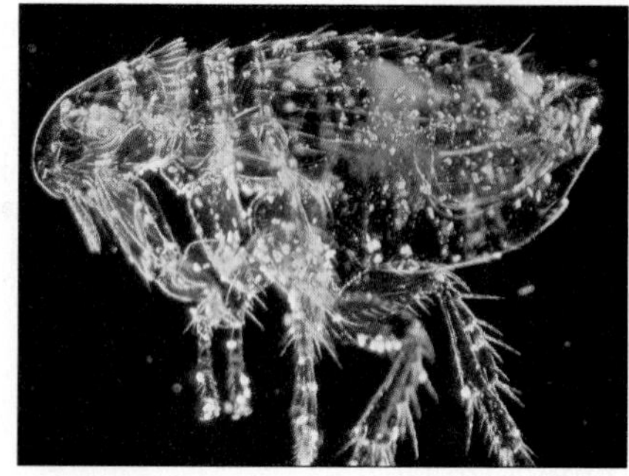

- A single drop of water contains approximately 3×10^{21} (3,000,000,000,000,000,000,000) water molecules. If you started to count these water molecules at the rate of one per second, it would take you . . . Well, you do the calculation. How many minutes? How many hours?

EXPLORATION 5

More Large Numbers!

You know that atoms are small. In fact, the aluminum atom that is part of aluminum foil has an approximate diameter of 0.00000001 cm (1×10^{-8} cm). How many atoms thick is the aluminum foil?

Here is the equation that you will need to get the answer:

$$\text{number of atoms} = \frac{\text{thickness of foil in centimeters}}{\text{thickness of one atom in centimeters}}$$

You still have a problem, though. You need to find the thickness of the aluminum foil in centimeters. So try this:

1. Cut out a piece of foil that measures exactly 10×10 cm.
2. Find its mass in grams.
3. Find its volume in cubic centimeters (cm^3), using its density of 2.7 g/cm^3. (Another way of stating this is to say that 2.7 g of aluminum has a volume of 1 cm^3. What would be the volume of 1.0 g of aluminum?)
4. Find its area in square centimeters (cm^2).

   (Area = length × width.)
5. Divide volume by area to find thickness:

   $$\frac{\text{V (length} \times \text{width} \times \text{thickness)}}{\text{A (length} \times \text{width)}}$$

   = thickness of the foil (cm)
6. Now use the first equation to calculate how many atoms thick a piece of aluminum foil is. Are you surprised?

The Miracle of Modern Fabrics

Is your jacket waterproof? If it is made of plastic, it is. Even some fabrics, such as nylon, can withstand some rain. Other fabrics are bonded or attached to a synthetic plastic. *Gore-Tex®* is one such fabric. The spaces in such fabrics are so small that raindrops cannot get through easily. The raindrops just run off the fabric. But even though water drops cannot pass through, individual water molecules can. As you perspire, these particles pass through the openings, from inside to outside. This keeps you dry. Can you think of some situations in which you'd most need such a fabric?

Tents made of fabrics bonded to plastic make great shelters from wet and cold weather.

6 Particles of Solids, Liquids, and Gases

From circumstantial evidence, you developed a model of the structure of matter. So far, your model includes the following ideas:

1. All matter consists of particles.
2. Particles are of different sizes, although all are small.
3. Elements are made up of particles called *atoms,* and compounds are made up of particles called *molecules.*

In Explorations 6 and 7, you will discover more ideas about the structure of matter to add to these three.

EXPLORATION 6

Compressing Gases, Liquids, and Solids

Fill a plastic syringe (without needle) with air. Then, with your finger over the end, push down the piston as hard as you can. What does this experiment tell you about the particles that make up a gas?

Now fill the syringe with water and again try to push down the piston. What is the difference between the particles that make up a gas and those that make up a liquid?

Finally, consider this. If you could get a solid (such as a piece of chalk) into the syringe, could you compress it? What differences are there between the particles of a solid and those of a liquid or a gas?

To conclude this Exploration, add a fourth statement to the particle model of matter. (See the list of ideas in the first column on this page.)

Compressed air in this rocket forces water through the opening at the bottom. As the water is forced downward, the rocket is propelled upward. Why do you think both air and water are needed to make the water rocket work?

EXPLORATION 7

Particles on the Move

Your model of the nature of matter is becoming more and more useful because it can explain more observations. Now you will make a few more observations of the behavior of matter. In each instance, explain your observations in terms of what the particles in the solid, liquid, or gas are doing.

You Will Need

- food coloring
- an eyedropper
- ice water
- hot water
- a balloon
- an ice chest with ice
- alcohol
- 2 microscope slides
- matches
- a beaker
- cotton batting
- a metal lid
- perfume
- a candle
- a watch or clock

Station 1

Place a drop of food coloring into very cold water and another drop into hot water. Explain what happens.

Station 2

Place a blown-up balloon into the ice chest. After half an hour, take it out and let it warm up in the room. Explain what happens.

Station 3  

Heat a microscope slide with a match. Then, after extinguishing the flame, place one drop of alcohol on the heated slide and one on an unheated slide. Using the particle model, explain the differences you observe.

Station 4

Pour ice water into a beaker. Now breathe on the beaker. What do you observe? Explain this observation in terms of what you think the water molecules in your breath are doing.

Station 5

Place a piece of cotton batting on a metal lid. Add a few drops of perfume to the cotton. How far away can you stand and still smell the perfume? What do you think the liquid particles that make up the perfume are doing?

Station 6 

Observe a burning candle. What is formed at the top of the candle (not the top of the flame)? What happens when the candle is blown out? Explain these observations in terms of what the particles of wax are doing.

Analysis, Please!

1. Now that you have finished Exploration 7, add at least one more statement to the three given on page 414.
2. Here are six words that help describe the processes you observed in Stations 1–6: *condensation, expansion, diffusion, evaporation, melting,* and *solidification.* Which word(s) would you associate with each station?

Expanding the Model

Mr. Chin's class expanded the particle model of matter, as described on page 414, by adding more ideas. If you agree with the statements they added, suggest at least one observation that supports each statement.

More Ideas

1. Particles in gases are far apart.
2. Particles making up liquids and solids must be as close together as possible.
3. Particles move.
4. Particles in a hot substance move faster than particles in a cold substance.
5. The faster gas particles move, the more pressure they exert on the sides of a balloon.
6. Liquid particles can become gas particles, and gas particles can become liquid particles.

Consolidating Your Knowledge

Here's a change of pace. The following four activities will help you consolidate what you have learned so far about the nature of matter. Work on your own or with fellow students.

Task 1

In your Journal, make a table like the one shown here. The first column lists some words that describe the ways particles may move. Which state of matter—solid, liquid, or gas—is most likely to exhibit each kind of movement? Suggest an everyday event that is similar to the way particles move. One has been done for you.

Word	State of Matter	Your Analogy
Wriggling	Solid	Like students wriggling while sitting in their seats.
Vibrating	?	?
Tumbling	?	?
Bouncing	?	?
Flying	?	?
Shaking	?	?
Whirling	?	?
Sliding	?	?

Task 2

The following pictures were drawn in a children's science book to illustrate the behavior of particles in solids, liquids, and gases. Write a sentence or two explaining to a 5th-grader what is happening in each picture.

Task 3

Here are some students' descriptions of solids, liquids, and gases. To which particular behavior of particles is each student referring?

Ésteban's description of a solid: "Solid—made up of the stay-at-home type of particles."

Corinne's description of a liquid: "Liquid—consists of particles slip-sliding away."

Brad's description of a gas: "Gas—particles with claustrophobia."

Now it's your turn to create an unusual definition. Share it with a friend, and see whether he or she can discover which term you are describing.

Task 4: Revisiting Kevin's Model of Air

In Exploration 1, you used ball bearings to make Kevin's model of air. You may wish to make this model again. For each observation Kevin made, explain how it describes the behavior of air and therefore supports the particle model of matter.

1. When the balloon was shaken gently, the ball bearings rattled around. I could feel and see them hitting the sides of the balloon.
2. Shaking the balloon harder caused more frequent and harder collisions.
3. When I doubled the number of ball bearings in the balloon, the number of collisions with the sides of the balloon increased.
4. When I stopped shaking the balloon, the ball bearings formed a pattern like this:

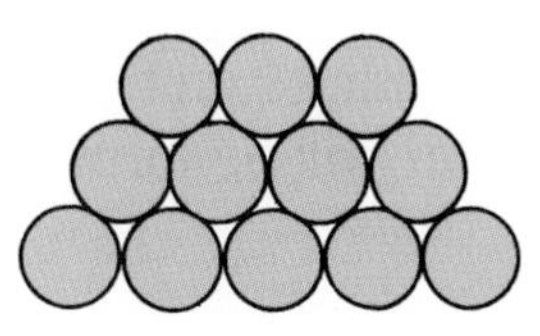

Challenge Your Thinking

1. Potassium permanganate was placed on the bottom of a graduated cylinder, and water was added carefully so that mixing did not occur. A record of what happened is shown by the illustrations. What inferences can be made about particles from this experiment? Write down two inferences.

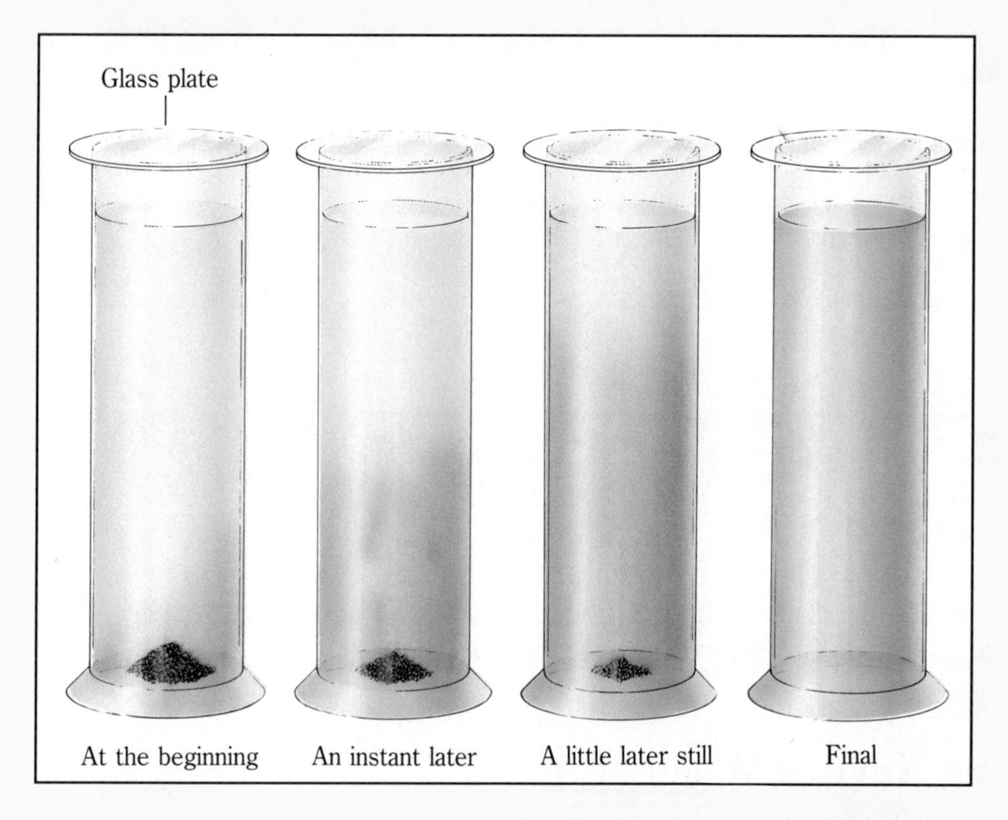

2. Fill in the blanks in the following paragraph so that it makes sense. The words you'll need are *molecules, atoms, compounds,* and *elements.*

   Gasoline and candle wax are two different ___(a)___. However, both are made up of the same ___(b)___, hydrogen and carbon. The reason gasoline and candle wax do not appear the same is that their molecules are made up of different numbers of hydrogen ___(c)___ and carbon ___(d)___. When both of these ___(e)___ burn, they produce ___(f)___ of water and carbon dioxide.

3. Brendan realized that in a given amount of water, the mass of oxygen atoms is always 8 times the mass of the hydrogen atoms.
   (a) Suppose his sample of water contains 4 g of hydrogen. How many grams of oxygen are in his sample?
   (b) Suppose he reacts hydrogen with 24 g of oxygen to form water. How many grams of water will be formed?

Temperature and Particles

You know that matter is made up of molecules that are in constant motion. The particle model of matter suggests that when heat is added to matter (solid, liquid, or gas), the molecules move faster and faster and farther and farther apart. Similarly, as matter cools, the molecules slow down and move closer together. The particle model suggests that this is true for all states of matter—solids, liquids, and gases.

EXPLORATION 8

Slowing Down and Speeding Up Particles

Sedrick's Experiment

Examine the notes jotted down by Sedrick and Leilani on the following two experiments, which test these inferences. You may want to verify that the results are correct. Then do the following:

1. For each experiment, devise a good title that's in the form of a question.
2. Draw an appropriate conclusion for each experiment.

Sedrick's Experiment

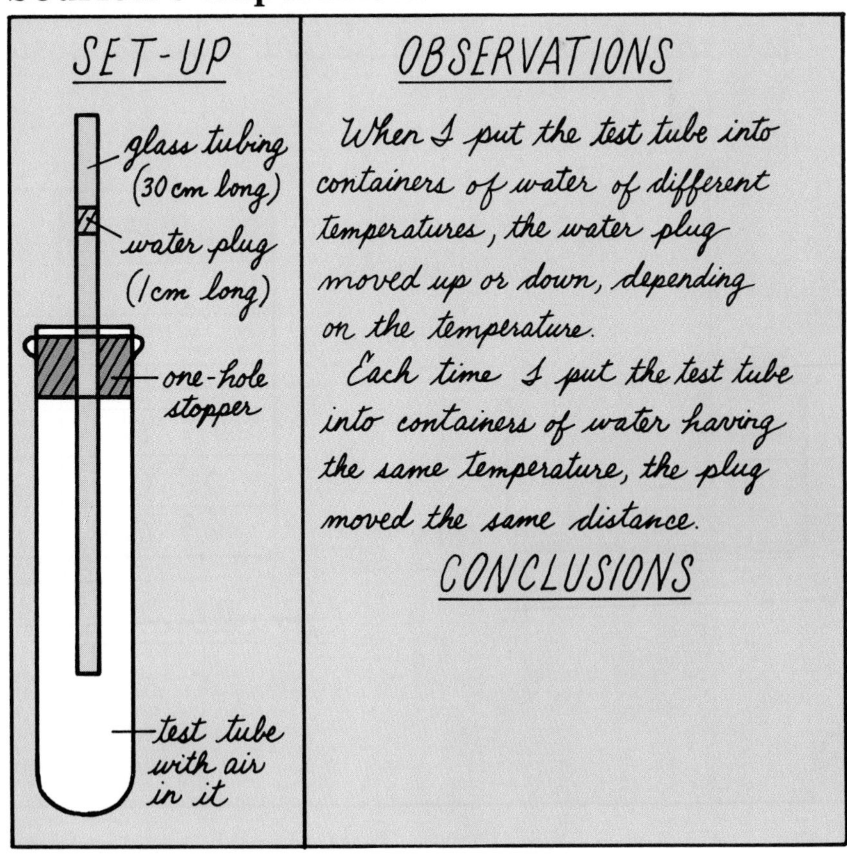

Leilani's Experiment

Set-Up

In this experiment, I filled two test tubes with two different liquids, glycerine and water.

Here are my two test tubes:

glass tubing (30 cm long)
one-hole stopper
water
glycerine

Procedure

When I put the stoppers in, I had to make sure that the liquids rose in the glass tubing about one-third of the way up from the top of the stopper to the top of the glass tubing. I also had to be careful that there was no air left in the test tubes. Then I put both test tubes in a beaker of ice and water. The water level had to almost cover the test tubes. After the liquids in the glass tubing stopped moving, I marked the level on each piece of glass tubing as "0 cm"—my baseline. I then put my test tubes into four more water baths of different temperatures. I measured how far the liquid moved up the glass tube each time.

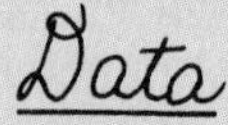

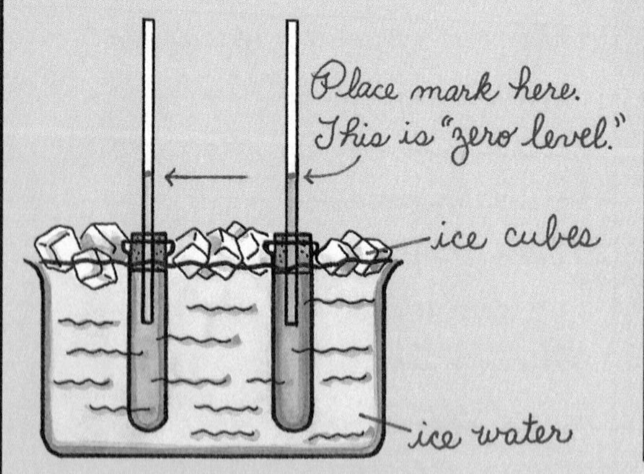

Temperature	Height of Liquid at Different Temperatures	
	Glycerine	Water
5 °C	0 cm	0 cm
24 °C	0.9 cm	0.4 cm
45 °C	3.8 cm	2.6 cm
50 °C	4.2 cm	3.1 cm
63 °C	5.5 cm	4.1 cm

Conclusions

Aisha's Secret

Use the particle model to explain why Aisha's technique works.

Egg Science

Why does an egg sometimes break when you boil it? To find an answer, try this activity at home.

Start with a pot of cold water, and heat an egg. Watch the large end. What do you observe? How does your observation explain why eggs sometimes crack?

Using the same egg (as long as it's not cracked), pierce the large end with a thumbtack. Start over again with cold water and heat the egg. Again, what do you observe?

Why do some eggs crack when boiled, while others don't? Can you suggest why it is not a good idea to put a raw egg in boiling water, even if you have pierced the large end?

Ask someone in your family how he or she boils eggs. Perhaps you can give a lesson on the science of cooking an egg!

8 Changes of State

"Does heating a material always cause a temperature change within the material?"

"Of course," said Derrick. "When you heat water, it gets hotter and the pot it's in gets hotter too." "And," added Wakenda, "because the particles making up the water and the pot are vibrating at a faster rate than they were before heating, the particles take up more space and the materials expand."

That explanation makes good sense. Now examine what happens when heat is *removed* from a substance. Believe it or not, this will help you answer the question above.

EXPLORATION 9

The Temperature Connection

In this Exploration, you will start with paradichlorobenzene at 60°–65°C and observe how its temperature changes as it cools for half an hour.

You Will Need

- paradichlorobenzene
- a beaker of hot water (60°–65°C)
- a hot plate
- a thermometer
- a test tube and stopper
- a test-tube rack
- a watch with second hand

Caution: *Handle paradichlorobenzene with care. Do not breathe the fumes!*

To melt paradichlorobenzene:

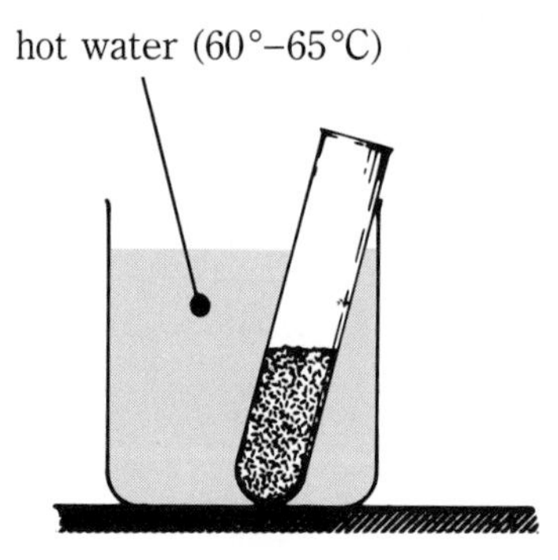

To cool paradichlorobenzene:

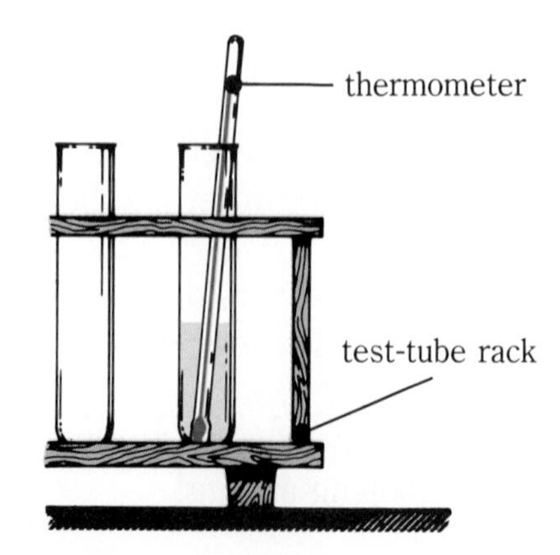

What to Do   

1. Half-fill a test tube with paradichlorobenzene and place it in 60°–65°C water.
2. When the paradichlorobenzene has melted, place the test tube in the test-tube rack and measure the temperature of the paradichlorobenzene.
3. Record the temperature of the molten liquid every half minute. Stir the liquid with a thermometer, taking care to keep the thermometer away from the sides of the test tube (to prevent breaking the glass). You can record your data in a table like this one.

Time (minutes)	Temperature (°C)
0	?
0.5	?
1.0	?
1.5	?
2.0	?
2.5	?
3.0	?
(etc.)	

4. Keep a record of the time and temperature until *all* the liquid has returned to the solid state. **Caution:** *Do not try to pull the thermometer out of the solid, as this could break it.*
5. If your thermometer is "frozen" in the solid, melt the solid again, as in step 1. Clean the thermometer if necessary, and stopper the test tube of paradichlorobenzene for future use.

Analysis, Please!

1. In your Journal, prepare a graph of your results. Place *time* on the horizontal axis and *temperature* on the vertical axis.

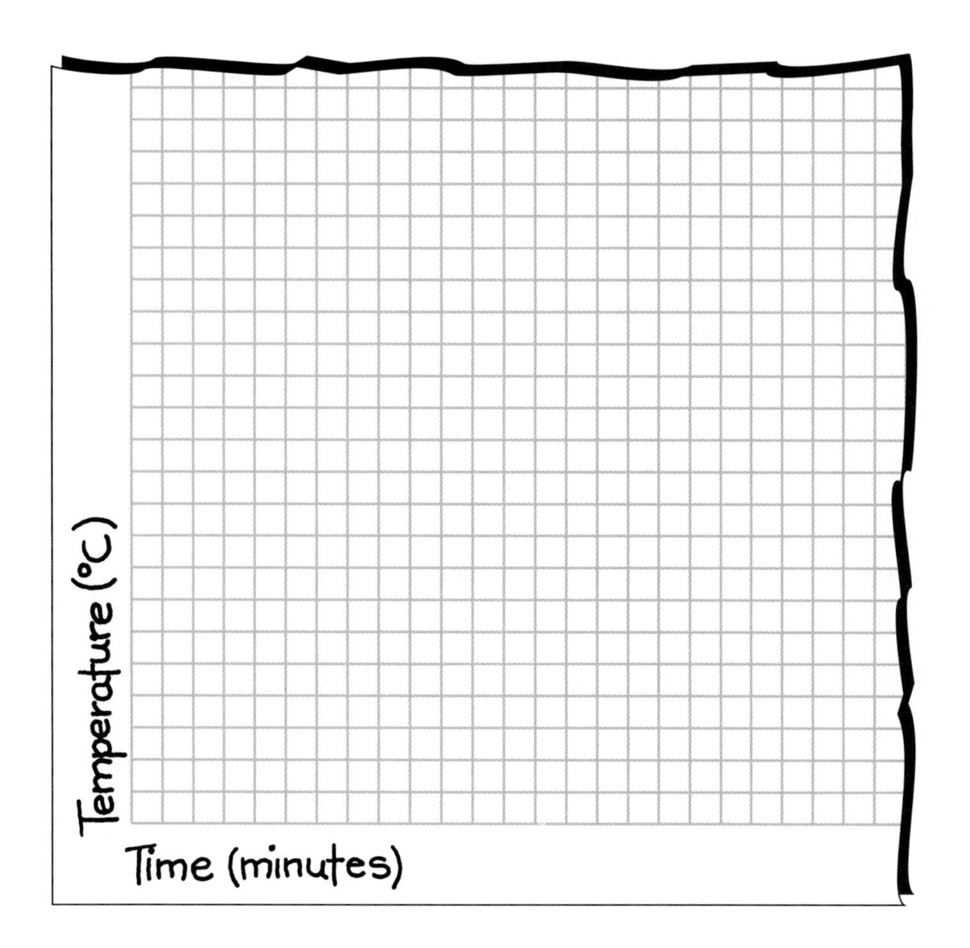

2. On the graph, locate and clearly label the points where the paradichlorobenzene is
   (a) liquid only.
   (b) solid only.
   (c) a mixture of solid and liquid.
3. How does your graph resemble those of other students in the class? How does it differ?
4. What is (a) the melting point and (b) the freezing point of paradichlorobenzene? Compare your answer with those of your classmates.
5. The graph should reveal that, at one point, the temperature does not drop as you might expect. Explain this observation.
6. Return to the question that began this lesson: Does adding heat to a material always cause a temperature change within the material? Have you changed your answer?

Derrick's Log Book

I was surprised by the results of the experiment, although I shouldn't have been. After all, I already knew that you can't cool liquid water below 0°C using only ice, and I knew that you can boil water all day without its temperature going over 100°C.

Here is what I learned by doing this experiment. I drew two graphs to illustrate. The first graph shows what I now predict will happen as paradichlorobenzene is heated. The second is what actually happened as the paradichlorobenzene cooled.

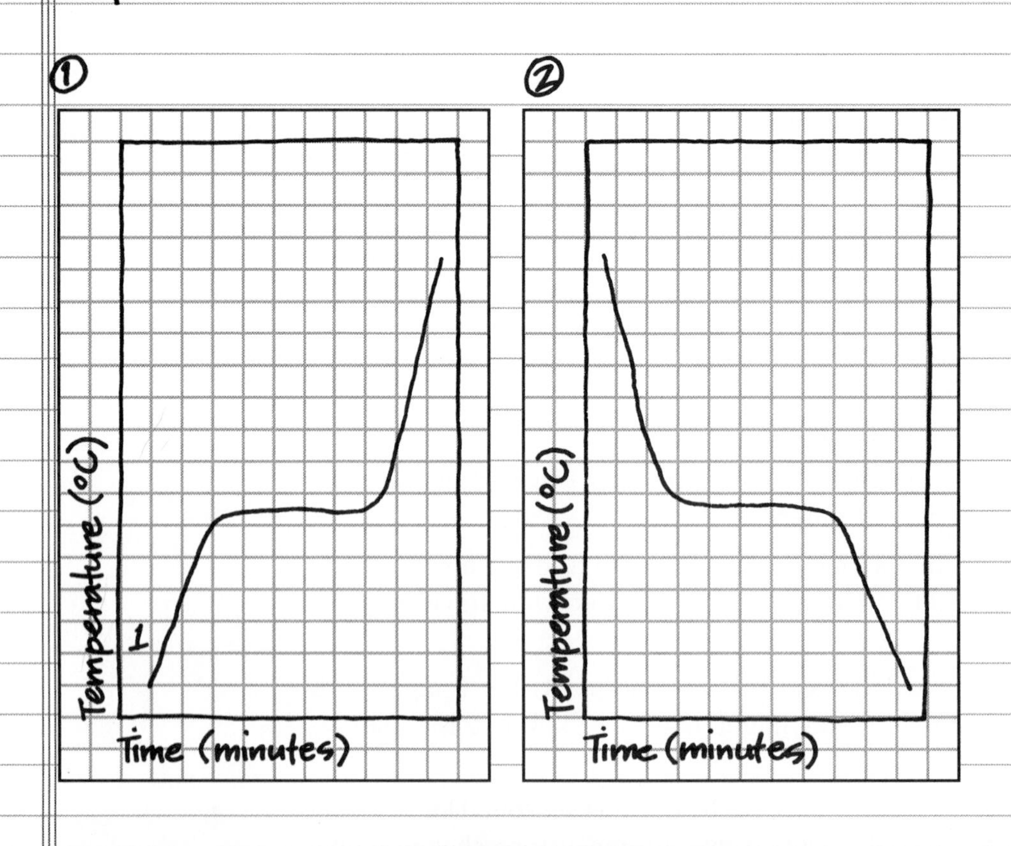

Derrick intends to label the parts of his graphs. The labels are listed below, but they are out of order. First copy his graphs (on a larger scale) into your Journal. Then, with a classmate, decide where each label belongs. Write the number of each label in the appropriate place on your graphs. Derrick has already added the first label.

1. Heat energy is added to solid paradichlorobenzene.
2. Paradichlorobenzene starts to form a solid.
3. The particles of solid paradichlorobenzene are moving more slowly.
4. The particles of solid paradichlorobenzene return to room temperature.
5. The particles of solid paradichlorobenzene are vibrating at a greater rate.
6. The temperature of liquid paradichlorobenzene goes up.
7. Melting temperature—no change in temperature
8. The temperature remains constant for a long time.
9. The paradichlorobenzene melts.
10. The particles of paradichlorobenzene release heat energy.
11. The temperature of liquid paradichlorobenzene goes down.
12. The freezing point of paradichlorobenzene

Follow Up

Ice melts at 0 °C. Paradichlorobenzene melts at 54 °C. In which substance is the force of attraction between molecules greater? How do you know?

Frozen water

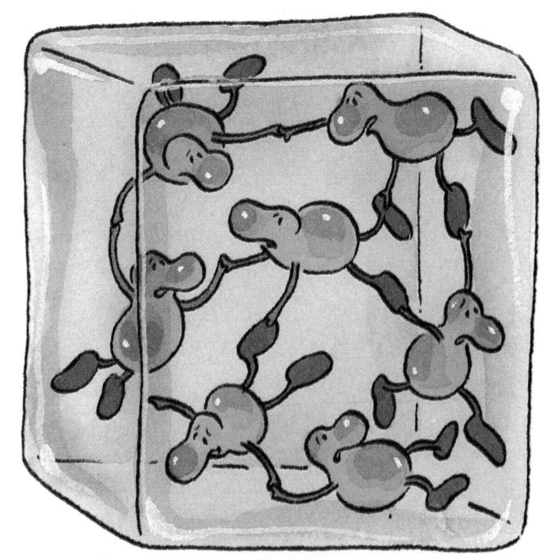

Frozen paradichlorobenzene

Absorbing and Releasing Heat

When enough heat energy is added to solids such as candle wax, ice, or paradichlorobenzene, the solid will increase in temperature and then start to melt. But during the melting process, the temperature of the solid-liquid mixture no longer increases. This phenomenon causes the plateau on your graph. For ice, the plateau is at 0 °C. For paradichlorobenzene, the plateau is at 54 °C. These are the melting points of ice and paradichlorobenzene. During melting, heat is *absorbed,* or taken *in,* by the substance being heated. *Endo* means "in." Melting is therefore an example of an **endothermic** change.

When a liquid solidifies, heat is released. During the solidification process, the temperature of the liquid-solid mixture remains constant. This time period of constant temperature corresponds to the plateau on your graph; it is the freezing point for the substance. For water, the plateau occurs at 0 °C, but for paradichlorobenzene the plateau occurs at 54 °C. Because heat energy is released, or given off, freezing is an **exothermic** change. *Exo* means "outside."

From solid to liquid. As the ice melts, heat energy is being *absorbed* by the ice. Melting is an *endothermic* change.

From liquid to solid. As the water freezes, heat energy is being *released* from the water, thus forming ice. Freezing is an *exothermic* change.

EXPLORATION 10

Salol and Changes of State

This Exploration involves changes of state, this time for a substance called *salol* (phenyl salicylate). As you follow the procedure, identify the points when:

- melting is occurring.
- freezing is occurring.
- energy is being absorbed.
- energy is being released.
- an exothermic change occurs.
- an endothermic change occurs.

You Will Need

- a glass microscope slide
- a match
- salol
- a magnifying glass or low-power microscope

What to Do

1. Add a small amount of salol to a glass slide. **Caution:** *Salol is poisonous.*

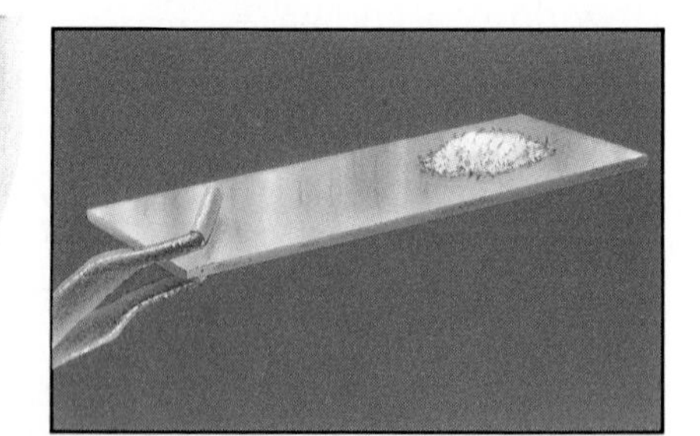

2. Heat the slide with a match until the salol just melts.

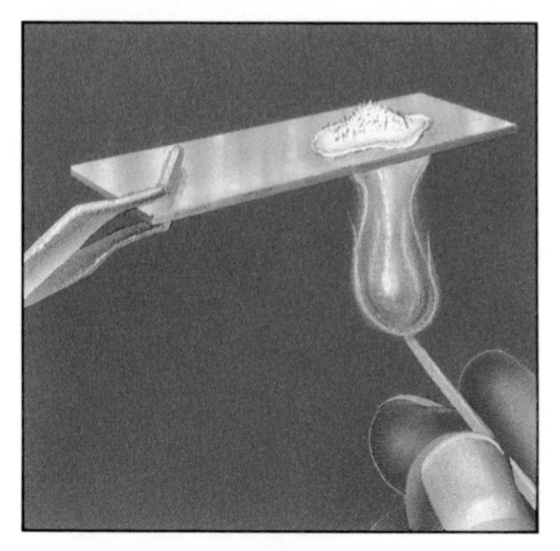

3. Extinguish the match, let the slide cool, and add a small crystal of salol to the slide. This is called a "seed crystal."

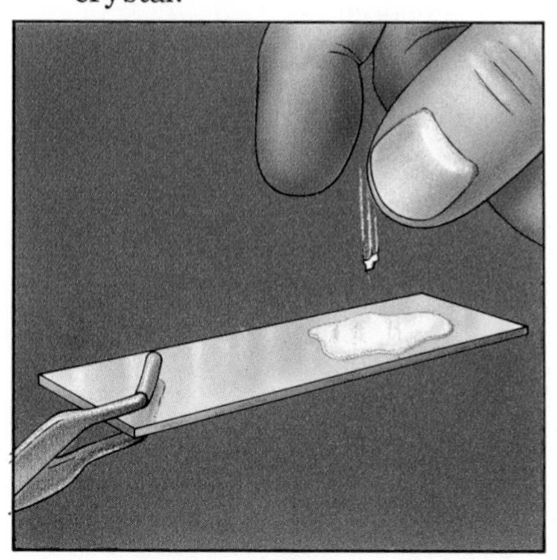

4. Using a magnifying glass or low-power microscope, observe what happens.

Panting Dogs and Other Phenomena

Panting dogs, a refrigerator, and even the Earth itself have something in common. They all use the same process to exchange heat. In other words, they all cool down by a similar process. Read on to find out how.

You've probably experienced the coolness that results when you dip your finger in alcohol, when you sweat, or when you lick your finger to test the wind direction. Why does this occur? What's happening involves a process called *evaporative cooling.* In each example above, a change of state occurs: a liquid evaporates to form a gas. When a liquid evaporates or a solid melts, heat energy is absorbed from the surroundings. As sweat on your skin absorbs nearby available heat (including heat from your body), the sweat changes into water vapor. Since heat has been removed from the skin to make this happen, you feel cooler. How does this process work for a panting dog?

A wet-bulb thermometer works by evaporative cooling too. (You made a wet-bulb thermometer in Unit 3 by covering the end of a thermometer with a sleeve cut from a piece of hollow shoelace.) When wet, a wet-bulb thermometer shows a lower temperature than a dry-bulb thermometer. And if you dip a wet-bulb thermometer into alcohol, you will obtain an even lower temperature. Using what you've learned about evaporative cooling, explain how this happens.

Here's an even greater challenge: why would a wet-bulb thermometer read much lower on some days than on others?

Endothermic or Exothermic?

Evaporation and melting are both endothermic changes because they involve putting heat *into* a liquid or a solid. The opposite is true when a gas condenses to form a liquid, or when a liquid solidifies. That's because heat is *released* in both of these cases. Condensation and solidification are therefore exothermic changes.

Explanations, Please!

1. (a) The Refrigerator

   Did you know that changes of state are involved in maintaining a colder temperature inside your refrigerator than outside? Examine the diagram of a refrigerator. Discuss with a classmate why it is colder inside the refrigerator than it is outside.

   (b) Discuss this with a friend: You can warm a room by turning on an oven and opening the oven door. Can you cool a room by opening the door of a refrigerator? (Notice that the back of your refrigerator has pipes containing a coolant. The coolant circulates in these pipes.)

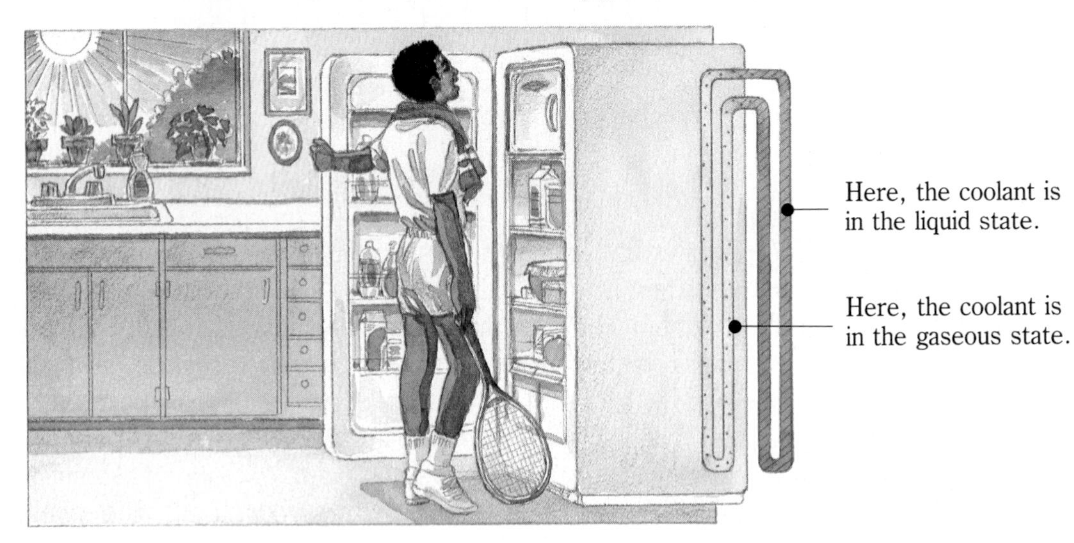

2. Earth's Air Conditioner

   The water cycle is described in diagram form below. Use the diagram to explain how heat is transferred over large distances by the processes of evaporation and condensation. Use the words *endothermic* and *exothermic* in your explanation.

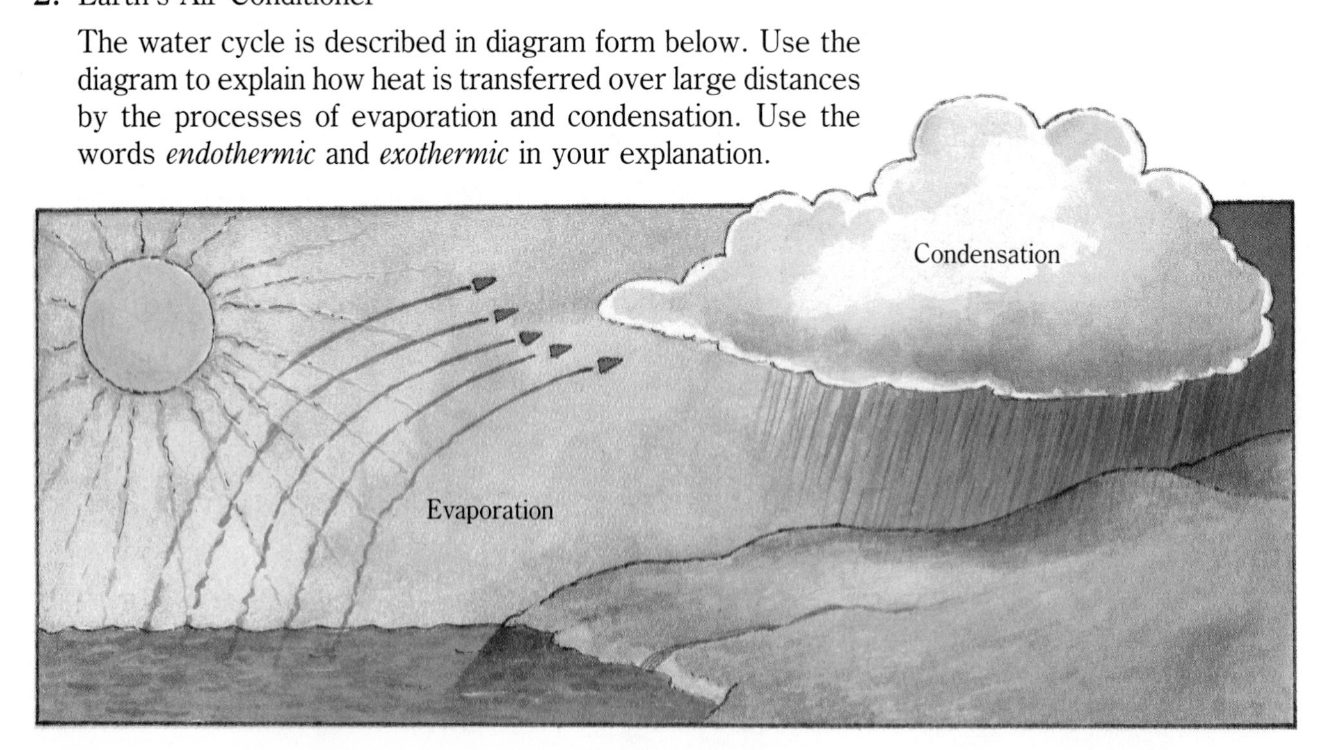

Challenge Your Thinking

1. Can the particle model of matter explain the phenomena below? If so, how?
   (a) When a sunbeam shines into a room, you can see dust particles dancing about in the path of light.
   (b) When making pancakes, you can test whether the pan is hot enough by dropping a bit of water onto it. If the drop bounces around and quickly disappears, then the pan is hot enough.
   (c) If you could trap a little smoke in a transparent box and put it under a microscope, you would see the particles of smoke moving in a haphazard pattern, darting in one direction and then in another.
2. Describe a phenomenon involving matter. Be sure to choose one that hasn't been discussed before. It could involve air, water, salt, or *any* form of matter. Then use the particle model of matter to help explain why the phenomenon occurs as it does.
3. Imagine that you are a particle of matter. Describe your experiences during one of the following happenings: melting, evaporating, diffusing, dissolving, or freezing. Remember: you belong to an entire community of particles, which may be quite different from you.
4. Three solids were heated, and their temperatures were plotted against the heating time.

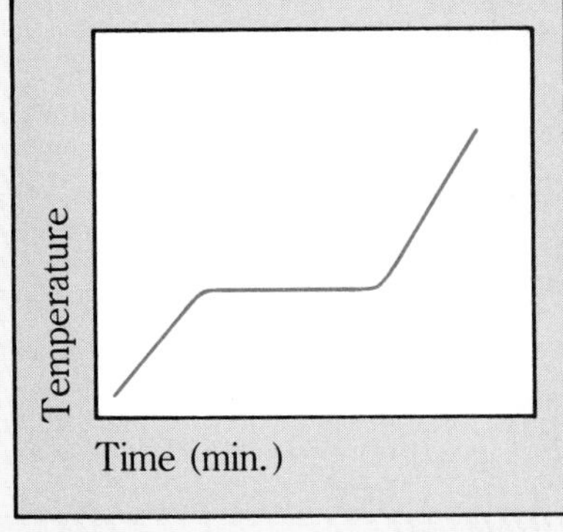

Solid 1

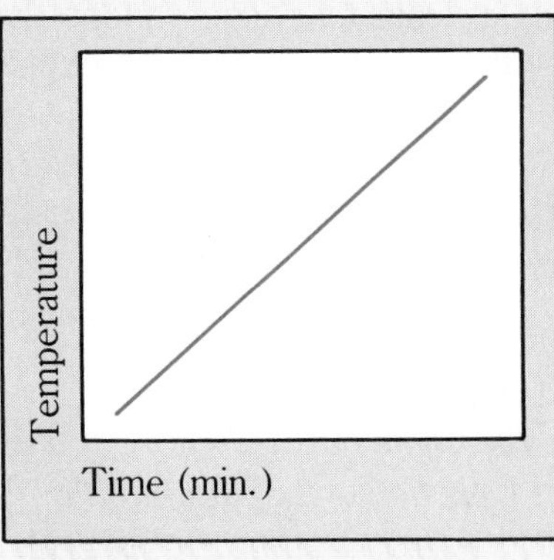

Solid 2

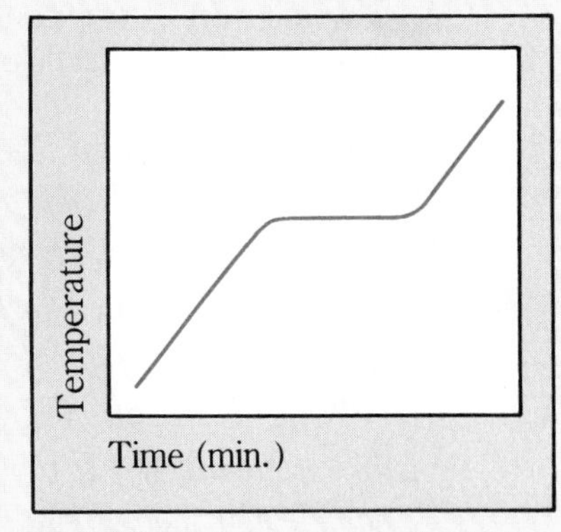

Solid 3

   (a) Which substance(s) melted when heated? How do you know?
   (b) Which substance has the highest melting point?
   (c) Which substance has the strongest forces of attraction between its particles?

Unit 7 Making Connections

The Big Ideas

In your Journal, write a summary of this unit, using the following questions as a guide.

- How do observations differ from inferences?
- What are models, and why are they useful?
- What evidence is there that matter is made up of particles?
- What everyday observations can be explained by the particle model of matter?
- Who was John Dalton, and what did he conclude about the makeup of matter?
- How does the particle model explain the properties of solids, liquids, and gases?
- What are changes of state, and how can they be explained by the particle model of matter?
- What is the effect of temperature changes on the particles making up matter?
- What are some examples of endothermic changes? exothermic changes?

Checking Your Understanding

1. Why do we feel hotter on hot, humid days than hot, dry days? Use the particle model of matter to explain your reasoning.

2. The following story contains at least five observations that can be explained with the particle model. List them and give explanations for each.

   "One more dive and then we gotta go. We can't be late for dinner again. Mom'll get mad." Ben and Josh each dove off the cliff and neatly split the water. "It shouldn't take too long to dry in this sun," said Ben.

   "Uh-oh," groaned Josh. "We're in trouble now. My front tire's flat. Guess I should've filled it before we left."

   With the tire pumped up, the brothers raced home to make up for lost time. The breeze felt cool on their half-wet skin and hair. When they got home, their mother said, "Put your wet things in the dryer and come eat. I want one of you to mow the grass before it gets dark, while it's still dry. There'll be too much dew on the grass to mow in the morning." "No prob, Mom," said Josh, "Ben'll do it. Dinner smells great. What're we having?"

3. Air fresheners are often placed in different areas of a home, such as kitchens, bathrooms, and basements. Over a period of weeks, the fragrant part of the air freshener simply disappears. What happens to it? Use the particle model to explain.

4. Gas particles are speedy. Under winter conditions, molecules zoom about at 1300 km/h. Under summer conditions, their speed is 1400 km/h—as fast as a bullet! Why does the speed of molecules in air differ from winter to summer?

5. An atom of silver in a spoon at room temperature vibrates about 4.6×10^{12} times per second! How can you decrease the number of vibrations per second?

6. Nathan's class decided that if they could see air in a flask, it might look like this:

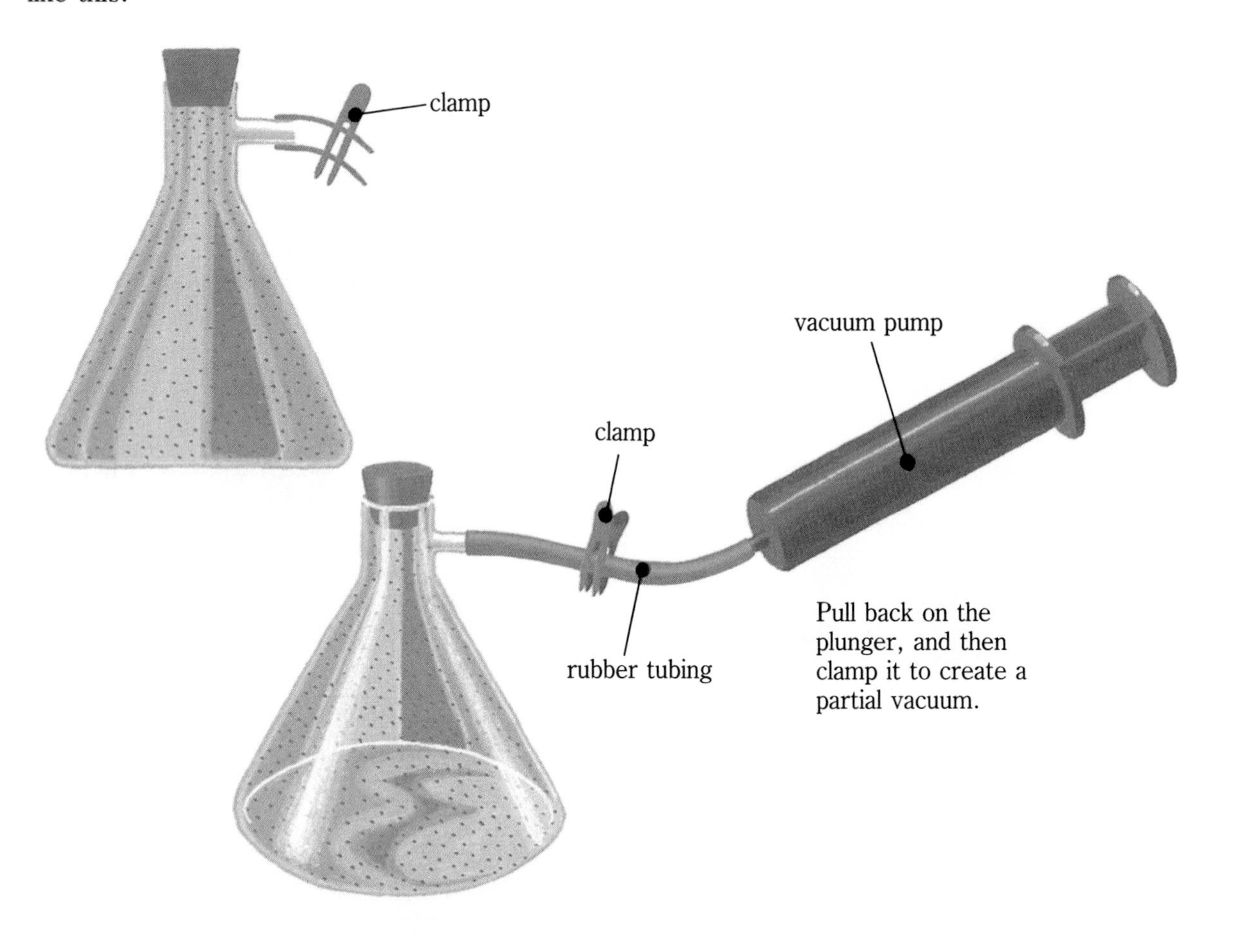

After some discussion of what would happen if half of the air were removed by using a vacuum pump, three suggestions were made.

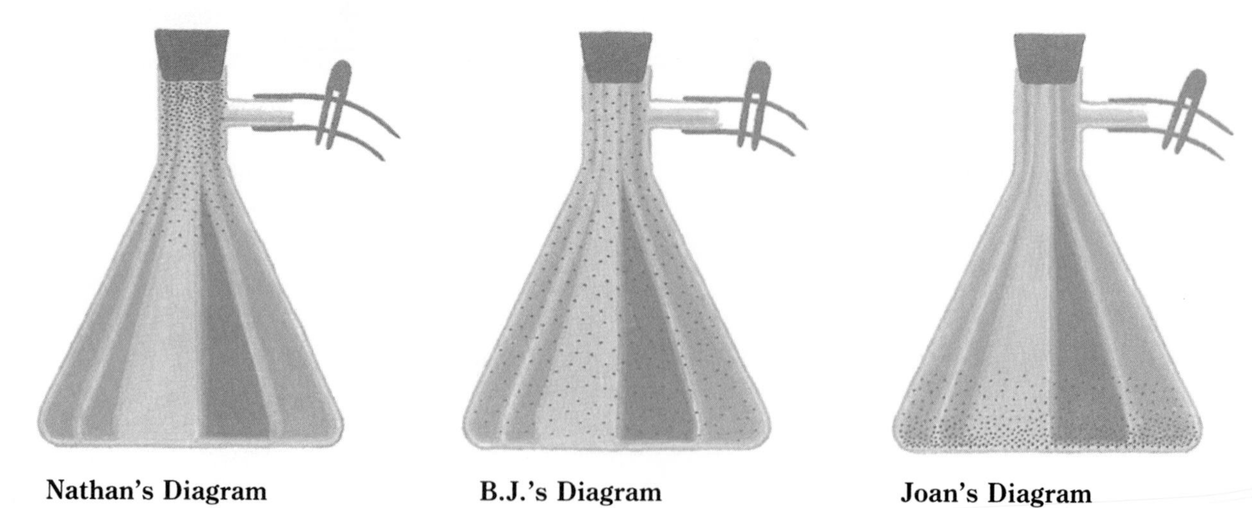

Nathan's Diagram **B.J.'s Diagram** **Joan's Diagram**

Which one do you think is the best model of air? Why?

7. When two different metals are bonded or welded together, the result is called a "bimetallic strip." When heated, a bimetallic strip bends. When it cools, it assumes its original shape. Why does it do this?

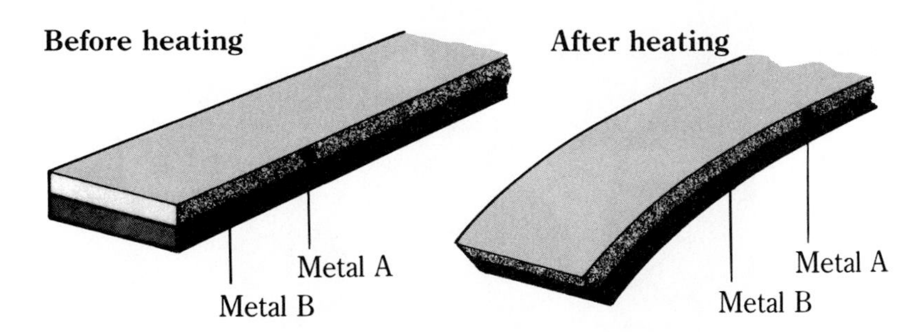

8. A balloon initially had a mass of 6.2 g and a volume of 2.3 L. What might have been done to the balloon to bring about the following changes?

	Mass	Volume
(a)	6.2 g	3.2 L
(b)	6.5 g	2.9 L
(c)	6.0 g	2.5 L

9. Scientists now theorize the following about the nature of particles:
   - All matter is made up of small particles.
   - These particles move in all directions.
   - Temperature affects the speed at which these particles move.

   Use any or all of these ideas to explain the occurrences below:

   (a) A soccer ball is pumped up in the morning. By evening, the ball does not feel as hard as it did earlier in the day.

   (b) A balloon was filled with hydrogen gas. An hour later, it was noticeably smaller, even though the temperature had not changed.

Reading *Plus*

Now that you have been introduced to particles, you can discover how scientists use them in measurement and theory. By reading pages S107–S124 in the *SourceBook,* you will learn about how particles help explain the kinetic theory of matter and about particles that are smaller than the atom.

Updating Your Journal

Reread the paragraphs you wrote for this unit's "For Your Journal" questions. Then rewrite them to reflect what you have learned in studying this unit.

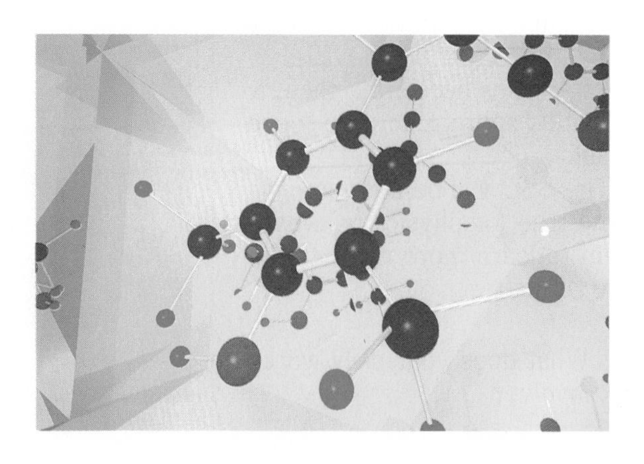

Science in Action

Spotlight on the Super Collider

Margaret Lutfey works on the ion source at the SSC.

When completed, the Superconducting Super Collider, or SSC, will be the largest scientific instrument in the world. The construction costs are expected to be about $8.2 billion, and yearly operating expenses will be around $270 million! Scientists look forward to using the SSC to help them learn more about the nature of matter and perhaps even to add insight into the origin of the universe.

Margaret Lutfey is a 24-year-old electrical engineer working on the SSC project. A graduate from the State University of New York at Stony Brook, Margaret has a bachelor's degree in Electrical Engineering.

Q: Exactly what is the SSC, and why is it important?

Margaret: When it is finished, the SSC will be the largest particle accelerator in the world. A particle accelerator is a device that increases the velocity of charged atomic particles called *ions*. This process gives the ions extremely high amounts of energy. The SSC will use superconducting magnets to guide the ions into circular paths while electric fields accelerate them to velocities approaching the speed of light.

The SSC will be a dream-come-true for physicists, allowing them to learn more about the nature of matter.

Q. What does your daily work involve?

Margaret: I work at the starting point of the SSC. The area is called the "ion source." It's where ions are generated in order to create the beam of particles that the SSC uses. My responsibilities include setting up a computer-based data analysis program. This program will be designed to monitor the status of the ion beam. As an operator, I'm also responsible for making sure everything in the lab runs smoothly. For example, if a power supply fails or something goes wrong, it's my job to fix it.

Q. You seem very young to have so much responsibility! How would you explain your success?

Margaret: I guess I'd attribute it to having self-confidence and motivation. I think it's also helpful to be personable and to get to know as many people as you can. I met some great people who helped me determine that I would enjoy research and development, and that the laboratory was the right place for me.

Margaret Lutfey and Ellie Lai work on computer models at the SSC.

The completed Super Collider ring, shown here under construction, will be over 87 km long.

Q: What was your main reason for choosing electrical engineering?

Margaret: The desire to be a problem-solver led me to engineering. My uncle, who was a chemical engineer, had me thinking about this field since I was 10 years old. Even though I really didn't know what an engineer did, I knew it would be challenging. My mother and father, with their strong emphasis on education, were also very encouraging and supportive.

Q: What were your favorite subjects in school?

Margaret: Physics and math were always my favorite subjects because they're so definite. With the principles these subjects provided me, I felt I had a strong foundation for understanding new technology. For me, applying the knowledge I acquired in school to create new technology is fascinating.

In this field you see a lot of theoretical concepts turn into realities. I think that's really neat! What it means is that I see the basic theories I've learned about in physics books, like Einstein's theory of relativity, actually being applied—and they work!

Q: What are the most exciting aspects of your job?

Margaret: I'm surrounded by world-renowned physicists and engineers, many of whom have 20 or more years of experience. It's exciting to be associated with such accomplished individuals.

The SSC itself is exciting technology. The collider ring, in which protons collide, is huge—it's over 87 km long. And right now, I'm recommending the best model for the actual ion source. That model will, in effect, get the whole SSC running. That's very exciting to me!

Q: Is there anything else you would like to add?

Margaret: My work is an important part of my life. At the same time, I think it's necessary to enrich my life with other activities and interests. At lunch I either jog, swim, or play the piano. On the weekends, I do something outdoors like canoeing, biking, or camping. I also participate in church activities. I love to cook, especially Arabic vegetarian food. Probably my favorite thing to do is just hang out with my friends!

A Project Idea

Did you know that you may have one or more particle accelerators in your home? Find out how a television or home computer works. Do some library research or interview an electronics technician. Which parts of a TV or computer are similar to the ion source, the linear accelerator, and the super-collider magnets of the SSC? Prepare a report showing your findings.

Science and Technology

Micromachines

The technology of making things smaller and smaller keeps growing and growing. Powerful computers can now be held in the palm of the hand. But what about motors smaller than grains of pepper? Or gnat-sized robots that can swim through blood vessels to fight disease? Are you ready for those? These are just a couple of possibilities for **micromachines.**

The new technology is well under way. Researchers have already built gears, motors, and other devices so small that you could accidentally inhale one!

The earliest working micromachine had a turning central rotor.

A researcher studies a wafer loaded with 7000 micromotors like the one shown behind him.

Working at the Micro Level

To work in the microscopic world, we need microscopic tools. Imagine, for example, the problems an eye surgeon faces. Today's surgical techniques are, for the most part, too clumsy for working on the tiny delicate structures of the eye. It's like a watchmaker who can't move his hands accurately enough to make a watch, as one eye surgeon explains it.

But micromachines may change all that. One day, eye surgeons may use microscopic power saws that are so small they can fit between the outer covering of the eye and its lens. Such saws could be used to grind away *cataracts,* which are a major cause of blindness.

New Uses for Old Technology

In the computer industry, scientists learned how to carve and etch silicon to create microscopic circuits in computer chips. Researchers are now using some of the same techniques to make micromachines from silicon. As it turns out, silicon is an excellent material for micromachines, since it is three times stronger than steel.

Moving Parts

Micromachines, unlike computer chips, have moving parts. For example, one engineer has devised a motor so small that five of them would fit on the period at the end of this sentence. This micromotor is powered by static electricity instead of electrical current, and it spins at 15,000 revolutions per minute. This is three times as fast as most automobile engines running at top speed.

So far, micromachines have been most useful as sensing devices. Micromechanical sensors can go in places too small for ordinary instruments—like inside blood vessels. Blood-pressure sensors, made from silicon, are shown in the illustration on the right. Each sensor has a patch on it that is so thin it bends when the pressure changes. About 16,000 of them can be produced from a single 10-cm silicon wafer.

Tweezers pick out one tiny sensor from thousands of identical devices.

Going to the Limit

The ultimate in micromachines would be machines that are built up one atom or molecule at a time. Already, scientists are able to manipulate individual atoms and molecules. For example, the tiny figure on the left is made of individual molecules. *Scanning tunneling microscope* technology, which first made it possible to take such pictures of atomic particles, is now being used to move the particles around.

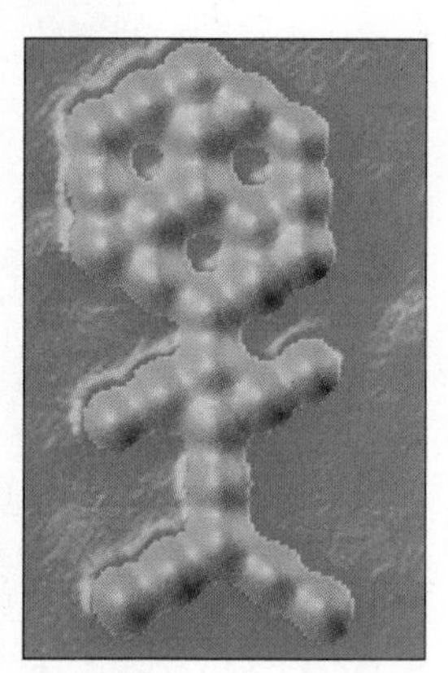

"Molecular Man," drawn with 28 carbon monoxide molecules

One team of scientists has used chemistry to create a molecular "on-off switch." With further development, a switch of this kind might be used for storing information in computers. It consists of a ring of atoms that moves back and forth between two different positions on a string of atoms.

Think About It

Quick—how many ideas can you think of for using micromachines? Perhaps you should write them down—they might even be worth a lot of money someday!

Do some research to find out more about micromachines that have already been built and about ideas people have for making them in the future.

Unit 8

Continuity of Life

For Your Journal

Write a paragraph summarizing your thoughts about each of the following:

1. Why don't we all look alike?
2. What scar do we all have, and how did we get it?
3. What instructions guide the development of new living things, and where are they located?
4. In this field of tulips, what makes the flower color uniform? What makes the individual flower stalks different heights?

A Family Likeness

You're the spitting image of your mom!
Your nose is just like Grandpa's.

You might have heard your relatives say things like this. Such similarities are called family "likenesses."

Look at yourself carefully in a mirror. Notice details such as the color and shape of your eyes and the shape and size of your nose, eyebrows, mouth, ears, and hands. Then look at other family members (your parents, sisters, brothers, cousins, aunts, uncles, and grandparents) and at family photos for resemblances.

How do the people in each family photo resemble one another?

Family Relations

The diagram at the right shows how everyone in Juan's immediate family is related. Use it to answer the questions that follow.

1. How is Juan related to each of the other people in the diagram?
2. How are Alicia and Laura related?
3. What is meant by a *generation* of a family?
4. How many generations are included in the diagram?
5. If Laura and Nathan have children, how will their children be related to Juan and Alicia?
6. From whom might Alicia get her looks? If she "has her father's ears," from whom did he get them?
7. Who has more *ancestors,* Juan or his father? Explain.

Four generations of a family

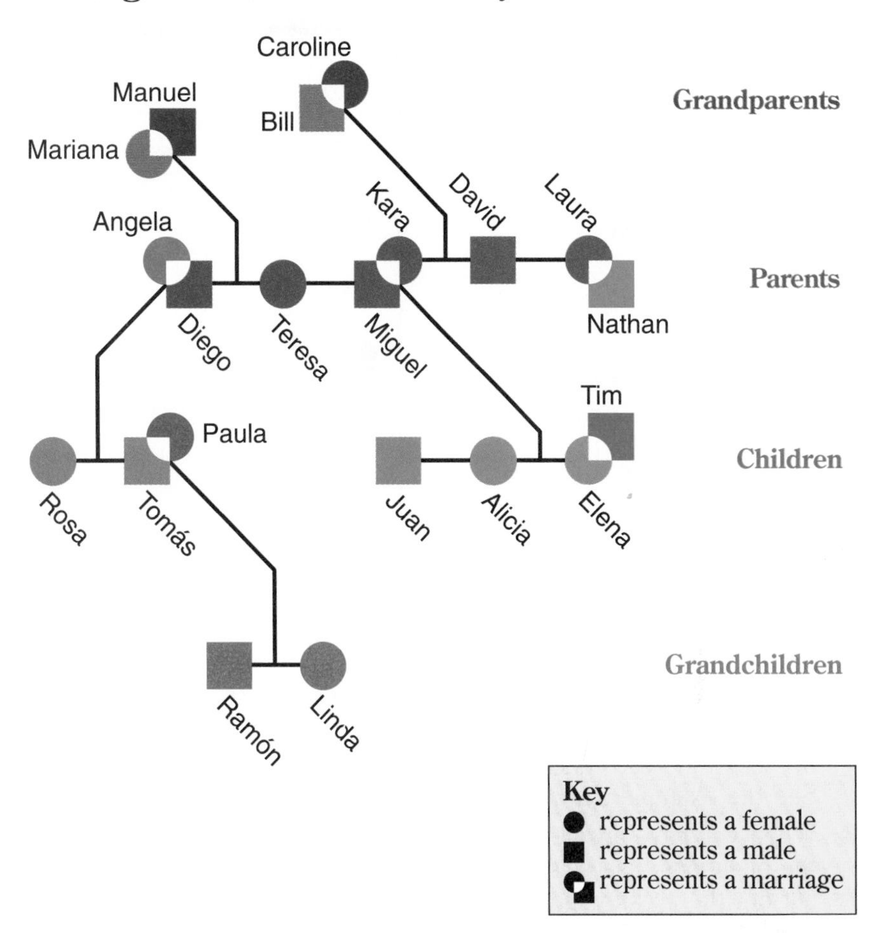

Family-Tree Project

Now is the time to begin a history of a family of your choice. You could choose your own family or one from which you can get the information you need. The family history will feature a diagram like the one drawn for Juan's family. Include as many generations as possible. Some families are fortunate enough to have four or five generations living at the same time. You may also include relatives who are no longer alive. Go back as many generations as you can by asking questions of family members and by looking through family records. The resulting **family tree,** as this type of diagram is often called, could become quite large.

Preparing a family tree can take quite a bit of time. You can add to it as you proceed through this unit. If you'd like, you could collect photographs to include on the family tree. The photos may help you observe family members' characteristics, or **traits,** more carefully than you could by relying on memory alone. The following Exploration will introduce you to some traits—some you know about and some you have probably never heard of.

Four generations of a family

Can You Roll Your Tongue?

You Will Need

- an index card
- a paper punch
- scissors
- a knitting needle

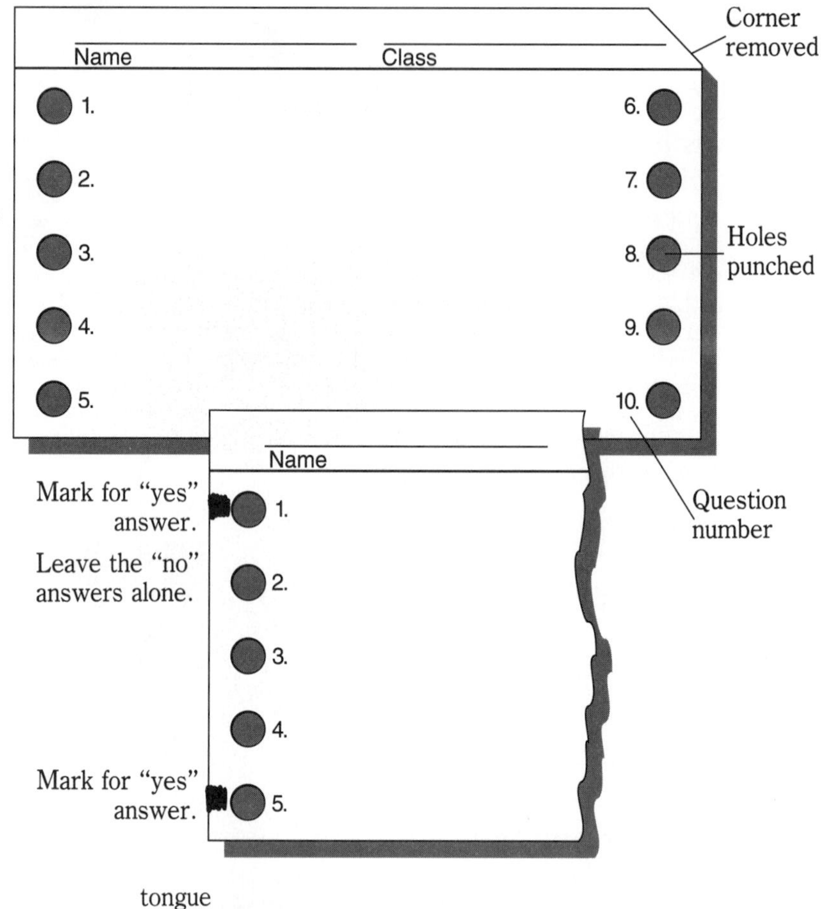

What to Do 

1. Prepare the index card as shown in the diagram.
2. Use the illustrations to help you answer the questions below. For each question, if your answer is *yes,* darken the area by the number, between the hole and the edge of the card. If your answer is *no,* leave the card unchanged. See the example shown.
3. When you have answered all the questions, use scissors to cut out the areas you darkened for step 2 above. This will open the holes for all the *yes* answers, as shown. Pass in your card.

Questions About Your Traits

1. Can you tell the difference between red and green? (Take the test shown.)

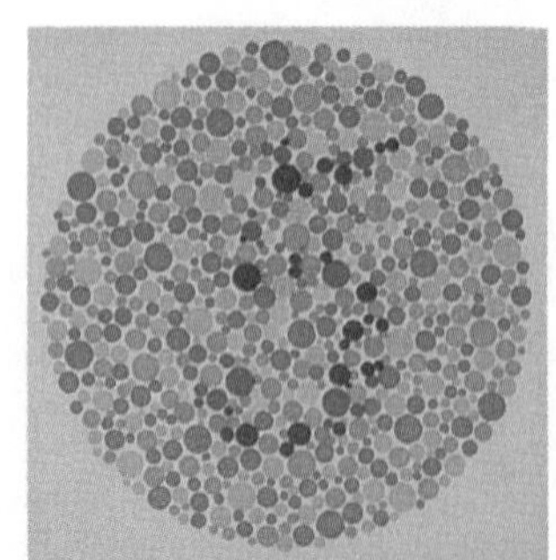

A test for colorblindness. A person with red-green colorblindness will not be able to read the number in the circle.

tongue

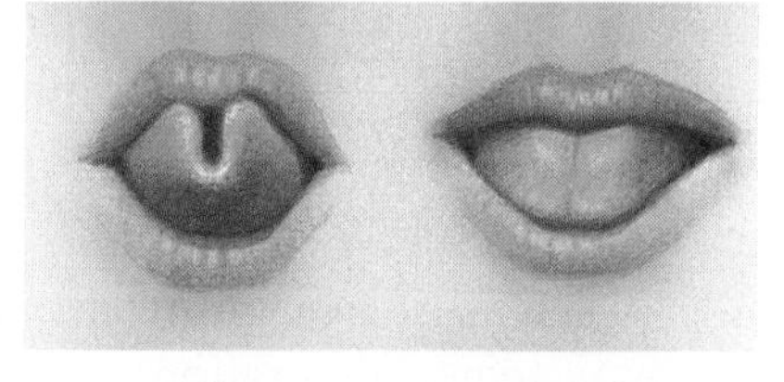

roller non-roller

2. Can you roll up the edges of your tongue?

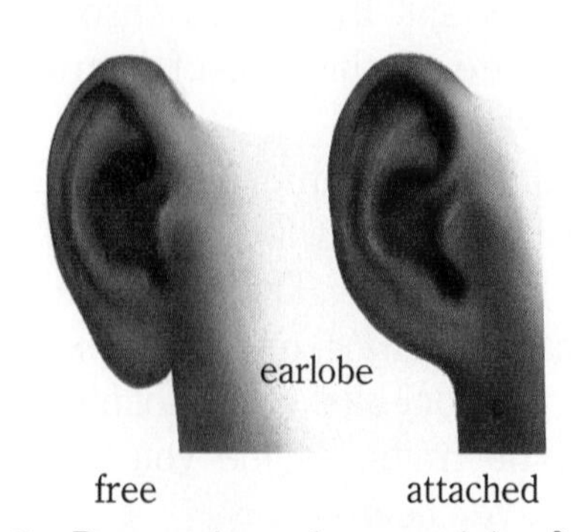

free attached

3. Do you have free ear lobes?

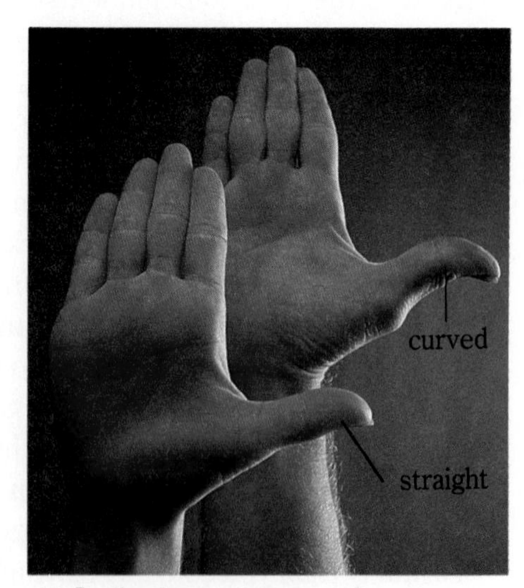

4. Is the last segment of your thumb straight?
5. Do you have dimples?

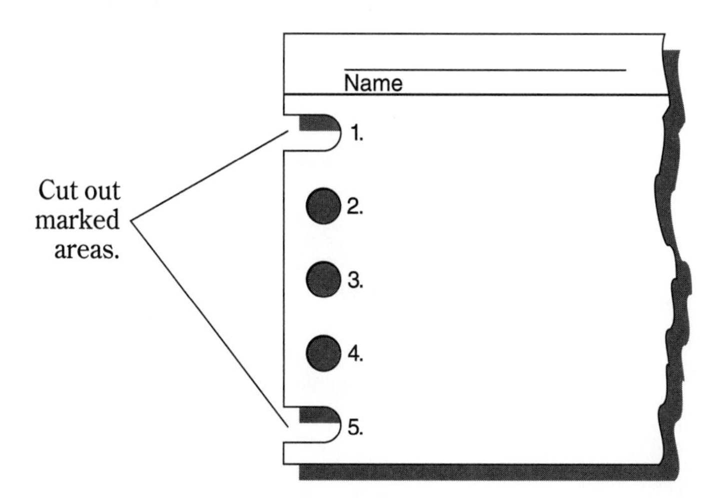

The Results

When all the cards have been collected, stack them together. Insert the knitting needle in each hole in turn and then lift the pile of cards. Gently shake the cards until some fall. Which ones will fall? For example, if you answered *yes* to question 1, will your card stay in or fall out of the group? Count how many cards fall and how many remain. Record the numbers in a chart. Repeat this process for each question.

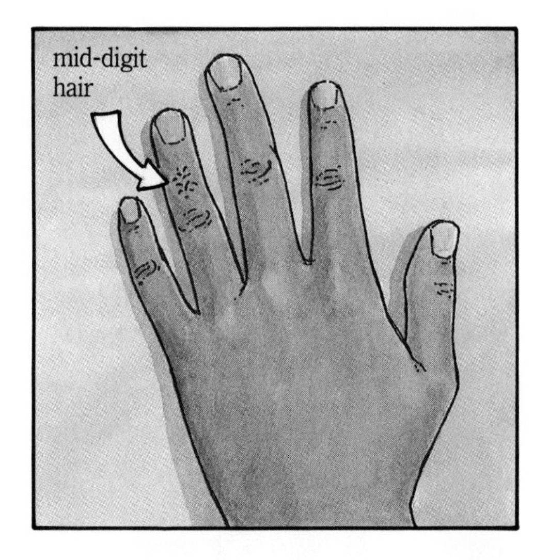

6. Do you have mid-digit hair?
7. When you fold your hands naturally, does your left thumb cross over your right thumb?

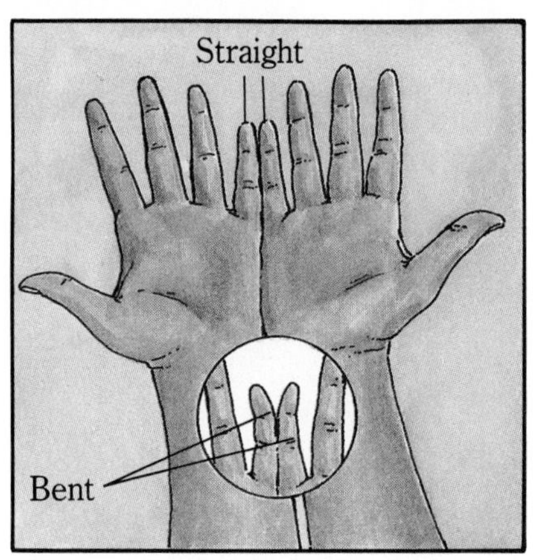

8. Is your fifth finger (pinky) straight?
9. Do you have an indentation in the middle of your chin (cleft chin)?

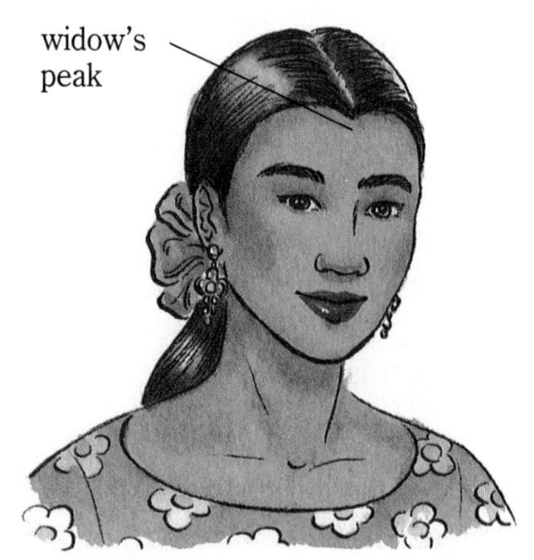

10. Do you have a widow's peak?

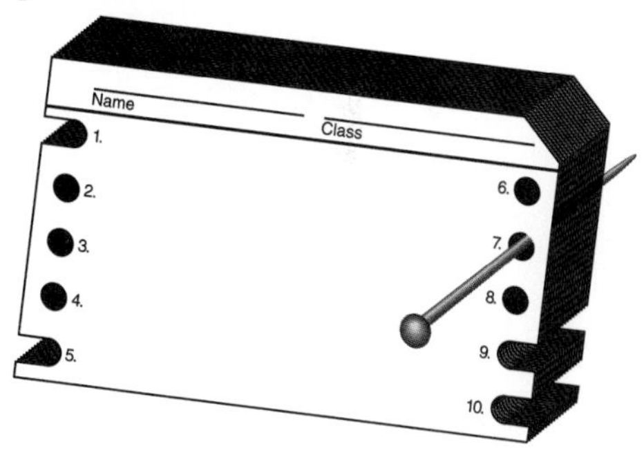

Analyzing the Data

1. Looking at the first four characteristics, how many people in your class answered *yes* for the same two characteristics? for three? for more?
2. Do any two people in your class have the same responses for all of these traits?
3. What do these results tell you about the characteristics of people?
4. What are some characteristics that all people have in common?

What Is Life?

Families share likenesses. So do all humans, all mammals, all vertebrates, and all animals. Some characteristics are shared by *all* living things. If you were asked to describe the characteristics of living things, what things would you include? The list of questions below suggests four essential properties that all living things exhibit. See if you can determine what the four properties are.

1. Why does a rabbit run away from a fox?
2. When you cut your finger, what does your body do?
3. Cats don't live forever. Why are there still so many of them around?
4. Why do we perspire?

Biologists call the four essential properties of life:

- Self-preservation—staying alive; obtaining energy to live and grow
- Self-regulation—the control of life processes, such as respiration
- Self-organization—forming, grouping, and repairing body cells
- Self-reproduction—making new life of the same kind (such that the young of monkeys are monkeys and the babies of whales are whales)

Now try matching the questions above with the four properties.

This unit is involved with the fourth essential property of life—self-reproduction—which deals with many interesting questions such as:

- Why can two relatives resemble each other in some ways but still look very different?
- Can two people look exactly alike?
- Who did you get your looks from? Where did they get their looks?
- If someone loses a toe in an accident, is it likely that his or her children will be born with a missing toe?
- How is life passed on?
- How does a human baby develop before birth?
- Is it possible to predict what a baby will look like before it's born?
- What is a "test-tube" baby?
- Could there ever be another "you"?

Under which property of life would you classify *perspiring?*

Recipes for Life

Throughout history, people have given many different explanations for how living things reproduce. As you read the following statements made centuries ago, suggest why each of the explanations might have seemed reasonable at the time it was offered.

1. "Flies come from rotting meat."
2. "Put one dirty shirt and some grains of corn in an old pot. In 21 days there will be a lively crop of mice."
3. "Some insects are born on dewy leaves; some are born from the hair and flesh of animals."
4. "Hairs from a horse's tail become horsehair worms, which are found in pools."
5. "The barnacle goose grows from goose barnacles found on rocks beside the ocean."
6. "The emanations rising from the bottom of the marshes bring forth frogs, snails, leeches, herbs, and a good many other things."—Jan Baptist van Helmont (16th century)

Jan Baptist van Helmont's recipe for mice

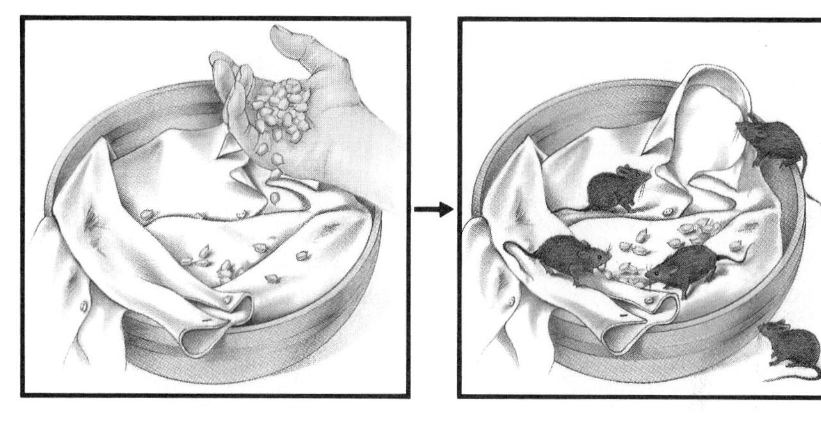

dirty shirt + grains of corn + old pot → a lively crop of mice

What do you think about these explanations? Choose any one of the statements listed above. What could you do to investigate that claim scientifically?

In van Helmont's day, ideas about the reproduction of living things were still being influenced by Aristotle, who lived in Greece 2000 years before van Helmont. Aristotle said that living things could spring from nonliving things by a process called *spontaneous generation*. For example, Aristotle said eels grow from mud and slime at the bottoms of rivers and oceans.

What happened to change people's ideas about how new generations of living things come about? Read the following story about Francesco Redi, whose experiment eventually changed people's ideas about reproduction.

Redi's Experiment

In 1628 an English doctor named William Harvey published a book in which he suggested that some living things might come from tiny eggs and seeds instead of from nonliving things (as had been thought previously).

When Francesco Redi, an Italian doctor, read Harvey's book, he decided to investigate whether flies grow from rotting meat. His experiment is pictured below.

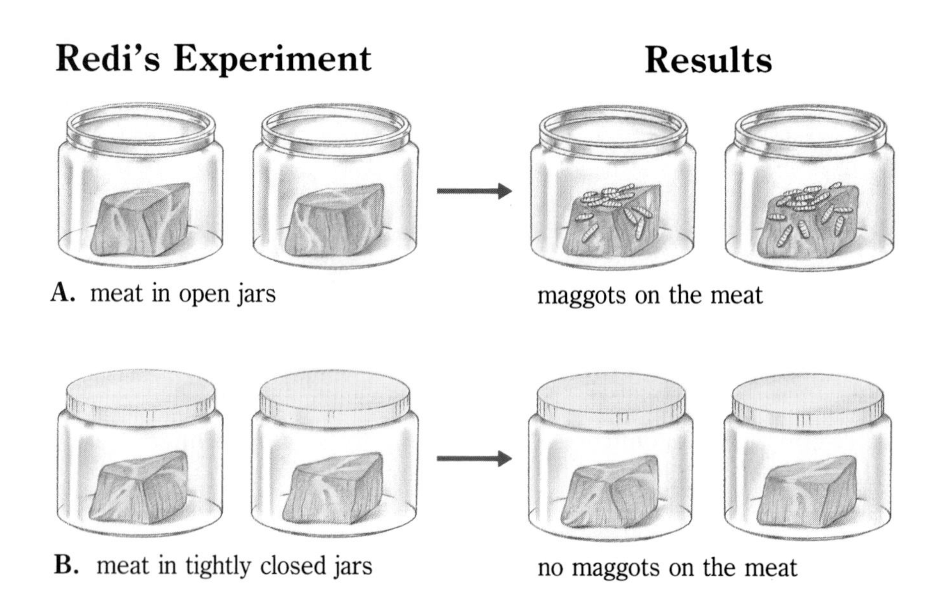

A. meat in open jars — maggots on the meat

B. meat in tightly closed jars — no maggots on the meat

Think About It

1. What must have happened in part A? in part B?
2. Where did the maggots in part A come from?
3. Redi repeated his experiment, making one important change: He substituted a gauze cloth for the solid lid he'd used in part B. Why do you think he did this?
4. Why was part A so important? What name is given to that part of the experiment?
5. What conclusion can be drawn about spontaneous generation?

Although people no longer believe in spontaneous generation, they remain deeply interested in the development of new living things. For example, when you were very young, you probably asked, "Where do babies come from?" In the next section, you'll begin your exploration of this and other questions about reproduction by studying single cells, the basic units of life.

Cells—The Basic Units of Life

You've probably used a microscope to look at single-celled living things like those below. They can be found in pond water.

In the following Exploration, you'll look at *Protococcus*—an alga that forms the greenish stain on tree trunks, wooden fences, flowerpots, and buildings.

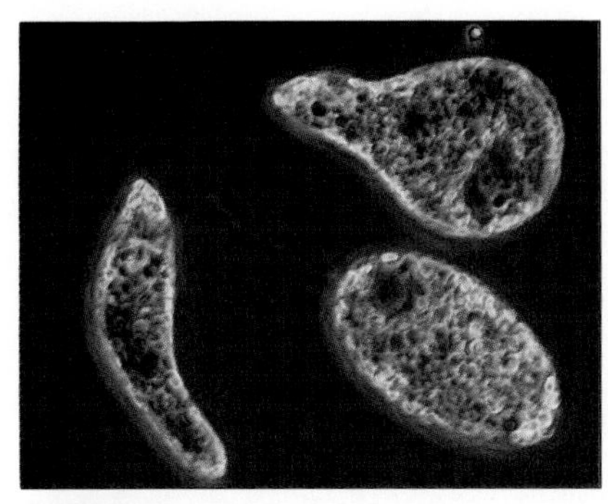

Euglena

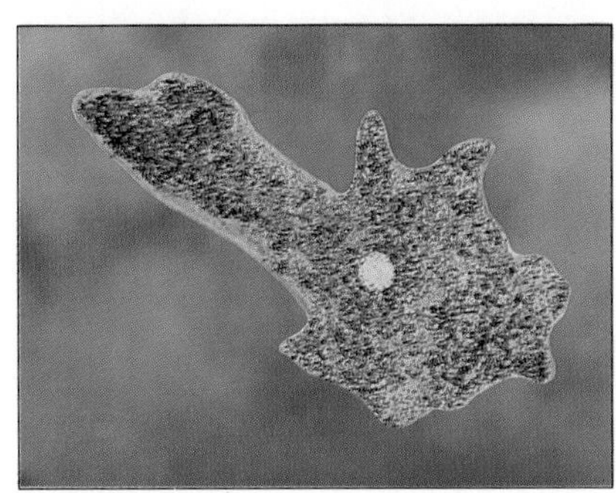

Amoeba

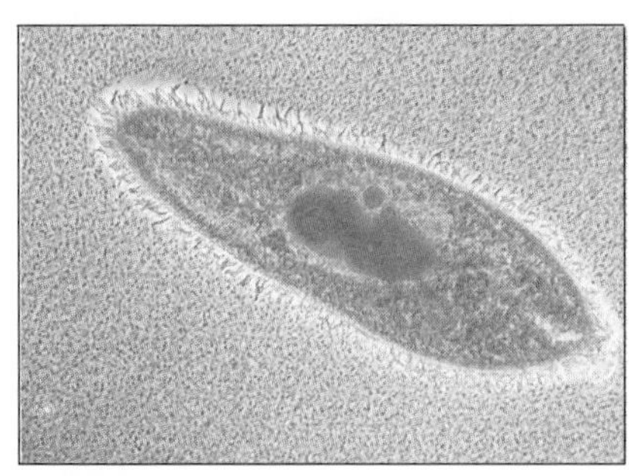

Paramecium

EXPLORATION 2

A Cell Has a Nucleus

You Will Need

- *Protococcus* (or other alga)
- a microscope
- water
- a microscope slide and coverslip
- an eyedropper
- a knife
- a plastic container with lid

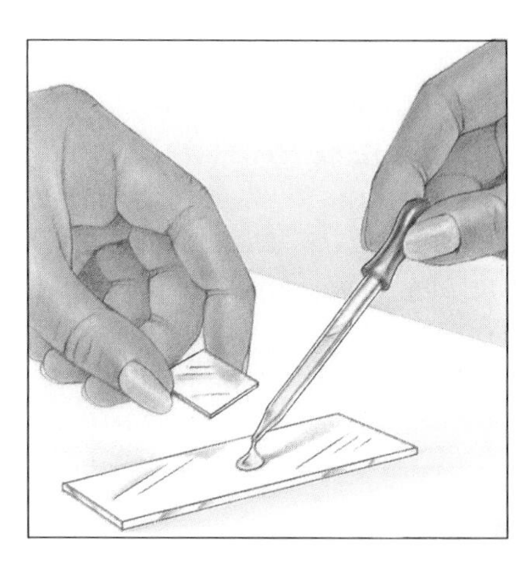

What to Do

1. Locate some of the green material that grows on tree trunks or one of the other places mentioned at left. Scrape a small sample of the material into a container. Bring it to the classroom and make a wet mount of it. If you can't find *Protococcus* outdoors, look for algal growth on the glass in an aquarium. Such algae may not be *Protococcus,* but it will serve your purposes here.

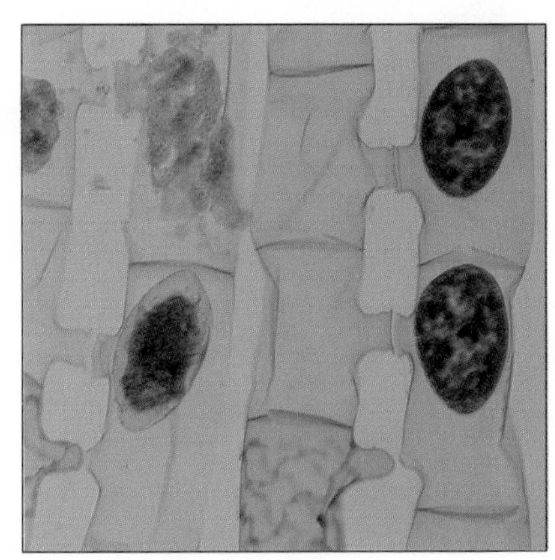

Can you find the nucleus in these algal cells?

2. Use low power and then high power to examine the algal cells.
3. Draw both a single cell and a group of cells.
4. You'll probably notice that each cell contains several chloroplasts—the parts of the cell that are responsible for photosynthesis. One other structure that should be clearly visible in all of the algal cells is the **nucleus.** The nucleus of a cell controls most of the activities that take place in that cell. Find and label the nucleus in one of your cells. As you will soon discover, the nucleus contains the "recipe" for life. In what process, then, must the nucleus play an important role?

❑

The Cell Is Like a Factory

For a moment, think about how the cell is like a factory. Both the cell and the factory put raw materials together to make finished products. In making those products, factory workers follow directions from the "head office." Where do you think the "head office" of a cell is, and what do you think it does?

To help you answer these questions, think about what goes on in a factory's head office: designers create blueprints for the products the factory will make, and managers make sure those products turn out correctly by giving clear instructions to the people assembling the product. Other blueprints in the head office contain the plans for the factory so that an identical factory could be built from them. What are the corresponding life processes that the "head office" of a cell would be in charge of?

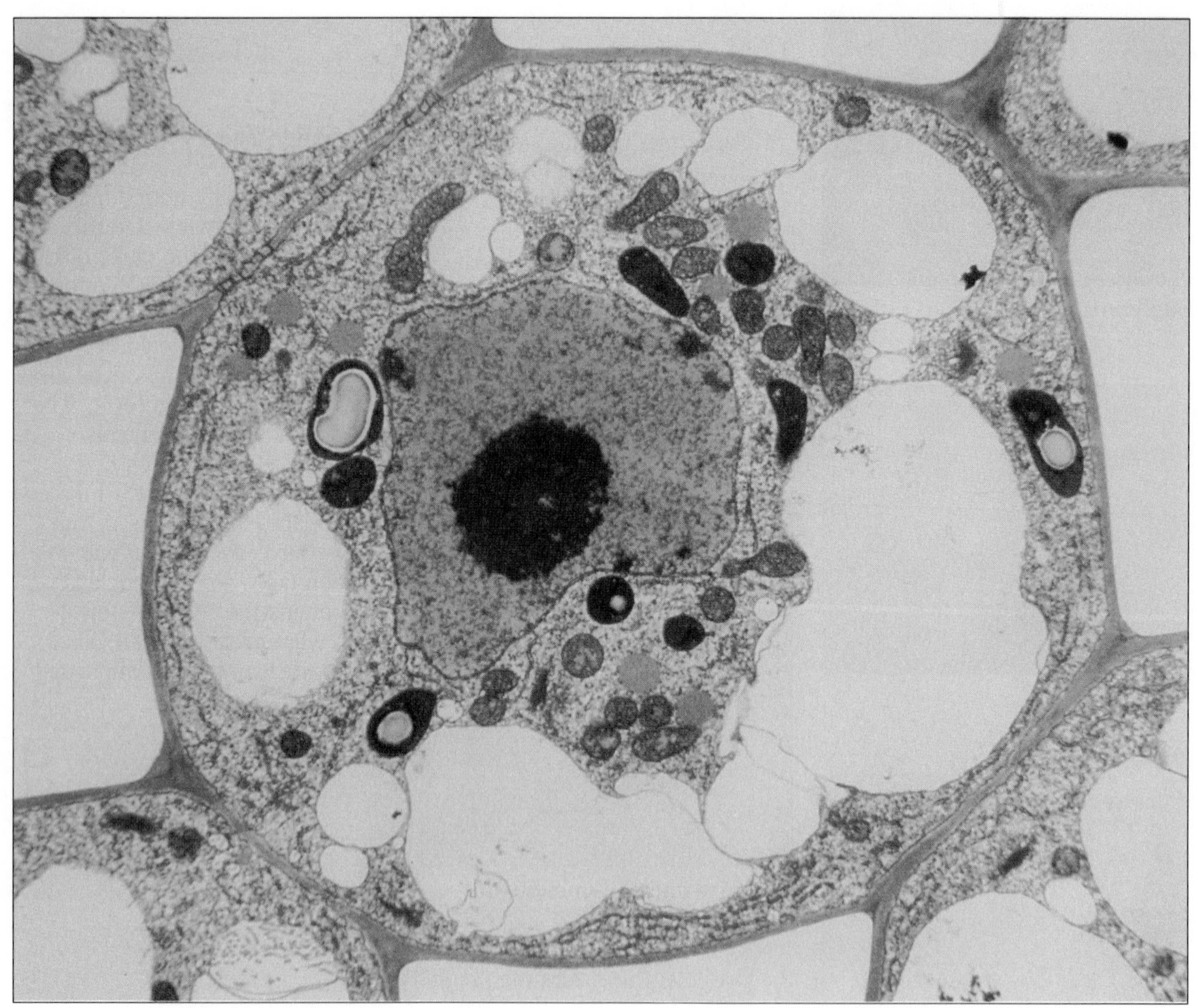

A root cell of an aquatic plant

The "head office" of a cell is the nucleus. It controls everything that goes on inside the cell, just as the head office controls a factory's activities. For example, the "head office" of a factory manages construction of the end product. Likewise, the nucleus of a cell directs the production of substances the cell needs. And just as the factory's head office holds the plans for duplicating the factory should another one need to be built, the nucleus contains the blueprints for creating new cells. Where are new cells needed? Look at how the creation of new cells works in a single-celled organism.

Multiplying by Dividing

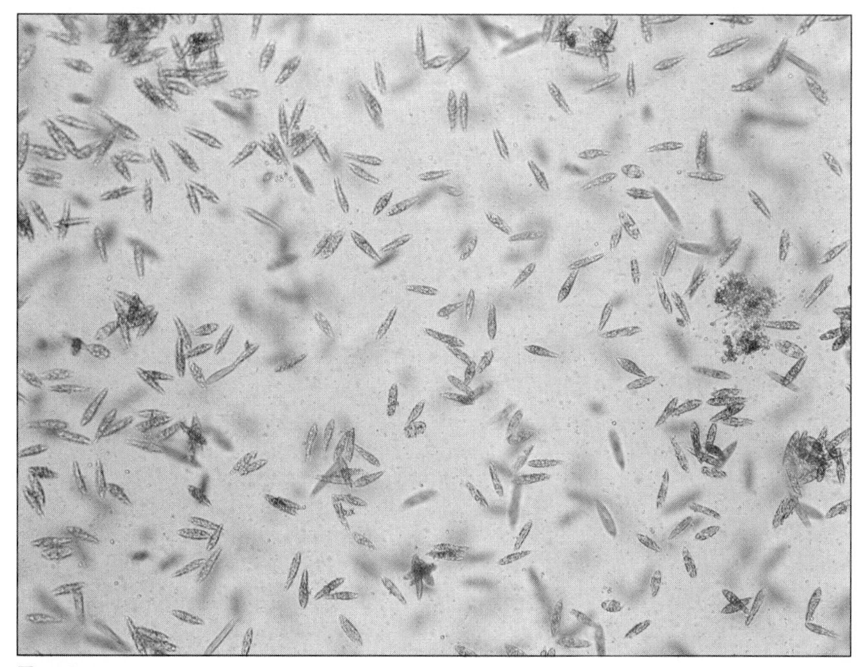

Euglenas

Each of the euglenas shown above is a single-celled living thing. Study the illustration closely. Are several generations of euglenas present here? If so, can you tell one generation from another?

Consider this strange-sounding fact: When euglenas multiply, they divide. When it is ready, one euglena splits into two identical cells, each of which contains all it requires for all of life's activities. Each new cell grows into a full-sized euglena. The new cells are commonly called **daughter cells.** What happened to their "parent"?

Many other single-celled organisms reproduce by this process, which is called **cell division.** Look again at the *Protococcus* cells you drew in Exploration 2 and at the *Protococcus* shown in the photo on page 447. Is there any evidence of cell division?

Cell division also takes place in your body when you need new cells for growth, repair, or other purposes. For example, skin cells are dividing continually to make new skin and heal cuts.

Not So Simple

Gloria and Yoon visited their town's new high-tech Science Discovery Center. Listen in as they talk to their guide, a microbiologist.

Guide: *During cell division, the materials within the cell, including the nucleus, organize themselves into two equal portions. As a result, the two new daughter cells are alike—and just like the "parent" cell.*

Gloria: *But wait! If you divide one apple into two equal parts, you end up with two half-apples. How can one cell produce two whole cells simply by dividing?*

Guide: *Good point! The split isn't at all "simple." And to be honest, the two new cells are actually a bit smaller than the original one. They'll gradually grow larger until it's time for each of them, in turn, to divide. Then there will be four new cells. For single-celled organisms, each of the resulting cells is a complete living thing that can carry out all of the activities necessary for life. To carry out these activities, they must take in substances such as food and release others such as wastes.*

Yoon: *So what does that have to do with cells dividing?*

Guide: *The new, smaller cells have a surface area that is large for their volume. This enables them to take in and release substances very well. When the cells grow in size, the increase in volume is greater than the increase in surface area. So their capability to take in and release substances lessens. Eventually, each cell must divide. But before the cell divides, an important process takes place: The material in the nucleus doubles!*

Gloria: *No way! You're saying matter can appear from nowhere. That's impossible!*

Guide: *You're right. To see how the material doubles, let's use some special equipment that will make you feel like you've been miniaturized—you'll be small enough to enter a cell. Are you up for it?*

Cell division

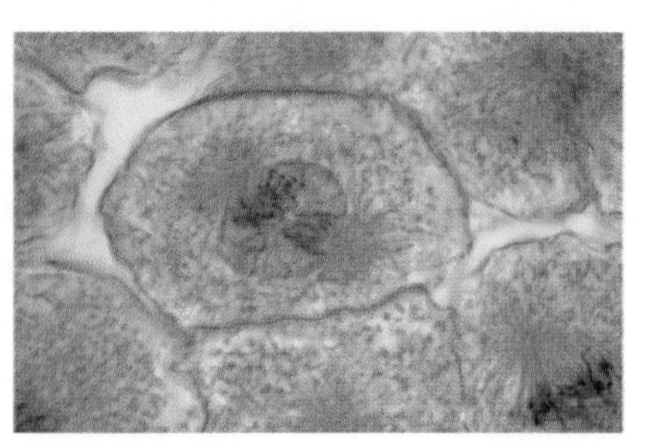

Parent cell

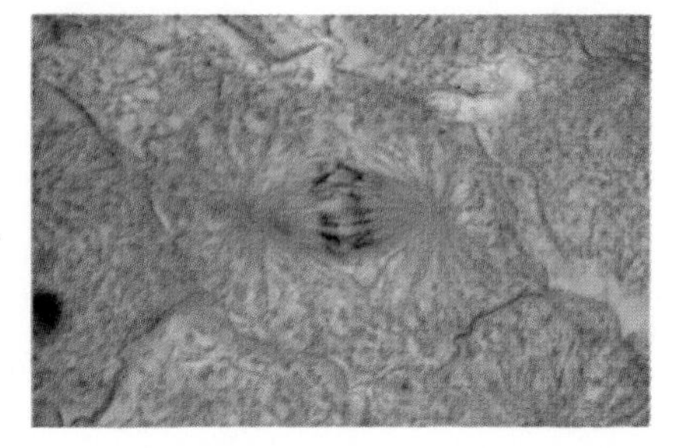

▲ Materials within the nucleus organizing themselves into two equal portions ▼

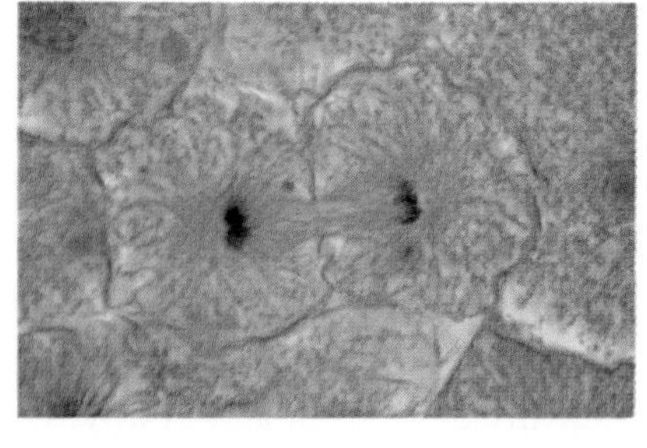

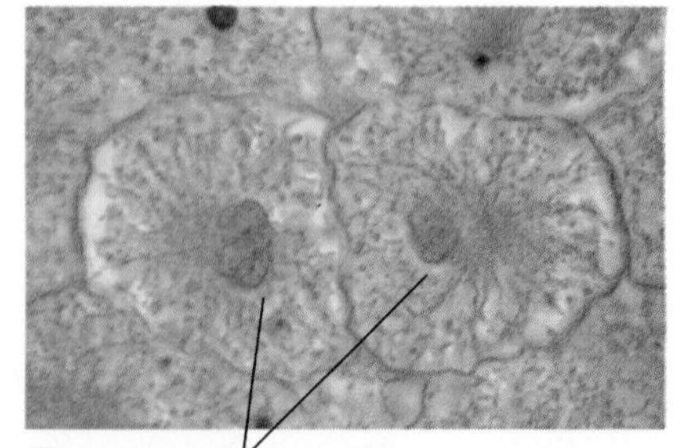

Two daughter cells

Yoon:	*You mean—become cell explorers?*
Guide:	*Right! You'll see what it's like to be a cytonaut. ("Cyto" means "cell.") You know, if we really could enter a cell, we'd have to wear a wet suit and carry a supply of air to breathe, since we'd be immersed in liquid. This special equipment lets us be cytonauts without all of that excess baggage. Let's go!*

Cytonauts approaching the nucleus

The miniaturizing equipment worked fast. Gloria and Yoon soon felt like they were under water, moving through a liquid containing many objects they'd never seen before. Dark masses loomed all around them. Gloria had the best view of the largest object in the cell.

Gloria:	*What is that?*
Guide:	*That's the nucleus. Watch it carefully. The cell is about to divide!*

Gradually, several long ladder-like objects came into view inside the nucleus. Each "ladder" was twisted in such a way that each structure looked like a spiral staircase. Gloria and Yoon gasped as each ladder rung began to break apart near its middle **(a)**. Right away, separate components that had been floating in the cell fluid moved toward the broken rungs and attached themselves **(b)**.

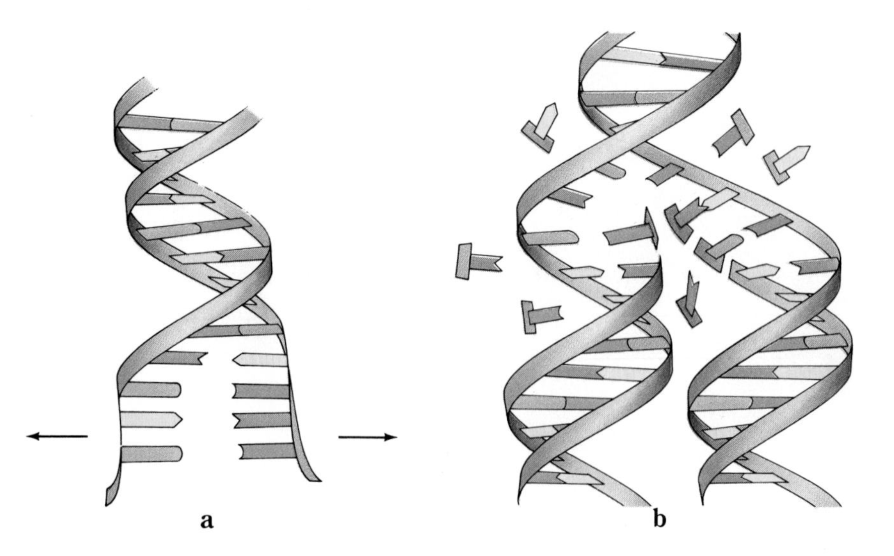

New, complete ladders took shape before their eyes. Now they saw twice as many ladders, but each spiral ladder was joined to its duplicate, similar to the way Siamese twins are joined **(c).**

Guide: *Look! The membrane around the nucleus is dissolving.*

Gloria and Yoon then watched as the spiral ladders lined up across the center of the cell. Suddenly, each "Siamese twin" separated, forming two sets of spiral ladders **(d).** Everyone watched as one set of spirals moved toward them, while the other set faded from view behind a film.

Guide: *Guess what? You're in one of the new nuclei now. A membrane has enclosed you in this new nucleus, keeping you separate from the other new nucleus. You just witnessed cell division from inside the nucleus! Pretty cool, huh? Let's return to normal size now.*

c

d

The photos below represent the stages of the formation of new nuclei that Yoon and Gloria saw—a process called **mitosis.** You'll notice that there aren't any spiral ladders here. That's because the photos show the way you'd see cell division through a microscope. Microscopes aren't powerful enough to show the ladders.

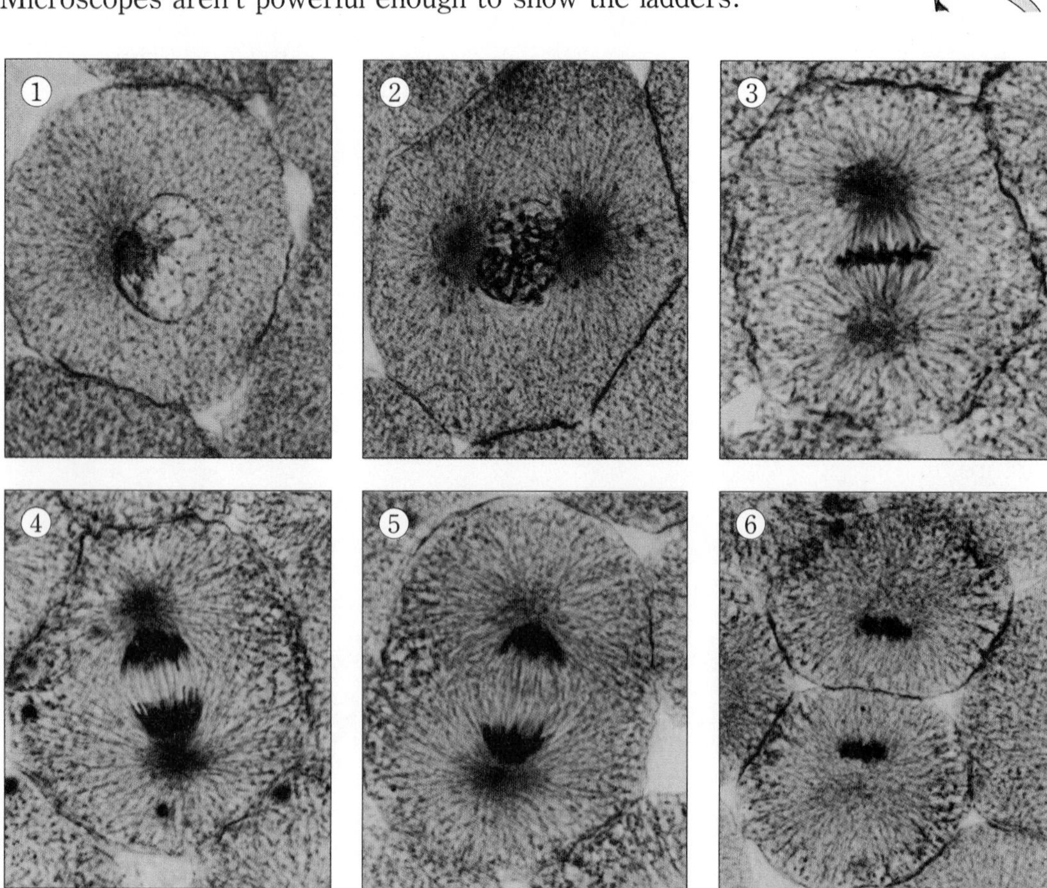

A Cytologist's Expertise

A **cytologist** studies cells and their structures. How well-informed a cytologist have you become so far? Check your expertise by doing the following matching exercise. Write eight sensible statements about cells in your Journal by matching the words in Column I with the most appropriate words in Column II.

Column I	Column II
(a) Cells are	growth and repair.
(b) Euglenas, paramecia, and amoebas are	produces new cells like the old.
(c) Cells can	control centers of life processes.
(d) Nuclei are	the material in the nucleus is equally divided.
(e) Cell division	the basic units of life.
(f) Before a cell divides, material in the nucleus	reproduce.
(g) Our skin cells undergo cell division for	single-celled living things.
(h) During cell division,	makes a copy of itself.

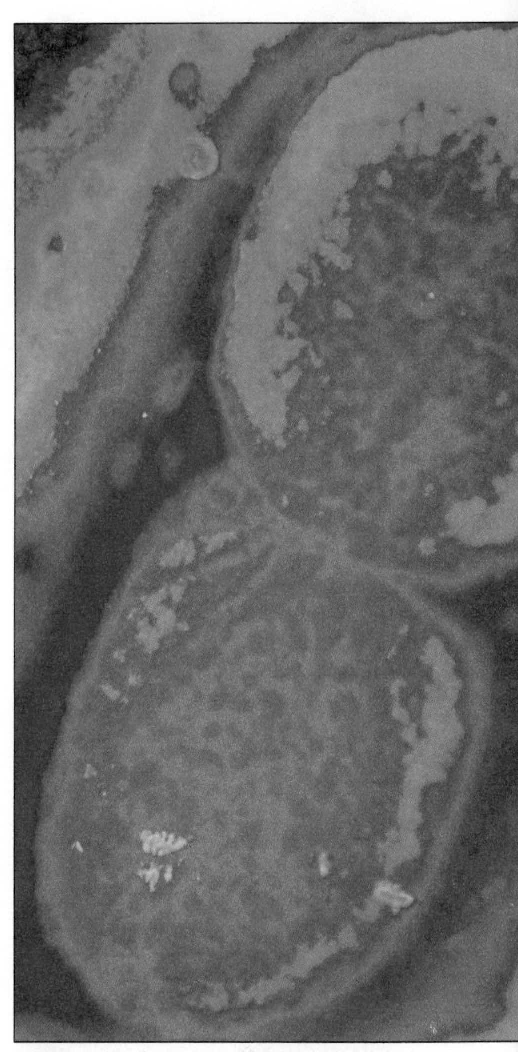

These *E. Coli* cells are dividing. *E. Coli* is a type of bacteria found in your digestive tract.

Twosomes

Cell division is accomplished by one cell alone. It is a successful form of reproduction for many one-celled organisms, such as bacteria. Multi-celled organisms (such as plants and animals) grow and repair damaged tissues by cell division, but they usually reproduce by a more complicated method. This method of reproduction requires two individuals, one male and one female. It is called **sexual reproduction.** In sexual reproduction, two cells unite—one from the male and one from the female. These two cells become parents of the next generation—the offspring, which inherit a mixture of traits from both parents. It's through this mixing of traits that diversity arises.

By contrast, the kind of reproduction that involves only a single parent is called **asexual reproduction.** Many single-celled organisms reproduce asexually by dividing into two cells. Each cell is then a new individual. Many plants also can reproduce asexually. For example, a new plant can be produced by planting a small stem that has been cut from another plant.

Going Further

Many single-celled organisms can reproduce sexually, but the process differs from that among multi-celled organisms. How do single-celled organisms reproduce sexually?

3 Mendel's Factors

What would you get if you crossed a snake plant with a trumpet vine?

If you've ever heard jokes like this, you probably already know what "crossed" means. In the case of plants, it means that the pollen from one plant is used to pollinate another plant. When pollen from a flower's stamens (a plant's male organs) contacts the ovules in a flower's pistil (the female organ), parts of the pollen and ovules unite. Sexual reproduction has occurred, and eventually seeds are formed. We cross plants to produce seeds that will grow into plants with a desirable mixture of characteristics inherited from the parent plants.

By the way, the answer to the joke above is: *a snake in the brass!*

Far-Reaching Labors

What would you get if you crossed a tall pea plant with a short pea plant? (This time, the question is real—not a joke.) The answer is that you would get a second generation of tall pea plants. Why not short plants or plants that fall somewhere in between? In the 1850s, Gregor Mendel, an Austrian monk with a serious gardening hobby, became intrigued by such questions. He had become curious about pea plants because he'd noticed that in some patches of his pea plants, all the plants grew tall. In other patches, there were only short plants, and in a third type of patch, there were mixtures of tall and short plants. He spent the next seven years growing pea plants and looking for answers to satisfy his curiosity. His discoveries became very important.

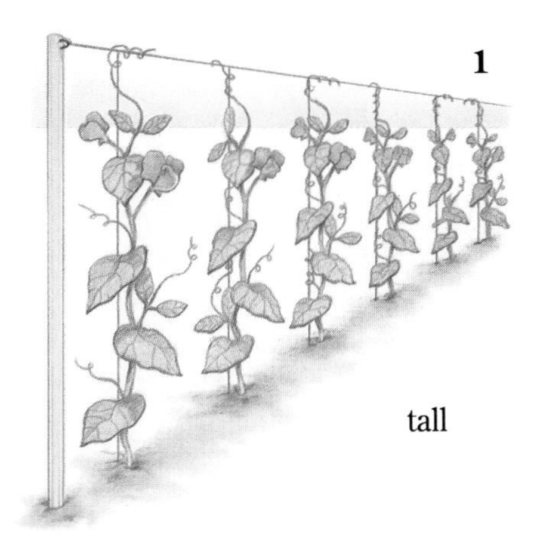

As you read about the stages of Mendel's investigations, search for clues to explain the results.

1. Mendel planted the seeds of tall pea plants in one garden area. He planted the seeds of short pea plants in another area. Why didn't he plant all the seeds in the same area of the garden?
2. He wrapped the flowers of each plant with pieces of calico cloth. Why?
3. He collected the seeds of each group of plants and planted them the next spring in separate beds. What was he trying to find out?
4. Mendel repeated this procedure many times until he was satisfied that the seeds from tall plants produced only tall plants, and the seeds from short plants produced only short plants. Why did this matter?
5. Mendel then planted the seeds from the pure tall plants, and he planted the seeds from the pure short plants. When they bloomed, he transferred pollen from tall plants to the flowers of short plants and later collected the seeds. From which plants (tall or short) did he collect the new seeds?

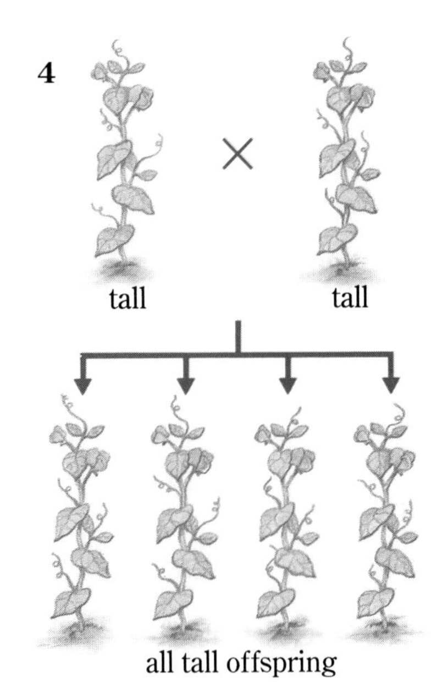

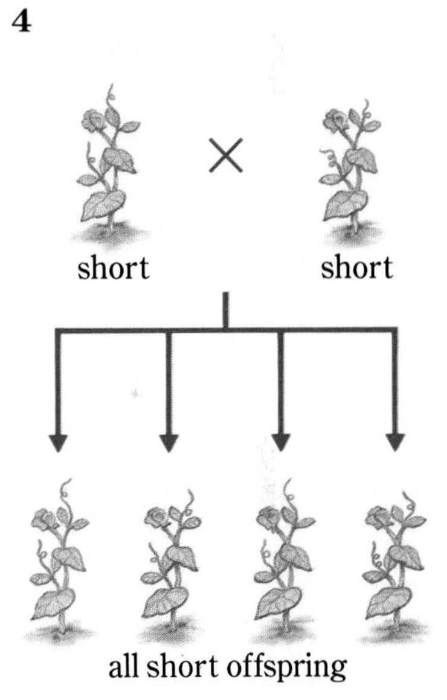

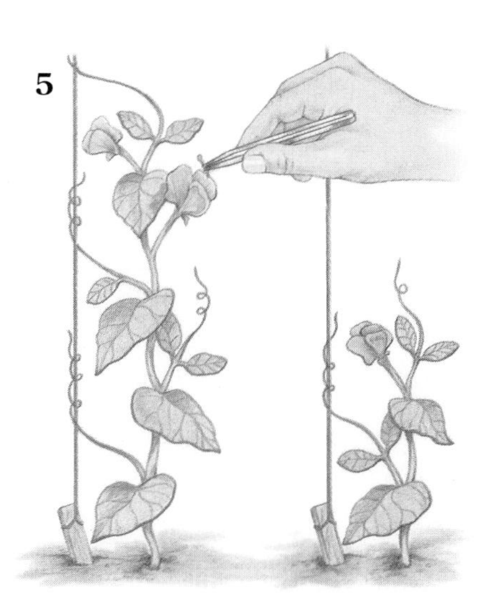

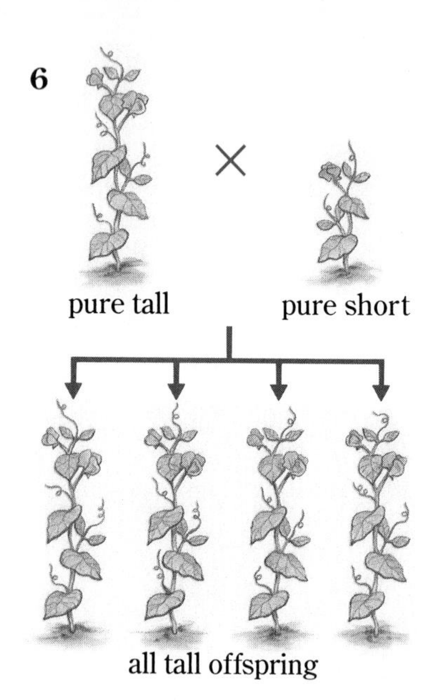

6. The next year, the seeds produced by **cross-pollination** were planted. All the resulting plants were tall. What might be inferred from this? How could it be tested?
7. Next, Mendel tried transferring pollen from short plants to tall plants, but the results were always the same—all the offspring were tall. How would you explain this?

To Mendel, it seemed that male and female each gave its own "instructions" for height to the seeds (and therefore to the next generation of peas). Why do you think this is what he inferred? He called the "instructions" *factors* and concluded that factors are inherited in pairs. For example, his pure tall parent plants contributed the factor for tallness, and his pure short parent plants contributed the factor for shortness. When two such plants were crossed, somehow tallness always "won." It seemed stronger, so Mendel called it **dominant.** He called the apparently weaker factor for height (shortness) **recessive** because it seemed to recede into the background. Even though the recessive factor didn't show up in the offspring, Mendel suspected that it was still present, just hidden.

8. Finally, Mendel proceeded one generation further. He crossed the offspring that resulted from crossing a pure tall plant with a pure short plant. The results were surprising. From a total of 1064 plants, Mendel counted 787 tall plants and 277 short plants. He noticed a definite pattern among the test plants. About what fraction was of each type? Why was this such an important result?

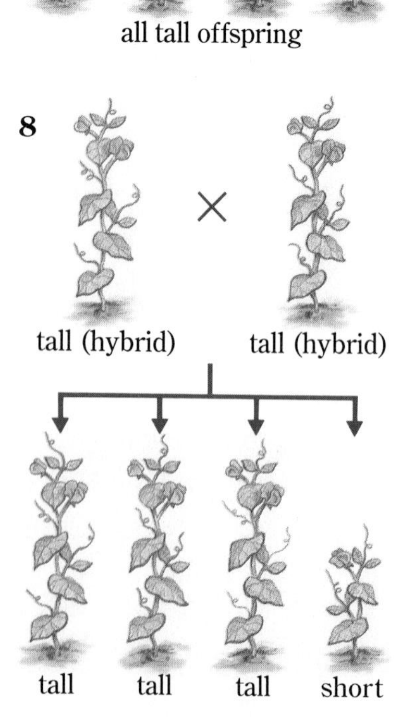

Mendel's hunch was right—the recessive factor was still there among the offspring that had come from crossing pure tall with pure short plants. Mendel gave the name **hybrid** to offspring that contained one dominant factor and one recessive factor. The diagram on the right can be used to show what Mendel did. Write a one-sentence description for each section of the diagram.

Mendel went on to investigate other characteristics of pea plants, such as seed color. He found that the same pattern prevailed. Again and again, 3⁄4 of the second generation of plants (produced from the hybrids) showed the dominant trait, and 1⁄4 showed the recessive trait. Mendel was right. The recessive factor was still there in the hybrids, but it was hidden. It reappeared in the next generation.

Mendel experimented for many years. He wrote in one of his reports, "It indeed required some courage to undertake such far-reaching labors."

Mendel's work did not get immediate notice. However, many years later, his results laid the groundwork for the basic ideas of **genetics,** which is the science that studies how traits are passed on from one generation to the next.

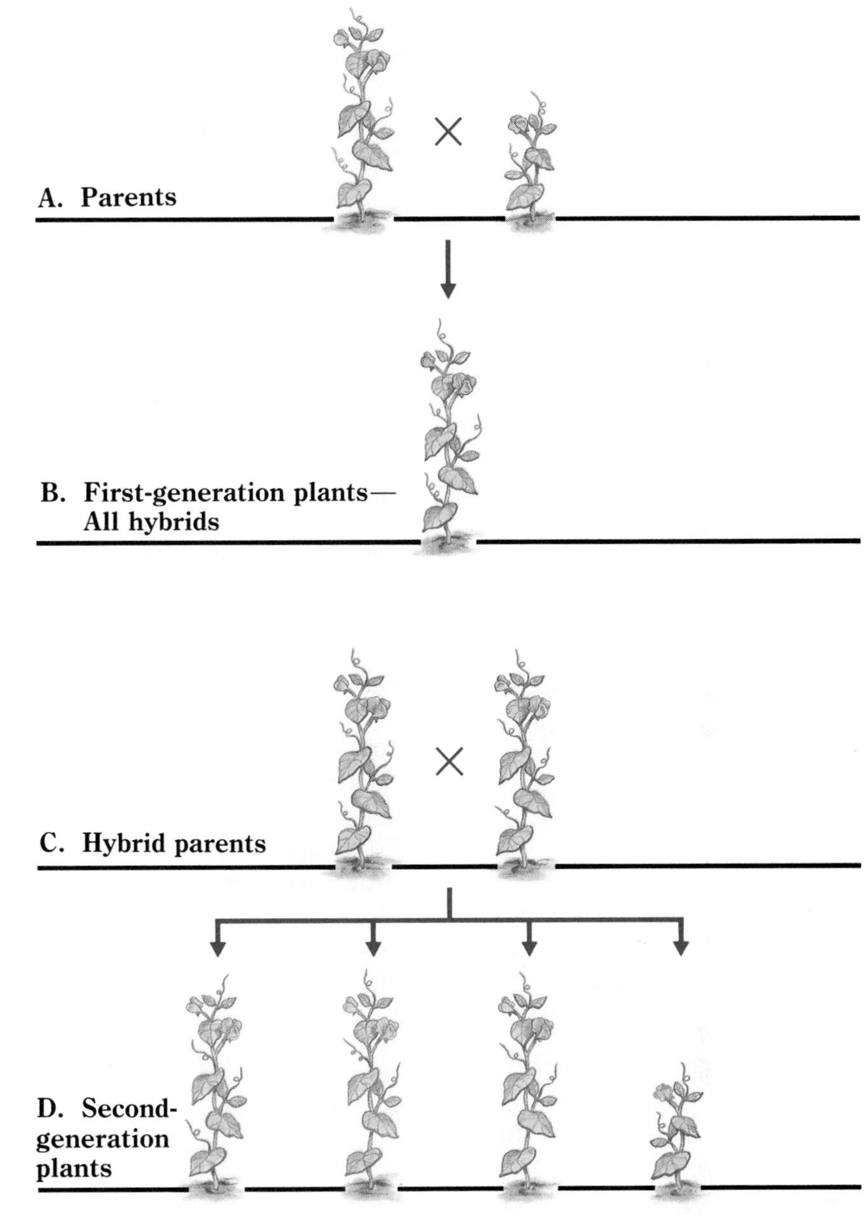

Dominant to Recessive: Why 3 to 1?

Mendel observed that when hybrids are crossed, the next generation shows the traits in about the same mathematical relationship each time: ¾ show the dominant trait, while ¼ show the recessive trait. That's a ratio of 3 to 1. This ratio occurred over and over again in Mendel's studies with other traits besides plant height. It didn't seem to appear by chance. The larger the number of plants involved in each test, the more noticeable the 3-to-1 ratio became. If only a few second-generation plants were grown, the ratio of 3 to 1 might not be so clear.

Using pieces of paper to represent Mendel's factors, the following Exploration shows how likely the 3-to-1 ratio is. As you do the Exploration, remember that Mendel was working with living things made of cells. What part of a living cell might the bag represent in Exploration 3?

EXPLORATION 3

Two Factors

You Will Need

- 200 squares of paper (about 2 × 2 cm)
- scissors
- a metric ruler
- two paper bags or other opaque containers

What to Do

1. Mark 100 squares with a capital *T*. Let this stand for the dominant factor for height—tallness.
2. Mark 100 squares with a small *t*. Let this stand for the recessive factor for height—shortness.
3. Drop 50 squares with a *T* and 50 squares with a *t* into a brown paper bag labeled "Male Hybrid." Then drop 50 squares with a *T* and 50 squares with a *t* into a brown paper bag labeled "Female Hybrid."
4. Without looking into the bags, select one square from the Male Hybrid bag and one from the Female Hybrid bag. Place them together on a table.

Combinations	Number drawn
TT	
Tt	
tt	

5. Repeat step 4 until all the squares have been used up and all possible combinations of the male and female factors have been made. Set each possible combination (TT, Tt, or tt) in its own area of the table.
6. Count the number of pairs of *TT* combinations, *Tt* combinations, and *tt* combinations. Record the number of each in your Journal, in a table like the one shown above.

Looking for Meaning

1. How many combinations of squares did you have?
2. In how many pairs did you have at least one *T*? What fraction is this of the total number of squares?
3. In how many pairs did you have only *t*'s? What fraction is this of the total number of pairs?
4. What is the ratio of pairs with at least one *T* to pairs with two *t*'s?

More Peas

Mendel thought each parent contributed, by chance, one of its factors for height to the peas. Suppose both parents supplied a *T*. The combination of *TT* would result in a tall plant in the next generation. The combination of a *T* from one parent and a *t* from the other parent would also result in a tall plant in the next generation, since *T* is dominant and *t* is recessive. If a plant inherited a *t* from each parent, however, the plant would be short. The diagram below illustrates this.

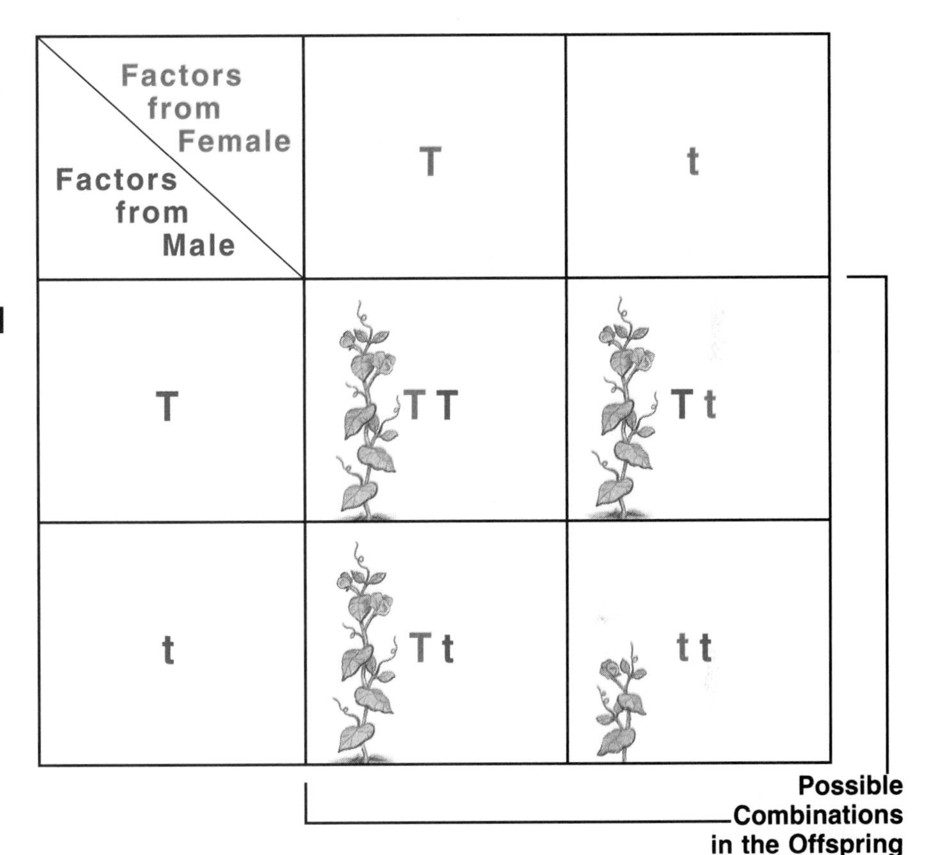

1. How many different combinations of factors are possible?
2. How does the number of combinations with at least one *T* compare with the number with only *t*'s?
3. For the cross shown in the above diagram, how many tall plants would you expect for every short plant?
4. What did Mendel actually find in his investigations for this particular cross?
5. Refer again to the diagram outlining Mendel's investigations (page 457). How would you label each plant shown, using combinations of *T* and *t*?

Making Punnett Squares

The diagram on page 459 shows all the possible combinations that can be passed on to the offspring of hybrid parent plants. Such diagrams are called **Punnett squares.** The passing on of traits from one generation to the next is called **heredity**. Punnett squares are commonly used in studying heredity.

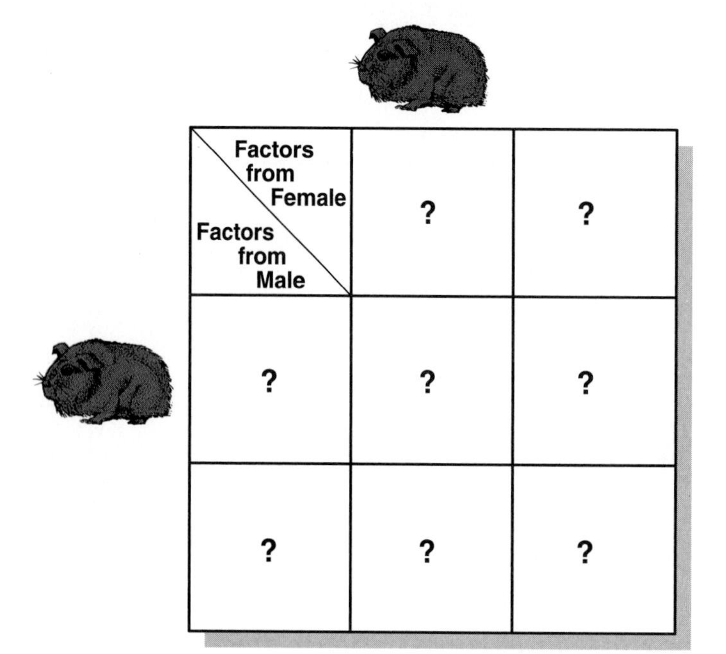

Now it's your turn to practice working with Punnett squares. Draw and complete a Punnett square that shows the possible results of mating a black guinea pig with a brown one. Black is the dominant color, while brown is the recessive. The black guinea pig carries two dominant factors for a black coat. To indicate the dominant factor for coat color, use the letter B. For the recessive factor, use the letter b.

What would be the results of crossing two hybrid guinea pigs? Draw and complete a Punnett square for such a cross. What ratio of black to brown coat color would you expect the offspring of these hybrids to show?

Factors from Female / Factors from Male	?	?
?	?	?
?	?	?

Predictable Ratios

In Exploration 3, you observed that there were three pairs with at least one T for every one pair that consisted only of t's. This is a ratio of 3 to 1. Had you chosen only a few pairs instead of 100, would you have gotten this 3-to-1 relationship? Why or why not?

To arrive at his results, Mendel experimented with many different inherited characteristics of pea plants, such as seed shape and color, pod shape and color, stem length, and flower position. He kept careful records of his experiments and used these records to calculate the ratios in which the traits appeared. In every case, the offspring of hybrid crosses produced the 3-to-1 ratio. One of Mendel's chief contributions was showing that many inherited traits would appear in predictable mathematical patterns.

Thinking Back

Now look over the data you recorded for Exploration 1. All the *yes* answers indicate dominant traits. That is, every *yes* answer you gave shows that you have the dominant expression of that characteristic. Every *no* answer shows that you have the recessive expression of that characteristic. If you show the recessive expression of a characteristic, what kind of factor did each of your parents contribute? Would the same be true if you show the dominant expression of a characteristic? Explain.

How did the daughter of tongue-rollers end up being a non-roller?

Going Further

1. Should you expect there to be a 3-to-1 ratio of dominant to recessive traits for the characteristics you observed in Exploration 1? Why or why not?
2. Suppose everyone in the school completed an index card. How do you think the number of dominant to recessive traits would compare?
3. Having a cleft chin is a dominant trait. So is having six fingers. Do you think that three-fourths of the students in your school have cleft chins? How about six fingers? How would you explain your observations?
4. Mendel's pea plants were either tall or short. How does this compare to the heights of your classmates? Can you think of a way to explain this?

As you can see, studying human inheritance factors is much more complex than studying Mendel's pea-plant factors.

And Where Did You Get That Nose?

Now use what you've learned to add to your family-tree project. Using the questions you answered in Exploration 1, interview or research as many members of the family you chose as possible. Keep a careful record of the information you find.

The diagram on the right looks similar to the family tree you drew earlier, but this one traces the inheritance of a particular *trait* in a family—tongue-rolling. Using a separate diagram for each trait, note as many different traits as you can for the family you are studying. Each of these diagrams is called a **pedigree.**

Are there any family members who have characteristics that neither of their parents possesses? What can pedigrees tell you about traits that are passed on from generation to generation? Pedigrees are not usually used for studying the inheritance of traits in plants. Why are pedigrees used for studying the inheritance of human traits?

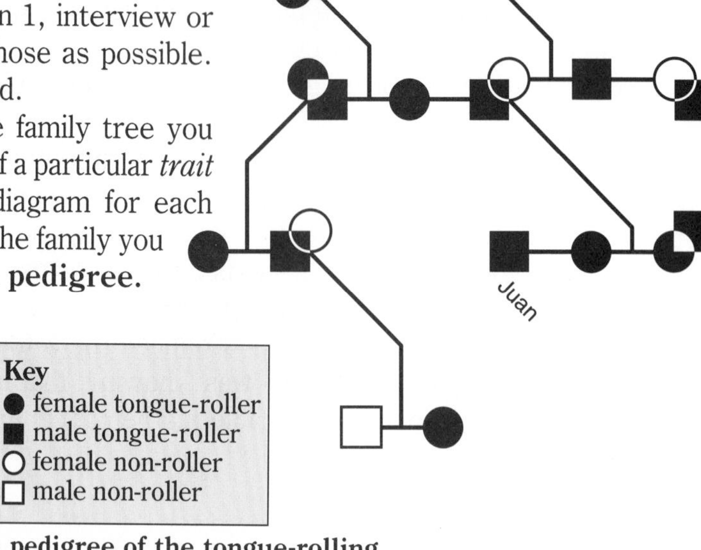

A pedigree of the tongue-rolling trait in Juan's family

A Modern Look at Mendel's Factors

You know a lot now about dominant and recessive factors. By now, too, you've seen how certain physical traits are *inherited,* or passed from one generation to the next. What are the mysterious *factors* that determine these traits? It took scientists quite a while before they discovered an answer to this question.

Toward the end of the 19th century, certain structures in the nucleus of a dividing cell were seen through a microscope. Because these structures easily absorbed dye or stain, they were called **chromosomes** (meaning "colored bodies"). Each type of living thing has a specific number of chromosomes in its cells. The chromosomes are in pairs, like Mendel's factors. Human body cells, for example, contain 23 pairs, or a total of 46 chromosomes.

By 1910, Mendel's factors were identified as individual sections of chromosomes. These components were named **genes** after a Greek word relating to "birth." Each chromosome in a pair carries one of the two genes that determine a trait. Altogether, human chromosomes carry at least 100,000 genes. Genes not only determine traits but also provide instructions to cells of an organism as it grows.

Human chromosomes

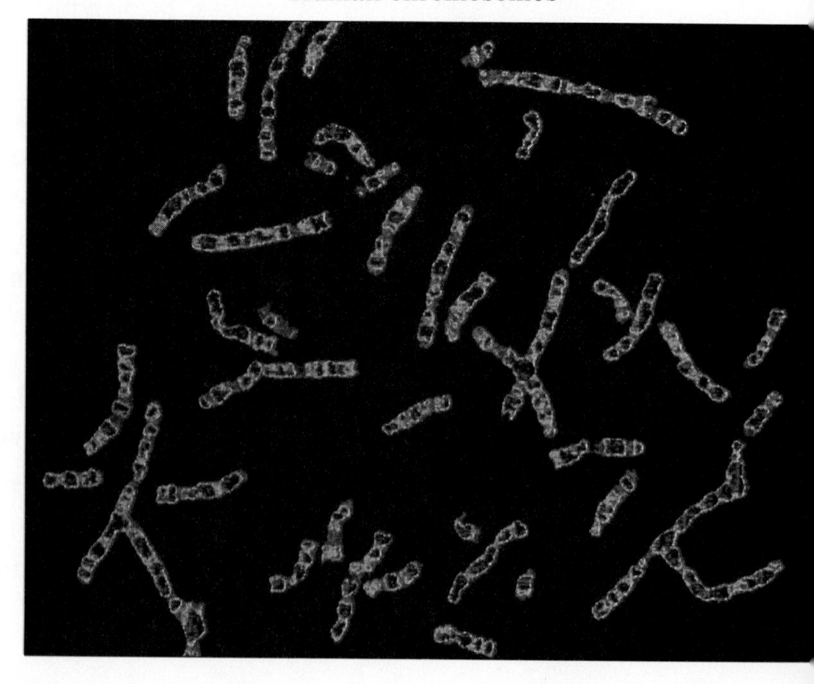

Sex Cells

When sexual reproduction takes place, a cell from each parent unites to form a new cell. This process is called **fertilization.** In flowering plants, pollen contains a sperm nucleus (the male contribution), which unites with an ovum, or egg cell (the female contribution), inside an ovule. In the animal kingdom, a sperm from a male unites with an ovum, or egg, from a female. Eggs and sperm are also called **sex cells.**

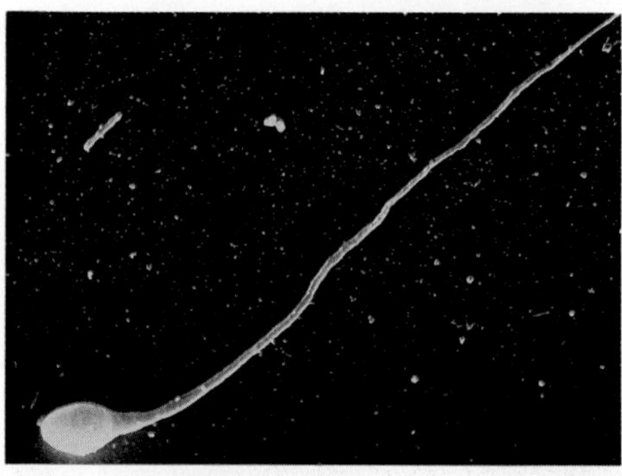

A human sperm cell

Now here is a problem! If each human sex cell supplied 46 chromosomes, the offspring would develop from a total of 92 chromosomes. But human cells have only 46 chromosomes. What happens to make this number remain constant, generation after generation? Through a complicated process, sex cells with only 23 *single* chromosomes (one chromosome of each pair) are produced. This way, when the two sex cells unite, the resulting cell has 23 pairs of chromosomes, or a total of 46 chromosomes—the correct number.

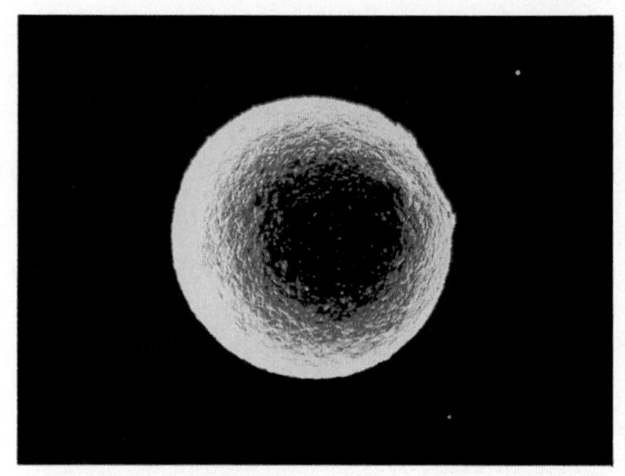

A human egg cell

The production of sex cells begins like ordinary cell division: the material in the nucleus doubles. Several further stages cause the chromosomes to be distributed among four cells, instead of between two.

Here is a simplified view of what happens when sex cells form:

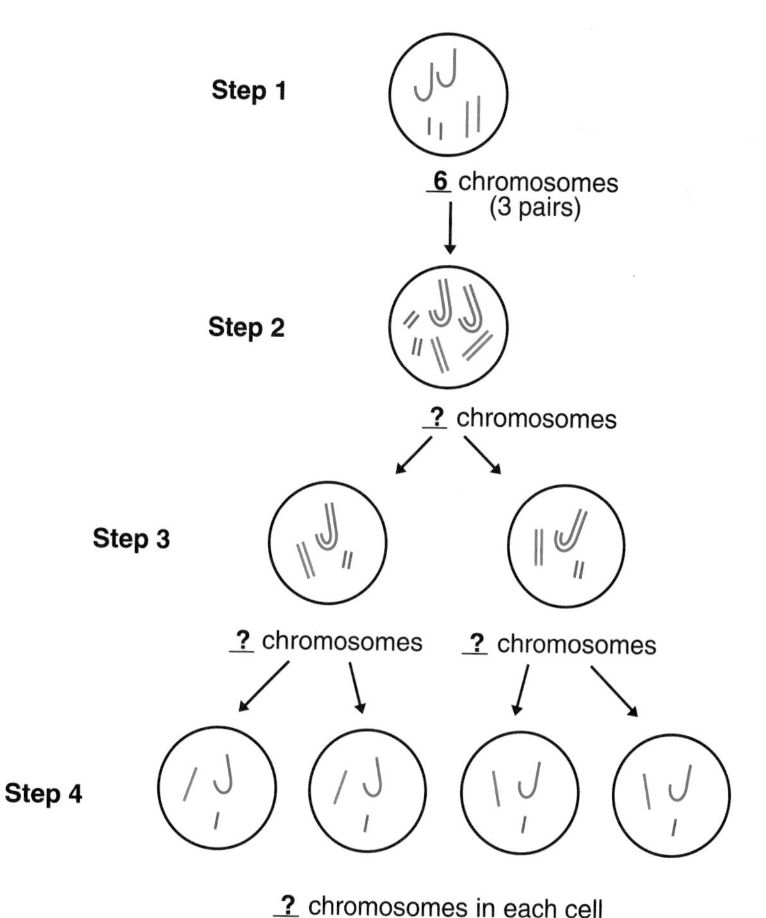

If this were a human cell, it would contain 23 pairs of chromosomes. Only three of these pairs are shown here. That way, you can more easily see what is happening. Look at the illustration for each step. What is happening to the number of chromosome strands at each step?

In a male, these 4 cells all become sperm. In a female, only 1 of the 4 becomes an egg cell. The other 3 gradually disappear.

You have just observed the process that results in the formation of sex cells. This process is called **meiosis.**

Life Story of the Unborn

The development of a new human life is a remarkable and fascinating nine-month process. At five different intervals in this lesson, you will pause to observe the development of an *embryo*. It all began with the fertilization of an egg cell by a sperm cell. Cell division then began, and from there, all kinds of things started happening. . . .

But first things first. It really begins with two sex cells, each formed through the process of meiosis. Female sex cells (egg cells) are produced in the female's ovaries, while male sex cells (sperm cells) are produced in the male's testes. Each sex cell has only 23 single chromosomes (as opposed to the 23 pairs, or 46 chromosomes, in a body cell). Do you remember why there's a difference? To examine the events that come before fertilization and the development of an embryo, read the following description, stopping to locate the numbers on the accompanying diagrams.

Before You Were You

The sperm cells are placed in the vagina (1) by the male reproductive organ, the penis (2). The sperm move upward through the uterus into the Fallopian tube (3), where one sperm cell (4) may fertilize an egg cell (5) that has been released from the ovary (6). The fertilized egg cell begins to divide (7) as it moves into the uterus (8). There, perhaps one week after the egg has been fertilized, the embryo embeds itself (9) in the wall of the uterus.

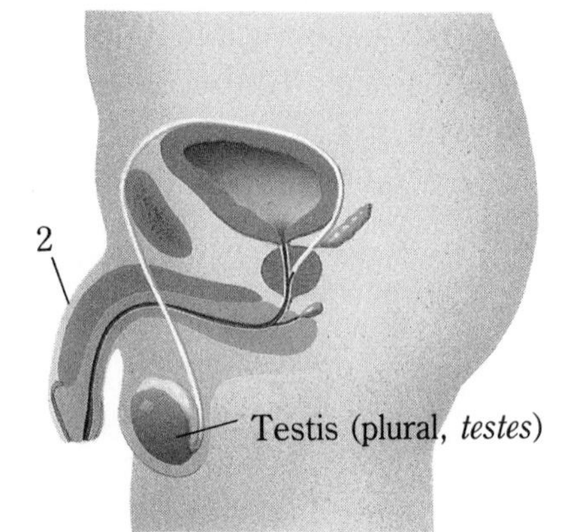

Side view of male reproductive system

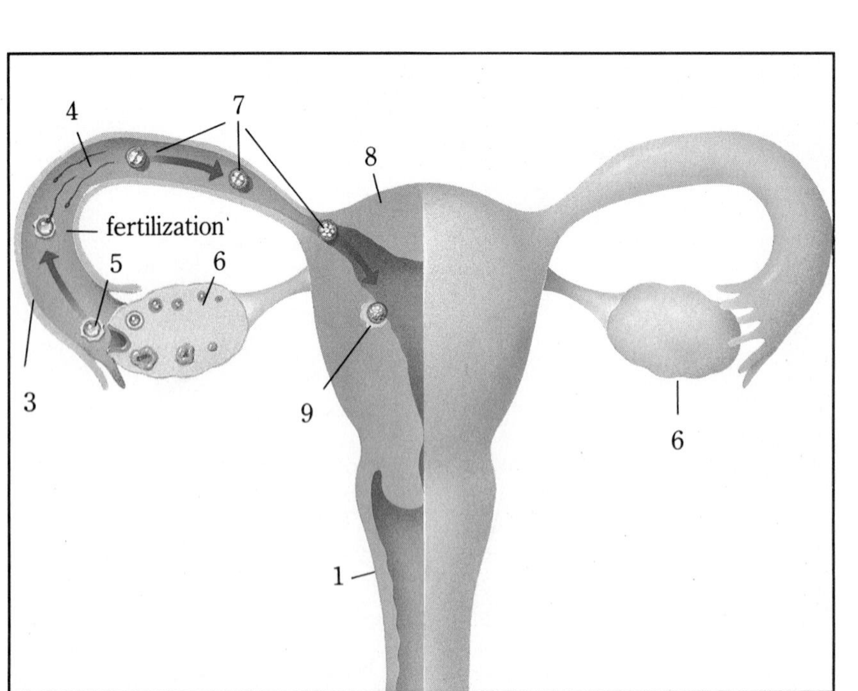

Front view of female reproductive system

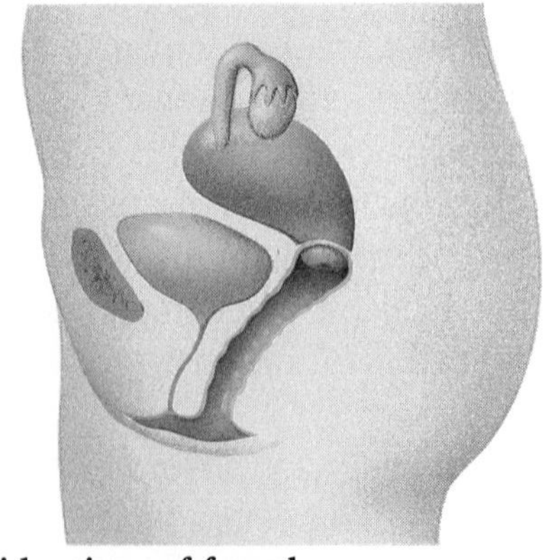

Side view of female reproductive system

This diagram shows some details of the embryo in the uterus wall.

1. What do you think happened at (10)?
2. What role might the amniotic cavity (11) play?
3. What observation can you make about the cells of the embryo (12) at this stage?
4. What part do you think the yolk sac (13) plays?
5. Why is it important that the mother's blood vessels (14) be near the embryo?

What necessary part of the life-support system has not yet developed in this diagram?

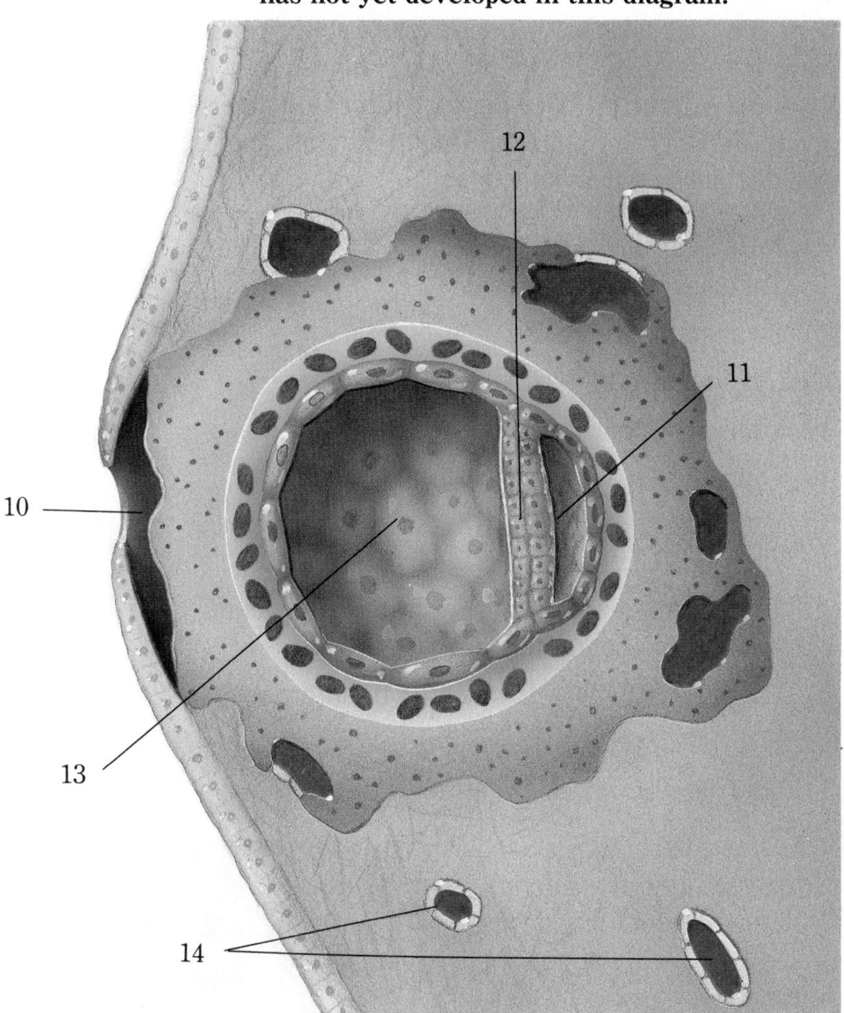

The life-support system for the embryo (15) is called the **placenta** (16). It is an organ that develops around the embryo and that attaches the embryo to the wall of the uterus. The placenta is rich with blood vessels that come from both the mother and the embryo. However, the blood of the mother does not mix directly with the blood of the embryo. Instead, substances are exchanged between the mother's blood and the embryo's blood across the walls of the blood vessels in the placenta.

As the embryo develops, it grows an umbilical cord (17) that connects the embryo to the placenta. It is through this cord that substances enter and leave the embryo. What substances are needed by the embryo? What substances must leave the embryo?

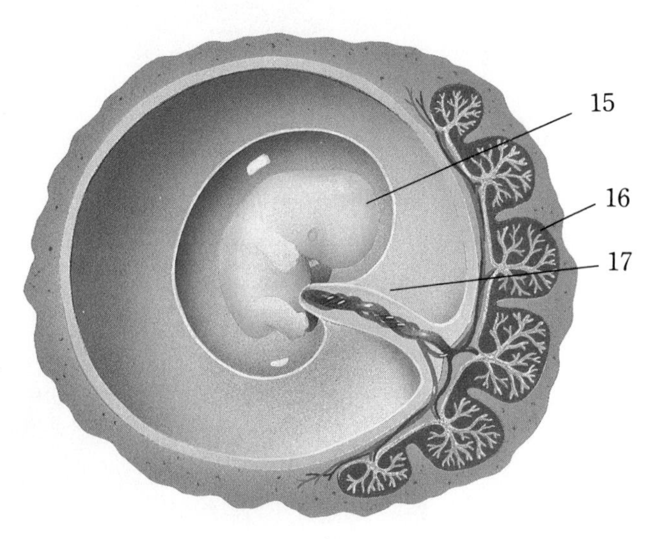

Your Early Development

Now consider what you already know about your own formation. In your Journal, place the following happenings in the order you think they occurred. Decide, for example, what might have happened during the first month in the womb, and so on. The list includes abilities and activities.

(a) Your fingers and toes took shape.

(b) Your heart began to beat.

(c) Your memory began.

(d) Leg and arm stumps appeared.

(e) Your eyes could respond to light.

(f) Your mother was aware of your movement.

(g) Your first teeth are starting to take shape, and tiny buds for your permanent teeth are beginning to develop.

(h) You began the cycle of sleeping and waking, and you developed the ability to dream.

(i) Your sex was clearly indicated by internal sex organs.

(j) You developed your own fingerprints.

(k) Your face and neck began to develop.

As you follow each episode in "Adventures of a Life in Progress," refer to your list to see how well you imagined your own beginning!

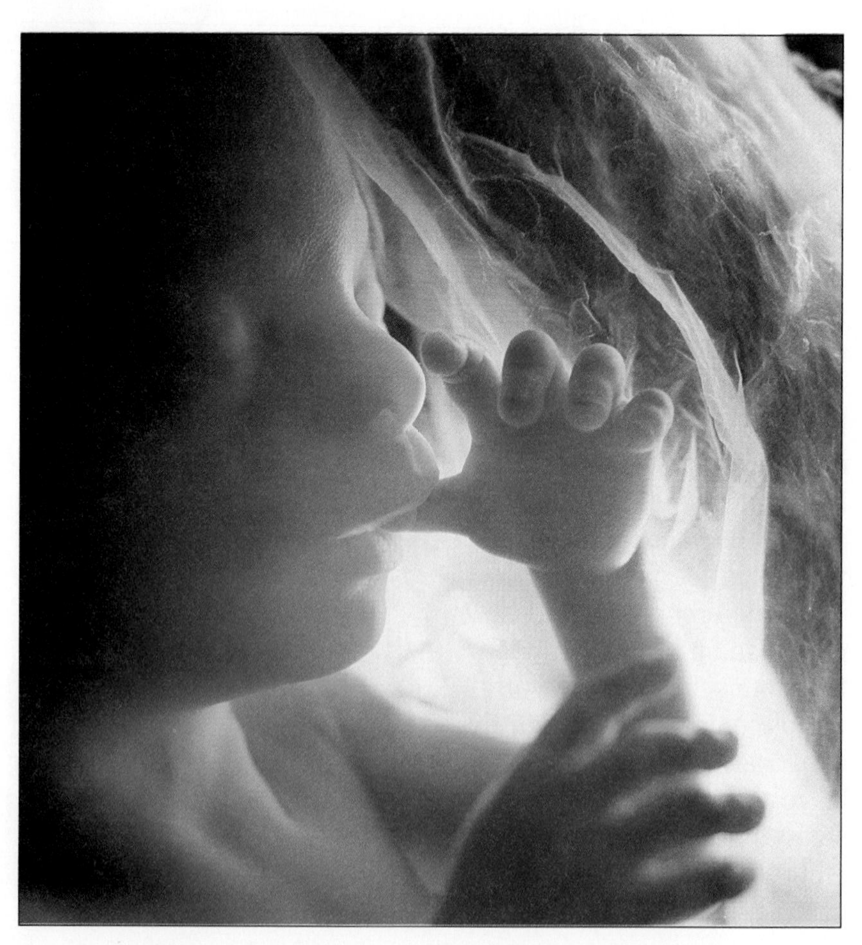

Adventures of a Life in Progress

Episode 1—The First Month

It started with the fertilization of an egg cell by a sperm cell. Then cell division began, and things happened like clockwork. You began as a human egg that was even smaller than the period at the end of this sentence—about 0.1 mm in diameter. The sperm that fertilized the egg was only one-thirtieth the size of the egg. After fertilization, you began to divide. Later, cells with a specific shape and function began to appear.

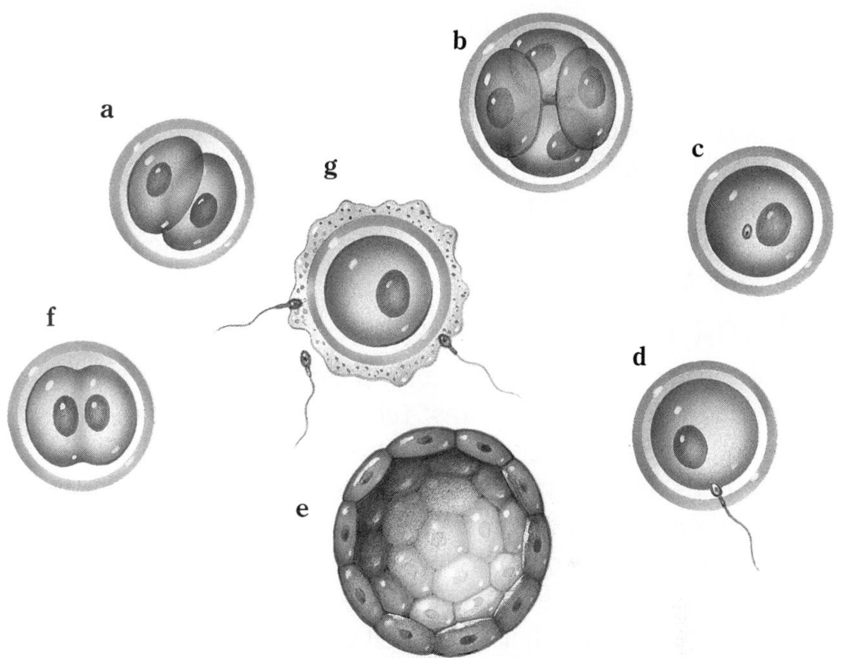

In what order would you place these drawings to illustrate the beginning of a new human life?

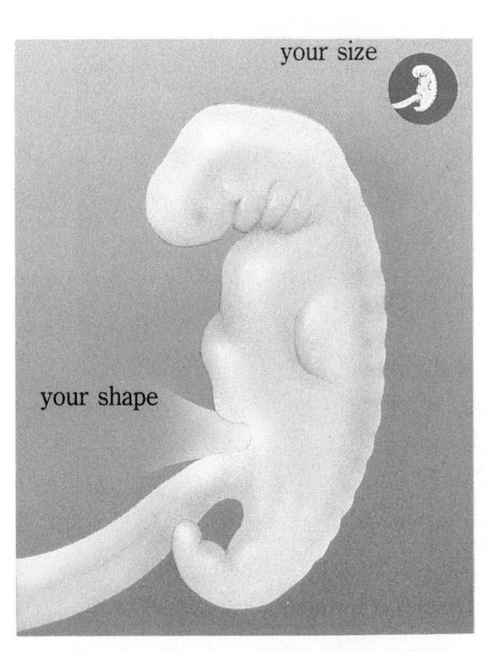

You, at one month of growth in the womb

By the age of 17 days, blood cells had formed, and shortly thereafter, you had a heart tube. Your heart began to twitch and started the beat that must continue until your life's end. By the end of the 3rd week, the cells that would eventually form your sperm or eggs were set aside. In the 4th week, a tube was formed in what became your mid-back region. At the front end, your brain later developed. The back part of the tube later formed your spine. Next came your food canal. By the 25th day, you had a head end and a tail end (and a bit of a tail also). With no face or neck, your heart lay close beside where your brain would eventually be.

By the end of the first month, you had tiny bumps where arms and legs would develop. Your lungs, liver, kidneys, and most other organs had begun to form. Your head had two pouches that would become eyes, two sunken patches of tissue where the nose would form, and a sensitive area behind each eye where the inner ear would form. A lot had happened in one month. Can you imagine what size and shape your embryo was by then?

Jean Hegland recorded in a journal the thoughts she and her husband had as they awaited the birth of their child. She published her thoughts in a book called *The Life Within*. Here is a part of her journal for the first month.

Then comes heart, and spinal cord, and gut, all primitive, all bulging and shifting like a rose opening in a time-lapsed film, all arising out of that speck of protein that was two cells, that became one, that doubled and divided, transforming from one thing to another as wildly and exactly as objects in a magician's show, where rabbits become scarves that become tulips and then coins, in a perfect, dizzying succession of change. And so these organs, these basic bits of human being, appear like rabbits out of nowhere—out of the intention of the universe—and a new creature takes shape.

What sort of instructions are being followed to make all of the embryo's body parts form at a certain time, in a certain way? Compare the events of the first month with the list of developments you wrote in your Journal. Were your predictions close?

Helping the Natural Process

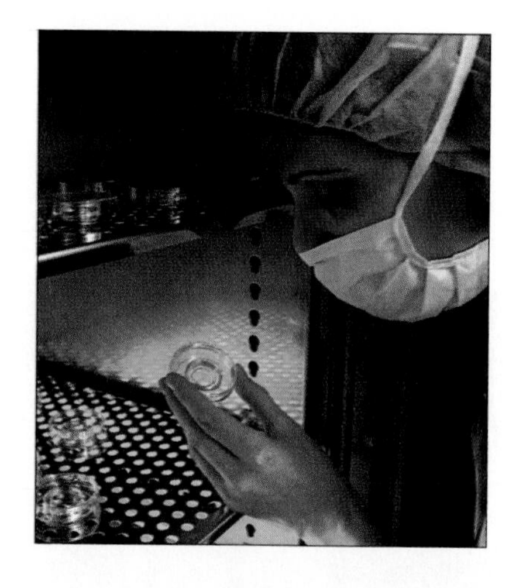

In 1978 a baby's birth made history. That baby's name is Louise Brown, and she was the first "test-tube baby." Did she really grow in a test tube? What does the term mean? Why was it done?

Sometimes, certain problems interfere with the natural process of fertilization. **In vitro fertilization** is a way for certain people to have children. Find out what this term means and why the procedure is performed. As you investigate this form of reproductive technology, as it is called, ask yourself some basic questions, such as:

1. How successful is in vitro fertilization likely to be?
2. What are the costs, in terms of
   (a) risks to parent or baby?
   (b) money?
   (c) time?
3. Why would someone choose to try (or not to try) this procedure?

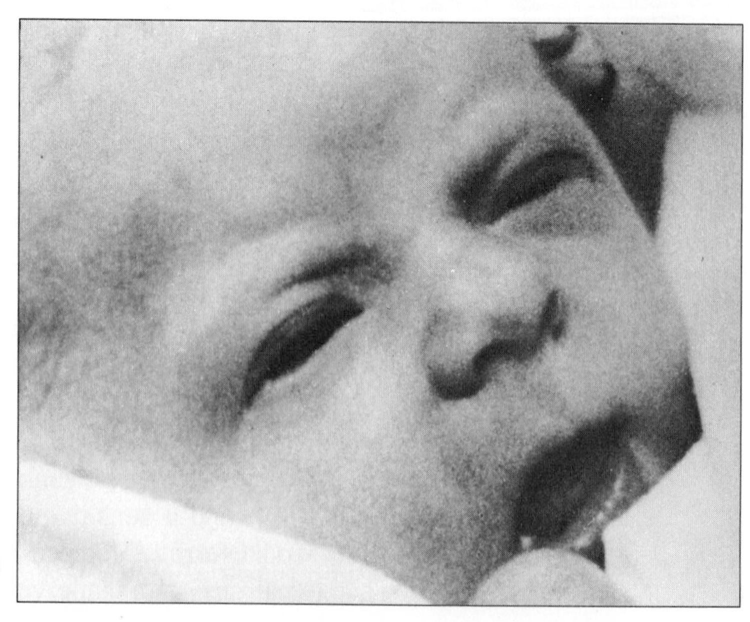

Louise Brown—the first "test-tube baby"

Outline a presentation (sometimes called a *brief*) that you could give at a public meeting. Include the pros and cons of in vitro fertilization and back up your statements with information gathered in your research. End by giving your recommendation for guidelines as to when in vitro fertilization is appropriate and when it is not.

Twin Heartbeats

Sometimes two embryos develop at the same time. The resulting children may look so much alike that even their parents may have trouble identifying them. They are called *identical twins*. Sometimes, though, twins look no more alike than any other children of the same parents (siblings). These are *fraternal twins*. Why are twins sometimes identical and sometimes fraternal? Consider two possibilities.

A. The mass of cells from a single fertilized egg separates into two halves early in development. Two babies result.

B. Two eggs are released by an ovary. They are fertilized by two different sperm cells. Each is implanted in the uterus and continues to grow. Two babies result.

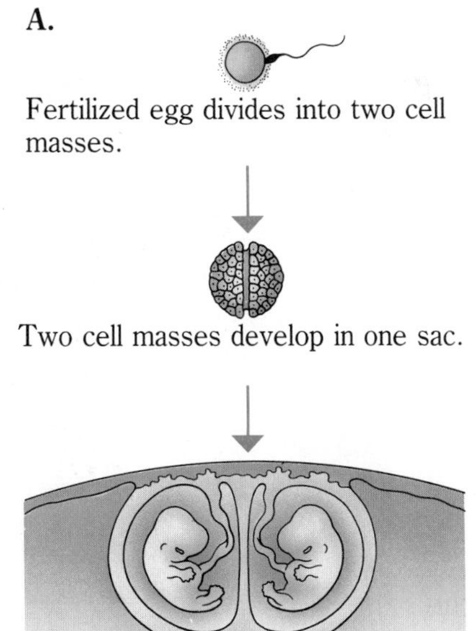

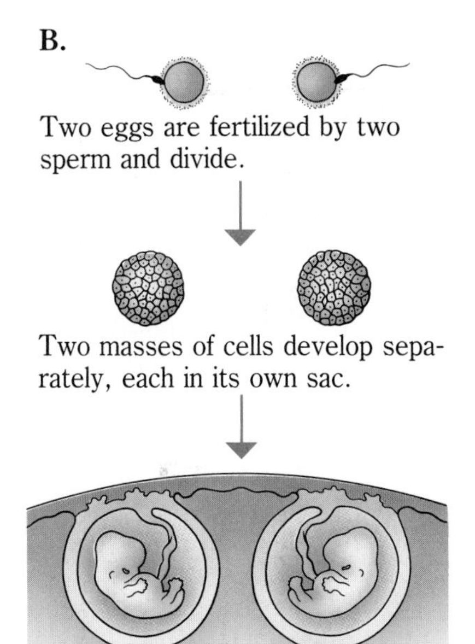

What Do You Think?

Think about the two possibilities above. Do they tell you why twins are sometimes identical and sometimes fraternal?

1. Which instance, A or B, do you think would produce two identical babies?
2. It is possible at birth to tell if twins are identical. How?
3. If you had identical twins, could they be:

   (a) both boys? (b) both girls? (c) one boy and one girl? Provide an explanation for your answers.
4. If you have fraternal twins, could they be:
   (a) both boys? (b) both girls? (c) one girl and one boy? Provide an explanation for your answers.

Episode 2—The Second Month

This month was a time of rapid growth. You grew about 5 times longer and about 500 times heavier. Your face and neck developed. Your brain developed at a faster pace than any other part of your body. Why do you think this happened?

By the end of the second month, your sex was clearly indicated by internal sex organs, which could probably have been seen externally too. Bones and muscles began to form. The arm and leg buds that formed last month began to grow into limbs, and your fingers and toes took shape. Gradually, constrictions formed to mark off elbow and wrist, knee and ankle.

Your bones first consisted of a soft substance called cartilage. Your bone structure has been changing throughout your prenatal (before birth) life so far, and it will continue to change throughout your postnatal (after birth) life. In fact, not until you become a mature adult will your skeleton be fully formed.

How big were you by the end of the second month? How many events did you predict correctly?

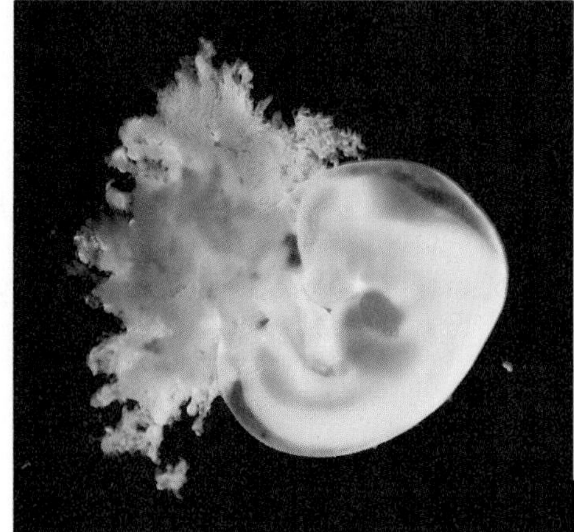

◄ **A 5-week-old embryo**

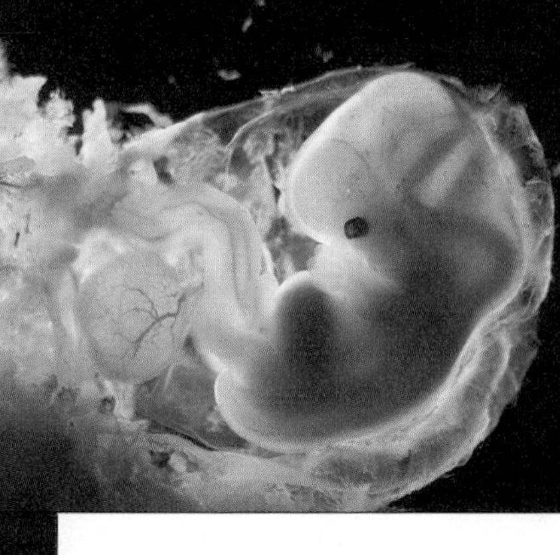

At 6 weeks, you were a little more than 1 cm long. The dark masses are this embryo's heart and liver. ►

◄ **A 6-week-old embryo from another angle**

Sometime during this month, this bit of a child, this baby grown to the size of a peanut, will begin to move, to flail its stubby hands and stretch its delicate legs. . . . I wonder what it must be like to bend a brand new elbow, to wiggle freshly made toes . . . .

And then there are the thousands of things that could go wrong . . . .

Each dividing cell, each growing organ must follow the four-million-year-old plan exactly—for the first time. There are no second chances. There is no going back, no catching up, no fixing mistakes. . . .

. . . we have been warned about Thalidomide, radiation, rubella. We have been told about nutrition and exercise, and what the Chinese knew a thousand years ago, that the child of a tranquil mother is calmer, brighter, and weighs more at birth. And so I drink milk for the sake of this creature's bones, eat protein for its mushrooming brain. I avoid alcohol, aspirin, and stress . . . .

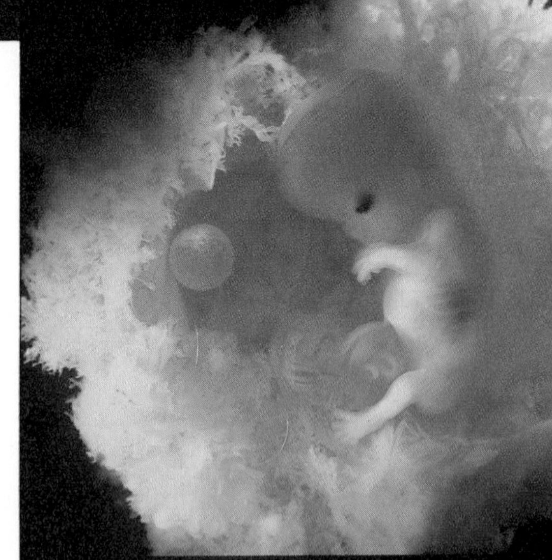

As an 8-week-old embryo, you were only the size of a walnut, but you had all of your basic organs and systems. Your head was almost half your total size.

Opinions and Viewpoints

As you have seen, the developing embryo is very delicate. What is the parents' responsibility to the developing embryo? This subject has prompted quite a bit of research and thought. Use the following questions to help you form your own informed opinions and viewpoints.

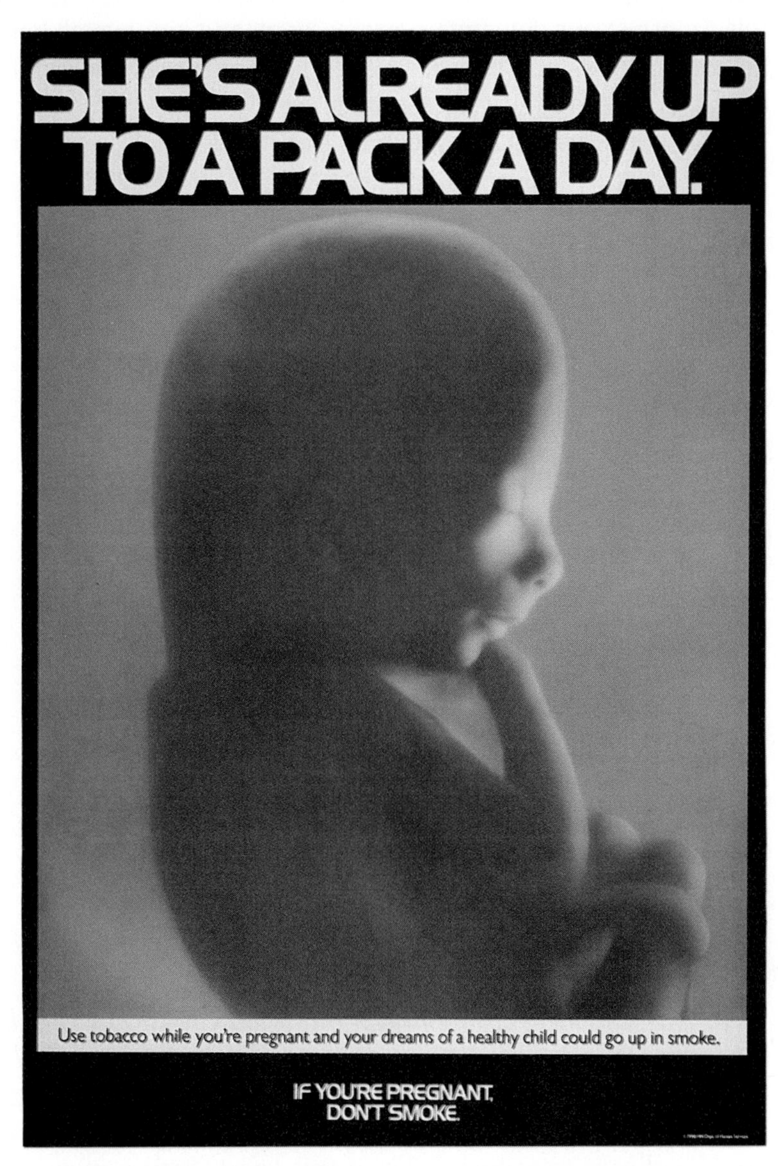

1. How dangerous is it for pregnant women to drink alcohol, to smoke, or to abuse drugs? What can happen to the developing embryo as a result of these activities? Find out all you can about this topic by consulting your library for books, magazines, and newspaper articles on the subject. You might start with *National Geographic's* article of January, 1992, on pages 36 to 39. For additional assistance, ask a nurse or doctor.
2. What is rubella? Investigate how a pregnant woman's exposure to this can affect the developing embryo.
3. Thalidomide was a drug used in some countries for a short time to help pregnant women who had nausea during the first months of pregnancy. There were disastrous results.
   (a) Do some research about this drug and the tragedy of its use.
   (b) What evidence indicates that it was used during the second month of pregnancy?
4. What is meant by radiation? How can it be harmful to the developing embryo?

Episode 3—The Third Month

From the third month on, the developing organism is called a **fetus.** A male fetus undergoes a lot of sexual development this month. This could be called the "tooth month" because your first teeth are starting to take shape, and tiny buds for your permanent teeth are beginning to develop. Your vocal cords formed during this month. Your digestive system, liver, and kidneys all began to function. Bones and muscles developed, and your face and body began to take on their human form.

After three months of growing, you were about 7 cm long.

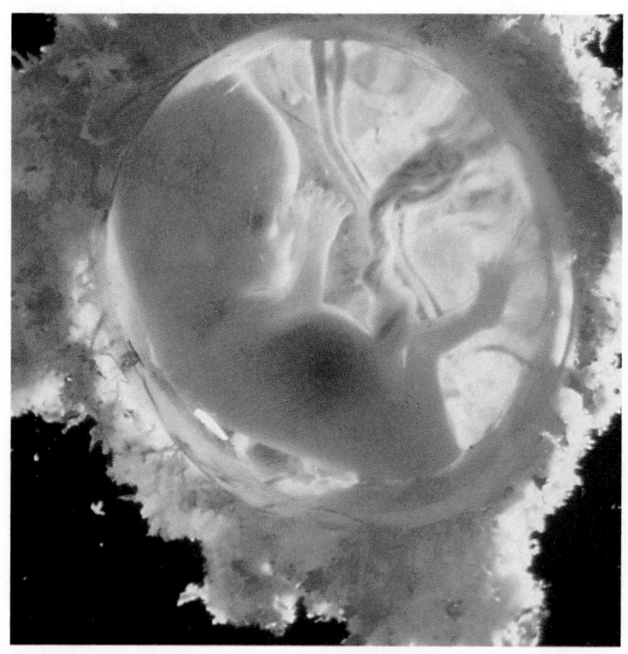

An 11-week-old fetus. Each hand was about the size of a teardrop, and your body weighed about what an envelope and a sheet of paper do.

. . . its umbilical cord connects our Riddle to its placenta like an astronaut to his spacecraft or a deep sea diver to her boat. A placenta is a weird, dense pudding of an organ, just thicker than a beefsteak and roughly round as a dinner plate. On one side, the pale cable of umbilical cord rises from it like a tree trunk from a rooty tangle of blood vessels, and on the underside, those veins have thinned to tiny capillaries that entwine with the capillaries in the wall of my uterus. There, this fetus and I communicate in . . . the language of diffusion, so that its needs are met, molecule by molecule, before it knows them. There, its blood rushes constantly to meet mine, to strip it of its lode of oxygen and nutrients, and give me in exchange carbon dioxide, uric acid, wastes.

. . . After it is born, and the cut stump of its umbilical cord sloughs off, this baby will be scarred forever with the pretty pucker of its navel, a remembrance of the communion it once took from my blood.

You have now reached the end of the first **trimester** of the developing life (first called an embryo, and later a fetus) inside a mother's body. Review what has happened. How does the order of events compare with what you predicted on page 466?

Considerations

1. (a) The writer of *The Life Within* calls her unborn child a "Riddle." Why do you suppose she does that?
   (b) Explain "the language of diffusion." Where are the membranes that permit diffusion? What substances diffuse? How is this like the process in a hen's egg?
   (c) You may have seen the "cut stump" of an umbilical cord on a newborn. What does "sloughs off" refer to?

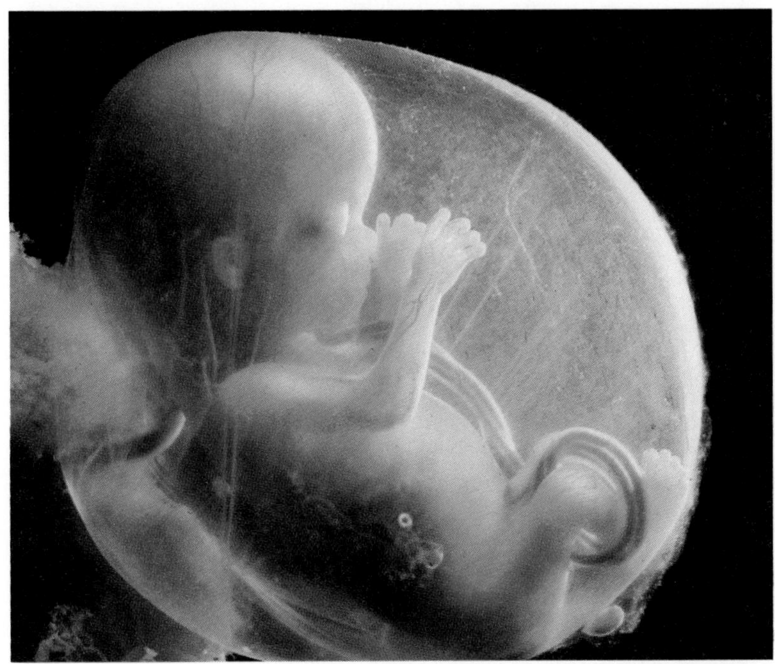

A 3-month-old fetus. By the 12th week, your eyelids closed over your eyes and remained shut for about 3 more months.

2. If a pregnant woman smokes cigarettes, do any of the substances from the smoke pass through the placenta into the fetus? Find out about this.
3. Suppose someone said, "A pregnant woman can do whatever she wants to do. In fact, she has the *right* to. This means she can smoke, drink, or take drugs, if that's what she decides to do." Would you agree or disagree with the statement? Gather information to support your opinion.

Episode 4—The Second Trimester

In this, the second trimester of development, your fingerprints and toeprints formed (month 4), followed by oil and sweat glands, scalp hair, nails, and tooth enamel (month 5). Your eyebrows and lashes continued to grow (month 6).

Your mother first became aware of your movement during the fourth month. You were ½ your birth height (⅓ of that was the length of your head), and you looked thin, wrinkled, and dark red.

During the fifth month, some of your skin cells died and fell off. They were replaced by new ones. A mixture of the dead cells and a substance from the skin glands formed a protective layer over your skin. Your mass was 0.45 kg. Your head was well-balanced, and your back grew straighter.

Your eyes (sealed at the age of 3 months) reopened during the sixth month. You had more taste buds on your tongue and in your mouth than you have now.

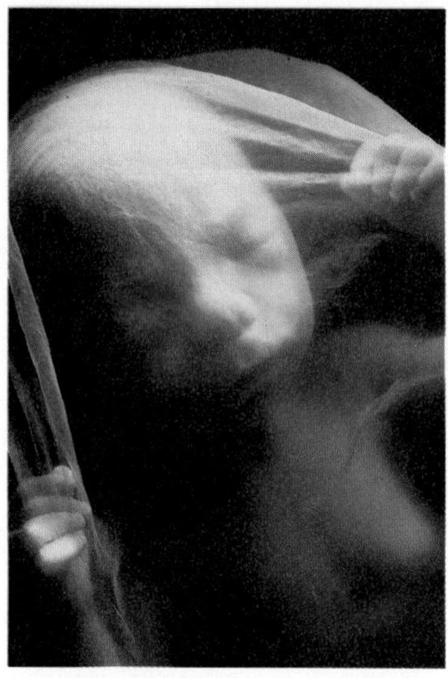

A 4-month-old fetus

Fetus at 4½ months

They say that already your face, with its newly finished lips, is like no one else's face.

. . . a twenty-week-old fetus is a porcelain doll of a child, complete with fingernails and eyebrows and a tender fringe of lashes sprouting from the lids of its still-sealed eyes. Fragile, luminous, it seems an unearthly being—more sprite than hearty human—through whose skin shines an intricate lace of veins.

. . . Always when it moves I know it now. It pummels my kidneys, kicks my stomach, punches my lungs. I know when it is resting, and when it wakes and stretches. I know it well by now. . . .

It knows me too. It dangles . . . inside me . . . listening to all my doings. . . . It hears all that I hear, the sounds intensified and distorted like the . . . racket one hears under the water in a swimming pool.

Thought-Provoking Questions

1. Why do you suppose the features of the fetus are already unique at such an early age?
2. Why are some mothers-to-be particular about the type of music they listen to?
3. Why are the sounds distorted?

Episode 5—The Third Trimester

In month 7, your eyes could respond to light, and the palms of your hands could respond to a light touch. Fat began to fill in the wrinkles during months 8 and 9. By the end of month 7, your mass was almost 1.4 kg. If you had been born at this point, you could have survived as a premature baby. Your activity increased, and you might have even changed position in the uterus. During this trimester, you were "practice-breathing," even though your lungs were filled with fluid. If your mother smoked just one cigarette, the practice-breathing stopped for up to half an hour.

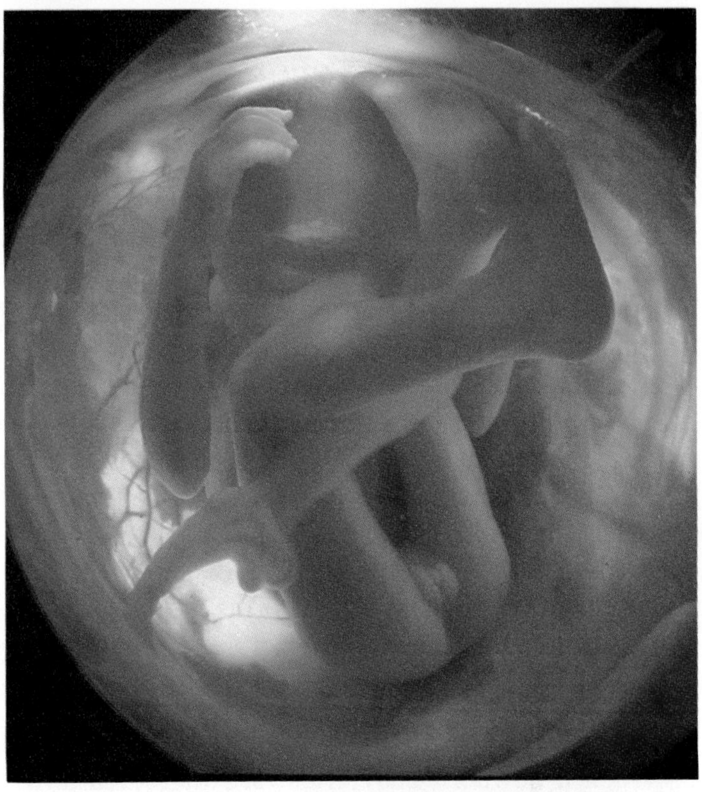

A 22-week-old female fetus. Notice the eyebrows. At this point, your mother could feel your movements, and she may have even been aware of your sleeping and waking cycles.

Medical researchers say that memory begins during the seventh month . . . . I wonder what this child will remember . . . .

. . . neurologists claim that by the eighth month this fetus will have begun a cycle of waking and sleeping whose rhythms are already influenced by the sun. And . . . when it sleeps, it dreams—or at least its brain waves resemble those of a dreamer.

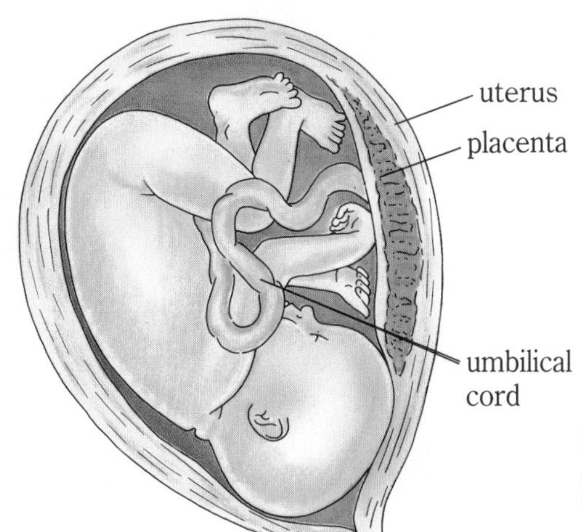

By the end of the ninth month, a fetus turns itself upside down, or into a head-down position. The umbilical cord supplies the fetus with food and oxygen and removes waste until after birth, when the cord is tied and cut.

Birth

At the end of pregnancy, the combined effects of several hormones and the stretching of the uterus cause the uterus to begin to contract with strong muscular movements. With these movements, the baby is pushed out of the uterus. Jean Hegland suggests, "For a baby, contractions may be caresses that express the fluid from its lungs and stimulate its skin in the same way that other mammals . . . do by licking their newborns."

At birth, you probably gasped, filled your lungs with air, and cried. The placenta and umbilical cord you no longer needed were cut away. When your wound healed, you were left with the scar of the navel. The navel has been called a "permanent reminder of our once parasitic mode of living."

People have very different life experiences, but everyone shares the prenatal phase, which forms the first personal history of each human being.

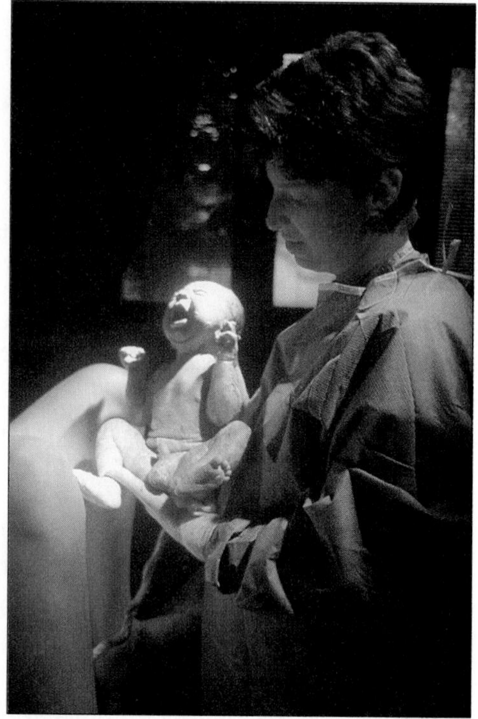

Within seconds after birth, you took your first breaths. Note that most newborns look wrinkled and red right after birth (regardless of their race). The cells that control skin color will begin to function soon after.

Going Further

1. When a newborn baby cries, are there tears? How do you explain this?
2. How old is a baby when starch can be added to its diet? Why?
3. What eye development has to take place after birth?
4. Explain what you think Jean Hegland meant when she wrote, "in the years to come, when we celebrate this Riddle's birthday, there is a way in which we will be commemorating little more than a change in its means of respiration and nutrition."
5. A sizable percentage of babies in the United States are delivered by *Caesarean section*. Find out what you can about this procedure. How is it performed? What are the advantages of delivery by Caesarean section? the disadvantages?

During pregnancy, the mother's breasts prepare for the production of milk. After the baby is born, breast milk can provide all the nutrients needed for almost the first year of a baby's life.

Be a Witness

Find out more about mammal birth by watching it or by talking to someone who has.

Here are some suggestions for what you might do.

1. Plan a visit to a farm, an agricultural fair, or a kennel to watch the birth of a mammal, such as a colt, a calf, or a puppy.

2. Interview a pregnant woman during her pregnancy and later, after her child's birth.
3. Interview a father, both before and after his child's birth. Many fathers are present during the birth of their children.
4. Watch a film of a human birth.

I Want to Know!

Patrick's class arranged for a visit from an obstetrician (a doctor specializing in human birth). Everyone was supposed to do some reading in advance and come up with some good questions. How might the obstetrician have answered them?

1. I read that a newborn has no bacteria living in its digestive system. Is that true? How do we get the bacteria?
2. One book said the fetus lives the life of a parasite. Do you agree? Explain.
3. Are the mother and fetus really separated, such that they don't touch? Would the mother's body really reject the fetus if they were touching?
4. If the mother's blood does not pass through the developing fetus, how does the fetus get what it needs?
5. I read that the fetus is floating and weightless, like an astronaut in space. Can that be so? Is there any advantage to that?
6. Someone told me that viruses and poisons can get into the fetus from the mother. Is that true? Can the AIDS virus do that?

What other questions would you like to ask? Make a list. Arrange to invite the doctor or other health professional to your class. You will likely make a lot of new discoveries.

Challenge Your Thinking

1. Could you draw a family tree for single-celled organisms? Could such organisms hold a family reunion? Why or why not?

2. Explain to a 9-year-old how this statement can be true: "Some living things multiply by division."

3. Miki told her mother this morning, "There are a bunch of fruit flies around the bananas in the cupboard, Mom. They just appeared—out of nowhere!" Is Miki right? Did the fruit flies grow from the bananas? Explain what must be happening, and suggest how she can keep fruit flies from becoming more numerous.

Where are these flies *coming* from?

4. The mother's blood does not flow through the body of the fetus. How, then, is the fetus harmed by chemicals from alcohol, tobacco, and other drugs taken in by the mother?

5. What characteristic of living things is shown when:
   (a) your body uses food to provide energy?
   (b) a dog pants?
   (c) a salamander grows a new tail?
   (d) a fly lays eggs on food?
   (e) a wound heals?
   (f) you leap out of the way of a car?

Living with Instructions

Miguel arrived home looking quite pleased.

"I got a model kit on sale—75 percent off the regular price," he announced, as he hurried to open a cardboard carton. "Look at this," he said, as he allowed dozens of parts to fall out onto the table.

"Oh no! There are no instructions—not even a diagram," he moaned. "I know it's a model of an airplane, but I'm not sure what it's supposed to look like. The illustration and model number on the outside of the box don't help me much."

Miguel called the store for help. The store manager said, "You'll have to send away for the instructions—to the head office."

What sort of instruction sheets should Miguel expect to get?

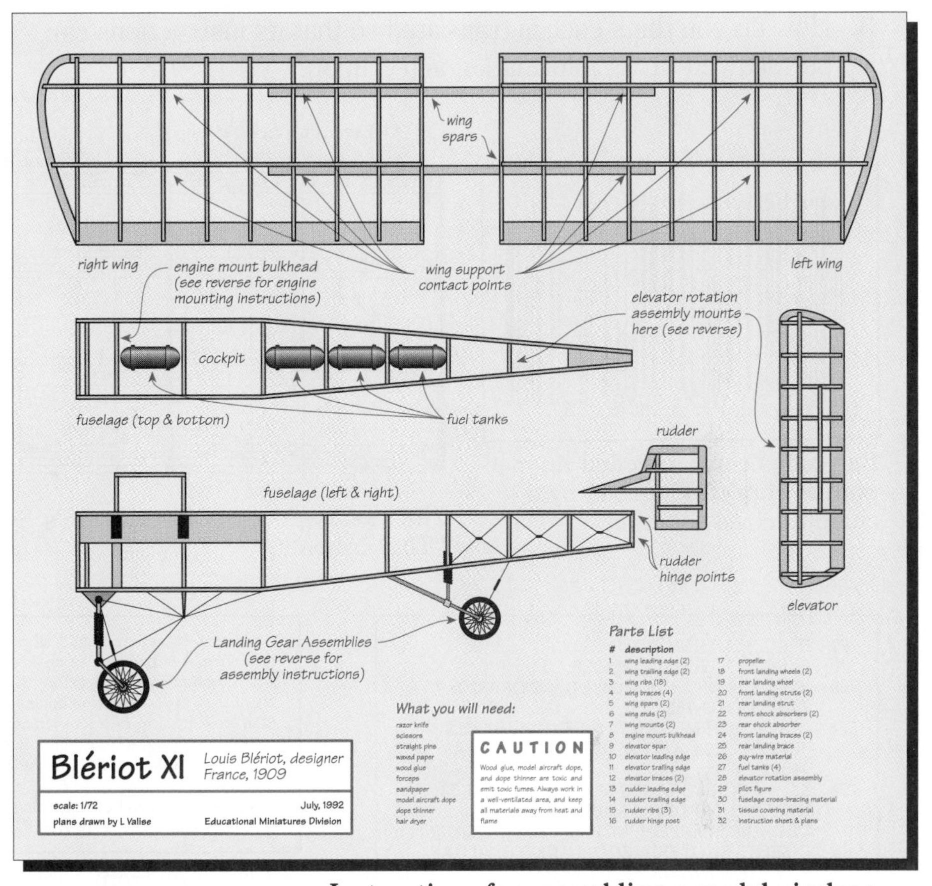

Instructions for assembling a model airplane

The pictures and diagrams show what the pieces look like, how to assemble them, and what the completed model should look like.

All Kinds of Instructions

However, not all information and instructions are given through pictures. Some instructions are in words alone, so they look nothing like the mental picture they produce. For example, a computer uses a series of codes for running a program. Earlier, you used holes in cards as a source of information about traits. The holes and cutaway sections did not resemble any of your traits, but they did serve to pass on the information. What mechanical means did you use to retrieve, or *translate,* the information? What information did you get? What other examples of instructions can you think of that do not use pictures?

Look at the following examples of sets of instructions and information. For each one, consider these questions:

1. What do you think the information or instructions are about? Do the instructions look like the end result?
2. What notation (set of characters) does each use? How is the notation arranged?
3. How do you think each is translated so that its instructions can be followed or its information acted upon?

Bar codes contain coded information that can be read by a computerized scanner.

The opening bars of Beethoven's *Symphony No. 3 in E flat, "The Eroica"*

```
190  IF UC = 4 THEN GOTO 220
200  IF UC = 7 THEN GOTO 250
210  IF UC < 4 OR UC > 4 OR UC < 7 OR UC > 7
     THEN GOTO 165
220  UL$ = "CORMORANTS & SEAGULLS":CL$ =
     "UPPER":TC$ = "WATERBIRDS":U = 5:AR =
     36521
230  X$ = "CORMORANTS SEAGULLS,BLUEFISH
     HERRING SARDINES SQUID,MOLLUSKS
     URCHINS STARFISH WORMS,CRUSTACEANS
     INSECTS,BACTERIA ZOOPLANKTON,ALGAE
     PHYTOPLANKTON PROTOZOANS,"
240  GOTO 280
250  UL$ = "WHALES & TUNA":CL$ =
     "FOURTH":TC$ = "FISH & MAMMALS":U =
     4:AR = 3521
```

This shows part of a computer program that keeps track of the food chains in a particular habitat.

dc in next sc, repeat from * to end of rnd. Then ch 1, and join with sl st to 3rd st of ch-4 first made. **9th rnd:** Ch 10, and complete cross st as before, skipping 2 dc between each leg of cross st and inserting hook under ch-1 sp. Ch 3 between each cross st. Skip 2 dc between each cross st. Repeat around (20 cross sts) and join last ch-3 with sl st to 7th st of ch-10 first made. **10th rnd:** Sl st in 1st 2 sts of 1st sp, ch 5, * dc in same sp, ch 3, sc in next sp, ch 3, dc in next sp, ch 2, repeat from * to end of rnd. Join last ch-3 with sl st to 3rd st of ch-5 first made. **11th rnd:** Ch 6, dc in same sp, * ch 4,

A few lines from some crochet instructions

In the examples above, the information is provided in a series of symbols or characters arranged in rows or lines. These sets of instructions are forms of *linear programming.* Which sets of instructions were developed for recent technology? Which have been used for centuries?

The mechanism of a music box

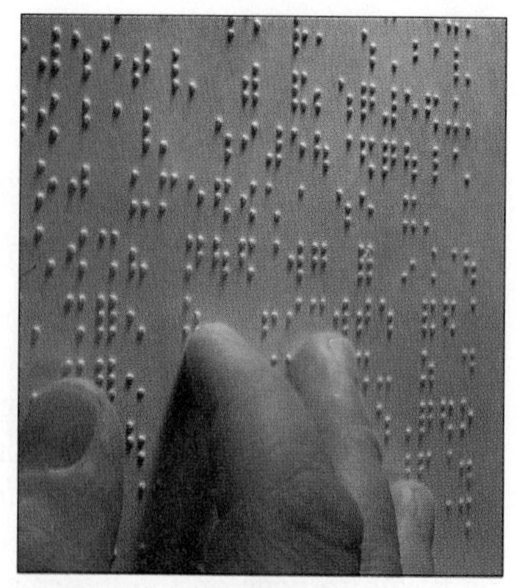

Braille words are printed as a series of raised dots. Blind people read by running their fingertips over the dots.

The Instructions That Shape Life

Like Miguel and his model airplane, scientists often have to work without instructions. For example, biologists looked for years for the instructions that guide the development of new living things. They needed to find out if there was a "head office" where the instructions were kept, what notation was used, how the notation was arranged, and how it was translated. Follow the story of how they found answers to some of these questions in the next Exploration.

EXPLORATION 4

Cell Search

What to Do

Read the descriptions below of some important findings made by scientists. Notice how technology helped them as they looked for answers to their questions about reproduction.

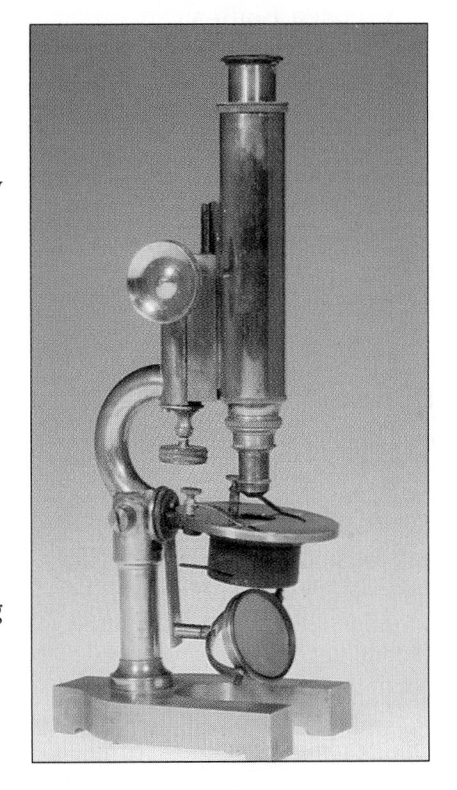

A. Body Fabrics
The tissues of the body (such as skin, bone, and muscle) were first described as woven materials, such as linen or rope. During the 19th century, the use of improved microscopes let biologists see the cells of living things for the first time.

B. Fertile Ground
Scientific techniques allowed biologists to observe living cells and some of their life processes in action. Using Petri dishes and microscopes, they watched fertilization of egg cells (ova) by sperm cells.

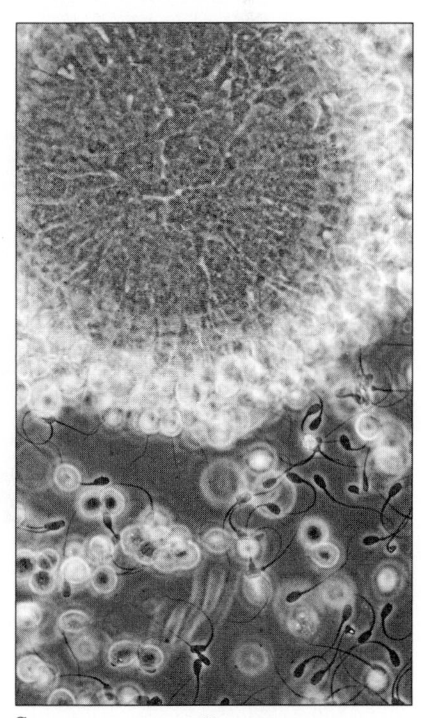

Sperm surrounding an egg

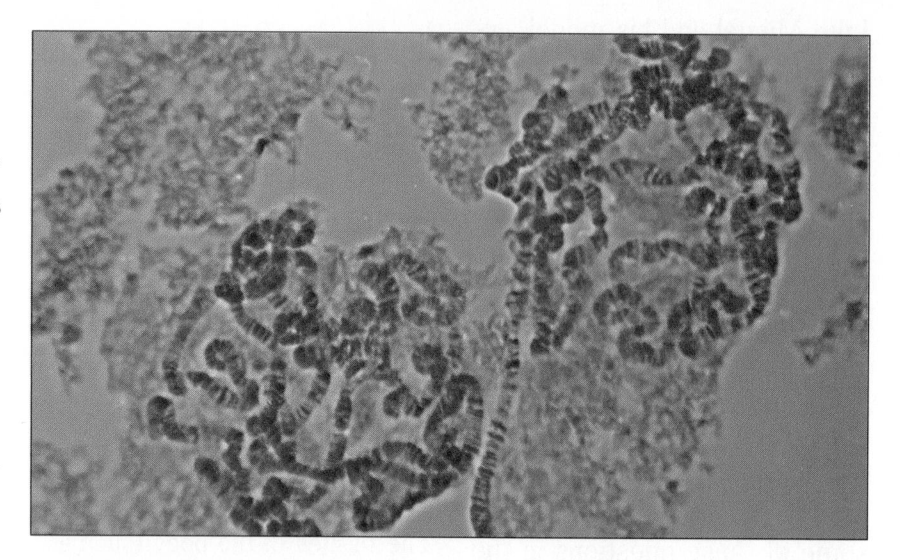

C. Colored Bodies

A special dye colored certain structures in the nucleus. These structures were named *chromosomes,* meaning "colored bodies." The function of the chromosomes was not known at first. The dye enabled scientists to observe the nucleus with much greater clarity.

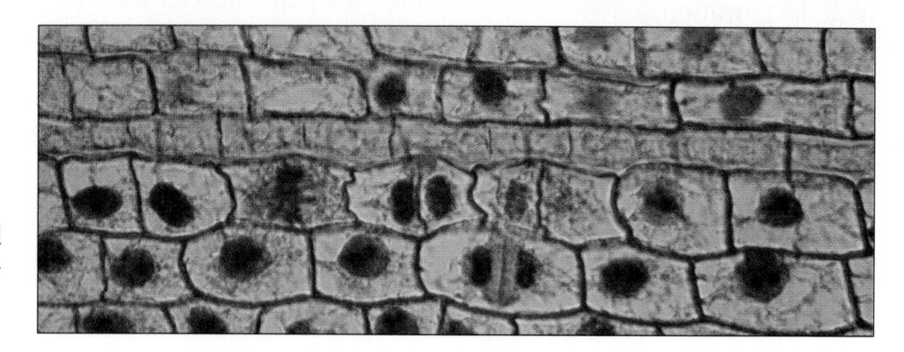

D. Inner Division

Biologists were keenly interested in observing the behavior of the chromosomes. Chromosomes seemed to appear right before cell division, and then a complex series of movements caused the chromosomes to be distributed equally between the two new cells. Because the same number of chromosomes appeared each time, they concluded that the chromosomes must be duplicated between each division. The careful way in which they were distributed to each new cell led them to believe that the chromosomes carried the hereditary instructions for the cells.

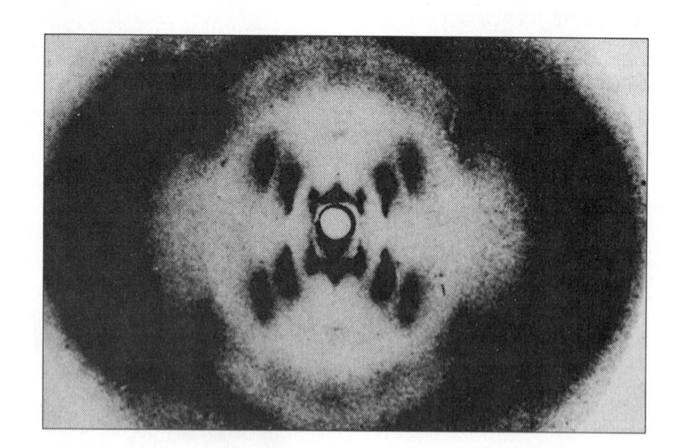

E. Shadowy Shapes

By the 1950s, scientists were using a photographic technique that uses X rays instead of light to study the structure of crystals. With this technique, Rosalind Franklin exposed crystallized molecules from the nucleus to X rays and then made photographs of the "shadows" cast by those molecules. James Watson and Francis Crick used her photographs to determine the structure of those molecules.

Inferences

Match each of the following inferences to observations A–E on pages 481 and 482.

(a) Two parents each contribute part of the instructions that lead to the development of offspring.

(b) A model could be constructed to resemble a molecule of material in the nucleus.

(c) The chromosomes contain instructions for offspring. These instructions are arranged in a form of linear programming.

(d) The nucleus is the "head office" of a cell—the location of the instructions for the development of new life.

(e) Body tissues are communities of similar cells joined together to form a structure.

Thinking It Over

What answers did biologists have so far about the mysteries of reproduction? As you read the following, think about the words that complete the sentences.

Biologists knew that two sex cells, __1__ and __2__, unite to start a new life. They called the combining of these cells __3__. The center of activity in the formation of the sex cells was always the __4__, where long, ribbon-like objects called __5__ took part in the process of cell division. The biologists believed these objects to be the sets of __6__ for the development of a new individual. They believed these sets were a form of __7__ programming. They also were able to identify the __8__ of a molecule of chromosome material.

Substance of Life

What was the material that biologists were viewing in the nucleus? Its shape and structure were identified in 1953 by James Watson and Francis Crick, who received the Nobel prize for their discovery. The material is called **DNA**—an abbreviation for the chemical that contains the instructions for cells. (*DNA* stands for *deoxyribonucleic acid*.) It is found in the chromosomes and is essential for all forms of life.

The models of DNA below may remind you of the spiral ladder-shaped structures that Gloria and Yoon saw in the nucleus at the Science Discovery Center. The spiral ladders they watched were molecules of DNA.

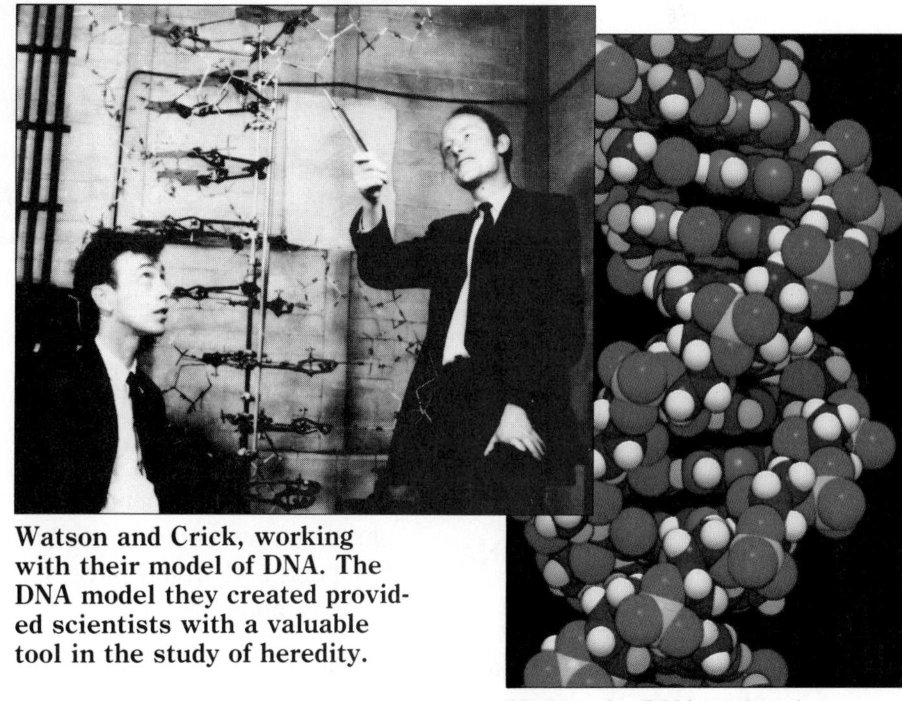

Watson and Crick, working with their model of DNA. The DNA model they created provided scientists with a valuable tool in the study of heredity.

Model of a DNA molecule

Further investigation showed how the DNA molecule stores the instructions for life. There are four chemicals that bond together, two at a time, to form the rungs of the DNA ladder. The instructions for life are based on specific sequences of these bonded pairs. A lot is known about how the chemical code is translated into action as new living things develop. However, much remains to be discovered.

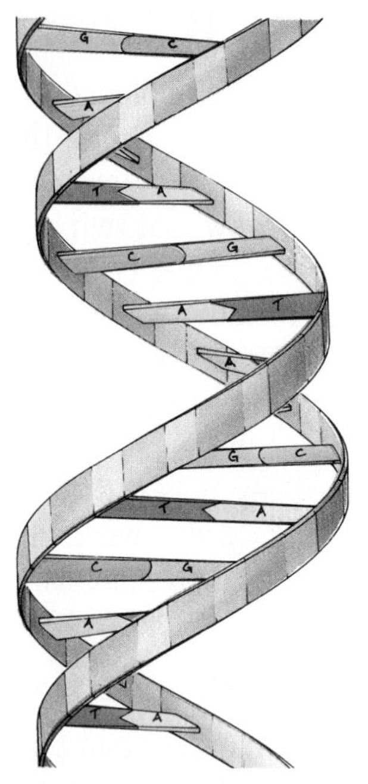

The structure of DNA is much like a twisted ladder or a spiral staircase.

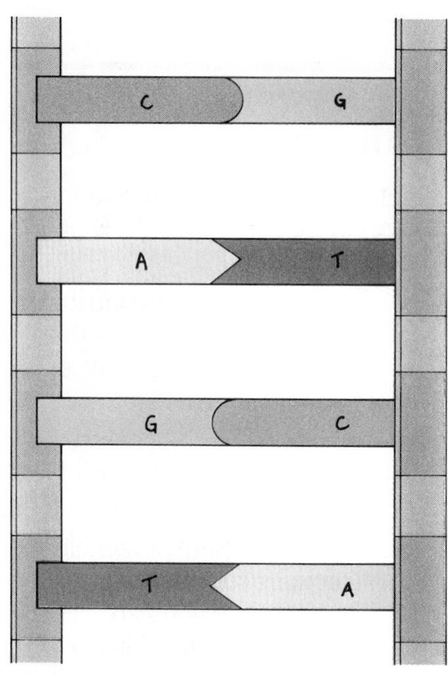

The structure of DNA, untwisted so that you can see the "ladder rungs" more clearly. The letters *C, G, T,* and *A* stand for the chemicals that combine to form the rungs of the "ladder."

Mind-Benders

Here are some mind-benders about the code of DNA—the language of life!

- DNA has an alphabet of four chemical "letters" that form the rungs of the spiral ladders. The DNA in a human contains more than a billion of these letters, which would fill 10,000 novels of 100,000 words each.
- If you could completely uncoil a DNA molecule in one of your cells, it would measure almost 2 m. Each cell contains many strands of DNA, and you have more than 10 trillion cells in your body. All the DNA they contain could stretch to the sun and back.
- Chromosomes contain long strands of DNA, which are "read" in segments called genes. A section of the DNA spiral ladder with 2,000 rungs might make up only one gene. Each gene is like a "sentence" in the language of life. Each of your cells contains about 100,000 genes, or "sentences" of instructions.
- The DNA of all 4 billion people alive now could fit on 1 teaspoon and would have a mass of about 1 gram.
- The long series of letters shown above is part of the instructions for a protein in human blood. You can see that it is in the form of linear programming. The whole code for the protein would fill up this page. To print the genetic code for one single cell from your body, however, would require about 1 million pages this size.

```
10        20        30        40        50        60
                              GACTCTAGAGGATCCCCGGGT
                              GACTCTAGAGGATCCCCGGGT
80        90        100       110       120       130
ATCATGGTCATAGCTGTTTCCTGTGTGAAATTGTTATCCGCTCACAATTCCA
AT?ATGGTCATAGCTGTTTCCTGTGTGAAATTGTTATCCGCTCACA?TTCCA
50        160       170       180       190       200
AGCATAAAGTGTAAAGCCTGGGGTGCCTAATGAGTGAGCTAACTCACATTAA
AGCATAAAGTGTAAAGCCTGGGGTGCCTAATGAGTGAGCTAACTCACATTAA
20        230       240       250       260       270
CCGCTTTCCAGTCGGGAAACCTGTCGTGCCAGCTGCATTAATGAATCGGCCA
CCGCTTT?CAGTCGGGAAA?CTGTCGTGCCAGCTGCATTAATGAATCGGCCA
90        300       310       320       330       340
TTTGCGTATTGGGCGCCAGGGTGGTTTTTCTTTTCACCAGTGAGACGGGCAA
```

Stop and Think

Maria: "So genes are found in chromosomes?"
Hugh: "Chromosomes are made of genes—aren't they?"
Kim: "And we have more than 100,000 genes?"
Hugh: "Well, where is this stuff called DNA?"
Carlos: "Someone said the genetic instructions have a kind of alphabet."
Maria: "There are 4 letters."
Kim: "Those are 4 chemicals—remember the ladder rungs?"
Carlos: "And groups of them make up the genes?"
Maria: "The genes must be like sentences of instructions . . ."
Hugh: "and those instructions tell cells how to develop . . ."

Help Maria and her friends start thinking clearly. Which of their comments do you agree with? Which do you think are wrong? Sort out and organize their thoughts about genes, chromosomes, DNA, and genetic instructions.

When you feel certain of the structure and organization of genetic material, reconsider the function of DNA. The instructions of the genetic code control the formation of a new life—from the first cell division, to the formation of cells of various sorts (such as nerve and muscle cells), to the organization of the cells into an organism. How do you think the instructions account for all the characteristic traits of an organism? Does this mean that every organism is completely programmed by its genes?

What other factors besides genes affect who you are? In the following lesson, you'll consider the importance of environment on the formation of life.

6 Environment or Heredity?

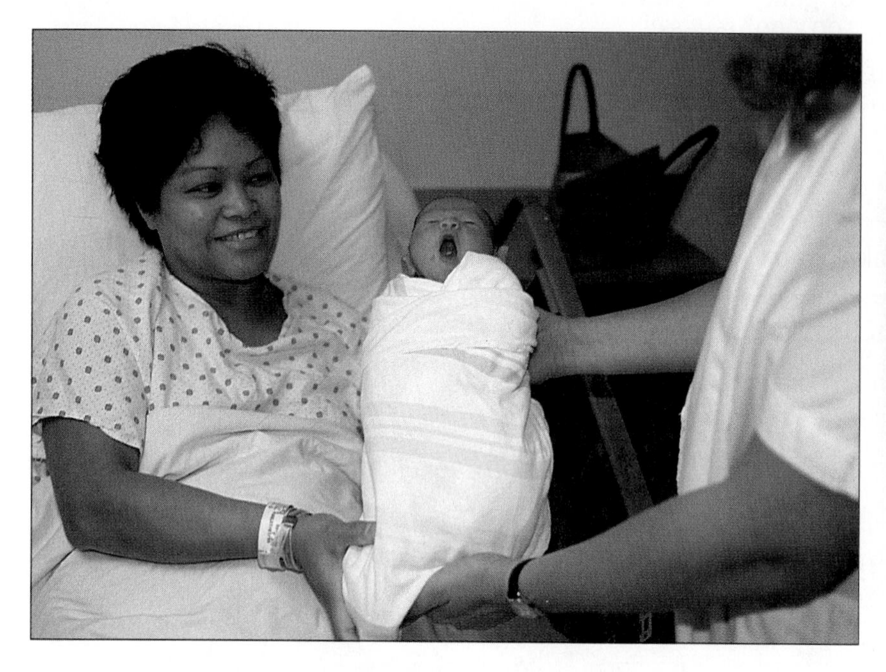

What are the phenotypes for several traits of the puppies shown here? Can you tell what their genotypes are for these traits?

The instructions have been given. All the pieces have been assembled properly. In other words, a baby has been born. The new human being has 100,000 gene pairs inherited in two sets—one from each parent. The huge number of traits that the gene pairs code for are assembled in a unique combination. With 100,000 gene pairs to juggle, you can see how no one else is exactly like you. But genes are not the only reason you are who you are. Factors from your environment also greatly affect your unique combination of characteristics.

Family-Tree Time

At this point you can add to your family-tree project. You will need to trace the family's traits back several generations. Talk to the family members you are studying to learn about those traits. In the project, include lists of the following:

1. All the observable traits
2. All the traits that may be inherited but that will not show up until later in life
3. Any recessive traits that may have skipped a generation (as short plant height did in Mendel's pea experiments)

Geneticists call the description of an organism's traits its **phenotype.** Your phenotype is the physical or chemical expression of your genes. A phenotype can be observed visually, as in flower color, or chemically, as in the test used to find out a person's blood type.

The actual pair of genes you have inherited for each of your traits is called your **genotype.**

For each individual, the traits you listed in 1, 2, and 3 above will fall under one or both of these categories—*phenotype* and *genotype*. See if you can identify which traits are part of your phenotype and which are part of your genotype. Can a trait be part of your genotype but not part of your phenotype? Can the reverse be true?

There's just one more piece of information to add to your family-tree project. Read "How Important Is Your Environment?" below to discover what that information is, and then complete your family tree. Of what value could it be to a family to have the information you have gathered for this family tree? In what ways do you feel more informed about heredity after doing this project? Record your thoughts about these questions in your Journal.

How Important Is Your Environment?

Which has a greater effect on what you are—the genes you inherit, or your environment? What does environment include? Food and nutrition are major factors. Consider a girl like Emily, shown on the right. In one environment she might eat a healthy, well-balanced diet. In a less fortunate environment, she might eat fewer meals and have less nutritious foods. In each environment, would her genotype be the same? Her phenotype? How might the food she eats affect her health and appearance?

Investigate, as far back as possible, the length of life of people in the family you're studying. Look for patterns. For example, many of the father's family may have lived to be 80 or 90 years old. Many people believe that the length of a person's life, or **longevity,** is influenced by heredity. In what ways could your environment affect your life span? Identify as many environmental factors as possible.

Add the information about the longevity of your ancestors (or the ancestors of the family you chose) to your family-tree project. Make a prediction about your own longevity. If the family is not your own, make a prediction about the longevity of one of the younger family members. How can people alter their environment to try to extend life?

"Environment can affect a person as much as heredity does."

Do you agree or disagree? Begin collecting information that illustrates why this statement is actively debated. Think about Emily, described above. Ask people you know about their opinions.

A Field Trip

Visit an animal breeding farm or an experimental nursery and explore the environment. Find out from someone there how they make use of heredity to produce healthy, productive animals or plants. Ask about environmental effects on living things. In your Journal, restate the previous quote for the plants or animals you investigated: *"Environment can (cannot) affect animals (plants) as much as heredity does."*

With continuous light and constant temperatures, these turnip plants can flower in only 3 weeks.

Surprising Evidence or Coincidence?

The twin brothers shown at right were separated at the age of 4 weeks and were not in touch with each other for 40 years. They grew up in different areas with different families. They were brought together again by a group doing an investigation about heredity and environment. The results surprised everyone. The two men were a lot alike—not just in appearance. They both liked math and hated spelling in school, had similar hobbies, went to the same beach areas in Florida each year, drove the same kind of cars, married women named Betty, named their firstborn sons James Alan and James Allen, had dogs named Toy, had high blood pressure, gained and lost weight at the same stages in their lives, and had the same type of recurring headaches. How does this study add to your understanding of the role of genes and environment in development?

EXPLORATION 5

How Alike?

1. If you know any identical twins, arrange an interview with each one separately. Make up a set of questions in advance about their likes, dislikes, habits, and abilities. Ask each person the same questions, and then compare their answers.
2. If you know any fraternal twins, arrange to ask them the same questions. Their inheritance is not identical, but they have been living in the same environment. Compare these figures with those of other mammals. What environmental factors do you think affect twins while they are still in the uterus? in their first year of life?
3. If you don't know any twins, you can do research to discover more about them. Many studies have been done on twins and other multiple births. What are some unusual similarities documented about twins, triplets, or quadruplets? What is the likelihood that a woman will have a multiple birth?
4. What do you think now about the effects of heredity and environment? Which one matters more?

When Heredity Goes Wrong

Most of the time, the instructions for life are followed precisely. But suppose something goes wrong. Although 99 percent of the babies born are the result of normal development, there are babies born with various birth defects. Some embryos cannot develop beyond the first stages of pregnancy and are rejected by the mother—often before she knows she is pregnant. It has been estimated that this may happen 500,000 times every year in the U.S. alone.

Why do these things happen? What goes wrong? Could it be that sometimes, in the process of cell division, instructions get improperly duplicated? Perhaps sections of instructions in the sex cells are lost, altered, turned upside down, or placed out of order. Perhaps the instructions are perfect, but there is an error in their translation in the developing embryo.

Study the flowchart, beginning at the top and working down. What is it saying? Write a script to accompany this chart so that someone else will understand it. What would be a good title for the flowchart?

A Disaster or a Wonderful Mistake?

A mutation is a permanent change in the genetic code. Mutations may produce changes that are disastrous to living things. On the other hand, without mutations, living things would not have changed, or evolved, to their present form. The remarkable diversity among the living things around you would not exist if mutations had not occurred over time.

What are some examples of nature's "wonderful mistakes"? This may be a subject you would like to investigate.

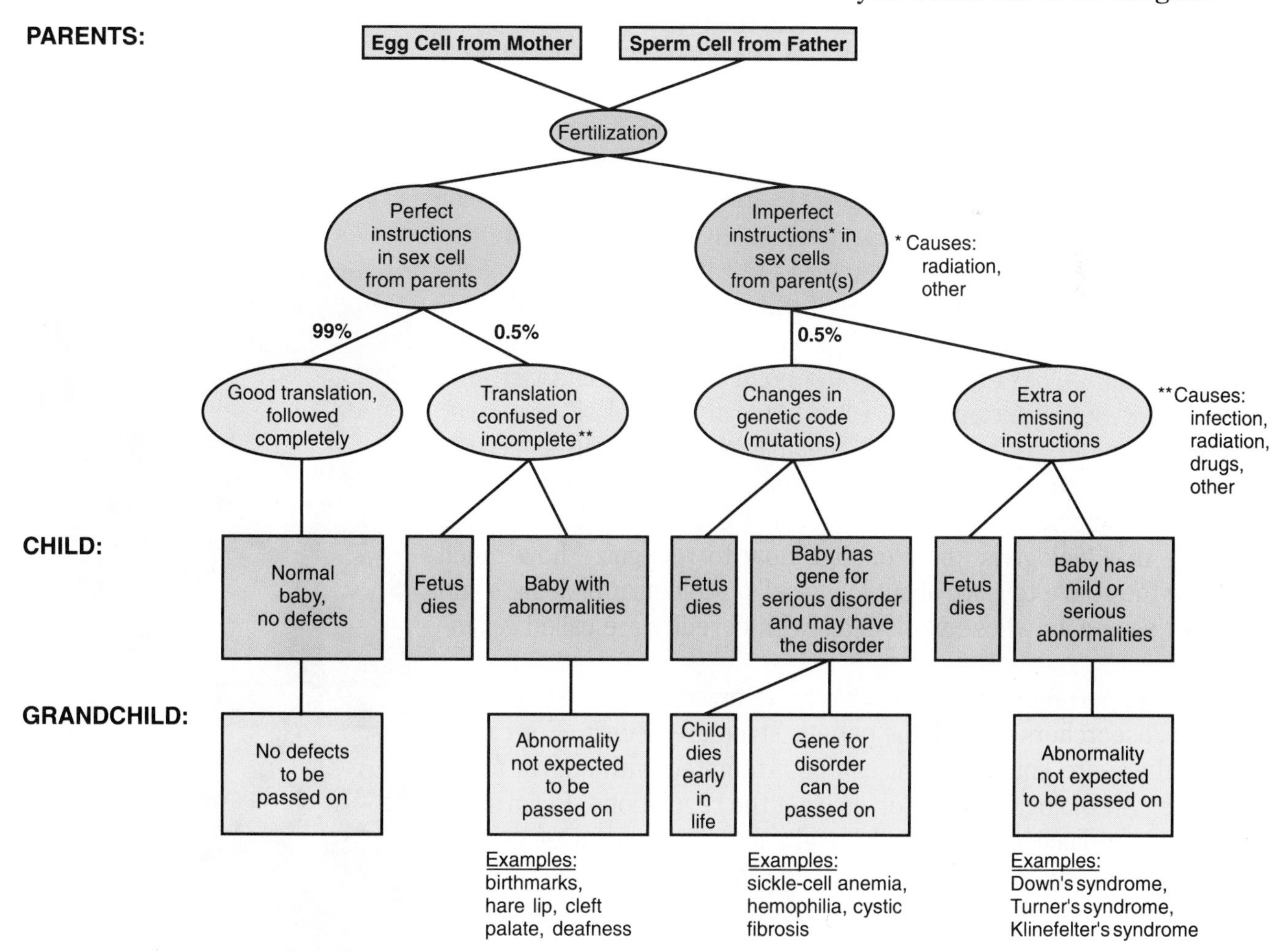

7 What the Future Holds

Since an early age, you have been able to examine life all around you. You have learned a lot about yourself, your environment, and other living things. You have learned something about the structure of living things and the use or function of their parts. You have become especially well-acquainted with the plants and animals you use for food. Your eyes alone have provided you with a lot of information about the world around you.

Your experience has mirrored that of biologists. Their examination of living things was limited, for a long time, to what their unaided eyes could tell them. After the invention and perfection of the microscope, their main approach changed from studying the whole animal to examining the animal's tiny details. Now biologists are able to study life at the level of the molecule of life—DNA. If you hear the term *molecular biology,* you will know what it refers to.

Technology has advanced to the point that biologists can now work within the nucleus of a cell. They can manipulate, or move, DNA from one cell to another. This manipulation of DNA is called **genetic engineering.** It has become possible to perform some remarkable procedures. However, the greatest realization seems to be that biologists know enough now to recognize how much more they have to learn about living cells. Molecular biologists and other scientists whose work deals with heredity are called *geneticists.*

It is important to note here that there are regulations to control what researchers do with the genetic structure of cells. For example, there are limits to what can be attempted with human fertilized eggs. Biologists themselves want the regulations. Form small groups to discuss why people might fear the results of genetic engineering involving humans. Do you think the possible dangers posed by genetic engineering outweigh its benefits to human life?

On the Horizon

What should you expect to hear about in the future? Several possibilities are listed below.

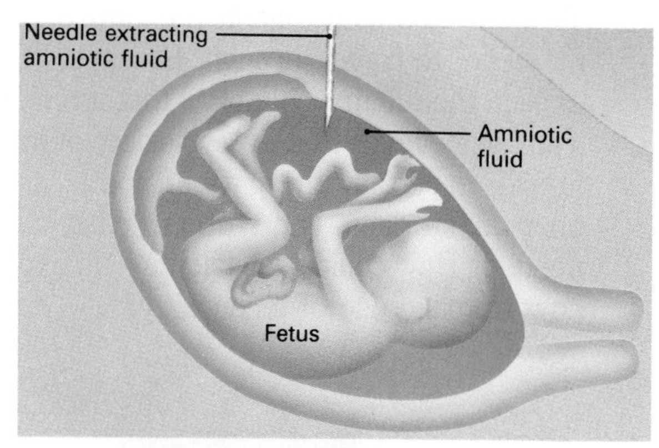

Amniocentesis involves withdrawing some of the amniotic fluid that surrounds a fetus and then analyzing cells from that fluid.

(a) *Improved ways of detecting genetic disorders early in pregnancy.* **Amniocentesis** is a procedure used to obtain fetal cells for examination. It involves the use of a long needle to withdraw some of the fluid surrounding the fetus. This fluid contains skin cells that have been shed by the fetus. Chemical tests of the fetal cells can detect genetic disorders.

There is a risk of death to the fetus in performing this procedure. A new technique is being developed that requires only a sample of the mother's blood. The technique involves finding and removing the very rare fetal blood cells that enter the mother's bloodstream through placental leaks. What do you think would be some advantages of using the new technology?

(b) *Discoveries about the location of genes that cause birth defects and diseases.* The genes associated with cystic fibrosis and a form of muscular dystrophy have already been located.

(c) *Gene therapy.* There is hope that hemophilia and other disorders can be corrected through gene therapy. In gene therapy, some cells are removed from the patient, healthy genes are inserted into them, and the cells are returned to the patient's body. The substance lacking in the person's body is then produced by the new genes. Would this procedure have any effect on the descendants of a person who had undergone gene therapy?

(d) *Attempts to change genes in a fertilized egg.* Why might this procedure have more far-reaching effects than the gene therapy described above? Explain.

(e) *Clones.* When you cut off parts of a mature plant (such as leaves or stems) and grow new plants from them, you are producing **clones.** Clones have the same genetic makeup as the mature plant. They are new plants produced without sexual reproduction.

It is now possible to clone certain animals too. Are human clones possible? This issue is being debated right now.

Keep in mind that human clones would not first appear as full-grown adults. Instead, they would have to mature as every other embryo does—and you have already seen the kind of effect that environment can have on development.

These plants are being cloned by placing cells from various plants in their own individual test tubes. The cells will produce plants that are genetically identical to the plants the cells came from.

What lies ahead is unknown. The noted scientist, Lewis Thomas, has said that in the future, there is a "wilderness of mystery" to travel through. He says, "Science will not be able to explain the full meaning of all that it makes possible." As a result, Thomas believes that every variety of talent will be needed for the future, especially the gifts of poets, artists, musicians, philosophers, historians, and writers. What talents do *you* think will be needed? How could such talents help explain the process of life and how to make use of technology?

In vitro fertilization allowed this eland to give birth to a bongo calf. Elands and bongos are different antelope species.

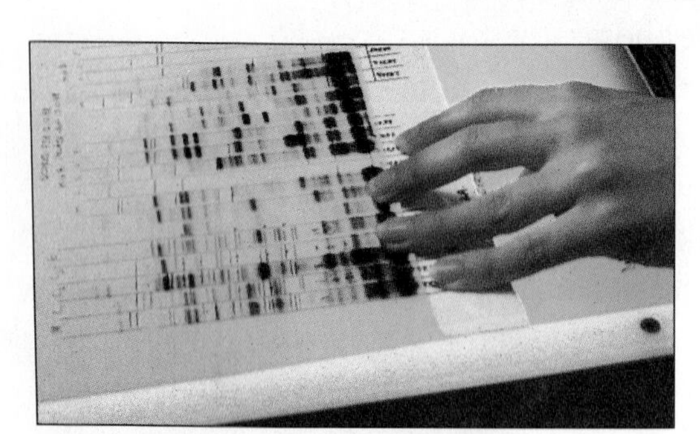

DNA fingerprinting is one of the newest tools used in solving crimes. When a small amount of body tissue or a strand of hair is discovered at a crime scene, scientists can make a "DNA fingerprint" from the body tissue. The DNA content can be analyzed and compared to the DNA of a suspect.

A genetic counselor advises couples about the probability of their passing on genes for a genetic disorder to their offspring.

Challenge Your Thinking

1. *The 16-year-old clone was nervous. He had just been reassigned as COPI-864 (Clone, On Parts Inventory), and was one of the next to be used if transplants were needed—a leg, a kidney, or perhaps his brain. Although most clones had been brainwashed into accepting their roles, COPI-864 plotted his escape.*

   His adventure came to a successful conclusion, and the movie ended.

   At home that evening, Tranh and Nu talked about the movie and the whole idea of human clones. They wanted to find out all they could about the possible results of cloning. They were interested not just in humans, but in all sorts of living things.

   They started writing down a list of experts who could contribute knowledge and valuable viewpoints to a discussion about cloning. Complete their list by adding at least five more people.

   - owner of prize cattle or other farm animals
   - director of a zoo for endangered wildlife

   What might each expert have to say?

2. Are there inherited characteristics that are not affected by the environment? Are there human characteristics that are not affected by genes? Support your answers with evidence.
3. Explain in your own words how each of the items grouped together are related.
   (a) linear programming, DNA, trait
   (b) double helix, Watson and Crick, X rays
   (c) heredity, genotype, phenotype

Unit 8 Making Connections

The Big Ideas

In your Journal, write a summary of this unit, using the following questions as a guide.

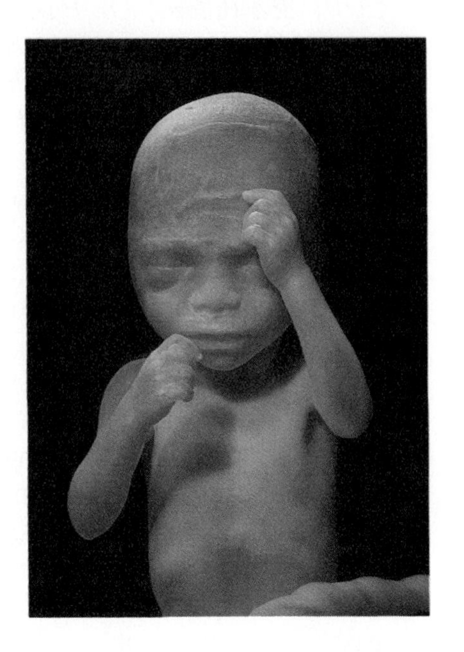

- What does it mean to "inherit" your grandfather's nose, eyes, or other feature?
- What are four abilities that all living things share? What does each involve?
- What part of a cell controls the formation of new cells? How is this accomplished?
- How do you distinguish asexual reproduction from sexual reproduction?
- What patterns did Mendel discover as he worked with pea plants?
- What parts of an embryo develop early in pregnancy? mid-way through? late?
- How is genetic information organized? What structures are involved?
- What is meant by "Chromosomes are a form of linear programming"?
- How can the environment affect development (both before and after birth)?

Checking Your Understanding

1. **Consider the Possibilities**
   Think about the situations presented below. What would be the possible effects on a couple's descendants if:
   (a) the father had lost a finger due to an accident at the plant where he worked?
   (b) the mother (unvaccinated) was exposed to rubella during the early weeks of pregnancy?
   (c) the mother abused drugs during pregnancy?
   (d) the mother went to a scary movie about vampires during the early weeks of pregnancy?
   (e) the mother was exposed to radiation during the early weeks of pregnancy?
   (f) the father was exposed to rubella during the early weeks of the mother's pregnancy?
   (g) a woman was vaccinated against rubella before she became pregnant?
   (h) a woman who abused drugs (of any sort, including alcohol and tobacco) gave them up before she decided to have a child?
   (i) the father had abused drugs known to cause chromosome damage?
   (j) the father worked around insecticides suspected of causing genetic mutations?

2. Tongue-Rolling
   (a) Draw a Punnett square to show the possible combinations among the children shown in the pedigree below.
   (b) Fill in the children's genotypes on the pedigree.
   (c) Who are the tongue-rollers and non-tongue-rollers in this family?

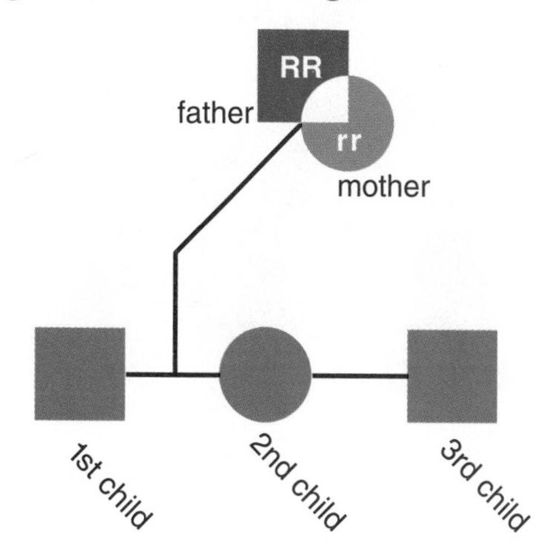

   (d) Use Punnett squares to show the possibility of tongue-rollers among the grandchildren.

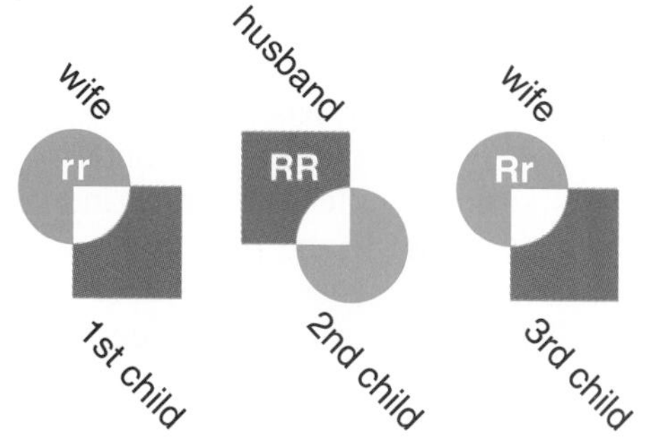

3. Develop a concept map using the following terms: *chromosome, dominant, fur color, gene, recessive,* and *trait.*

Reading *Plus*

You have been introduced to how hereditary factors are responsible for traits. By reading pages S125–S142 in the *SourceBook,* you will take a closer look at how hereditary patterns are determined.

Updating Your Journal

Reread the paragraphs you wrote for this unit's "For Your Journal" questions. Then rewrite them to reflect what you have learned in studying this unit.

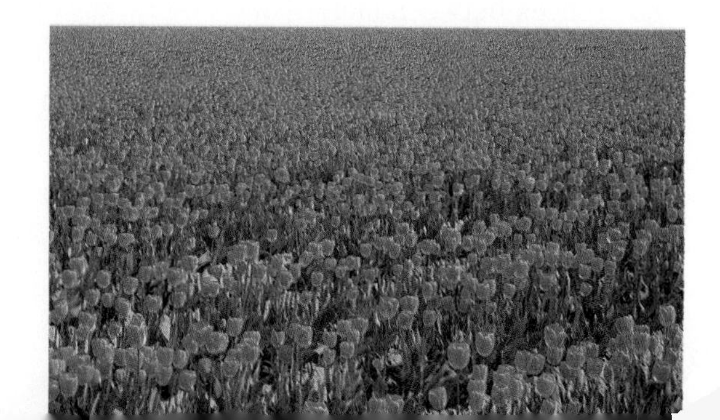

Science in Action

Spotlight on Molecular Genetics

Nora Lapitan examines experimental plants in her laboratory.

One branch of genetic research is offering exciting possibilities for farmers of the future. Geneticists are working to increase the disease resistance and stress tolerance of plants, and to improve the yield and nutritional quality of major crop plants. They do this by manipulating the genes of plant cells and then growing these cells under sterile laboratory conditions.

Nora Lapitan is a geneticist and assistant professor of genetics at Colorado State University. She received her bachelor's, master's, and doctoral degrees in Genetics.

Q: What do you do on a daily basis?

Nora: My position involves both research and teaching. I usually teach one semester out of the year and spend the second semester and the summer doing research.

Q: What does your research involve?

Nora: My overall goal is to use molecular techniques for plant improvement. I do this with a team of students and researchers by first isolating DNA from an organism and cutting it into fragments for study. When there are genetic differences between individuals, differences in sizes of DNA fragments are visible. We look for these differences in order to build what we call a molecular map. Once we've built the map, we use it to find genes that affect the economically important traits in crops.

For example, we're currently looking for genes that provide resistance to the Russian wheat aphid. This pest has recently come to the United States and has seriously affected barley and wheat crops. In the past five years, it has caused millions of dollars' worth of crop damage to farmers.

With a molecular map, we hope to find the gene that makes barley and wheat resistant to the Russian wheat aphid. Then with regular plant-breeding procedures, we'll cross a susceptible plant with a resistant one. The offspring will include both susceptible and resistant plants. We can then use those that pick up the resistant genes and discard the others.

Q: When did you first become interested in the field of genetics?

Nora: Good question! The high school I went to was strongly focused in the liberal arts. My favorite subjects were writing and public speaking. As a result, when I went to college I was originally a mass communications major. The university was very strong in the sciences, however. As I became more exposed to math, physics, biology, and genetics, I soon realized that I excelled in the sciences. So, I changed my major to genetics, and here I am!

This scientist studies a field of barley infected with Russian wheat aphids.

A special dye helps these scientists examine the roots of experimental soybean plants.

Q: What are the most exciting aspects of your work?

Nora: What I enjoy most about my job is the research. I think asking questions, doing experiments, and finally getting results is very exciting—especially when you get the unexpected! These results are exciting because they can really have an impact—they may even change the way people think about what they can and cannot do with genetics.

Q: Are environmental issues involved in your research or activities?

Nora: By producing crops that are resistant to some pests, farmers can reduce the amounts of pesticides they use. Since pesticides can be harmful to the environment, I feel like we're not only helping farmers cut their costs, but we're also doing something positive for the environment.

Q: What qualities do you feel are essential for a person to be successful in genetics?

Nora: I think having a positive attitude and being persistent in achieving goals are the most essential qualities. I've seen really smart people who haven't gone far in this field because when things go wrong, they get discouraged and give up. On the other hand, I've seen people who are of only average intelligence become successful in genetics because they really apply themselves and believe in what they're doing.

Q: How would you describe the importance of your work to high school students?

Nora: My work is important because by understanding how organisms inherit characteristics, we have a way of controlling how those characteristics are passed on. Similarly, the more we understand about human genetics, the more likely we are to find cures for inherited diseases.

A Project Idea

As human populations grow, deforestation has become a widespread concern. How do you think genetic engineering could help solve the problem? Look into how plant clones are being used to produce new trees and to improve forests. For example, examine how scientists have been using "plantlets" from a single piece of redwood tissue to produce seedlings that grow twice as fast as regular seedlings. Prepare a poster or a report showing your findings.

Science and Technology

Rain Forest Treasures

Some of the world's greatest treasures are not under the ground or at the bottom of the sea. They are not precious stones, or minerals, or works of art. They can be found all over the world, but they are especially abundant in the Earth's rain forests.

Tropical rain forest

Genetic Resources

The treasures are genetic resources—that is, unique collections of genes, which allow living things to pass their characteristics to their offspring. These collections of genes are unique to each species of life. They are irreplaceable if a species becomes extinct.

The danger of losing valuable genetic resources has caused scientists to increase their efforts to learn all they can about life in the rain forests. They must act fast. Every minute of every hour, 20 hectares of trees are cut down for timber or to make room for agriculture or mining.

The race is on, and the rewards can be great. For example, a disease-resistant strain of corn, now worth billions of dollars, was bred from a type of maize discovered in a Mexican rain forest.

Setting up camp in the treetops

Floating on the Treetops

Like other treasures, the genetic resources of the rain forests can be difficult to reach. There are no roads in the most remote regions of the rain forests.

A group of French scientists has found an unusual way to get to the rain forest species they want to study. They use a large "raft" that they lower into place from a dirigible. The raft settles down onto the treetops and "floats" there, forming a base of operation high above the ground. From this base, the scientists study the plant and animal life of the *canopy* formed by the treetops, which is where many unidentified rain forest species are found.

The "Indiana Jones" of Botanists

Getting to the places where tropical species live can be dangerous. For example, botanist Michael Doyle, who does his field work in remote forests in Hawaii, sometimes has to climb steep cliffs. He has even dangled from a helicopter by a 50-meter line to collect plant samples.

Doyle has devoted years of his life to studying a single rare plant, the *gunnera.* This plant grows in Hawaiian mountain forests covered by clouds. Its huge, brilliant green leaves, which can be up to 3 meters wide, may be spotted through the mist, on cliffs surrounded by waterfalls.

Michael Doyle on location, with gunnera plants

The gunnera plant is worth the trouble, Doyle will tell you. It may turn out to be extremely valuable as a genetic resource. Most importantly, the plant has a unique way of producing its own fertilizer. A type of blue-green bacteria lives in the cells of its stems. These bacteria convert nitrogen from the air into plant food.

According to Doyle, if the self-fertilizing ability of the gunnera plant could be transferred to common food plants, major progress could be made against the problem of world hunger. People could grow food in soil that would otherwise be too poor to support plant growth without costly fertilizers.

The non-improved soybeans were killed by the herbicide.

The New Engineering

Doyle's idea is not far-fetched. Genetic engineering technology makes it possible to create new varieties of plants and animals by transferring genes from one species to another. For example, scientists have used gene transfer to develop plants, like the soybeans in the photo on the left, that are not harmed by glyphosate. Glyphosate is a weed-killing product, or herbicide, that is harmless to humans and is readily broken down in the environment. If farmers plant glyphosate-resistant crops, they can spray their fields with glyphosate to kill weeds without harming the crops.

Find Out for Yourself

The gunnera plant is the largest herb in the world. But some varieties of the plant are very small, with tiny leaves. This indicates that there may be a single gene that controls the growth of the plant. What engineering possibilities does this suggest?

Do some research to find out ways in which different species have been altered by using genes from other species.

What Is a Concept Map?

Have you ever tried to tell someone about a book or a chapter you've just read, and you find that you can remember only a few isolated words and ideas? Or maybe you've memorized facts for a test and then weeks later, you're not even sure what topic those facts are related to.

In both cases, you may have understood the ideas or concepts *by themselves,* but not in relation to one another. If you could somehow link the ideas together, you would probably understand them better and remember them longer. This is something a concept map can help you do. A concept map is a visual way of showing how ideas or concepts fit together. It can help you see the "big picture."

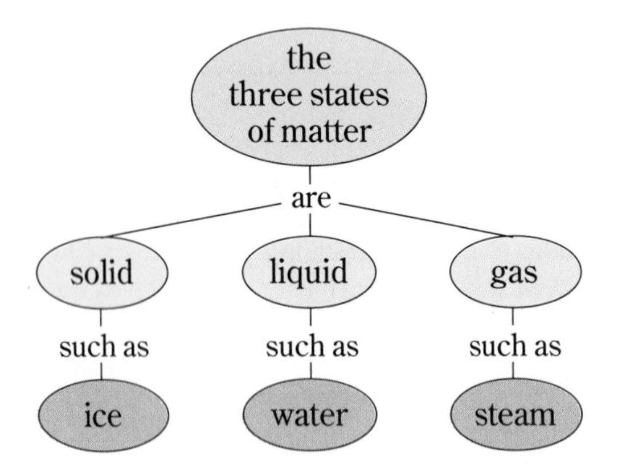

How to Make a Concept Map

1. **Make a list of the main ideas or concepts.**
   It might help to write each concept on its own slip of paper. This will make it easier to rearrange the concepts as many times as you need to before you've made sense of how the concepts are connected. After you've made a few concept maps this way, you can go directly from writing your list to actually making the map.
2. **Spread out the slips on a sheet of paper and arrange the concepts in order from the most general to the most specific.**
   Put the most general concept at the top and circle it. Ask yourself, "How does this concept relate to the remaining concepts?" As you see the relationships, arrange the concepts in order from general to specific.
3. **Connect the related concepts with lines.**
4. **On each line, write an action word or short phrase that shows how the concepts are related.**

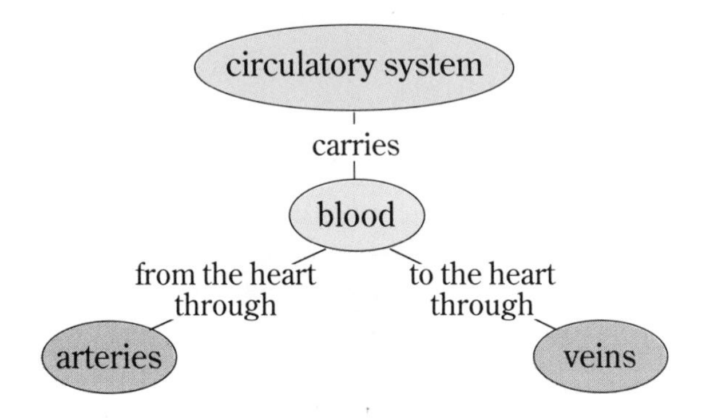

Look at the concept maps on this page and then see if you can make one for the following terms: *plants, water, photosynthesis, carbon dioxide,* and *sun's energy.* The answer is provided below, but don't look at it until you try the concept map yourself.

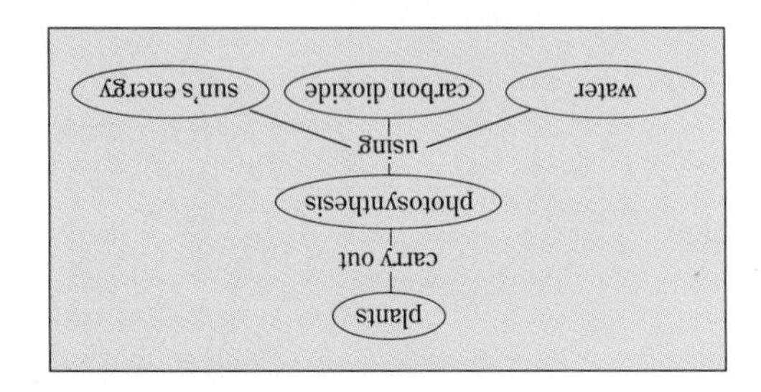

SourceBook™

This *SourceBook* is designed as a handy reference so you may learn more about many of the concepts you have discovered in your reading of Science*Plus*. For each unit of Science*Plus*, you will find a corresponding unit in the *SourceBook*. Questions on the first page of each unit will help direct your thinking as you read the material.

Contents

Unit 1

Reading *Plus*

Now that you have been introduced to some of the life processes, let's explore the chemical basis of life and then take a closer look at the basic processes that are characteristic of all living things. Read pages S2 to S14 and write a report about the nature and activities of living things. Your report should include answers to the following questions.

1. What are the basic chemicals of life, and what are their primary functions?
2. In what ways is water important to the processes of life?
3. In what ways are the processes of metabolism and growth related?

In This Unit

1.1 THE CHEMISTRY OF LIFE

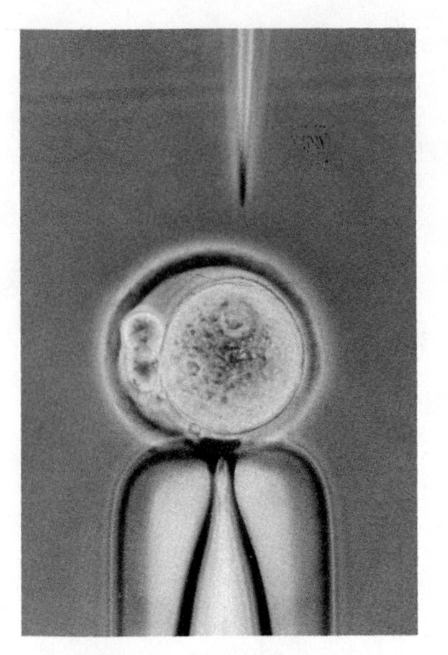

Figure 1-1 Using an extremely fine syringe, a biologist injects nucleic acid into this mouse cell. This procedure allows scientists to study the biochemistry of the nucleus.

Everything that happens in a living thing is the result of many coordinated chemical reactions. Just flexing your index finger, for example, requires thousands of individual chemical changes. Think about this as you bend your finger. Chemical changes cause nerve cells to send the "bend your finger" message from your brain to your finger. Then other chemical changes cause some of the muscle cells in your finger to contract and others to relax. As a result, your finger bends. However, this action also involves many other chemical processes that take place within a living organism.

A better understanding of the chemical processes of life may be the key to solving many human problems, from producing enough food for the people of the world to curing diseases such as cancer, heart disease, and AIDS. Currently, scientists are doing extensive research to help them further understand the chemical basis of life as well as the pathways through which the processes of life are organized. See Figure 1-1.

The Elements of Life

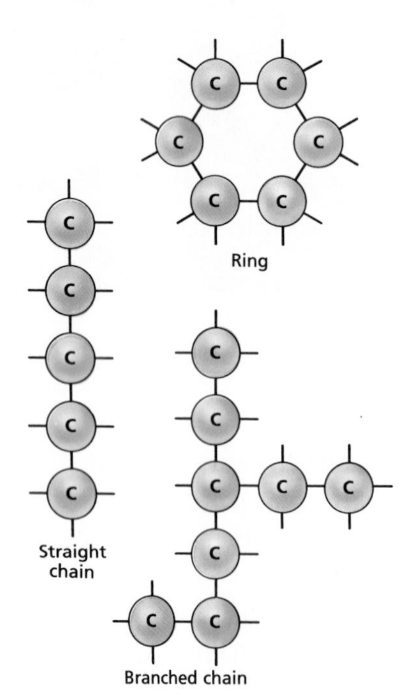

Figure 1-2 A carbon atom readily bonds with other carbon atoms in various formations.

Organic Compound A compound that contains carbon and usually is produced by living things.

As you know, all matter—living and nonliving—is made up of particles called *atoms*. Atoms are the smallest particles of elements. But of the 92 naturally occurring elements, just four of them—carbon, hydrogen, oxygen, and nitrogen—make up over 99 percent of the living matter on Earth. Atoms of these four elements combine with each other by forming very strong and stable chemical bonds. Carbon forms the backbone of most of these compounds. In fact, two unique abilities of carbon make life as we know it possible. First, carbon atoms can chemically bond with up to four different atoms. Second, carbon atoms can bond to other carbon atoms, making long chains or rings. See Figure 1-2. In doing so, they produce millions of different chemical compounds. The variety of life forms on Earth is a direct result of the tremendous variety of carbon-containing compounds. Over 4 million compounds of carbon are now known, and the list continues to grow.

Chemists classify carbon-containing compounds as **organic compounds**. *Organic* means "coming from life," and indeed, most of the compounds produced by living things do contain carbon. Many organic compounds, particularly the ones that are important to living things, consist of very large molecules. Some of these molecules may have thousands or even millions of individual atoms. These atoms are arranged in units called *monomers* that repeat over and over throughout the length of the molecule. Since these molecules are made of many monomers, they are called *polymers.* The prefix *mono* means "one," and the prefix *poly* means "many." Polyethylene, a type of plastic, is a polymer (as its name suggests) of repeated ethylene molecules. Nylon, which is a very long fiber used to make fabric, is another common polymer. See Figure 1-3.

Figure 1-3 *Individual strands of nylon, shown here being wound onto a stirring rod, are very long polymers of two alternating monomers.*

Basic Biochemicals

The study of organic chemistry is a very complex science because there are so many different kinds of organic compounds. But although most organic compounds originate from the activities of living things, only certain types of organic compounds actually occur in living organisms. These compounds are of four basic types: *carbohydrates, lipids, proteins,* and *nucleic acids.* As you read about each one, keep in mind that plants synthesize all of these compounds from the inorganic materials carbon dioxide (CO_2), water (H_2O), and a variety of minerals. Animals, on the other hand, must eat foods that contain the monomers they use to build the organic compounds they need.

Carbohydrates Sugars, starch, and cellulose are complex organic compounds that are known as **carbohydrates**. Figure 1-4 shows some of the sources of these substances in your diet. All carbohydrates are energy-storing substances. Carbohydrates are made entirely of carbon, hydrogen, and oxygen. Most carbohydrates are polymers of smaller repeating units called *monosaccharides,* or simple sugars.

Carbohydrates Organic molecules composed of monosaccharides that store energy for use by cells.

Glucose is one type of monosaccharide. It is the main product of photosynthesis. When two monosaccharides join together, they form a *disaccharide* (or double sugar).

Figure 1-4 *Fruits and honey* (left) *contain the simple sugars glucose and fructose. Complex carbohydrates, such as starches and cellulose, are found in bread, cereal, and pasta* (right).

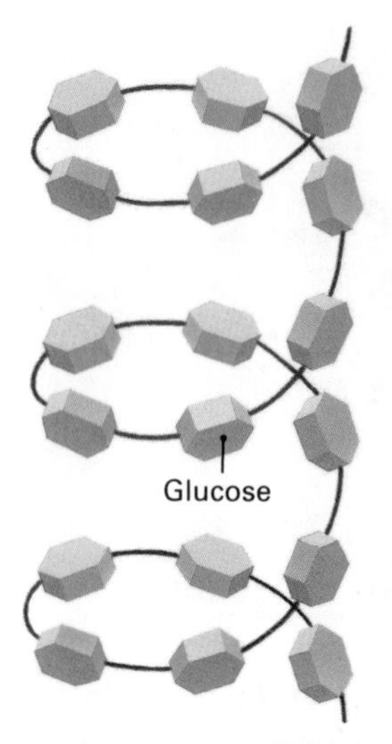

Figure 1-5 Starch is a polysaccharide, a polymer formed from many glucose units.

Common table sugar is a disaccharide. It is made by joining two different monosaccharides: glucose and fructose. Many glucose molecules joined together form the *polysaccharides* starch and cellulose. See Figure 1-5.

Lipids Fats, oils, waxes, and steroids are examples of the group of organic compounds called **lipids**. They consist mostly of carbon, hydrogen, and oxygen. Lipids are primarily energy-storage molecules. Fats store excess food energy in animals. Oils store excess food energy in plants. Fats and oils are made by joining three *fatty-acid* molecules to a *glycerol* molecule.

Lipids Organic molecules composed of fatty acids and glycerol that store energy and make up cell membranes in living things.

In addition to their energy-storage function, lipids play other important roles in living things. Cell membranes, for example, are made of two layers of lipids. Waxes, like the ones found covering the leaves of plants and lining the inside of your ears, protect the tissues they cover from dehydration and attack by microorganisms.

Cholesterol is another type of lipid, which you have probably heard much about. Although excess cholesterol can lead to heart disease, a small amount of it is necessary in animals. Cholesterol is used for making cell membranes, nerve and brain tissues, and for producing certain *hormones*, the chemicals that regulate life processes such as growth and reproduction. See Figure 1-6 for some of the sources of lipids in your diet.

Figure 1-6 Lipids are found in each of the items shown here.

Proteins Most of the organic molecules in living things are **proteins**. Figure 1-7 shows some of the sources of protein in your diet. All proteins contain carbon, hydrogen, oxygen, and nitrogen, but they may also contain some other elements as well. Proteins are very complex polymers of smaller molecules called *amino acids*.

Proteins Large organic molecules composed of amino acids that either act as structural materials in an organism or regulate the chemical activities of life.

There are twenty-three different kinds of amino acids. These amino acids make up all the different proteins of

Figure 1-7 *Proteins are found in dairy products, nuts, meat, and fish.*

living things in much the same way that the twenty-six letters of the alphabet make up the different words of the English language. In the same way that words with different meanings are spelled with different letters, proteins (with different chemical properties and functions) are made by varying the kinds, the number, and the order of amino acids. For example, two proteins that differ by only one amino acid can have completely different properties, just as the words "live" and "love" have different meanings. The longest word in the English language has only a few dozen letters, but the longest proteins have over 1000 amino acids. Just think how many different words you could make if words could have more than 1000 letters! It is because of this that there are so many different kinds of proteins.

Proteins are very important to the structure and function of living things. Some proteins, such as *collagen* in human skin and *myosin* in muscle cells, form strong, elastic fibers. See Figure 1-8. These fibers make up the primary support structures of living tissue. Other proteins are important to the chemical activities of cells. *Enzymes*, for example, are proteins that are involved in nearly every chemical reaction in a living cell. Still other proteins, such as the hemoglobin in the blood, carry substances to places in the body where they are needed. Cell membranes also contain many carrier proteins in their structure.

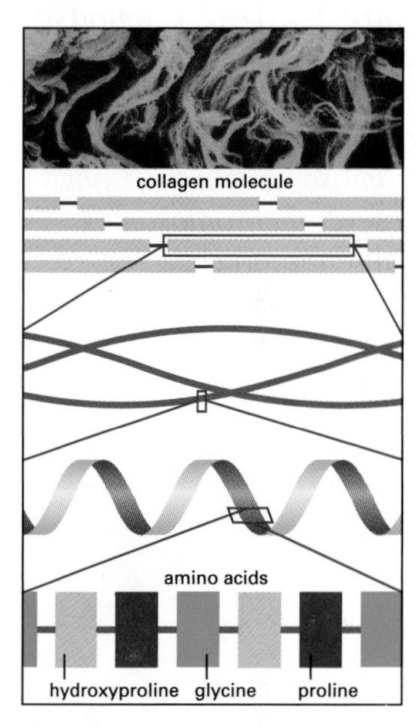

Figure 1-8 *Collagen fibers from human skin* (top) *are composed of many collagen molecules.*

Nucleic Acids The largest and most complex organic molecules in living things are the **nucleic acids**—DNA (deoxyribonucleic acid) and RNA (ribonucleic acid). Nucleic acids are very long chains of monomers called *nucleotides*. Nucleotides are composed of the elements carbon, hydrogen, oxygen, nitrogen, and phosphorous. To give you an idea of how long nucleic acids are, your body contains about 25 billion kilometers (more than 2 trips to Pluto and back) of DNA molecules alone!

Nucleic Acids Large organic molecules composed of nucleotides that contain the instructions for the structure and function of organisms.

Nucleic acids contain the instructions for putting the amino acids of a protein together. These instructions constitute a code, called the *genetic code*. Each group of three nucleotides forms a "code word" for a specific amino acid. A string of these code words causes amino acids to be put together in the proper sequences to form a particular protein.

DNA, which is located in the nucleus of the cell, contains the genetic code. See Figure 1-9. DNA has a double structure, which allows it to split and make copies of itself. In this way, it is able to transmit genetic instructions from one generation to the next during reproduction. RNA, on the other hand, actually does the physical work of synthesizing proteins. RNA copies information from the DNA and moves out of the nucleus into the cytoplasm of the cell. There its nucleotides act as a pattern for lining up the amino acids, thereby making all the proteins required for building and operating cells.

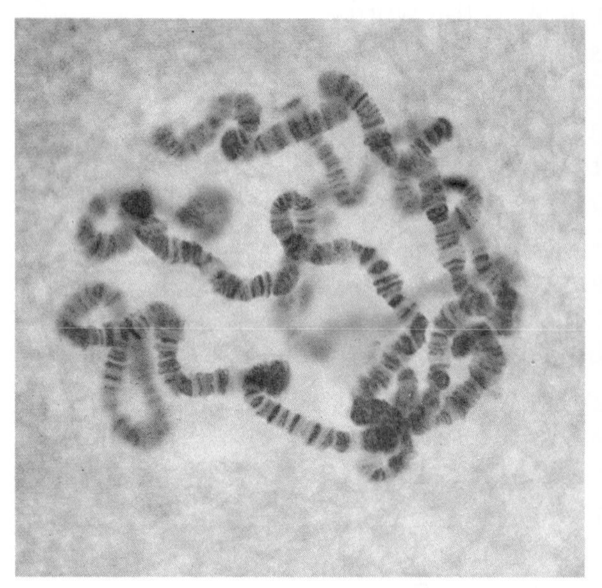

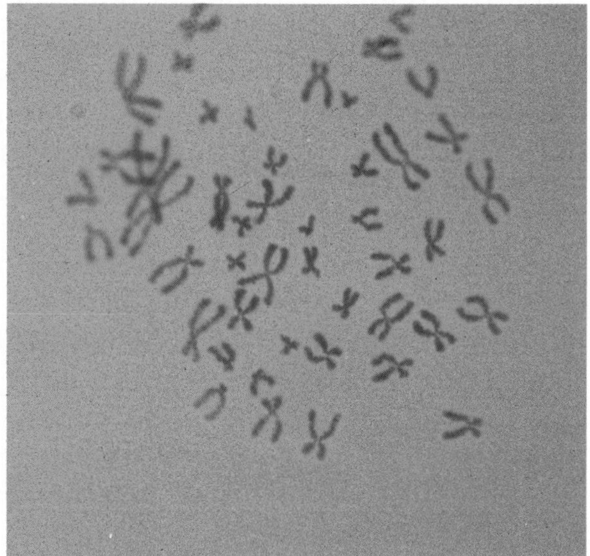

Figure 1-9 *DNA is found in the chromosomes of a cell's nucleus. On the left is a close-up of a fruit fly chromosome. On the right are several chromosomes from a human cell.*

Water

You could live about five weeks without eating a bite of food. However, you could live only about five days without drinking water. Water is the most important inorganic substance in living things, making up from 50 to 98 percent of the mass of a living cell. About 65 percent of the total volume of your body is water, and the loss of only 15 percent of your body's normal amount of water could be fatal.

We share our need for water with all living things. For one thing, water dissolves and transports materials necessary for life. See Figure 1-10. Among its many unique properties, water has a high capacity to absorb and give off heat without great changes of temperature. Water's *heat capacity*, therefore, keeps living things from both freezing and overheating. As the only common substance that is a liquid at most of the temperatures found on Earth, water is also the habitat of millions of living species.

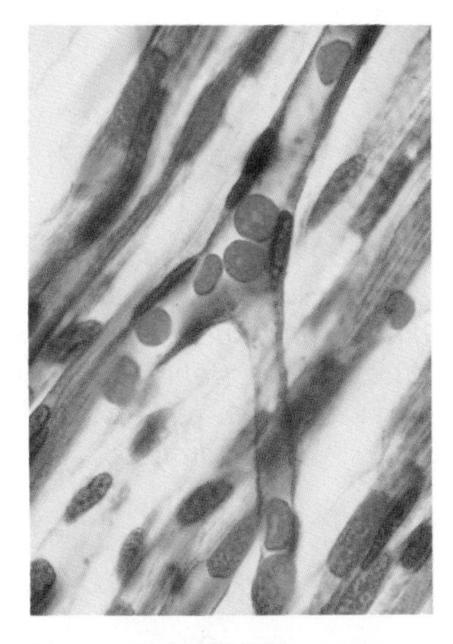

Figure 1-10 Without water, there can be no life. These blood and muscle cells exist in a solution of salts and water.

A Polar Substance Some of the unique properties of water that make it so valuable to life are due to its its structure. Two hydrogen atoms covalently bond to an oxygen atom to form a water molecule. Because it has a bent shape, the electrons of the water molecule spend more time close to the oxygen nucleus than the hydrogen nuclei. As a result, the oxygen side of a water molecule has a slight negative charge, and the hydrogen side has a slight positive charge. See Figure 1-11. Due to its shape and its positive and negative poles, a water molecule is said to be a **polar molecule**.

Polar Molecule A molecule that carries unevenly distributed electrical charges.

Polar molecules are attracted to other objects that have electric charges. This makes water an excellent solvent. For example, when crystals of common table salt (NaCl) are dropped into water, the water molecules are attracted to the individual sodium and chloride ions that make up the salt. The sodium ions (Na^+) are positive and the chloride ions (Cl^-) are negative. The attraction of the water molecules pulls the ions away from the salt crystals and causes each ion to be surrounded by water molecules. The Na^+ ions and the Cl^- ions separate and spread throughout the solution. As a result, the salt disappears into the water, or dissolves.

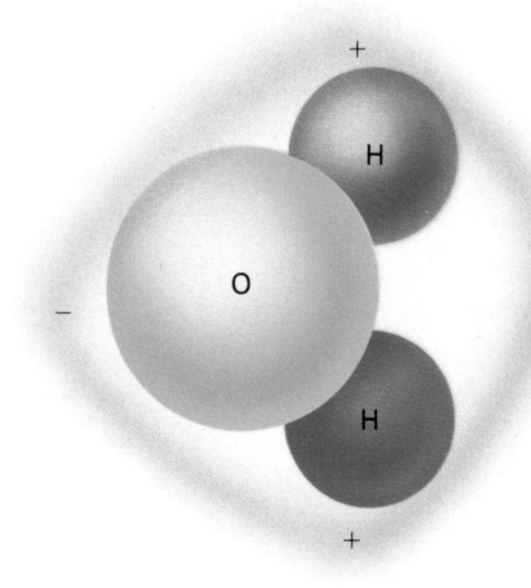

Figure 1-11 The unequal distribution of positive and negative charges around a water molecule results in its bent shape.

Adhesion and Cohesion Water molecules also stick to things they cannot dissolve, such as glass, soil particles, and cells. This property is called *adhesion*. As you know, adhesion is partly responsible for water's ability to rise inside of small tubes, like the vessels in plants that conduct water from the roots to the leaves. This type of movement, called *capillary action*, also helps to move blood through the arteries, veins, and capillaries of your circulatory system. Water molecules are attracted to each other as well, due to their polarity. This property, called *cohesion*, keeps water columns from breaking as they rise inside the vascular cells of plants.

Water's Chemical Role In addition to its other roles, water also plays an important part in the chemical reactions that occur in living things. First, dissolved substances react much more rapidly than undissolved substances. In fact, many substances will not react at all unless they are dissolved in water. Water dissolves most of the substances that participate in the chemical reactions of life. Second, water is also a participant in many chemical reactions. For example, splitting apart large organic molecules requires the addition of water. This type of reaction is called *hydrolysis*, which means "splitting with water." On the other hand, large organic molecules are made by a reaction known as a *condensation* reaction. In a condensation reaction, water is produced when monomers are joined to form polymers. See Figure 1-12.

Figure 1-12 *A condensation reaction between glucose and fructose produces sucrose (table sugar) and water. This reaction is reversible by hydrolysis.*

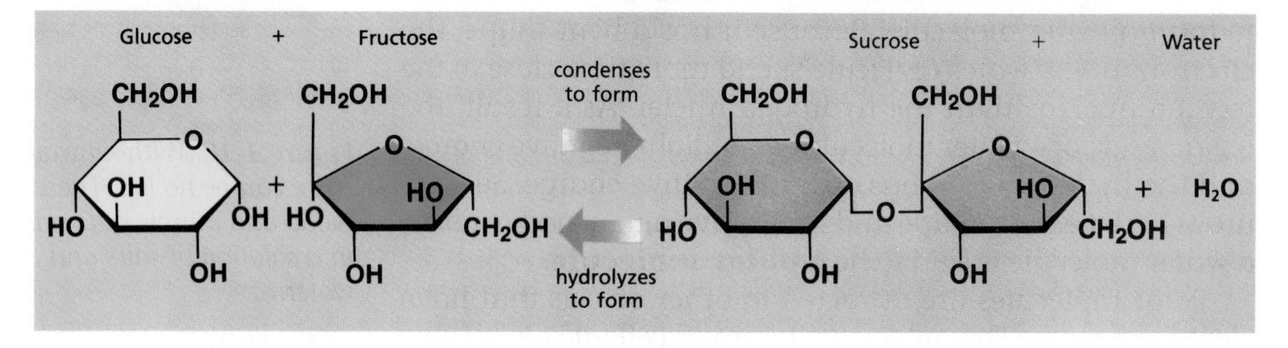

Summary

Carbon is the element that forms the backbone of most of the important compounds of living things. These organic compounds include carbohydrates, lipids, proteins, and nucleic acids. Carbohydrates are made from monosaccharides, and lipids are made from fatty acids and glycerol molecules. They both are used as energy-storage compounds. Proteins are made by joining amino acid molecules together. Some proteins are structural materials, while others make the chemical reactions necessary for life possible. Nucleic acids are made of long chains of nucleotides. Nucleic acids contain the instructions for making proteins. They also transmit these instructions during reproduction. Water is the most abundant inorganic substance in living things. It dissolves and transports materials and provides the medium in which life's chemical reactions take place.

1.2 BASIC LIFE PROCESSES

The chemical activities of living things are organized into several different processes. Among these processes are *obtaining raw materials, metabolism, excretion,* and *regulation.* You may have observed processes that resemble these in non-living things. For example, streams gather and transport sediments, while acids seem to digest substances. However, all organisms—and the cells of which they are made—perform not just one, but all of these processes. As you read about them, notice that each process is related to and depends upon the others. See Figure 1-13.

Figure 1-13 *Even though a computer has complex organization, responds to stimuli, and has moving parts, it is not alive. What can this mouse do that the computer cannot?*

Obtaining Raw Materials

To remain alive, all living things must take in raw materials such as food, water, and mineral nutrients. Some organisms, such as plants, algae, and certain bacteria, take in only inorganic substances. From these, they make the organic substances they need by utilizing the sun's energy. These organisms, called *producers*, form the basis of all food chains. Other organisms, called *consumers*, must eat other organisms to obtain organic substances.

The materials obtained by both producers and consumers must be transported to the individual cells of the organism, where they are used. However, in order for cells to take in materials, the materials must pass through the cell membrane. The movement of materials into and out of cells is regulated by the cell membrane, which allows certain substances to enter and keeps others out. Two processes by which movement across a cell membrane is accomplished are diffusion and osmosis.

Diffusion and Osmosis The movement of molecules or ions that occurs when the concentration of a dissolved substance differs in two neighboring areas is called *diffusion*. A difference in concentration can be pictured as a hill. Just as an object (like a ball, for instance) rolls down the slope of a hill, the overall movement of molecules and ions in diffusion also

Concentration Gradient The difference in the concentrations of a substance between two areas.

occurs from high concentration to low concentration. In other words, they move down (or with) the **concentration gradient**. The steeper the slope of the hill is, the faster the ball rolls. See Figure 1-14. The same is true for diffusion—the greater the concentration gradient is, the faster the process of diffusion occurs.

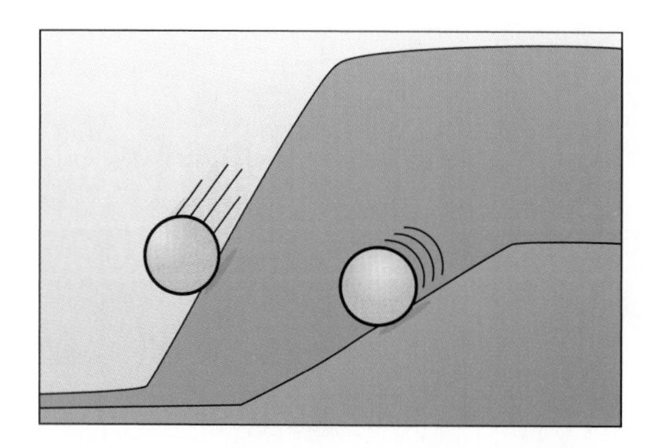

Figure 1-14 *A concentration gradient can be compared to the slope of a hill.*

Of course, molecules and ions do not really "roll down hill." However, they are constantly moving back and forth between areas of different concentration. Since there are more molecules or ions to move from high to low, and fewer to move in the opposite direction, an area that is less concentrated gains more molecules or ions than it loses. As a result, the less concentrated area experiences an overall increase in molecules or ions of that kind.

Osmosis follows the same principle, but usually refers only to the diffusion of water through a cell membrane. The concentration of water molecules is related to the amount of solutes contained in the water. When there are more water molecules outside of a cell (fresh water), they tend to move into the cell in order to equalize the concentration. If there are less water molecules outside the cell (salt water), water tends to leave the cell. When the concentration of water molecules is equal on both sides of the membrane, there is no net movement in either direction. In this case, the rate of osmosis is the same both into and out of the cell. See Figure 1-15.

Figure 1-15 *Three types of cell environments are shown here. Which one would allow net water movement into the cell, out of the cell, and equally in both directions?*

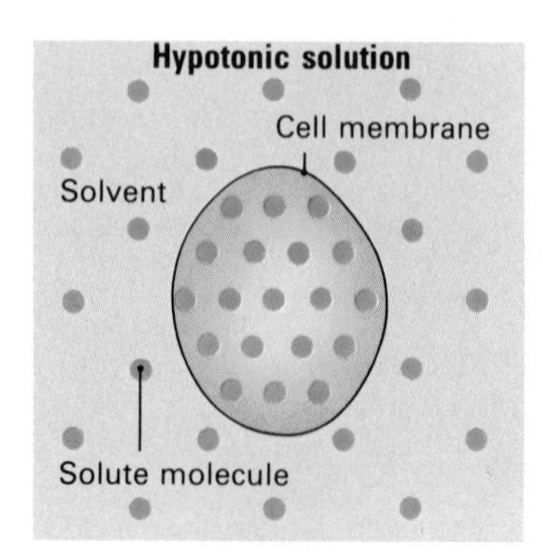

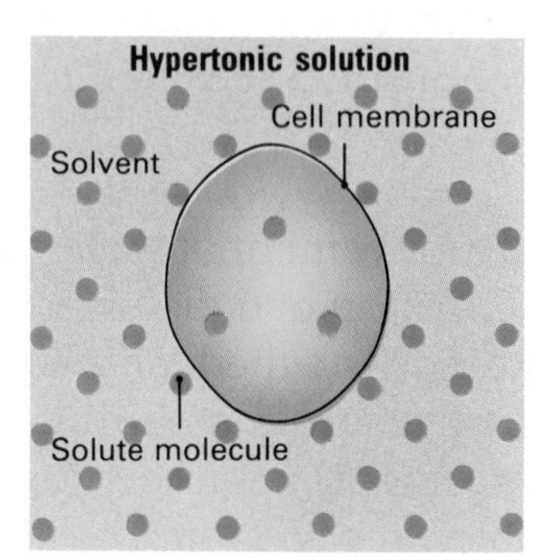

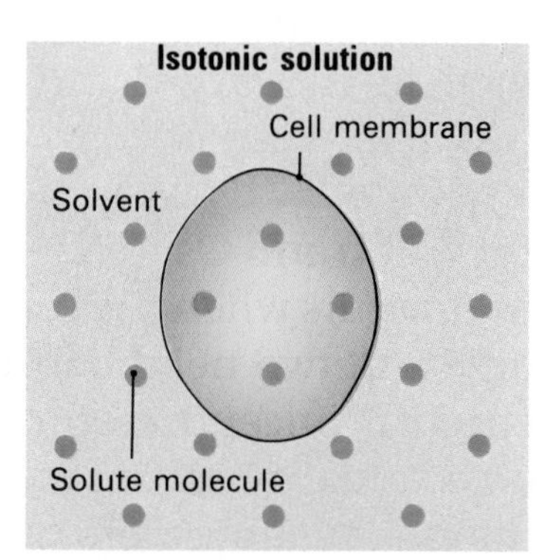

Diffusion and osmosis are things that, you might say, "just happen" to a cell. The cells themselves do not take an active part in obtaining materials by either process. However, only certain materials can enter a cell by these means. First of all, the material entering a cell must be in higher concentration outside the cell. Second, the materials must be soluble in water and must have molecules or ions small enough to pass through the openings in a cell membrane.

Active Transport Sometimes a cell may need a substance that is already more concentrated inside the cell than it is outside the cell. Such substances cannot get in by diffusion. However, there *is* a method by which these substances can enter. This method is called **active transport**. Unlike diffusion, active transport requires energy to operate.

Active Transport The movement that requires energy to carry materials into and out of cells against a concentration gradient.

To understand why active transport uses energy while diffusion and osmosis do not, imagine that you are riding your bicycle and you come to a hill. If the hill goes down, you can relax because all you have to do is steer. You don't get tired because you don't need to use energy just to roll downhill. But if you come to an uphill climb, you have to use energy to pedal. As long as the hill goes up, you must continue to use energy. See Figure 1-16.

Figure 1-16 Coasting downhill is similar to diffusion and osmosis. Stand-up pedaling is more like active transport.

Substances entering a cell by diffusion are like your bicycle rolling downhill. Since both movements occur with (or down) the gradient, no energy is needed. On the other hand, active transport moves materials *against* the concentration gradient. Just as you must expend energy to ride your bicycle uphill (against the gradient), a cell must use energy to move materials against a concentration gradient.

Figure 1-17 The cell membrane is composed of a double layer of lipids in which proteins are embedded.

The secret to active transport is in the structure of the cell membrane. See Figure 1-17. Large protein molecules called *carrier molecules* are positioned in the lipid layers of the membrane. These molecules move ions or molecules in and out

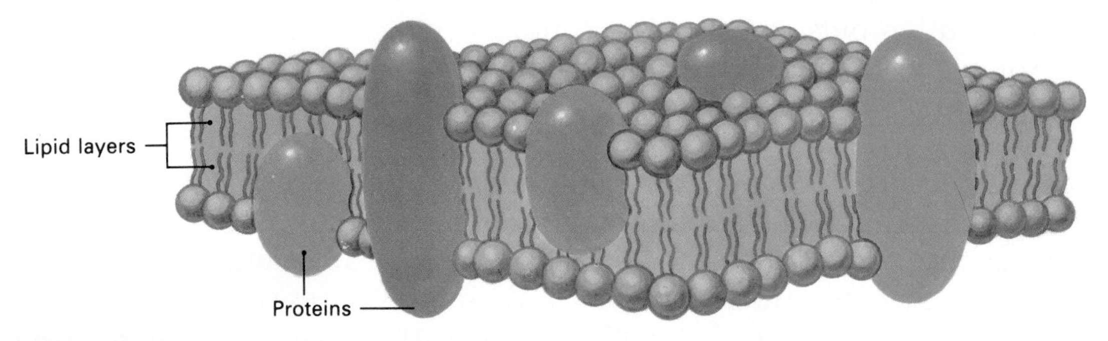

of the cell by changing their shape. In this way, they act like gates. When an ion has to go against its concentration gradient through one of these gates, a chemical reaction inside the cell provides the energy that makes this possible. By a similar process, cells are also able to take in molecules that are too large to fit through the small openings in the cell membrane. See Figure 1-18.

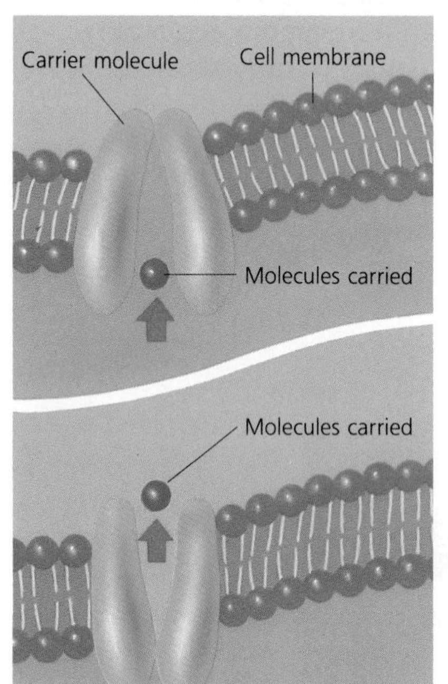

Figure 1-18 Some of the proteins of the cell membrane act as carrier molecules that allow materials to pass from one side of the membrane to the other by changing shape.

Metabolism

Living things are able to organize and reorganize the substances they take in for specific purposes. This involves breaking down large organic molecules into smaller molecules, as well as joining small molecules together to make larger ones. Organisms are also able to convert energy into more usable forms. Together, all the chemical breaking-down, building-up, and energy-conversion processes of organisms are referred to as **metabolism**.

There are three basic chemical processes that make up metabolism: digestion, respiration, and synthesis. In digestion, large food molecules are broken down into simpler compounds. Some of these compounds are then used in the process of respiration, which supplies the energy for cell activities. The simple sugar glucose is the primary fuel for respiration, but fatty acids and amino acids can be used as well. In fact, if an organism is suffering from starvation, it will break down any basic organic compound to obtain energy.

Some of the products of digestion are combined in new ways to make the specific carbohydrates, lipids, proteins, and nucleic acids an organism needs for its activities. This process of putting smaller molecules together to make larger molecules is called **synthesis**. *Photosynthesis* uses light energy to put CO_2 and H_2O molecules together to

Metabolism All of the chemical reactions of an organism.

Synthesis The chemical combination of small molecules to make larger molecules.

form glucose molecules. The combining of amino acids to make proteins is called *protein synthesis.* From the raw materials obtained from its environment, and organism can synthesize the substances that are necessary for life. See Figure 1-19.

Figure 1-19 *These mushrooms turn materials they obtain from dead wood into the chemical compounds of their own living cells. In other words, they turn dead wood into living mushrooms.*

Excretion

Building things up and breaking things down usually produces some materials that are unusable. These unusable materials are called *wastes.* Wastes that are produced by the chemical building up and breaking down processes of metabolism are called *metabolic wastes.* They include carbon dioxide, water, nitrogen compounds (ammonia and urea), inorganic salts, and heat. If these wastes build up in a cell or an organism, they may harm it by blocking other processes. For example, a build up of too many wastes in a cell prevents food and oxygen from entering. As a result, the cell may eventually die from lack of energy and fresh supplies of building and repair materials if its wastes are not removed.

The removal of metabolic wastes is called **excretion**. Individual cells and unicellular organisms eliminate these wastes primarily by diffusion. See Figure 1-20. In large multicellular animals, wastes are usually gathered from the area around individual cells by the fluids of a *circulatory system.* These wastes are then filtered out of the fluid for removal from the body. In humans, organs such as the lungs, skin, and kidneys play an important role in excretion. For example, the lungs remove carbon dioxide and water from the body. Skin excretes water, nitrogen compounds, and inorganic salts in the form of perspiration. The kidneys remove nitrogen compounds, inorganic salts, water, and other unnecessary or harmful substances from the blood, which they excrete in the form of urine. In plants, the primary metabolic waste product is oxygen, which diffuses out of the leaves through the stomata.

Excretion The removal of metabolic wastes.

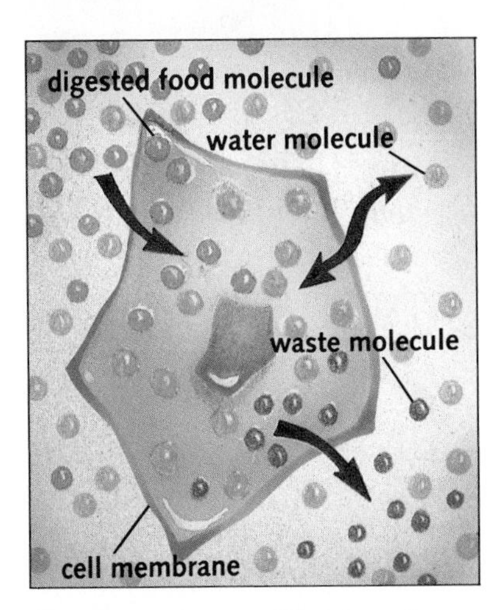

Figure 1-20 *Diffusion of food into the cell and waste out of the cell take place at the same time.*

Regulation

To remain alive, all organisms must continually monitor the activities of their cells, tissues, organs, and organ systems and keep them operating in a coordinated manner. This life process is called **regulation**. In order to regulate its activities, the individual parts of an organism must be able to send and receive messages. Plants, for example, regulate their activities by producing chemical substances called *growth regulators.* These chemicals initiate such changes as cell specialization, growth, tropisms, and flower formation.

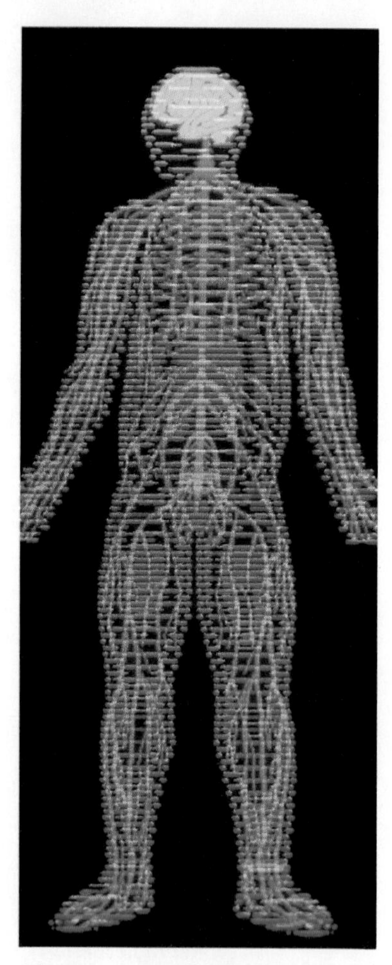

Figure 1-21 *The nervous system is controlled by the brain. Its messages are sent through the spinal cord and a network of nerves.*

Animals, on the other hand, have two highly-developed systems that work individually and together to regulate the activities of their bodies.

One of these, the *nervous system*, acts very quickly, controlling adjustments that must be made immediately. For example, fleeing a predator is a response that requires quick action. The nervous system also monitors and controls such things as muscle tone and heart beat. The messages of the nervous system are sent through a network of nerves that branches from the brain and spinal cord to all the tissues of the body. See Figure 1-21. The electrochemical messages of the nervous system are passed along by transporting ions into and out of nerve cells. Such messages are transmitted along the length of a nerve cell to the next cell in the network at 100 m/s.

The *endocrine system* acts more slowly, keeping things running smoothly and controlling gradual changes, such as growth and development. The organs of the endocrine system, called endocrine glands, produce a variety of chemicals. These chemicals, called *hormones*, act as messengers that travel through the blood stream to affect certain parts of the organism. For example, insulin is a hormone produced by the pancreas that controls the uptake of glucose by cells. Insulin molecules chemically bind to proteins in a cell's membrane, thus helping glucose to diffuse into the cell.

The nervous system assists the endocrine system by monitoring the level of glucose in the blood stream. It causes the pancreas to produce more insulin as the glucose level rises and less insulin as the glucose level falls. Your body produces dozens of other hormones that work with your nervous system to keep your body operating properly.

Summary

All living things must carry out several processes that are necessary for maintaining life. First, organisms must take in raw materials. The materials are then transported to the cells, where they pass through membranes by diffusion, osmosis, or active transport. These materials are processed by the various chemical reactions of metabolism to produce the energy and the many different molecules that organisms need to maintain themselves. Then, by the process of excretion, metabolic wastes and excess water are removed from cells and organisms before they become harmful. The various functions of life are regulated by slow-acting chemical messengers and (in animals) a quicker-acting nervous system.

Unit 2

In This Unit

Reading *Plus*

Now that you have been introduced to machines and how they relate to work and energy, let's take a closer look at the physics behind machines. Read pages S16 to S30 and then do a survey of your own home. The survey should list all the mechanical systems you can find. For each item listed, answer the following questions.

1. What simple machines make up the mechanical system?
2. How is the input energy supplied to each system? What does the output energy do?
3. What is the mechanical advantage and efficiency of each mechanical system?

2.1 FORCE, WORK, AND POWER

Do you realize that you are constantly under the influence of an all-pervasive force? Even though you might not feel it, this force is actually pulling on you right now. It is, of course, the force of gravity. The size of this gravitational force, as you have learned, is equal to your weight. Fortunately, your weight is usually balanced by another force. When you are standing or sitting, for example, there is an equal force pushing up on you from below. This force acts in the opposite direction from the gravitational force and is supplied by the ground or the seat of the chair. In these situations, the forces are balanced, and so there is no motion. See Figure 2-1.

Figure 2-1 *When forces are balanced, there is no motion.*

When a force is not balanced, there is always some kind of change. For example, if you push hard enough on the side of a chair, the chair will probably move. This is because the force you supply is not balanced by an opposing force.

Work: Force over a Distance

Work The result of using a force to move an object through a distance.

To a scientist, **work** means using a *force* to move an *object* through a *distance*. Thus, work is related to motion. In everyday language we might say that sitting at a desk reading this book is "work." But in a scientific sense, we could not call this work. In order for work to be done, some object must be moved through a distance. See Figure 2-2.

Think about the last time you played tug of war. When you win a game of tug of war, you do work in the scientific sense. The purpose of such a contest is, of course, to move an object (the other team) through a distance (across the center line). Your team has to apply a force (your pulling on the rope) to move the other team. From your team's perspective, we call this force the *effort*. At the same time your

Figure 2-2 *In the scientific sense, climbing stairs is a lot more work than holding a bar bell above your head.*

team is pulling on the rope, the opposing team is resisting your force, so we call their force the *resistance*.

As long as effort force is equal to resistance force, both forces are balanced and nothing moves. See Figure 2-3. Since neither team is moving, nobody has done any work. You might be pulling and straining, yet you are not doing any *work* in the scientific sense. By using more force, however, your team can move the opposing team across the line. Your effort overcomes the resistance and causes the other team to *move*. Now work is being done!

Figure 2-3 *Work is defined as force applied through a distance. For all their efforts, these people are not doing any work unless one team is moved through a distance.*

Would you like to know how much work your team did in winning the game? To find out, we have to know how much force your team used to move its opponents and how far you moved them. You could have used a force meter attached to the rope to measure the actual force and a meterstick to measure the distance. Any work that was done is equal to the *force* applied multiplied by the *distance* over which it was applied. This can be expressed by the equation:

$$\text{Work} = \text{Force} \times \text{Distance}$$

or

$$W = Fd$$

Since force is measured in newtons, and distance in meters, the unit of work is the *newton-meter* (N·m). This unit is also given the name *joule* (J). The term *joule* is used to describe a force of 1 N acting through a distance of 1 m, which is 1 N·m. See Figure 2-4.

Let's assume for the moment that your team applied a force of 10,000 N to overcome the resistance of the other team and that your team pulled the other team a distance of 3 m. The work your team did can be calculated as follows:

$$W = F \times d$$
$$W = 10{,}000\text{ N} \times 3\text{ m}$$
$$W = 30{,}000\text{ N·m or } 30{,}000\text{ J}$$

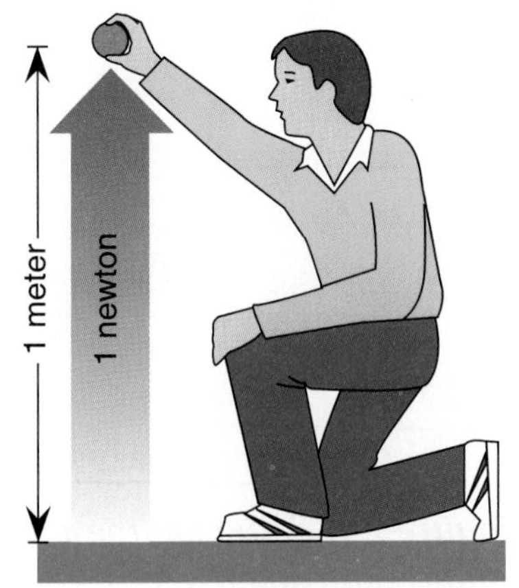

Figure 2-4 *This person is doing 1 J of work, an amount that is not very large.*

Figure 2-5 *In most situations, the force applied in doing work is not constant. In this situation, average force is used to calculate work.*

Therefore, your team did a total of 30,000 J of work to win the tug of war.

In the example above, it was assumed that the force stayed the same throughout the contest. In most situations, however, the applied force is not constant. For example, suppose you are pulling on a spring. The farther you stretch the spring, the greater the force you must apply. If you are using a bow, as shown in Figure 2-5, you must also apply different amounts of force. You apply a small force when you start pulling back on the bow string, but you must apply more force as you continue to pull. Other situations might have even more complex force patterns. In cases where the amount of the force changes, *average force* is used to calculate the work accomplished.

Power: Work Over Time

Power The rate at which work is done.

Suppose your team is challenged to a tug-of-war rematch. This time you beat them even faster than the first time. Before, it took your team 5 minutes to win. Now it takes you only 2 minutes to win. Assuming the force your team used was the same in both games, how do you think this time difference affected the amount of work your team does? Actually, the work was exactly the same in both cases. The amount of work you do does not depend on how long it takes to do the work. Notice that in the equation for work, there is no variable for time. Another term, **power**, is used to describe *how long* it takes to do a certain amount of work. The faster work is done, the greater the power. See Figure 2-6. Power equals *work* divided by *time* (in seconds). This can be written in an equation as

$$P = \frac{W}{t}$$

Figure 2-6 *Although these runners are doing approximately the same amount of work, they do not all have the same power. Which runner has more power? Why?*

Since work equals force times distance, the power equation can also be written as

$$P = \frac{(F \times d)}{t}$$

If you exert a force of 1 N over a distance of 1 m for 1 second, the power would be calculated as

$$P = \frac{(1\text{ N} \times 1\text{ m})}{1\text{ s}}$$

$$P = 1\frac{\text{N}\cdot\text{m}}{\text{s}} \text{ or } 1\frac{\text{J}}{\text{s}}$$

You can see from the equation that power is measured in *joules per second*, which has been given as special name, the *watt* (W). The watt is the SI unit of power. It was named after James Watt, the inventor of the sliding-valve steam engine. See Figure 2-7. Because the watt is rather small, however, a more commonly used unit is the *kilowatt*. One kilowatt is equal to 1000 W. If 1 J of work is done in 1 second, 1 W of power is generated. If 1000 J of work is done in 1 second, then 1000 W or 1 kW of power is generated.

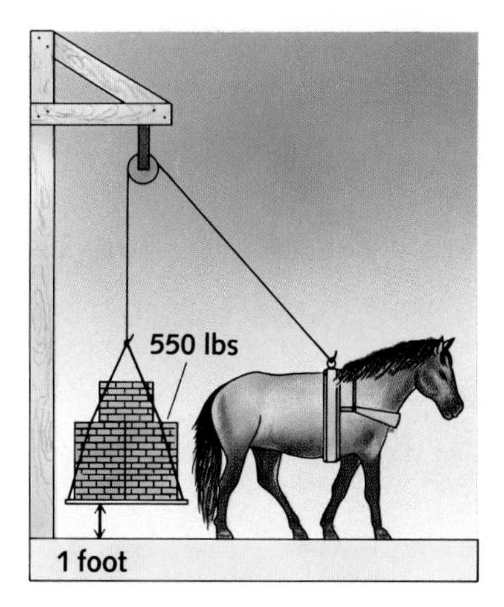

Figure 2-7 *James Watt actually defined power in terms of a workhorse. One horsepower (hp) was the amount of power it takes a horse to lift 550 lbs. 1 ft. in 1 second. One hp = 746 W.*

The power of your tug-of-war team in each of the two matches would be figured as follows.

$$P = \frac{W}{t} = \frac{30{,}000\text{ J}}{5\text{ min}} \times \frac{\text{min}}{60\text{ s}}$$

$$P = 100\text{ J/s or } 100\text{ W}$$

$$P = \frac{30{,}000\text{ J}}{2\text{ min}} \times \frac{\text{min}}{60\text{ s}}$$

$$P = 250\text{ W}$$

Note: The term "$\frac{\text{min}}{60\text{ s}}$" is a conversion factor for changing minutes into seconds.

What common household appliance do you use every day that has a similar power rating?

Power and Energy Consumption The rate at which a light bulb transforms energy is usually expressed in watts. For example, light bulbs are labeled 40W, 60W, 100W, and so forth. See Figure 2-8. However, the total amount of energy

Figure 2-8 *Light bulbs come in a variety of power ratings.*

used by a bulb depends on how long it operates. The units for total energy usage are *watt-hours* (Wh) or *kilowatt-hours* (kWh). For example, a 60-W light bulb burning for 1 hour uses 60 W × 1 h = 60 Wh of electricity.

The electrical energy used in a home is usually measured in kilowatt-hours (where 1 kWh = 1000 Wh). The kilowatt-hour, you must realize, is a unit of energy, not power.

$$\begin{aligned} 1\ \text{kWh} &= 1000\ \text{W} \times 1\ \text{h} \\ &= \frac{1000\ \text{J}}{\text{s}} \times 3600\ \text{s} \\ &= 3{,}600{,}000\ \text{J} \end{aligned}$$

You are actually buying energy when you pay your electric bill. For example, a 60-W light bulb consumes 216,000 J of energy in 1 hour:

$$60\ \text{J/s} \times 3600\ \text{s} = 216{,}000\ \text{J}$$

A meter such as that shown in Figure 2-9 measures electrical energy use. On an average day, a house may use between 60 and 100 kWh of electrical energy.

Figure 2-9 *An electric watt-hour meter measures the number of kilowatt-hours of electricity used.*

Summary

When an unbalanced force causes motion, work is done. Work means using a force to move an object through a distance. In mathematical terms, work is equal to force multiplied by distance and is measured in joules. Average force is used to calculate work when force is not constant. Work, however, does not involve time at all. Power gives the rate at which work is done. Mathematically, power equals work divided by time and is measured in joules per second, or watts.

Electrical energy is usually measured in kilowatt-hours, which is a unit of energy.

2.2 SIMPLE MACHINES

Can you imagine the amount of work that was done in building the ancient Egyptian pyramids? Each of the great stones used to build the pyramids has a mass of thousands of kilograms. See Figure 2-10. Yet because it took many years to build a pyramid, the power involved was not very great. You have already seen that not much power is generated in a game of tug of war. If it took your team even 1 minute to move the other team a distance of 3 m, your team would have had an output of only 500 W of power. This small amount is only enough power to light five 100-W light bulbs. It should be obvious that human power is not very effective in doing large amounts of work quickly. This is why machines were invented.

Figure 2-10 *Because few machines were used, it took over 100,000 workers nearly 20 years to build some of the pyramids.*

Simple machines are used to change the size or direction of a force. For example, moving a heavy boulder probably requires more force than you could normally supply alone. You might not be able to move it at all by simply pushing on it. However, by applying the same force to a pole or bar placed under the boulder, you may find that you can move it. See Figure 2-11. By using a simple lever, you can produce a large force to move a heavy load.

Simple Machine A device that changes the size or direction of a force.

Figure 2-11 *With a simple machine, such as this metal bar, you can apply more force.*

You already know that complex mechanical systems are made from simpler machines. Among the simple machines are wheel-and-axles, pulleys, gears, screws, and wedges. But all simple machines can be reduced to just two basic machines—the *lever* and the *inclined plane.*

The Lever

It might seem like a lever can supply more work than you put into it. However, you know this is not true. The work done in moving one end of a lever is always equal to the work done on the load at the other end. This means that a lever only *transfers* work. It does not increase the amount of work done. However, by using a lever, you can do the same amount of work with less force. Remember, work depends only on the *force* applied and the *distance* moved. Of course, the distance over which you must supply that force is increased. This is a fundamental feature of all machines. To decrease the force needed to accomplish work, you must increase distance. On the other hand, if you want to decrease the distance over which that work is done, you must increase the force.

A lever always turns on a fixed point called a *fulcrum*. The fulcrum is the place where the lever is supported. The part of the lever between the fulcrum and the load is called the *resistance arm*. From the fulcrum to the end of the lever where the force is applied is the part called the *effort arm*. See Figure 2-12. When the effort arm is longer than the resistance arm, applying a small force to the end of the effort arm results in a larger force at the end of the resistance arm. This is because the effort force acts over a greater distance than the resistance force.

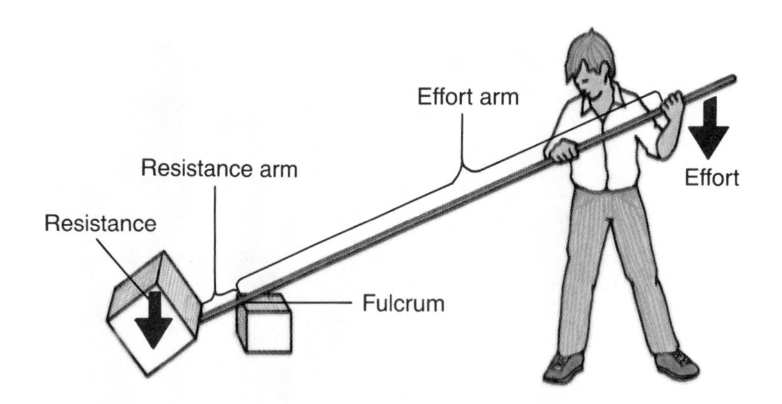

Figure 2-12 *This diagram shows the parts of a lever. It converts a small force applied over a large distance into a large force applied over a small distance.*

So far, we have been discussing a first-class lever—one in which the fulcrum is *between* the resistance and the effort. A first-class lever can change the size of the applied force and reverse its direction as well. There are two other methods of moving a resistance with a lever, depending on the position of the fulcrum. As shown in Figure 2-13, second- and third-class levers can change the size but not the direction of the applied force. You probably have seen and used many examples of all three kinds of levers.

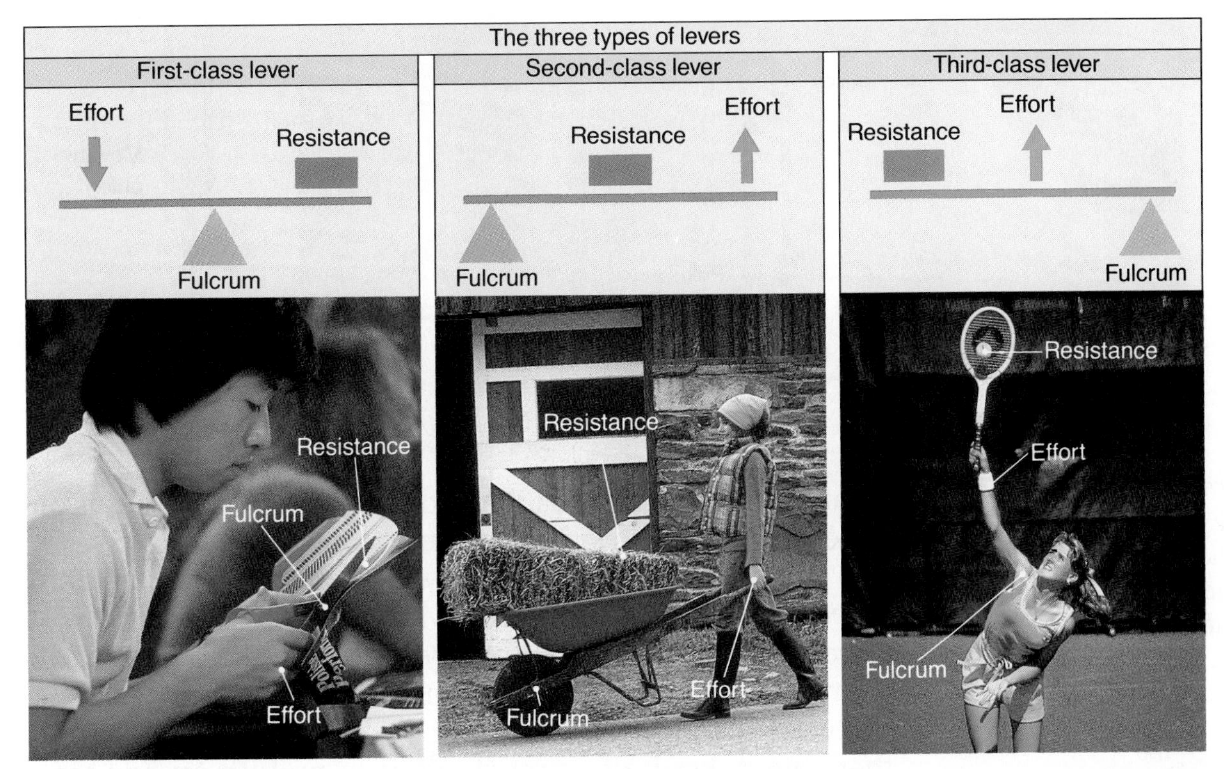

Figure 2-13 *There are three classes of levers, as shown here. Notice the position of the fulcrum in each type.*

The Wheel and Axle Did you ever try to turn the shaft of a doorknob when the handle was missing? If so, you know how much harder it was to open the door. Could you imagine trying to make an automobile turn a corner without a wheel on the steering column? A doorknob and a steering wheel are common examples of the *wheel and axle*, a machine that is similar to the lever. A wheel and axle has a fulcrum at the center of the axle. Its resistance arm is the radius of the axle, and the effort arm is the radius of the wheel. See Figure 2-14. A small effort applied to the wheel results in a large force turning the axle.

Have you noticed that large trucks and tractors have big steering wheels? It is the size of the wheel that determines the amount of force applied to the axle. The larger the wheel, the more distance it covers in turning the axle. The force needed to turn that distance is therefore reduced, while the force applied at the turning axle is increased. This makes it much easier to turn a heavy truck.

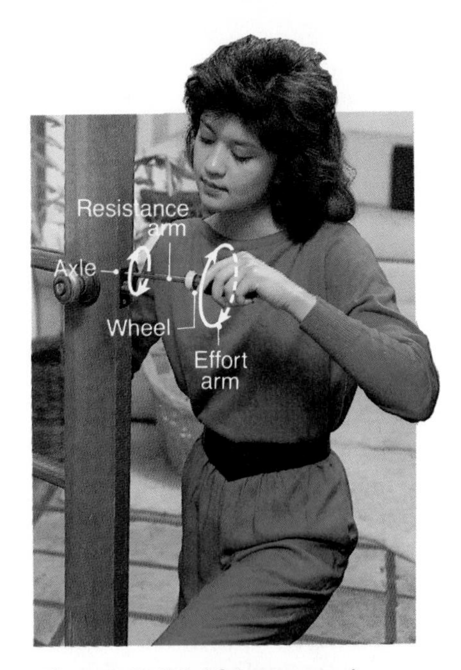

Figure 2-14 *This person is using a wheel and axle in the form of a screwdriver to apply a force to a screw.*

The Pulley A pulley is similar to a wheel and axle. But the axle on a pulley does not turn. A rope runs over the pulley with the resistance hanging down on one side and the effort applied to the other side. Because the effort and resistance distances are the same, there is no gain in force. A single *fixed* pulley simply changes the direction of the force. See

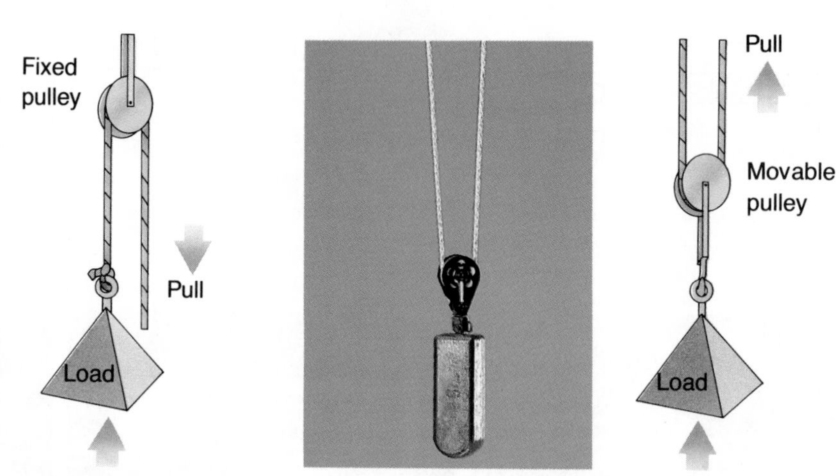

Figure 2-15 *The fixed pulley is used to change the direction of a force, while the movable pulley is used to multiply force.*

Figure 2-15 *(left)*. Such pulleys can be useful because it is often easier to pull downward than to lift upward.

With a single *moveable* pulley, like the one in Figure 2-15 *(right)*, you can lift a resistance with half the effort you would have to supply to a fixed pulley. With the fixed pulley, all the resistance hangs on one rope, while the resistance is supported by two ropes in the case of the moveable pulley. It, therefore, takes half as much effort to pull up on one rope using a moveable pulley.

As you know, pulleys can be combined to multiply force even more. Moveable pulleys and fixed pulleys work together in a *block and tackle.* By moving the end of the rope a long distance with a small force, a very large force can be applied to lift a heavy object a small distance.

The Inclined Plane

Figure 2-16 *Ramps are examples of inclined planes connecting one level with a higher level.*

An *inclined plane* is a simple machine with no moving parts and works on a basic principle. An inclined plane simply increases the distance over which an applied force acts. Since it forms a sloping surface, such as the ramp shown in Figure 2-16, you can use an inclined plane to move a load with less force than would be needed to raise it straight up to the same height. Remember the relationship between force and distance that you saw in the lever and its related machines? It also holds true for the inclined plane. As the length of the inclined plane increases, the force needed to move the load decreases. If you shorten the length of the inclined plane, more force is required. Again, the work done in either case is the same. It is only a matter of which is more important in getting the job done—the amount of force you must use or the distance you must travel.

Screws Though they are considered simple machines in their own right, screws are actually forms of the inclined plane. By looking at Figure 2-17, you can see that a screw is basically an inclined plane wrapped around a cylinder. Screws are very useful because the length of the inclined plane can be spread over a large distance. The more tightly wound the screw is, the larger the distance. It will take more turns to fully tighten such a screw, but it will take less force to do it.

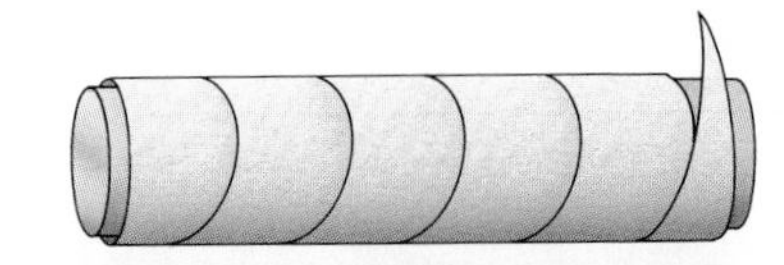

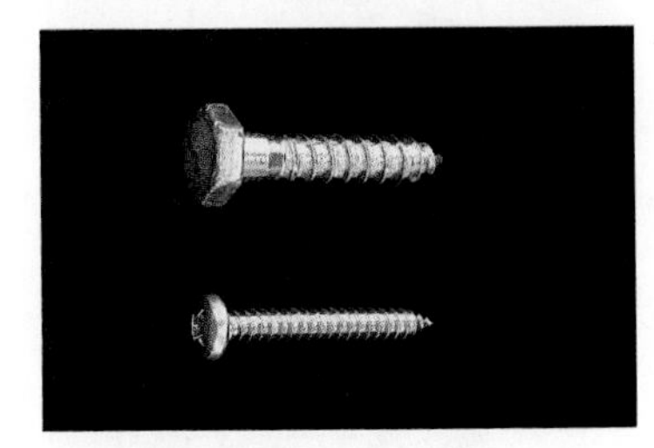

Figure 2-17 A screw is an inclined plane wrapped around a cylinder.

Wedges Two inclined planes placed back-to-back make a wedge. When a wedge is used to split a log, as shown in Figure 2-18, the two inclined planes move the resistances apart on each side. You could imagine that a short, wide wedge would split a log without being driven very far. However, the force with which you pound the wedge would have to be much greater. A long, narrow wedge, though it would need to be driven further, would be a lot easier to pound. There are probably many wedges on your kitchen counter. A knife is actually a vary narrow wedge. Can you explain the function of a knife in terms of it being a double inclined plane?

Figure 2-18 A wedge is two inclined planes placed back to back.

Summary

Machines were invented to overcome the limitations of human power. The lever and the inclined plane are the most basic machines, from which all other simple and complex machines are built. Simple machines are used to change the size or direction of a force. They do not increase the amount of work done. To decrease the force needed to accomplish work, distance must be increased. In contrast, if distance decreases, force must be increased.

2.3 MACHINES AT WORK

Any tool that helps to do work is a machine. Pencil sharpeners, drills, screwdrivers, and crowbars are machines. See Figure 2-19. Machines help do work in three ways. First, a

Figure 2-19 Machines can even be used to play ball.

machine can increase the force applied to an object. Second, a machine can change the direction of a force. Third, a machine can increase the distance through which a force moves. You already know that work is equal to force multiplied by distance. Because of this relationship, machines make it possible to reduce the force needed to do a given amount of work by increasing the distance. Nevertheless, the product of force and distance—work—remains the same.

Figure 2-20 *It would be impossible to lift this elephant without a machine. Yet the machine does not save you any work.*

Ideal Machine An imaginary machine in which there is no friction.

Work Is Work

Although machines make it easier to perform a task, they do not reduce the amount of work needed to do the job. For example, it takes the same amount of work to lift a heavy crate regardless of how the crate is lifted. You might lift the crate without the aid of a machine, or you might use a simple machine such as an inclined plane or a pulley to lift it. If you use a machine, you may exert less force, but you will be exerting that force over a longer distance. No matter how you do it, the amount of work required to lift the crate to a certain height is always the same. In other words, the amount of work you put into the machine can never be less than the work that comes out. See Figure 2-20.

In an **ideal machine**, *input work* is always equal to *output work*. Since work also equals force times distance, we can set up the following equations.

$$\text{input work} = \text{output work}$$

$$\text{(effort force) (effort distance)} = \text{(resistance force) (resistance distance)}$$

$$F_e \times d_e = F_r \times d_r$$

Now suppose you have a task in which it is important to minimize the force you must supply. For example, you decide to use a lever to lift a 360-N load 0.5 m off the ground. If you can only apply a force of 90 N, how far would you have to push the opposite end of the lever?

$$F_e \times d_e = F_r \times d_r$$

$$90\text{ N} \times d_e = 360\text{ N} \times 0.5\text{ m}$$

$$d_e = \frac{360\text{ N} \times 0.5\text{ m}}{90\text{ N}}$$

$$d_e = 2\text{ m}$$

You can see from this example that in order for the *effort force* to be less than the *resistance force*, the *effort distance* must be greater than the *resistance distance*. In this case, the resistance force (F_r) is four times larger than the effort force (F_e) you applied. However, your effort force moved four times farther than the load was lifted. See Figure 2-21. This clearly illustrates the fact that a machine can change the size of a force, but it cannot supply more work than is put into it.

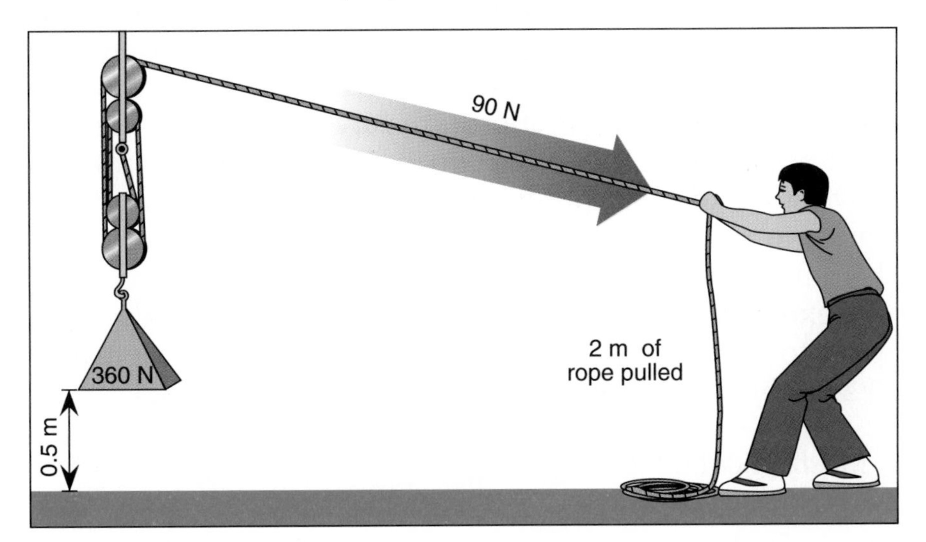

Figure 2-21 The work equation applies to all machines. In this case, the rope being pulled must move four times farther than the load.

Mechanical Advantage

If machines do not increase the amount of work you can accomplish, what would make any one machine better than another? Besides the fact that you can choose whether to increase force or distance, what else might help you decide to use a particular machine? The usefulness of any machine can be rated by comparing the amount of force you put in with the amount of force you get out.

Mechanical Advantage
The amount by which a machine increases the effort force.

You can find the amount by which an ideal machine increases force by simply dividing the output (resistance) force by the input (effort) force. The resulting number is called the **mechanical advantage** of the machine. Mechanical advantage is a ratio of the output force of a machine compaired with the input force. For example, suppose an automobile mechanic uses an engine hoist to pull an engine from a car. See Figure 2-22. If the engine weighs 4000 N, and the mechanic used a force of only 50 N to lift it, the mechanical advantage of the hoist could be figured as follows.

$$\text{Mechanical Advantage} = \frac{\text{Output force}}{\text{Input force}} = \frac{4000\text{ N}}{50\text{ N}} = 80$$

The mechanical advantage of the hoist is 80. This means the mechanic was able to apply 80 times more force to raise the engine with the machine than without it.

Figure 2-22 Without the mechanical advantage of a machine, it would be impossible to do this task alone.

Mechanical advantage can also be *less* than one. In this case, a machine would increase distance (or speed) rather than force. For example, when you turn the handle of an

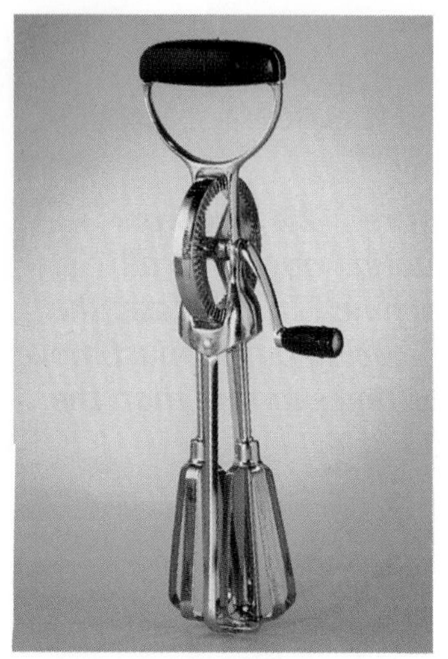

Figure 2-23 The mechanical advantage of an eggbeater is less than 1. The blades turn faster than the handle, exchanging speed (or distance covered) for force.

eggbeater, the blades turn much faster than the handle. See Figure 2-23.

Computing Mechanical Advantage Mechanical advantage can sometimes be calculated without knowing the input and output forces. Levers and their related machines increase effort force because their effort arms are longer than their resistance arms. Thus, you can compute the mechanical advantage of a simple lever by simply relating effort distance to resistance distance. This is done by dividing the length of the effort arm by the length of the resistance arm. For example, if the length of the effort arm of the lever you plan to use is 2 m and the length of the resistance arm is 0.2 m, the mechanical advantage found as follows.

$$\text{Mechanical advantage} = \frac{\text{Effort arm}}{\text{Resistance arm}} = \frac{2\text{ m}}{0.2\text{ m}} = 10$$

Your lever has a mechanical advantage of 10. In other words, this lever multiplies your effort force by 10.

The mechanical advantage of a pulley system can be easily determined by counting the ropes that support the resistance force, or the load. See Figure 2-24.

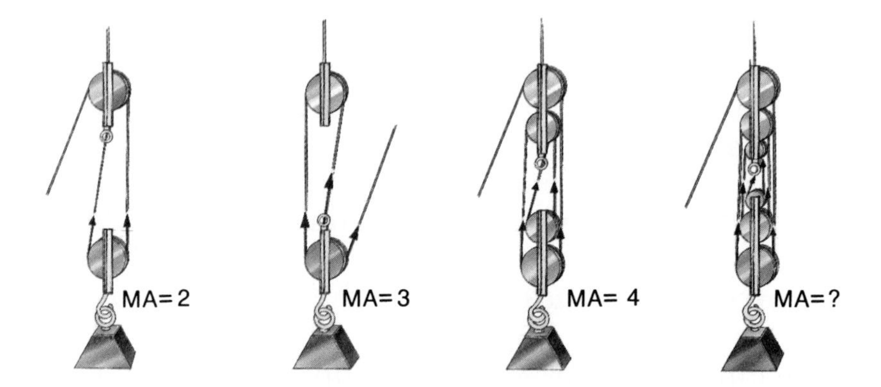

Figure 2-24 What is the mechanical advantage of the pulley system on the far right?

The mechanical advantage of an inclined plane can be found by dividing its length by its height. For example, if the length of an inclined plane is 3 m and its height is 0.6 m, its mechanical advantage is:

$$\text{Mechanical advantage} = \frac{\text{length}}{\text{height}} = \frac{3\text{m}}{0.6\text{ m}} = 5$$

So the force needed to push a box up this inclined plane would ideally be only one-fifth of the force you would need to lift the box directly. The same equation can be used to calculate the mechanical advantage of a wedge or screw. In the case of a screw, the more tightly wound the screw is, the greater its mechanical advantage.

The Effect of Friction Friction, as you have learned, is a force caused by two surfaces rubbing together. It opposes the motion of two objects in contact with each other. For example, suppose that you want to use an inclined plane to help lift a heavy box, as shown in Figure 2-25. Dividing its length by its height will give its mechanical advantage. However, since there is friction between the box and the surface of the inclined plane, some of the force you use is needed to overcome this friction. Therefore, the mechanical advantage of the inclined plane is actually less than what you originally calculated.

Figure 2-25 *The actual mechanical advantage of this inclined plane is less than its ideal mechanical advantage because some force is used to overcome friction.*

Imagine you and some friends are helping to roll a piano up an inclined plane, or ramp, into a truck. The ideal mechanical advantage of the ramp is 10. This means that for every newton of input force you and your friends apply to the piano, 10 N of force (ideally) are applied to move the piano upward. Unfortunately, friction between the wheels of the piano and the ramp increases the input force you must exert to move the piano. This is true in every real situation when work is done with a machine. The greater the friction, the smaller the actual mechanical advantage.

Efficiency The ratio of the work done by a machine to the work put into it.

Efficiency

Since all machines have some friction, the actual mechanical advantage of a machine is always less than its ideal mechanical advantage. A real machine always puts out less work than is put in. The difference between these two values is the measure of the machine's **efficiency**, which is expressed as a percentage. The smaller the difference, the greater the efficiency. Since all machines involve at least two surfaces in contact, friction is a major consideration in the design and application of machines. See Figure 2-26.

Figure 2-26 *Friction must be minimized for efficient operation af this fan.*

One way to find the efficiency of a system is to divide the work that comes out of the system by the work that is put

Figure 2-27 Every new appliance sold in the United States now carries a notice indicating the efficiency of the machine.

into it. For example, suppose you use 16 kJ (kilojoules) of energy to pedal a bicycle. Of this amount, 5 kJ are used to overcome friction. The efficiency of the bicycle would be calculated as follows:

$$\text{Efficiency} = \frac{\text{Work output}}{\text{Work input}} = \frac{16\text{ kJ} - 5\text{ kJ}}{16\text{ kJ}} = \frac{11\text{ kJ}}{16\text{ kJ}} = 0.69 = 69\%$$

Therefore, your bicycle has an efficiency of 69 percent. This means that only 69 percent of the work you do in pedaling is used to actually move the bicycle forward. Efficiency, then, is a measure of how well machines overcome friction. See Figure 2-27.

Increasing Efficiency One way to improve the efficiency of a machine is to reduce friction. Friction can be reduced in many ways. For example, oil or grease makes the surfaces of rotating parts turn more easily. See Figure 2-28. Bearings, such as the ball bearings inside bicycle wheels, are also often used to reduce friction between rotating parts.

Figure 2-28 The oil on this rotating spindle helps reduce friction as well as increase efficiency.

As you know, many people throughout history have tried to invent machines that would be 100 percent efficient. These machines were called "perpetual motion" machines. They were designed to make at least as much energy as they used. In other words, a perpetual motion machine is one that would continue operating forever with no additional input of energy.

However, no one has ever succeeded because no one has ever been able to eliminate the loss of energy due to friction. Even complex systems—such as the solar system—that appear to be running without any energy input have been found to be running down very slowly.

Summary

Machines make work easier to do, but they do not lessen the amount of work needed for any task. They can change the size of a force, but they cannot supply more work than is put into them. The mechanical advantage of any machine relates its output force to the input force. Some of the work put into a machine is always used to overcome friction, therefore, the work that comes out of a machine can never equal the work that is put in. Efficiency is a comparison of the work output with work input. Despite many efforts to build a perpetual motion machine, no one has been able to make output work equal input work because of the presence of friction.

In This Unit

Reading *Plus*

Now that you have been introduced to oceans and climates, let's take a closer look at the atmosphere and its structure, as well as the resources of the ocean. Read pages S32 to S50 and write a script for a TV special about the importance of the atmosphere and the oceans. Your script should include answers to the following questions.

1. How does the composition of the air in the atmosphere affect life on Earth as we know it? How has this changed over time?
2. In what ways do heat and moisture control the weather patterns in the atmosphere?
3. How are the resources of the ocean utilized, and in what ways must these resources be protected?

3.1 THE ATMOSPHERE

You could, perhaps, live about five weeks without food and maybe about five days without water. But you could last less than five minutes without air. In a sense, then, Earth's blanket of air, called the **atmosphere**, could be considered the most important of all its resources.

Atmosphere The layer of gases that surrounds the Earth.

The atmosphere is a vital part of our planet for many reasons. In addition to supplying the gases required for life processes, it protects Earth's organisms from exposure to harmful radiation. It also protects the Earth itself from bombardment by meteors. See Figure 3-1. Without an atmosphere, the Earth would have extremely hot days and terribly cold nights. There would be no wind, no clouds, and no liquid water. Without an atmosphere, our planet would be nothing like it is today. However, Earth's atmosphere was once very different from what it is today. In fact, you would not have been able to breathe the air of Earth's early atmosphere. How, then, did the present atmosphere come to be able to support life as we know it?

Figure 3-1 *The surface of Mercury, shown in this photograph taken by* Mariner 10, *is unprotected by an atmosphere.*

Evolution of the Atmosphere

The atmosphere, like the life forms and physical features of the Earth, is thought to have evolved slowly into its present form. Scientists theorize that the sun and all the planets formed about 4.5 billion years ago from a spinning cloud of dust and gas. Most of the dust and gas gathered into a large central body that became the sun. The nine planets eventually formed from the remaining material. Earth's first atmosphere probably consisted of gases that were part of the original cloud—mainly hydrogen, ammonia, methane, and water. But once the sun began to shine, the high-energy radiation streaming from it, known as the *solar wind*, scattered the gases of Earth's first atmosphere into space.

The Primitive Atmosphere Evidence suggests that after the solar wind scattered Earth's original atmosphere, a new atmosphere formed from gases that were trapped inside the rocks of Earth's crust when it first solidified. Most of these gases—which included hydrogen, water vapor, carbon dioxide, and nitrogen—were released during the volcanic eruptions that were common during Earth's early history. See Figure 3-2. But while this new atmosphere was forming, it began to change rapidly. Hydrogen, which is too light to be held by gravity, escaped into space. As Earth and its atmosphere cooled, much of the water vapor condensed and fell as rain. As a result, the first oceans were formed when rainwater collected in low places on the Earth's surface. Much of the carbon dioxide in the atmosphere dissolved in the rainwater and was washed out of the atmosphere.

Figure 3-2 *Many of the gases of the primitive Earth's atmosphere probably came from volcanoes.*

Conditions in this primitive atmosphere allowed certain important chemical reactions to occur. For example, energy from ultraviolet solar radiation and the electrical discharges of lightning caused some of the gases in the atmosphere to combine into a variety of complex organic compounds. These compounds fell to the Earth's surface and collected in the oceans. From these organic compounds, the first life forms on Earth are thought to have developed over 3.5 billion years ago.

Oxygen Enters the Picture Oxygen, a gas that was not abundant in the Earth's early atmosphere, was introduced to the atmosphere by two processes. A small amount of oxygen entered the atmosphere when water molecules were split into hydrogen and oxygen as they absorbed energy from the sun. However, most of the oxygen that is now part of the atmosphere is the result of photosynthesis. During photosynthesis, carbon dioxide and water are used to make sugar. Oxygen is a byproduct of that process.

Figure 3-3 *Scientists have discovered a hole in the ozone layer of the upper atmosphere over Antarctica. If similar holes should develop over populated areas, large amounts of dangerous ultraviolet radiation could cause an increase in genetic disorders and skin cancers.*

Some of the free oxygen in the atmosphere formed the gas ozone (O_3). This important form of oxygen protects us from excess exposure to ultraviolet radiation, which is harmful to living things. See Figure 3-3. However, when enough ozone formed to shield the Earth's surface from ultraviolet radiation, the rapid evolution of early life forms began. Increasing numbers of photosynthetic bacteria and algae began to release a steady supply of oxygen gas into the atmosphere and the ocean.

Figure 3-4 *Early corals and worms, from which these fossils formed, used much of the carbon dioxide dissolved in ocean water to make their skeletons and tubes.*

By about 600 million years ago, enough oxygen had dissolved in ocean water to allow larger multicellular animals to evolve there. Most of these primitive animals had shells made of *calcium carbonate*—a compound formed by the reaction of carbon dioxide and the element calcium, which dissolves in water. Thus much of the carbon dioxide that had been part of Earth's early atmosphere was used by these animals to make their shells. See Figure 3-4. Over millions of years, the shells of dead organisms collected on the ocean floor and were compressed and cemented together to form limestone. Today, most of the carbon dioxide from the primitive atmosphere remains chemically bound in these massive deposits of limestone. Only a small amount of carbon dioxide now remains in the present atmosphere.

Characteristics of the Atmosphere

With the removal of hydrogen, water vapor, and carbon dioxide from the early atmosphere, nitrogen became the most abundant atmospheric gas. Due to a rapid increase in the numbers of photosynthetic algae and plants, oxygen became the second most abundant gas. Today the relative amounts of gases in the atmosphere remain fairly stable. An equilibrium is maintained by the cycles through which gases are exchanged among the living and nonliving parts of the environment. See Figure 3-5.

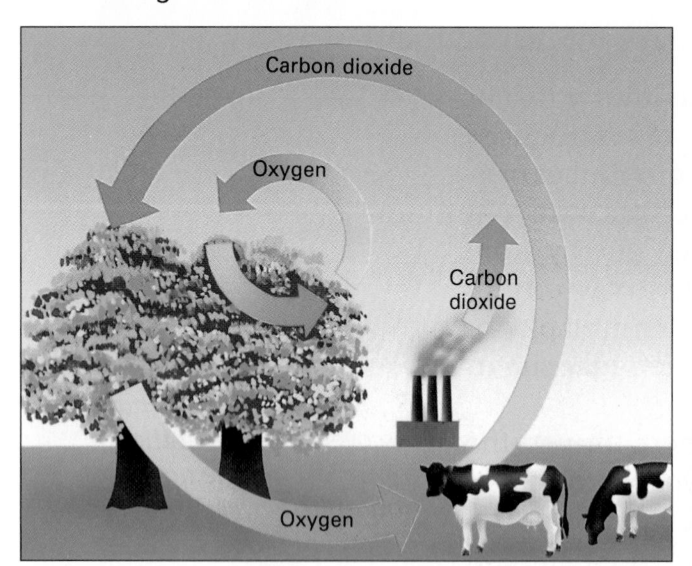

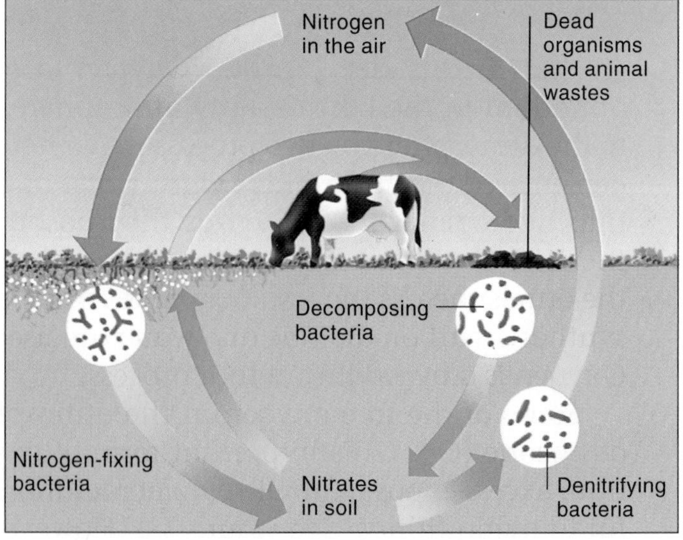

Figure 3-5 *The carbon dioxide–oxygen cycle* (left) *maintains a stable amount of oxygen in the atmosphere. The nitrogen cycle* (right) *maintains a stable amount of nitrogen in the air.*

Air The mixture of gases in the atmosphere.

Currently the atmosphere consists of about 78 percent nitrogen, 21 percent oxygen, and 1 percent argon, carbon dioxide, and water vapor. This mixture of gases is called **air**. Earth's atmosphere, however, is not a uniform layer of air.

To illustrate this fact, let's take an imaginary trip from the beach to a nearby mountain.

Pressure Changes When you travel up a mountain, you might notice that your ears pop. This happens because the air pressure outside your ears gradually becomes lower the higher you climb, while the air pressure inside your ears remains the same. You may also hear a popping sound when the air pressures inside and outside of your ears quickly become equal as you yawn or swallow. But why does air pressure change as you go up a mountain?

Gravity pulls everything, including atmospheric gases, toward the Earth's surface. Air, in turn, presses down on the objects under it, creating air pressure. At sea level, all of the atmosphere is above you, making the air pressure greatest there. See Figure 3-6. On the mountain, you have less air above you, so the air pressure is less. Air also pushes down on itself. Thus the atmosphere's density is also greatest at lower altitudes.

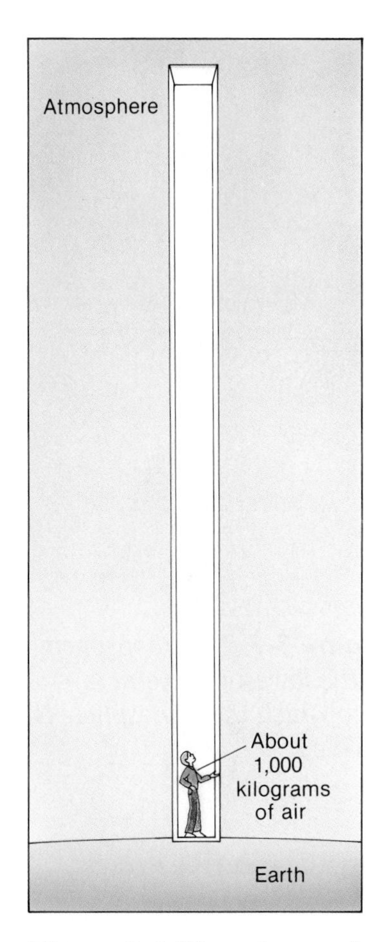

Figure 3-6 *The column of air pushing down on you at sea level weighs more than 1000 kg! At higher altitudes, there is less air above you, so the air pressure is lower.*

Temperature Changes In addition to the change in air pressure, you will notice that the temperature drops as you go up the mountain. If you measured this drop, you would find that the temperature falls about 6.5 °C for each kilometer you ascend. For example, at the top of a mountain 3 km high, you would find that the temperature is about 20 °C colder than it is at sea level.

The farther you travel above Earth's surface, the colder it gets, at least until you reach an altitude of about 10 km above sea level. There, the temperature remains steady at about –55 °C for several more kilometers. Then a surprising thing happens. At an altitude of about 20 km, the temperature of the atmosphere begins to rise. This change marks the boundary between two separate layers of the atmosphere.

Layers of the Atmosphere The Earth's atmosphere consists of several layers. Each layer is characterized by either a rise or a fall in air temperature. Closest to the Earth is the densest layer of the atmosphere, the *troposphere*. The troposphere contains about 90 percent of the atmospheric gases. It is also the layer in which almost all of Earth's weather occurs. The air in the troposphere is not heated directly by the sun. Instead, the Earth's surface is warmed by the sun and it, in turn, heats the air near the surface. Warm air near the surface of the Earth rises because it is less dense than the colder air at higher altitudes. Cold air then sinks and replaces the rising warm air. This constant movement of air in the troposphere is what drives the Earth's wind and weather systems. See Figure 3-7.

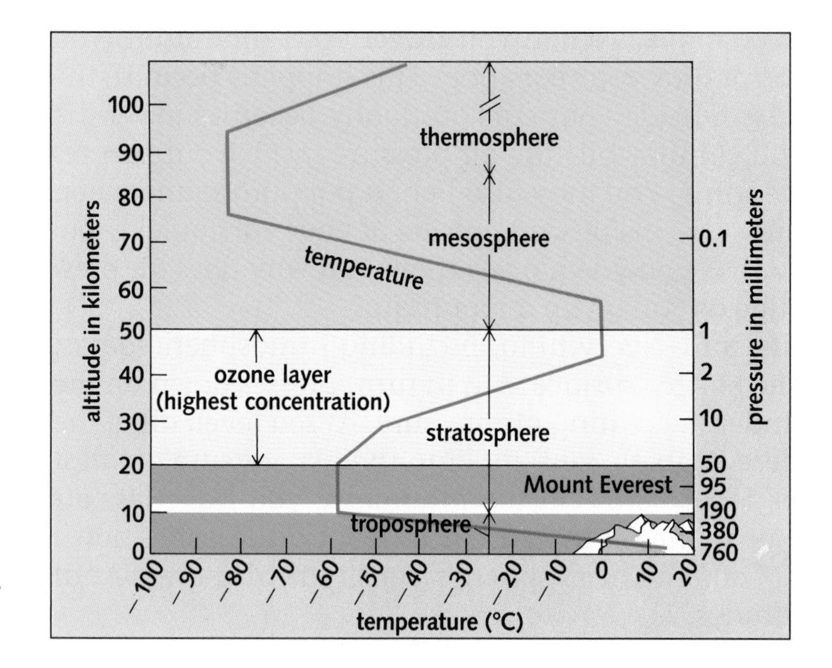

Figure 3-7 The troposphere is the lowest of several layers into which the atmosphere is divided.

Beginning at an altitude of about 20 km, and extending upward to an altitude of about 50 km, is a clear, cold layer of air called the *stratosphere*. In the stratosphere, the air is very thin and contains little moisture or dust. As a result, practically no weather occurs in the stratosphere. However, at the base of the stratosphere are broad, fast-moving "rivers" of air, called **jet streams**. See Figure 3-8. These jet

Jet Stream A belt of high-speed wind circling the Earth between the troposphere and the stratosphere.

Figure 3-8 Fast-moving jet streams form where warm air from the tropics meets cold air from the poles. Jet streams in the Northern Hemisphere flow from west to east.

streams circle the planet and affect weather patterns in the troposphere below. The stratosphere also contains most of the ozone in the atmosphere. The concentration of ozone rises with altitude, reaching a maximum at about 50 km. Ozone molecules absorb some of the sun's energy, causing the temperature in the stratosphere to steadily rise until it reaches about 0 °C.

Above the stratosphere is the coldest layer of the atmosphere—the *mesosphere*, which extends from about 50 to 85 km above the Earth's surface. Again, the temperature drops steadily with increasing altitude, due to a decreasing concentration of ozone and the decreasing density of gases.

The uppermost layer of the atmosphere, which is characterized by very thin air and rising temperatures, is called the *thermosphere*. Reaching to an altitude of about 600 km, the thermosphere blends gradually into the near vacuum of outer space. Due to the absorption of high-energy solar radiation, temperatures in the thermosphere are high. Special instruments have recorded temperatures in the thermosphere as high as 2000 °C.

The Ionosphere Within the thermosphere is a region called the *ionosphere*. Beginning at a height of about 80 km and continuing up to about 400 km, the ionosphere also includes the uppermost part of the mesosphere. The ionosphere is so named because it contains electrically charged particles called *ions*. These ions are formed when atoms and molecules in the atmosphere absorb solar radiation and lose some of their electrons. Colorful light displays, known as *auroras*, result from the recapture of electrons by ions in the ionosphere. See Figure 3-9. Auroras, however, are usually only visible near the poles. Ions, which are attracted by Earth's magnetic field, are concentrated there because the magnetic field is much stronger at the poles.

Figure 3-9 *The Aurora borealis (northern lights) as seen in Alaska.*

The ionosphere also reflects AM radio waves back toward the Earth. These radio waves are able to travel long distances by bouncing back and forth between the Earth and the ionosphere. But every time a radio wave is reflected, some of its energy is lost so it eventually fades out. Have you noticed that radio signals often come from greater distances at night than during the day? This is because at night the lower layers of the ionosphere disappear because ions cannot form in the absence of the sun's rays. Thus, radio waves are able to travel farther since they are reflected fewer times. See Figure 3-10.

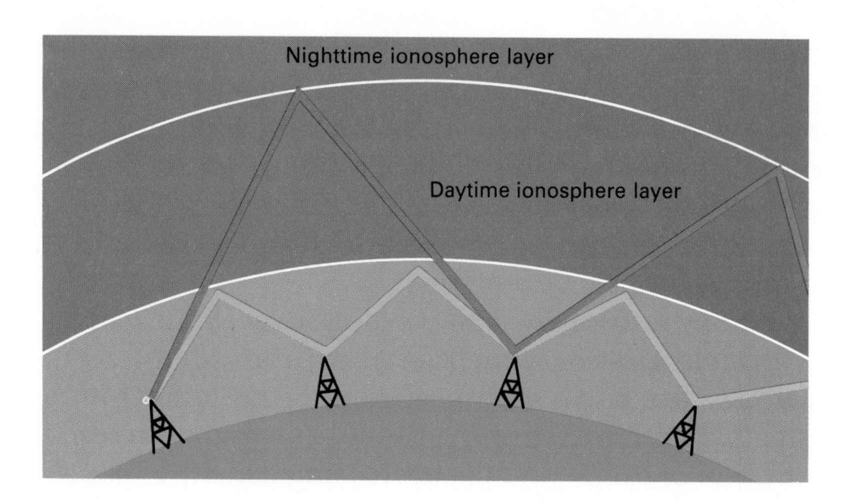

Figure 3-10 *Some radio waves can be transmitted around the world by reflecting them off the ionosphere. At night, the radio waves can travel farther because the lowest ion layer disappears and the waves are reflected off a higher ion layer.*

Summary

To air-breathing organisms, Earth's atmosphere is one of its most vital resources. According to current theory, the gases of Earth's original atmosphere were scattered by the solar wind and eventually replaced by gases that escaped from inside the planet through the action of volcanoes. The makeup of the Earth's second atmosphere slowly changed as cooling caused water vapor to condense into rain, and chemical reactions involving other gases formed organic compounds. Living organisms were primarily responsible for producing the oxygen in the atmosphere of today. The atmosphere is divided into several invisible yet distinct layers. These layers are characterized by air pressure that decreases with increasing altitude as well as periodic increases and decreases in temperature.

3.2 WEATHER, WIND, AND WATER

If you live along the Gulf Coast, in Hawaii, or in Puerto Rico, you know it will be warm and humid most of the year. If you live in the Northeast or the Midwest, your summers will be warm and humid, but winters will be cold and snowy. If you live in the Southwest, it will be warm and dry most of the year. If you live on the Northwest Coast, it will be cool and rainy most of the year.

Weather The condition of the atmosphere at a given time and place.

Weather is the condition of the atmosphere at a particular time and place. Specific weather conditions, such as temperature, humidity, clouds, winds, and precipitation, usually change from day to day and often from hour to hour. The average weather conditions over a long period of time constitute the **climate** of that region. Weather and climate both are created by the sun's uneven heating of Earth's atmosphere.

Climate The average weather for a particular place.

Heating the Atmosphere

The sun releases a huge amount of radiant energy, only about two-billionths of which is received by the Earth. Even this tiny fraction of the sun's radiation contains a very large amount of energy. In fact, the total amount of energy that reaches the Earth is 13,000 times greater than all the energy used by everyone on Earth. So what happens to all this energy? About 20 percent of the solar radiation that reaches Earth is absorbed by the gases of the atmosphere. Another 33 percent is reflected back into space by clouds, snow, and ice. The Earth's surface absorbs an additional 47 percent. See Figure 3-11.

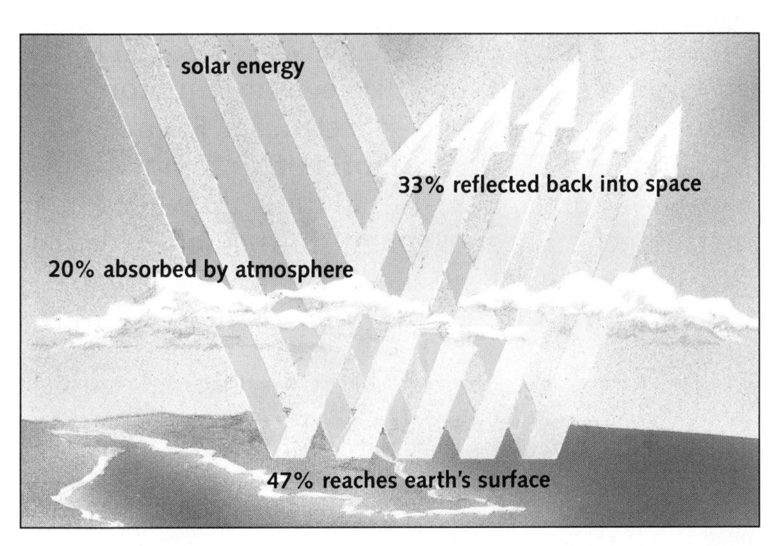

Figure 3-11 The majority of the sun's energy that reaches Earth is reflected back into space or absorbed by the atmosphere.

However, the solar radiation striking the Earth is not evenly distributed. Several factors affect the amount of radiation that strikes at different locations. First, because Earth is a sphere, the sun's rays strike different places at different angles. See Figure 3-12. This uneven distribution of solar radiation causes unequal heating of the surface. Since the troposphere is heated by the Earth's surface, it too is heated unevenly. For example, air over the equator is heated more than the air over the poles. This causes the air over the equator to be less dense and have a lower pressure than the air over the poles.

Figure 3-12 Since the sun's rays are most direct at the equator, the surface of the Earth is warmed the most there.

Air Circulation

In general, air flows from a region of high pressure toward a region of low pressure. Thus, cold dense air from the poles

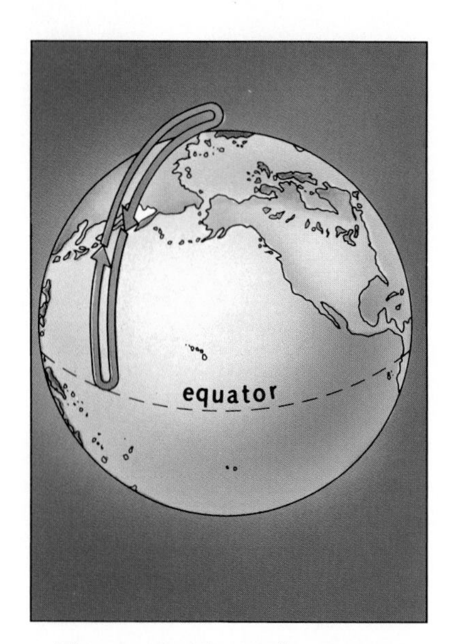

Figure 3-13 At the equator, air is warmed and rises upward. At the poles, air is cooled and sinks.

flows toward the equator along the Earth's surface. At the equator, the air becomes warmer, expands, and rises. This warmed air is also pushed upward by the cold air moving in behind it. The warm air flows back toward the poles along the top of the troposphere. At the poles, the air cools, becomes denser, sinks, and starts a new cycle as it flows back toward the equator. Figure 3-13 shows this general pattern of air circulation.

However, the Earth's pattern of air flow also consists of several smaller *wind belts.* As warm air from the equator flows toward the poles along the top of the troposphere, it gradually cools. About one fourth of the way to the poles, most of this air sinks back to the surface. When it hits the surface, the sinking air divides. Some of it flows back toward the equator, while the rest flows toward the poles. The air finally reaching the poles sinks even more and then flows back toward the equator.

Coriolis Effect The apparent bending of the path of a moving object due to Earth's rotation.

Because the Earth rotates, the path taken by the moving air is not a straight line between the equator and the poles. Rather, it appears to be deflected. See Figure 3-14. This phenomenon, called the **Coriolis effect**, makes objects in the Northern Hemisphere drift to the right when you are facing their direction of movement. In the Southern Hemisphere, they drift to the left. In the Northern Hemisphere, the Coriolis effect causes air that is moving back toward the equator

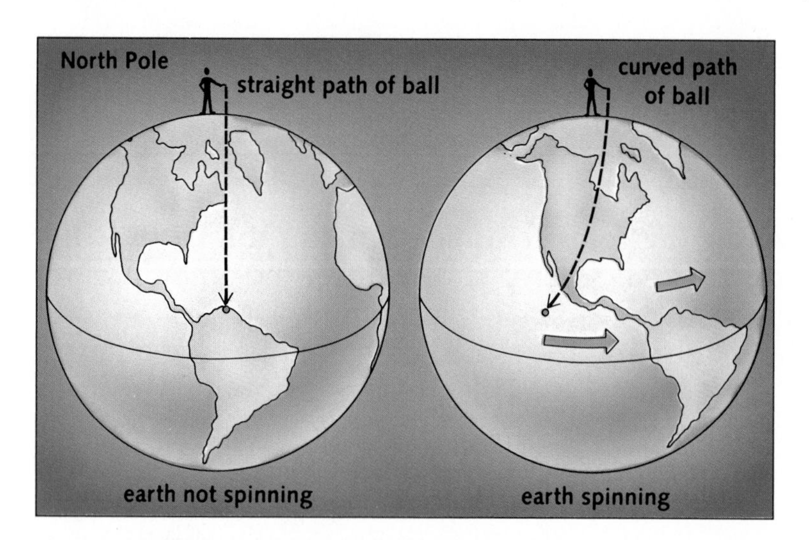

Figure 3-14 Because of the Coriolis effect, objects traveling in a southerly direction in the Northern Hemisphere appear to curve to the right.

to create winds that tend to blow from northeast to southwest. Such is the case with the wind belts known as the *tradewinds* and the *polar easterlies.* See Figure 3-15. Air moving toward the poles creates winds that blow from southwest to northeast. These winds are called *prevailing westerlies.* Notice how the direction of the winds in the wind belts differs between the Northern and Southern Hemispheres.

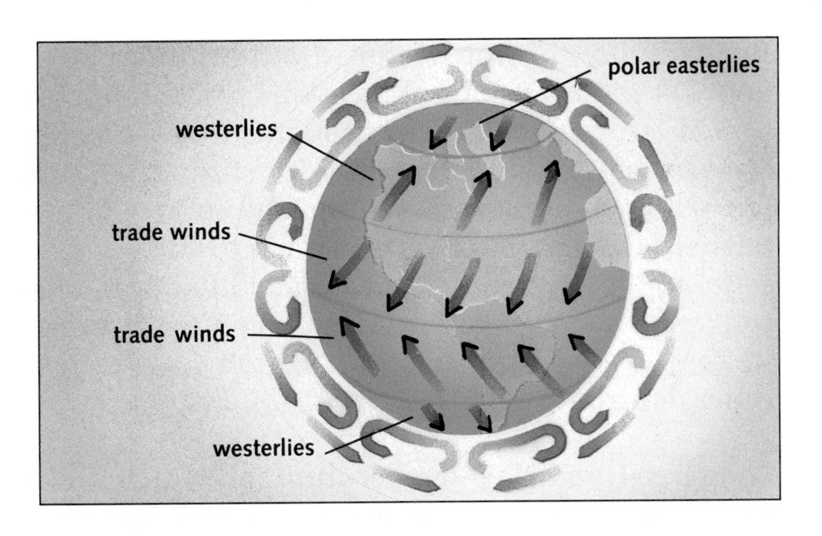

Figure 3-15 A combination of the Coriolis effect and the circulation of air to and from the poles causes the global wind belts.

Water Enters the Atmosphere

When water is heated by energy from the sun, some of it evaporates and enters the atmosphere as water vapor. Lakes, streams, rivers, and plants contribute water vapor to the atmosphere, but the oceans are the source of most of the water vapor that enters the air. But air can only hold a certain amount of water vapor. The amount of water vapor that air can hold is determined by its temperature. Warm air can hold more water vapor than cold air. Air that is holding as much water vapor as possible is said to be *saturated.*

The amount of water vapor in the air is called *humidity.* But air does not always contain as much water vapor as it could hold. The term **relative humidity** refers to the amount of water vapor that *is* in the air compared to the amount it *can* hold. In other words, relative humidity is the percent to which air is saturated with water vapor. For example, if a certain volume of air could hold 40 g of water but has only 20 g, the relative humidity would be 50 percent. If the same mass of air had 30 g of moisture, the relative humidity would be 75 percent. When the relative humidity is 100 percent, the air is saturated. See Figure 3-16.

Relative humidity also changes as air temperature changes. The amount of water vapor itself does not change, but the amount of water vapor that air can hold does change. For example, as air temperature drops, the air can hold less water vapor. Therefore the relative humidity rises. When air is warmed, it can hold more water vapor. Therefore the relative humidity drops.

Relative Humidity The amount of water vapor in air compared to the amount of water vapor that air could hold at that temperature.

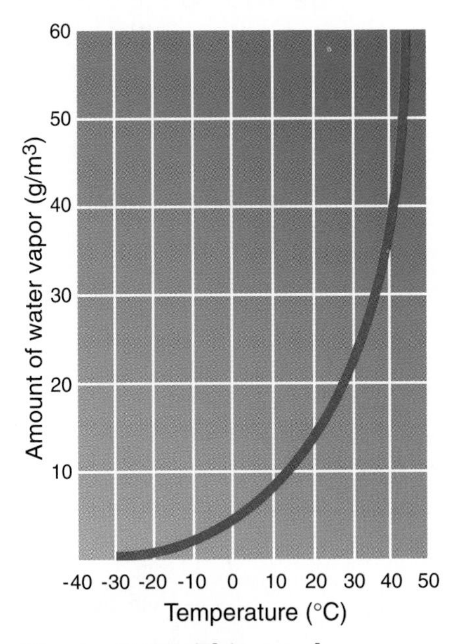

Figure 3-16 This graph shows the amount of water vapor that a given volume of air can hold at various air temperatures.

Water Leaves the Atmosphere

Each day about a trillion metric tons of water move from the Earth's surface into the atmosphere. Each day about the same amount of water returns to the surface as precipitation. Yet many regions of the world, including parts of the United States, face serious droughts and water shortages. Meteorologists, the scientists who study the atmosphere, know how moisture moves from the surface to the atmosphere. But what causes it to return to the surface only in certain places? Weather forecasters attempt to answer this challenging question every day, often with rather limited success. A discussion of some of the factors involved in water returning to Earth's surface follows.

Figure 3-17 *At the dew point, moisture in the warm, damp air from the lungs of the horse and its rider condenses.*

Condensation The changing of a gas into a liquid.

Condensation The temperature at which the relative humidity reaches 100 percent is called the *dew point*. If the temperature continues to drop, water vapor in the air will change back into a liquid. This process is called **condensation**. See Figure 3-17. Water vapor may condense directly on cold objects, such as rocks, metal, windows, or plants. This moisture is called *dew*. If the dew point is below 0 °C, *frost* will form. Frost is solid ice crystals that form directly on an object. It is not frozen dew.

There are several ways that water can be cooled to the dew point. Air may be cooled as it is pushed upward to rise over mountains. It may also be pushed upward by denser air moving under it. Visible water droplets will form when the air is cooled to the dew point, provided there are solid particles such as dust or salt crystals in the air. The solid particles give water vapor a surface on which to condense. The visible droplets that result form clouds.

Precipitation Cloud droplets are so small that even slight air movements keep them floating. When cloud droplets collide and merge, however, larger, heavier drops form and may fall as *rain* or *drizzle*. See Figure 3-18. Whenever the temperature in a cloud is well below 0 °C, ice crystals may form. Falling ice crystals collect into snowflakes that may fall as *snow* if the air is cold enough all the way to the ground. However, if snowflakes pass through a layer of warm air, they will melt and become rain. Sometimes, rain passing through a cold layer of air freezes as it falls. It is then called *sleet*. In violent storms, raindrops may be tossed high enough in a cloud to freeze. As they fall again, a new layer of water freezes onto them. If they are tossed up and down several times, multiple layers of ice accumulate, forming *hail*. See Figure 3-19. Sometimes hail can grow large enough to cause damage to objects on which they finally land. The largest recorded hailstone fell in Kansas and was larger than a grapefruit.

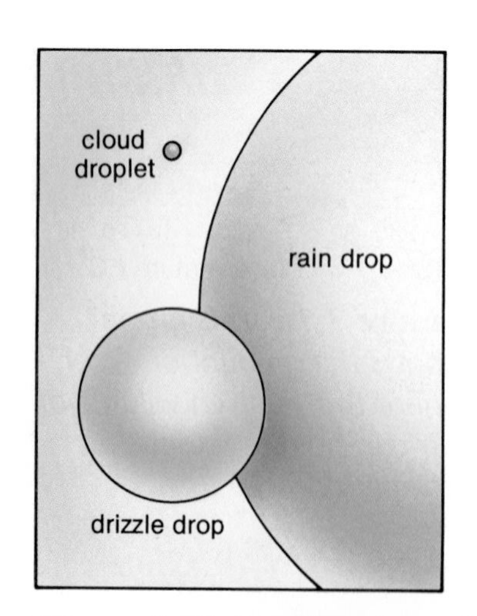

Figure 3-18 *Relative sizes of condensed water vapor.*

Figure 3-19 *Hail is precipitation in the form of layered lumps of ice.*

Clouds In 1802 an Englishman named Luke Howard developed a classification system to describe various types of clouds that form when water vapor condenses. Cloud formation depends on certain conditions in the atmosphere, such as air temperature, air pressure, humidity, and air currents (wind). On a fair day, for example, warm air rising from the Earth's sun-heated surface cools and forms puffy white clouds. Howard called the puffy clouds that form rounded tops *cumulus clouds. Cumulus* is a Latin word meaning "heap." Another type of cloud, which consists of thin, white wisps, he named *cirrus*—the Latin word meaning "a lock of hair." Cirrus clouds are formed high in the troposphere where the temperature is very cold. Water vapor there forms ice crystals, which are blown by the high winds of the troposphere into patterns resembling delicate strands of hair. A third type of cloud is formed where warm air moves over colder air. A solid layer of flat clouds forms when the warm air is cooled past its dew point. These broad, flat clouds often form a solid blanket over the entire sky. Howard called clouds of this type *stratus clouds. Stratus* comes from the Latin word meaning "layered" or "wide spread." Figure 3-20 shows these three basic cloud types.

If you have done much cloud gazing, you know that clouds do not always fit into one of the three categories above but appear to be combinations of the three basic types. Howard suggested names for these combinations as well. For example, sometimes cumulus clouds become so crowded together that they almost form a layer across the sky. Howard called these clouds *stratocumulus.* The modern classification system for clouds also takes into account the altitude at which clouds form, and whether or not they produce rain. Clouds that form at altitudes above 7 km are given the prefix "cirro," and those that form between 2 and 7 km above the ground are given the prefix "alto." Clouds that produce rain or snow are called *nimbus clouds. Nimbus* comes from the Latin word meaning "shower." Figure 3-21 shows some of the positions and characteristics of cloud types.

Figure 3-20 *Three types of clouds are* (from top to bottom) *cumulus, cirrus, and stratus.*

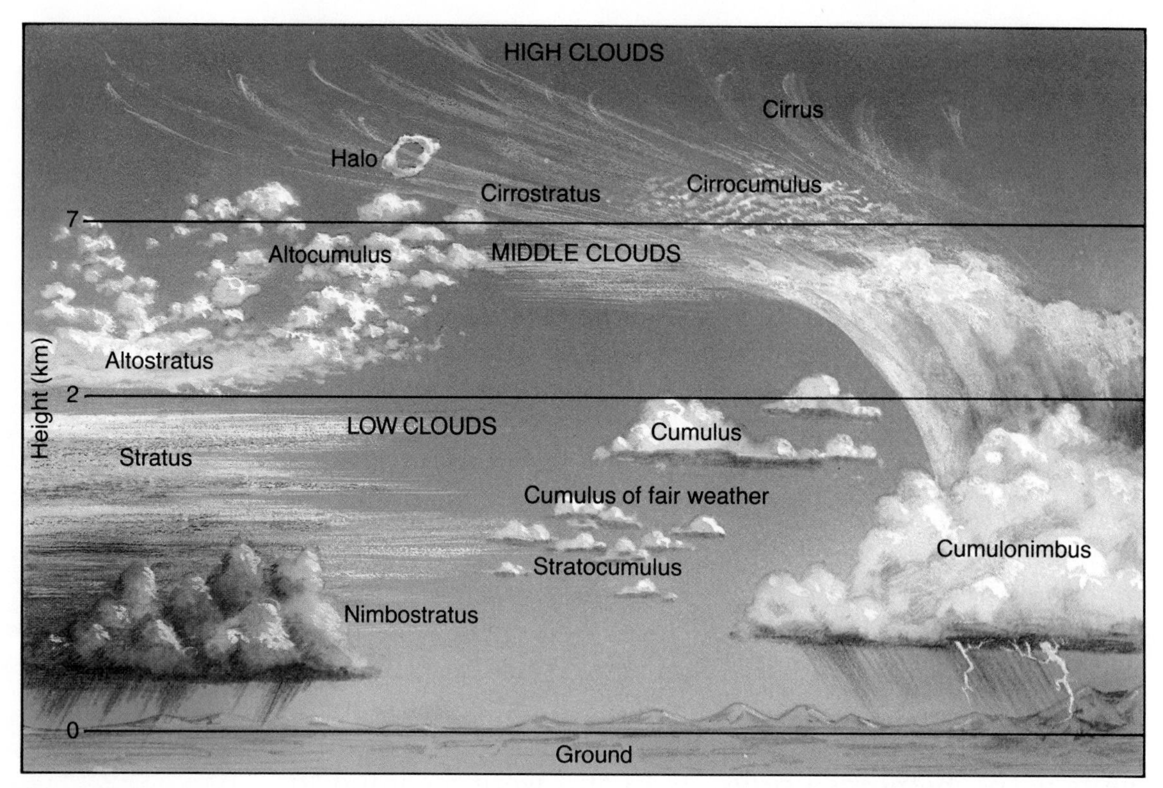

Figure 3-21 *Classification of clouds according to height and form.*

Air Masses

Air Mass A large body of air with uniform temperature and humidity.

When air remains over one part of Earth's surface for a long period of time, it takes on the characteristic temperature and humidity of that region. Such a body of air is called an **air mass**. In general, air masses that form over land are dry, and those that form over oceans are humid. Air masses that form in northern regions are cold, and those that form in regions close to the equator are warm. See Figure 3-22.

Only regions that have light winds and consistent surface features over a large area can form air masses. For example, the Earth's surface at the equator is mostly covered by oceans. At the poles, vast areas are covered by the polar ice caps. Regions such as these are called *source areas*.

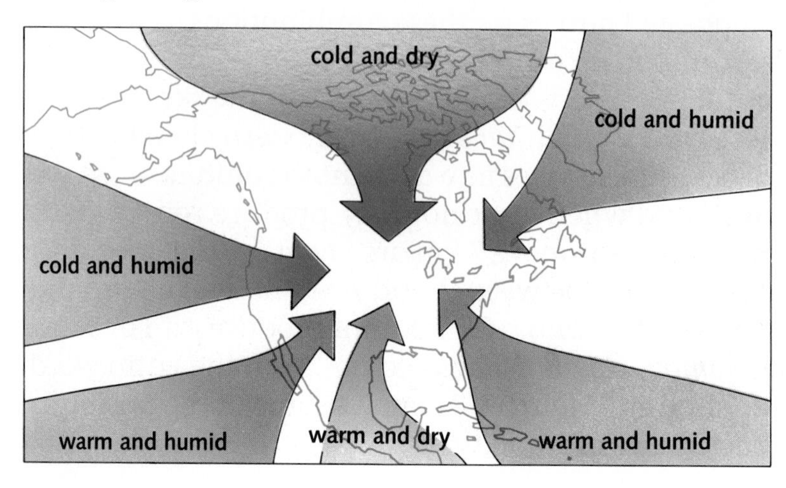

Figure 3-22 *Air masses that affect the weather in North America.*

Fronts

As air masses move, they meet each other. Where they meet, the winds bring air from the different masses together. Even though some of the air at the boundary mixes, it is always replaced by fresh air from farther back in each air mass, so the boundary between the air masses stays sharp. Such a boundary is called a **front**. Temperature, pressure, and wind direction usually change significantly with the passage of a front. Most fronts form when one air mass moves into a region occupied by another air mass.

Front The boundary between two air masses.

Warm Fronts When a warm air mass pushes into a cold air mass, a warm front is formed. See Figure 3-23. Because it is less dense, the invading warm air rides up over the colder air. As the warm air is lifted, cirrus clouds are formed high in the troposphere. As time passes, the cloud cover becomes lower and thicker until a solid sheet of stratus clouds covers the sky. A long period of gentle rain may occur, followed by slow clearing and rising temperatures. The weather changes that follow a warm front are not as noticeable as the changes that follow a cold front.

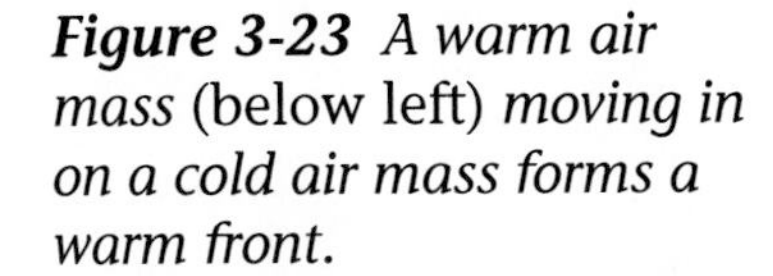

Figure 3-23 *A warm air mass* (below left) *moving in on a cold air mass forms a warm front.*

Figure 3-24 *A cold air mass* (below right) *moving in on a warm air mass forms a cold front.*

Cold Fronts If a cold air mass invades warmer air, a *cold front* is created. See Figure 3-24. Since colder air is denser, it

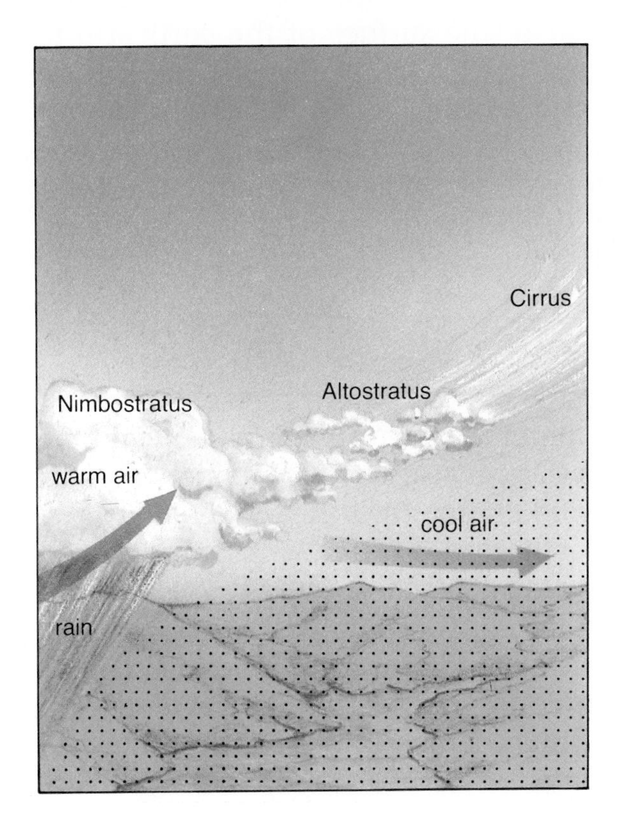

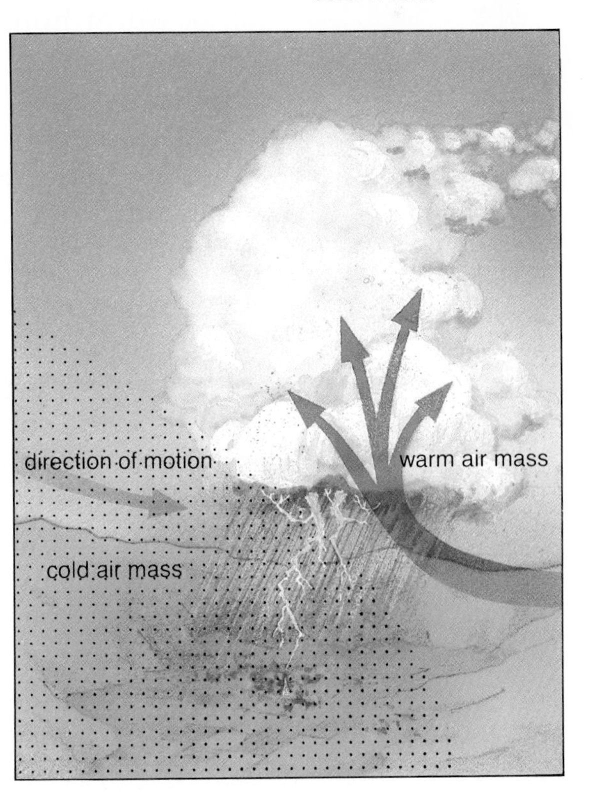

pushes under warmer air like a wedge. The lifting of a warm air mass usually causes cumulus-like clouds to form along the cold front. These clouds often bring stormy weather and heavy rain. Cold fronts tend to move quickly because dense, cold air easily pushes into warmer air. The turbulent weather associated with a cold front usually passes quickly. The air mass behind a cold front normally brings cool, dry weather.

Figure 3-25 *A thunderhead.*

Thunderstorms are a common occurrence during the approach and passage of a cold front. They are produced when warm, humid air is lifted very rapidly by the cold air mass. Towering *cumulonimbus* clouds develop as a result. These clouds are commonly called *thunderheads.* See Figure 3-25. Sometimes, cold fronts are preceded by a solid line of thunderstorms and strong winds. Such a line is called a *squall line.* Local conditions can also give rise to thunderstorms. For example, warm, moist air blown up the side of a mountain, or heated by an area of warm ground, may form thunderstorms.

Figure 3-26 *The funnel cloud of a tornado.*

Occasionally, very small but extremely violent storms called *tornadoes* form in the warm, moist air ahead of a cold front. See Figure 3-26. A tornado usually drops from a thunderstorm as a thin, funnel-shaped cloud. Scientists are not really sure how a tornado forms. But they do know the kinds of weather conditions that are likely to produce tornadoes—when very cool and dry air in the upper atmosphere is on top of warm humid air at the surface of the earth. Up to 1000 tornadoes occur in the United States each year. Most of these occur in a corridor from Texas to Michigan appropriately called "Tornado Alley." Although they usually last only a few minutes, tornadoes can be very destructive. Their winds, sometimes up to 800 km/h, can destroy almost everything in their path.

Summary

Weather is the condition of the atmosphere at a given time and place. Climate is the general weather tendencies of a region over a long period of time. Weather is created by the sun's uneven heating of the atmosphere. Heated air rises and is replaced by sinking cold air from the poles, creating a general pattern of air circulation. This pattern is broken up into several wind belts. Water vapor in the atmosphere condenses when it cools, forming clouds and precipitation. Clouds and precipitation commonly occur along fronts, which are boundaries between different kinds of air masses.

3.3 OCEAN RESOURCES

One outcome of the evolution of the atmosphere was the formation of the oceans. As water vapor condensed out of the early atmosphere and fell as rain, it gathered in the lower depressions on the Earth's surface. The ocean became home to a vast number of organisms, past and present. Today, the oceans provide a wealth of important natural resources, including food, valuable minerals, and energy sources.

Food From the Sea

Oceans cover almost two-thirds of the Earth's surface. All over the ocean, living things have established a niche. In fact, the average amount of plant growth for each square meter of ocean surface is about the same as it is for land surfaces. In theory then, oceans could produce as much food as the land. Then why does less than 10 percent of the world's food supply come from the ocean? Part of the answer is the type of food we collect there.

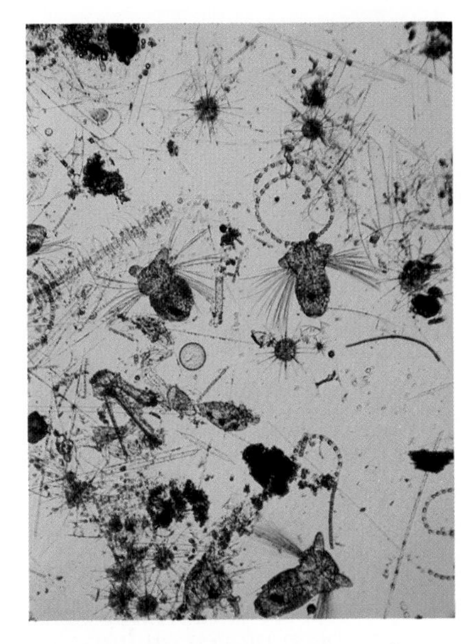

Figure 3-27 *Many parts of the ocean contain large amounts of microscopic plankton.*

Plankton The small plants and animals floating near the surface of the ocean.

Plankton Most of the mass of the organisms (biomass) in the ocean is in the form of **plankton**. Most plankton are so small that they can be seen only with a microscope. See Figure 3-27. Microscopic plants called *phytoplankton* are found mostly in the upper few meters of the sea where there is enough light for photosynthesis. Tiny animals called *zooplankton* feed on the phytoplankton. Both kinds of plankton are eaten by even larger animals that, in turn, become food for larger fish and other marine animals. Thus phytoplankton are the beginning of a food chain that supports the animal life of the ocean. See Figure 3-28.

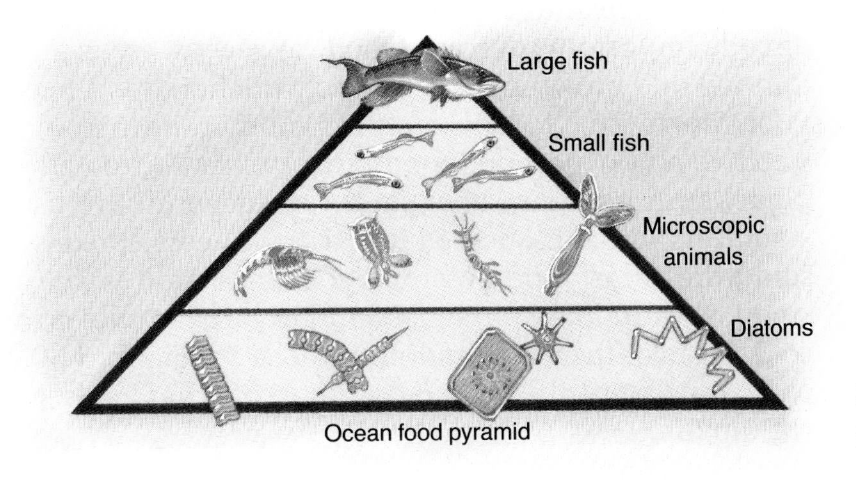

Figure 3-28 *A typical food pyramid in the ocean.*

It has been suggested that if there were a way to harvest the plankton in the ocean, many more people could be fed by the ocean than it currently feeds. But even if we could eat plankton, most of the upper ocean (where plankton are found) is poor in nutrients. As a result, most of the ocean could not produce much more than it does now.

Fish Fish are the main food we take from the ocean. They are caught primarily in areas of the ocean that lie over continental shelves. There the water is rich in the phytoplankton that are the basis of most ocean food chains. Some areas near the coasts act like natural fish farms. This is partly because of *upwelling,* which is the rising of cold water from below to replace surface waters removed by steady winds and evaporation. Upwelling brings up nutrients from the ocean bottom that are needed for the growth of phytoplankton. Upwelling also occurs in parts of the open ocean.

Figure 3-29 *Redfish were almost eliminated from the Gulf of Mexico because of the popularity of these fish in cooking. Strict limits have now been placed on harvesting redfish.*

Today, humans catch nearly one-third of all the fish large enough to be taken in nets. But there is a limit to the number of fish that can be taken without destroying the source. Some kinds of fish that were once plentiful, such as cod, are now much less abundant. Laws currently protect some endangered fish populations in an attempt to prevent their disappearance from the oceans. See Figure 3-29. In addition, scientists are trying to determine all of the conditions in the ocean that affect fish populations. This information may help researchers to learn how many fish can be harvested without affecting the population size. Using that information could help to ensure that there would always be a supply of fish.

Other Ocean Food Sources Shrimp and shellfish, two marine organisms that are lower in the ocean food chain, are already important sources of food. Seaweeds, especially certain red and brown algae that grow anchored to the ocean floor, are another food resource. In many countries, the seaweed *kelp* is a popular food. Substances extracted from algae, such as algin, agar, and carrageenan, are used as stabilizers in foods such as ice cream, cheese spreads, and salad dressings. One potential source of food is *krill.* This small, shrimp-like animal is found in large numbers in the cold waters of the Southern Hemisphere. See Figure 3-30. Krill can be used directly for food or as part of other foods, just as shrimp are.

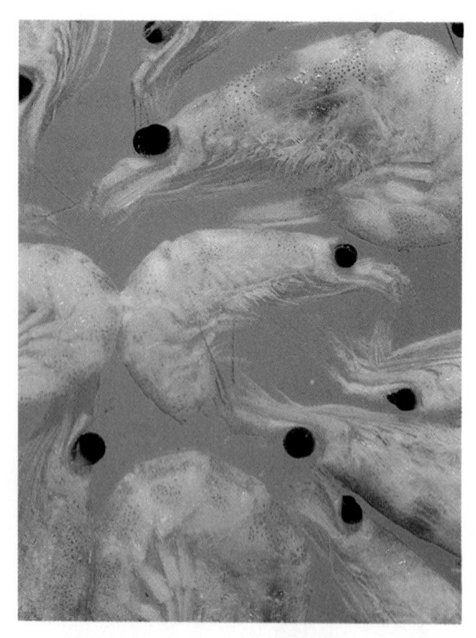

Figure 3-30 *Krill is a small animal found in huge numbers in some parts of the ocean.*

Minerals from the Ocean

Because of the constant inflow of deposits carried by rivers and streams, the ocean contains huge amounts of dissolved minerals. However, the mineral resources of the ocean are distributed throughout a vast amount of water. To recover useful amounts of these minerals, large quantities of seawater must be processed. This comes at great expense. It is even too expensive to take a mineral as valuable as gold from seawater because it costs more to process than it is worth. At the present time, sodium, bromine, and magnesium salts are the only minerals extracted from seawater commercially.

Ocean deposits, called **nodules**, are another possible source of minerals from the ocean. These egg-sized to potato-sized lumps litter large areas of the deep ocean floor. Scientists do not know exactly how nodules are formed, but they do know that they are a vast reserve of important metal ores. Nodules are composed mostly of iron, manganese, nickel, and cobalt. They also contain smaller amounts of copper, zinc, silver, gold, and lead. See Figure 3-31. However, taking nodules from the ocean bottom is not currently economical. Since nodules are only found far beneath the surface, special mining methods must be used to gather them.

Nodules Lumps of sediment found on the ocean floor that contain valuable minerals.

Figure 3-31 *Iron, manganese, nickel, and other elements can be obtained from these nodules.*

Energy from the Sea

One of the most important sources of energy from the sea is *petroleum.* When marine organisms die, their remains fall to the ocean floor along with other debris. Since the water at the ocean bottom does not contain much oxygen, organic matter does not readily decay. Instead, it accumulates along with other, inorganic sediments. As more sediments accumulate, the buried organic matter is eventually chemically converted into petroleum. The largest known petroleum deposits on Earth formed when the African continent drifted northward and pushed organic-rich sediments against Asia.

These sediments ultimately became the large petroleum reserves of the Middle East. Other large reserves are found on the north shore of Alaska, beneath the North Sea, and along the coast of the Gulf of Mexico. See Figure 3-32.

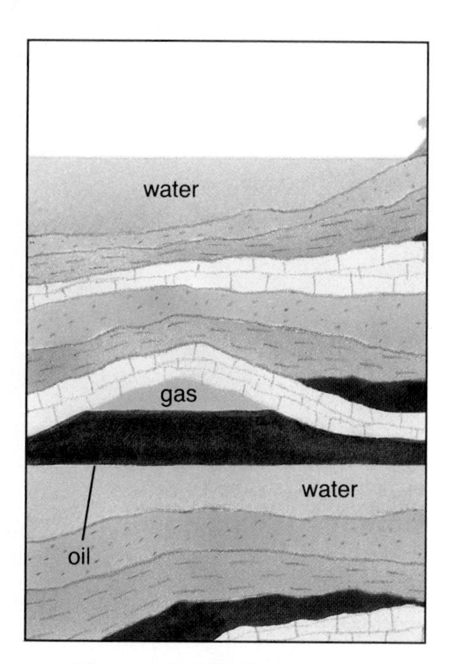

Figure 3-32 *Structure beneath the ocean showing the layers in which oil is trapped.*

Protecting Ocean Resources

The technical difficulty of obtaining ocean resources may not be the biggest problem we must overcome. Deciding who owns these resources and how they should be divided may be an even bigger problem. Countries have not yet agreed on answers to questions about using the resources of the sea. How far into the ocean do the borders of a country extend? Do nations have the right to collect these resources from the ocean beyond their borders? Do countries without a coastline have any rights to ocean resources? Until there is agreement among nations, there cannot be full or fair use of the ocean's resources.

But even if these political questions could be solved, the pollution of the ocean is a present and growing problem that must be addressed immediately. The chief sources of ocean pollution are sewage and garbage, industrial waste, and agricultural chemicals and wastes. These pollutants are not only detrimental to ocean life in the areas where they are dumped, but they can be spread by currents to affect the entire ocean. See Figure 3-33. Oil spills from tankers and drilling platforms must be cleaned up more effectively or, better yet, prevented. One accident can damage the ecosystems within the water and on the shores that surround it for many years.

Figure 3-33 *Raw sewage flushed directly into the ocean is one practice that can cause serious pollution problems.*

The resources of the ocean can only be protected if all the nations of the world work together to solve the problem. You can assist in this effort by doing your part to see that effective laws are passed by your local, state, and national governments.

Summary

The Earth's oceans contain many valuable natural resources. Food is taken from the ocean, mostly in the form of fish. Other sources of food, such as plankton, krill, and algae, may one day be of greater importance. The ocean is also a source for many valuable metals, which are dissolved in the water or found in nodules that litter the sea floor. It is also a source of petroleum. Ocean resources, however, are threatened by pollution and overuse. They must be protected and shared equitably through worldwide cooperative efforts.

Unit 4

In This Unit

Reading *Plus*

Now that you have been introduced to electromagnetic systems, let's take a closer look at the basics of these systems. Read pages S52 to S68 and write a proposal for a research grant to support your further investigations into the nature of electromagnetism. Your proposal should raise questions similar to those that follow.

1. How can the nature of static electricity be distinguished from current electricity?
2. In what way are the observed properties of magnets explained by the characteristics of their atoms.
3. What is the relationship between magnetic effects and electric effects?

4.1 TWO KINDS OF ELECTRICITY

When you think of electricity, you probably think of something that comes from a wall outlet and operates our machines and appliances. That certainly is electricity—electricity in motion. But have you ever walked across a carpet and then felt a shock when you touched a metal doorknob? Do your clothes sometimes stick together after being in the dryer? Have you ever rubbed a balloon along a cat's fur? Does your hair ever crackle and cling when you comb it? If it is dark enough, you might see tiny sparks jumping between the comb and your hair. Such experiences are the result of *static electricity*. See Figure 4-1.

Figure 4-1 *The effects of static electricity can be demonstrated by this simple activity.*

Static Electricity

Since ancient times people have been very curious about a natural substance called *amber*. Amber is the sap of ancient trees that has hardened over the centuries into a solid. Amber ranges in color from yellow to brown and sometimes contains the remains of insects. The ancient Greeks, however, noticed another interesting property of amber. When it is rubbed with cloth, it attracts objects, such as bits of straw and paper. See Figure 4-2. The Greek word for amber is ἤλεκτρον (*electron*), and its attraction for small objects is due to what we now call *electric charges*.

Figure 4-2 *When amber is rubbed with a cloth, objects such as paper are attracted to it.*

Electric Charges Benjamin Franklin, who was a scientist as well as a statesman, studied electric charges. See Figure 4-3. In his day, it was known that there are different types of

The Granger Collection, New York

Figure 4-3 *Franklin's experiment showed that lightning is a form of electricity. Why was this experiment dangerous?*

electric charges. It was also known that these charges can be created by rubbing two materials together. Franklin hypothesized that there are only two kinds of electric charges—positive (+) and negative (–). What Franklin called a *negative charge* is defined as the type of charge given to a rubber rod when it is rubbed with fur or wool. On the other hand, a *positive charge* is defined as the charge given to a glass rod that has been rubbed with a silk cloth. An object that has neither a positive nor a negative charge is considered *neutral.* But how do objects become electrically charged?

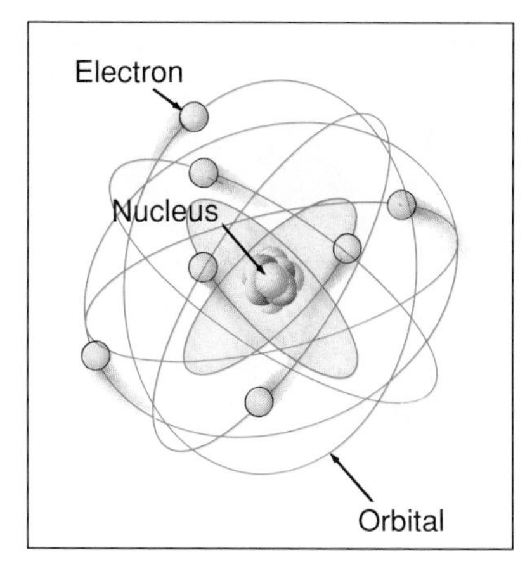

Figure 4-4 *Bohr's atomic model indicated that electrons travel around a nucleus that contains protons and neutrons.*

You have learned that all substances are made of atoms. According to one scientific model, atoms are composed of three basic particles—negatively charged electrons, positively charged protons, and neutrons that have no charge. See Figure 4-4. Atoms are usually electrically neutral because they have equal numbers of electrons and protons. Normally, you are also neutral because the negative electrons in your body are balanced by the positive protons.

Producing Charges Rubbing two materials together causes electrons to be torn away from one material and added to the other. For example, clothes taken from a dryer often cling together because they are electrically charged. This charge results from the different kinds of cloth rubbing together as the clothes are tumbled in the dryer. Clothes that have picked up electrons take on a negative charge. Those that have lost electrons take on a positive charge. See Figure 4-5. Likewise, running a comb through your hair may cause electrons to move from your hair to the comb. In this case, the comb takes on a negative charge because it has extra electrons. Your hair is left with a positive charge because it has lost electrons. When you walk across a carpet, the friction of your shoes on the carpet may cause you to pick up extra

Figure 4-5 *Has this ever happened to you? Static electricity makes the clothes cling together after they are removed from the dryer.*

Figure 4-6 Electricity can make your hair stand on end. The electric charge here was produced by a Van de Graaff generator.

electrons, giving your body an overall negative electric charge. See Figure 4-6.

An electric charge results when an object either gains or looses electrons. This type of electric charge, because it builds up on objects and does not move as it would through a conductor, is called **static electricity**. The word static refers to staying in one place. Any substance, conductor or insulator alike, can be given a static electric charge.

Static Electricity Electrical charges that accumulate on an object.

Electric Forces and Fields When two charged objects are brought close together, they either attract each other or they repel each other. The behavior of objects with electric charges can be described by a simple rule: *Like charges repel, and unlike charges attract.* This is called the ***law of electric charges***. See Figure 4-7.

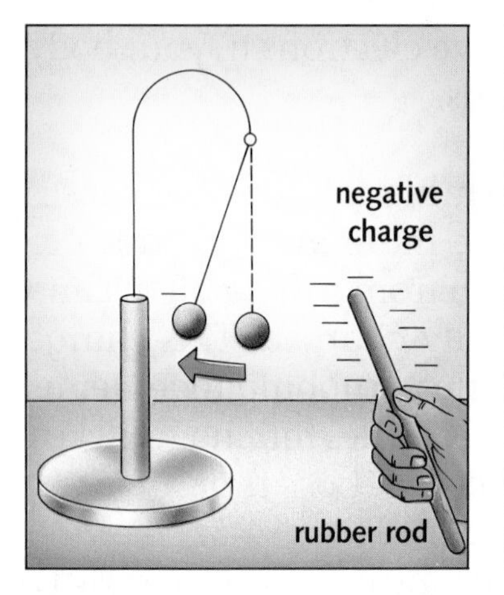

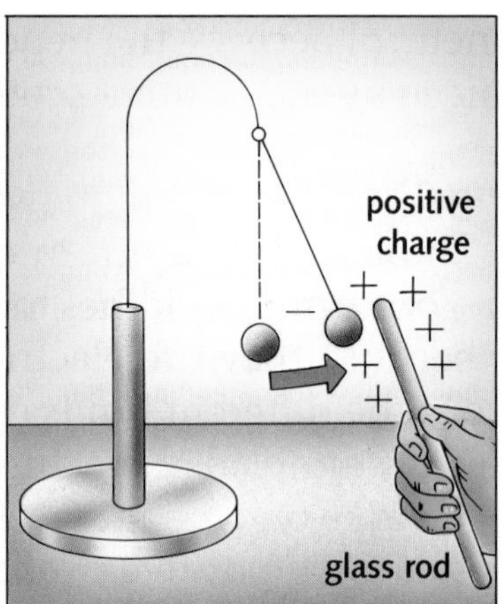

Figure 4-7 A rubber rod that has been rubbed repels a negatively charged ball (left). *A glass rod that has been rubbed attracts the ball* (right).

Recall that force is always needed to cause an object to move. Thus, when two electrically charged objects cause each other to move, a force must be involved. This force is called **electric force**. Electric forces cause charged objects to move apart or come together according to the kind of charge they have.

Electric Force The force that causes two like-charged objects to repel each other or two unlike-charged objects to attract each other.

Two things seem to affect the strength of an electric force. First, the more an object is rubbed to give it an electric charge, the stronger the electric force that is produced. Furthermore, the strength of the electric force between two objects increases as the amount of electric charge on those objects increases. The distance between two charged objects also affects the strength of an electric force. See Figure 4-8. Charged objects have a greater effect on each other as they come closer together. For example, if the distance between two charged objects is cut in half, the strength of the force becomes four times greater. In the same way, moving the charged objects twice as far apart reduces the force to one-fourth of what it

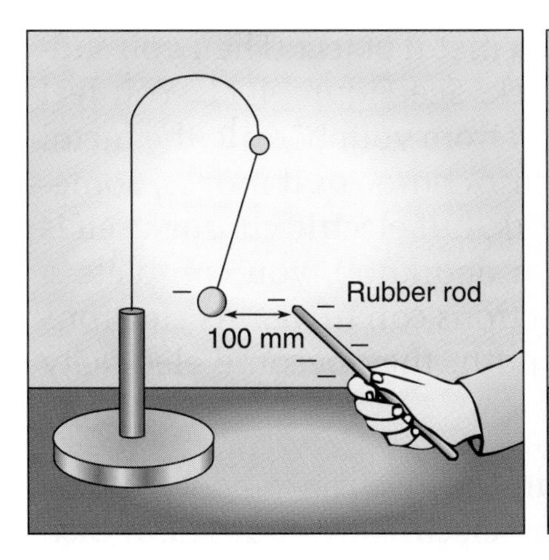

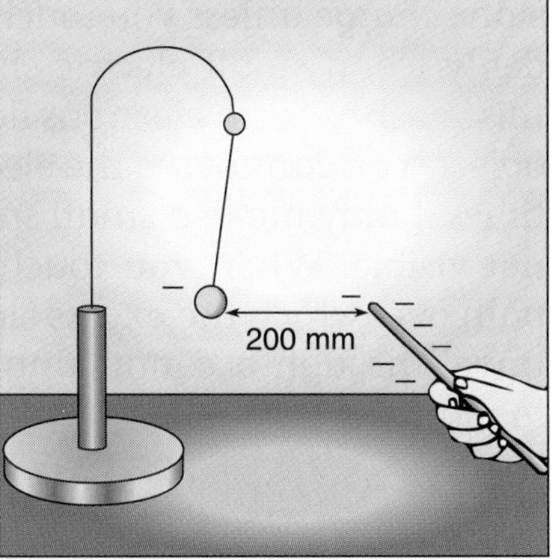

Figure 4-8 The electric force between objects depends on their charges and the distance between the objects. The precise relationship is given by Coulomb's law.

was. These two relationships are expressed in ***Coulomb's law:*** *The force between two charged objects is directly proportional to the product of their charges and inversely proportional to the square of the distance between them.* How is this law similar to Newton's Law of Gravitation?

Around every charged object there is a space in which the effects of the electric force may be observed. Such a region of space contains an **electric field**, which surrounds the charged object in all directions. See Figure 4-9. As the

Electric Field The region of space around an electrically charged object in which an electric force is noticeable.

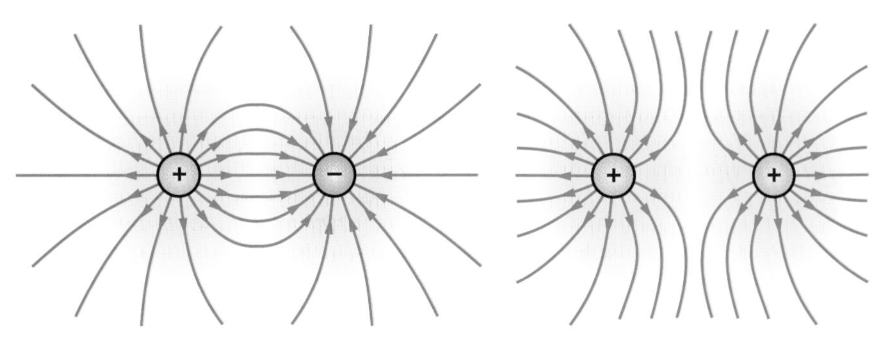

Figure 4-9 The lines indicate the nature of the electric field near two opposite charges (left) *and two similar charges* (right).

distance from a charged object increases, the strength of its field becomes weaker. At a far enough distance, the electric field is too weak to be noticeable.

Electricity in Motion

A bolt of lightning has tremendous power. In fact, a single flash of lightning can heat the air to a temperature higher than the surface of the sun. See Figure 4-10. Lightning has melted holes in church bells and welded chains into iron bars. Potatoes in a field struck by lightning have been cooked in the ground. But for all its great power, lightning is basically a spark—very much like the spark that jumps from your hand to a doorknob—only much bigger.

Scuffing your feet across a carpet may give your body an overall electric charge. However, you are not aware of this

Figure 4-10 Every year, many people are killed or seriously injured by being struck by lightning.

Figure 4-11 Electric charges can build up on your body when you walk across a rug. When you touch a doorknob, the charges can jump from your hand to the metal knob.

static charge unless you touch a metal object. Then you suddenly feel a small electric shock. See Figure 4-11. You feel this shock as extra electrons flow from your body to the metal object. You cannot see the electric charge as it moves. Sometimes it may make a small spark, but electric charge itself is not visible. When you touch a doorknob, you complete a path over which the excess electrons can travel. As electrons travel through a conducting path, they become electricity in motion.

Electric Current Not only can the electron model be used to explain how objects become electrically charged, it also helps to explain how electrons move from one place to another through a conductor. In some ways, electrons behave like water. Usually, gravity causes water to flow downhill. In a similar way, electrons also move or "flow" from one place to another. See Figure 4-12. It is not gravity, however, that causes electrons to move. Electrons will move from an area where they are in excess to an area where they are lacking. For example, when you pick up extra electrons from a carpet and then touch something that has no extra electrons, such as a doorknob, extra electrons will flow from you to the doorknob. This flow of electrons from one place to another is called an **electric current**.

Electric Current The result of electrons moving from one place to another.

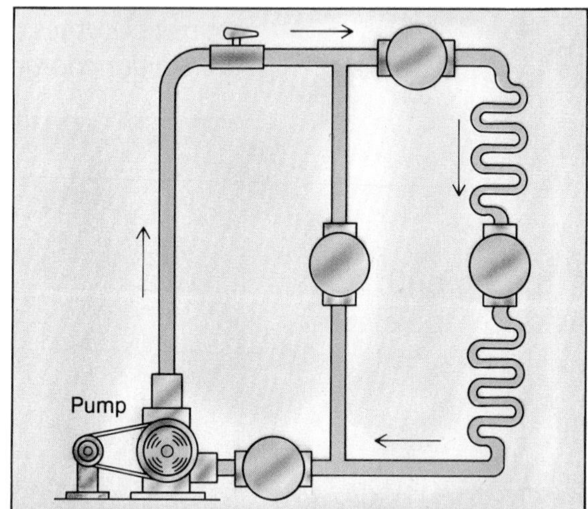

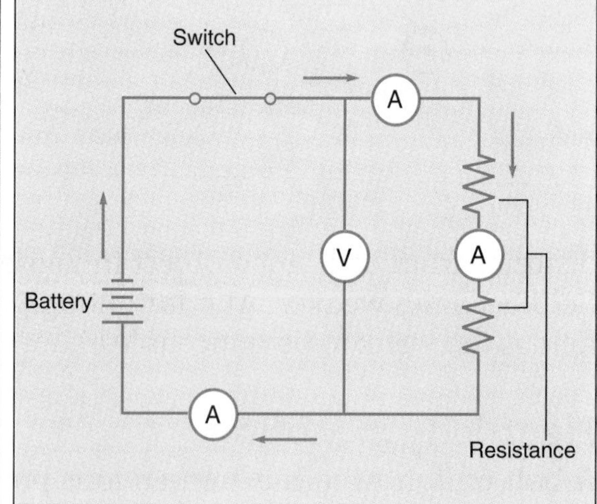

Figure 4-12 The flow of water in a system of pipes (left) *can be compared to the flow of electricity through a circuit* (right).

The movement of electrons through a conductor is also an electric current. For example, electrons can travel from the negative pole to the positive pole of a battery. However, in order for electrons to flow, there must be a path. If a conductor touches only the negative pole of the battery, the electrons will pile up on the conductor. For electrons to flow, the other end of the conductor must touch something that has a positive charge. If a conductor such as a wire is touched to both the negative and positive poles of a battery, electrons will flow through the conductor. This constitutes an *electric circuit*. See Figure 4-13. An electric circuit allows movement of electrons from a place where there are many electrons to a place where there are fewer.

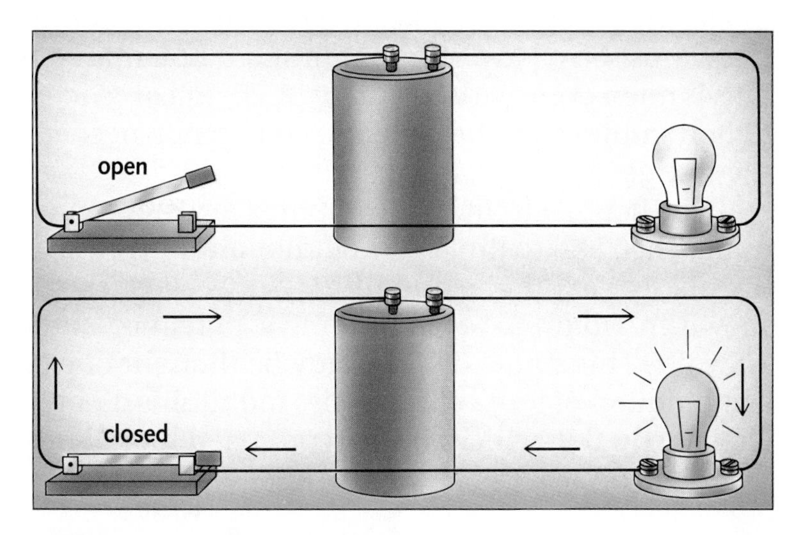

Figure 4-13 *An electric circuit must be a continuous path in order for electrons to travel.*

Describing Electric Current

The flow of electrons in an electric circuit can be compared to the flow of water in a pipe. Water will not flow in a pipe unless a force pushes or pulls it along. That force is usually gravity, which causes the water to flow downhill. Pumps are used to lift the water to the top of a water tower or reservoir. The pumps provide the water with *potential* energy. When the water is allowed to flow from the tower or reservoir, its potential energy is converted into the kinetic energy of motion, which can do work.

In the same way, potential energy can be stored by electric charges. Due to the electrical force of attraction between opposite charges, it takes work to pull them apart. Once these charges are separated, they possess a form of potential energy called *electric potential energy*. When the separated charges are allowed to recombine, this electric potential energy is converted into other forms of energy—heat, light, and sound for example.

Potential Difference When work is involved in moving a charge from one point to another in an electric field, these two points have a different electric potential. The amount of work done on a charge as it is moved between these points is called **potential difference**. Potential difference is measured in volts (V) and is, therefore, also called voltage. A *volt* is a measure of the amount of work done while moving a unit of charge between two points in an electric circuit.

Potential Difference The amount of work done to move a charge between two points in an electric field.

Potential difference can be compared to water pressure. For example, water will not flow through a hose unless there is pressure. This pressure is supplied by either a pump or by the fact the the water source is higher than the outlet. In a like manner, charge will not flow unless there is a difference in electric potential. To keep charge flowing continuously, there must be a way of maintaining a potential difference. This is often done with a battery. A chemical reaction in the battery maintains the potential difference needed to keep the charges flowing.

If the flow of electrons is compared to water running down a hill, then the voltage can be thought of as the height of the hill. For example, an ordinary dry cell, such as the one shown in Figure 4-14, maintains a potential difference of 1.5 volts. This single dry cell may be thought of as a hill. A combination of four cells in series maintains a potential difference of 6.0 V, which would represent a higher hill.

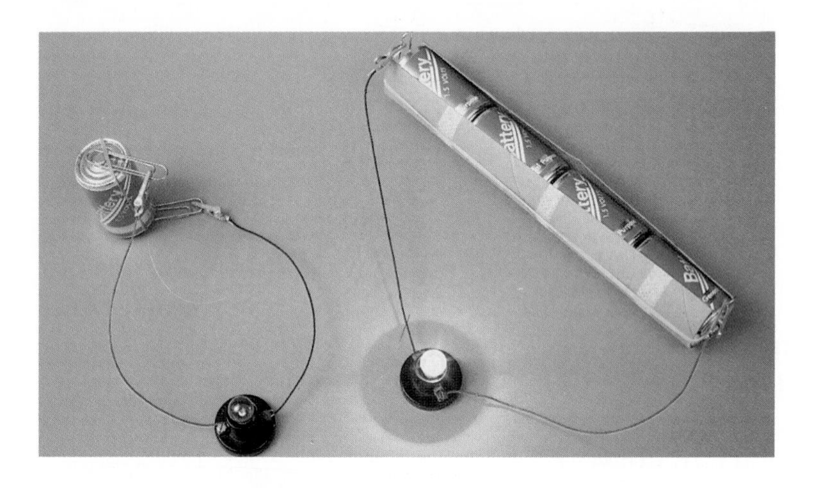

Figure 4-14 *A single battery* (left) *does less work than four similar batteries* (right). *The 6-V combination has a potential difference four times greater than the 1.5-V battery.*

Amperage The amount of current passing a point in one second.

Amperage Since the flow of electric charge through a circuit constitutes an electric current, the amount of charge passing a point in one second is a measure of the current, or **amperage**. The unit of electric current is the ampere (A). One *ampere* equals the charge of 6.25×10^{18} electrons moving past a point in one second. It is similar to "liters per minute" when referring to flowing water. In your home, currents may range from a few amperes in electrical appliances, such as space heaters and hair dryers, to a few milliamperes (or thousandths of an ampere) in electronic calculators and computers.

Resistance The opposition to current in an electric circuit.

Resistance If you pinch a water hose, you will restrict the flow of water through it, and the amount of water coming out the end will be reduced. When electrons move through a material, they meet **resistance**. Resistance is an opposition to the current that limits the flow of electrons in an electric circuit. Resistance is measured in ohms (Ω). A circuit has a resistance of one *ohm* when a potential difference of 1 V produces a current of 1 A.

As you may recall, conductors have low resistances; insulators have high resistances. For example, the wires in your home have very little resistance to electric current, while a rubber hose or a piece of glass has a large resistance. The amount of current in any electric circuit is determined by *both* the voltage and the resistance in the circuit. Think again of water flowing down a hill through a pipe. The amount of water that will pass through the pipe is determined by both the height of the hill and the size of the pipe. A narrow pipe offers a great deal of resistance to the flow of water, and so the amount of water that can pass through the pipe is smaller than that in a larger pipe. See Figure 4-15. The narrow pipe is similar to an electric circuit with high resistance. When an electric circuit has high resistance, the current in the circuit is less for any given voltage.

Figure 4-15 *The resistance to flow in these pipes is a function of their diameter.*

Resistance can also be thought of as the *friction* of electricity. Consider the wires in a toaster oven. Even though they are made of metal (a conductor), they have a relatively high resistance to current. When electric current meets resistance, heat is produced. You know that the harder you rub your hands together, the warmer they get. As friction (resistance) increases, more heat is generated. Since there is a lot of resistance in the wires of the toaster, they glow red hot when electric current passes through them.

Ohm's Law The voltage, current, and resistance in an electric circuit are related to each other by a rule known as ***Ohm's law***. This relationship was discovered by a German schoolteacher, Georg Ohm, in the early 1800s. Ohm experimented with electric circuits made of wires having different amounts of resistance. By doing so, he discovered a general rule that describes the relationship among voltage, current, and resistance in a circuit. Ohm's law can be written in equation form as follows.

$$\text{amperes} = \frac{\text{volts}}{\text{ohms}}$$

$$I = \frac{E}{R}$$

For example, an automobile with a 12-V battery has headlights whose resistance is 4 Ω. When the lights are on, the current needed is

$$I = \frac{E}{R} = \frac{12\text{ V}}{4\ \Omega} = 3\text{ A}$$

Figure 4-16 *Using this illustration may help you recall the alternative forms of Ohm's law.*

By rearranging the terms, Ohm's law can also be written as:

$$\text{volts} = \text{amperes} \times \text{ohms}$$

$$E = IR$$

or

$$\text{ohms} = \frac{\text{volts}}{\text{amperes}}$$

$$R = \frac{E}{I}$$

In other words, if you know two of the values for a circuit, you can always find the third. See Figure 4-16.

Transforming Electricity

In getting electricity to do what you want it to do, it is often necessary to transform it—raise or lower its voltage. Since voltage is a measure of the amount of "push" that an electric current has, it is very much like pressure. The higher the voltage, the stronger the push.

Alternating current is easy to transform. The device used to do this is called a *transformer*. See Figure 4-17. Notice that there are two sets of coils around an iron core. Alternating current from the source flows through the first, or *primary coil*. With each surge of current, the iron core becomes an electromagnet. Every time the current alternates, or switches direction, the electromagnet changes polarity.

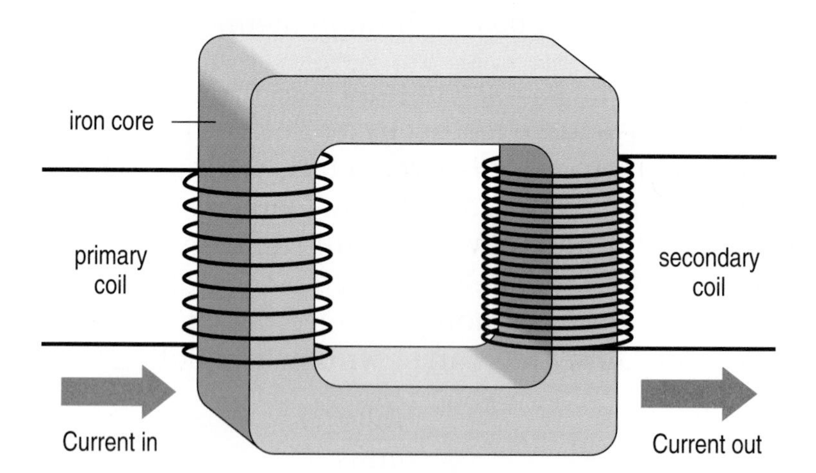

Figure 4-17 *As shown here, a transformer consists of an iron core and two coils.*

The other coil of the transformer responds as though it were rotating between the poles of a magnet. Current is therefore "induced" in the *secondary coil.*

Wind-Up Voltage The amount of voltage induced in the secondary coil depends on the number of turns of wire in each of the two coils. If the primary coil has more turns than the secondary, the voltage will drop across the circuit. This type of transformer is called a step-down transformer. If, on the other hand, the secondary coil has more turns, the voltage will increase. In this case, it would be a step-up transformer. See Figure 4-18.

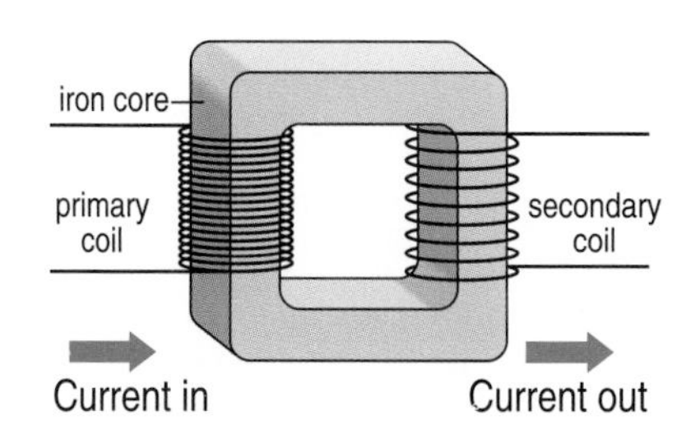

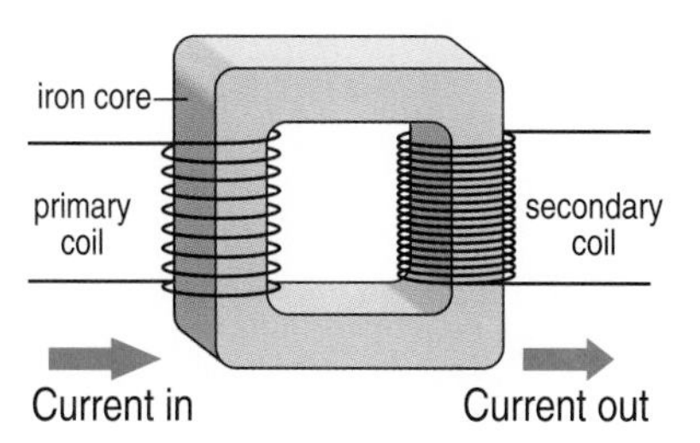

Figure 4-18 *A step-down transformer* (top) *and a step-up transformer* (bottom).

You can figure out the voltage in either of the two coils if you know the voltage in one and the number of turns in both. The relationship between voltage and turns in a transformer is given by the following equation.

$$\frac{E_P}{E_S} = \frac{T_P}{T_S}$$

The value E is the voltage and T is the number of turns in the coil. The subscripts P and S refer to the primary and secondary coils respectively. If you know three values, by cross-multiplying, you can find the fourth. For example, if the primary coil had 100 turns and the secondary coil had only 50 turns, the voltage would be transformed from high to low voltage. If this transformer were attached to a house circuit with 120 V, the voltage in the secondary coil would be 60 V.

Effects on Current If you transform the electricity by increasing or decreasing the voltage, there is a corresponding effect on current. A transformer can change the voltage (force) in a circuit, but the total electrical energy (work) must remain the same. In other words, when voltage is stepped up, current is stepped down. Remember that current depends on both voltage and resistance. If the same type of wire is used in both the primary and secondary coils, the change in current can be calculated as follows.

$$\frac{E_P}{E_S} = \frac{I_S}{I_P}$$

If, for example, the voltage was boosted from 100 V to 1000 V and the current in the primary was 2 A, the current in the secondary would be 0.2 A.

Electricity is usually transmitted over power lines at very high voltages of over 200,000 V. This high voltage produces very low currents. In this way, there is a smaller loss of energy as the electricity travels through the lines. Once the electricity reaches your neighborhood, the voltage is reduced by step-down transformers to the 120 V that are used by home appliances. See Figure 4-19.

Figure 4-19 *Power transformers convert high-voltage electricity into low-voltage electricity for home use.*

Summary

Benjamin Franklin proposed the idea that all electric phenomena could be explained in terms of positive and negative electric charges. An electric charge results when something either gains or loses electrons. Objects with like charges repel each other and those with unlike charges attract each other. An electric charge that does not move is called static electricity. The force between two charged objects depends on their charges and the distance between them. An electric current is the flow of electrons as they move from a place where there are more of them to a place where there are fewer. Ohm's law states that the electric current increases with increasing voltage but decreases with increasing resistance. Voltage can be increased or decreased by using a transformer. Current is affected by changes in voltage, a fact that is used to save energy as electricity is transmitted.

4.2 MAGNETS AND MAGNETIC EFFECTS

Figure 4-20 *Lodestone is the only naturally occurring magnetic material.*

You can see the effects of electricity all around you, and you have learned that electricity is related to magnetism. But what is magnetism, and what produces magnetic forces? Over 2000 years ago, the Greeks discovered that a material called *lodestone* attracts pieces of iron. See Figure 4-20. This discovery occurred in a region of Greece known as *Magnesia*. As a result, the term *magnet* was given to any material that attracted iron and other metals. Common magnets are specially treated pieces of steel that can attract iron. See Figure 4-21. They also attract steel (because steel contains iron), as well as nickel and cobalt.

Figure 4-21 *Magnets can have many different shapes and sizes.*

Describing Magnetism

Magnetism is not evenly distributed throughout a magnet. If you dip a bar magnet into a box of small nails, it will come out with nails clinging to the end of the bar. The ends of a bar magnet are called *poles*. Poles are the points on a magnet at which its magnetic attraction is the strongest. See Figure 4-22.

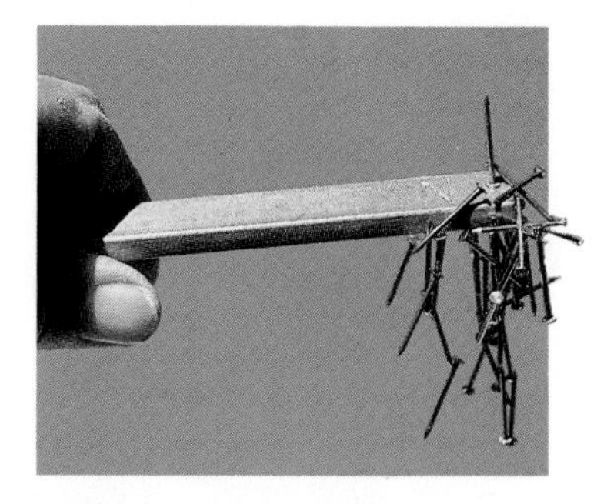

Figure 4-22 *Attractive force is strongest at the poles of a magnet.*

Magnetic Poles A simple demonstration illustrates an important fact about the poles of a magnet. If you allow a bar-shaped magnet to swing freely from a string, it will always point in a north–south direction. One pole points north; the other pole points south. If you happen to disturb the magnet, it will return to this same orientation. Because magnets always act this way, the pole of the magnet pointing north is called its north-seeking pole or *north pole*. The opposite end is called the *south pole* of the magnet. See Figure 4-23. A compass is simply a magnet that is free to turn so that the north pole always points north.

Figure 4-23 *Because a magnet always aligns itself in a north–south direction, its two poles are called its north pole and south pole.*

What do you suppose would happen if you cut a bar magnet in half? You might think that the north pole would be in one piece and the south pole would be in the other piece. The actual result is shown in Figure 4-24. Each piece becomes a complete magnet with both north and south poles. If the pieces are cut into even smaller ones, the smallest piece will still be a complete magnet with a north and a south pole.

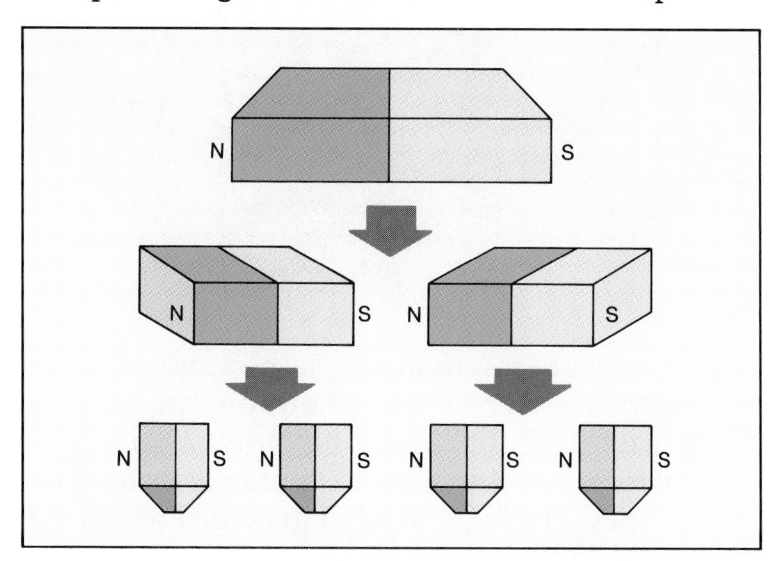

Figure 4-24 *If you continue cutting a magnet, each piece will still have two poles.*

Magnetic Forces You know what happens when you try to put two magnets together—like poles repel and unlike poles attract. This attraction and repulsion is called **magnetic force**. If the north pole of one magnet is brought near the north pole of another magnet, there is a repulsive force between them. If the north pole is brought near the south pole of a second magnet, there is an attractive force between them. In this way, magnetic force is similar to electric force. Magnetic force is similar to electric force in other ways as

Magnetic Force The attraction or repulsion between the poles of a magnet.

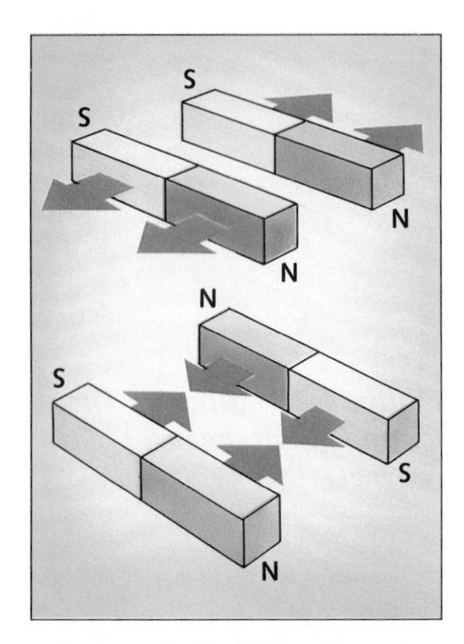

Figure 4-25 Opposite poles attract while similar poles repel each other, even at a distance.

well. First, both kinds of force work at a distance. Two magnets do not need to touch each other in order to attract or repel. See Figure 4-25. Furthermore, both magnetic force and electric force become weaker as the distance between two magnets or two electrically charged objects increases. As two magnets are brought closer together, the push or pull between them becomes much greater.

Magnetic Fields You can feel the force between two magnets as they either attract or repel one another. A **magnetic field** exists in the space around each magnet in which magnetic forces are noticeable. You can see the shape of a magnetic field by sprinkling iron filings around a magnet. See Figure 4-26. The filings line up along curving lines running from one end of the magnet to the other. These lines are known as *lines of force*.

Magnetic Field A region of space around a magnet in which magnetic forces are noticeable.

As you move a compass around a magnet, the compass needle turns. This happens because the magnet within the compass is also affected by the magnetic field produced by the larger magnet. At any location in the field, the compass

Figure 4-26 This horseshoe magnet has a circular magnetic field. Like all magnets, the field is strongest at the magnetic poles.

needle comes to rest in line with the magnetic lines of force. See Figure 4-27. These lines, however, are only imaginary. Like the lines of latitude and longitude on a map, magnetic lines of force are "inventions" that help us understand how things work. They can be used to predict which way a compass needle will point.

Figure 4-27 Shown here are compasses positioned around a bar magnet. The needle of each compass shows the direction of the magnetic field around the magnet.

Magnetic lines of force never cross each other. When two or more magnets produce fields that overlap, the result is a combined field. See Figure 4-28. If opposite poles are aligned, most of the lines run from the north pole of one magnet to the south pole of the other.

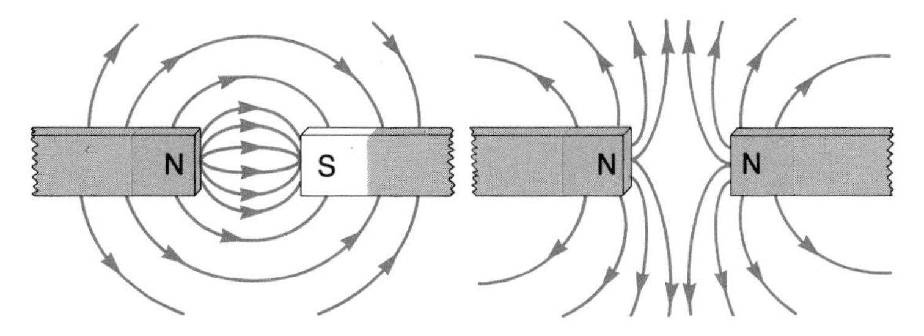

Figure 4-28 *Magnetic lines of force never overlap, even when the poles of two magnets are brought close to one another, as shown here.*

The Source of Magnetism

Although people have known about magnets for centuries, it was not until the early 19th century that theories were developed to explain magnetism. The currently accepted theory suggests that magnetism is actually a property of electric charges in motion. You already know that the electrons of an atom are in constant motion. Not only do they orbit the nucleus, but they also "spin" about their own axis. This motion of electrons creates opposite magnetic poles. Because electrons may spin in either direction, the magnetic effects of a pair of electrons spinning in opposite directions cancel each other. Therefore, the magnetic properties of an atom are related to how many unpaired electrons it has. See Figure 4-29. The motion of unpaired electrons can give an atom north and south poles, making it act like a tiny magnet.

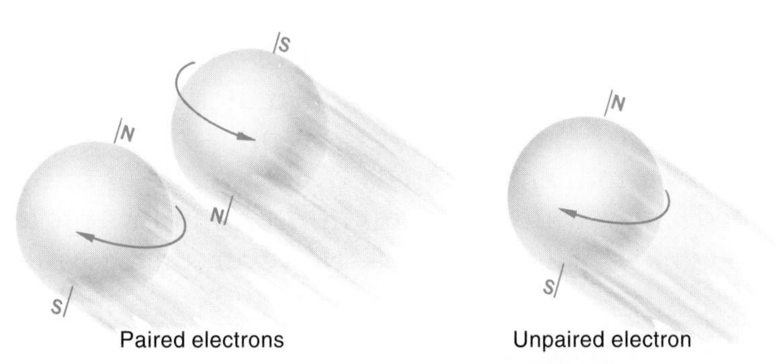

Figure 4-29 *Spinning electrons act like tiny magnets. While paired electrons cancel each other out, unpaired electrons cause the atom to be magnetic.*

Within materials that are affected by magnets, such as iron, atoms are grouped together in clusters called **domains**. The atoms within each domain have the same magnetic alignment. In other words, all of their individual north and south poles point in the same direction. In an unmagnetized piece of iron, for example, these small domains are arranged in a helter-skelter manner. Because of this, their magnetic properties cancel each other. These domains are like a crowd of people sitting in a park, facing in all directions. When

Domains Clusters of atoms in magnetic materials in which the magnetic fields of most atoms are aligned in a similar direction.

Figure 4-30 *In most materials* (left), *magnetic domains are randomly arranged. In magnets, however, the domains are aligned and produce a magnetic effect* (right).

you bring the south pole of a magnet near the iron bar, however, the magnet attracts the north poles of the atoms in the iron bar and causes most of the domains to line up in the same general direction. Now the domains are more like people sitting in a theater in rows. When most of the domains are aligned, it makes the whole iron bar into a single large magnet. See Figure 4-30.

Types of Magnets

The early Chinese discovered that a piece of lodestone will always point in the same direction when allowed to move freely. As a result, lodestone was used as the first compass. The Chinese also found that iron can be magnetized by rubbing it with lodestone. Scientists now know that lodestone is actually a magnetized form of iron oxide called *magnetite.* It is thus a natural magnet. However, since iron and other metals, such as cobalt and nickel, possess natural magnetic properties, they can be made into artificial magnets.

Magnets that lose their magnetism easily are called *temporary magnets.* Some harder metals, such as steel, are harder to magnetize but tend to keep their magnetism better. A magnet made of such material is called a *permanent magnet.*

Artificial Magnets There are only five elements that can be made into magnets—iron, cobalt, nickel, gadolinium, and dysprosium. In their pure forms, however, these metals can only be magnetized temporarily. Excess heat or a sudden blow may cause their domains to become disorganized. To make a permanent magnet, you need an alloy. An alloy is a mixture of two or more metals, or a mixture of metals and nonmetals. The most common material used for making permanent magnets is steel, an alloy of iron and carbon. Strong permanent magnets can be made of another alloy called *alnico,* which contains iron, aluminum, nickel, cobalt, and copper. See Figure 4-31. However, the best material for

Figure 4-31 *Some alnico magnets are strong enough to support your weight.*

permanent magnets is called *magnequench*. It was invented in 1985 and consists mostly of iron, with small amounts of neodymium and boron.

Magnets can also be made of a ceramic that has been mixed with oxides of iron and other metals. Ceramic magnets are light in weight but break easily. Loadstone, a natural ceramic, is a mineral made of various oxides of iron. Soft, flexible magnets, such as the ones you may have on your refrigerator, are made of powdered magnetic ceramics embedded in rubber.

Electromagnets As you know, electromagnets are made by using an electric current. Due to the close relationship between electricity and magnetism, magnets can be used not only to produce electricity (as in a generator), but electricity can be used to make non-permanent magnets. These magnets can be turned on and off simply by throwing a switch. Powerful electromagnets have been constructed for a variety of uses. See Figure 4-32.

Figure 4-32 *An electromagnet attached to this crane is used to lift scrap iron and steel.*

The Earth as a Magnet

In 1600 William Gilbert, who was the physician of Queen Elizabeth I of England, proposed a reason for why a suspended loadstone lines up in a north–south direction. He believed that the Earth itself was a giant lodestone! To test his theory, Gilbert performed an experiment in which he carved a natural loadstone into the shape of a sphere. He found that a compass placed on the surface of the loadstone acts just as a compass does at different places on the Earth's surface.

In fact, the Earth behaves as if a giant bar magnet were running through its center. See Figure 4-33. Because the north

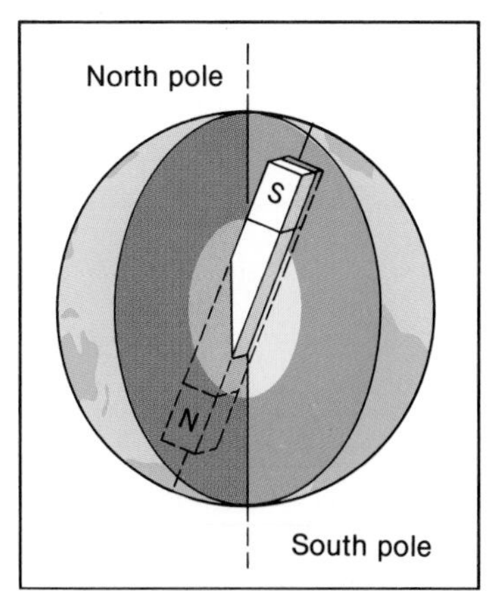

Figure 4-33 *The Earth itself is a magnet, with north and south poles. The North Pole of the Earth is actually a magnetic south pole.*

pole of a magnet (by definition) is the pole that points northward, the magnetic pole at that location on Earth must in fact be a magnetic south pole. This follows from the behavior of magnetic poles—unlike poles attract. So when you use an ordinary compass, you are actually making use of two magnets. One of these magnets is small—the needle of the compass. The other magnet is very large. That magnet is the Earth itself.

Like all magnets, the Earth is surrounded by a magnetic field. Recall that you can see evidence of a field around a small magnet if you sprinkle small pieces of iron on a piece of paper over the magnet. If there were some way to do this for the Earth, you would have a picture of this magnetic field. See Figure 4-34.

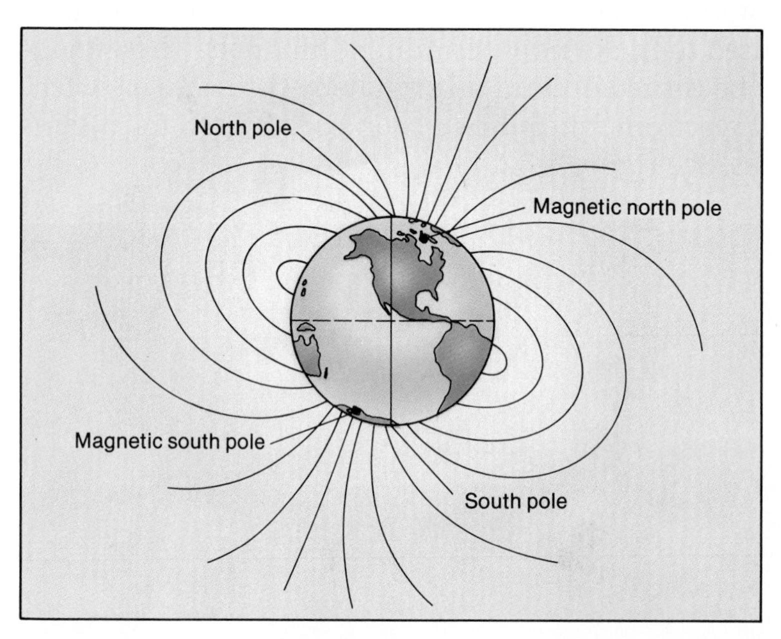

Figure 4-34 *Like other magnets, the Earth has a magnetic field that is strongest at the poles. Many stars and other planets also have magnetic fields.*

Summary

The term magnet refers to any material that attracts iron and certain other metals. A magnet always has both a south and a north pole. As with electric charges, like magnetic poles repel and unlike poles attract. Magnetic fields surround magnets much like electric fields surround electrically charged objects. The magnetic properties of an atom are related to the movement of its electrons and to the number of unpaired electrons it has. In a magnetized bar of iron, the magnetic domains line up with like poles pointing in the same direction. Artificial magnets are made from alloys or ceramics, and electromagnets are produced by the action of an electric current. The Earth also has a magnetic field, as if a giant bar magnet were running through its center.

Unit 5

Reading *Plus*

Now that you have been introduced to sound and the nature of its qualities, let's take a closer look at the physical basis for sound—waves. Read pages S70 to S86 and write a letter you might send to the editor of your local newspaper explaining how sound is transmitted and the effects it might have on the general public. Your letter should raise concerns highlighted by the following questions.

1. How do waves transmit energy from one place to another? How can this be observed with sound waves?
2. In what ways does the wave characteristics of sound affect how we hear sound. Give examples.
3. In what ways do people use sound, and how can they be protected from sound levels that are too high?

In This Unit

Figure 5-1 *Energy from a stone radiates outward in the form of waves.*

Wave A disturbance that transmits energy as it travels through matter or space.

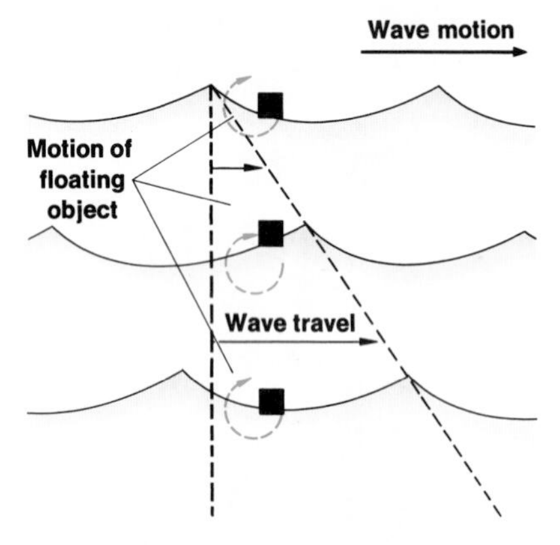

Figure 5-2 *Notice that although the wave continues traveling to the right, the floating object (as well as the water itself) remains in one place, making small circular motions.*

Transverse Wave A wave in which the motion of the particles is perpendicular to the path of the wave.

5.1 WAVES

You have been learning about sound, and the waves of which it is composed. But what are waves? There are many kinds of waves in addition to sound waves. You may have heard about water waves, radio waves, microwaves, earthquake waves, and electromagnetic waves, as well as others. Since sound is made of waves, it shares some of the properties of all waves. So let's take a step back and investigate the phenomenon of waves in general.

If you have ever dropped a stone into a quite pond, you produced many waves, or disturbances, in the surface of the water. A **wave** is a disturbance that transmits energy as it travels through matter or space. The material through which a wave travels is called a *medium*. The waves on the pond are produced when kinetic energy from the stone is transferred to the medium, the water. See Figure 5-1. The waves then carry the energy outward through the medium, from the point where the stone hit the water.

It is important to understand that the energy that causes a wave moves *through* the medium; the medium itself remains in one place. See Figure 5-2. It is like a long line of people standing close together. Imagine that the person at the end of the line pushes the next person in line, and that one pushes the next, and so on. Can you see how the disturbance can pass to the head of the line without any person taking a step forward?

Types of Waves

A liquid, such as water, is one medium for waves. But gases, like those in the atmosphere, and solids, such as iron, can also be mediums for waves. Therefore, we can study the action of waves without "getting wet." For example, the waves in a rope act much like waves in water. The rope acts as the medium for the waves. To study waves in more detail, look at Figure 5-3. When the student moves her hand up and down, she sets the rope in motion and creates a wave in the rope that travels along the rope to the other end. A wave carries energy through a rope because each point on the rope passes its energy on to the next point.

Transverse Waves The waves produced by the up-and-down motion of a medium are called **transverse waves**. *Transverse* means "moving across." In a transverse wave, the material of the medium moves *across* (at right angles to) the direction in which the wave moves. Look at Figure 5-3 again. Notice that although the wave is moving to the right, no

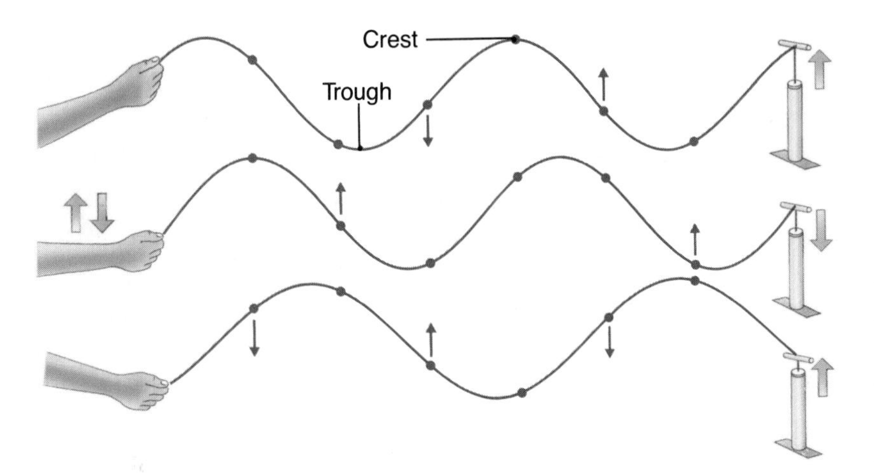

Figure 5-3 *Here a wave in a rope is being used to operate a bicycle pump. The handle is raised and lowered each time a wave reaches the pump.*

part of the rope is moving to the right. Look at the colored dots marked on the rope. If you follow the motion of any one of these dots from one drawing to the next, you will see that each dot is moving up and down. Every point on the rope follows the motion of the student's hand.

The top of a transverse wave is called a *crest*. Each crest is followed by a depression called a *trough*. Transverse waves consist of a series of crests and troughs that follow each other through a medium. The motion of the particles of the rope is perpendicular to the direction in which the wave is traveling. If the rope were shaken from side to side, would the wave still be transverse?

Longitudinal Wave A wave in which the particles move back and forth, parallel to the direction of motion of the wave.

Longitudinal Waves A different kind of wave motion can be demonstrated by using a coiled spring like the one shown in Figure 5-4. In this case, a student is producing a wave by moving her hand toward and away from a bicycle pump with a motion that is parallel to the length of the spring. The particles of the spring also move toward and away from the pump. Such waves, in which the particles of the medium move back and forth parallel to the path of the wave, are called **longitudinal waves**.

Figure 5-4 *Here a spring is being used to operate a bicycle pump. Each time a complete wave reaches the pump, it moves the pump handle in and out.*

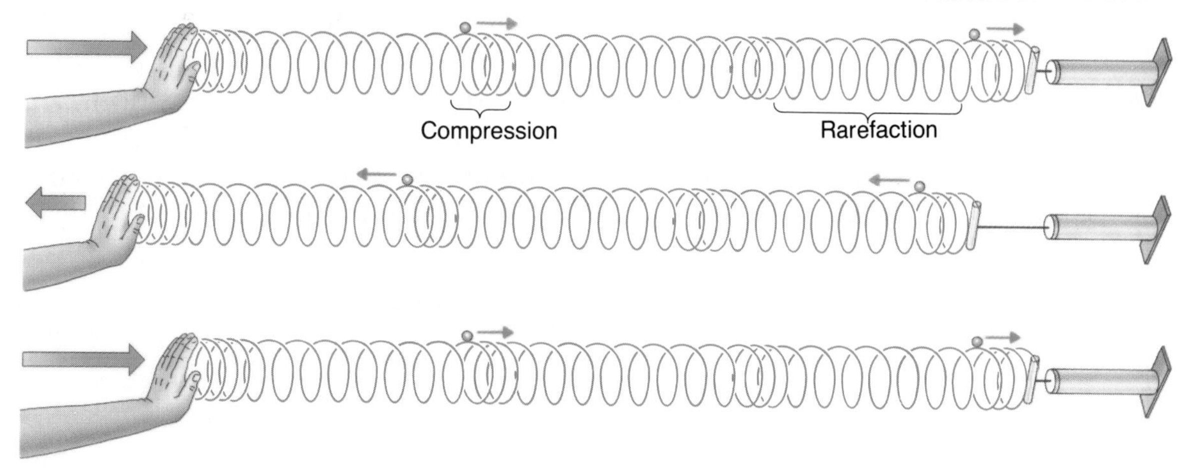

Notice that longitudinal waves do not have crests and troughs. Every time the student pushes the spring forward, she pushes some of the coils of the spring closer together. In other words, the student causes a *compression* in the spring. In each compression of a longitudinal wave, the particles of the medium are more dense than they are in the surrounding material. When the student pulls her hand back, she separates the coils. This causes an expansion in the spring. This part of a longitudinal wave, where the particles are the least dense, is called a *rarefaction*. See Figure 5-5. Longitudinal waves consist of a series of compressions and rarefactions following each other through a medium.

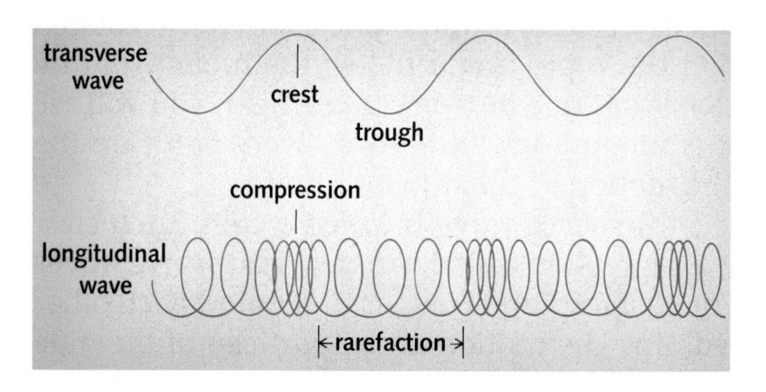

Figure 5-5 *The crest of a transverse wave can be compared to the compression of a longitudinal wave, while a trough can be compared to a rarefaction.*

Properties of Waves

Just as you can describe matter by various physical properties, such as color, size, mass, and density, waves also have certain properties by which they can be described. Although they are produced in many different ways and cause many different effects, all waves can be described by four basic properties.

Amplitude The distance a wave rises or falls from a normal rest position.

Amplitude You can make a small wave in a rope by flicking your wrist while holding the end of the rope with your hand. You could make a much larger wave by moving your whole arm to shake the rope. Scientists use the term *amplitude* to refer to the sizes of waves. **Amplitude** is the distance a certain point on a wave moves from a rest position. See Figure 5-6. The amplitude of a wave depends on the amount of energy in the wave. The greater the energy the wave carries, the greater the wave's amplitude. See Figure 5-7.

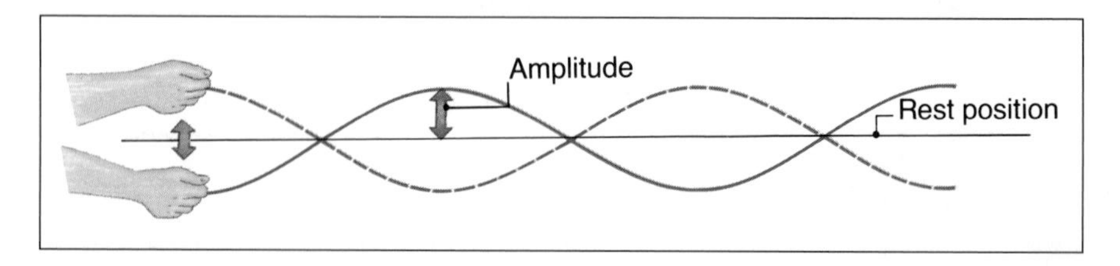

Figure 5-6 *The amplitude, or height, of a wave is measured from the rest position as shown.*

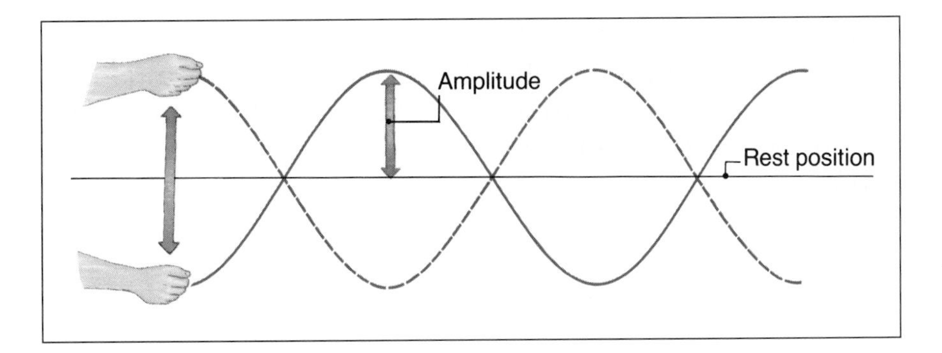

Figure 5-7 *As energy input increases, amplitude increases.*

Wavelength Waves are characterized not only by their amplitude, but also by *wavelength*. **Wavelength** is the distance between a particular point on one wave and the identical point on the next wave. The easiest way to find the wavelength of a transverse wave is to measure the distance from one crest to the next. The easiest way to find the wavelength of a longitudinal wave is to measure the distance between two compressions. See Figure 5-8.

Wavelength The distance between two identical points on neighboring waves.

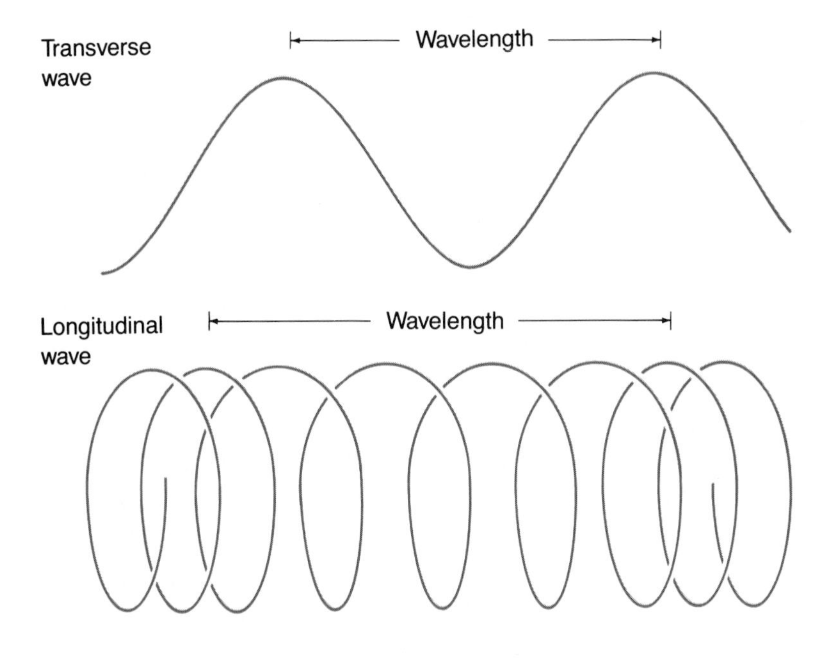

Figure 5-8 *The wavelengths of transverse and longitudinal waves are illustrated here.*

Frequency If you shake one end of a rope very slowly, you will not produce many vibrations in the rope. However, if you shake the rope quickly, you will create a large number of vibrations. The number of vibrations produced in a given amount of time is called **frequency**. Frequency tells the rate at which vibrations are produced. It is usually measured in hertz (Hz), which is equal to the number of vibrations per second.

Frequency The number of complete waves that pass a point each second.

Objects that vibrate with constant frequencies can be used to measure time. For example, a pendulum can be used to operate the mechanism of a grandfather clock. See Figure 5-9. Such a pendulum might vibrate, or swing back and forth, one time per second—a frequency of 1 Hz. On the other hand, the quartz crystal used in most digital watches vibrates at a frequency of 32,768 Hz. Frequencies also determine phenomena such as the pitch of musical tones and the colors of light.

There is an important relationship between the frequency of a wave and its wavelength. As the frequency increases, the wavelength decreases. You can see this by watching the vibrations in a rope as you shake it at different rates. The faster you shake the rope, the closer together the crests will be. See Figure 5-10.

Figure 5-9 The pendulum of this clock operates at a constant frequency.

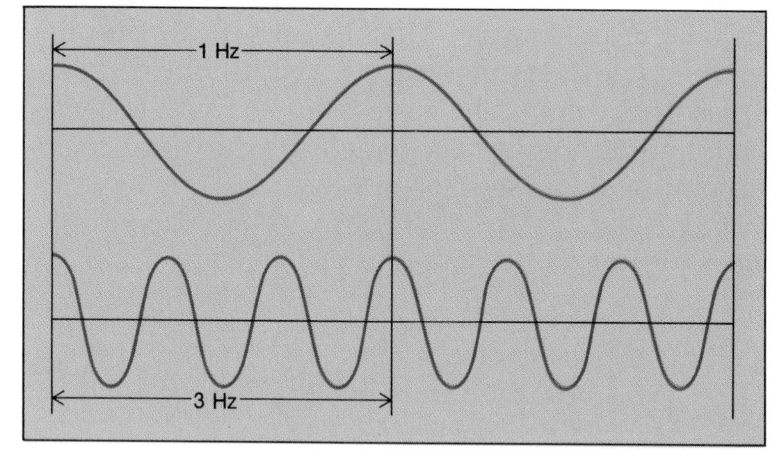

Figure 5-10 The wave on the bottom has a wave-length 1/3 that of the wave on the top, while it has a frequency 3 times as great.

Wave Speed Another basic property of waves is the speed at which they travel. The speed of a wave can be determined by observing a certain point on the wave as it moves. In longitudinal waves, you could observe a single compression or rarefaction as it travels through the medium. In transverse waves, you could observe a single crest or trough. For example, the speed of a wave on water can be found by observing one crest and measuring the distance it moves in a certain amount of time. Wave speed is usually given in meters per second (m/s). See Figure 5-11.

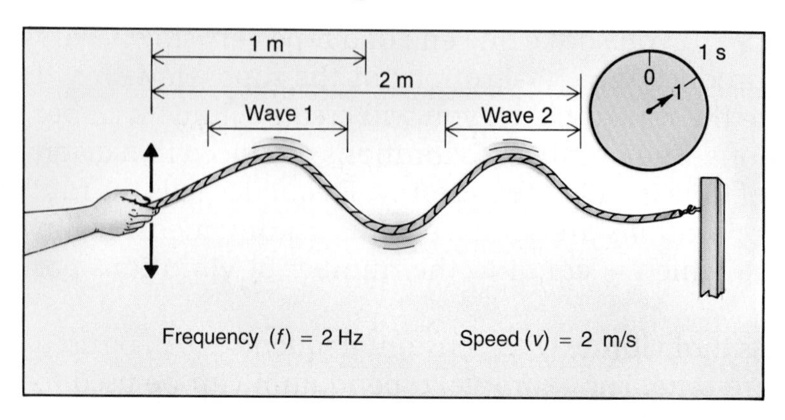

Figure 5-11 This diagram shows how speed, frequency, and wavelength are related.

Wave speed can be calculated if you know the wavelength and frequency of a wave. The relationship between speed, frequency, and wavelength is expressed in the following formula:

$$\text{speed} = \text{frequency} \times \text{wavelength}$$

or

$$v = f\lambda$$

For example, suppose that you are fishing from a stationary boat. Several waves pass by the boat. You could find the speed of those waves by calculating the number that pass in 1 second (frequency) and measuring the length of 1 wave. If the frequency is 2 waves/s (2 Hz) and the wavelength is 0.5 m, the speed is:

$$v = f\lambda$$
$$v = 2 \text{ waves/s} \times 0.5 \text{ m/wave}$$
$$v = 1.0 \text{ m/s}$$

The same formula can be used for sound waves. The speed of sound (v) is known to be about 346 m/s. So, if you can determine either the frequency (f) or the wavelength (λ) of a sound, you can calculate the other.

Characteristics of Waves

There are certain characteristics of behavior that all waves have in common. *Reflection,* for example, is displayed by waves when they encounter obstacles. Other characteristics are *refraction, diffraction,* and *interference.*

Figure 5-12 *A skillful billiard player can cause the balls to bounce at unusual angles.*

Reflection The process in which a wave bounces back after striking a barrier that does not absorb energy.

Reflection Have you ever watched or played a game of billiards? A skillful player knows how to use the side rail of the table to his or her advantage. Success often depends on getting the balls to bounce at the desired angles. Normally, a ball will bounce off the side rail of the table at the same angle at which it strikes it. See Figure 5-12.

Waves bounce off barriers in the same manner that a billiard ball does. This is called **reflection**. To demonstrate that waves are reflected, try these experiments. Tie one end of a rope to a doorknob. From the opposite end, send a transverse wave through the rope. You will see that when the wave reaches the doorknob, it bounces back. Longitudinal waves are also reflected by barriers. Hold a coiled spring at one end. If you send a longitudinal wave down the spring, the wave will come back when it reaches the end of the spring. To demonstrate the reflection of water waves, try making a wave in a sink or tub full of water by tapping your finger on

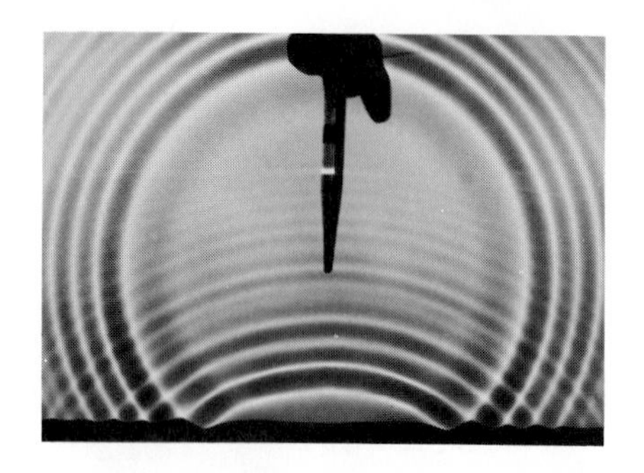

Figure 5-13 *This circular wave reflects from a straight barrier.*

Refraction The process in which a wave changes direction because its speed changes.

the surface of the water. When the wave reaches the side of the sink or tub, you will see it bounce back. See Figure 5-13. For an example of the reflection of sound waves, recall the behavior of sound as it reflects from walls or the sides of buildings to produce echoes.

Refraction The bending of a wave when it passes from one medium to another is called **refraction**. Refraction is due to a change in the speed of a wave. You can observe the refraction of waves at the ocean shore. As waves in the ocean approach the shore, they enter shallow water. The wave troughs drag on the bottom and slow down. If the wave approaches the shore at an angle, the end closest to the shore slows down first. As a result, the wave bends toward the beach. See Figure 5-14.

Figure 5-14 *If waves approach the shore at an angle, the ends closest to the shore slow down first, and the waves are refracted.*

Refraction can be compared to the column of a marching band that marches at an angle from hard pavement into a patch of soft mud. The band members at one side of the front row will encounter the mud first and, finding it harder to march, will slow down. The others in the front row will continue at their initial speed until, in turn, each enters the mud. In effect, each row in the column will turn toward the mud. The opposite effect will occur when the column returns from the mud to the pavement.

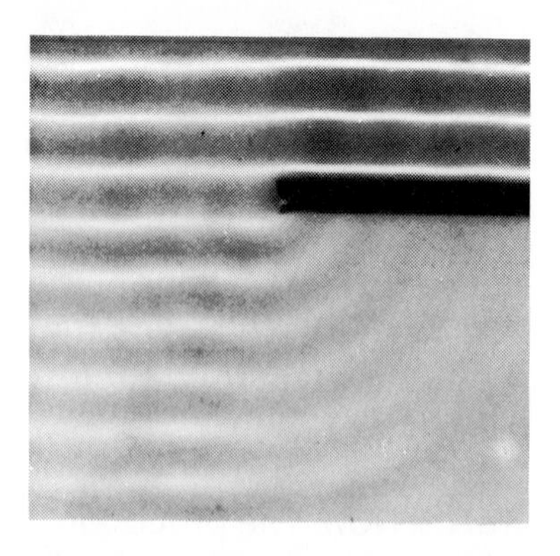

Figure 5-15 *Straight waves can be bent around the edges of a barrier.*

Diffraction The bending of waves as they pass the edges of objects.

Diffraction Waves can also go around corners. Water waves clearly illustrate this behavior. Figure 5-15 shows ripples from a vibrating source touching the water surface in a "ripple tank." Straight water waves (coming from the top of the picture) bend around the edge of the barrier. This is called **diffraction**. All waves, including sound waves, exhibit diffraction. For example, you can hear a sound from around a corner, even if there is nothing to reflect the sound to you.

Interference Have you ever thrown two stones into a still pond at the same time and seen the ripples from the two splashes pass through each other? See Figure 5-16. Waves that meet in this manner affect each other by a process called **interference**. When two troughs meet, they combine to make a single trough. Likewise, when two crests meet, they combine to make a single crest. The result is lower troughs and higher crests. This phenomenon is called *constructive interference*. On the other hand, when a crest and a trough meet, they cancel each other out. This is called *destructive interference*. See Figure 5-17. Interference is a characteristic of all waves, including sound waves.

Figure 5-16 *When two waves cross each other's path, they affect each other's amplitude.*

Interference The results of two or more waves overlapping.

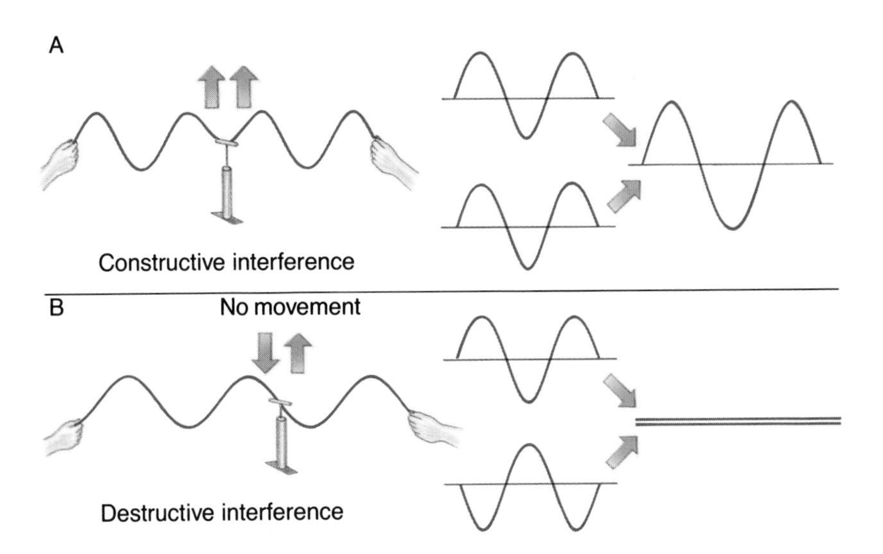

Figure 5-17 *Waves that undergo constructive interference* (top) *produce a single, stronger wave. Waves that undergo destructive interference* (bottom) *cancel each other out.*

Summary

A wave transmits energy through a medium. Both transverse waves and longitudinal waves exhibit the properties of amplitude, wavelength, frequency, and speed. Reflection—like refraction, diffraction, and interference—is a characteristic of all waves. Reflection is the bouncing back of a wave when it meets a barrier. Refraction is the bending of a wave when it passes at an angle from one medium to another, either slowing down or speeding up. Diffraction is the bending of straight waves around the edge of a barrier. When two or more waves meet, they interfere with each other. If two troughs or two crests meet, they combine to form a single lower trough or higher crest. When a crest and a trough meet, they cancel each other out.

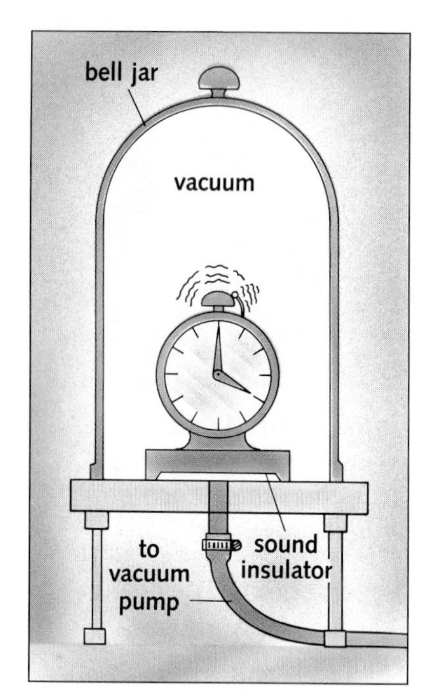

Figure 5-18 When air is removed from the bell jar, the alarm clock will be silent. No air molecules are there to transfer the energy as a wave.

5.2 SOUND WAVES

Without air, there would be no familiar sounds. In fact, we would hear no sounds at all. This is because sound moves through air the way ripples move through water. See Figure 5-18. A sound begins when a vibrating object, such as a guitar string, pushes on the air particles around it. These particles then push on other particles, causing a disturbance that is transmitted outward from the source. As sound waves move through the air, the air particles are alternately crowded together and spread apart. This back-and-forth motion of the particles of air is similar to the way longitudinal waves move through a spring. Therefore, sound consists of longitudinal waves. See Figure 5-19.

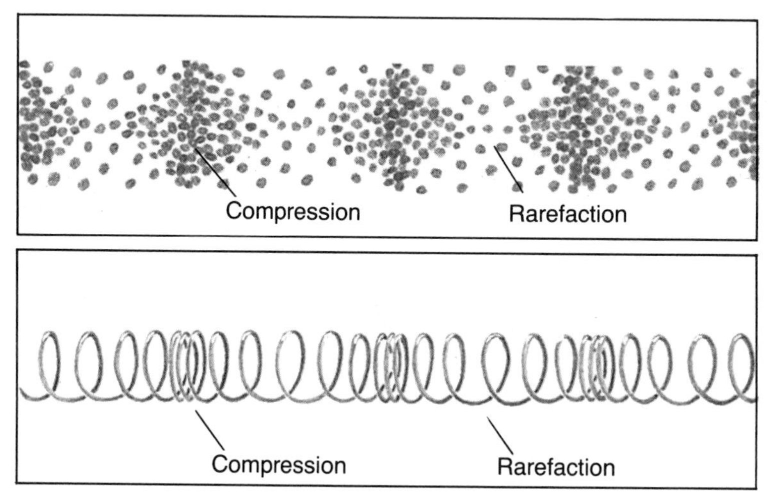

Figure 5-19 This diagram compares the compression and rarefaction of sound waves with the movement of waves in a spring.

Vibrating Strings

Pythagoras was a Greek philosopher and mathematician who lived in the 6th century B.C. He is most remembered for his description of the relationship among the three sides of a right triangle. However, Pythagoras also pursued many other areas of study. For example, he was very interested in the sounds produced by vibrating strings. He discovered that pleasant-sounding chords could be produced if the lengths of the strings were in the ratio of two small numbers. For example, consider two similar strings that are under the same tension. If one is two times the length of the other, they will produce two notes that are one octave apart (for example, from C to C). If the strings are in the ratio of two to three, then a fifth chord is produced (for example, from C to G). These chords are considered to be very pleasing to the ear.

Pythagoras was so impressed by his discovery that he made it the basis for a whole school of thought. He believed that the beauty of the universe must be related to the beauty of these chords. For example, he thought that the movements of the planets and stars were based on the same ratios that he discovered in vibrating strings. The phrase "music of the spheres" comes from this idea. Later scientists disproved Pythagoras' theory concerning the movements of stars and planets. However, his law of vibrating strings, which came from direct observation, has withstood the test of time.

Standing Waves

Another important characteristic of vibrating strings is the patterns of waves that they form. Suppose you send a wave through a rope and the wave is reflected by a barrier. If you send only one crest, it will travel to the barrier and then be reflected back toward yourself. The crest will encounter no interference. However, if you continue to make waves, the waves traveling toward the barrier will soon encounter waves reflected from the barrier. In this situation, the waves will interfere with one another. Since the strings on a musical instrument like the guitar are connected at both ends, they reflect from both ends with the same amplitude, frequency, and speed. If waves traveling in opposite directions have matching characteristics, **standing waves** will be formed. See Figure 5-20.

Standing Waves Waves that form a pattern in which portions of the waves do not move and other portions move with increased amplitude.

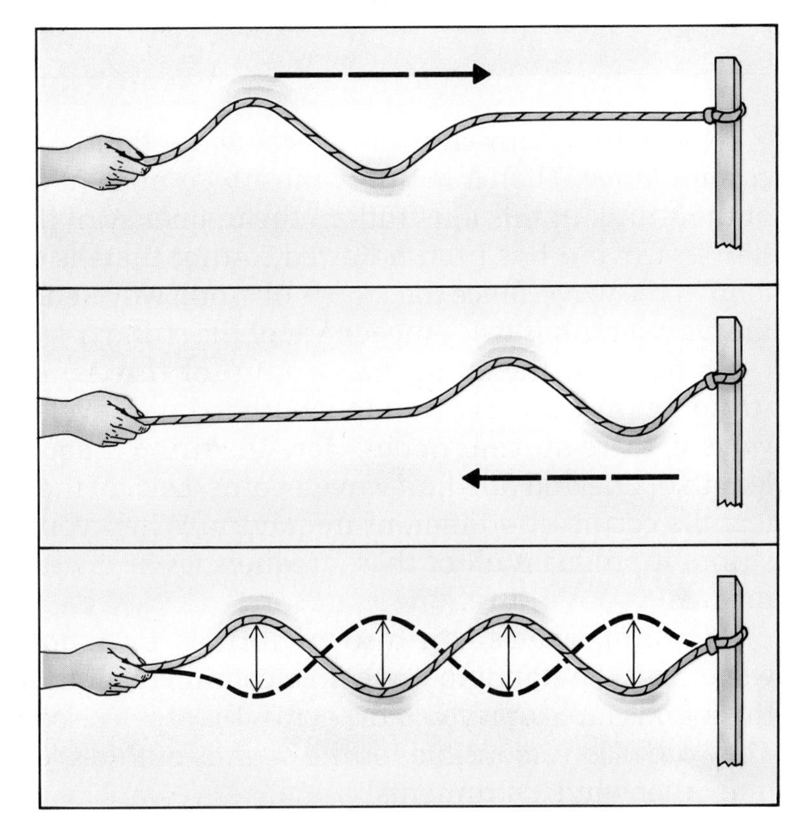

Figure 5-20 *A single wave* (top) *is reflected from a post* (middle). *When reflected waves meet waves moving in the opposite direction, standing waves are formed* (bottom).

The points on standing waves that have no vibration are called *nodes*. Nodes are caused by destructive interference between waves moving in opposite directions. The points on standing waves that vibrate with the greatest amplitude are called *antinodes*. Antinodes are the result of constructive interference between waves moving in opposite directions.

Standing waves look very different from traveling waves. Standing waves appear not to move through the medium. Instead, the waves cause the medium to vibrate in a series of stationary *loops*. Figure 5-21 shows standing waves in a vibrating string. Each loop is separated from the next by a node, or point of no vibration. The distance from one node to the next is always one-half wavelength.

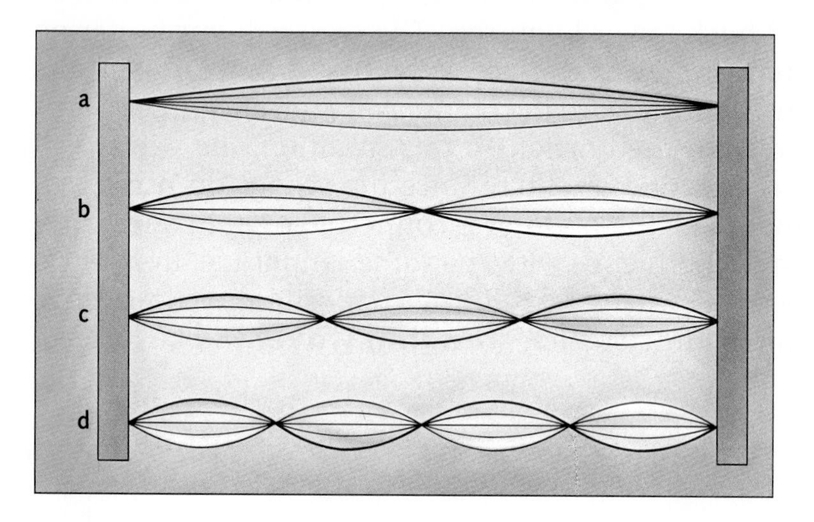

Figure 5-21 *This diagram shows the vibrations of a string. In each example, a standing wave has been produced.*

Notice that both ends of the string are tightly held and cannot move. Therefore, there must be a node at each end of the string. In this illustration, the frequency of the vibration in wave **a** has been adjusted so that there is only one loop in the wave. Since the loop is one-half wavelength long, the wavelength is twice the length of the string.

In wave **b**, the string has a total of three nodes. The string, in this case, is equal in length to one wavelength. In wave **c**, there are four nodes. Here the string is equal to the length of one and one-half wavelengths. Each of these wavelengths occurs at a different frequency. When you pluck a string, it produces all of these frequencies at once. See Figure 5-22.

Standing waves can also be formed by longitudinal waves. It is the standing waves in a column of air that causes the sound you hear if you blow across the top of a soda bottle. They are also responsible for the sounds of flutes, clarinets, and other wind instruments.

Figure 5-22 *When guitar strings are plucked, a variety of standing waves are produced, each blending to form a guitar's distinctive sound quality.*

The Doppler Effect

Have you ever heard the blast of the warning horns as a train passed through a railroad crossing? Did you notice how the sound of the horns changed as the engine passed by? You probably heard a sudden drop in the pitch of the sound. But why?

Figure 5-23 shows the pattern of sound waves produced by a moving train. Notice that as the train moves forward, the sound waves in front of the train are closer together, giving these waves a shorter wavelength and thus a higher frequency. The sound waves in back of the train are spread farther apart, so they have a longer wavelength and a lower frequency. This apparent change in the frequency of the waves from a moving wave source is called the **Doppler effect**.

Doppler Effect An apparent change in the frequency of waves caused by the motion of either the observer or the source of the waves.

Figure 5-23 *The pitch of the horn on this train seems to get lower as the train passes by.*

As sound waves are crowded toward you, the frequency at which they reach your ears would be increased. Remember that higher frequency sound waves have a higher pitch. Therefore, as a train approaches you, you hear a higher pitch than you would if the train were standing still. Once the train has passed by, the apparent pitch of each horn decreases. However, the actual pitch does not change. The engineer on board always hears the same sound.

The Doppler effect also applies to situations in which the observer, rather than the wave source, is moving. Think of watching water waves from a moving boat. As the boat heads into the waves, the waves will hit the boat more often. The frequency of the waves will appear to increase. As the boat moves in the same direction as the waves, the waves will hit the boat less often. The frequency of the waves will appear to decrease. To someone watching the waves from the shore, the actual frequency of the waves will not change at all.

The Doppler effect is a property of all waves, not just sound and water waves. Light waves from distant galaxies also display the Doppler effect in the form of a shift toward the red end (longer wavelengths) of the spectrum of light. This *red shift* indicates that these galaxies are moving away from us at high speed. See Figure 5-24.

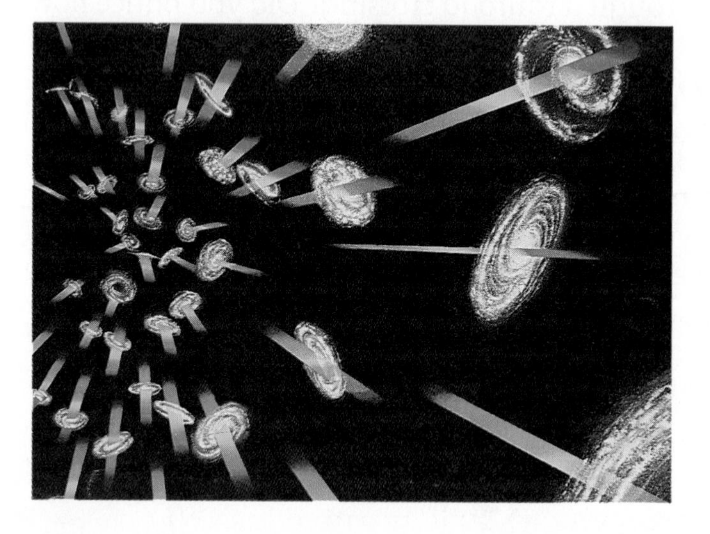

Figure 5-25 *As indicated by this artists conception, light from faster-moving galaxies would appear shifted more toward the red end of the spectrum.*

Summary

There would be no sound without air or some other transmitting medium. Sounds begin when vibrating objects push on air particles, causing a back-and-forth motion of these particles, or longitudinal waves. Standing waves are formed when similar waves traveling in opposite directions interfere with each other. They form a pattern in which the medium vibrates with nodes and antinodes. The Doppler effect—a property of all waves—is an apparent change in the frequency of waves caused by the movement of the wave source or its observer.

5.3 SOUND EFFECTS

In addition to the sounds we can hear, there are sounds we cannot hear. The effects of the sounds we cannot hear can be used in a variety of ways, from locating earthquakes and measuring the depth of the ocean, to ensuring good health. Whether produced by natural phenomena or by artificial means, audible and inaudible sound is an important aspect of our daily lives.

Infrasonic Waves

Sounds with frequencies below 20 Hz consist of **infrasonic waves**. These frequencies are too low for human ears to detect. Earthquakes produce infrasonic waves that move through the solid material of the Earth. Two types of waves originate from the focus of an earthquake. See Figure 5-25. The faster-moving type of wave is a longitudinal wave. The longitudinal waves of an earthquake travel by compressing rock and soil in front of them and stretching it behind them. Waves of this type are the first ones to reach earthquake-recording stations. For this reason, they are called *primary waves*, or P waves.

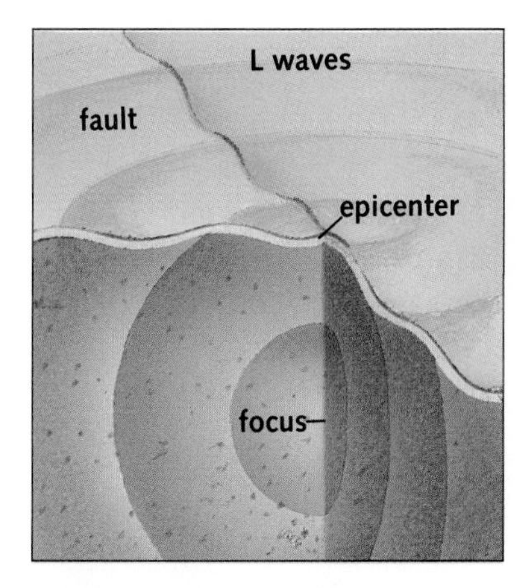

Figure 5-25 *As this diagram shows, the focus of an earthquake is often far below the epicenter.*

Infrasonic Waves Sound waves at a frequency below that which humans can hear.

The second type of earthquake wave is a transverse wave, which travels more slowly than P waves. These slower waves are called *secondary waves*, or S waves. They are similar to the movement of a rope shaken from side to side. The interaction of P and S waves produces a third type of wave, called a surface wave. These waves cause the surface of the Earth to shake and roll. Surface waves are unique in that they originate from the epicenter, not the focus of an earthquake. Their rolling action, which is like the rising and falling motion of ocean waves, causes many buildings to collapse.

The instrument used to record earthquake waves is called a *seismograph*. A seismograph consists of a rotating drum wrapped with paper, and a pen attached to a suspended weight. The pen presses gently against the paper-wrapped drum. The recording from a seismograph is called a seismogram. See Figure 5-26.

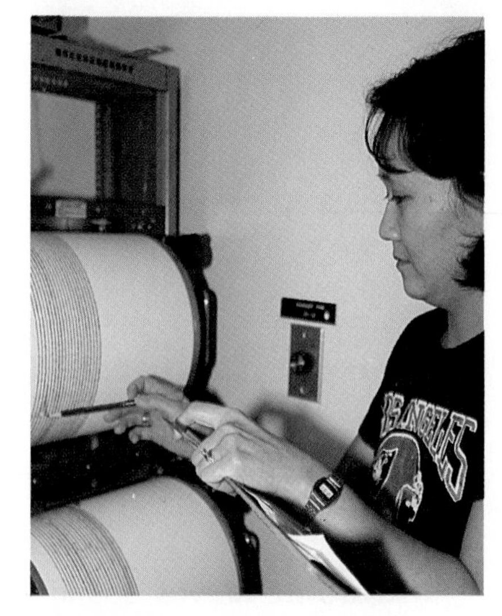

Figure 5-26 *Seismologists keep detailed records of earthquake activity.*

Since P waves travel fastest, they are the first waves recorded on the seismogram. The difference in travel time between the arrival of P and S waves is used to determine the distance between the recording station and the epicenter of the earthquake. See Figure 5-27.

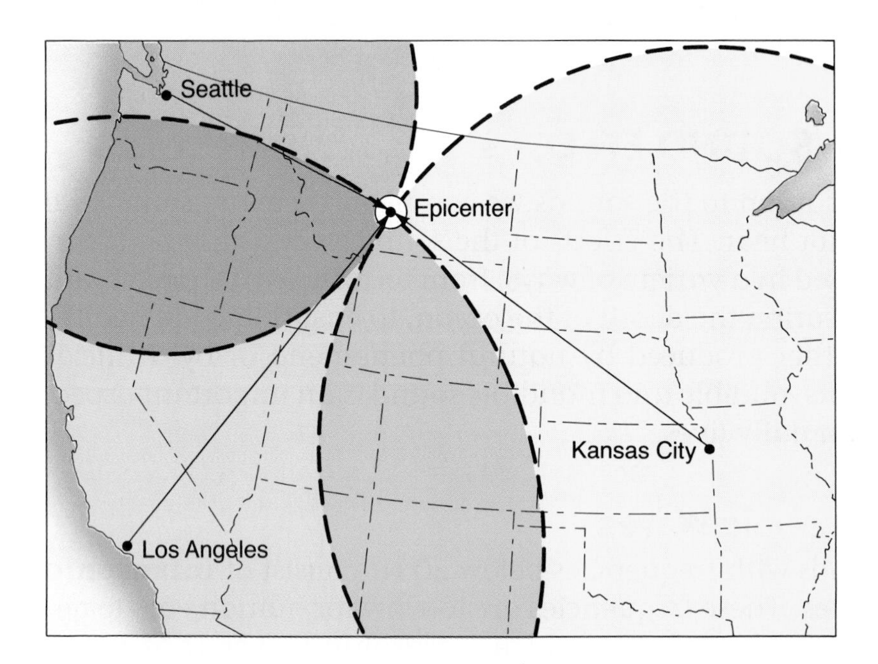

Figure 5-27 *Triangulation, diagrammed here, is the process used to determine the location of an earthquake from the recordings of three seismic reporting stations.*

Volcanic eruptions and tornadoes also produce infrasonic waves that travel for thousands of kilometers through the atmosphere. When Krakatoa erupted in the largest volcanic event in recorded history, the infrasonic waves it produced could still be detected after they had circled the Earth several times.

Infrasonic waves are not all bad. Scientists are using infrasonic waves to investigate the inner structure of the sun. *Helioseismology*, the study of solar "earthquakes," is based on the assumption that sound waves generated within the sun cause a pattern of peaks and valleys on the sun's surface. Because the movement of waves through a medium depends on factors such as elasticity and density, helioseismologists can learn about processes inside the sun by analyzing the surface wave patterns.

Figure 5-28 *Bats use ultrasonic waves to locate prey in the dark.*

Ultrasonic Waves

Insects and other animals can detect sounds that are much higher than those detected by the human ear. See Figure 5-28. Sounds in this upper range are called ultra-high frequency sounds, or **ultrasonic waves**. Ultrasonic waves have been put to use in a variety of ways, from industrial cleaners to medical diagnostic and treatment instruments.

Ultrasonic Waves Sound waves at a frequency above that which humans can hear.

Many ships have a navigation system that uses echoes of ultrasonic waves to find the depth of the water. This system is called *sonar*, which stands for *so*und *na*vigation ranging. This system works in very much the same way as echolocation does in bats. A sonar device sends short pulses of ultrasonic waves through the water. When the sound waves hit the sea floor or other underwater objects, some of the waves are reflected back to the ship as an echo. The echo is then detected by a receiver, and the time delay indicates the distance. See Figure 5-29. Some autofocus cameras operate in the same way.

Figure 5-29 *Submarine pilots rely on sonar completely in order to navigate underwater.*

Ultrasonic waves can also be used to clean jewelry, machine parts, and electronic components. A device called an *ultrasonic cleaner* consists of a container that holds a bath of water and a mild detergent. Sound waves are sent through the bath, causing intense vibrations in the water that remove dirt from the items placed in the bath. See Figure 5-30. The major advantage of ultrasonic cleaners is that they are nonabrasive, so they do not harm the items being cleaned.

Figure 5-30 *An ultrasonic cleaner, such as the one shown here, uses high-frequency sound waves to clean jewelry.*

Ultrasonic waves used for medical applications are referred to as *ultrasound.* In some cases, ultrasound is used to remove kidney stones—crystals that form in the kidneys—without surgery. Ultrasound shatters these crystals without damaging the soft tissues nearby. The tiny fragments can then easily pass out of the body in the urine.

Ultrasound also provides a way to "see" inside the body. Ultrasonic waves bounce off high-density tissues and are converted into electrical signals that are fed into a computer. The computer uses these signals to form an actual picture, called a *sonogram.* Using this technique, doctors can locate tumors and gallstones. They can also examine a developing baby inside its mother to determine whether it is forming normally and if it is in the proper position.

Sounds of Silence

Most people can hear a wide range of sounds. This is not the case for hearing-impaired individuals. However, there are ways that even a person who is completely deaf can detect sound waves in the environment. Since sound waves cause vibrations in solid materials, these vibrations can be felt through the sense of touch. Low-frequency sounds, especially, can be felt this way. Deaf people can dance to the beat of a loud band by sensing the strong driving beat of the bass player. For partially hearing-impaired people, hearing aids can sometimes help to amplify frequencies that are hard to hear. These devices have been made so small by modern electronic methods that they can be inserted completely into the ear canal.

If you have normal, healthy ears, you should protect them. Loud sounds and loud music can cause damage to the ears that cannot be repaired. Sound is measured in *decibels* (dB). The higher the decibel level, the more intense the sound. See Table 5-1. Above a certain level of intensity, noise

SOUND INTENSITY LEVELS		
Type of Sound	**Intensity (dB)**	**Hearing Damage**
Whisper	10–20	
Soft Music	30	None
Conversation	60–70	
Vacuum Cleaner	70	After long
Heavy Street Traffic	70–80	Exposure
Construction Equipment	100	Progressive
Thunder	110–120	
Loud Rock Concert	110–130	Immediate and
Jet Engine at takeoff	120–150	Irreversable

Table 5-1

can harm your health. A quiet conversation, for example, is only about 40 dB, a harmless level of sound. Hearing damage may begin, however, when a person is exposed to noise levels of about 75 dB for 8 hours a day. A noise of more than 115 dB also causes immediate pain. Loud music and crowds at rock concerts have been recorded at levels reaching 100 to 130 dB. Repeated exposure to high noise levels at rock concerts and the long-term use of personal stereos may cause permanent hearing impairment. See Figure 5-31. Do not let the world of sound fade away due to carelessness!

Figure 5-31 *Hearing loss can result from long-term exposure to loud noise or the music that comes from radios played at high volumes or personal stereos that block out all other sound.*

Summary

Sound waves, both audible and inaudible, can be used in many ways. Infrasonic waves are very low-frequency sound waves. They are produced by several natural phenomena, such as earthquakes, and are even used in the study of the sun's interior. Ultrasonic waves are used by sonar devices, ultrasonic cleaners, and medical instruments. Deaf and hearing-impaired people cannot hear the sounds that are constantly around us, but they can sense strong vibrations. Loud sounds or music can damage the ears and cause permanent hearing loss.

Unit 6

Reading *Plus*

Now that you have been introduced to light and its relationship to heat, color, and images, let's take a closer look at the composition of light and some of the ways technology uses it. Read pages S88 to S106 and prepare a consumer report on an electronic appliance of your choice that describes how the properties of light (or another form of electromagnetic radiation) affect its operation. Your report should provide information concerning the following questions.

1. In what ways does light behave in this appliance—as a stream of particles, a series of waves, or both? How can you tell?
2. At what frequencies does this appliance operate, and what might happen if the frequency were adjusted up or down?
3. How is the operation of your appliance an improvement over similar appliances that do not use light technology?

In This Unit

6.1 LIGHT: A SPLIT PERSONALITY?

Isaac Newton, whose laws of motion and gravity are the basis of traditional physics, wrote his great work on light in 1672. Besides using prisms to study color, he also studied how light affects, and is affected by, objects in its path. Throughout the history of the scientific investigation of light, there have been conflicting ideas, not only about what gives it color, but also about the very makeup of light itself. Light seems to act in different ways under different circumstances. Various theories of light have been advanced to explain some of these properties, but as you will see, none of the theories can explain all of the known properties.

The Particle Theory

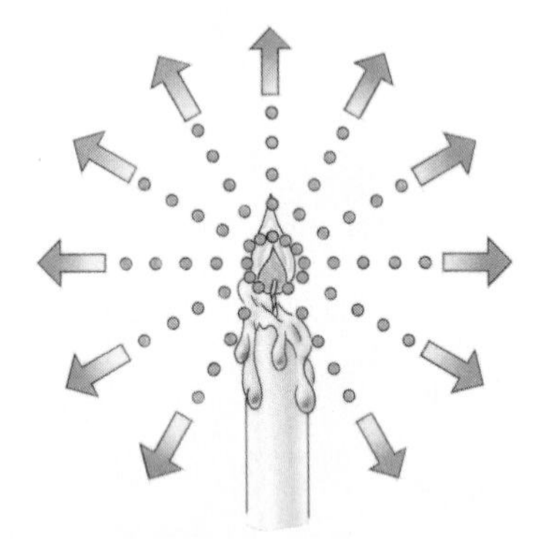

Figure 6-1 *Newton thought light consisted of streams of tiny particles.*

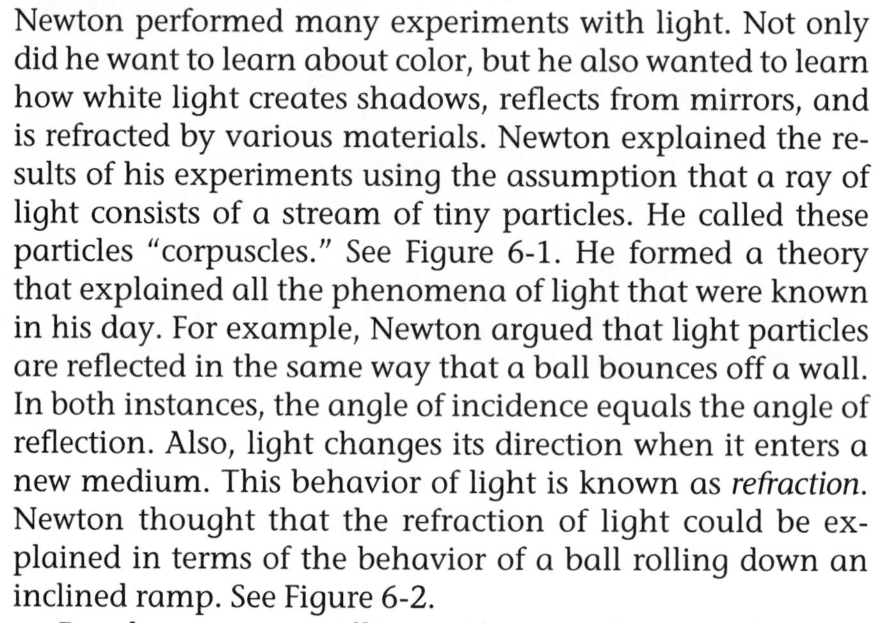

Newton performed many experiments with light. Not only did he want to learn about color, but he also wanted to learn how white light creates shadows, reflects from mirrors, and is refracted by various materials. Newton explained the results of his experiments using the assumption that a ray of light consists of a stream of tiny particles. He called these particles "corpuscles." See Figure 6-1. He formed a theory that explained all the phenomena of light that were known in his day. For example, Newton argued that light particles are reflected in the same way that a ball bounces off a wall. In both instances, the angle of incidence equals the angle of reflection. Also, light changes its direction when it enters a new medium. This behavior of light is known as *refraction*. Newton thought that the refraction of light could be explained in terms of the behavior of a ball rolling down an inclined ramp. See Figure 6-2.

But the most compelling evidence in favor of Newton's particle theory of light was the fact that light seems to travel in straight lines. For example, if an object is placed in the path of light, it casts a shadow; in other words, the light is

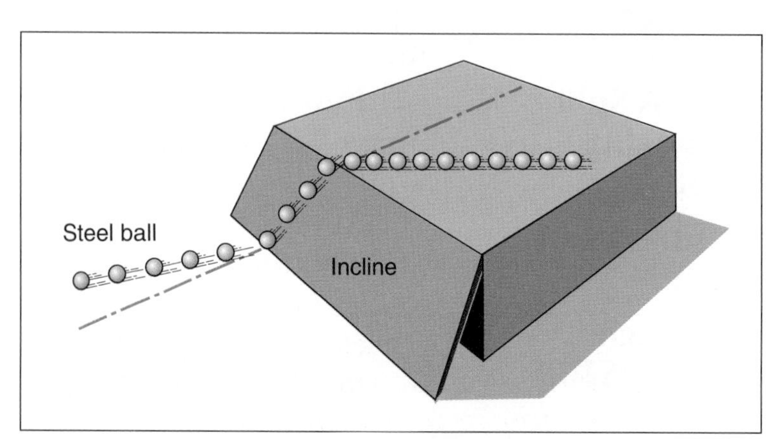

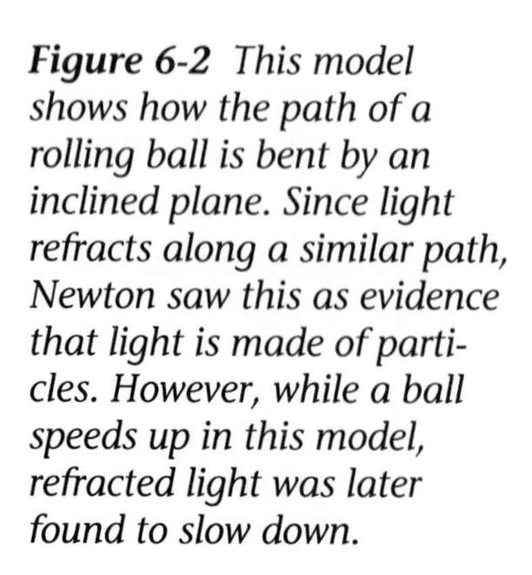

Figure 6-2 *This model shows how the path of a rolling ball is bent by an inclined plane. Since light refracts along a similar path, Newton saw this as evidence that light is made of particles. However, while a ball speeds up in this model, refracted light was later found to slow down.*

stopped. In the same way, a rolling ball is stopped by a barrier placed in its path.

Meanwhile, the Dutch physicist Christiaan Huygens theorized that light travels as waves. See Figure 6-3. To support this idea, he pointed out that one beam of light can pass through another without either beam being disturbed. Waves can do this; streams of particles presumably cannot. A wave theory could also explain reflection and refraction of light, since both phenomena are common to wave action, as demonstrated by water waves. But Huygens could not explain why light seemed to travel only in straight lines. At that time, it was known that sound waves could travel around barriers, a behavior known as *diffraction*. If light is made of waves, it should also be diffracted around barriers, but observations seemed to indicate otherwise. For example, if both sound and light are carried by waves, why can you hear, but not see, around corners? Since there was not enough evidence in support of Huygens' wave theory, Newton's particle theory of light became the accepted theory for the next 100 years.

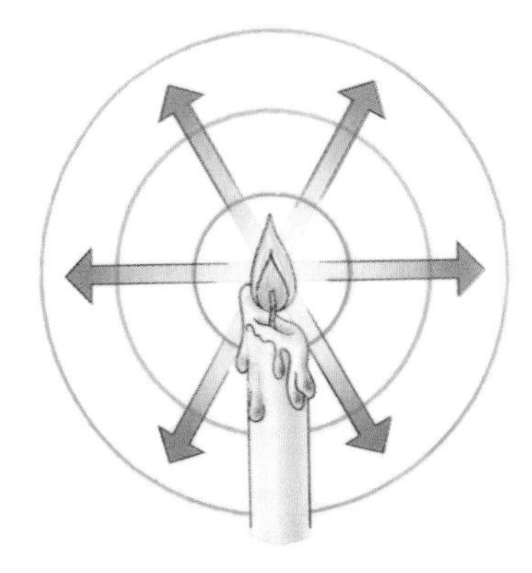

Figure 6-3 *Huygens thought light consisted of a series of waves.*

The Wave Theory

In 1801, however, an English scientist named Thomas Young discovered new evidence supporting the wave theory of light. He discovered that light, like sound, *could* be diffracted. This was a powerful argument in favor of the wave theory of light, since diffraction is a property of waves but not of particles. An example of the diffraction of waves in water is shown in Figure 6-4. Notice that when water waves pass through sufficiently small openings, they are diffracted, or bent, around the edges of the openings.

Young did a simple experiment to see whether light would behave as particles or as waves. He made two narrow, parallel slits in a card. He then placed a light source on one side of the card and a screen on the other. Young reasoned that if light is made of particles, two bright bands of light should appear on the screen. But if light is made of waves, something very different should happen. The light would be diffracted as it passed through the slits, so each slit would act as a separate new wave source. Then the waves from these two new sources would *interfere* with each other, resulting in a pattern of bright and dark bands on the screen. At places on the screen where the crests or troughs of the waves from one source matched the crests or troughs of the waves from the other source, constructive interference would occur and bright bands would appear. At places where the wave crests from one source coincided with wave troughs

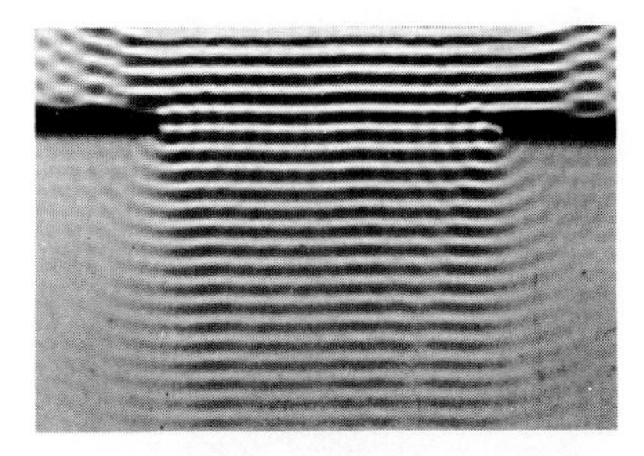

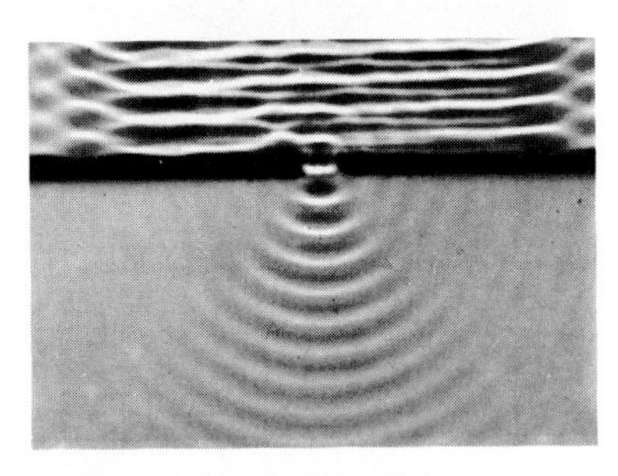

Figure 6-4 *When an opening is large* (top), *little or no diffraction takes place. When it is small* (bottom), *waves diffract around the edges of the barrier.*

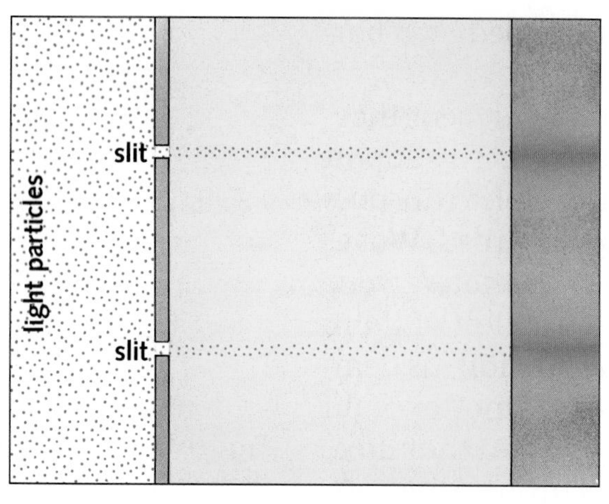

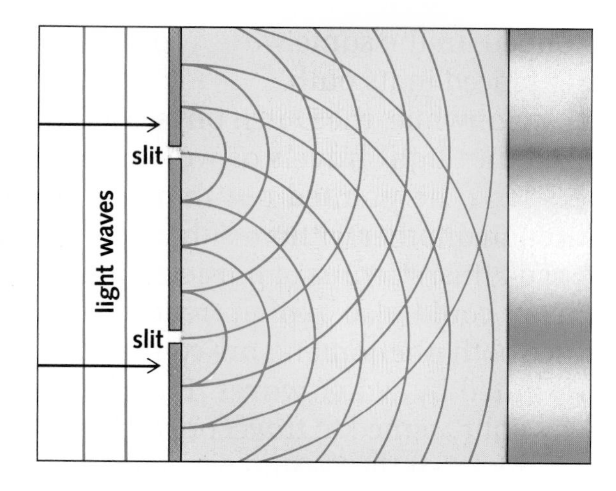

Figure 6-5 *If light is made of particles* (left), *no interference will occur. If light is made of waves* (right), *the diffracted waves will interfere with one another.*

from the other source, destructive interference would occur, and so there would be dark bands. This is exactly what happened in Young's experiment. See Figure 6-5.

Huygens may not have detected the diffraction of light because, in order for a wave to be diffracted, the opening through which it travels must not be much larger than its wavelength. See Figure 6-6. Because the wavelengths of visible light are very small (~0.0005 mm), the opening must be very narrow. The slits Young used were narrow enough to cause diffraction and interference of light. Following Young's experiments, the wave theory eventually replaced Newton's particle theory as the accepted model of light.

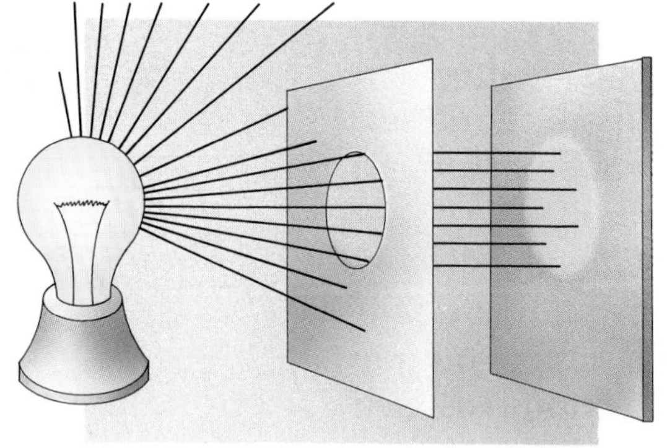

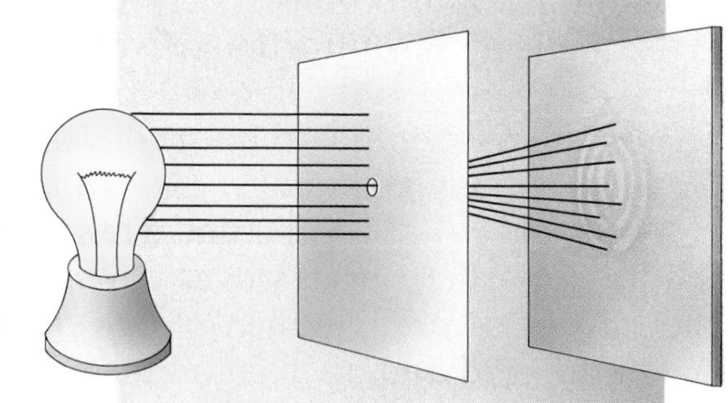

Figure 6-6 *When light passes through a large hole, the light follows the particle model. When it passes through a small hole, it follows the wave model.*

Broadening the Spectrum

Electromagnetic waves Waves that carry both electric and magnetic energy and move through empty space at the speed of light.

The wave theory of light received a major boost when, in 1864, the Scottish physicist James Clerk Maxwell showed that light consisted of **electromagnetic waves**. Maxwell was studying the relationship between electricity and magnetism. Using mathematics, he showed that an electromagnetic wave consisting of an electric field and a magnetic field at right angles to each other would travel at a constant velocity. See Figure 6-7. What surprised him was the fact that this velocity was the same as the speed of light! He was not

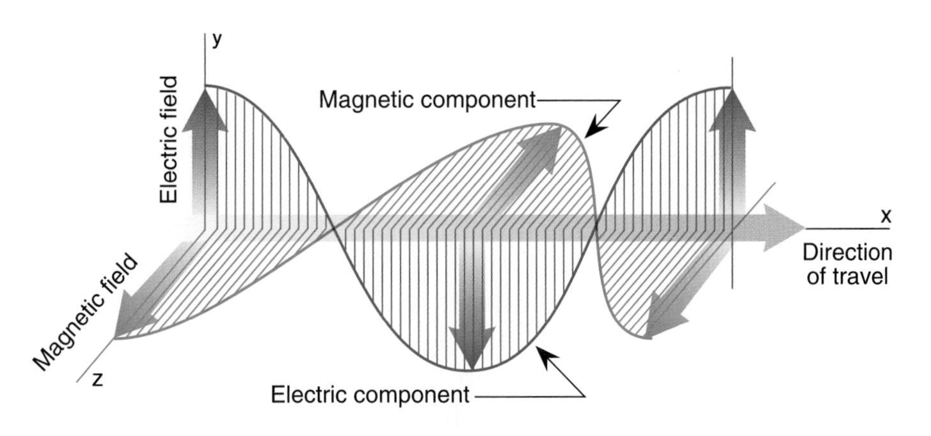

Figure 6-7 In this diagram of an electromagnetic wave, the electric and magnetic fields are at right angles to each other and to the direction of travel.

looking for a theory of light, but the coincidence was striking. He suggested that light is actually an electromagnetic wave, generated by electric and magnetic fields.

The German scientist Heinrich Hertz verified Maxwell's theory experimentally. He was able to generate electromagnetic waves with a wavelength of approximately 1 meter. Later experiments showed that a wider range of electromagnetic waves could be generated, with wavelengths from thousands of kilometers to millionths of a centimeter. These waves are called *electromagnetic radiation*. Light, whose wavelength (and frequency) is approximately in the middle of this range, is only part of a much broader spectrum of electromagnetic radiation.

Figure 6-8 Some metals are sensitive to visible light frequencies, which can be used as a switch to operate street lights along the highway.

Particle Theory Revisited

After the discovery of electromagnetic waves, scientists thought they had learned all there was to know about light. However, in 1887, Heinrich Hertz discovered a phenomenon called the **photoelectric effect**. You may have encountered this phenomenon when riding in a car. See Figure 6-8. Hertz found that when light hits a thin metal plate, electrons are sometimes ejected. See Figure 6-9. However, the light must be above a certain frequency for each kind of metal plate. If the frequency of the light is a little below this threshold frequency, no electrons are emitted, even if the intensity of the light is great. On the other hand, if the frequency of the light is even a little above the threshold, electrons are emitted, no matter how weak the intensity of the light is. Once the light is above this threshold frequency, the number of electrons emitted increases as light intensity increases.

Photoelectric Effect The emission of electrons by a substance when illuminated by light of a sufficient frequency.

This discovery, however, presented a problem. The wave theory of light could not explain the photoelectric effect. If light is composed of waves, then both low- and high-frequency light should give electrons enough energy to escape from any metal plate if the intensity of the light was great enough. According to the wave theory, the energy of

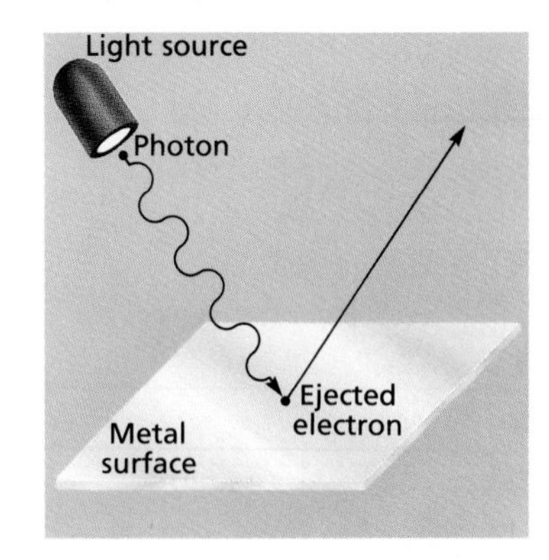

Figure 6-9 The photoelectric effect occurs when some metals emit electrons as light shines on them.

light would be spread out over the entire wave front. Thus the higher the intensity of the light, the greater the energy distributed over the wave front. But the characteristic threshold frequency of each material showed that this was not the case.

Then in 1905 Albert Einstein offered an explanation of the photoelectric effect. Using the ideas of German physicist Max Planck, he showed that light is actually composed of particles. These particles were later called **photons**—tiny packets of light energy. According to Einstein's explanation, each photon of light has a definite amount of energy that relates to its frequency. When light hits the atoms of a metal plate, each photon gives up its energy to a single electron, and the electron escapes from the metal plate. See Figure 6-10. Low-frequency photons do not have enough energy to provide an electron with the energy needed to escape, whereas high-frequency photons do. Above the threshold frequency, the number of electrons emitted increases with the intensity of the light, since there are more photons to eject electrons.

Photon A tiny package of light energy.

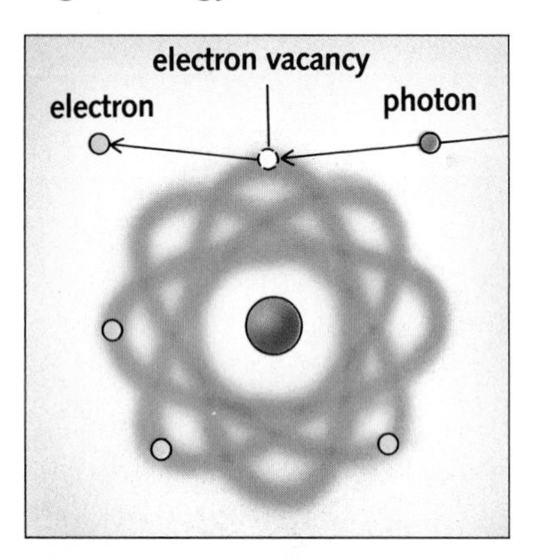

Figure 6-10 *When a photon strikes the proper electron, the electron is emitted. Many photons striking many electrons can initiate an electric current.*

Packages of Energy Each photon carries a specific amount of energy. In order to cause the photoelectric effect, the amount in each package must be great enough. To illustrate, imagine that you and your class decide to visit the zoo. The price of an admission ticket is $7. If there are 30 students in your class, and each student has $5, the class would have a total of $150. There are two things the class might do with that money. On the one hand, the money might be pooled, and students would then be admitted to the zoo until the $150 was used up. (This would, however, be unfortunate for the last 9 students in line.) On the other hand, each student might hold onto his or her own $5 bill. In other words, no student would get in because no one had enough to pay the admission price. But if the teacher then gave $2 to each student, all of you would get into the zoo. As long as each student has $7, the number of students admitted would then only be limited by the class size.

According to the wave theory, the photoelectric effect should follow the first plan, where money is shared. Energy (money) is distributed along the wave front (class), and the number of electrons (tickets) emitted would be related to the total amount of energy (class funds). But according to the particle theory, the photoelectric effect would follow the second plan, where each student keeps his or her own money. The energy is carried by individual photons (students), each of which must have sufficient energy (admission price) in order to eject an electron (ticket). As long as each photon meets the minimum requirements, the number of electrons emitted is related to the intensity of the light (class size). In the case of the photoelectric effect, the particle theory matches what is actually seen in experiments.

The Particle–Wave Theory

The wave model of light that replaced the particle model lasted until 1905 when Albert Einstein published his paper explaining the photoelectric effect. Einstein proposed that light *does* sometimes behave as particles. In order to explain this behavior of light, he suggested that the energy of the light comes out in tiny packets, or photons. On the other hand, Young had shown that light behaves as waves. It seems that neither waves nor particles alone can explain the behavior of light. What is light then—a wave or a stream of photons?

Today scientists realize that in some experiments the behavior of light is best explained by a wave model, while in other experiments the photon model is more useful. See Figure 6-11. This has led to what is called the *particle–wave theory* of light. In other words, scientists have yet to determine the true nature of light. How can light exhibit the qualities of both particles and waves, yet not be one or the other? Perhaps a third alternative is still awaiting discovery.

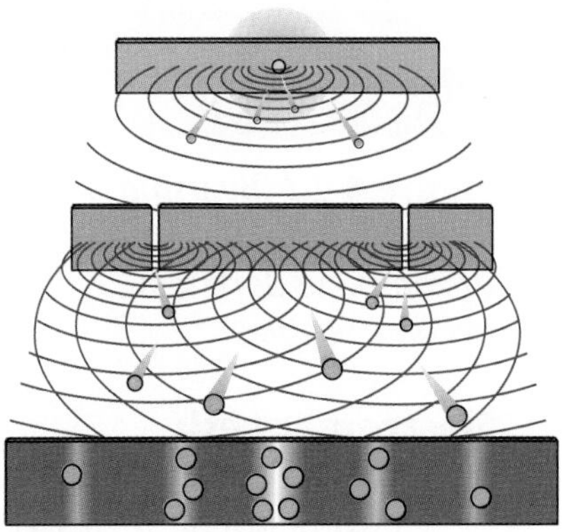

Figure 6-11 *Light shows both particle and wave characteristics. Individual photons can be detected as they hit the screen, but a wave-like interference pattern is also evident.*

Summary

Newton explained light in terms of a stream of tiny particles. His particle theory of light was largely based on the fact that light travels in straight lines. Christiaan Huygens' theory that light travels as a wave was supported by Young's discovery that light could be diffracted and undergo interference. Light was found to be part of a broader spectrum of electromagnetic waves. Heinrich Hertz discovered the photoelectric effect, which Einstein explained in terms of light particles called photons. Today the properties of light are explained by a combination particle–wave theory.

6.2 INVISIBLE LIGHT

Much of the behavior of light can, nevertheless, be discussed in terms of waves alone. As you know, visible light waves are not the only form of electromagnetic radiation. Though we usually think of light in terms of the waves that we *can*

Figure 6-12 By using infrared binoculars, objects in complete darkness become visible.

Electromagnetic Spectrum The entire range of electromagnetic waves, of which visible light is a part.

see, the total number of electromagnetic waves far exceeds our eyes' abilities to discern it. See Figure 6-12. These visible and invisible waves together constitute a spectrum of electromagnetic radiation.

The Electromagnetic Spectrum

Visible light represents only a small portion of the known range of all electromagnetic waves. The color of light within this range varies with its frequency from red to violet. Similarly, the properties of all electromagnetic waves vary according to their frequency. The total range of electromagnetic wave frequencies is known as the **electromagnetic spectrum**. See Figure 6-13. It is divided into several different regions based on the differing characteristics of the waves at different frequency levels. (Remember, frequency is the number of complete waves passing a point in 1 second.) Electromagnetic waves have frequencies between 10^1 and 10^{21} hertz (Hz). Visible light has a frequency of about 10^{14} Hz, while the frequency of microwaves is about 10^9 Hz. The major parts of the electromagnetic spectrum are discussed below.

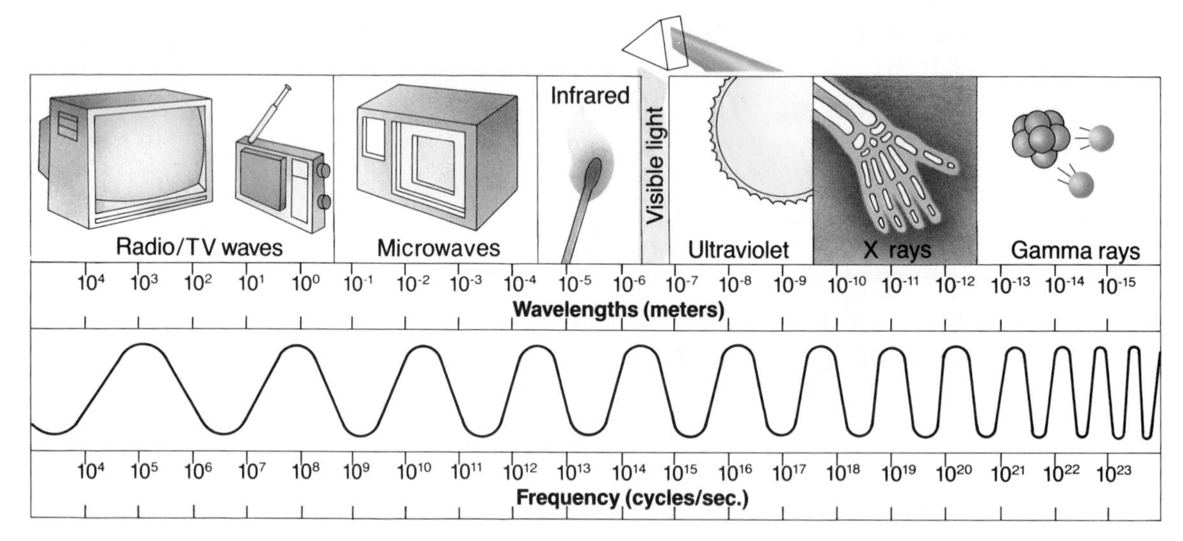

Figure 6-13 Notice that visible light represents only a small portion of the electromagnetic spectrum.

Power Waves Power waves are very low frequency waves that are produced by electrical generators. These waves are used to carry electrical power through transmission lines. They cause static on your car radio when you drive near them. Their frequencies range between 10 Hz to 100 Hz, with wavelengths of thousands of kilometers.

Radio Waves When you turn on your stereo, you are receiving the effects of another type of electromagnetic wave. Radio waves include a wide range of electromagnetic waves with frequencies between 10^4 and 10^{12} Hz. Ordinary AM radio broadcasts use the lower-frequency waves between 535 kilohertz (kHz) and 1605 kHz. (A kilohertz is 1000 waves per second.) One characteristic of AM radio waves is that they are reflected back to the Earth's surface by the layer of the atmosphere called the *ionosphere*. Thus AM radio broadcasts can be received far away from the station. See Figure 6-14. At night, you might be able to pick up stations from half-way across the country. FM radio waves have much higher frequencies of between 88.1 megahertz (MHz) and 107.9 MHz. (One megahertz is 1,000,000 waves per second.) FM radio waves are not reflected by the ionosphere, so they do not travel as far as AM waves. Like FM, television broadcasts also use high-frequency radio waves that can only travel short distances. By using satellites, however, radio and television signals can be relayed half-way around the world.

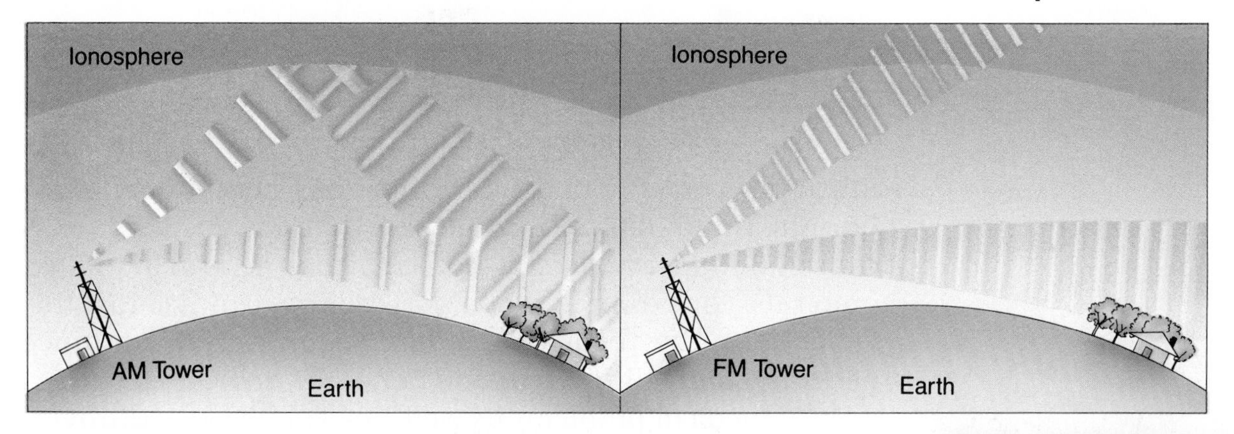

Figure 6-14 *AM radio waves have a longer wave-length and are reflected by the ionosphere. FM waves have a shorter wavelength, which allows them to pass through the ionosphere.*

Radar waves are radio waves that have higher frequencies than FM or TV. Waves at these frequencies are reflected by many materials, especially metals. Reflected radar waves are like reflected sound waves, which are heard as an echo. This radar "echo" can be used to "see" objects through darkness, fog, or at a great distance. For this reason, radar is used for navigation on ships and planes. See Figure 6-15.

Microwaves, unlike radar, are not reflected by most materials. They either pass through the materials or are easily absorbed. For example, microwaves pass through glass but are absorbed by water. When a material absorbs microwaves, its molecules move faster and it becomes hotter. Thus moist food can be heated in a microwave oven in a glass dish. The food gets hot, but the glass dish does not.

Figure 6-15 *This radar specialist uses radar to scan the sky for incoming aircraft.*

Infrared Waves Objects become warmer when they absorb *infrared waves*. For example, your skin feels warm in sunlight because it absorbs infrared waves produced by the sun.

Figure 6-16 The lighter colors in this infrared photograph indicate higher temperatures.

Dark surfaces absorb infrared waves better than light-colored surfaces. For this reason, light-colored clothing is cooler than dark clothing in hot summer sunlight. Objects also give off infrared waves as shown in Figure 6-16.

Visible Light Waves *Visible light* refers to the very narrow band of frequencies that can be seen with the human eye. We see different frequencies within this narrow band as different colors. As you know, the lowest frequencies of visible light are seen as red; the highest frequencies appear as blue or violet. All of the frequencies of visible light together are seen as white light. Because each frequency is refracted at a slightly different angle, a visible spectrum can be produced by shining white light through a prism. Sunlight shining through water droplets produces a natural spectrum we call a rainbow.

Ultraviolet Rays If you spend a lot of time outdoors, you should be aware of ultraviolet rays. The waves in this part of the electromagnetic spectrum have frequencies just above visible light. Ultraviolet rays in sunlight cause sunburn. Too much ultraviolet light can be dangerous since it may cause harmful mutations and serious skin damage, including skin cancer. But ultraviolet light can also be beneficial in limited amounts. For example, when skin cells are exposed to ultraviolet light, they begin to produce vitamin D, a substance necessary for healthy teeth and bones. It can be used to treat certain medical problems as well. See Figure 6-17. Ultraviolet light also kills germs and is often used for this purpose in hospitals.

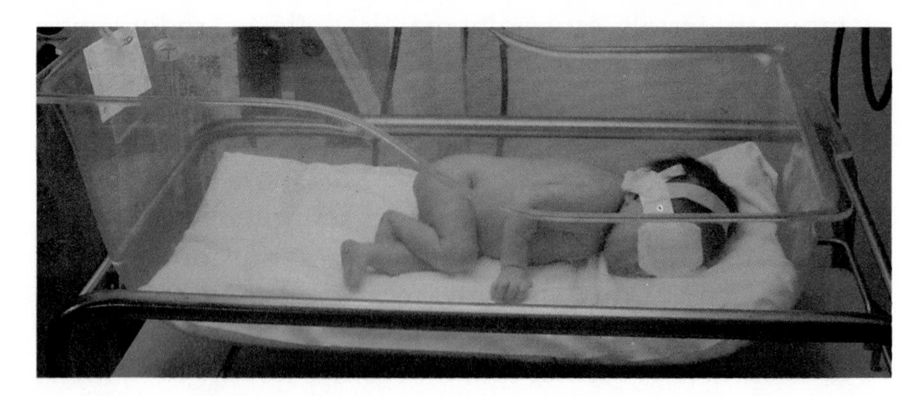

Figure 6-17 Ultraviolet light can be used to treat jaundice in newborn babies.

X Rays *X rays* are high-frequency electromagnetic waves that are very useful in the field of medicine. They can easily pass through skin and other tissues, but not through bone. X rays can produce an image on film or a picture on a screen. This means that X rays can be used to see inside your body.

You have probably had X-ray photographs made of parts of your body by a doctor or dentist. See Figure 6-18. However, too much exposure to X rays can kill living cells. Therefore, machines that produce X rays must be used only by trained personnel.

Figure 6-18 *X rays are produced in a special tube when electrons strike a tungsten target. The reflected X rays pass through the object to form an image on film.*

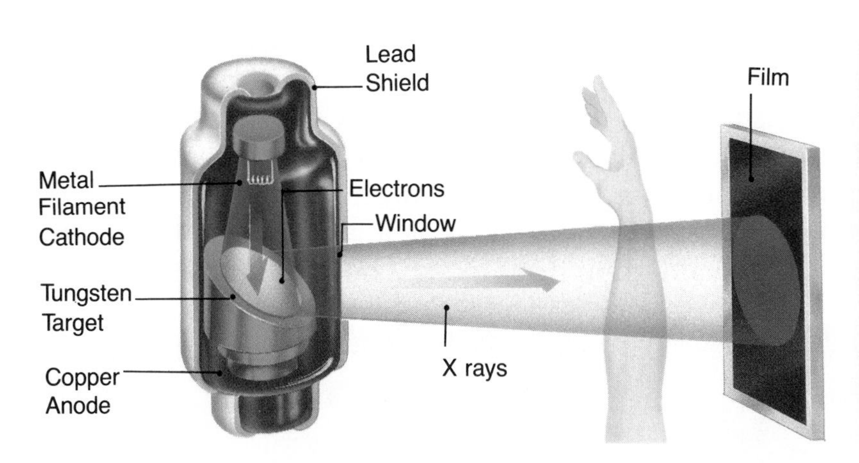

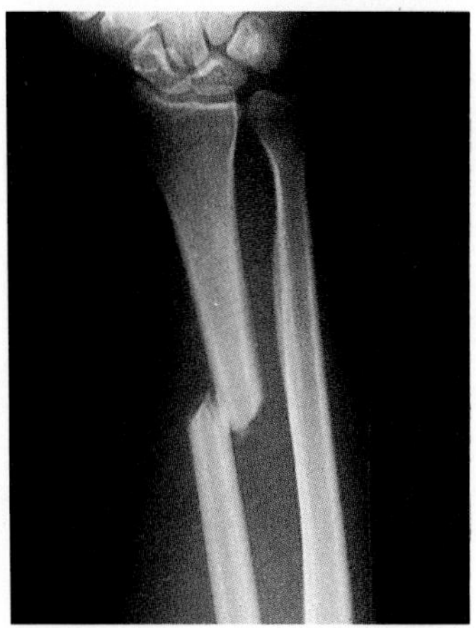

Gamma Rays *Gamma rays* are very high-frequency waves similar to X rays. They are much more dangerous, however, because they pass through matter more easily. Nevertheless, gamma rays are useful in fighting cancer. Beams of gamma rays can be aimed at the location of the cancer. The rays can pass deep into the body to reach and kill the cancer cells. See Figure 6-19.

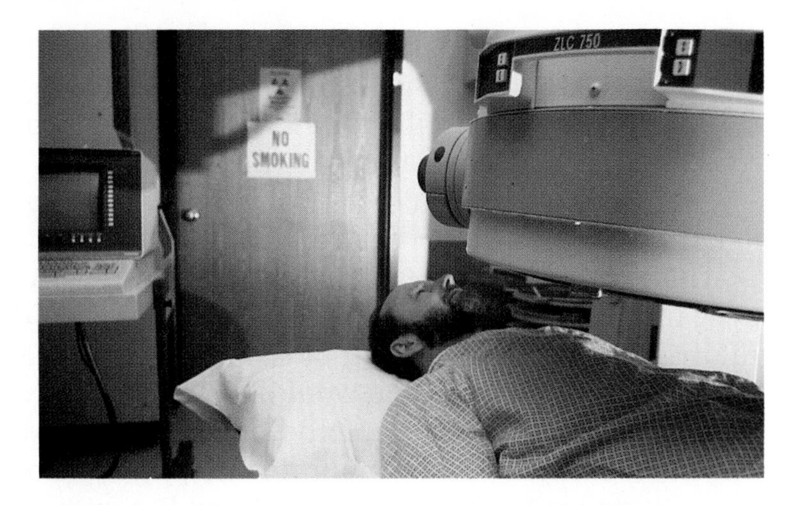

Figure 6-19 *This device uses gamma rays to kill cancer cells.*

Quicker than the Eye

Since the most easily detected electromagnetic radiation is visible light, most early investigations were done with light. However, many of the results of experiments with light can be applied to other forms of electromagnetic radiation. One of the most impressive characteristics of light, and all other electromagnetic radiation, is its incredible speed.

Speed of Light Just how long does it take light from a lamp to get to your eyes? The first scientists who tried to measure the speed of light found that it is far too fast to measure by ordinary means. In the 16th century, the Italian scientist Galileo attempted to measure the time it took light to travel about 2 kilometers. However, light covers this distance almost instantly. Galileo had no way to measure such a short time. About 1676, however, the Danish astronomer Olaus Roemer succeeded in finding out how fast light travels. By studying the eclipses of the moons of Jupiter, he was able to figure out how long it takes for light to travel to the Earth. See Figure 6-20. He found a value for the speed of light that was not very different from today's best measurement.

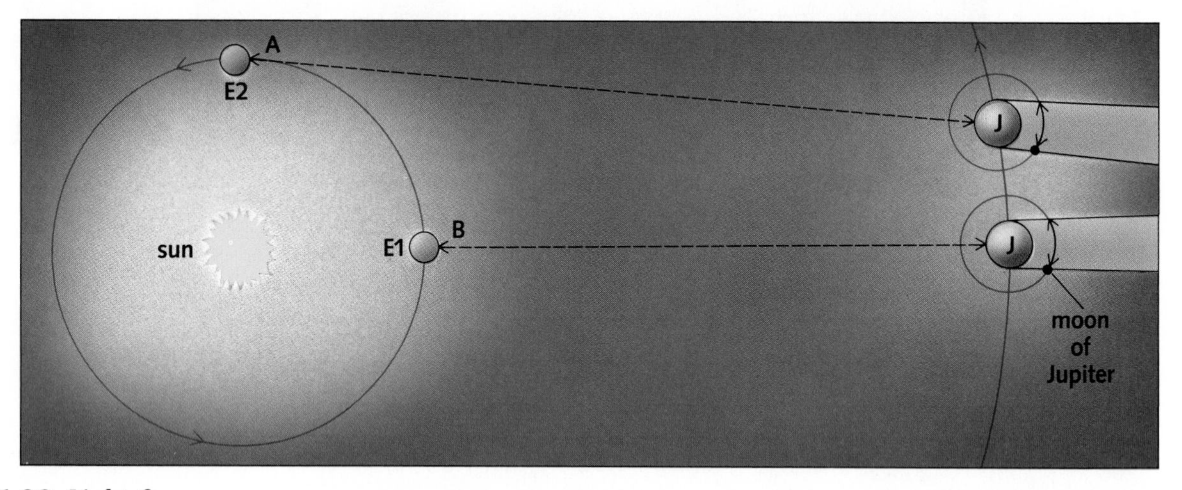

Figure 6-20 *Light from an eclipse of one of Jupiter's moons takes longer to reach the Earth when it is farther away from Jupiter. By comparing this difference in time, Roemer estimated the speed of light.*

Modern measurements, of course, are more accurate than Roemer's. Today the value used for the speed of light is accurate to 1 part in 10 billion. In a vacuum, light travels at a speed of 2.99792458×10^8 m/s, or about 3×10^8 m/s. At this speed, a light beam could get from Los Angeles to Atlanta in less than a hundredth of a second.

This same speed holds for all other forms of electromagnetic radiation. Radio waves from Los Angeles would reach Atlanta in the same amount of time as light waves. Radio listeners in Atlanta, in fact, would hear the music of a concert broadcast live from the Hollywood Bowl before those sitting in the back row of the amphitheater, since sound travels so much slower than electromagnetic radiation.

Speed Limits The speed of light as given above is only valid for light in a *vacuum*. In air, the speed of light is slightly slower. Also, when light moves from one material into another, its speed changes. For example, when light moves from air into water, it slows down. In fact, light waves travel about 25 percent slower in water than in air. On the other hand, light moving from water to air speeds up. This change of speed causes light entering or leaving the water to be refracted into a new path. See Figure 6-21. As you know, the refraction of light makes objects underwater appear to be closer to the surface than they really are.

Figure 6-21 *Light is refracted when it travels from one medium to another because it has a different speed in each medium.*

Some materials slow light more than water. A diamond, for example, slows light to less than 50 percent of its speed in air, or only 1.24×10^8 m/s. As a result, light rays entering a diamond are refracted and separate into different colors just as they are by a prism.

The amount by which a material can bend light is given by its **index of refraction**. The index of refraction for any material is found by dividing the speed of light in a vacuum by its speed in that material. For example, the index of refraction for a diamond is given by the following calculation:

Index of Refraction The ratio of the speed of light in a vacuum to its speed in another medium.

$$\text{index of refraction} = \frac{\text{speed of light in a vacuum}}{\text{speed of light in diamond}}$$

$$= \frac{3.00 \times 10^8 \text{ m/s}}{1.24 \times 10^8 \text{ m/s}} = 2.42$$

The index of refraction of some common materials is given in Table 6-1. In general, the larger the index of refraction, the greater the change in the path of the light rays.

Infrared binoculars demonstrate that electromagnetic waves other than visible light are also affected by the materials through which they travel. Infrared waves are refracted by the lenses of the binoculars in much the same way that light is refracted through the lenses of a normal set of binoculars. Radio waves can also undergo refraction. Since the atmosphere changes density at different altitudes, radio waves are gradually refracted. This refraction sometimes enables them to travel past the horizon.

INDEX OF REFRACTION OF SOME COMMON SUBSTANCES

Substance	Index of Refraction
Air	1.00
Ice	1.31
Water	1.33
Quartz	1.46
Glass	1.52
Amber	1.54
Ruby	1.76
Diamond	2.42

Table 6-1

Summary

Visible light is only one part of a larger electromagnetic spectrum. Electromagnetic waves travel through empty space and can be classified by their frequencies. All electromagnetic waves known to exist are part of the electromagnetic spectrum, which includes power waves, radio waves, infrared waves, visible light, ultraviolet rays, X rays, and gamma rays. The speed of light (and other electromagnetic radiation) through a vacuum is about 3.00×10^8 m/s. This speed slows down as light enters another medium. This change of speed causes light to be refracted by an amount determined by each material's index of refraction.

6.3 HIGH-TECH LIGHT

Several other unique properties of light have led to innovations that affect our daily lives and improve our understanding of the universe around us. Both wave and particle models of light have inspired advances in technology. From lasers that scan prices at the grocery store to sunglasses you wear at the beach and CD players you listen to after school, the technology of light has greatly affected our lives. We will look at a few of these technologies and discover how some of the properties of light are used.

Polarization

Picture yourself at a lake shore on a beautifully clear day. Sunlight shimmers from the sparkling lake, making you squint. You put on a pair of polarizing sunglasses, and the glare disappears. Now you can even see into the water. How can a pair of sunglasses make such a difference?

Sunglasses that are able to eliminate glare contain polarizing filters. Figure 6-22 shows what polarizing filters can

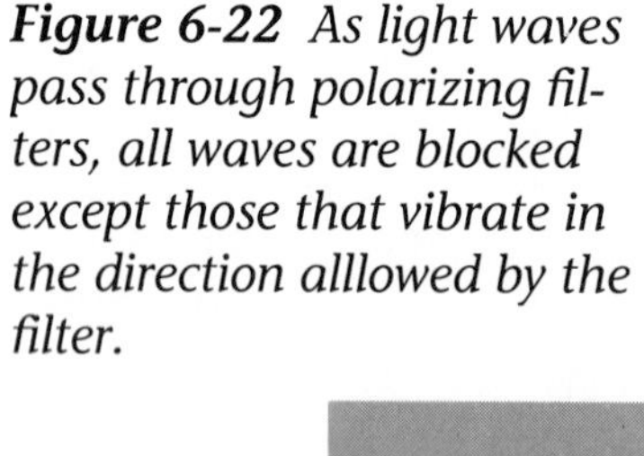

Figure 6-22 *As light waves pass through polarizing filters, all waves are blocked except those that vibrate in the direction alllowed by the filter.*

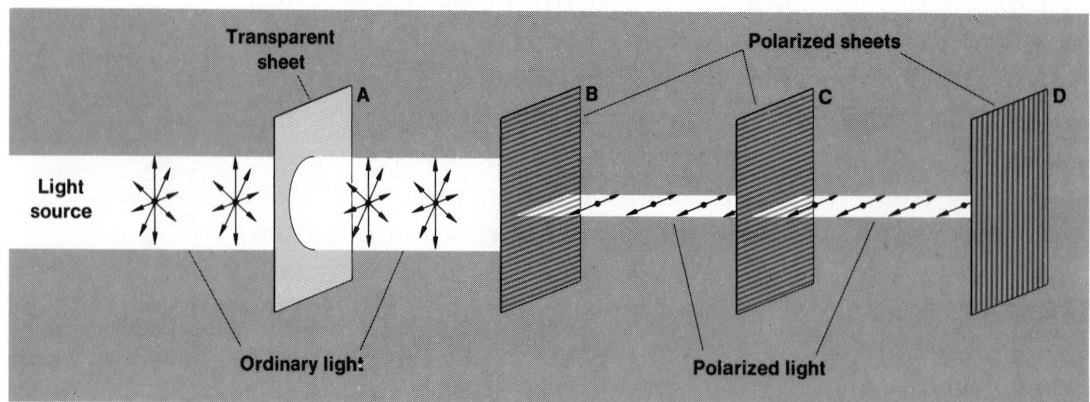

do to light. Light waves, like all electromagnetic waves, are transverse waves. However, some of the waves travel up-and-down, some side-to-side, and others at all different angles in between. The first filter blocks all waves but those that vibrate in a horizontal, or side-to-side, direction. The light passing through the filter is said to be **polarized light** because its waves vibrate in only one plane. After being polarized by the first filter, the light then travels to the second filter. This filter is vertical, so it blocks all waves but those with vertical vibrations. Since only light waves with horizontal vibrations could pass through the first filter, the second filter allows no light to pass.

Polarized Light Light consisting of light waves that vibrate in only one plane.

The glare coming from the surface of a lake is polarized horizontally when it reflects from the surface of the water. Your sunglasses have a vertical polarizer that blocks the glare from the lake so it does not pass through to your eyes. Since the rest of the light in the scene is not polarized, the light waves with vertical vibrations pass through unobstructed. See Figure 6-23.

You cannot detect polarized light with your eyes alone. You can find out if light is polarized by using a polarizing filter. While light passes through the filter to your eyes, rotate the filter 90 degrees. If the light is polarized, it will get dimmer or brighter as you rotate the filter.

Figure 6-23 The photograph on the right does not show the glare because it was taken through a polarizing filter.

Lasers

American industry spends billions of dollars a year making and using laser light for hundreds of purposes. From medical applications or space research to home entertainment, lasers have changed the direction of modern technology. But how is laser light different from ordinary light?

Properties of Laser Light You can see one difference between laser light and ordinary light by examining the two light beams shown in Figure 6-24. The flashlight beam is about 3 cm across when it leaves the flashlight. The beam enlarges to about 40 cm by the time it reaches the other end of the room. By contrast, the beam of light produced by the laser is only about 2 mm across at the laser. When the beam reaches the other end of the room, it makes a 2-mm spot on the opposite wall. In fact, laser light spreads out so little, it can be reflected off of mirrors that were left on the moon by American astronauts. See Figure 6-25.

Figure 6-24 *Laser light remains in a compact beam, while ordinary light expands quickly.*

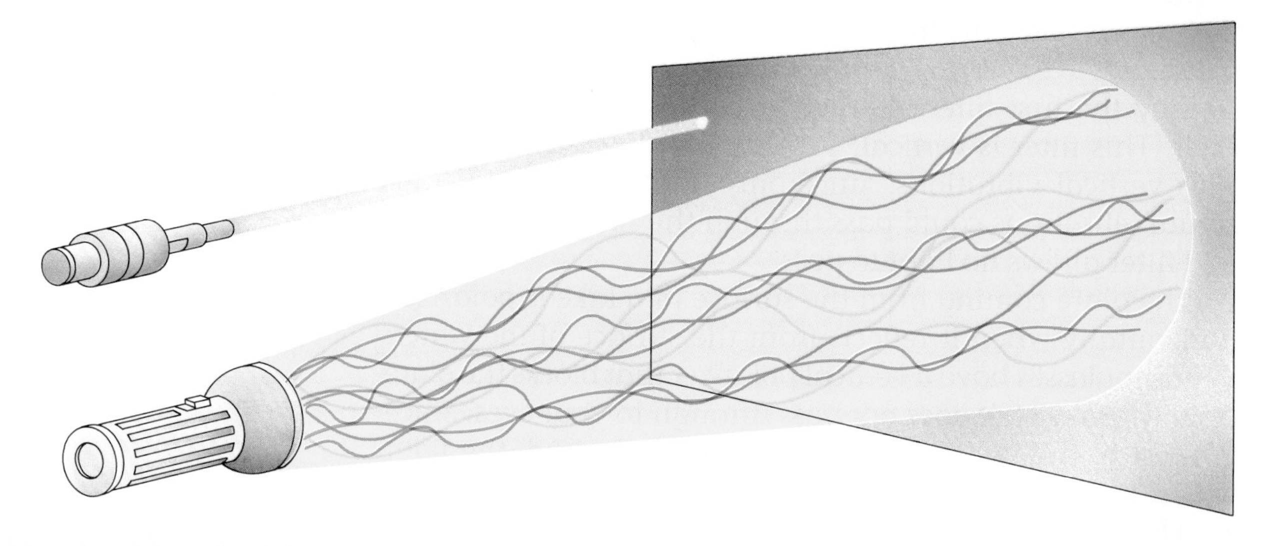

Figure 6-25 *Lasers can be used to measure the distance to satellites in orbit around the Earth.*

The beams of light from the flashlight and the laser have about the same amount of total energy, but there is a big difference between them—the energy of the laser beam is concentrated into a tiny spot. In fact, the energy of the laser beam can be concentrated even more. If a convex lens is used to focus the beam, all of its energy can be forced into a spot no more than a micrometer (0.001 mm) in diameter.

In a laser beam, all of the light has the same wavelength. For instance, if you are using a helium–neon laser, the wavelength is in the long-wave part of the visible spectrum and the light therefore looks red. Other types of lasers produce different wavelengths (and colors) of light. Although a flashlight with a colored filter may also produce pure single-wavelength light, laser light is unique because all the light waves are *in step*. That is, crests travel with crests and troughs travel with troughs. In a beam of ordinary light, many waves overlap and there is a jumble of crests and troughs. When all of the light waves in a beam are in step, the light is called **coherent light**. A laser is a device that produces coherent light.

Coherent Light Light consisting of waves that are in step.

How Lasers Work The most commonly used laser is the low-power, helium–neon laser. It produces a thin, straight beam of red light. The light has the same wavelength as the light from the red neon signs that you see in the windows of stores. Both the neon tube and the laser tube emit radiation in the form of red light.

The tube that is used to make neon signs is filled with neon gas at low pressure. As an electric current passes through the tube, it causes the gas to glow. Each atom of neon gas acts like a spring that can absorb and store energy when it is compressed. When an atom contains stored energy, it is said to be in an *excited state*. However, there is no mechanism to hold the neon atoms in their excited state, and they quickly release their energy. When atoms give off the energy resulting from an excited state, they return to their *ground state* and release the extra energy by emitting a photon of light. See Figure 6-26. When neon releases its energy, every released photon has the same wavelength. However, in the neon tube the photons are released in all directions, at all times.

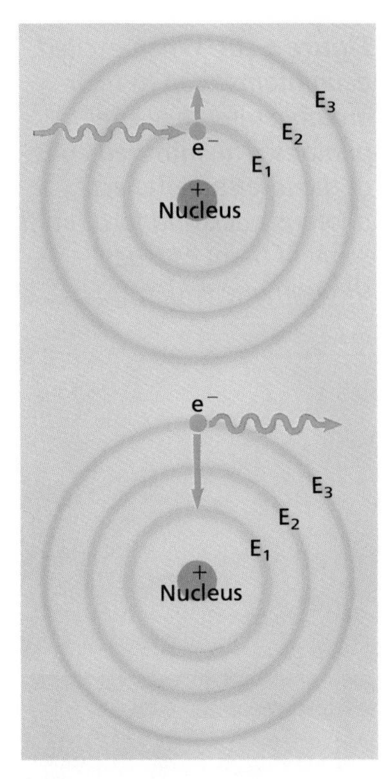

Figure 6-26 *When an electron absorbs the energy of a photon, it rises to a higher energy level* (top). *As if falls* (bottom), *it emits a photon of light.*

A neon tube can be converted into a laser because of a unique property of neon atoms. When an excited neon atom is struck by a photon of red light, the atom immediately releases another photon of red light. The two photons are identical, and they leave the atom together in the same direction and *in step*. The process of forcing identical photons into step is called **stimulated emission**. See Figure 6-27. It is from this process that the laser gets its name. The word *laser* is an acronym for *l*ight *a*mplification by *s*timulated *e*mission of *r*adiation.

Stimulated Emission A process of forcing identical photons into step in order to produce coherent light.

Stimulated emission from one photon produces two photons. If each of these photons strikes another excited neon atom, four photons are produced. Soon there are eight, sixteen, and so on. In a microsecond, there are so many photons produced that the neon tube glows brightly. The two ends of the tube in which the gas is contained are coated with a reflecting surface, which causes any photons traveling down the tube to bounce back and forth between the ends. The bouncing photons will stimulate emission from other atoms as they pass by. These newly emitted photons go in the same direction, in step. Soon there is an enormous flow of photons back and forth in the tube, all in step with one another. One of the mirrored ends has a thinner coating, which allows some of the light to pass through. From this end, five to ten percent of the coherent light escapes. This light becomes the laser beam. See Figure 6-28.

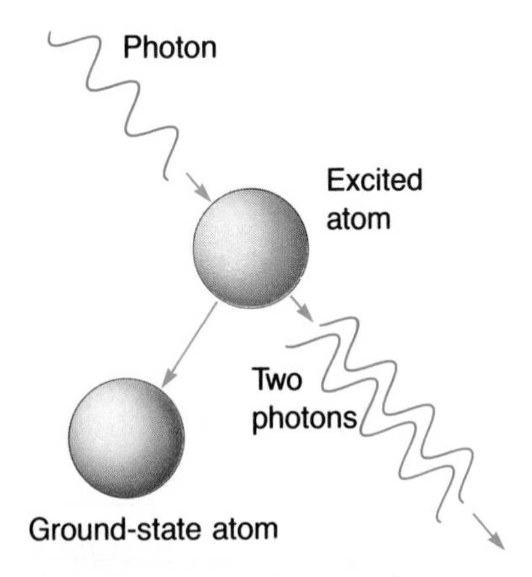

Figure 6-27 *When a photon strikes an excited atom, the excited atom will emit a photon. The two photons will then leave the atom together in step with one another.*

Figure 6-28 *When excited neon atoms drop to their ground state, they emit photons. The mirrored ends of the laser tube reflect photons back and forth until some of them escape through one end.*

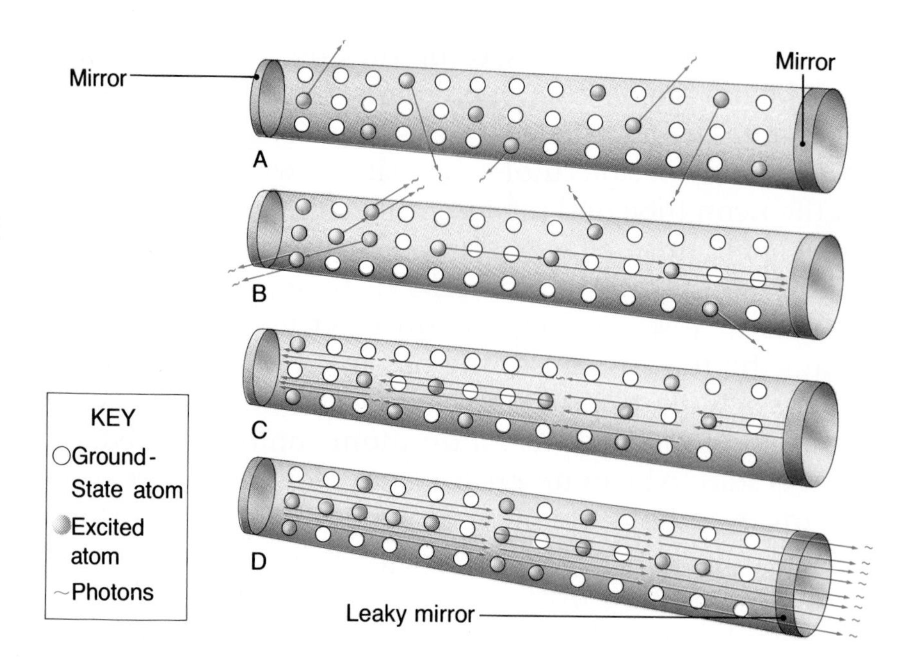

***Figure 6-29** The "coins" in this hologram are actually images on a flat piece of film.*

Holography One of the most unusual uses of laser light is in the making of *holograms.* A hologram is a picture that often looks three-dimensional, sometimes with images that seem to protrude out toward the observer. See Figure 6-29. Looking at some holograms is like looking through an open window. By looking through different parts of the window, you can see the same scene from different angles. Other holograms can be viewed from both front and back as well as all sides.

In order to make a hologram, the light from a laser is split into two beams, an *object beam* and a *reference beam.* See Figure 6-30. The object beam is reflected from a mirror, is spread out by a lens, and is then reflected from another mirror to the object being photographed. The reference beam is also spread by a lens and reflected by a mirror. However, it hits the film without ever striking the object. The two beams of light form a microscopic interference pattern of dark and light areas on the film. Light waves from every spot on the object interact with the reference beam and are recorded on

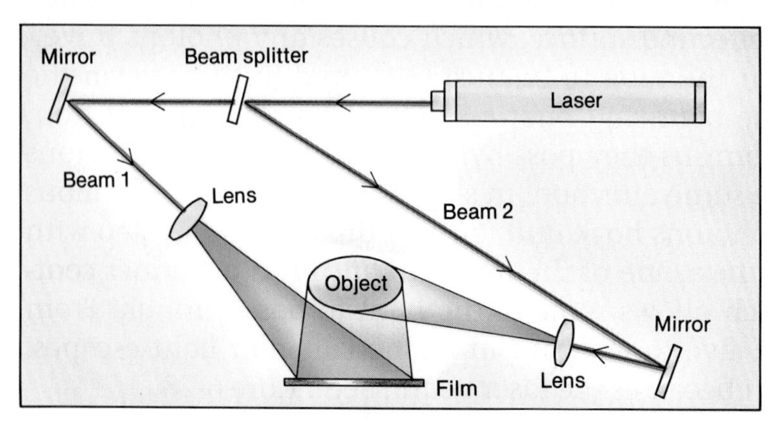

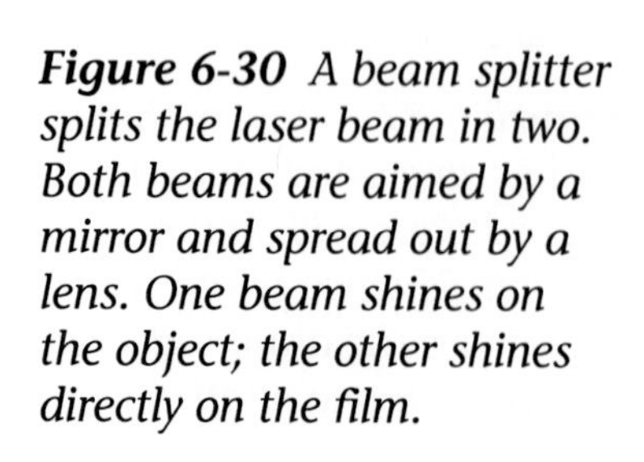

***Figure 6-30** A beam splitter splits the laser beam in two. Both beams are aimed by a mirror and spread out by a lens. One beam shines on the object; the other shines directly on the film.*

the film. When properly illuminated, this interference pattern "reconstructs" the three-dimensional image of the object.

Engineers use holograms to study how things deform under pressure. Biologists have made holograms of bacteria to study their structure. Artists use holograms to create unusual images. The largest use of holograms, however, is by the banking industry. Banks use holograms as security devices on the credit cards they issue. See Figure 6-31. In the future, it may be possible to make holographic television and motion pictures that will let you feel as if you are actually in the picture.

Figure 6-31 *This hologram would be difficult to reproduce, making it hard to counterfeit the card.*

Fiber Optics

Have you ever seen telephone-company workers installing new cables along the highway? Chances are that the cables are made not of copper wire but of fine glass fibers. What flows through these cables is not electricity, but light. See Figure 6-32. New long-distance lines transmit conversations in the form of tiny pulses of light.

Figure 6-32 *Even though the optical-fiber cable* (right) *is much smaller, it can carry much more information than the copper-wire cable* (left).

The principle that allows glass fibers to carry light has been known for a long time. When light travels through a medium in which the light travels slower than it would through air, the light may not be able to get out. If the angle of incidence at the surface is too great, the light is reflected inward. This reflection from the inside surface is called *total internal reflection.* See Figure 6-33. The phenomenon of total internal reflection is used in many kinds of optical instruments, including binoculars, telescopes, and the viewfinders of cameras.

Total internal reflection is also used in the operation of the light pipe. A *light pipe* is a clear plastic rod that can be bent into a curved shape to carry light into unreachable places. As long as the curvature is not too great, the light always strikes the surface at a high angle of incidence and

Figure 6-33 *Once the angle of the light beam becomes great enough, it is totally reflected at the surface of the water.*

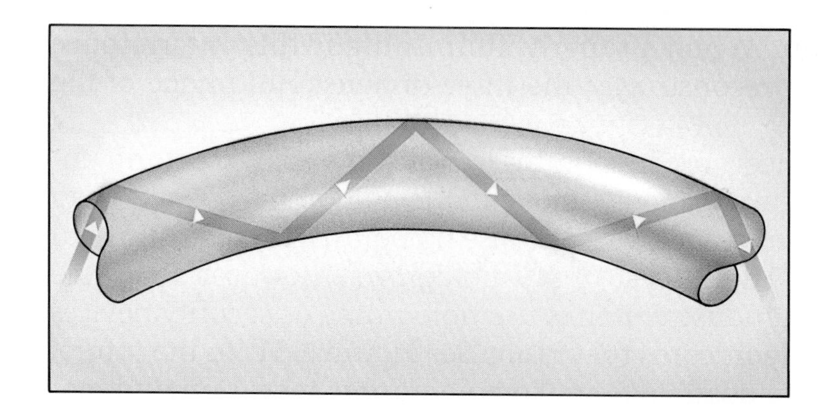

Figure 6-34 A beam of light is totally internally reflected inside a thin glass fiber.

Optical Fiber A long, thin light pipe, which uses total internal reflection to carry light.

is reflected inward. See Figure 6-34. A long, thin light pipe is called an **optical fiber**. Each fiber, made of glass and encased in plastic, is considerably thinner than a human hair. Light travels slower in glass than it does in plastic, making total internal reflection possible. Each tiny glass fiber picks up one small portion of the light from the object and carries it to the other end of the fiber. The fiber can be bent and twisted into all sorts of shapes without losing the light.

Fiber-optic telephone cables also consist of fine glass fibers encased in plastic. When you talk on the telephone, your voice is converted into an electrical signal. At a local telephone station, this signal is changed into digital code. The electric bits of the code are used to trigger a tiny laser, no bigger than a grain of salt. See Figure 6-35. The laser translates this code into a series of flashes of infrared light, which travel through the glass fiber. As you speak, the code takes up only a minute portion of the signal, leaving spaces between the flashes for your conversation. In the fiber, these intervals are filled in with bits from someone else's conversation. Since each fiber can carry 46 million bits per second, a cable of 12 fibers can carry nearly 50,000 conversations at once.

Figure 6-35 The laser used to send light in an optical fiber is very small. One such laser is shown here—lying on the eye of George Washington on this quarter.

Summary

The unique properties of light have led to innovations that affect our daily lives. Polarized glasses are able to eliminate glare because they contain filters that allow light waves vibrating in only one plane to pass. A laser is a device that can produce coherent light by forcing identical photons into step through a process of stimulated emission. Laser light can be used in the making of holograms. Optical fibers carry light and information by total internal reflection. Telephone systems use both optical fibers and lasers to greatly enhance our communications ability.

Unit 7

IN THIS UNIT

Reading *Plus*

Now that you have been introduced to the concept of particles, let's take a closer look at how scientists use particles in measurement and theory. Read pages S108 to S124 and write a story for a popular science magazine about how particles form the basis for our present understanding of matter. Your story should include answers to the following questions.

1. What is the meaning of the term *mole*? How is it used in measuring matter?
2. How do particles fit into the kinetic theory of matter?
3. How small can a substance be divided and what are the pieces called? In what way does the particle model of matter fail? How could you account for this failure?

7.1 PARTICLES OF MATTER

Particle Any small unit (of matter or energy) that, with others, forms a larger whole.

Scientists often use the term *particle* when discussing subjects such as matter, light, and energy. But just what do they mean by *particle*? A **particle** is any small unit that forms a larger whole. An atom is a particle of an element; a proton is a particle of an atomic nucleus; and a *quark* is a particle of a proton. Light consists of particles called photons. According to theory, even gravity may be composed of particles called *gravitons*. See Figure 7-1.

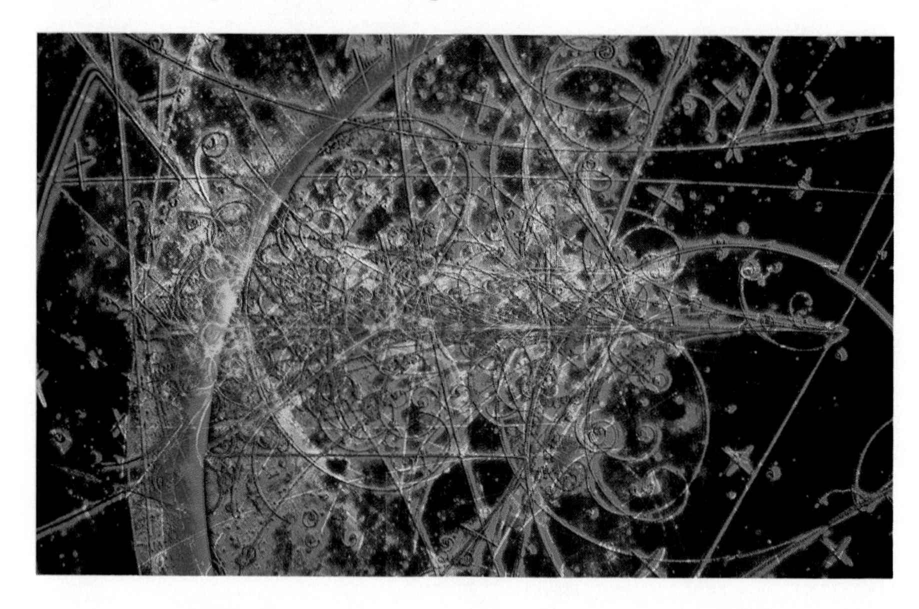

Figure 7-1 *Artificially colored bubble-chamber photograph showing tracks of subatomic particles.*

Molecules: What Are They?

Molecule A particle of matter composed of two or more atoms joined together by covalent bonds.

The word "molecule" is frequently used to describe any particle composed of two or more atoms. We often say that all compounds are made of "molecules." This, however, is not quite accurate. The term *molecule*, in the scientific sense, refers only to groups of atoms that are joined by one type of bond. You may remember that there are two distinct types of chemical bonds—ionic and covalent. It is the covalent bonds that form the particles we call **molecules**. Covalent bonds are formed when atoms share electrons. For example, water is composed of H_2O molecules in which the hydrogen and oxygen atoms share electrons. See Figure 7-2.

On the other hand, a substance such as sodium chloride (table salt) is not composed of molecules. Rather it contains particles called *ions*. Ions are formed when electrons are transferred between atoms. Atoms that lose or gain electrons take on either a positive or a negative charge. Ions with opposite charges are then attracted to each other. The positive ions of sodium (Na^+) and negative ions of chlorine (Cl^-) join together with ionic bonds to form a crystal. Each sodium ion

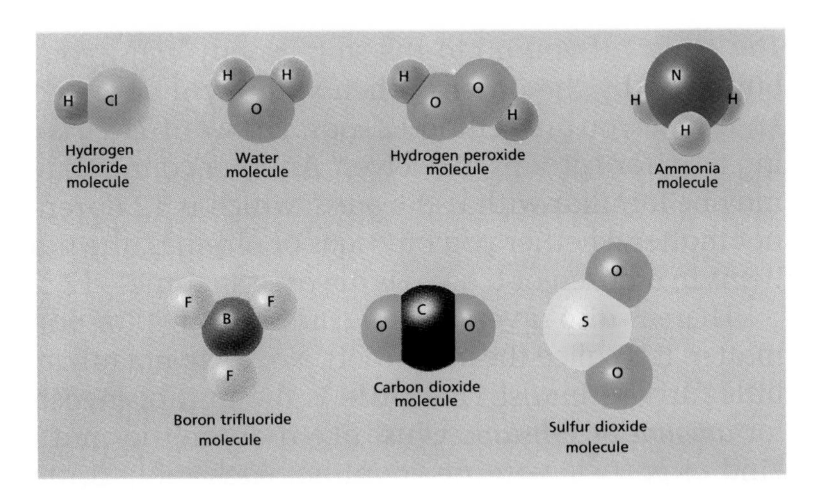

Figure 7-2 *Models of molecules of various chemical compounds.*

is surrounded by chloride ions, which in turn are surrounded by sodium ions, as shown in Figure 7-3. Since these ions do not "share" electrons with any particular neighbor, their groupings are not considered true molecules. The simplest group consisting of one Na^+ ion and one Cl^- ion is therefore called a **formula unit**. Since we cannot isolate one "molecule" of salt, the designation NaCl is simply a means of writing formulas and equations in which sodium chloride takes part.

Formula Unit The simplest unit indicated by the formula of an ionic compound.

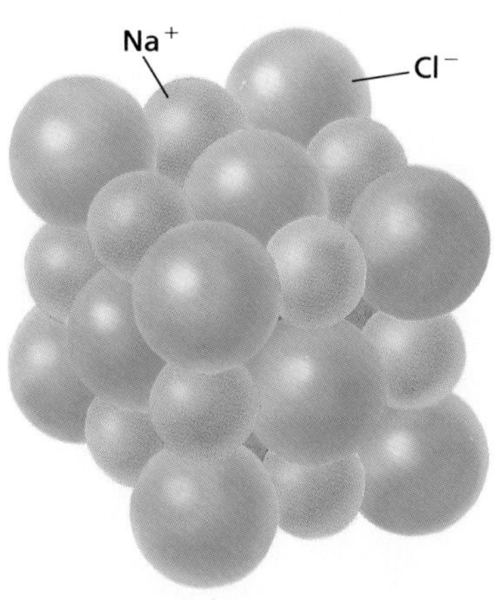

Figure 7-3 *The shape of salt crystals displays the arrangement of its ions.*

You should now be able to understand the usefulness of the term "particle." We can talk about particles of sugar (sugar molecules) and particles of salt (Na^+ and Cl^- ions) as well as the properties they share, without the necessity of considering the details. This comes in handy in a number of areas of study.

Counting Particles

No matter what the identity of the particular particles under study, certain characteristics are shared among them. One of these is the ability to be counted. Particles can either be counted individually or in groups. But since particles of matter are so small, we need a special way to count them.

The Mole If you go to the store to buy eggs, you usually buy them by the dozen. When you buy a dozen eggs, you know that you are getting 12 eggs. The word *dozen* is a counting term for "groups of twelve." Another counting term you may be familiar with is the *gross*, which is 12 dozen. It does not matter whether you buy eggs or oranges, the number of items in a dozen or a gross is always the same—12 or 144.

Mole The SI unit for the amount of a substance.

There is also a very useful counting term for particles of matter. It is called the *mole* (mol). No, we're not talking about little furry animals! The **mole** is the unit of measurement for *amount of substance*. Just like a dozen, no matter what kind of particle you are counting, each mole contains the same number of particles.

Avogadro's Number Since the particles of matter are so small, the number of particles in a mole is very large. But just how many particles is this? The number of particles in one mole of any substance has been determined to be 6.022137×10^{23}. This number is called **Avogadro's number** in honor of the Italian scientist Amedeo Avogadro.

Avogadro's Number The number of particles in one mole of a substance (6.022137×10^{23}).

While one dozen is only a small value, the size of Avogadro's number is quite large. Written out, it comes to:

602,213,700,000,000,000,000,000

If you could imagine getting the help of all 5 billion people on Earth to count the number of atoms in only 1 mole, at the rate of 1 atom per second, it would still take everyone almost 4 million years! Put another way: If you made $40,000 every *second* at your job and you had worked since the world was formed 4.5 billion years ago, you would not yet have earned Avogadro's number of *pennies*. See Figure 7-4.

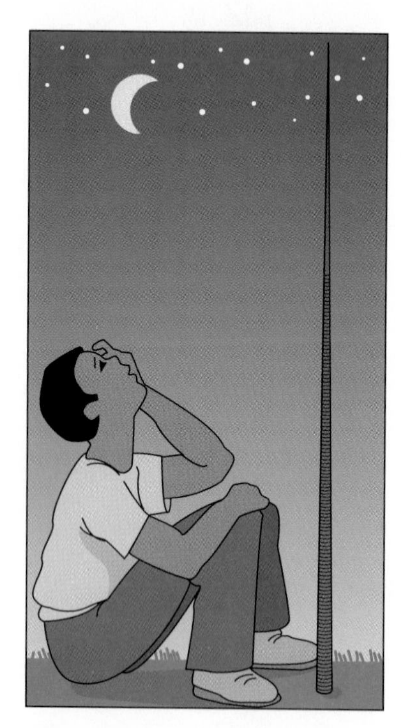

Figure 7-4 *If 6.02×10^{23} pennies were stacked, one on top of another, the stack would reach to the other side of the Milky Way.*

Using the Mole Using the mole is like using any other standard quantity, such as a dozen or a gross. Suppose you were in charge of breakfast for a class campout and you wanted to serve breakfasts with two eggs and one sausage patty on each plate. Thus you could order, say, 6 dozen eggs and 3 dozen sausage patties. This way, there would be just enough eggs and just enough sausages so that none would be left over when all the breakfast plates were served.

In a similar way, chemists generally determine the amounts of chemical substances they will use in terms of moles. For example, an experiment might call for 1 mol of copper or 2 mol of potassium. Just as you were sure the number of eggs to sausages would match, chemists know the number of atoms will match by using moles. But how would you translate these instructions into units you are already familiar with? To understand the answer to this question, you will need to refer to the Periodic Table.

Moles to Grams If you look on the Periodic Table for copper (Cu), you will find that its atomic mass is 64. This value

Atomic Mass

5 **B** Boron 11	6 **C** Carbon 12	7 **N** Nitrogen 14	8 **O** Oxygen 16	9 **F** Fluorine 19
13 **Al** Aluminum 27	14 **Si** Silicon 28	15 **P** Phosphorus 31	16 **S** Sulfur 32	17 **Cl** Chlorine 35

Figure 7-5 You can determine the number of grams in 1 mol of any element by finding its atomic mass on the Periodic Table.

tells you the number of grams in 1 mol of that element. In other words, there are 64 g of copper in 1 mol of copper. Likewise, the atomic mass of potassium (K) is 39, so 1 mol of potassium has a mass of 39 g. As you might expect, 2 mol of potassium would be twice this amount, or 78 g. To find out the number of grams in 1 mol of any particular element, just look up its atomic mass and use it as the number of grams. See Figure 7-5.

Let's use eggs again to illustrate how moles relate to grams. Suppose you have three sizes of eggs—small (30 g), medium (40 g), and large (50 g). If you have a dozen of the small eggs, you would have a total of 360 g. A dozen medium eggs would be 480 g, and a dozen large eggs would be 600 g. If it was true that all small eggs had a mass of 30 g, you could order 360 g of small eggs and expect to get 1 dozen.

In the same way, scientists use grams to count out the number of moles they would like to use. For example, if you had three elements—carbon, iron, and copper—each type of atom would have a different mass. Since the atomic mass of carbon is 12, a mole of C atoms equals 12 g. In the case of iron, which has an atomic mass of 56, it would take 56 g to equal 1 mol (or 6.02×10^{23} atoms) of Fe. In other words, if you know the atomic mass of the atoms of an element, you can determine how many grams of that element are needed to get 1 mol of its atoms. See Figure 7-6.

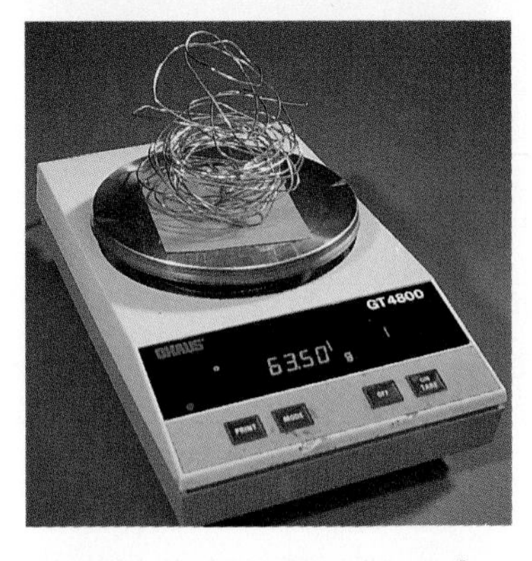

Figure 7-6 Approximately one mole of each of three substances: carbon (top), *iron* (middle), *and copper* (bottom).

So far, we have only talked about moles of *elements*. But *compounds* can also be measured in moles. Remember that atoms are not the only particles that can be counted with moles or Avogadro's number. You can also have a mole of molecules, a mole of ions, or a mole of electrons. Recall that a molecule of water (H_2O) has two atoms of hydrogen and one of oxygen, as indicated by its formula. Therefore, 1 mol of water molecules would contain 2 mol of hydrogen *atoms* and 1 mol of oxygen *atoms*. Nevertheless, there are only 6.02×10^{23} *molecules* in a mole of water.

To determine the number of grams in 1 mol of the compound water, for example, we would first find the atomic mass for each atom, and then add the atomic mass of each of the atoms in one molecule together: $1 + 1 + 16 = 18$. Therefore, by applying the same rule as with moles of elements, there are 18 g in 1 mol of water.

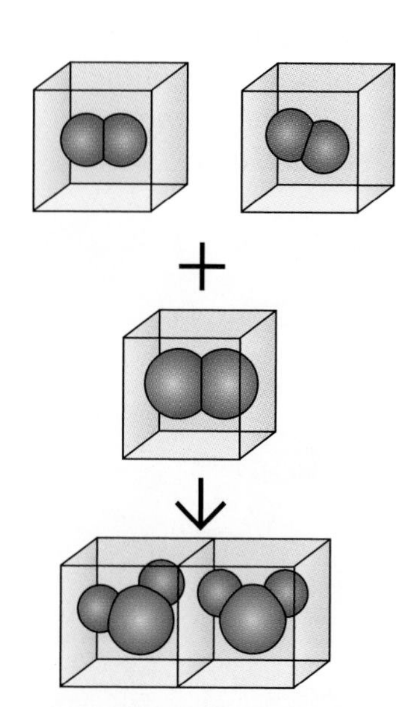

Figure 7-7 *Two molecules (or moles) of H_2 plus 1 molecule (or mole) of O_2 yields 2 molecules (or moles) of water.*

The Mole in Chemistry Knowing the mass of a mole of any substance is very important to chemists. Suppose, for example, you wanted to react hydrogen with oxygen to get water according to the following balanced chemical equation.

$$2H_2 + O_2 \rightarrow 2H_2O$$

The equation states that two *molecules* of H_2 react with every *molecule* of O_2. That is, you would need twice as many H_2 molecules as O_2 molecules in your experiment. See Figure 7-7. This is where moles come in. If you use twice as many moles of H_2 as you do of O_2, you will have the right proportions of the two chemicals. Suppose you use 10 mol of H_2 and 5 mol of O_2; how much would these be in grams?

Atomic mass of H	=	1	Atomic mass of O	=	16
So, 1 mol of H_2	=	2 g	So, 1 mol of O_2	=	32 g
and 10 mol of H_2	=	20 g	and 5 mol of O_2	=	160 g

Therefore, 20 g of H_2 will completely react with 160 g of O_2 to give you 180 g (10 mol) of water.

Summary

Scientists often use the term "particle" when discussing various phenomena in nature. The particles called molecules are formed by covalent bonds between atoms, while formula units describe the simplest grouping of ions in a compound formed by ionic bonds. A mole of any one substance has exactly the same number of particles as a mole of any other substance, which is equal to Avogadro's number. The mole is a very useful unit of measurement for counting these particles, whether they be molecules, atoms, or ions.

7.2 PARTICLES IN MOTION

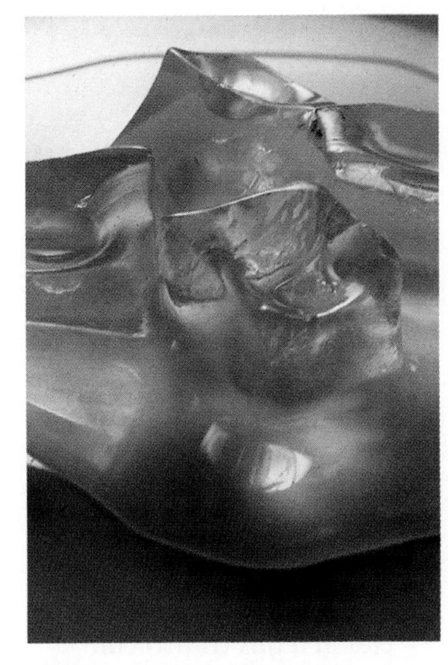

Figure 7-8 Ice, like most solids, has a definite melting point.

Joseph Black, the Scottish chemist who first identified carbon dioxide gas, devoted much of his life to the investigation of how heat affects matter. Among the phenomena he studied were the freezing, melting, and boiling points of different substances. See Figure 7-8. In one experiment, Black measured the temperature of water as it was heated to a boil. Of this experiment he wrote:

> The liquid gradually warms, and at last attains the temperature, which it cannot pass without assuming the form of vapor. In these circumstances, we always observe that the water (or other liquid) is thrown into violent agitation, which we call boiling.... Another peculiarity attends this boiling of liquids, which, when first observed, was thought to be very surprising. However long and violently we boil a liquid, we cannot make it hotter than when it began to boil. The thermometer always points to the same degree, namely, the vapor point of that liquid. Hence the vapor point of a liquid is often called its boiling point.

The Kinetic Theory of Matter

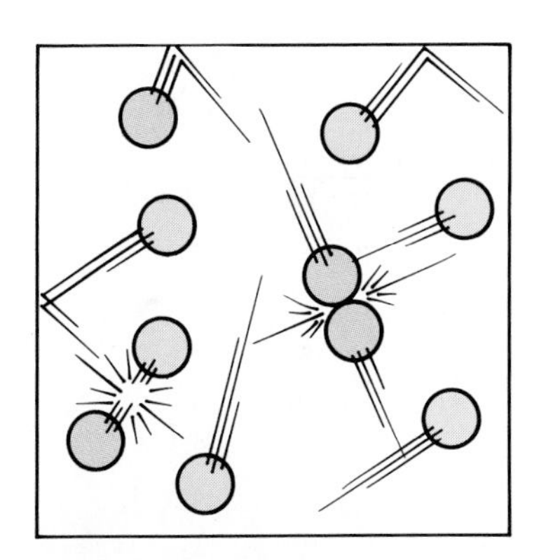

Figure 7-9 Particles with a large amount of kinetic energy may even collide with one another.

The kinetic theory of matter is a scientific model that helps to explain many of the physical properties of matter. It is one of the most important theories in modern science. The basic concept of the kinetic theory of matter is that *matter is composed of very tiny particles that are in constant motion.* These particles are, of course, atoms and molecules. As you have learned, kinetic energy is energy of motion. Atoms and molecules have kinetic energy because they are constantly moving. The amount of kinetic energy an atom or molecule possesses is related to how fast it moves. See Figure 7-9. The kinetic theory of matter, then, is about particles in motion.

Cohesion What holds a piece of paper, a steel beam, or a raindrop together? As you know, the property of matter that holds solids and liquids together is called *cohesion.* **Cohesion** is the tendency for similar particles of matter to stick to each other. Cohesion is a force of attraction between particles. Therefore, the properties of cohesion can be discussed in terms of the kinetic theory of matter.

Cohesion The force of attraction between like particles.

In a solid, for example, the atoms or molecules cannot move about freely. They are only free enough to vibrate back and forth about fixed positions. They do not have enough kinetic energy to overcome the cohesive forces that hold them together. For this reason, solids have a definite shape and a definite volume. The arrangement of molecules in their fixed positions is often very orderly, which gives the solid a crystal shape.

In a liquid, the strength of cohesion between its particles is generally less than that in a solid. While the particles of a liquid are still relatively close together, they are free to move around or roll over each other, much like marbles in a plastic bag would roll around if you squeezed the bag. Thus, because of this characteristic, liquids have a definite volume but no definite shape. Their structure is dictated by the shape of the container in which they are held.

Gas particles, on the other hand, move so quickly that they are able to overcome cohesion completely. They fly off in straight lines until they collide with other particles. For this reason, gases do not have definite shapes—they expand to fill whatever container they occupy. Gas particles are separated by an average distance of about 10 times the diameter of the particles themselves. Figure 7-10 shows how particles move in a solid, a liquid, and a gas.

Figure 7-10 *Molecules in a* solid (left) *vibrate in a fixed position. Liquid molecules* (center) *can move around but are always in contact with other molecules. Particles in a gas* (right) *can freely move from place to place.*

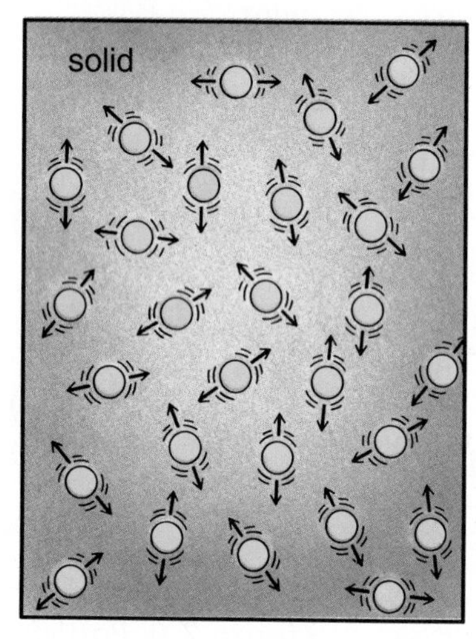

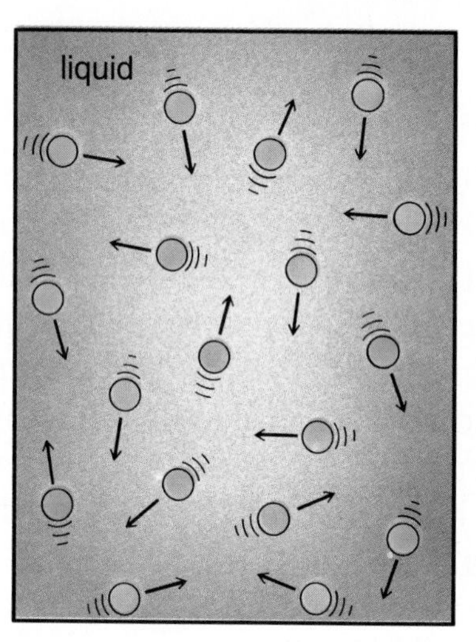

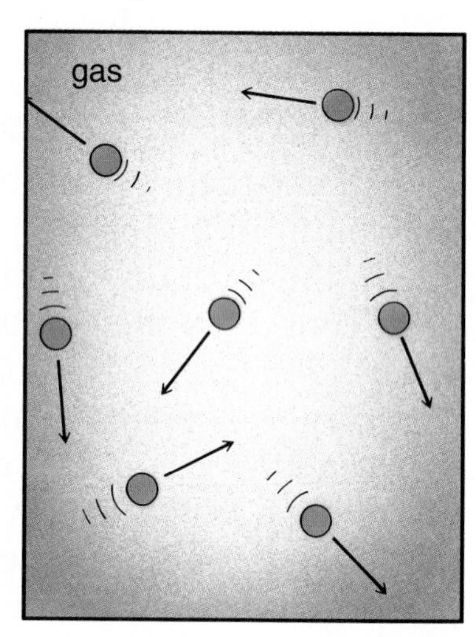

Changes of State As described at the beginning of this section, Joseph Black discovered that the temperature of a liquid at its boiling point remains constant even when it is heated further. The additional heat energy increases the kinetic energy of the individual particles, thus changing the liquid into a gas. Likewise, adding heat to a solid increases the kinetic energy of its particles. When the particles acquire enough energy to break the forces holding them, the solid melts. In other words, heating overcomes the cohesive forces between the particles of liquids and solids. Thus, the kinetic theory of matter explains why all liquids have definite boiling points and why all solids have definite melting points.

What happens when a gas or liquid cools? As heat energy leaves a substance, its particles lose kinetic energy and slow down. Because they have less energy, the attractive forces between the particles draw them closer together. In this situation, we say that gases *condense* and liquids freeze. The condensation point of a gas is the same as its boiling point. Likewise, the freezing point of a liquid is the same as its melting point.

The kinetic theory can illustrate such properties as the *sublimation* of dry ice. Molecules on the surface of dry ice (frozen carbon dioxide) acquire enough heat energy from the surrounding air to break loose and enter the gaseous state. These molecules go directly from the solid to the gaseous state without becoming a liquid. See Figure 7-11.

Figure 7-11 *Dry ice goes directly from its solid state to a gas by the process of sublimation.*

Diffusion The kinetic theory of matter can also explain the property of matter called **diffusion**, which is the mixing of molecules of two or more substances as a result of their motion. You have already learned about how diffusion operates in living systems. The process of diffusion indicates that particles are in constant motion. Because gas molecules move very quickly, diffusion occurs quickly in gases. For example, if you open a bottle of strong perfume inside a room, the perfume molecules will quickly diffuse through the air. The fast-moving air molecules help to distribute the perfume molecules, and the odor of perfume soon becomes apparent all over the room.

Diffusion The mixing of particles of one type of substance with the particles of a second type of substance due to the motions of their respective particles.

If you place a drop of food coloring into a glass of water, the particles of the food coloring will move between the particles of the water until they are equally distributed throughout. See Figure 7-12. The kinetic theory also explains why the temperature of the water influences the rate of diffusion. Since particles of warmer water move faster than those of cooler water, diffusion should occur faster at higher temperatures. This is indeed the case.

Figure 7-12 *Food coloring and water gradually diffuse to form a uniform mixture.*

Figure 7-13 *The surface tension of water particles causes water to form spherical drops.*

Diffusion even occurs between the molecules of solids. For example, if you leave a sheet of lead on top of a sheet of gold for several months, a very small amount of diffusion will occur, and the particles of one solid can be detected in the other. However, because molecules of solids are not free to move around quickly, diffusion occurs very slowly in solids.

Particles of Liquids

In liquids, as you know, the molecules are able to move around each other, while still remaining close together. Thus, liquids may change their shape but still take up a certain volume. Liquid water, for example, can be poured from a tall, narrow glass into a short, wide glass, but will occupy the same volume in both glasses. However, certain aspects of the shape of a liquid are still due to the attractive forces of the individual particles of the liquid itself.

Surface Tension Have you ever watched water dripping from a leaky faucet? See Figure 7-13. The water does not drip molecule by molecule. Rather, cohesion between the water molecules keeps them together until a large drop forms. Just as the drop falls, it becomes a sphere. Raindrops also tend to be spherical. But without a container, what keeps these liquid drops in this shape? The spherical shape of falling water drops results from a property of liquids called **surface tension**. Surface tension is caused by the cohesive forces between molecules at the surface of a liquid. Look at the diagram of a container of water in Figure 7-14. A

Surface Tension The tendency of the particles at the surface of a liquid to pull together.

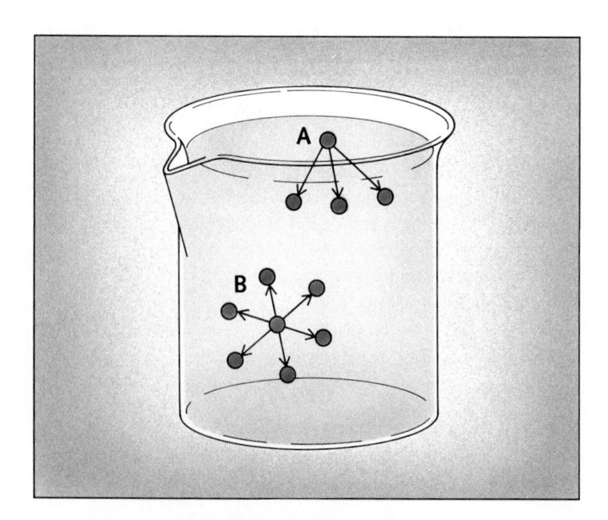

Figure 7-14 *Surface particles have a net downward attraction that draws them together.*

molecule at **B** is attracted to other water molecules on all sides. A molecule on the surface of the water (position **A**) is attracted to other water molecules beneath it as well as next to it. There are no water molecules to attract it upwards. Thus, the molecules at the surface pull close together, causing them to act like a thin elastic film. It is surface tension that allows some insects to stride across the surface of a pond without sinking into the water. See Figure 7-15.

Surface tension also explains why falling water drops are spheres. Look at the diagram of the water droplet shown in Figure 7-16. Molecules at the surface of the droplet are pulled inwards by the cohesive forces of the water molecules inside the drop. The cohesive forces between surface molecules and those within the drop are what causes the droplet to take on a spherical shape.

Figure 7-15 *Surface tension allows this water strider to walk on water.*

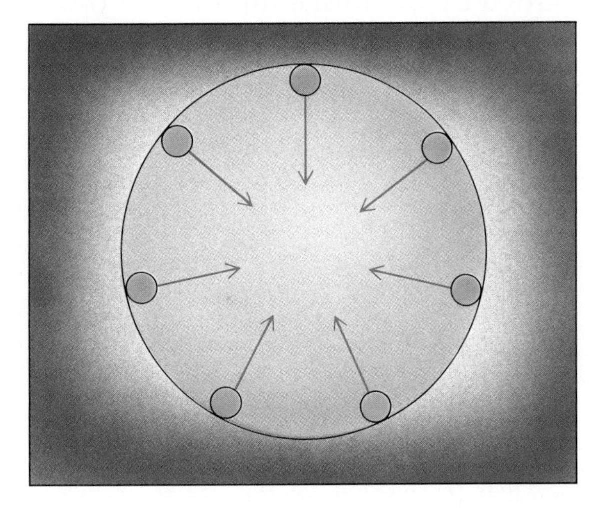

Figure 7-16 *In a drop of water, all surface particles are drawn toward the center of the drop, thus giving it a spherical shape.*

Adhesion When you place a glass tube in water, the water will flow up the tube, against the force of gravity. If you look closely, you will see that the surface of the water in the tube is not level. The water is drawn up the side of the tube slightly so that the surface of the water forms a curve. This curve is called a *meniscus*.

Adhesion The force of attraction between unlike particles.

Water and other liquids behave in this way because of two forces—*cohesion* and *adhesion*. You know that cohesion is the attractive force between particles of the same substance. **Adhesion**, on the other hand, is the attractive force between the particles of two different substances. For example, when water comes in contact with glass, the water and glass particles attract one another. See Figure 7-17.

Figure 7-17 *Water molecules stick to glass, as well as each other, forming drops on the window.*

Pressure in a Liquid Blaise Pascal was a French philosopher, mathematician, and scientist who lived during the 17th century. Among other things, he is known for his work on pressure in liquids. ***Pascal's law*** describes the effects of applying pressure on a liquid in a closed container. This law states that *whenever pressure is increased at any point, the change*

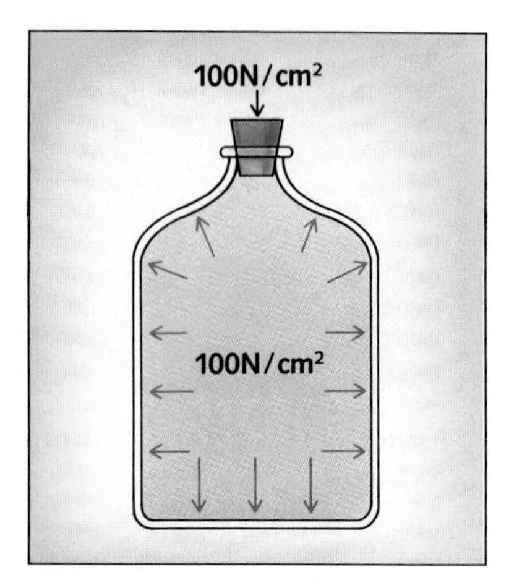

***Figure 7-18** Pascal determined that when pressure is increased at one point in a contained liquid, pressure is dispersed equally throughout the liquid.*

in pressure takes place equally throughout the liquid. Since the particles of liquids are already close together, they cannot be compressed in the same way that gases can. This explains why a corked bottle, when completely filled with water, breaks if the cork is suddenly pushed down. See Figure 7-18.

A hydraulic lift works on this same principle. See Figure 7-19. Since the output piston has a larger area than the input piston, a small force exerted on the input piston creates a larger force on the output piston. If the area of the output piston is 20 times that of the input piston, the force is multiplied by 20. The brake system of an automobile also uses this principle. A small pressure on the brake pedal is transmitted by the brake fluid to the brake pads where the force is multiplied many times, enabling the driver to bring the car to a stop.

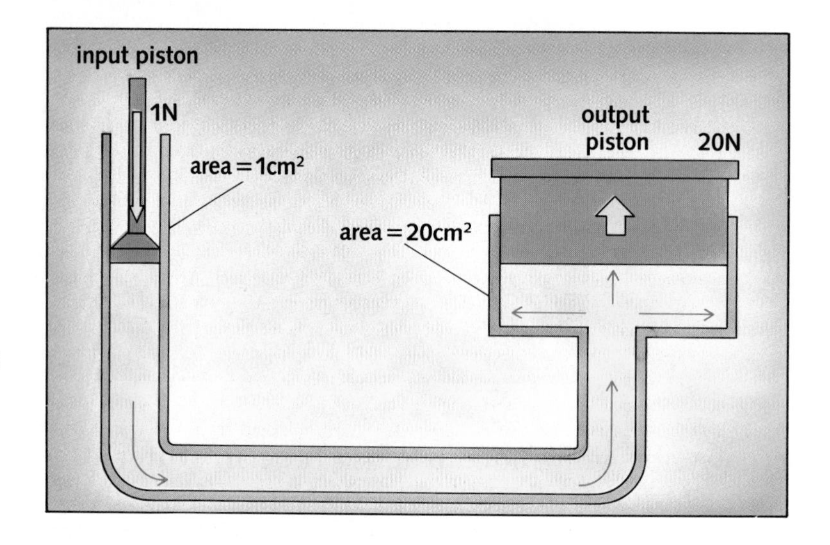

***Figure 7-19** The incompressibility of liquids is the basis of a hydraulic lift. How do the initial and resulting forces compare in this lift?*

Gases: A Special Case

Unlike solids and liquids, gases do not have definite volume. The volume of a gas is related to its temperature and pressure. For example, if the handle of a bicycle pump is pushed down, the volume of the gas inside the pump is decreased and the pressure is increased. If a balloon full of air is left in the bright sunlight, the balloon expands as the air inside warms up.

The behavior of gases can be described by general laws because gas particles are, on the average, very far apart. Each particle is independent of its neighbors. Only conditions that affect the way the particles move—such as temperature and pressure—will change the volume occupied by gases. During the 17th and 18th centuries, scientists began to question exactly how and why the volume of a gas changes.

Boyle's Law Robert Boyle was an Irish scientist who lived during the 17th century, around the same time as Newton. He studied how gases behave when the pressure on them changes. Robert Boyle experimented with several gases, such as oxygen and carbon dioxide. In each case, he found that *doubling* the pressure always reduces the volume by *one-half.* See Figure 7-20. His observations led to what is now called ***Boyle's law***, which can be stated as follows: *The original volume (V_1) occupied by a certain amount of gas multiplied by its pressure (p_1) is equal to its new volume (V_2) times its new pressure (p_2)*, or

$$V_1 \times p_1 = V_2 \times p_2$$

For example, suppose the air trapped in an air piston has a volume of 17.5 mL (V_1) when under a pressure equal to 4.7 atm (p_1). If the total pressure is increased to 7.7 atm (p_2), its new volume (V_2) is 11.7 mL.

$$17.5 \text{ mL} \times 4.7 \text{ atm} = V_2 \times 7.7 \text{ atm}$$

$$V_2 = \frac{82.25 \text{ mL·atm}}{7.7 \text{ atm}}$$

$$V_2 = 10.7 \text{ mL}$$

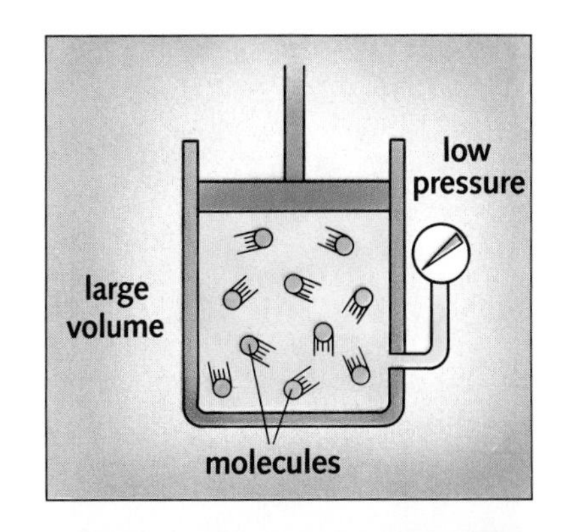

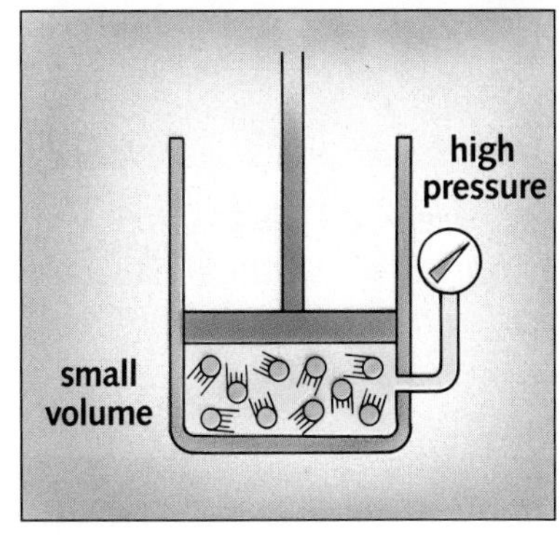

Figure 7-20 As pressure increases, volume decreases.

The kinetic theory of matter can be used to explain Boyle's law in the following way. As the force causing the pressure pushes down on the air inside the piston, the air inside the piston pushes upwards with an equal force. Otherwise, the piston would go all the way down to the bottom of the cylinder. The pressure exerted on the bottom of the piston to hold it up is a result of the constant bombardment by the trapped air particles inside. Each time the volume occupied by the air inside the piston is reduced by *half,* the particles of air are *two times* closer together. Thus, twice as many particles strike the piston per second. See Figure 7-21.

1 atm = atmosphere pressure at sea level (101.3 kPa)

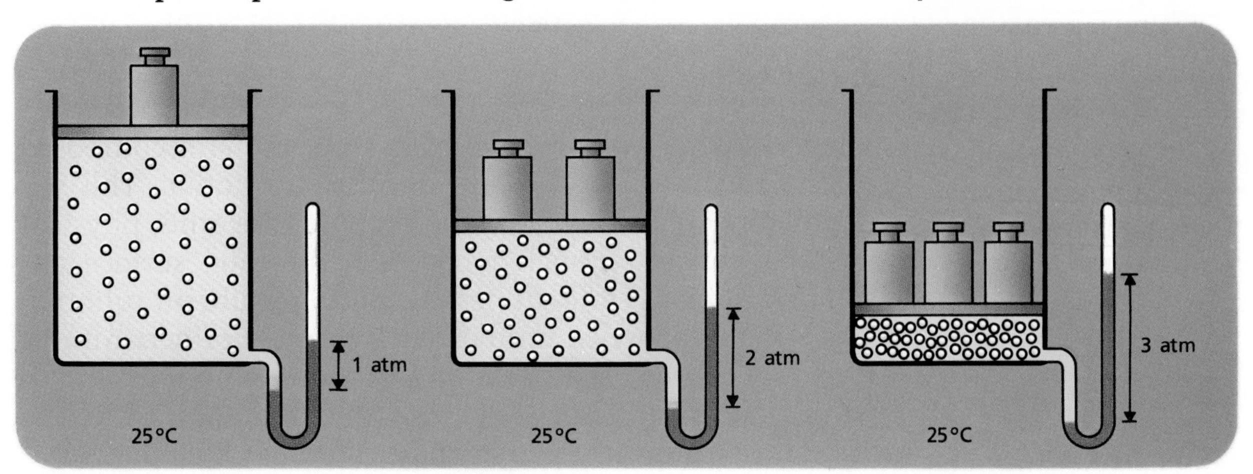

Figure 7-21 At constant temperature, the volume of a gas will be reduced in proportion to the increase in pressure.

Figure 7-22 By using Charles' law, balloonists can calculate the amount of air necessary for lift.

Charles' Law Boyle's law applies only when the temperature of a gas is constant. Now suppose you warm or cool the gas while keeping it under constant pressure. How is volume affected? In the late 18th century, a French scientist named Jacques Charles observed that the volume of air increases steadily as the temperature goes up. See Figure 7-22. **Charles' law** states that: *The volume of a gas increases regularly as the temperature increases if the pressure remains the same.*

Most gases differ very little in the percentage of increase in volume per degree of temperature rise, however. For example, when heated from 0 °C to 1 °C, almost all gases expand about 1/273 (0.37%) in volume. Suppose that you raise the temperature twice this amount, or 2 °C, from 0 °C to 2 °C. How will volume change? If the volume changes by 1/273 with a 1 °C change in temperature, then according to Charles' law, the volume must change twice as much, by 2/273, with a 2 °C change.

When using the Kelvin temperature scale, this relationship is expressed as a direct proportion. In other words, if you double the Kelvin temperature, you double the volume. With temperature in kelvins, Charles' law can be written as:

$$\frac{V_1}{T_1} = \frac{V_2}{T_2}$$

Just as in using Boyle's law, if you know three values, you can calculate the fourth.

How can the kinetic theory be applied to Charles' law? If the temperature of a gas is increased, the gas particles move faster and collide with one another more often. Because of this, they are exerting more force. In order to keep the overall pressure constant, gas particles must travel a greater distance, and therefore the volume they occupy increases. See Figure 7-23.

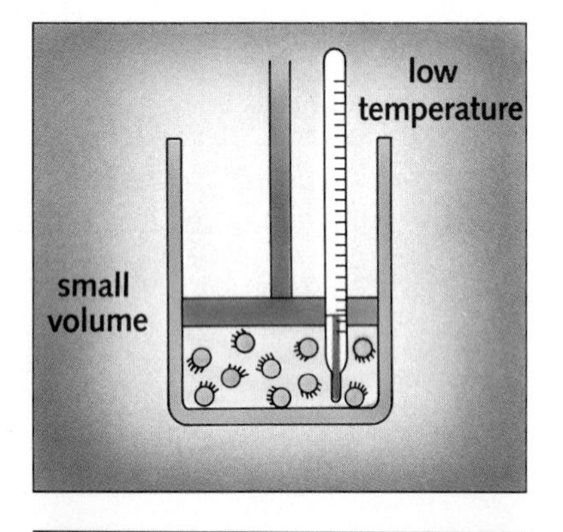

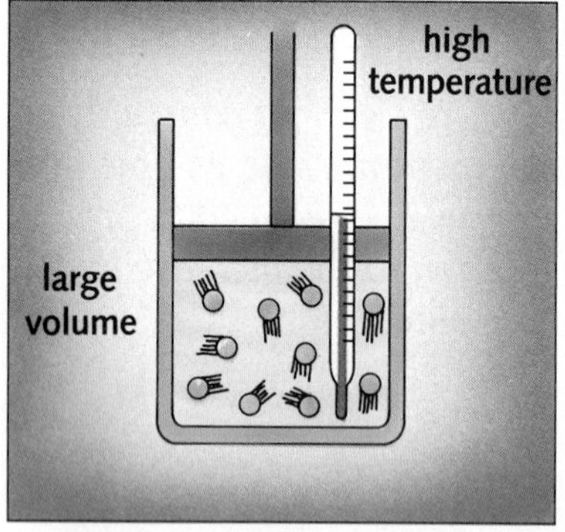

Figure 7-23 At constant pressure, as the temperature increases, the volume of a gas also increases.

Summary

The kinetic theory of matter states that matter is composed of very tiny particles that are in constant motion. This scientific model helps to illustrate many of the other physical properties of matter, such as cohesion, diffusion, surface tension, changes of state, adhesion, and pressure on a liquid. The kinetic theory of matter leads to a set of general gas laws that can be used to explain how doubling pressure always reduces the volume of a gas by one-half and why the volume of a gas increases steadily as temperature goes up.

7.3 PARTICLES OF PARTICLES

Each element consists of its own unique atoms—the smallest particles of matter that we usually deal with in the field of chemistry. But are atoms the smallest particles to be found in nature? The answer is no. See Figure 7-24. Among the other particles that you have come across in your investigations of matter are protons, neutrons, and electrons. It is in the realm of physics that these particles are usually considered.

The current view of the universe considers *matter* to be the most basic entity. This means that every substance in existence must be composed of fundamental particles of matter. Until the late 19th century, matter was thought to be formed of indivisible atoms. Then, due to the work of J.J. Thomson, Ernest Rutherford, and James Chadwick, a new model of the atom—composed of protons, neutrons, and electrons—was formed.

More recently, smaller particles have been identified. A proton, for example, is now considered to be made of three particles called *quarks*. Although no quarks have ever been isolated, their existence has been inferred from the interactions of atoms in huge machines called particle accelerators. See Figure 7-25.

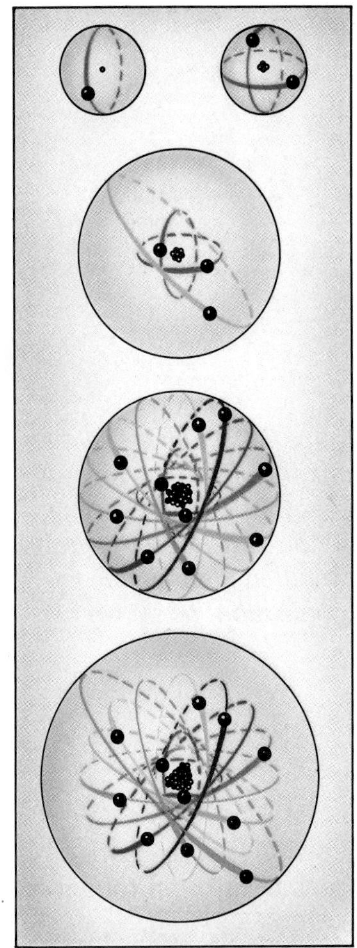

Figure 7-24 *Atoms consist of smaller pieces called subatomic particles.*

Figure 7-25 *This particle accelerator, at Fermilab in Batavia, Ill., is used to explore the interior of atoms.*

Particle Interactions

Scientists have used their knowledge of subatomic particles and the forces involved in their interactions to explain the behavior of matter and the particles of which it is composed. The primary force affecting stars, planets, and most of the visible objects around you is *gravity*. It is this force of gravity that pulls everything toward the Earth and keeps you from flying off into space. According to present theories, however, there are three other forces that also help to hold everything together. See Figure 7-26.

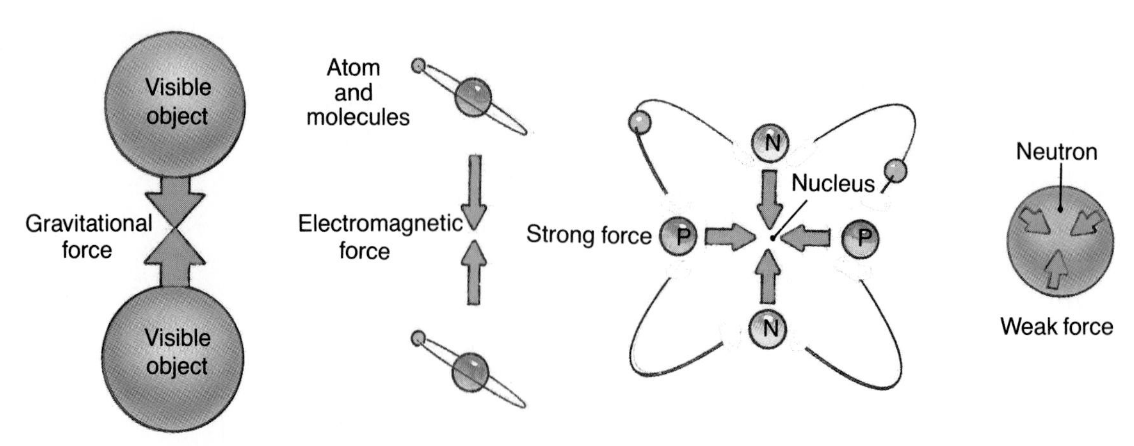

Figure 7-26 Some scientists think that there are four forces that hold everything together. Each force operates at a different level of organization, as shown here.

At the level of atoms and molecules, the *electromagnetic force* is more important than gravity. Atoms stick together to form molecules because of electric attraction between negatively charged electrons and positively charged nuclei. Recall that like charges repel and unlike charges attract. The theory explaining this electromagnetic force is called *quantum electrodynamics*, or QED.

The force that holds an atomic nucleus together is called the *strong force*. As you know, atomic nuclei are made up of the small particles called protons and neutrons. However, these particles are made of still smaller particles called **quarks**. Scientists think that protons and neutrons each consist of three different quarks. See Figure 7-27. The strong force holds the quarks inside each proton and neutron together. The strong force also holds the nucleus itself together. This means that the strong force is stronger than the electromagnetic force that tends to push protons, which have like charges, away from each other.

Quark A theoretical particle that forms protons and neutrons.

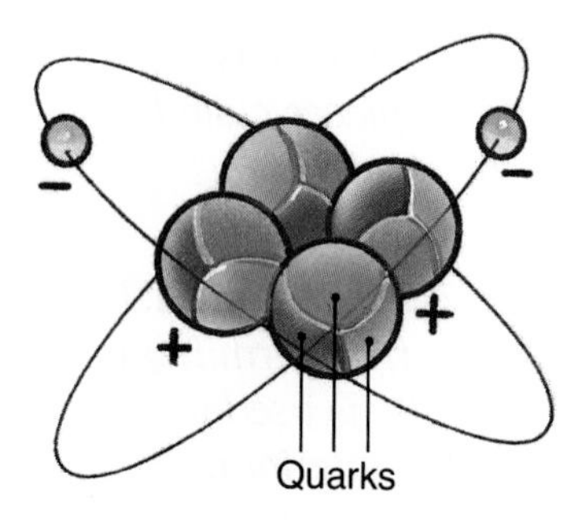

Figure 7-27 This model of a helium atom shows that protons and neutrons are made of smaller particles called quarks.

The fourth force is called the *weak force*. The weak force is responsible for several types of radioactive decay. This force also helps to hold particles such as neutrons together. In weak-force interactions, particles called *neutrinos* are either produced or absorbed. Neutrinos are very small. In fact, physicists are still trying to determine whether they have any mass at all. As indicated by their name, neutrinos do not have any charge.

Quantum Theories

The theories explaining the electromagnetic, strong, and weak forces are all called *quantum* theories. All quantum theories assume that only certain-sized packages, or **quanta**, of energy can be transferred when particles interact. This transfer of energy can be compared to buying eggs at the grocery store. You can buy one egg or a dozen eggs, but you

Quanta Tiny packages of energy.

cannot buy 2.7 eggs. See Figure 7-28. You have already learned about one type of quanta, the photon, which is a quantum of light energy.

The fact that the electromagnetic, strong, and weak forces can all be described by quantum theories has led many scientists to try to combine the theories into a single theory that explains them all. One of these theories, called the *electroweak theory*, combines electromagnetic and weak forces. The electroweak theory explains everything that is explained by the two separate theories and has even successfully predicted the existence of some previously unknown particles.

Since the time that the electroweak theory was proposed, a number of theories that describe the strong force and the electroweak force in a single theory have also appeared. As a group, these are called *grand unified theories*, or GUTs. All of the GUTs share some features, including the prediction that electroweak and strong forces become a single force at extremely high energies when particles are very close together. Eventually, all four fundamental forces may be explained by a single "theory of everything." See Figure 7-29.

Figure 7-28 *An egg can be divided only by breaking the shell. Quanta, however, are indivisible.*

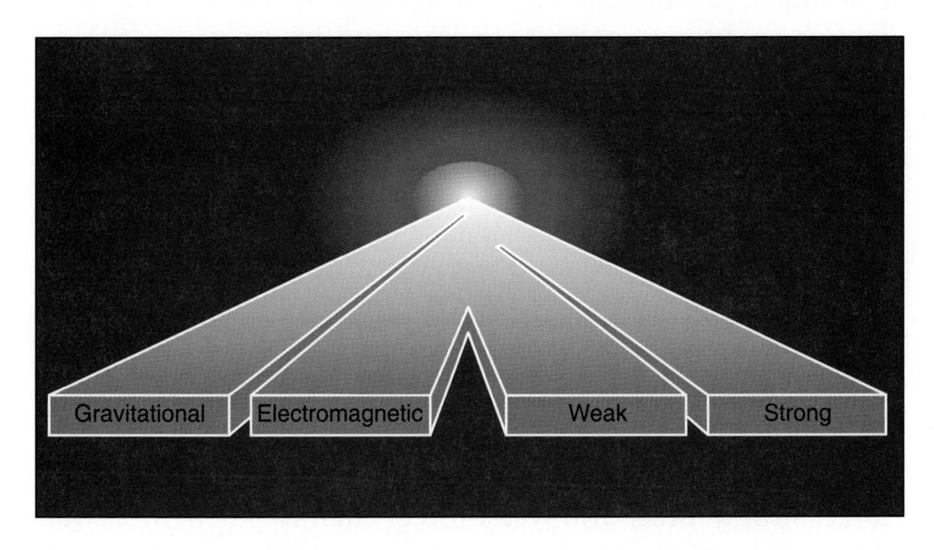

Figure 7-29 *Scientists theorize that in the first few seconds after the big bang, the four forces separated from one original unifying force.*

Where Do We Go From Here?

Before the mid-20th century, the *law of conservation of mass* and the *law of conservation of energy* were thought to hold true for all phenomena. But it is now known that matter can be transformed into energy. For example, it is estimated that the sun converts 657 million tons of hydrogen into 653 million tons of helium every second. The "missing" 4 million tons of matter are converted into radiant energy, some of which we receive as sunlight. See Figure 7-30. The

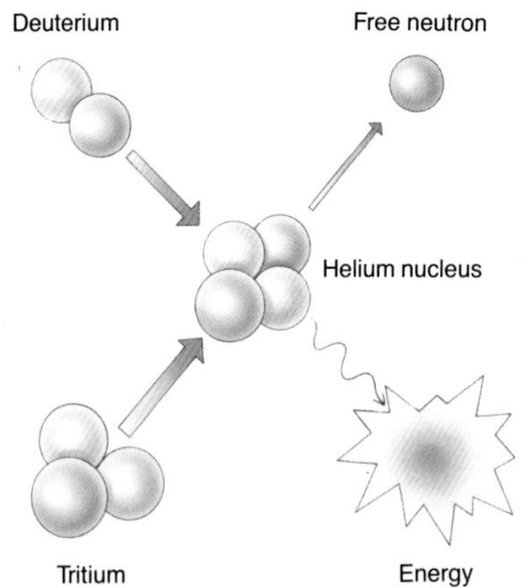

Figure 7-30 *Nuclear fusion results when two hydrogen atoms combine to form helium. A free neutron and a great deal of energy are released from this reaction.*

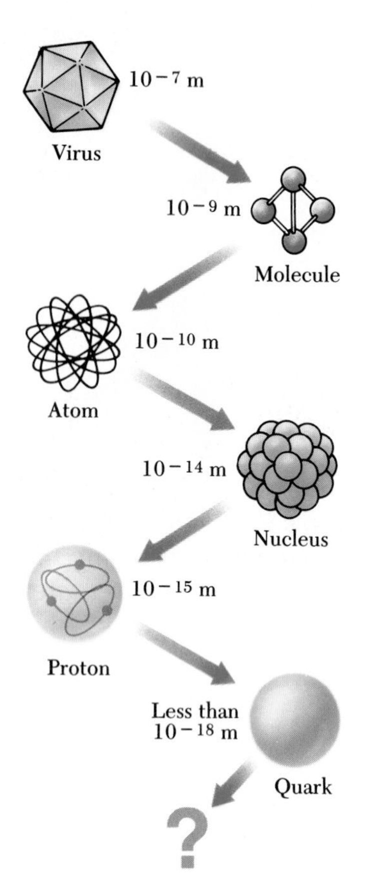

Figure 7-31 *As observational techniques have improved, smaller and smaller particles have been discovered. Although the quark itself has not yet been observed, it may be composed of something even smaller.*

formula expressing the mathematical relationship between mass and energy was proposed by Albert Einstein with his equation: $E = mc^2$. This famous equation states that *energy* is equal to *mass* multiplied by the *speed of light* squared. According to this equation, mass can be changed into energy and energy can be changed into mass.

The law of conservation of mass (which states that matter can neither be created nor destroyed) and the law of conservation of energy (which states that the total energy in the universe is constant) were therefore affected by this discovery. To account for the creation of energy from mass, a new law had to be established: *The total amount of matter* AND *energy in the universe does not change.* This combination of the two previous laws has become known as the ***law of conservation of mass and energy***. Whether mass is converted into energy or energy is converted into mass, the total amount of mass and energy remain the same.

The original "indivisible" atom has been subdivided into protons, neutrons, and electrons. It has also been further broken into quarks, leptons, and bosons, as well as many other pieces. In fact, over 200 subatomic particles have now been cataloged. Just how far can this continue? See Figure 7-31. If matter can be changed into energy, and energy can be turned into mass, can we consider any particle of matter as fundamental? Perhaps there is there an entity common to both matter and energy that may answer such questions.

Summary

According to current understanding of the universe, every substance in existence is composed of particles of matter. These particles, in turn, are affected by four forces that hold everything together—the force of gravity, electromagnetic force, strong force, and weak force. Each force operates at a different level of organization, from galaxies to subatomic particles. Quantum theories involve all forces but gravity. All quantum theories assume that only discrete quanta of energy can be transferred when particles interact. Grand Unified Theories attempt to combine three forces, while a "theory of everything" would combine all four fundamental forces. The relationship between mass and energy is given by the equation $E = mc^2$. According to the law of conservation of mass and energy, the total amount of matter and energy in the universe does not change. Nevertheless, the basic nature of matter and energy is still very much undecided.

IN THIS UNIT

Reading *Plus*

Now that you have been introduced to heredity and the processes of reproduction and development, let's take a closer look at some patterns of inheritance and DNA—the molecule of heredity. Read pages S126 to S142 and write a short story about the role of DNA and heredity in the continuing history of life on Earth. Your story should provide answers to the following questions.

1. What evidence did the discoveries of mitosis and meiosis provide in support of Mendel's laws of heredity?
2. How does our modern understanding of genetics differ from Mendel's?
3. In what ways does the structure of DNA make it an ideal molecule for the transmission of inherited traits?

8.1 FOUNDATIONS OF HEREDITY

Figure 8-1 *What principle of heredity does this litter seem to violate?*

Life continually renews itself by means of the process of *reproduction*. Not only are new individuals made by the reproductive process, but they are similar to their parents in the most fundamental of ways. For example, live oak trees produce offspring that are also live oak trees, and human beings produce other human beings. In other words, each species produces offspring that are of that same species. See Figure 8-1.

At the same time, offspring often differ from their parents in many ways. External differences are the easiest to see. Such differences include height, coloration, and shape of body parts. But, there are internal differences as well. For instance, most people have type O or type A blood, while others have type B or type AB blood. Such differences, external and internal, are called *variations*.

Figure 8-2 *Santa Gertrudis cattle are the result of thousands of years of selective breeding. They are resistant to heat and insects and also produce high-quality meat.*

Long ago, people realized that many characteristics are passed from parents to offspring. With this knowledge, people began selecting and breeding certain types of plants and animals that had desired characteristics. See Figure 8-2. Yet it was not known how traits pass from parents to offspring until the 20th century.

Mendel's Discoveries

Heredity The passing of traits from parents to offspring.

Gregor Mendel was an Austrian botanist who lived from 1822 to 1884. His claim to fame is that he discovered the basic principles of **heredity**. Mendel became interested in plants while growing up on his family's farm. At the age of 21, he entered a monastery to become a priest and a teacher. Later, the monastery sent Mendel to the University of Vienna where he studied science and mathematics. After his return, Mendel taught natural science at a nearby high school.

Mendel also applied what he had learned at the university to his interest in plants. In the monastery garden, he conducted the research that led to his discoveries about heredity. Other scientists had previously tried to determine how traits are transmitted from parents to offspring, but had been unsuccessful. Mendel, on the other hand, discovered the basic principles of heredity primarily because of the scientific methods he used. To begin with, Mendel tested only one trait, or variable, at a time. He also used many samples, repeated his experiments several times, and kept accurate records of all his methods and results. Unlike other researchers, Mendel analyzed his results by using the mathematical principles he had learned at the University of Vienna. He was the first to use such analysis in biological experiments.

Mendel's Procedure Mendel studied seven pea-plant traits that have two contrasting forms. See Figure 8-3. He began by *cross-pollinating* (crossing) plants with contrasting forms of a trait. For example, he cross-pollinated tall plants with pollen from short plants and short plants with pollen from tall plants. The offspring of such crosses are called **hybrids**. Mendel found that both types of crosses produced only tall plants. The contrasting form, shortness, had disappeared. In crosses involving the six other traits he studied, the same thing happened. Only one form of each trait appeared among the hybrid offspring.

Hybrids The offspring of crosses between individuals that breed true for the alternative forms of a trait.

	Seed Shape	Seed Color	SeedCoatColor	Pod Shape	Pod Color	Flower Position	Stem Length
Dominant	Round	Yellow	Colored	Inflated	Green	Axial	Tall
Recessive	Wrinkled	Green	White	Constricted	Yellow	Terminal	Short

Figure 8-3 *Of the many characteristics in the garden pea plant, Mendel chose to study these seven contrasting pairs of traits.*

In the next stage of his experiments, Mendel allowed the hybrid offspring to *self-pollinate*. When he grew the seeds that resulted from these self-pollinations, both forms of each trait appeared among the resulting plants. For example, tall hybrid pea plants produced both tall and short offspring. In every case, the form of the trait that had disappeared in the hybrid offspring reappeared in the next generation.

Mendel also noticed that among the offspring of the hybrids, the contrasting forms of each trait appeared in the same ratio. In every case, three-fourths of the plants showed the form of the trait that appeared in the hybrid offspring. Only one-fourth showed the form that had disappeared and then reappeared.

Mendel's Factors Seeing that one form of each trait had disappeared in his hybrids but reappeared in the next generation, Mendel inferred that some distinct element must be responsible for each of the traits he observed. He called these distinct elements *factors*. He also came to several important conclusions about how traits are inherited.

1. Each parent contributes one factor for each trait to its offspring. Thus individuals have a *pair* of factors for each trait.

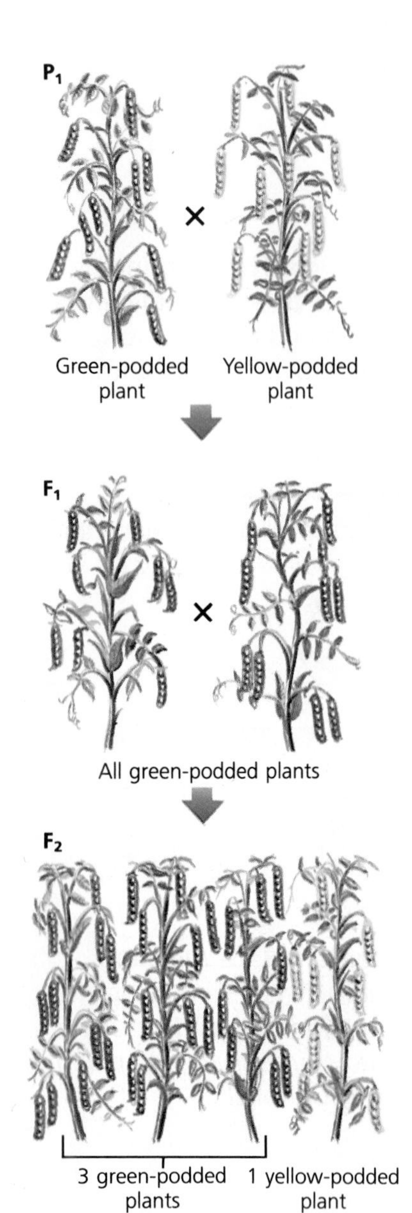

Figure 8-4 *Green-podded plants crossed with yellow-podded plants produce only green-podded plants. Yet when this second generation is permitted to self-pollinate, some yellow-podded plants appear.*

2. Since only one factor for each trait is contributed by a parent, the pair of factors for a trait must *separate* when reproductive cells are formed. As a result, reproductive cells receive and carry only one factor for each trait. When two reproductive cells combine to produce a new individual, there is again a pair of factors for each trait. This principle is called ***Mendel's law of segregation***.
3. There must be a different form of the factor for each expression of a trait. In peas, for example, there are two kinds of factors for the trait of plant height—one for tallness and the other for shortness. Likewise, there are two kinds of factors for seed shape (round and wrinkled), and two kinds of factors for seed color (yellow and green), and so forth.
4. One of the two factors for a trait prevents the expression of the other. This explains the disappearance of one form of each trait from the hybrid generation. For example, the factor for tallness prevents the expression of the factor for shortness. The factor for yellow seeds prevents the expression of the factor for green seeds. Mendel called the factor that prevented the expression of the other a *dominant* factor. The factor whose expression was prevented he called a *recessive* factor.

Mendel's hybrids displayed only the dominant form of each trait because each had received one dominant factor and one recessive factor. The recessive factors reappeared in the offspring of the hybrids because some of them received the factor for shortness from both the male and female reproductive cells. Only when two recessive factors were paired together was the recessive factor expressed. Figure 8-4 shows the results of another of Mendel's crosses.

Mendel published the results of his work in 1865. However, no one paid much attention to them at the time. Perhaps this was because of the mathematics he used in his work, which other scientists may not have understood. But over the next 35 years, important discoveries about cells were made. Those discoveries prepared the scientific community for the rediscovery of Mendel's work in 1900. Unfortunately, this was after Mendel had died, so he never knew of the impact that his work had on our understanding of heredity.

Cell Division

During the 1860s, several scientists began to examine cells of living tissues in the process of dividing. Certain structures, found in the nucleus of each cell, were observed during the division process. These structures were named *chromosomes*, because they accepted a colored dye that made them more visible under the microscope. *Chrome* means "color" in Greek. See Figure 8-5. Scientists noticed that when cells divided, equal numbers of chromosomes were separated into two groups and distributed to each new cell. Each cell became two new cells that were exactly like the original cell.

Later, scientists who were studying the formation of reproductive cells observed a second type of cell division. During this process, two successive divisions occurred. The result was four new cells, each having only one-half of the original cell's number of chromosomes. Let's take a closer look at these two types of cell division.

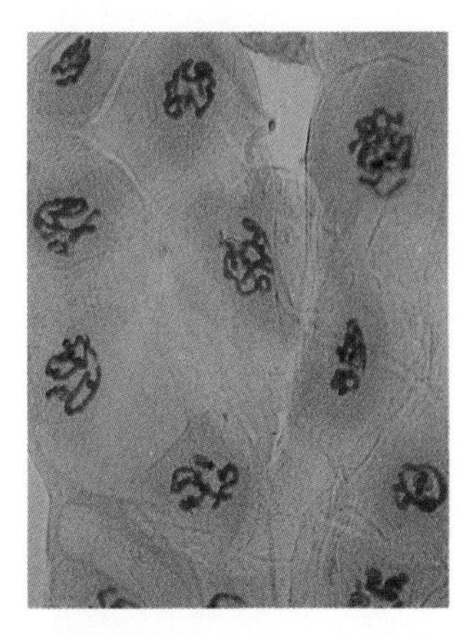

Figure 8-5 *The darkly stained objects in this photograph are chromosomes in the cells of a fruit fly.*

Mitosis Look at your hand. Each square centimeter of skin is made of more than 150,000 cells. Yet most of these cells will be gone by tomorrow. During the next 24 hours, two complete generations of skin cells will live, reproduce, and disappear. In other parts of your body, cells are also reproducing rapidly. For example, new red blood cells are being made at the rate of about 100 million cells a minute. If you are still getting taller, your bones are growing in length by adding new bone cells at the ends of the bones.

Cells reproduce by simply dividing into two new cells. These new cells are called *daughter cells*. See Figure 8-6. The original cell is called the *parent cell*. Daughter cells are exactly alike, and both are exactly like their parent cell.

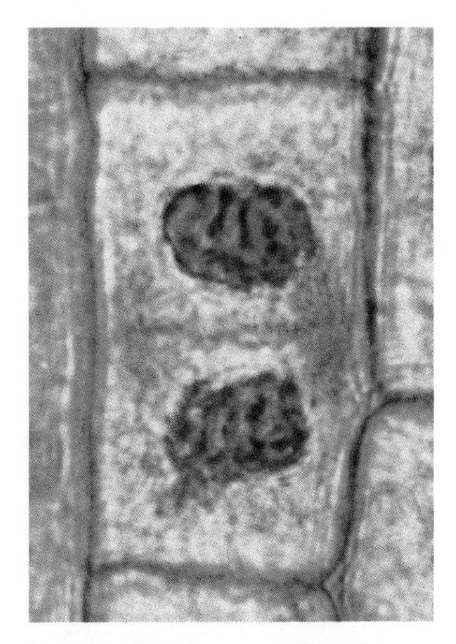

Figure 8-6 *A new cell wall can be seen beginning to form between these two daughter cells in an onion root tip.*

The chromosomes in the nucleus of a cell are the "blueprints" for that cell. They contain all the information needed to build new cell materials and to control the cell's activities. Suppose you wanted to have two identical houses built at the same time. You would not cut the blueprints in two and give one half to each builder. Instead, you would make an exact copy, or duplicate, of the blueprints so that each builder could have a full set of instructions. The same is true for the blueprints of a cell. Each new cell must carry a full set of chromosomes. Therefore, chromosomes must be duplicated before the nucleus of a cell divides.

This division of the nucleus is called **mitosis**. During mitosis, the division of the nucleus occurs in four consecutive stages that allow each daughter cell to receive one copy of each chromosome. Scientists call these stages—from first to last—*prophase*, *metaphase*, *anaphase*, and *telophase*. After cell division, the daughter cells go through a period of growth and activity called *interphase*. Refer to Figure 8-7 as you read about each stage in the division of a cell's nucleus.

Mitosis A process by which cells produce exact copies of their nuclei.

1. During prophase, each of the chromosomes inside the nucleus begin to twist and thicken. As the chromosomes become visible, the membrane around the nucleus disappears. Since the chromosomes have been duplicated, they appear as two strands joined together by a structure called a *centromere*. The two strands are complete and identical chromosomes. Slender fibers, called *spindle fibers*, extend from opposite sides of the cell. A spindle fiber from each side attaches to the centromere of each doubled chromosome.
2. During metaphase, the duplicated chromosomes line up across the middle of the cell. Then the centromeres divide, separating the duplicate chromosomes.
3. During anaphase, the spindle fibers shorten and pull the duplicate chromosomes through the cytoplasm toward opposite sides of the cell.
4. During telophase, the separated chromosomes cluster at the opposite sides of the cell. A nuclear membrane forms around each group of chromosomes. At the same time, the chromosomes untwist, becoming longer and thinner. At the end of telophase, the cytoplasm is divided between the two new cells and they separate.

Figure 8-7 *From left to right, the four phases of mitosis are prophase metaphase, anaphase, and telophase.*

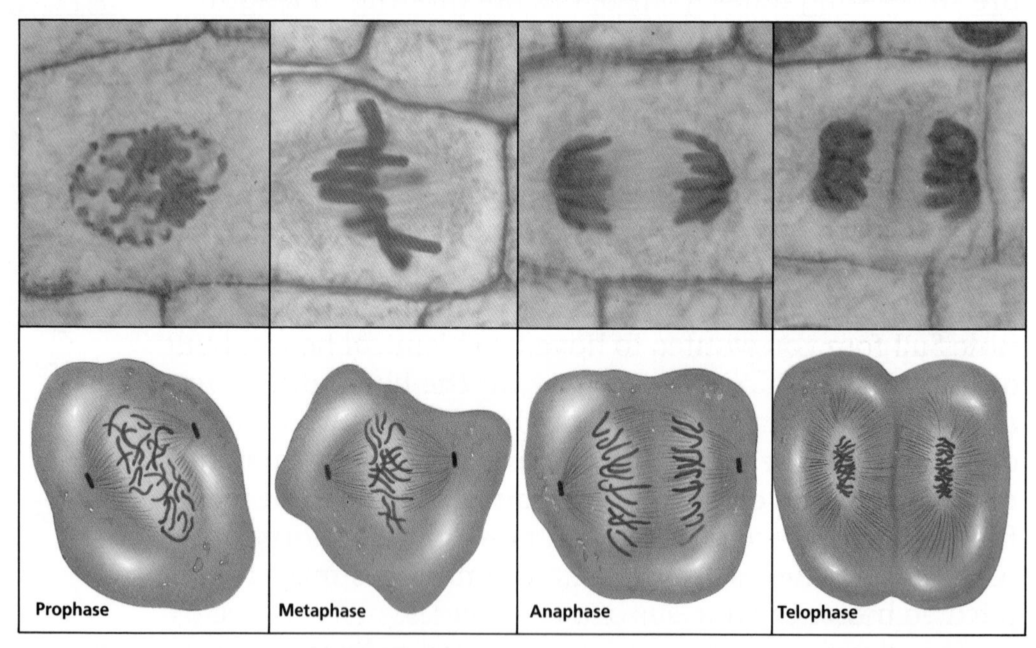

Meiosis Many organisms reproduce by means of sexual reproduction. This type of reproduction involves the uniting of two sex cells—one male and one female—in a process called *fertilization*. If sex cells (egg and sperm) contained the same number of chromosomes as other body cells, the number of chromosomes would be doubled whenever fertilization occurred. After a few generations of doubling, the cells produced by the uniting of sex cells would be bursting with chromosomes. This, of course, does not happen. Each species has a characteristic number of chromosomes in each of its cells, which remains stable from one generation to the next. See Table 8-1. Scientists realized that for this to be the case, the number of chromosomes in sex cells would have to be exactly half the number of chromosomes in body cells.

CHROMOSOME NUMBERS OF VARIOUS SPECIES			
Species	**Number**	**Species**	**Number**
Alligator	32	Fruit fly	8
Amoeba	50	Garden pea	14
Brown bat	44	Goldfish	94
Bullfrog	26	Grasshopper	24
Carrot	18	Horse	64
Cat	32	Human	46
Chicken	78	Lettuce	18
Chimpanzee	48	Onion	16
Corn	20	Redwood	22
Earthworm	36	Sand dollar	52

Table 8-1

Meiosis Cell-division process that produces sex cells with half the chromosome number of the parent cell.

While studying the formation of sex cells, scientists discovered a cell-division process that results in daughter cells with half as many chromosomes as the parent cell had. This process is called **meiosis**. In meiosis, there are two successive divisions. See Figure 8-8. Four daughter cells are produced instead of two, and each has exactly one half the number of chromosomes in the organism's body cells. In males all four daughter cells become sperm. In females the cytoplasm is divided unequally, and only one of the four daughter cells becomes an *ovum* (egg cell).

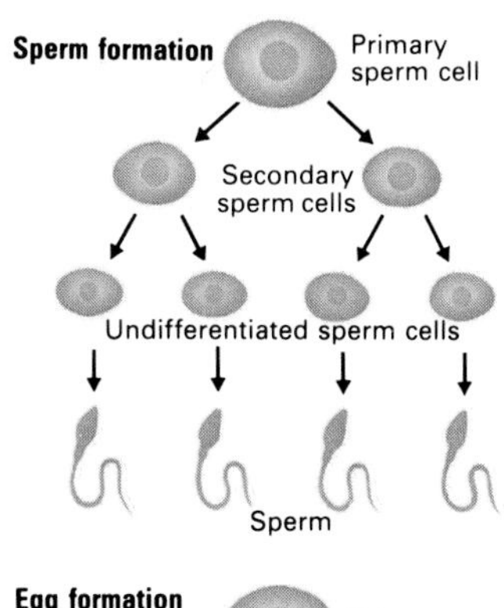

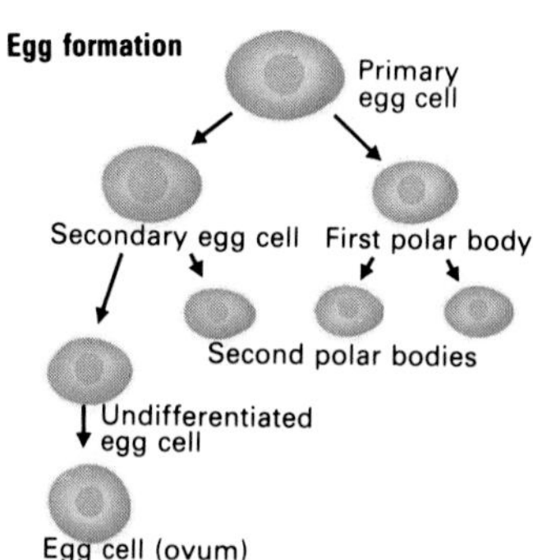

Figure 8-8 Meiosis produces four sperm cells (top) *or one egg cell* (bottom).

Scientists made another important discovery while studying the formation of sex cells. In meiosis, the nucleus divides in stages that are very similar to those of mitosis, with one exception. In meiosis, the duplicated chromosomes do not line up randomly during metaphase for the first division as they do in meiosis. Instead, each chromosome lines up across from another chromosome of the same size and shape. Seeing this, researchers realized that chromosomes occur *in pairs*. These pairs separate during the first division of meiosis, and each daughter cell carries only one chromosome from each pair. Mendel had described this same pattern in explaining the inheritance of factors. Traits are determined by a pair of factors, the pairs separate during reproduction, and only one factor for each trait is transmitted by each parent. See Figure 8-9.

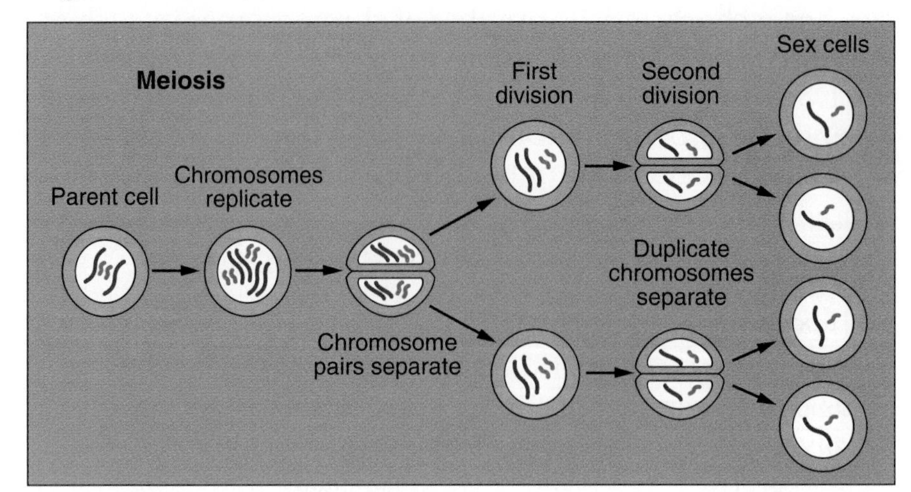

Figure 8-9 *During meiosis, pairs of duplicated chromosomes separate during the first division. These duplicates separate during the second division. Each resulting sex cell has only one of each kind of chromosome.*

Summary

Life continues by means of reproduction. During reproduction, characteristics are passed from parents to their offspring. Gregor Mendel was the first to describe the basic principles by which traits are inherited. He theorized that a trait is determined by a pair of factors. These factors separate during reproduction. Each parent transmits one factor for a trait to its offspring. Factors occur in two forms. The dominant factor prevents the expression of the recessive factor. Cells reproduce themselves through a process called cell division. During mitosis, the chromosomes in the nucleus of a parent cell are doubled and then divided so that two daughter cells receive the same kind and number of chromosomes as the parent cell had. During meiosis, pairs of like chromosomes separate so that four daughter cells are produced, each having half as many chromosomes as the parent cell had.

8.2 GENES, CHROMOSOMES, AND HEREDITY

Without knowing what they were or where they were located, Mendel had inferred that "factors" inherited from parents are responsible for the characteristics of an organism. Many years later, the behavior of chromosomes was observed during the formation of reproductive cells. However, the significance of this behavior was not recognized at first. When the work that Mendel had done previously was rediscovered in 1900, a new science, called **genetics**, was born. This name, which comes from a Greek word meaning "beginning," was first suggested at a scientific meeting on heredity in 1906.

Genetics The science that studies heredity.

The Chromosome Theory

In 1903 Walter S. Sutton, a researcher at Columbia University, was conducting microscopic studies of meiosis in grasshoppers. Having read Mendel's paper about the inheritance of traits in pea plants, Sutton realized that the behavior of Mendel's factors was the same as the behavior of chromosomes during meiosis. He inferred that, because their behavior was the same, factors must be related to chromosomes. He also realized that for every organism, there are many more traits than there are chromosomes. See Figure 8-10. As a result, Sutton also inferred that there must be several hereditary factors located on each chromosome. Sutton's ideas became known as the **Chromosome Theory**, which simply states that *hereditary factors, or genes, are found on chromosomes.*

Chromosome Theory
Genes are found on chromosomes.

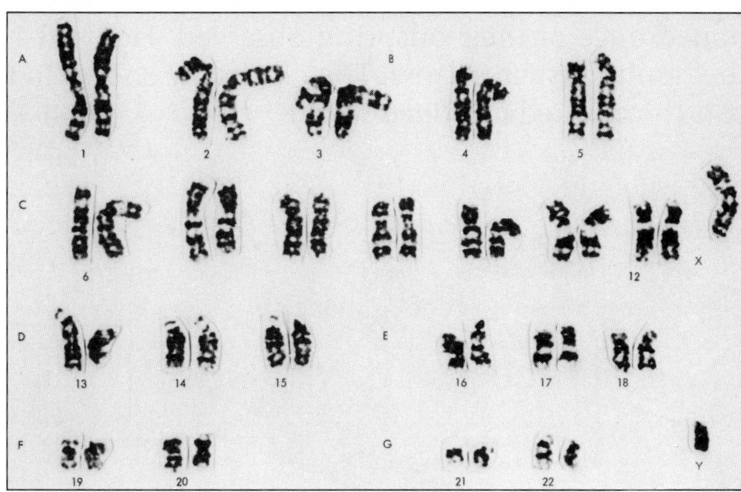

Figure 8-10 *Humans have 23 pairs of chromosomes. Yet they have many more inherited traits.*

The term *gene* was first used to refer to a hereditary factor in 1909. Over the next fifty years, *geneticists* studied the way many genes are inherited. They even identified the locations of genes on the chromosomes of many organisms. However, the exact nature of a gene and how it works remained a mystery until the 1950s. Even to this day, there is still much more to be learned about genes and chromosomes.

Working With Genes

Geneticists represent individual genes with letters. Capital letters are used to indicate dominant genes. For example, T is used to represent the gene for tallness in pea plants, and R is used to represent the gene for round seed shape in pea plants. The lowercase version of the *same* letter indicates the recessive gene for the same trait. The gene for shortness, then, is t and the gene for wrinkled seed shape is r. See Figure 8-11.

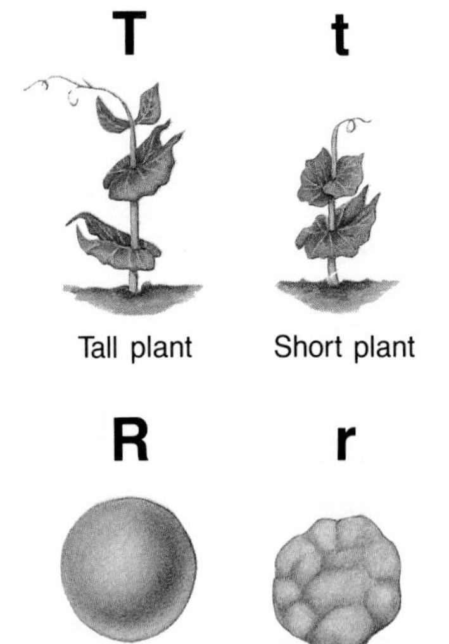

Figure 8-11 *The symbol for a dominant gene is a capital letter, and the symbol for a recessive gene is a lowercase letter.*

Genotype The actual pair of genes an organism has for a trait.

Phenotype The physical appearance of a trait.

Homozygous Condition where the pair of genes for a trait are identical.

Heterozygous Condition where the pair of genes for a trait are of contrasting forms.

Since traits are produced by pairs of genes, a pair of letters can be used to represent the genes that a particular individual has for a trait. For example, pea plant height is determined by one of three possible gene pairs—TT, Tt, or tt. Pea seed shape can be determined by RR, Rr, or rr. The specific pair of genes that an organism has for a trait is called its **genotype**. The physical expression of a genotype, such as round seeds, wrinkled seeds, tall plants, or short plants, is called a **phenotype**. If both genes for a trait are identical (TT or tt) the genotype is said to be **homozygous**, or pure, for that trait. If the two genes are different (Tt), the genotype is said to be **heterozygous**, or hybrid.

Geneticists determine how a trait is inherited by using methods similar to those Mendel used. First they choose two individuals that are homozygous for the contrasting forms of a particular trait. These individuals are crossed and the appearance of their offspring observed. Figure 8-12 shows the results of such a cross. The results are shown in the form of a chart called a *Punnett square*.

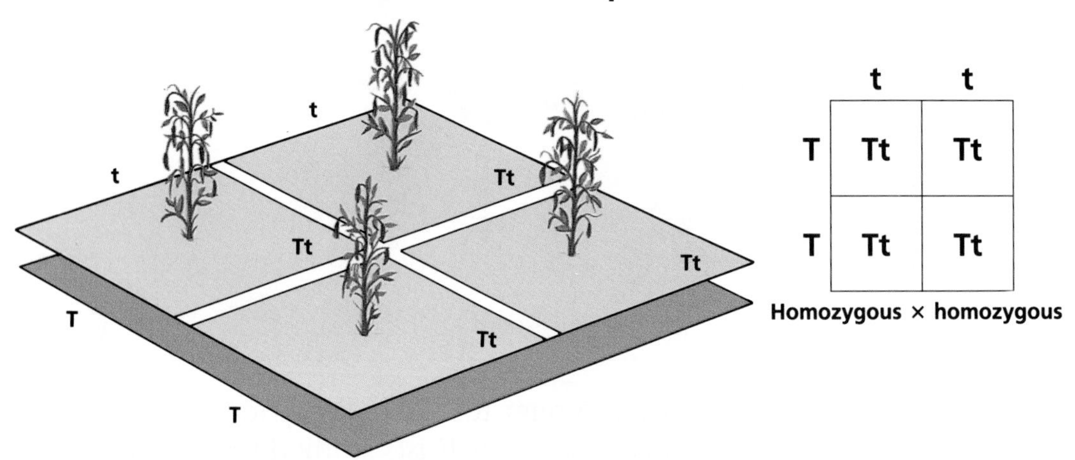

Figure 8-12 *A pea plant that is homozygous for tallness crossed with a pea plant homozygous for shortness will produce only offspring that are heterozygous for tallness.*

Punnett's Square The Punnett-square method for displaying the possible results of a genetic cross was developed by British scientist Reginald Punnett. To make a Punnett square, first draw a square and divide it into two rows and two columns. Now write the two genes of one parent's genotype at the top of the square, one above each column. Then write the two genes of the other parent's genotype down the left side of the square, one beside each row. Finally, copy the gene symbol from the top of each column into the boxes below and the gene symbol from the side of each row into the boxes of that row. All of the possible gene combinations from the cross now appear in the boxes. Look again at Figure 8-12. Notice that all the genotypes in the square are heterozygous. Since tallness (T) is dominant to shortness (t), all of the offspring of this cross will be tall.

Pedigrees Suppose you decide to raise rabbits with black fur (B), which is dominant to brown fur (b). You go to a pet store to buy several pairs of black rabbits. But how will you know if the rabbits you buy will produce only black young? You cannot tell just by looking at them since a black rabbit can have either one of two different genotypes—homozygous BB or heterozygous Bb. See Figure 8-13. If some of the rabbits you buy have the Bb genotype, your rabbit herd will eventually contain some brown rabbits. To be sure that you buy only homozygous black rabbits, you must obtain a **pedigree** for each rabbit.

A pedigree is a chart of the genetic history of a particular individual. To be useful, a pedigree usually must go back several generations and show as many family members as possible. For example, if there are no brown rabbits among the family members of a particular rabbit for several previous generations, you could be fairly certain that the rabbit is homozygous for black fur. However, if the rabbit has a brown parent, you would know that the rabbit is heterozygous for the trait. A brown parent can only pass the gene for brown fur to its offspring.

Figure 8-13 *What is the genotype of the brown rabbit on the top? the black rabbit on the bottom?*

Pedigree Chart that traces the inheritance of a genetic trait through several generations of a family.

Sex Determination

What makes an individual male or female? In the early 1900s, Thomas Hunt Morgan found one answer to this question. While conducting genetic experiments with fruit flies, Morgan noticed that female flies have four pairs of chromosomes that are alike in shape and size. In male flies, however, only three of the pairs match, the fourth pair does not.

Morgan believed that this difference in chromosomes was responsible for the sex of a fruit fly. He called the nonmatching chromosome in the male the *Y chromosome*. The other chromosome in the pair he called the *X chromosome*. Further research revealed that sex is determined by the inheritance of X and Y chromosomes in most animals, including humans. Females have two X chromosomes, and males have one X and one Y chromosome.

Morgan's discovery also explained why certain traits appear more frequently in males than in females. For example, red–green colorblindness is more common in human males than in females. Such traits are controlled by genes found on X chromosomes. The X chromosome is a typical chromosome with many genes. The Y chromosome, however, is much smaller and has very few genes. As a result, most of the genes on an X chromosome are not matched by corresponding genes when paired with a Y chromosome. For conditions that are determined by a recessive gene, such as colorblindness, females must have two recessive genes to show the condition. Males, on the other hand, only need to receive one recessive gene to be colorblind. Such traits are called *sex-linked traits*.

Other Patterns of Inheritance

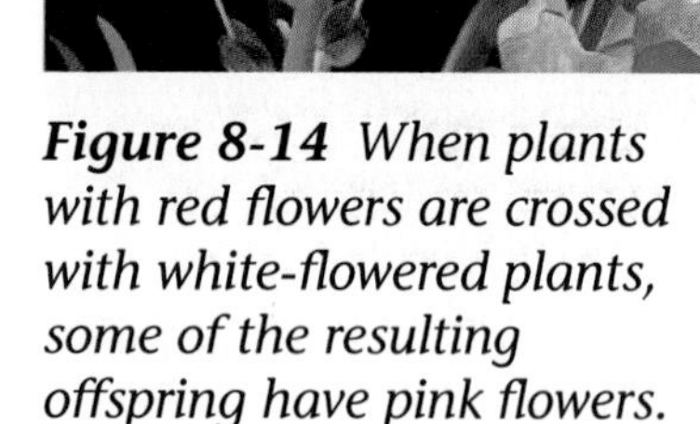

Figure 8-14 *When plants with red flowers are crossed with white-flowered plants, some of the resulting offspring have pink flowers.*

Mendel described only one pattern of inheritance. The traits he studied in peas are all controlled by one pair of genes. Each trait also has two different kinds of genes—one dominant and one recessive. Modern geneticists refer to this pattern of inheritance as *simple dominance*. However, simple dominance is only one way that genetic traits are inherited. For many traits, there are no dominant or recessive genes. Instead, both genes in a heterozygous pair are expressed. This pattern of inheritance is called *incomplete dominance*. Flower color in snapdragons, for example, shows incomplete dominance. See Figure 8-14. Crosses made between two pink-flowered plants produce three different phenotypes among the offspring in a ratio of 1:2:1—one-fourth red, two-fourths pink, and one-fourth white.

Some traits are determined by several pairs of genes. This pattern of inheritance is called *polygenic inheritance*. The effect of each pair of genes is added together to produce the phenotype of the individual. Human skin color, for example, is a trait produced by the action of several pairs of genes. See Figure 8-15.

Figure 8-15 *These high school students exhibit a wide range of skin colors.*

Summary

Hereditary factors called genes are located on chromosomes. Dominant genes are indicated by capital letters, while recessive genes are indicated by lowercase letters. A Punnett square is used to determine the possible genotypes and the resulting phenotypes of the offspring from genetic crosses. A pedigree is used to study the inheritance of traits when controlled crosses cannot be made or to trace an individual's genetic history. Sex is determined by the inheritance of X and Y chromosomes. In addition to simple dominance, there are other patterns of inheritance, including incomplete dominance, polygenic inheritance, and sex linkage.

8.3 DNA—THE MATERIAL OF HEREDITY

DNA was discovered in 1869. At that time, all that was known about DNA was that it is found in the nucleus of cells. Thus, it was dubbed a "nucleic acid." In 1924, DNA (deoxyribonucleic acid) was found to be a part of chromosomes. These structures were already known to transmit hereditary traits. But it was not until 1944 that scientists found evidence identifying DNA as *the* material responsible for heredity. Yet, how does DNA work? How is it made?

The Watson and Crick Model

Have you ever put together a model of an airplane or an automobile? Would you be able to put all the tiny pieces of the model together correctly if you lost the instructions? You might if you know what the model is supposed to do. In a similar fashion, scientists must often work with little or no instructions. However, by using what they do know, scientists can construct models of the structures and processes they are studying. Scientists then test their models to find out if they are consistent with what is known about the actual structure or process.

Using a three-dimensional model, American biologist James Watson and English physicist Francis Crick worked out the structure of DNA in 1953. Their model was based on information that many scientists had collected. In working out the structure of the molecule, they also discovered how DNA makes exact copies of itself and how it acts as the hereditary material. Watson and Crick won the Nobel Prize in physiology and medicine in 1962 for their work.

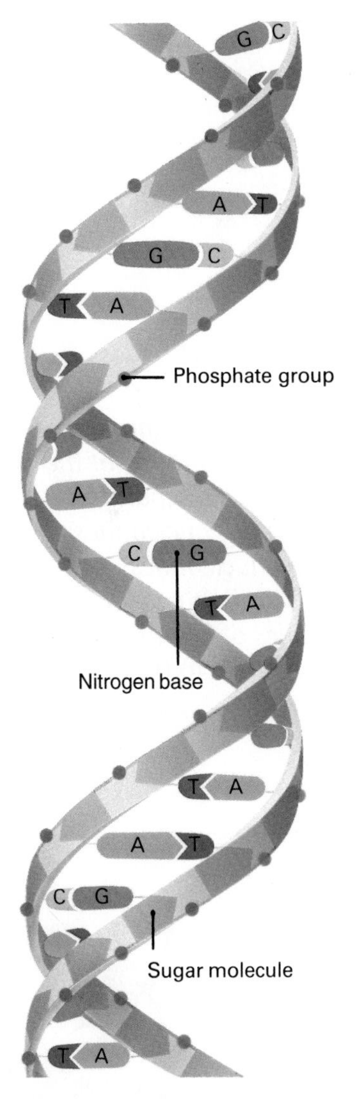

Figure 8-16 *A DNA molecule resembles a twisted ladder. Its sides are formed from long chains of sugars and phosphates. The "rungs" are pairs of nitrogen bases.*

Structure of DNA Each DNA molecule consists of two very long chains of smaller units called *nucleotides*. Stretched out, a DNA molecule from one of your cells would be about two meters long. DNA's two chains are connected by crosspieces, or "rungs," that give the molecule a "ladder-like" appearance. See Figure 8-16. The molecule is also twisted into a

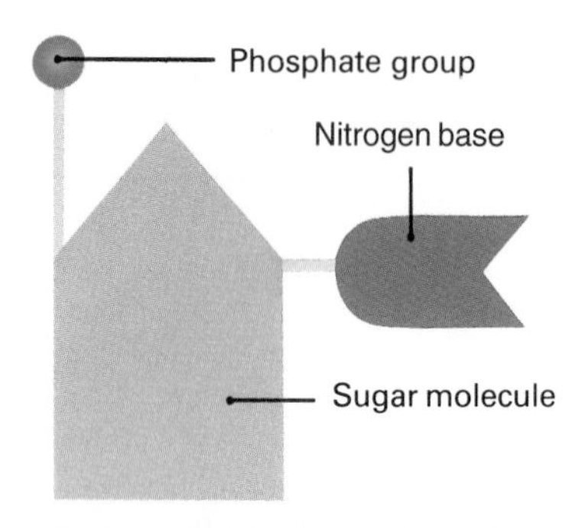

***Figure 8-17** A nucleotide consists of a phosphate, a sugar, and a nitrogen base.*

spiral, or *helix*. This twisted ladder shape is called a *double helix*.

Each nucleotide in a DNA molecule is made of three parts—a sugar molecule, a phosphate ion, and a molecule called a *nitrogen base*. See Figure 8-17. Four different nitrogen bases are found in the nucleotides of DNA. They are *adenine, guanine, cytosine*, and *thymine*. Respectively, the nitrogen bases are often abbreviated as A, G, C, and T. They can be thought of as the "letters of the genetic alphabet."

The key to DNA's functioning, however, is in the way the nucleotides are arranged to form the "sides" and the "rungs" of the "ladder." Nucleotides are bonded together in a specific way to form the spiral helix. The sides are formed by bonding the sugar of one nucleotide to the phosphate of the next in a continuous chain. The nitrogen bases stick out from the sugar and phosphate sides. Each of the nitrogen bases of one side is bonded to one from the other side. Look again at Figure 8-16. The resulting nitrogen-base pairs form the rungs of the ladder. But each base can be paired with only one other. Adenine (A) always pairs with thymine (T), and guanine (G) always pairs with cytosine (C).

DNA Replication *Replication* is the making of an exact copy of a DNA molecule. The process of replication begins when the two chains of a DNA molecule begin to separate at one end, as shown in Figure 8-18. This happens because the nitrogen-base pairs are pulled apart by a type of protein called an *enzyme*. The separation of the two sides of the DNA molecule is often compared to the unzipping of a zipper. When the chains have separated, individual nucleotides

***Figure 8-18** During replication, the two strands of the DNA molecule separate. Free-floating nucleotides pair with the nucleotides of each chain, forming two identical molecules of DNA that are also identical to the original.*

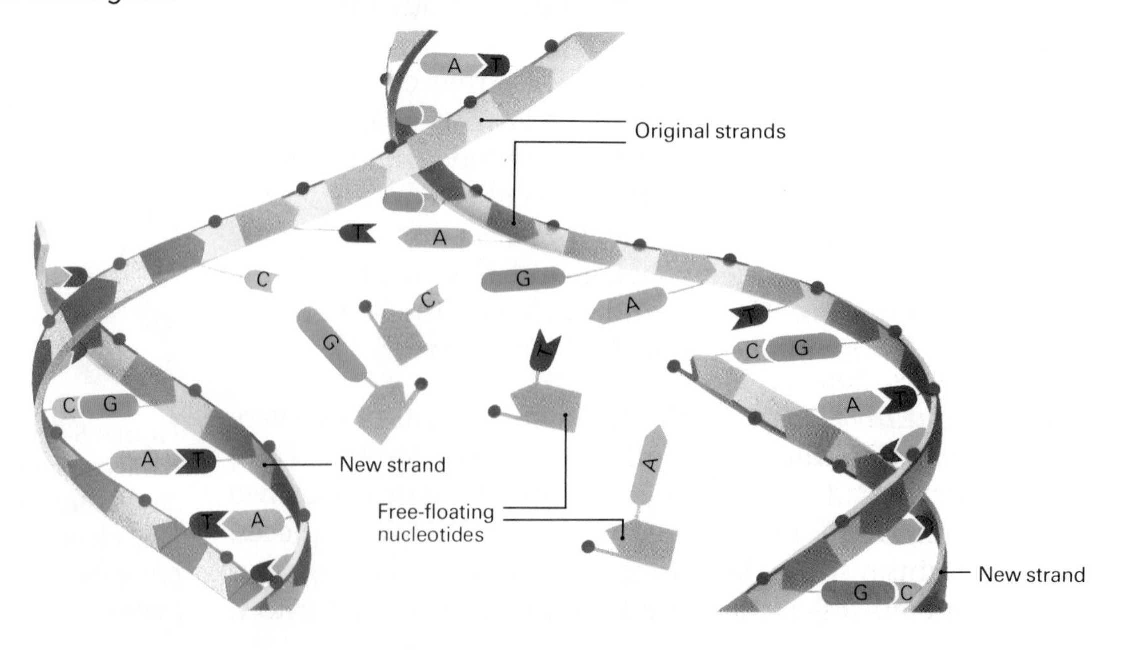

floating freely in the nucleus line up across from the nitrogen bases of the separated chains. New base pairs form as the nitrogen bases of the new nucleotides bond to the nitrogen bases on each half of the old molecule. But remember, there are only two kinds of nitrogen base pairs that can be made—C always pairs with G, and A always pairs with T. As a result, the base-pair sequence of the original DNA reappears. Once the process of replication is complete, each side of the original DNA has produced a new DNA molecule. The two new DNA molecules are identical to each other and to the original molecule.

Translating DNA

The DNA in each of the chromosomes of an organism has a different sequence of nitrogen base pairs. Instructions for the making of that organism are contained in this sequence. Genes are actually sections of a DNA molecule. One gene may consist of thousands of nitrogen-base pairs.

The individual genes of a DNA molecule contain instructions for making specific proteins. This is an extremely vital role since the primary structural and regulatory materials of living things are proteins. Some of the many different proteins made by an organism are used to build its various parts. Other proteins, called enzymes, assist in the making of the variety of other substances that an organism produces.

The Genetic Code The instructions for proteins are not written in words that you are familiar with. Instead, they are written in a "language" that uses the nitrogen bases A, T, G, and C as "letters." This language is called the **genetic code**. Each "word" in the genetic code is composed of three "letters," or nitrogen bases. Sixty-four such 3-letter "words", or triplets, can be made. Each triplet attracts a particular amino acid to itself. See Table 8-2. Recall that amino acids are the small molecules that make up proteins. Therefore, the string of code words in a gene causes many amino acids to come together in a particular order to make a particular protein.

Proteins are made in the cytoplasm of a cell, and DNA is found only in the nucleus. For this reason, another substance called RNA is required for the making of proteins. RNA, or *ribonucleic acid*, is similar to DNA. Both kinds of nucleic acids are made of nucleotides. RNA, however, is only a single chain of nucleotides. It also has two chemical parts that are different from DNA. RNA contains a sugar called *ribose*, while DNA contains a similar sugar called *deoxyribose*. As you can see, the molecules' names are based on their sugar parts.

GENETIC CODES AND ASSOCIATED AMINO ACIDS

Triplet	Amino Acid
AAA	Lysine
GUG	Valine
UUA	Leucine
AGA	Arginine
GUU	Valine
CAG	Glutamine
UGU	Cysteine
CAC	Histidine
AUU	Isoleucine
UUU	Phenylalanine
CCC	Proline
AUG	Methionine

Table 8-2

Genetic code The language in which the instructions for proteins are written in the nucleic acids.

RNA also has a nitrogen base called *uracil* (U) instead of thymine. As with DNA, the nitrogen bases of RNA also form pairs, but there is one difference. The U of RNA pairs with A, since U takes the place of T in RNA.

Protein Synthesis The formation of proteins using the information coded on RNA.

Codon A sequence of three nitrogen bases in a molecule of mRNA that codes for a particular amino acid.

Anticodon A sequence of three nitrogen bases in a molecule of tRNA that is the opposite of a codon.

Protein Synthesis The information written in the genetic code is "translated" by the cell during **protein synthesis**. Some RNA molecules, called *messenger RNAs*, copy the instructions from DNA in the nucleus and take them into the cytoplasm. The instructions in a messenger RNA (mRNA) molecule are written in three-base code words called **codons**.

In the cytoplasm, there is another kind of RNA, known as *transfer RNA* (tRNA). Each tRNA molecule, which has a 3-looped or cloverleaf shape, carries its own particular amino acid. See Figure 8-19 (inset). For example, one kind of tRNA carries the amino acid *leucine*, while another carries *valine*. When a number of tRNA molecules come together in a certain way, the amino acids they carry line up end to end. Then they are joined together to form proteins.

The correct lining up of the tRNA molecules, with their amino acids, is brought about by the messenger RNA from the nucleus. Each kind of tRNA has a certain 3-letter code word, or triplet, known as an **anticodon**. Each anticodon will attach to a single codon on an mRNA chain. For example, an AAA anticodon on a tRNA molecule will attach to a UUU codon on the mRNA. Similarly, a GGG anticodon will pair with a CCC codon, and a UAC anticodon will pair

Figure 8-19 mRNA and tRNA work together in the process of protein synthesis. Ribosomes in the cytoplasm of a cell are the sites where proteins are made.

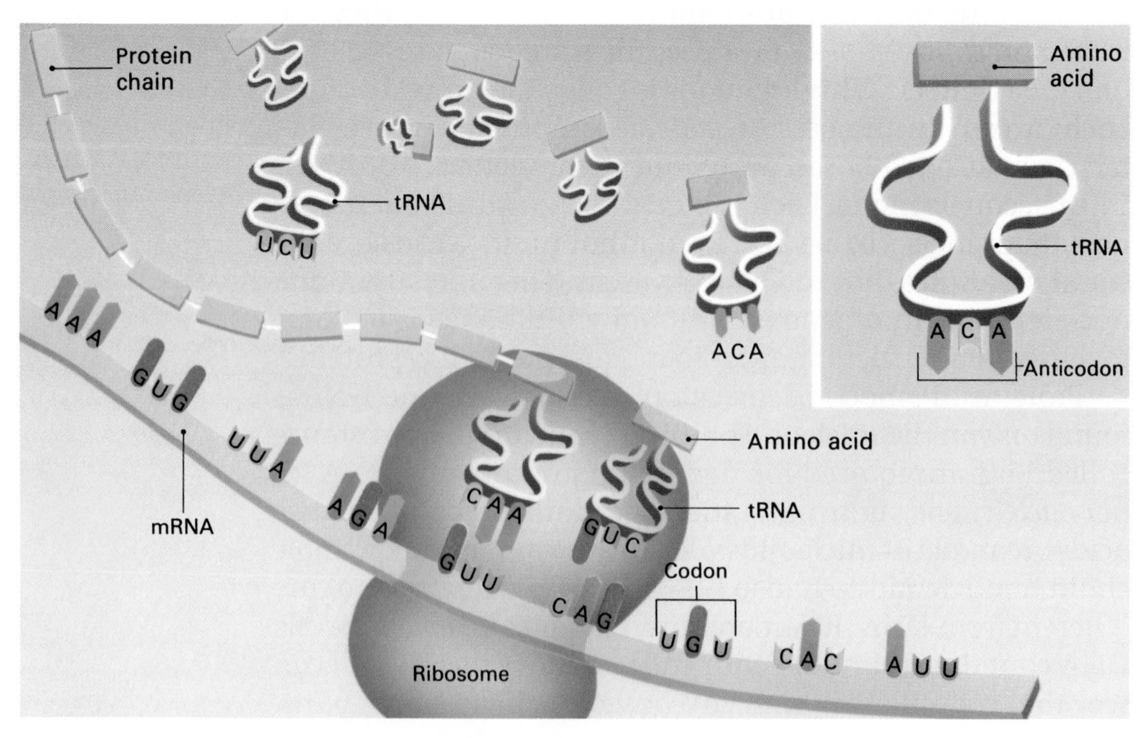

with an AUG codon. When anticodons have been attached to all the codons on an mRNA molecule, all the amino acids for a single protein will be lined up. Special enzymes then join them together to form the protein. Figure 8-19 illustrates this process.

Changes in DNA

Every time a cell divides, its DNA is replicated so that each new cell receives a copy of the complete set of genetic instructions for an organism. Since the nitrogen bases of DNA pair only in certain ways, the copies are normally exact. However, copying mistakes are sometimes made. Also, factors from the environment, such as chemicals and radiation, can cause accidents that destroy parts of a DNA molecule. In both cases, the result is a change in the base sequence of a DNA molecule. Thus its genetic instructions may change as well. Such a change is called a *mutation*.

Figure 8-20 *This koala is white because a gene mutation has resulted in the inability to produce normal pigments.*

Mutations are important because they are the source of new genetic variations. As such, they have played a major role in the evolution of life on Earth. Although mutations can be beneficial, they are more frequently harmful. Fortunately, a mutation usually affects only one cell. In most cases, only mutations that occur in reproductive cells affect an entire organism. This occurs only when they are transmitted to the next generation. There are two basic types of mutations—*gene mutations* and *chromosome mutations*.

Gene Mutations Gene mutations are changes in the structure of DNA molecules. They usually affect only one section of a DNA molecule, or gene, at a time. The order in which the nitrogen bases occur in a gene determines the chemical the gene will produce. A change as small as even one base in the sequence of bases that makes a gene can cause the gene to produce an altered chemical, or no chemical at all. See Figure 8-20.

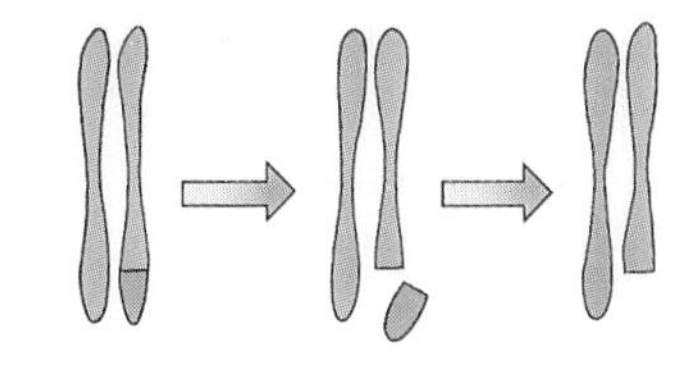

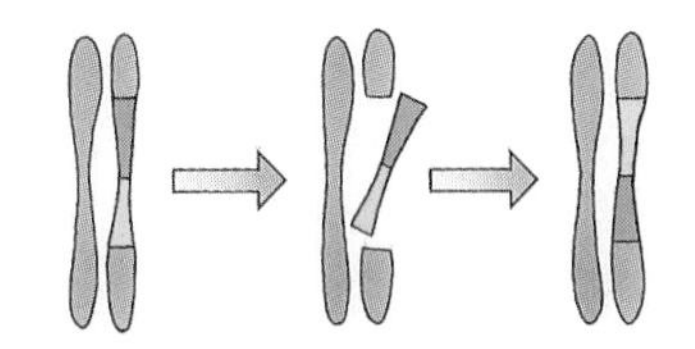

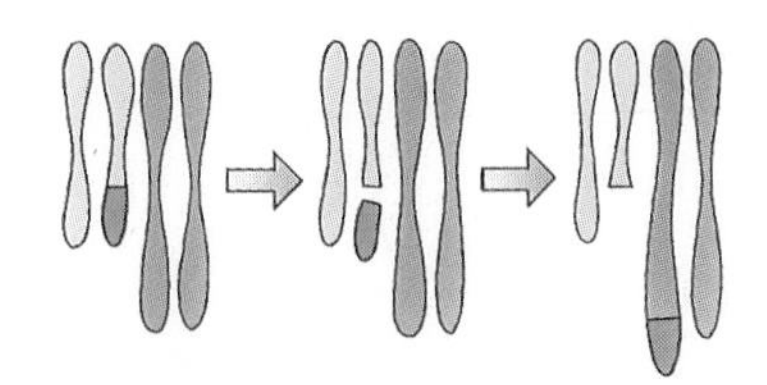

Figure 8-21 *Various types of mutations can occur in chromosomes.*

Chromosome Mutations Chromosome mutations are changes caused by the breaking of chromosomes. Normally, chromosome breaks are repaired. But if the parts of a broken chromosome are improperly rejoined or if some parts are lost, major changes in the genetic instructions of a cell can occur. Figure 8-21 shows three ways that chromosome mutations occur. These changes affect many genes. As a result, chromosome mutations are usually far more serious than gene mutations. In fact, most are fatal, often even before the offspring is born.

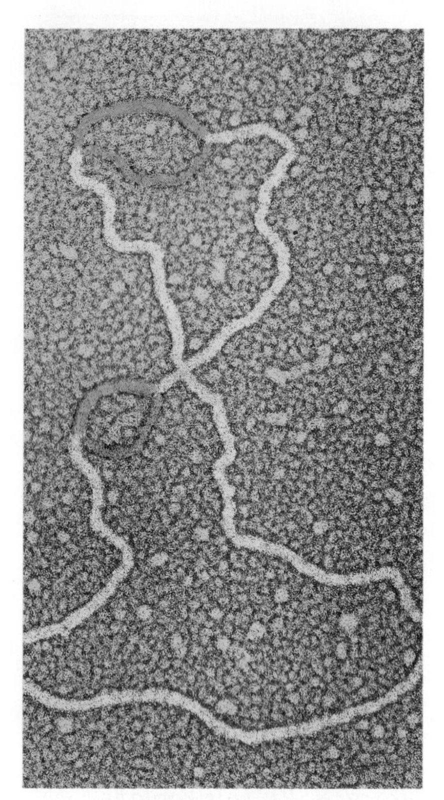

Figure 8-22 *This electron micrograph shows recombinant DNA in a bacterium. The ordinary double-stranded DNA is yellow. The red portions are foreign DNA that have been added. The blue portions are individual genes that have been isolated.*

Engineering DNA

Humans have learned to engineer their own changes in DNA. Early in the 1970s, scientists discovered several chemicals that could cut DNA molecules into pieces. As a result, scientists can take a specific section of DNA (a gene) out of a cell from one organism and place it into a cell from another organism. The section that was moved combines with the DNA that was already there. That cell then has some new hereditary instructions. The new DNA formed by adding DNA from one organism to the DNA of another organism is called *recombinant DNA.* See Figure 8-22.

Bacteria containing recombinant DNA are now being used to make substances that are vital to human beings. One such substance is insulin, which is needed by many people who suffer from a genetic disease called diabetes. Insulin was once extracted from the organs of cows or pigs. Now *human* insulin is synthesized by bacteria—and at a much lower cost. *Interferon,* a human protein that prevents the multiplication of viruses, is another chemical that is made by bacteria containing recombinant DNA. Certain types of cancer are treated with interferon.

Recombinant DNA may also be used to replace incomplete or undesirable DNA instructions in plants and animals. Scientists plan to improve food crops and farm animals by adding genes for higher nutritional value and disease resistance. There is hope that recombinant DNA techniques may someday be used to cure human genetic diseases.

Summary

DNA is the material responsible for heredity. A molecule of DNA is a double strand of nucleotides that is twisted into a spiral shape. This structure is called a double helix. The hereditary information carried by DNA is in a code that is contained in the sequence of the four nitrogen bases adenine, thymine, cytosine, and guanine. Because only certain of these bases can bond together in pairs, DNA can be copied exactly by the process of replication. The instructions of the genetic code are for putting amino acids together to make proteins. RNA molecules take the instructions for proteins from the nucleus into the cytoplasm, where proteins are made. A change in the sequence of bases in a DNA molecule is called a mutation. A single gene or an entire chromosome may be changed by a mutation. Humans can alter DNA by taking DNA pieces from one cell and placing them into another cell, thus making recombinant DNA.

A

ADP (adenosine diphoshate) one of the molecules that is involved in the process by which the body obtains energy from the chemical breakdown of food. Energy is stored when a phosphate group is added to ADP, forming ATP **(55)**

ATP (adenosine triphosphate) one of the molecules that is involved in the process by which the body obtains energy from the chemical breakdown of food; energy is released when a phosphate group is removed from ATP, leaving ADP **(55)**

Absorption taking light in but not reflecting or transmitting it, as when a filter absorbs certain colors of light but transmits others **(345)**

Absorption spectrum a diagram that shows which frequencies of light are absorbed by a particular substance **(23)**

Abyssal plains extensive featureless areas, largely lifeless, that make up much of the deep ocean bottom **(154)**

Adhesion a force of attraction between molecules of different substances whose surfaces are in contact with each other **(43)**

Air mass a body of air covering a large area and having fairly uniform temperature and humidity characteristics **(166)**

Alternating current an electric current that constantly changes direction **(226)**

Amniocentesis a procedure used to obtain human fetal cells for examination. A sample is drawn from the amniotic fluid surrounding the fetus **(491)**

Ampere (amp) the basic unit of measure for the flow of electric current, equal to 1 coulomb of charge per second **(249)**

Amplitude the greatest distance a wave moves from its neutral position **(277)**

Asexual reproduction reproduction that involves only the division of single cells into new cells **(453)**

Atom the smallest particle into which an element may be divided and still be the same substance **(403)**

B

Barometer an instrument for measuring atmospheric pressure **(178)**

Battery a group of connected cells that store electric charge and can produce an electric current **(207)**

Bimetallic strip a strip, composed of two different metals bonded together, used for detecting temperature changes **(242)**

C

Capillary action the rising of a fluid in a thin tube, caused by the force of attraction between the liquid and the walls of the tube **(41)**

Chemical cell a device, consisting of an electrolyte and two electrodes, that stores electric charge and can produce an electric current **(213)**

Chemical energy a form of potential energy, stored in fuels, food, and other chemicals, that is released by chemical reactions **(82)**

Chlorophyll the green pigment, found in plant cells, that is necessary for the process of photosynthesis to occur **(20)**

Chromosomes microscopic bodies that contain the genes that pass hereditary traits from parent to offspring **(462)**

Circuit a path for the flow of electrons **(231)**

Circuit breaker a mechanical device that breaks a circuit when too much current flows through it **(257)**

Clone an individual produced asexually from a single parent, as by cuttings, bulbs, or cell division, etc. **(491)**

Cohesion the attractive force of the particles of a substance for one another **(42)**

Cold front the leading edge of an advancing cold air mass **(167)**

Compound a substance that consists of two or more elements chemically combined **(406)**

Concave mirror a mirror in which the reflective surface curves inward **(371)**

Condensation the process by which a gas or a vapor changes to a liquid **(415)**

Condense to change from a gas or a vapor to a liquid, as when water vapor in the air becomes visible moisture **(146)**

Conductor a material through which electric current can pass **(209)**

Conservation of energy the principle which states that energy can be transferred and transformed, but its total amount remains the same **(101)**

Continental climate a climate typical of large land masses, having relatively large daily and seasonal temperature changes **(170)**

Continental shelf former coastal plains, now submerged due to the melting of ice-age glaciers **(155)**

Continuous spectrum a characteristic pattern of colored lines produced by the light an element gives off under special circumstances **(343)**

Converging lens a lens that is thicker in the center than at its edges; causes beams of light passing through it to come together **(366)**

Convex lens a lens in which both of the surfaces are convex (curve outward), or in which one one side is convex and the other is flat **(366)**

Convex mirror a mirror in which the reflective surface curves outward **(364)**

Coriolis effect the effect, caused by the rotation of the Earth, in which winds and ocean currents veer to the right in the Northern Hemisphere and to the left in the Southern Hemisphere **(184)**

Coulomb a unit of measure of electrical charge, equal to the charge on 6.24×10^{18} electrons (6.28 billion billion) **(249)**

Cross-breeding producing offspring from two individuals with different genetic characteristics **(456)**

Current a flow of electric charge through a conductor **(249)**

D

DNA an abbreviation for *deoxyribonucleic acid;* the chemical in chromosomes that contains instructions for cells **(483)**

Daughter cells the two cells that result from the division of a single "parent" cell **(449)**

Decibel a unit of measure of the loudness of sound **(307)**

Density current currents caused by density differences in liquids or gases; the denser substance flows downhill underneath the less dense substance **(165)**

Diffuse reflection scattered reflection; that is, reflection in which the light is reflected in more or less random directions **(358)**

Diffusion the intermingling of the particles of one substance with those of another because of the motion of both types of particles. As a result of this process, particles move from regions of greater concentration to regions of lesser concentration **(29) (415)**

Digestion the process by which an organism breaks down food into substances usable by individual cells **(46)**

Direct current an electric current that flows one direction only, without alternating **(227)**

Dominant describing a genetic factor that is always expressed; prevents a recessive trait from showing up in offspring **(456)**

Dry cell a type of chemical cell in which the electrolyte is in paste form **(216)**

E

Echolocation the determination of the positions of objects by the use of sound that is reflected back to the sender as echoes. Bats use echolocation to navigate **(293)**

Efficiency the ratio of the work output to the work input of a machine **(96)**

El Niño effect a recurring climatic phenomenon caused by a disruption of the usual pattern of winds and ocean currents in the Pacific Ocean **(187)**

Electrical energy the energy carried by charged particles **(86)**

Electrode a terminal that conducts electrons in a circuit into or away from the electrolyte in a battery **(213)**

Electrolyte a substance that conducts electricity when in solution or in liquid form **(213)**

Electromagnet a magnet produced by electric current flowing through a coil of wire wrapped around a soft iron core **(245)**

Electron a particle that is found in all atoms and has a negative electrical charge **(210)**

Element a substance that cannot be separated into other substances by ordinary chemical changes. The smallest division of an element is an atom **(406)**

Endothermic describes processes that absorb heat **(426)**

Energy the ability to cause change **(73)**

Evaporation the process by which particles of a nonboiling liquid escape from the surface of the liquid and enter the gaseous state **(415)**

Exothermic describes processes that give off heat **(426)**

Exponent a small number, placed at the upper right of another number, that tells how many times the lower number is to be multiplied by itself. For example, $10^3 = 10 \times 10 \times 10$ **(392)**

F

Family tree a diagram for a family, showing parents and offspring for a number of generations **(441)**

Fertilize to cause the union of a sperm cell with an egg, resulting in a new organism beginning to develop **(463)**

Fetus a developing organism, still in the womb or egg, in its later stages of development **(472)**

Filter a piece of glass or other material that absorbs certain colors of light while permitting other colors of light to pass through **(336)**

Flywheel a heavy rotating wheel that stabilizes the speed of the device to which it is attached **(107)**

Focal length the distance of the focal point of a lens from the center of the lens **(368)**

Focal point the point at which light passing through a converging lens comes together **(368)**

Forced vibration the vibration produced in an obect when it comes in contact with another object that is already vibrating **(305)**

Fraternal twins two individuals who are born at the same time from the same mother but have different genetic makeup **(469)**

Frequency the number of repetitions of something in a given interval of time, such as the number of vibrations of an object or the number of waves passing by a point per second **(277)**

Front a boundary between two different air masses **(167)**

Fuse an electrical device that contains a strip of metal that melts if too much current passes through a circuit **(254)**

G

Genes individual components of chromosomes which carry the factors that pass hereditary traits from parents to offspring **(462)**

Genetic engineering the manipulation, by splicing and recombining, of the DNA of living organisms in order to produce species with new characteristics **(490)**

Genetics the study of heredity **(457)**

Genotype the makeup of an organism in terms of its genes **(486)**

Global warming the apparent overall warming trend of the Earth's atmosphere **(134)**

Glucose a simple sugar that can be absorbed by the cells of living things **(46)**

Greenhouse effect an effect, similar to the warming that occurs in a greenhouse, which some scientists hypothesize is taking place in the Earth's atmosphere **(136)**

Gulf Stream a warm current that flows from the tropical Atlantic Ocean northward along the eastern coast of North America **(156)**

H

Heat energy the form of energy that results when, among other ways, surfaces are rubbed together **(86)**

Heredity the passing of traits from one generation to the next by means of genes **(460)**

Hertz the basic unit of measure for frequency, equal to 1 cycle per second **(227) (300)**

Hybrid an offspring produced from parents with different genetic makeup **(457)**

Hydrothermal vent springs of hot, mineral-rich water commonly found along the mid-ocean ridges **(154)**

Hypertonic solution a solution having a higher concentration of a dissolved substance than the concentration within a cell **(45)**

Hypotonic solution a solution having a lower concentration of a dissolved substance than the concentration within a cell **(45)**

I

Identical twins two individuals born from the same fertilized egg. Such individuals have identical genetic makeup **(469)**

Image the visual impression of an object, produced by reflection in a mirror or refraction by a lens **(360)**

Impermeable not allowing anything to pass, especially fluids **(34)**

Incident beam a beam of light that falls upon or strikes something, such as a beam of light falling upon a mirror **(354)**

Inclined plane a simple machine consisting of a plane set at an angle to a horizontal surface, forming a ramp **(94)**

Insulator a nonconducting substance **(209)**

Insulin the chemical substance that is released into the blood by the pancreas; enables the body to use sugar as a fuel in the process of respiration **(56)**

***In vitro* fertilization** fertilization that takes place outside of a living organism. The fertilized egg is maintained in an artificial environment until it can be placed into the uterus of a female for normal embryonic development **(468)**

Isotonic solution a solution that has the same concentration of a dissolved substance as another solution. Usually refers to a solution of salt (NaCl) and water with the same salt concentration as blood **(38)**

J

Joule the amount of work done when a force of 1 newton is exerted through a distance of 1 meter **(77)**

K

Kinetic energy the energy of moving objects **(86)**

L

Lens a piece of glass or other transparent material, curved on one or both sides, used for refracting light **(366)**

Lever a simple machine consisting of a bar that pivots about a fixed point **(98)**

M

Machine a device that increases force at the expense of distance or that increases distance at the expense of force **(70)**

Magnetic lines of force invisible curved paths along which magnets exert their influence **(223)**

Maritime climate a climate typical of the ocean, with relatively small daily and seasonal temperature changes **(170)**

Mechanical advantage the ratio of the output force of a machine to the input force **(97)**

Mechanical energy another name for kinetic energy **(86)**

Mechanical system a machine composed of more than one simple machine **(72)**

Medium a substance, such as air or water, through which a wave travels **(284)**

Meiosis the process, occurring within the nucleus during cell division, that leads to the production of sex cells with half of the organism's usual number of chromosomes **(463)**

Mid-ocean ridges undersea mountain ranges that form at the boundary between oceanic plates that are moving apart **(154)**

Mitosis the process, occurring in the nucleus during cell division, in which the chromosomes in the nucleus are distributed into two regions before the nucleus divides **(452)**

Model a representation of an idea or phenomenon that simulates the structure, function, or effect of the phenomenon **(396)**

Molecule particles made of two or more atoms that are chemically united **(406)**

N

Negative charge the type of charge possessed by electrons **(209)**

Noise a non-musical sound; appears on an oscilloscope as a jagged, haphazard, irregular display **(265)**

Nucleus the central mass of protoplasm, usually spherical in shape, found in most plant and animal cells. The nucleus controls the activities of the cell and contains genes **(447)**

O

Octave the musical distance, or *interval*, between a musical tone and one of twice the frequency, such as from middle C to the C just above it **(302)**

Offspring the child or children of a parent or parents. In the case of cell division of one-celled organisms, the new cells are the offspring of the original cell **(449)**

Oscilloscope an electronic instrument that shows sound waves as a curve on a "television screen," or cathode-ray tube **(311)**

Osmosis the passage of water by diffusion through a semipermeable membrane; the direction of flow is from the region of greater concentration of water to the region of lesser concentration **(36)**

P

Pancreas the gland that enables the body to use sugar as a fuel in the process of respiration; it does this by releasing insulin into the blood **(56)**

Parallel circuit a type of circuit in which a current is split before passing into different components **(235)**

Pascal a unit of measure of pressure, equal to 1 newton per square meter **(178)**

Pedigree a diagram that shows the inheritance of a trait through several generations **(462)**

Permeable allowing passage, especially by fluids **(34)**

Phenotype the makeup of an organism in terms of its inherited traits **(486)**

Photosynthesis the process by which green plants and plant-like organisms use energy from sunlight to convert water and carbon dioxide into sugars that the organisms can use for food **(6)**

Piezoelectric effect the production of electric currents by certain crystals when they are squeezed or stretched **(229)**

Pitch the "highness" or "lowness" of a sound, determined by the sound's frequency **(266)**

Plane mirror a flat mirror, in which images appear to be the same size as the objects **(360)**

Positive charge the type of charge possessed by protons **(209)**

Potential energy stored energy, such as the energy in a stretched or compressed spring or in an object raised above the surface of the Earth **(81)**

Power work done per unit of time **(79)**

Pressure the amount of force exerted per unit of area **(173)**

Prism a piece of glass or other transparent material used to separate light beams into spectrum patterns. A prism, as used in optics, has three rectangular sides and triangular ends **(336)**

Proton a type of particle that is found in all atoms and has a positive charge **(210)**

Punnett squares a diagram showing all possible genetic combinations that can be passed on to the offspring of hybrid parents **(460)**

R

Radiometer a windmill-like device consisting of a set of vanes in a vacuum. The vanes, which are black on one side and white on the other, rotate when exposed to light **(330)**

Real image a type of optical image of an object which can be projected onto a screen **(363)**

Recessive describing a genetic factor that is not always expressed; prevented from showing up in offspring when a dominant trait is present **(456)**

Reflected beam a beam of light after it bounces off of a mirror or other reflecting surface **(354)**

Refraction the bending of a beam of light, or other wave form, as it passes from one medium into another **(376)**

Resistance the property of a conductor to oppose the flow of electricity through it, causing the production of heat **(233)**

Resistor a device placed in an electric circuit to provide resistance **(234)**

Resonance vibration in an object or medium produced by a vibrating source having the same frequency **(306)**

Respiration the process by which organisms combine oxygen from air or water with digested food in order to release chemical energy for use by the cells **(48)**

Root hair a hair-like extension of the epidermal, or outer, cells of a root **(28)**

Scattering sending light in random directions by a reflecting surface or by the particles of a medium through which the light is passing **(345)**

Semipermeable allowing some substances to pass through, but not others **(34)**

Series circuit a type of circuit in which the parts are joined end to end, with the same current passing though each part **(235)**

Sex cells egg or sperm cells, containing half the organism's usual number of chromosomes **(463)**

Sexual reproduction reproduction that involves the union of an egg and a sperm, usually from two separate parents **(453)**

Simple machine one of the basic machines, such as a lever, pulley, wheel and axle, wedge, screw, or inclined plane **(98)**

Simulation a type of experiment that is intended to duplicate certain actual conditions **(133)**

Solidification the process by which a substance becomes solid or firm **(415)**

Sonic boom a loud noise produced by the shock wave of an object traveling faster than the speed of sound **(288)**

Spectrometer an instrument that breaks light into its characteristic pattern of colored bands or lines **(343)**

Spectrum (plural: spectra) a continuous pattern of colored bands or a series of colored lines, produced by passing light through a prism or similar device **(336)**

Specular reflection reflection, such as reflection by a mirror, that occurs without scattering **(356)**

Starch a white, tasteless, odorless food substance found in rice, potatoes, and cereal products such as wheat **(6)**

Stationary front a nearly motionless boundary between two air masses **(168)**

Sublimation the process by which a gas becomes a solid without passing through the liquid state **(400)**

Subsystem a system that is a part of some larger system **(72)**

Switch a device for opening and closing a circuit to start or stop the flow of electric current **(206)**

T

Thermocouple a device that converts heat energy into electric energy. Thermocouples are often used to measure temperature **(229)**

Thermoelectric device a device that produces heat when electric current flows through it **(234)**

Thermostat a device that controls temperature by automatically switching a heating or cooling source on and off **(243)**

Thyroid gland the gland that releases the hormone *thyroxin* into the blood to regulate the rate of respiration in body cells **(56)**

Total internal reflection the complete reflection of a beam of light from the inside surface of the medium, so that none of the light leaves the medium **(377)**

Trade winds relatively permanent, steady winds that blow toward the equator from about 35 °N and 35 °S latitudes **(166)**

Trait a distinguishing characteristic, such as hair color or height **(441)**

Translucent allowing light to pass through, but causing scattering in the process, so that images of objects are not clearly visible through the material **(350)**

Transmission the passing of light through a medium **(345)**

Transparent allowing light to pass through without scattering, so that images are clearly visible through the material **(349)**

Transpiration the movement of water out of a plant by evaporation from the leaves **(27)**

Trench a place in the ocean where the Earth's crust is slowly being pushed into the interior of the Earth; the ocean floor dips downward, forming a steep-walled valley **(153)**

Trimester one of the three 3-month periods into which a human pregnancy is divided **(473)**

Tuning fork a metal device consisting of two prongs that vibrate when struck, producing a tone of exact pitch **(268)**

V

Variable resistor a device in which the resistance can be changed to control the flow of current in an electrical circuit **(234)**

Vibration a complete back-and-forth movement of a rhythmically moving object **(276)**

Virtual image an image such as that seen in a mirror or when looking through a lens at an object. Such an image cannot be projected onto a screen **(363)**

Volt the basic unit of measure for electrical pressure **(249)**

Voltage the pressure, or electromotive force, of an electric current **(250)**

W

Warm front the leading edge of an advancing warm air mass **(168)**

Wavelength the distance from one point on a wave, such as a crest or a trough, to the corresponding point on the next wave **(283)**

Wet-bulb thermometer a type of thermometer used to measure atmospheric humidity **(145)**

Wet cell a type of chemical cell that uses a liquid for the electrolyte **(217)**

Wheel and axle a simple machine consisting of a wheel attached to a smaller wheel or shaft to increase force or speed **(102)**

Work force applied on an object to move it through a distance **(73)**

Work input the work that is put into a machine **(93)**

Work output the work that is done by a machine **(93)**

Boldface numbers refer to an illustration on that page.

A

B

C

D

E

M

P

T

Z

Abbreviated as follows: (t) top; (b) bottom; (l) left; (r) right; (c) center.

Table of Contents: iii(tr), Nathan Bilow/Stock Imagery; iii(b), HRW Photo by John Langford; iii(inset), Ken McGraw/Stock Imagery; iv(tl), Paul Franz Moore; iv(tr), Oliver Strewe/Tony Stone Worldwide; iv(bl), Shiki/The Stock Market; v(t), Naoki Okamoto/The Stock Market; v(b), HRW Photo by John Langford; vi(l), E.R. Degginger; vi(b), HRW Photo by John Langford; vii(t), Myron J. Dorf/The Stock Market; vii(b), HRW Photo by John Langford; viii, HRW Photo; ix, John Pinderhughes/The Stock Market.

Unit 1: 2, David Muench; 4, HRW Photo by Eric Beggs; 5(cl), Ontario Ministry of Agriculture and Food; 5(cr), W.H. Hodge/Peter Arnold, Inc.; 5(bl), Ontario Ministry of Agriculture and Food; 6, L.M. Biedler, University of Florida; 10, Grant Heilman; 11, Grant Heilman; 16(tl), Jay Maisel/The Image Bank; 16(tr), Pamela J. Zilly/The Image Bank; 16(bl), Yuri Dojc/The Image Bank; 16(br), TV Ontario; 17, Ed Reschke/Peter Arnold, Inc.; 20, Ruth Dixon; 24, NASA; 25(tl), HRW Photo by Frank Wing; 25(tr), Walter Chandoha/Gartman Agency; 25(c), Phillip Leonian; 25(b), David Muench; 26, HRW Photo by Richard Haynes; 27, Randall Hyman; 38, Michael Mitchell; 39(t), Ontario Ministry of Agriculture and Food; 39(b), Michael Mitchell; 40, David Muench; 51, Michael Mitchell; 58, NASA; 60(t), Jack Zehrt/FPG International; 60(b), Mark Barnes/Vandivier Stock; 63, David Muench; 64(b), Steve Whalen/Nawrocki Stock Photo; 65(t), HRW Photo by Michelle Bridwell; 65(b), Charles C. Place/The Image Bank; 66(t), NASA; 66(b), NASA; 67, Grant Heilman Photography.

Unit 2: 68, T. Rosenthal/Superstock; 70(tl), Dorothy & Fred Bush/Peter Arnold, Inc.; 70(tr), Chris Michaels/FPG International; 70(bl), Tom Tracy/FPG International; 70(br), Richard Francis/Wheeler Pictures; 71(t), Peter Grumann/The Image Bank; 71(b), Camerique; 72(tl), HRW Photo by Bruce Buck; 72(tr), David R. Frazier Photolibrary; 72(cl), Leonard Lessin/Peter Arnold, Inc.; 72(bl), NASA; 72(br), Barry Erickson; 73(t), HRW Photo by Eric Beggs; 73(c), HRW photo by Richard Haynes; 73(b), P.R. Dunn; 74(t), ©Barrera/TexaStock; 74(bl), Hanson Carroll/Peter Arnold, Inc.; 74(br), Ken Lax/Stock Shop; 75(t), Paul E. Loven/The Image Bank; 75(b), Jerry Wachter/Focus on Sports; 82(tr), Michael Mitchell; 82(cr), David R. Frazier Photolibrary; 82(cl), Michael Mitchell; 82(bl), Ben Rose/The Image Bank; 82(br), H.E. Edgerton; 83(cl), David Young-Wolff/PhotoEdit; 83(c), Steve Smith/Wheeler Pictures; 83(cr), Garry Gay/The Image Bank; 83(bl), Howell/Leo de Wys, Inc.; 83(br), Terrence Moore/Tom Stack & Associates; 85, HRW Photo by Eric Beggs; 88(tr), E.R. Degginger; 88(cl), HBJ Photo/Richard Haynes; 88(c), Lee Foster/Bruce Coleman, Inc.; 88(cr), E.R. Degginger; 88(bl), HRW photo by Richard Haynes; 88(bc), HRW Photo by Richard Haynes; 88(br), HBJ Photo; 98, Michael Mitchell; 99(tl), The Image Bank; 99(tr), P.R. Dunn; 99(lc), HRW Photo by Russell Dian; 99(cc), Michael Mitchell; 99(lcb), Michael Mitchell; 99(cb), Michael Mitchell; 99(b), Mark E. Gibson; 101(t), The Bettmann Archive; 101(b), The Bettmann Archive; 106, Richard Megna/Fundamental Photographs; 107(t), Jim Pokarchak; 107(b), G.S. Chapman/The Image Bank; 113, Jim Pokorchak; 114(t), Amy C. Etra/PhotoEdit; 114(b), David R. Frazier Photolibrary; 115(t), Jim Pokorchak; 115(c), Norco Products, Ltd.; 115(b), Jim Pokorchak; 123, T. Rosenthal/Superstock; 124(t), Michael Lyon; 124(b), HRW Photo by Michelle Bridwell; 125(t), John McGrail; 125(b), A.M. Rosario/The Image Bank; 126(b), Tate Gallery, London/Art Resource, New York; 127(t), Giraudon/Art Resource.

Unit 3: 128, Richard Harrington/FPG International; 130, Jack Zehrt/FPG International; 131(t), NASA; 131(b), NASA Photo/Research by Grant Heilman Photography; 132(t), Telegraph Colour Library/FPG International; 132(b), NASA/Peter Arnold, Inc.; 140(tl), R. Hamilton Smith/FPG International; 140(tr), Otto Hahn/Peter Arnold, Inc.; 140(bl), Lou Jacobs, Jr./Grant Heilman Photography; 140(br), Kent & Donna Dannen/Photo Researchers; 142, NASA's Goddard Space Flight Center; 149(t), H. Van/Superstock; 149(c), Marcel ISY-Schwart/The Image Bank; 149(cb), Marc Romanelli/The Image Bank; 149(b), Mike Schneps/The Image Bank; 152, World Ocean Floor by Bruce C. Heezen and Marie Tharp©, 1977. Reproduced by permission of Marie Tharp, 1 Washington Avenue, South Nyack, NY 10960; 154, Dudley Foster/Woods Hole Oceanographic Institute; 155, Lincoln Pratson/Lamont Doherty Geological Observ.; 156, National Oceanic and Atmospheric Administration; 157, E.R. Degginger; 162, E.R. Degginger; 163, E.R. Degginger; 165, NASA; 170, Weather Graphics Courtesy of Accu-Weather, Inc., 619 West College Avenue, State College, Pa. 16801. Other Educational Weather Products Available© 1992 ; 171, Bruno P. Zehnder/Peter Arnold, Inc.; 177, The Bettmann Archive; 178, TV Ontario; 182, HRW Photo; 184, Runk/Schoenberger/Grant Heilman Photography; 188, NASA; 192, Warren Faidley/Weatherstock; 193(b), Richard Harrington/FPG International; 194(t), Michael Lyon; 194(b), HRW Photo by Michelle Bridwell; 195(t), National Weather Service, Austin, Texas/HRW Photo by Michelle Bridwell; 195(b), Cliff Feulner/The Image Bank; 196(t), NASA; 196(b), National Oceanic and Atmospheric Administration; 196(inset), Earth Satellite Corporation/Science Photo Library/Photo Researchers; 197, C. Oricco/Superstock.

Unit 4: 198, D. Luria/FPG International; 200(t), Tom Mareschal/The Image Bank; 200(b), Jawitz/The Image Bank; 201, Vandivier/Vandivier Stock; 211, Schomburg Center for Research in Black Culture. The New York Public Library. Astor, Lenox & Tilden Foundations.; 228(t), Grant Heilman/Grant Heilman Photography; 228(b), F. Reginato/The Image Bank; 245, ©Tom Tracy; 257, D. Luria/FPG International; 258(t), Michael Lyon; 258(b), Howard Sochurek; 259, Howard Sochurek; 260(t), Cindy Lewis; 260(b), Martyn Cowley; 261, Martyn Cowley.

Unit 5: 262, Francisco Manrrique/Universidad de Venezuela, Caracas; 264, Grant Heilman/Grant Heilman Photography; 265(t), Dick Luria/Science Source/Photo Researchers; 265(b), John Coletti/Stock Boston; 274(t), Ocean Images/The Image Bank; 274(b), Marcia W. Griffen/Animals Animals; 275, Zig Leszczynski/Animals Animals; 280, Runk/Schoenberger/Grant Heilman Photography; 287, Woods Honda-Kawasaki Fun Center and Longhorn Speedway/HRW Photo by Michelle Bridwell; 291, Bob Daemmrich; 294(t), Rexford Lord/Photo Researchers; 300, Hans Wendler/The Image Bank; 303, Austin Chamber Music Center/HRW Photo by Michelle Bridwell; 307, Barry L. Runk/Grant Heilman Photography; 309, Charles S. Allen/The Image Bank; 310, Lennart Nilsson/*The Incredible Machine*/National Geographic Society; 311, Bill Ivy; 312, Douglas Mesney/Leo de Wys Inc.; 316, HRW Photo by Eric Beggs; 321, Francisco Manrrique/Universidad de Venezuela Caracus; 322(t), Michael Lyon; 322(b), John C. Beatty/Silverwing Photography; 323(t), Steve Dunwell/The Image Bank; 324, HRW Photo by Jim Newberry; 325, HRW Photo by Jim Newberry.

Unit 6: 326, Kelvin Aitken/Peter Arnold, Inc.; 328(tl), Janeart/The Image Bank; 328(tr), M. Roessler/Superstock; 328(b), Paul Slaughter/The Image Bank; 331, Dominique Sarraute/The Image Bank; 333, Larry Chiger/Superstock; 334, Barry L. Runk/Grant Heilman Photography, Inc.; 335(l), NASA; 335(r), Richard Hamilton Smith/FPG International; 335(b), Clyde H. Smith/Peter Arnold, Inc.; 343(t), Welch Scientific Company; 343(c), Runk/Schoenberger/Grant Heilman Photography, Inc.; 343(b), Runk/Schoenberger/Grant Heilman Photography, Inc.; 349, Bill Ivy; 350, HRW Photo by Eric Beggs; 352(t), Canapress; 352(b), Ontario Hydro; 353, Steve Satushek/The Image Bank; 360, Fox and Jacobs Homes/HRW Photo by Michelle Bridwell; 364, Michael Mitchell; 365, Courtesy of Stein Mart of Austin/HRW Photo by Michelle Bridwell; 366(tl), TV Ontario; 366, Michael Mitchell; 371, HRW Photo by Eric Beggs; 374, HRW Photo by Michelle Bridwell; 375, Runk/Schoenberger/Grant Heilman Photography; 382, Icon Comm/FPG International; 383(t), Obremski/The Image Bank; 383(b), Kelvin Aitken/Peter Arnold, Inc.; 384(t), Richard Spencer/University of Central Florida; 384(b), Howard Sochurek; 385(t), Lawrence Manning/Westlight; 385(c), Philippe Plailly/Science Photo Library/Photo Researchers; 385(b), Howard Sochurek; 386(t), P.R. Dunn; 386(b), Rickey Schum; 387(t), HRW Photo by Eric Beggs; 387(cl), E.R. Degginger; 387(cr), E.R. Degginger; 387(b), Rob Atkins/The Image Bank.

Unit 7: 88, Yagi Studio II/Superstock; 390(t), California Institute of Technology and Carnegie Institution of Washington; 390(br), NASA; 390(inset), J. Stubblefield/Superstock; 391(t), J. Stubblefield/Superstock; 391(cl), J. Stubblefield/Superstock; 391(cr), HRW photo by WARD'S Natural Science Establishment, Inc.; 391(b), Blair Seitz/Photo Researchers; 394, CDC/Science Source/Photo Researchers; 396, HRW Photo by Michelle Bridwell; 398, V. Sigl/Superstock; 405(t), E.R. Degginger; 405(b), Warren Jacobi/Berg & Assoc.; 406(l), Helga Lade/Peter Arnold, Inc.; 406(r), Comstock; 412(t), L. Fitzgerald/Superstock; 412(b), Michel Tcherevkoff/The Image Bank; 413, Bill O'Connor/Peter Arnold, Inc.; 414, South Austin Recreation Center/HRW Photo by Michelle Bridwell; 415, Leonard Lessin/Peter Arnold, Inc.; 427, R. Ramaekers/Superstock; 433, Yagi Studio II/Superstock; 434, Michael Lyon; 435, Michael Lyon; 436, Peter Menzel; 437(t), Peter Menzel; 437(b), IBM Corporation, Research Division, Almaden Research Center.

Unit 8: 438, Alan Kearney/FPG International; 440(tl), Elyse Lewin/The Image Bank; 440(tr), Helene Tremblay/Peter Arnold, Inc.; 440(bl), Maria Taglienti/The Image Bank; 440(br), Regine Mahaux/The Image Bank; 441, Cody/FPG International; 442(l), Grace Moore/Medichrome; 442(r), William Hopkins; 443, Daniel P. Mass, MD; 444(l), Renate & Gerd Wustig/Okapia/Photo Researchers; 444(r), Austin Rowing Club/HRW Photo by Michelle Bridwell; 447(tl), Manfred Kage/Peter Arnold, Inc.; 447(cl), Runk/Schoenberger/Grant Heilman Photography; 447(bl), Michael Abbey/Photo Researchers; 447(c), Ken Wagner/Phototake; 448, Biophoto Associates/Science Source/Photo Researchers; 449, Runk/Schoenberger/Grant Heilman Photography; 450, Michael Abbey/Science Source/Photo Researchers; 452, Philip Coleman; 453, T.R. Broker/Phototake; 454, The Bettmann Archive; 462, Custom Medical Stock Photo; 463(t), Dr. G. Schatten/Photo Researchers; 463(b), John Giannicchi/Science Source/Photo Researchers; 466, Lennart Nilsson/*A Child is Born*/Dell Publishing Company; 468(t), Hank Morgan/Rainbow; 468(b), UPI/Bettmann Newsphotos; 470(t), Lennart Nilsson/*Behold Man*/Little Brown and Company; 470(l), Lennart Nilsson/*A Child Is Born*/Dell Publishing Company; 470(r), Lennart Nilsson/*The Incredible Machine*/National Geographic Society; 470(b), Lennart Nilsson/*A Child is Born*/Dell Publishing Company; 471, Custom Medical Stock Photo; 472, Lennart Nilsson/*The Incredible Machine*/National Geographic Society; 473, Lennart Nilsson/*Behold Man*/Little Brown & Company; 474(l), Lennart Nilsson/*Behold Man*/Little Brown and Company; 474(r), Lennart Nilsson/*A Child is Born*/Dell Publishing Company; 475, Lennart Nilsson/*The Incredible Machine*/National Geographic Society; 476(t), Custom Medical Stock Photo; 476(b), David R. Frazier Photolibrary; 477, John Colwell/Grant Heilman Photography; 480(l), Leonard Lessin/Peter Arnold, Inc.; 480(r), Michael Tamborrino/Medichrome/The Stock Shop; 481(t), Photo Researchers; 481(b), Photo Researchers; 482(t), Biology Media/Science Source/Photo Researchers; 482(c), Runk/Schoenberger/Grant Heilman Photography, Inc.; 482(b), J.D. Watson/THE DOUBLE HELIX; 483(l), The Bettman Archive; 483(r), Tom Stack & Associates; 484, Philippe Plailly/Science Photo Library/Photo Researchers; 486(l), Hans Halberstadt/Photo Researchers; 486(r), John Colwell/Grant Heilman Photography, Inc.; 488(t), Dr. Jeremy Burgess/Science Photo Library/Photo Researchers; 488(c), Enrico Ferorelli/DOT; 488(b), Erika Stone/Peter Arnold, Inc.; 490(t), HRW Photo by Michelle Bridwell; 490(b), Martin Rogers/FPG International; 491(b), Runk/Schoenberger/Grant Heilman Photography, Inc.; 492(tr), Martin Dohrn/Science Source/Photo Researchers; 492(c), Dr. Betsy Dresser/Cincinnati Zoo; 492(bl), David Parker/Science Photo Library/Photo Researchers; 492(br), Texas Genetic Screening and Counseling Center/HRW Photo by Michelle Bridwell; 494, Lennart Nilsson/*A Child is Born*/Dell Publishing Company; 495, Alan Kearney/FPG International; 496, HRW Photo by

Perry Conway; 497(t), HRW Photo by J. Victor Espinoza; 497(b), Frank Fotex/ Nawrocki Stock Photo; 498(t), M. Fogdem/Bruce Coleman, Inc.; 498(b), Raphael Gaillarde/Gamma Liaison; 499(tl), Michael F. Doyle; 499(tr), Michael F. Doyle; 499(b), Monsanto Company; 501, Tim Fuller.

SOURCEBOOK

Abbreviated as follows: (t)top; (tl)top left; (tc)top center; (tr)top right; (c)center; (b)bottom; (bl)bottom left; (bc)bottom center; (br)bottom right; (l)left; (r)right.

Unit 1: S1, David Muench; S2, Jon Gordan/Phototake; S3(t), Charles D. Winters; S3(bl), HRW Photo by Richard Haynes; S3(br), HRW Photo by Russell Dian; S4, HRW Photo by Russell Dian; S5(tl), HRW Photo by Russell Dian; S5(br), Fred E. Hossler/Visuals Unlimited; S6(l), Manfred Kage/Peter Arnold, Inc.; S6(r), Manfred Kage/Peter Arnold, Inc.; S7, Lennart Nilsson/*The Incredible Machine*/National Geographic Society; S9(l), Joe McDonald/Animals Animals; S9(r), David Madison/Bruce Coleman, Inc.; S11(l), David Madison; S11, Marc Romanelli/The Image Bank; S13, E.R. Degginger; S14, Frank Levy/Stock Market.

Unit 2: S15, T.Rosenthal/Superstock; S16(t), Vandystadt/Allsport USA; S16(bl), HRW Photo by Yoav Levy/ Phototake; S16(br), Walter Iooss/The Image Bank; S17(t), Lou Jones/The Image Bank; S18(tl), Steve Smith/Wheeler Pictures; S18(b), Bill Ross/Westlight; S19, HRW Photo by Eric Beggs; S20, Jules Bucher/Photo Researchers; S21(tr), Allan Seiden/The Image Bank; S21(b), HRW Photo by Bruce Buck; S23(l), Richard Hutchings/Photo Researchers; S23(c), Whitney L. Lane/The Image Bank; S23(r), Manny Milian/ Sports Illustrated; S23(br), HBJ Photo by Rodney Jones; S24, HRW Photo by Richard Haynes; S25(tr), HBJ Photo by Rodney Jones; S25(c), HRW Photo by Richard Haynes; S25(br), Grant Faint/ The Image Bank; S27, S. Barrow/ Superstock; S28, HRW Photo by Richard Haynes; S29(t), HRW Photo by Russell Dian; S29(br), Steve Drexler Photography/Stockphotos; S30(t), HBJ Photo by Sam Joostan; S30(b), W. Beermann/Superstock.

Unit 3: S31, Richard Harrington/FPG International; S32, Jet Propulsion Laboratory; S33(tl), Catherine Ursillo/Photo Researchers; S33(br), Michael Gilbert/ Science Photo Library/Photo Researchers; S34, E.R. Degginger; S37, Shostal Associates; S42, Jonathan Wright/ Bruce Coleman, Inc.; S43(tl), Nuridsany & Perennou/Photo Researchers; S43(tr), Donald Deitz/Stock Shop; S43(c), Jeff Foott/Bruce Coleman, Inc.; S43(b), Warren Faidley/Weatherstock; S46(t), Doug Millar/Photo Researchers; S46(b), Frederic Lewis; S47, Runk/ Schoenberger/Grant Heilman; S48(t), Jeff Rotman/Peter Arnold, Inc.; S48(b), Tom McHugh/Steinhart Aquarium/ Photo Researchers; S49, Steinhart Aquarium/Tom McHugh/Photo Researchers; S50, Mark Sherman/Bruce Coleman, Inc.

Unit 4: S51, D. Luria/FPG International; S52(t), Roger Ressmeyer/Starlight; S52(c), HRW Photo by Yoav Levy; S52(br), The Granger Collection, New York; S53, HRW Photo by Russell Dian; S54, Ontario Science Center; S55, Thomas Ives/Black Star; S58, HRW Photo by Yoav Levy; S59, Walter Bibikow/The Image Bank; S61, C. Peulner/The Image Bank; S62(l), E.R. Degginger; S63(tr), William E. Ferguson; S63(c), HRW Photo by Richard Haynes; S64(c), HRW Photo by Richard Haynes; S64(b), William E. Ferguson; S66, Provided Courtesy of Science Kit & Laboria Laboratories; S67, E.R. Degginger.

Unit 5: S69, Francisco Manrrique/ Universidad de Venezuela, Caracas; S70, Fred Anderson/Photo Researchers; S74, HBJ Photo by Ed McDonald; S75, Jean-Marc Loubat/Photo Researchers; S76(tl), PSSC Physics, D.C. Heath & Company; S76(c), Dohrn/Science Photo Library/Photo Researchers; S76(b), Educational Development Center, Courtesy Film Studio; S77, Educational Development Center, Courtesy Film Studio; S81(t), HBJ Photo by Richard Haynes; S81(l), HBJ Photo by Sam Joosten; S81(br), HBJ Photo by Sam Joosten; S82, George V. Kelvin; S83, David Frazier; S84, Stephen Dalton/Animals Animals; S85(t), Steve Kaufman/ Peter Arnold, Inc.; S85(b), E.R. Degginger; S86(t), Mike Mazzachi/ Stock Boston; S86(b), Bill Gallery/ Stock Boston.

Unit 6: S87, Kelvin Aitken/Peter Arnold, Inc.; S89(t), Educational Development Center, Courtesy Film Studio; S89(b), Educational Development Center, Courtesy Film Studio; S91(c), HRW Photo by Eric Beggs; S94, ©Alon Reininger Contact/Woodfin Camp; S95, B. Campbell/FPG International; S96(t), Dr. R.P. Clark & M. Goff/Photo Researchers; S96(b), Eric Kroll/Taurus Photos; S97(t), J. Stevenson/Photo Researchers; S99, Fundamental Photographs; S101(l) (r), Ken Lax; S102, Royal Greenwich Observatory/Science Photo Library/Photo Researchers; S104, Paul Siverman, ©1991/Fundamental Photographs; S105(t), Ray Ellis/Photo Researchers; S105(c), HBJ Photo by Rodney Jones; S105(b), Michaud Grapes/Photo Researchers; S106, Alexander Tsiaras/Photo Researchers.

Unit 7: S107, Yagi Studio II/ Superstock; S108, Science Photo Library/Photo Researchers; S109, Runk/Schoenberger/Grant Heilman; S111(t), HRW Photo by Dennis Fagan; S111(c), HRW Photo by Dennis Fagan; S111(b), HRW Photo by Dennis Fagan; S113, HRW Photo by Richard Haynes; S115(t), HRW Photo by Yoav Levy; S115(l), HRW Photo by Dennis Fagan; S115(c), HRW Photo by Dennis Fagan; S115(r), HRW Photo by Dennis Fagan; S116, David W. Hamilton/The Image Bank; S117(t), Hermann Eisenbeiss/Photo Researchers; S117(b), R. Carr/Bruce Coleman, Inc.; S120, Susan Van Etten/The Stock Shop; S121(l), Dan McCoy/Rainbow; S121(br), Hank Morgan/Photo Researchers; S123, Tom Kelly©1990/ FPG International.

Unit 8: S125, Alan Kearney/FPG International; S126(t), Jen-Claude Carton/Bruce Coleman, Inc.; S126(b), E.R. Degginger; S129(t), Manfred Kage/Peter Arnold, Inc.; S129(b), Runk/Schoenberger/Grant Heilman; S130(a), Lester V. Bergman & Associates; S130(b), Lester V. Bergman & Associates; S130(c), Lester V. Bergman & Associates; S130(d), Lester V. Bergman & Associates; S133, Martin M. Rotker/Taurus Photos; S135(t) (b), Karl Maslowski/Photo Researchers; S136(tl), Luis Villota/The Stock Market; S136(b), Peter Vandermark/Stock Boston; S141, Tom McHugh/Photo Researchers; S142, P.A. McTurk, University of Leicester & David Parker/Science Photo Library/ Photo Researchers.

ART CREDITS

Accu-Weather, Inc. 169

Scott Thorn Barrows 46 (left), 410

Boston Graphics, Inc. 106 (a & b), 110 (top), 133, 138 (top), 143, 144, 145, 146 (bottom), 148 (top), 153, 155, 158, 159 (right), 162, 164, 167, 168, 176 (right), 268, 269, 270, 271, 276, 277 (left), 278, 280 (right), 289 (center), 295 (top-right), 296 (all center), 303, 304 (top & left), 305 (top), 306, 315 (right), 317 (top), 318, 331, 343, 345, 346, 347, 351, 355 (top), 365, 376, 377, 378, 379, 397 (left), 445 (top), 446, 447, 455, 456, 457, 459

Warren Budd & Associates 193, 359 (bottom), 368 (bottom), 372, 400, 402 (right), 418, 426, 427, 451 (bottom), 452, 484

Les Case 135 (top), 138 (bottom), 139, 151 (top), 190, 252, 254 (top), 255 (top), 266 (top), 273, 285, 314 (bottom), 338 (center), 354 (bottom), 423, 424, 441 442 (top), 443 (top & center-right), 462, 479, 495

Heather Collins/Glyphics 12 (bottom), 14, 17, 23, 29 (top), 37, 62, 63, 76, 77, 208 (top), 215, 218 (left), 220, 255 (bottom), 275, 301 (top), 304 (bottom-right), 305 (bottom), 311, 312, 313, 314 (top & center), 315 (top), 316 (right), 317(bottom), 319, 341 (bottom), 350, 368 (top & center), 371, 373, 379 (top), 397 (right)

Holly Cooper 12 (top), 29 (middle & bottom), 45 (bottom), 86, 90, 102, 137, 147, 161, 171, 204 (right), 239 (top), 264, 272, 279, 290 (top & center-left), 291 (center), 292 (center), 293 (center), 295 (top), 301 (center & bottom), 316 (bottom), 320, 359 (top), 380, 393, 398, 407, 411, 421, 425, 443 (bottom), 451 (top), 461, 478, 485, 487, 493

Sam Daniel 212, 222, 232, 233 (top) 234 (right), 236, 237, 238, 250 (bottom), 251, 253 (center), 281, 282, 283, 284

James Dowdalls 18

Barry Erickson 72

Erickson/Dillon Art Services, Inc. 460

Mike Faison 186

David Fischer 1, 21, 28, 34, 46 (right), 48 (lower left), 56, 96, 174, 175, 176 (left), 177, 178, 187, 189, 223, 226, 230, 361, 369, 442 (center-bottom), 464, 465, 467

Helen Fox 78, 79 (left & right), 98 a & b

David Griffin 50, 120, 141, 150, 151 (center), 159 (left), 165, 179, 183, 206, 207, 216, 218 (right), 219 (top), 233 (bottom), 249, 250 (top), 287-288, 299, 289-300 (bottom), 302, 342, 443 (center-left), 444, 445 (bottom), 458

Linda Hendry 428 (bottom)

Andrew Hickson 204 (left), 229 (bottom), 234 (left), 240, 241 (top), 242 (center), 244 (bottom), 253 (bottom), 256

Mike Krone 135 (bottom), 200-201

Vesna Krstanovich 266 (except top), 267, 290 (center-right), 352, 353

Narda Lebo 49

Jock MacRae 79 (center), 105c, 113, 117, 119

Michael Martchenk 22, 36, 396, 416, 417, 431

Bernard Martin 105 (right), 110 (bottom), 208 (bottom), 209, 273 (top), 227,319 (top), 241 (bottom), 242 (top & bottom), 243, 244 (top), 247, 257

Paul McCusker 80 (left), 122-123, 202, 203, 210, 214, 239 (right)

Leonard Morgan 491

Julian Mulock 8, 9, 11, 32, 33, 35, 41, 42, 54, 55, 91, 100, 103, 104, 108, 109, 111, 112, 121 (top), 205, 217, 221, 231, 245, 246, 248, 332, 333, 370, 381, 401, 402 (left), 432

Precision Graphics, Inc. vii, 170, 402

Joan Rivers S10, S17, S27, S41, S55, S60, S88, S91, S93, S110, S111, S112, S119, S123, S132

Den Schofield 285, 289 (left), 286

Gary Undercuffler 15, 19, 52, 53, 80 (right), 81, 148 (bottom), 173, 184 (top), 277 (top), 280 (left), 344, 348, 349, 358 (top), 363, 367, 395, 415, 422 (left), 428 (top)

Angela Vaculik 97, 98d, 105 (left), 121 (bottom)

Peter Van Gulik 31, 40, 43, 44, 48, 224, 225, 253 (top), 254 (bottom), 297 (right), 298 (right), 329, 330, 336, 337, 338 (top & bottom), 339, 340, 341 (top), 355 (bottom), 356, 357, 358 (bottom), 366, 374, 375, 382, 412, 422 (right)

Pascale Vial 92-93, 94-95, 419, 420

Martin Williams 89 (copied by permission of Stanley Thames Publishers from *GCSE Physics for You* by Keith Johnson © Keith & Ann Johnson, 1978, 1980)

Sarah Forbes Woodward 146 (top), 469

ACKNOWLEDGMENTS

For permission to reprint copyrighted material, grateful acknowledgment is made to the following sources:

Accu-Weather, Inc.: Weather Graphics. Copyright © 1992 by Accu-Weather, Inc., 619 West College Avenue, State College, PA 16801, (814) 237-0309. Other educational weather products available.

Harcourt Brace Jovanovich, Inc.: From *HBJ Earth Science* by Cesare Emiliani, et al. Copyright © 1989 by Harcourt Brace Jovanovich, Inc. From *HBJ General Science* by Patricia A. Watkins, et al. Copyright © 1989 by Harcourt Brace Jovanovich, Inc. From *HBJ Life Science* by Patricia A. Watkins, et al. Copyright © 1989 by Harcourt Brace Jovanovich, Inc. From *HBJ Physical Science* by William G. Lamb, et al. Copyright © 1989 by Harcourt Brace Jovanovich, Inc.

Holt, Rinehart and Winston, Inc.: From *Holt Earth Science* by William L. Ramsey, et al. Copyright © 1978, 1982, 1986 by Holt, Rinehart and Winston, Inc. From *Holt General Science* by William L. Ramsey, et al. Copyright © 1979, 1983, 1988 by Holt, Rinehart and Winston, Inc. From *Holt Life Science* by William L. Ramsey, et al. Copyright © 1978, 1982, 1986 by Holt, Rinehart and Winston, Inc. From *Holt Physical Science* by William L. Ramsey, et al. Copyright © 1978, 1982, 1986, 1988 by Holt, Rinehart and Winston, Inc.

Humana Press Inc.: From *The Life Within: Celebration of a Pregnancy* by Jean Hegland. Copyright © 1991 by The Humana Press Inc.

Lucent Books, Inc.: From "A Project Idea" by Isaac Asimov from *Computers: Mechanical Minds* by Don Nardo. Copyright © 1990 by Lucent Books, Inc.